Modern Glacial Environments

Processes, dynamics and sediments

Editor:
John Menzies

Glacial Environments: **Volume 1**

BUTTERWORTH
HEINEMANN

Butterworth-Heinemann Ltd
Linacre House, Jordan Hill, Oxford OX2 8DP

℞ A member of the Reed Elsevier plc group

OXFORD LONDON BOSTON

MUNICH NEW DELHI SINGAPORE SYDNEY

TOKYO TORONTO WELLINGTON

First published 1995

© Butterworth-Heinemann Ltd 1995

British Library Cataloguing in Publication Data
A catalogue record for this book is available from the British Library.

ISBN 0 7506 2351 9

Library of Congress Cataloguing in Publication Data
A catalogue record for this book is available from the Library of Congress.

Printed in Great Britain

CONTENTS

Preface   xi
Authors   xiii
Acknowledgements   xv
List of Symbols   xix

Chapter 1   Glacial Environments   **1**
J. Menzies

1.1.   Introduction   1
1.2.   Impact of Ice Masses on Global Habitats and Earth Systems   2
1.3.   Research in Modern Glacial Environments   2
1.4.   Research Issues in Glacial Studies   3
1.5.   Synopsis of Chapters   5

Chapter 2   Global Glacial Chronologies and Causes of Glaciation   **9**
P. E. Calkin (With contributions by G. M. Young)

2.1.   Introduction   9
2.2.   Pre-Cenozoic Glaciation   11
2.3.   Late Cenozoic Glaciation and the Classic Subdivisions   22
2.4.   Oxygen Isotope Stratigraphy and the Marine Record of Glaciation 28
2.5.   Long Terrestrial Records of Climate and Glacial Fluctuations for the Quaternary   36
2.6.   The Late Pleistocene Climatic Record Revealed by Deep Ice Cores 44
2.7.   Correlation of Late Cenozoic Glaciations in the Northern and Southern Hemispheres   51
2.8.   The Last 25,000 Years and Synchroneity of Northern and Southern Hemisphere
       Glacial Cycles   59
2.9.   Causes of Climatic Change for Late Cenozoic Glaciation   65
2.10. Future Climate Change and Glaciation   73

Chapter 3 Ice Sheet Modelling and the Reconstruction of Former Ice Sheets from Glacial Geo(Morpho)logical Field Data   77
T. J. Hughes

3.1.	Introduction	77
3.2.	Modelling Strategies	78
3.3.	Specifying Boundary Conditions	82
3.4.	First-Order Glacial Terrains	86
3.5.	Computing Flowline Profiles	90
3.6.	Basal Shear Stress Variations	92
3.7.	Applications	97
3.8.	Summary	99

Chapter 4 Glaciers and Ice Sheets   101
J. Menzies

4.1.	Introduction	101
4.2.	Ice Mass Types	102
4.3.	Formation of Glacier Ice	103
4.4.	Mass Balance and Glacier Sensitivity	105
4.5.	Structure and Thermal Characteristics of Ice Masses	113
4.6.	Ice Structures	124
4.7.	Concluding Remarks	138

Chapter 5 The Dynamics of Ice Flow   139
J. Menzies

5.1.	Introduction	139
5.2.	Ice Mechanics and the Thermo-Mechanical Problem	139
5.3.	Rheological Behaviour of Ice	139
5.4.	The Flow of Ice	146
5.5.	The Flow of Ice Sheets	147
5.6.	The Flow of Valley Glaciers	165
5.7.	The Flow of Ice Shelves	166
5.8.	Velocities of Ice Sheets and Glaciers	171
5.9.	Ice Velocities and Basal Ice Conditions	176
5.10.	Glacier Surges	179
5.11.	Concluding Remarks	195

Chapter 6 Hydrology of Glaciers   197
J. Menzies

6.1.	Introduction	197
6.2.	Meltwater Sources	197
6.3.	Meltwater Systems	197

6.4.   Hydraulic Systems and Deformable Beds   226
6.5.   Subglacial Lakes   231
6.6.   High-Magnitude Meltwater Discharges   233
6.7.   Concluding Remarks   239

Chapter 7 Processes of Erosion   **241**
N. R. Iverson

7.1.   Introduction   241
7.2.   Abrasion   242
7.3.   Quarrying   251
7.4.   Rates of Erosion   255
7.5.   Large-Scale Processes   256
7.6.   Summary   257

Chapter 8 Processes of Transportation   **261**
M. P. Kirkbride

8.1.   Introduction   261
8.2.   Ice Properties Affecting Sediment Transport   263
8.3.   Debris Sources   264
8.4.   Transport Pathways through Ice Masses   270
8.5.   Modification of Sediment During Transport   285
8.6.   Glaciological Effects of Debris in Transport   289
8.7.   Summary   291

Chapter 9 Processes of Terrestrial Deposition   **293**
C. A. Whiteman

9.1.   Introduction   293
9.2.   Factors Contributing to Terrestrial Glaciogenic Deposition   293
9.3.   Mechanisms of Till Deposition   295
9.4.   Mechanisms of Glaciofluvial Deposition   305
9.5.   Tills of Eastern England—Debates about Mechanisms of Till Deposition   306

Chapter 10 Processes of Glaciotectonism   **309**
F. M. van der Wateren

10.1.   Introduction   309
10.2.   Soft Sediment Deformation   310
10.3.   Glaciotectonic Regimes   313
10.4.   Shear Zones   315
10.5.   Marginal Compressive Belts   322
10.6.   Passive Deformations   332
10.7.   Conclusions   333

Chapter 11 Sedimentary and Hydrologic Processes within Modern Terrestrial Valley Glaciers   337
D. E. Lawson

11.1.	Introduction	337
11.2.	Sediment Erosion, Entrainment and Transport	337
11.3.	Water Entrainment and Transport	350
11.4.	Summary	362

Chapter 12 Sediments and Landforms of Modern Proglacial Terrestrial Environments   365
J. Maizels

12.1.	Distinctiveness of Proglacial Environments	365
12.2.	Morphology and Landforms of Proglacial Environments	374
12.3.	Proglacial Meltwater Channel Systems	383
12.4.	Characteristics of Proglacial Outwash Sediments	395
12.5.	Current Issues and Future Prospects	414

Chapter 13 Glaciolacustrine Environments   417
G. M. Ashley

13.1.	Introduction	417
13.2.	Limnology of Glacier-Fed Lakes	420
13.3.	Ice-Contact Glacier-Fed Lakes	427
13.4.	Distal Glacier-Fed Lakes	436
13.5.	Stratigraphy and Landforms	442
13.6.	Closing Remarks	444

Chapter 14 Modern Glaciomarine Environments   445
R. Powell and E. Domack

14.1.	Introduction	445
14.2.	Subglacial Processes	447
14.3.	Marine-Ending Termini	448
14.4.	Types of Grounding-Lines	459
14.5.	Proglacial and Paraglacial Environments	465
14.6.	Other Non-Glacial and Modifying Processes	477
14.7.	Sedimentation Rates and Fluxes	480
14.8.	Advance and Retreat of Marine-Ending Glaciers	484

Chapter 15 Glacial Crushing, Weathering and Diagenetic Histories of Quartz Grains Inferred from Scanning Electron Microscopy

487

W. C. Mahaney

15.1.	Introduction	487
15.2.	The Sediments	488
15.3.	Microtextures vs. Ice Thickness	505
15.4.	Conclusions	505

References **507**
Subject Index **591**

PREFACE

The purpose of this text is to provide a current and comprehensive survey of the glaciology, geomorphology and sedimentology of modern glaciers and ice sheets through an appreciation of the processes, dynamics and sediments found in these environments.

For many of us glaciers hold an almost hypnotic fascination whether it is due to their sheer beauty and grandeur, their relentless movement and awe-inspiring existence, or the interaction of these ice masses with the surrounding and underlying terrain. Glaciers and ice sheets are no longer the esoteric and exclusive realm of explorers and scientists but rather affect all humanity in terms, for example, of their interaction with global warming, sea level rise and predicted climate change.

This book began as part of a larger text on modern and past glacial environments. However, the scope and size of such a book, encompassing modern, Pleistocene and pre-Pleistocene glacial environments with attendant chapters, for example, on dating techniques, stratigraphy and drift prospecting was too large. This present text, therefore, covers modern ice masses and terrestrial and marine glacial environments; past environments being contained in a companion volume: *Past Glacial Environments—Sediments, Forms and Techniques.*

It would be reasonable to describe the study of glacial environments and the development of glacially-related disciplines as not rapidly expanding over the past 40 years but 'exploding'. Like any 'explosion' the effect has scattered widely, enveloping diverse scientific fields such as paleomagnetism, evolutionary biology, astronomical physics and scanning electron microscopy. This 'explosion' has had an attendant vast burgeoning of data and literature.

This book was conceived in order to present university students with a comprehensive survey of modern research and ideas on glacial environments incorporating glaciology, glacial geo(morpho)logy and sedimentology. The basic thesis behind the text originated from a course delivered to my third year students over the past decade. The *raison d'être* behind the course was to understand the complexity of glacial environments from a detailed understanding of their resultant sediments and landforms. In this it is essential to comprehend the fundamentals of ice physics and dynamics coupled with an appreciation of process geomorphology and sedimentology. This conceptualization of glacial environments within Earth Science might be termed a *glacio-sedimentological paradigm* as distinct from the past *morpho-sedimentological* methodological approach where descriptive terminology and scant regard for glaciology and glacial sedimentology merged to render narrow or self-limiting explanations. The approach presented in this book is open-ended permitting a broad view of modern glacial environments highlighting those areas and problems that remain in our, as yet, incomplete understanding of complex glacial conditions and environments. Emphasis is placed on an understanding of primary processes, their physics and impact upon the various interrelated glaciological, geological and sedimentological conditions under which glaciers and ice sheets operate.

Perhaps central to this text and its underlying thesis is a phrase I came across many years ago in which Yatsu (1966) incisively pointed out that "geomorphologists had asked the questions where, when and what but had rarely asked the questions how and why". Within the context of glacial environments, this text strives to redress this imbalance by emphasizing these two, all too often, rarely asked questions. As no text can completely address the many aspects and topics encompassed under the heading of glacial environments, I have attempted to touch upon those that seem most relevant at this stage in our understanding of these complicated and, at times, perplexing environments.

The production of this book has been made possible by the remarkable and tireless efforts of the authors of each chapter. I cannot adequately sum up my thanks or calculate the incredible efforts and sacrifices the authors have made in giving of their valuable time, in delaying important research, in giving up quality time and taking on such a huge undertaking. To Gail Ashley, Parker Calkin, Gene Domack, Terry Hughes, Neal Iverson, Martin Kirkbride, Dan Lawson, Judith Maizels, Bill Mahaney, Ross Powell, Dick van Wateren, Colin Whiteman and Grant Young, I extend my sincere thanks. The authors have revised chapters and answered what must, at times, have seemed interminable requests and have, along with me, watched and waited for the arrival of this book. Numerous people have kindly and willingly lent their time to review chapters, find photographs, answer queries and generally encourage the creation process—without their help this book would still be unfinished. I am indebted to numerous members of the academic and library faculty and staff at Brock University for answering queries, reviewing chapters, finding references and generally assisting me. I am especially grateful to Kamil Zaniewski for his incredible feats in checking references and in computer graphic design of diagrams. I owe thanks to Loris Gasparotto for his superb skills in cartography and to Colleen Catling who so often acted as my unofficial executive secretary. Eilleen Kraatz is to be thanked and praised for wading through so much terse and over-academic prose to edit and refine the chapters in this book. I am particularly grateful to colleagues throughout the world who reviewed chapters and in many cases encouraged me to see this project to its conclusion. I also have a debt of gratitude to Peter Henn, the senior publishing editor at Elsevier Science, who has kept faith with this project throughout and has never ceased to support and encourage my endeavours. I must acknowledge an enormous personal debt to David Sugden and Brian Sissons who were inspirational in their teaching and research and to Don Coates who was instrumental in the initiation of this project. Finally, I am cognisant of and humbled by the immense gratitude I owe to so many fellow geomorphologists, geoscientists, glaciologists and sedimentologists over the past twenty years who have wittingly and unwittingly provided me with so much help, advice and guidance, and whose work is embodied in this text. Whatever innovative ideas and concepts myself and fellow authors have brought to this text we have only built upon the findings and insights provided by these scientists.

No-one, however, has shared and lived with this book in quite the manner that my wife, Teresa, and daughters, Erica, Fiona and Rebecca have; to them I dedicate this book. They have endured, helped, suffered but never given up on me. They have been my inspiration and solace. My wife has given unceasing support, read innumerable chapters, introduced me to the marvels of e-mail and the internet, provided space and time when I needed it most, given prudent and sound advice, and above all else given me a sustaining perspective and devoted encouragement.

John Menzies
Brock University
Ontario
Canada

AUTHORS

G.M. Ashley, Department of Geological Sciences, Rutgers University, New Brunswick, New Jersey, U.S.A.

P.E. Calkin, Department of Geological Sciences, SUNY at Buffalo, Amherst, New York, U.S.A.

E.W. Domack, Department of Geology, Hamilton College, Clinton, New York, U.S.A.

T.J. Hughes, Institute of Quaternary Studies, University of Maine, Orono, Maine, U.S.A.

N.R. Iverson, Department of Geology and Geophysics, University of Minnesota, Minneapolis, Minnesota, U.S.A.

M.P. Kirkbride, Department of Geography, University of Dundee, Dundee, Scotland

D.E. Lawson, Cold Regions Research and Engineering Laboratories, Hanover, New Hampshire, U.S.A.

W.C. Mahaney, Geomorphology and Pedology Laboratory, York University, Toronto, Ontario, Canada

J.K. Maizels, Department of Geography, University of Aberdeen, Aberdeen, Scotland

J. Menzies, Departments of Geography and Earth Sciences, Brock University, St Catharines, Ontario, Canada

R.D. Powell, Department of Geology, Northern Illinois University, DeKalb, Illinois, U.S.A.

F.M. van der Wateren, Rijks Geologishe Dienst, Haarlem and the Institute for Earth Sciences, Vrije Universiteit, Amsterdam, The Netherlands

C.A. Whiteman, Department of Building, University of Brighton, Brighton, Sussex, England

G.M. Young, Department of Earth Sciences, University of Western Ontario, London, Ontario, Canada

ACKNOWLEDGEMENTS

The editor, authors and publisher wish to thank the following for their kind permission to reproduce copyright material. Full citation can be found by consulting figure and plate captions and references.

The authors wish to thank the following individuals and publishing companies for their kind permission to use figures and plates.

Figures: J.C. Crowell for Fig. 2.2; W. Ruddiman and H.E. Wright for Figs 2.7d and 2.23a; D.G. Martinson *et al.* for Fig. 2.9; W. Dansgaard *et al.* for Fig. 2.16; J. Imbrie and K.P. Imbrie for Fig. 2.23b; R.Powell for Fig. 4.1; G. deQ Robin for Figs 4.11 and 4.13; C.Raymond for Figs 5.8 and 5.15; G.S. Boulton for Figs 4.17 and 8.6; R. LeBHooke for Fig. 6.4; B. Hallet for Figs 7.4 and 7.5; J. Harbor for Fig. 7.11; N. Eyles for Fig. 8.2; D. Lawson for Figs 8.5 and 8.13b; R.J. Small for Figs 8.9a and 8.9b; T. Gustavson and J. Boothroyd for Fig. 13.6; J. Teller for Fig. 13.12; E. Cowan and R. Powell for Fig. 14.11; R. Powell for Fig. 14.12; and E.Domack and C.R. Williams for Fig. 14.13.

Plates: C.C. Langway for Plate 2.1; B. Lucchitta for Plate 4.1; G. Clarke for Plate 4.2; E. Evenson for Plates 4.3, 5.4a, 5.4b, 6.1a, 6.1b, 6.3, 6.5, 6.6a and 6.6b; G. Holdsworth for Plates 4.5, 4.8, 4.9, 4.10a, 4.10b, 5.2b and 6.2; H. Gayler for Plate 5.1; J. Shaw for Plate 6.8; H. Björnsson for Plate 6.9; D.E. Sugden for Plate 8.1; L. Homer for Plate 8.2; J.J.M. van der Meer for Plates 10.1 and 10.4 ; W. Mitchell for Plate 12.1; A. Russell for Plate 12.8; T. Gustavson for Plate 13.1c; J. Hartshorn for Plate 13.2a; and Elsevier Science Publishers BV for Plate 14.11.

Publishers: *Academic Press* for Figs 2.9a, 5.8, 5.15 and 5.32; *American Meteorological Society* for Fig. 2.21; *American Association for the Advancement of Science* for Figs 2.13b and 2.16; *American Geophysical Union* for Figs 2.7a, 2.7b, 2.7c, 2.25, 5.22, 5.24a, 5.24b, 5.25, 5.26, 5.27a, 5.27b, 5.28a, 5.28b, 5.28c, 5.29, 5.30, 5.31, 6.17 and 14.3; *A.A. Balkema Publishers* for Figs 2.11a, 2.11b, 6.11, 8.2, 9.2, 9.3, 10.10 and 10.16; *C.R.C. Press* for Fig. 2.3; *Cambridge University Press* for Figs 4.11, 4.13 and 4.17; *Canadian Society of Petroleum Geologists* for Figs 12.24 and 12.25; *Elsevier Science Publishers BV* for Figs 2.12a, 2.12b, 2.12c, 5.5, 5.13, 5.14, 6.19, 14.4 and 14.6; *Elsevier Science Ltd* for Figs 2.8b, 2.9b, 2.11c, 2.18a, 2.18b, 2.19, 2.20, 2.24, 4.12, 4.16, 7.1, 7.2, 9.5 and 9.8; *Edward Arnold Publishers* for Figs 8.1 and 12.9; *Geological Society of London* for Fig. 14.8; *Geological Survey of Canada* for Fig. 12.7; *IAHS* for Fig. 3.1; *Icelandic Glaciological Society* for Figs 6.20, 6.22 and 12.8; *International Glaciological Society* for Figs 4.5b, 4.7, 4.8, 4.9a, 4.9b, 4.9c, 4.10, 4.19, 4.20, 4.22, 4.23, 4.25, 5.9, 5.10, 5.12a, 5.12b, 5.15, 5.17a, 5.17b, 5.18, 5.20, 5.21a, 5.21b, 6.5, 6.7, 6.8, 6.9, 6.10, 6.12a, 6.12b, 6.13, 6.18a, 6.18b, 6.21, 7.9b, 7.10b, 8.9a, 8.9b, 8.11a, 8.11b, 8.15a, 8.15b and 8.15c; *International Association of Sedimentologists* for Figs 8.4b, 8.13a, 8.14b and 13.9; *Kluwer Academic Publishers* for Figs 2.4, 2.5, 2.8a, 2.13b, 2.26b, 5.1, 5.7, 5.11 and 12.27; *Laval Université Presse* for Figs 4.3 and 5.2; *Lingua Terrae* for Fig. 10.14; *MIT Press* for Fig. 4.2; *National Academy of Sciences* for Fig. 2.1; *Nature* for Figs 2.10a, 2.10b, 2.10c, 2.13a, 2.14, 2.15, 2.17 and 5.19; *Ohio State University Press* for Fig. 8.13b; *Oxford University Press* for Fig. 5.23; *Routledge Publishers* for Fig.

2.22; *Royal Society of Edinburgh* for Fig. 12.21; *Scandinavian University Press* for Figs 4.6, 4.24, 6.2, 9.4, 9.6, 10.12, 12.15, 12.17, 12.26 and 13.7b; *Society of Economic Paleontologists and Mineralogists* for Figs 12.12, 12.18 and 13.6b; *Arctic and Alpine Research* and the *University of Colorado* for Fig. 4.15; and *John Wiley and Sons* for Figs 6.1, 6.3, 6.6, 6.14, 6.15, 6.16, 12.2, 12.4, 12.10, 12.22 and 12.23.

The authors wish to acknowledge the following individuals and organisations.

Chapter 2: (P.C.) No one other than myself is responsible for any errors or misrepresentations in this chapter. However, I acknowledge William Ruddiman for his review of an early version of the text; he also supplied the long, 3 million year ^{18}O isotope record for Fig. 2.7. Chester Langway provided important suggestions on the ice core section. Grant Young's contribution on the Pre–Late Cenozoic ice ages is substantive and much appreciated.

(G.Y.) I would like to acknowledge the great educational benefit derived from participation in numerous IGCP projects related to glacial topics and to thank individuals such as Georges Choubert, Anne Faure-Muret, Kalervo Rankama, W.B. Harland, Janine Sarfati, Max Deynoux and Julia Miller who initiated and administered these valuable projects.

Chapter 4: The author is grateful to several colleagues who have given advice and reviewed early versions of this chapter. In particular, I wish to thank Simon Ommanney for his review.

Chapter 5: The author thanks several individuals for advice and valuable criticism of early versions of this chapter.

Chapter 6: The author appreciates the advice of many individuals but especially Roger Hooke who read an early version of the chapter and gave constructive and incisive advice.

Chapter 7: I am indebted to B. Hallet, who either formulated or contributed to most ideas in this chapter and who provided a preprint of his latest work on erosion. I also thank R. LeB. Hooke for his helpful criticism and suggestion during my studies of erosion.

Chapter 9: The writer is grateful to Peter Allen, Allen Cheshire, Jurgen Ehlers, Philip Gibbard, Jane Hart, Hilton Johnson, Juha-Pekka Lunnka, Jim Rose and Martin Sharp for stimulating discussion of these tills in the field. Both Jim Rose and John Menzies made helpful suggestions for improving the manuscript.

Chapter 10: The concept of glaciotectonic regimes evolved during a course in glacial geology, geomorphology and glaciotectonics at the University of Amsterdam by Professor G.S. Boulton, Dr J.J.M. van der Meer and myself. Dr J.J.M. van der Meer kindly gave permission to use the thin sections shown in Plates 10.1 and 10.4. Figure 10.14 is reproduced by permission of Lingua Terrae, Amsterdam.

Chapter 13: Most of the concepts summarized here result from discussions carried out over many years with colleagues, Jon Boothroyd, Tom Gustavson, Norman Smith, John Shaw and Mike Sturm. I am grateful to Bob Gilbert for his incisive and constructive review of an earlier version.

Chapter 14: (E.D.) I would like to thank the Office of Polar Programs of the U.S. National Science Foundation for support of my Antarctic research program over the past few years.

(R.P.) Much of my past work has been supported by grants from the U.S. National Science Foundation and the ideas expressed in this chapter have been an outgrowth from discussion with other colleagues over several years.

Chapter 15: I thank Klaus Fecher (Institute of Geology and Paleontology, Philipps University, Marburg, Germany) for assistance with the SEM and for many helpful suggestions. I also thank

Walter Vortisch (Institute of Geology and Paleontology, Philipps University) and Helmut Schloessin (Department of Geophysics, University of Western Ontario, London, Ontario, Canada) for information on quartz and glacial-crushing processes. Use of SEM facilities at the University of Toronto is also gratefully acknowledged. Samples were prepared with the assistance of Rick Boyer, Drew Clarke, David Hinbest and Julie Oakley. This research was supported by Minor Research Grants from York University and by the Deutsche Forschungsgemeinschaft in cooperation with Walter Vortisch and Wolfgang Andres (Institute of Geography, Philipps University).

I gratefully acknowledge W. Blake Jr (Geological Survey of Canada), Graeme Claridge (Days Bay, New Zealand) and Rein Vaikmae (Estonian Academy of Sciences, Tallinn) for submitting samples for SEM analyses.

LIST OF SYMBOLS

a   Ablation

$\dot{a}$ Ablation rate

a_g Geometric factor in clast 'ploughing'

a_p Sinusoidal bed amplitude

a_s Summer ablation

a_{sp} Bed asperity roughness

a_{sr} Measure of surface roughness

a_t Total ablation

a_w Winter ablation

A   Ice hardness coefficient

b   Mass balance

b_L Balance rate at the ice terminus

b_n Net balance

b_o Stream channel width

b_s Summer balance

b_w Winter balance

B   Ice sliding coefficient

B_e Energy balance on ice mass surface

B_f Buoyancy flux

B_p Back pressure within an ice shelf

c   Accumulation

$\dot{c}$ Accumulation rate

c_b Debris concentration in ice

c_m Debris concentration in moraine septa

c_r Constant related to regelation process

c_s Summer accumulation

c_t Total accumulation

c_w Winter accumulation

C   Cohesion

C_r Areal concentration of clasts in contact with the bed

C_3 Empirical factor related to the mode of water flow in a conduit, the conduit geometry and ice flow conditions

d   Obstacle size

d_c Controlling obstacle size

d_o Mean particle size diameter

d_w Depth of water film at the ice/bed interface

D   Constant related to ice and water density

D_c Conduit diameter

D_e Longitudinal strain or extension

E_i Young's modulus of ice

E^* Activation energy of the creep of ice

f   Area of ice/bed interface occupied by meltwater

f_{sh} Shape factor affecting frictional drag in valley glaciers and ice streams

f_t Fraction of the bed that is thawed

F_r Froude number

F_c Effective contact force between an indenting clast and the bed

F_w Longitudinal normal force within a sediment wedge

F_{wd} Channel width/depth ratio

g Acceleration due to gravity

G Hydraulic gradient

G_{eo} Geometric factor relating mean basal normal stress to basal shear stress and that fraction of the ice/bed interface covered in meltwater

h Ice elevation above grounding line of floating ice margin

h_b Depth of deforming layer

h_{bc} Height of cavity backwall

h_d Thickness of debris-rich ice layer

h_i Ice thickness

h_m Thickness of debris septa taken as equal to ice thickness

h_R Present-day elevation of bedrock above or below the margin of the ice sheet

h_s Sediment thickness in a wedge

h_{sp} Thickness of supraglacial debris cover

h_{sw} Height of submerged ice-cliff wall

h_w Water depth

H Maximum surface elevation of an ice mass

H_o Channel depth at stream mouth

i Hydraulic gradient

i_c Critical hydraulic gradient

I_p^* Ploughing index on a clast in a deforming sediment

J_s Approximation of sediment flux along a channel

k Permeability of sediment

k_m Reciprocal of the Manning roughness parameter

k_o Critical wave number

k_r Hydraulic roughness parameter for conduits

k_{tc} Thermal conductivity

k_w Wave number $= 2\pi/\lambda$

K Hydraulic conductivity

K_i Fracture toughness of ice

K_* Material constant

K_b Constant related to deforming sediment

K_f Constant related to ice shelf ice fabric, temperature and ratio of components of strain rate

K_p Apparent conductivity of a particle array in relation to ice

l Glacier length

L Distance from ice divide to ice margin

L_c Length of main channel

L_h Latent heat of fusion of ice

L_{ic} Length of ice cliff face

L_m Latent heat of melting

L_v Length of valley axis

m Constant of ice sliding related to ice creep

$\dot{m}$ Rate of melting of conduit walls

$\dot{m}_b$ Rate of basal ice melting

m_* Constant related to spectral density of the bed in terms of mean quadratic slope in the direction of the ice bed slip

$\dot{m}_c$ Rate of conduit closure by volume

M Manning roughness coefficient

M Annual mass balance

M_1 Melting rate of ice cliffs

M_2 Mean melting rate of ice cliffs

M_f Momentum flux of a meltwater jet

n Visco-plastic component for ice creep deformation

n_K Number of cavities/passageways across the width of a glacier

n_p Porosity of sediment

P Solid precipitation

P_c Cavity roof perimeter

P_{cs} Channel sinuosity

$P_{cs(tot)}$ Total channel sinuosity

P_o Normal stress distribution across a glacier bed

q_s Proportion of sand-size particles (0.063–0.5 mm diameter) to total sediment weight

q_z Proportion of quartz particles to all other mineral particles in the 0.063–0.125 mm size range

Q Volumetric flux (discharge) of meltwater

Q_{max} Peak discharge from a glacier during a jökulhlaup

Q_o Stream discharge

Q_s Suspended load carried by stream

Q_x Water influx along a section of channel x metres in length

$\dot{r}$ Rate of closure of a conduit

r Radius of a conduit

r_c Radius of curvature of a clast

r_p Particle radius

r_s Grain shape

r_{sr} Isostatic ratio variable

R Universal gas constant

R_e Reynolds number

R_I The Cailleaux–Tricart roundness index

R_n Richardson number

R_u Bed roughness parameter

s Areal fraction of the bed cavity/passageway uncoupled

s_r Bed roughness slope

S Cross-sectional area of channel

S_g Stream gradient

S_j Area of clast imposed upon by ploughing pressure

t_h Time in years for an ice mass to equilibrate

T_m Response time of an ice mass to changes in equilibrium status

T_s Surface water temperature

T_∞ Far-field ambient water temperature

TCL Total length of all braid channels

u Spreading ice velocity

$\bar{u}$ Average ice velocity in ice column

u_+ Sliding velocity of a surging ice mass

u_b Basal sliding velocity of ice

u_{by} Vertical ice flow vector

u_{bz} Transverse ice flow vector

u_c Velocity of a clast

u_{cs} Creep velocity of sediment

$\dot{u}_\varepsilon$ Effective rate of cavity closure

u_i Overall ice mass velocity where $u_i = u_s + u_b$

u_L Terminus ice velocity

u_m Sliding velocity of mobile sediment at its upper interface

u_n Component of ice velocity normal to the bed

u_o Component of u_n due to melting of ice caused by geothermal heat and sliding friction

u_s Surface velocity of ice

u_t Component of ice velocity parallel to the bed interface

u_w Dynamic viscosity of water

U_m Mean centreline velocity of stream at conduit mouth

U_{mo} Initial near-centreline stream velocity

U_w Water velocity

v_p Radius of a particle

$\bar{v}_w$ Average water velocity within a meltwater film at the ice/bed interface

V_m Mean flow velocity of sediment in a channel

V_{max} Total maximum volume of meltwater drained from an ice mass due to a jökulhlaup

V_w Specific discharge along a channel where inflow and outflow are in equilibrium

w Width of ice column

W Width of deforming sediment layer, transverse to maximum ice flow

W_b Width of channel bed

w_m Width of debris septa

W_s Stream power

w_{sp} Width of supraglacial debris cover

z Some vertical thickness of ice

z_{cs} Depth within a deforming sediment

z_s Thickness of deforming sediment

$z_{s(crit)}$ Depth within a deforming sediment at which point basal shear stress = r_s

Z_0 Constant related to sediment flux along a channel

α Ice surface slope

α_b Slope at the basal ice/sediment (bed) interface

α_c Slope of a conduit

α_d Constant dependent upon sediment layer thickness and water pressure

α_{hc} Constant dependant on hardness of clasts, the bed and the geometry of the striator point

α_i Incremental ice surface slope

β Slope of glacier bed

Γ Constant related to ice regelation processes

Γ_{pg} Pressure gradient of meltwater in a channel in the direction of the mean slope

γ Shear strain

γ_f Finite longitudinal strain

γ_g Density of sediment

γ_s Shear strain rate of sediment in plane of shearing

γ_w Specific weight of water

Δ Dilation within a sediment

Δ_h Incremental ice thickness steps

Δ_x Incremental length steps from ice divide to margin

ΔP Difference in pressure in ice adjacent to a conduit and the water pressure within a conduit

ΔT Temperature difference between ice and water

$\Delta\sigma_p$ Stress fluctuation

δ Parameter in Fowler's surging theory

δ_f Loss of water pressure in a conduit due to friction

δ_h Incremental steps of ice thickness

δ_p	Change in local water pressure in a conduit	Ξ	Melting-stability parameter	
δ_s	Length of a straight conduit	ξ	Balance velocity	
δ_u	Incremental lateral spreading of ice due to internal deformation	ξ_j	Controlling obstacle factor	
		ρ_a	Average compressive stress	
δ_z	Incremental vertical motion of ice due to internal deformation	ρ_{ad}	Ambient fluid density	
		ρ_i	Density of ice	
γ_i	Incremental longitudinal strain	$\bar{\rho}_i$	Average density of ice	
$\dot{\varepsilon}_d$	Strain rate for dilatant deforming sediment	ρ_o	Initial discharge density	
$\dot{\varepsilon}_{sh}$	Strain rate within an ice shelf	ρ_R	Rock density	
η	Ice viscosity	ρ_s	Density of sediment	
η_s	Newtonian viscosity	ρ_{sw}	Density of sea water	
η_{sed}	Sediment viscosity	ρ_w	Porewater pressure	
η_w	Viscosity of water	ρ_{wt}	Density of water	
Φ	Angle of internal friction of sediment	σ	Standard deviation of the Gaussian distribution for stream velocity across plume width	
Φ_s	Slope of conduit			
κ	Thermal diffusivity of ice			
κ_b	Bulk density of deforming sediment	σ_b	Ice overburden normal pressure at the base of an ice mass	
κ_s	Fluid density			
λ	Bed wavelength	$\bar{\sigma}_b$	Mean stress level at ice bed	
λ_f	'Frictional parameter'	σ_{cj}	Local effective stress upon a clast of jth size class	
λ_{rs}	Removal rate constant for suspended particles			
		σ_ε	Effective stress	
λ_t	Transition wavelength at ice/bed interface where regelation and ice deformation are equally efficient	σ_i	Normal stress at a point within an ice mass	
		σ_{zs}	Effective stress within deforming sediment	
		τ	Shear stress	
λ_{wr}	Characteristic bed wavelength— 'white roughness'	τ_0	Shear stress of ice along flow line	
		τ_b	Basal shear stress of ice	
λ_1	Principal direction of finite extension of sediment under shear	τ_c	Yield strength of sediment	
		τ_D	Shear stress for a dry bed	
λ_3	Shortest direction of finite extension of sediment under shear	τ_g	Gravitational shear stress in sediment wedge	
		τ_i	Yield strength of ice	
λ_R	Bed roughness term	τ_j	Basal shear stress applied to a clast of jth size class	
μ	Coefficient of rock friction			
μ_{cs}	Coefficient of friction of sediment	τ_s	Shear strength of deforming sediment	

LIST OF SYMBOLS

τ_w Shear stress for a wet bed

τ_{xy} Internal ice shear stress

τ_{xz} Applied shear stress in z-plane

Ψ Drag coefficient applied to a clast at the ice/bed interface

Ψ_c Rate of channel closure due to ice creep

Ψ_e Rate of erosion in evacuating a till channel

Ψ_{el} Rate of erosion by laminar flow evacuation in conduit

Ψ_{et} Rate of erosion by turbulent flow evacuation in conduit

Ω Conduit shape factor

Ω_w Constant between 0 and 1 that depends on λ and the transition wavelength (λ_t)

ϖ Angle made by equipotential lines with the ice surface

Chapter 1

GLACIAL ENVIRONMENTS

J. Menzies

1.1. INTRODUCTION

Glaciers and ice sheets are found in remote and wild terrains far from centres of population. Yet this physical and mental distance should not betray their fundamental significance and importance to all our lives. Modern glacial environments hold an essential key to our knowledge of the present, as well as past and future global environmental conditions. The Huttonian principle of understanding the present to permit a grasp of the past (and of the future) is never better exemplified than in our attempts to comprehend and explain contemporary dynamic processes in modern glacial environments. At first sight glacial environments are chaotic, complex and geologically ephemeral, yet beneath this apparent chaos exists a process-driven pattern. Perhaps no other sedimentological environment exhibits such rapid, dynamic and spatially-variable changes of processes which when viewed at the small scale appear chaotic yet at other larger scales begin to reveal a precision in the spatial and temporal organisation and execution of individual processes. The underlying theme of this book is to bring these environmental patterns into better focus.

Glaciers and ice sheets have today, and in the past, had an enormous influence, both beneficial and detrimental, upon all aspects of earth systems. At present, understanding of many aspects of glaciers remains poor. For example, the complexities of ice dynamics and the relevant variables and influences they have on ice mass balance, and on the specific reasons for ice front and meltwater discharge fluctuations, remain imprecisely understood. The processes of ice basal movement and the intricate relationship between ice motion and subglacial hydrology continues to be poorly understood, as do thermal and sedimentological states within subglacial environments. Study of processes of erosion, transport and deposition in modern glacial environments furnish evidence as to how these processes operate. Modern glaciers can also be viewed as active analogues of past glaciers and ice sheets. Modelling of modern ice masses and mass balance studies advance our knowledge in explaining past global ice sheet development and expansion. The diverse subenvironments of modern glaciers, both on land and subaquatically, provide an active field laboratory for studying present glacial sedimentological processes that can be employed to understand and interpret glacial sediments of the geological past. However, in utilising modern ice masses as analogues of the past, care must be exercised since past ice masses may have been significantly different in many critical aspects. For example, ice sheets in the Quaternary would appear to have carried substantially greater debris loads than present day ice sheets and had significantly different subglacial marginal environments.

1.2. IMPACT OF ICE MASSES ON GLOBAL HABITATS AND EARTH SYSTEMS

The effect of modern glaciers, particularly on global habitats and earth systems, can be viewed at two levels of impact: first their influence upon humans and habitats within their immediate locality and second, on their much more pervasive influence on all global habitats due to the effect of modern ice masses on global climate and sea-level. The effect of ice masses in the immediate proximity to humans is well documented (Ladurie, 1971; Lamb, 1977, 1985; Tufnell, 1984; Grove, 1988; Bell and Walker, 1992; Hambrey and Alean, 1992) in terms, for example, of meltwater outbursts and rapid ice advances resulting in the loss of pasture lands, property and in some cases human fatality. These detrimental aspects, of course, need to be balanced with the beneficial resources of water for hydro-electric projects, irrigation and fresh domestic water supplies. Less obvious, but actively researched today, is the more insidious and pervasive influence of present day ice masses on global climate and oceanic currents. As predictions of global warming increase, so knowledge of modern glacial conditions need to be amplified if we are to cope with and predict sea-level rise in the coming century when this knowledge will become acutely significant.

Related to potential sea-level rise due to ice sheet melting in Greenland and Antarctica is the question of future ice sheet stability (Hughes, 1987c). If these ice sheets melt at an increased rate vast plumes of cold fresh water may be injected into the polar oceans affecting future oceanic habitats, currents, surface ocean water temperatures (e.g. El Niño–Southern Oscillation events) and consequently global weather patterns (Sarmiento, 1993). However, whether ice sheet accelerated melting may or may not be occurring due to global warming remains the subject of considerable controversy (cf. Thomas *et al.*, 1979; CO2RC, 1983; Meier *et al.*, 1985; Hughes, 1987; Meier, 1987; Kasting, 1989, 1993).

A further aspect of modern glacial environments, as noted above, is in providing analogues to past glacial conditions as a key to understanding glacial sedimentation processes. By studying present day glacial environments and sedimentological processes,

considerable knowledge can be gleaned as to how sediments of past glacial events have been derived, transported and finally deposited both on land and in water. The impact of past glacial environments influences the earth's habitats: the soils, vegetation and ultimately animal life. Pleistocene glacial sediments cover at least 30% of the Earth's continental land masses and an even greater area must be included when Pre-Pleistocene sediments ranging over vast areas of India, Australia, Africa and South America are considered (Harland and Hambrey, 1981a; Eyles, 1993). These sediments affect almost every aspect of human life from establishing foundations and footings for buildings, roads and runways, to the nutrient content of soils, to the nature of groundwater supplies and to the potential soil-routes for contaminant waste disposal and the location of landfill sites. These few examples illustrate the vital need to understand glacial processes ongoing in modern glacial environments (cf. De Mulder and Hageman, 1989). Until relatively recently this figure of 30% for Pleistocene sediments was accepted, yet today perhaps approximately 60% would be a more accurate figure if glaciomarine sediments are included. These thick sediments lie over the ocean floors covering enormous parts of the northern and southern areas of the Atlantic and Pacific oceans. The impact of glaciomarine sediments on land-based habitats is certainly limited, affecting only fisheries to a little-known degree, but if future utilisation of oceanic basins occurs the influence of these vast areas of glacial sediments may become increasingly meaningful.

1.3. RESEARCH IN MODERN GLACIAL ENVIRONMENTS

Research in a wide range of modern glacial environments has only become significant since the 1950s. Prior to then, work in often remote areas was restricted to occasional explorers and adventurers and to sporadic scientific studies. In Europe, especially in the Alps, inaccessibility was not a problem but scientific interest was fostered by only a few individuals (Nilsson, 1983). In the nineteenth century, the work of the Swiss from the time of Charpentier and Agassiz is well known as are the travels and observations of James Forbes (Forbes, 1842, 1843; Cunningham, 1990). Likewise, considerable work on both modern and past glacial environments began in Germany and Scandinavia (Nilsson, 1983). The tragic contretemps between Forbes and Agassiz, concerning a misunderstanding as to who had first publishing rights on certain scientific observations, may have engendered what was an already increasing interest in glaciers and ice mechanics in the scientific community by the mid-nineteenth century. By the early 1900s the observations of Nansen in Greenland, of Amudsen and Sverdrup in other parts of the Arctic and many others increased the scientific interest in modern glaciers and ice sheets; and without doubt the expeditions of Shackelton and the doomed trek of Scott's desperate expedition in Antarctica only heightened the curiosity and wonderment of the general public to these environments. Other equally famous expeditions occurred in the Arctic to Nova Zemlya, Spitsbergen and the Canadian Arctic islands. In continental North America the famous expeditions, for example, of Harriman in the Canadian Rockies and the area of Glacier Bay in Alaska brought forth the beginnings of scientific writings on modern glacier environments (Gilbert, 1910). Following this period of intense interest and the sending off of expeditions, a phase of scientific development began that is still very much in existence today. Many influences beyond natural curiosity have engendered and encouraged this research work. In particular, the conflicts of the Second World War and the potential invasion of Alaska and continental North America by Japanese forces brought a renewed focus on polar areas and research problems. The Cold War that followed spurred on research in Alaska, and the Canadian and Soviet north to a degree never before (or since) witnessed. Perhaps in the interests of petty nationalism and territorial rights, Antarctica by the early 1950s was sectorised into areas where different nations strove to establish research bases, under United Nations auspices, to further scientific research. Whatever the motives may or may not have been, the establishment of so many research bases in Antarctica has led to a vast outpouring of scientific literature and an expanding

appreciation of modern glacial environments (Boulton, 1987b; Clarke, 1987a; Robin and Swithinbank, 1987).

The sparsity of texts on modern glacial environments attest to the past limited interest. Only recently have texts such as those by Gurnell and Clark (1987) and Maizels and Caseldine (1991), directly and exclusively addressed these environments with any scientific vigour. Textbooks on glacial geo(morpho)logy often utilise modern glacial environment research findings but rarely focus exclusively on them. In direct contrast, research by individuals has and continues to appear in many journals devoted to modern glacial environments such as Jökull and the Antarctic Journal of the United States; while most other geological/geographical journals contain many papers pertaining to this environment. It is one of the intentions of this book to focus on this, at times, disparate and scattered literature and to provide a synthesis and synopsis of fundamental research on modern glacial environments.

It is remarkable to learn that in the late 1800s James Geikie and his brother, Archibald, in the company of Benjamin Peach ventured to Spitsbergen to examine and appreciate modern glacial environments and processes. Their expedition was aimed solely to aid their efforts in explaining the Quaternary glacial sediment sequences and events they had mapped in Scotland. Their investigations remain beacons of accurate scientific observations and interpretations (Geikie, 1863; Geikie, 1894; Peach and Horne, 1930; P.A. Peach, *pers. commun.*). Likewise, Gilbert's work on glacial erosion, derived and enhanced by his observations in Glacier Bay, Alaska, remains a supreme illustration of early scientific glacial research (Gilbert, 1906). The scientific merit of research, as exemplified by these early researchers, in modern glacier environments therefore cannot be underestimated in comprehending and interpreting past glacial environments.

1.4. RESEARCH ISSUES IN GLACIAL STUDIES

When considering the vast array of research on glacier environments some general trends can be observed that have had an enormous impact on various facets of glacier studies. Over the past several decades the techniques of studying glaciers have vastly improved, for example, in the accuracy of measurements of ice movement and mass balance due to the use of Landsat and other satellite imaging methods. From the first ice cores extracted from near Byrd Station, Antarctica (Gow, 1963) to the present recently announced finds of the Greenland Ice-Core Project (GRIP) (cf. Dansgaard *et al.*, 1993; GRIP, 1993), the accuracy, detail and outpouring of data on ice cores, their geochemistry, geochronology and climatic implications have been astounding. The precision of techniques in geochemical analyses of ice cores has reached a level whereby details of atmospheric chemistry can be now be revealed that were hitherto unknown (cf. Souchez and Lorrain, 1991). As understanding of modern ice sheets has grown, the intricate relationships of feedbacks and counter-feedback between the atmosphere, oceans and ice mass modifications have been translated into first-order ice sheet models (Chapter 3). Recently, sophisticated computer generated ice sheet models have increased our understanding of ice sheet growth and potential stability/instability permitting future predictions of ice sheet behaviour (e.g. Fisher *et al.*, 1985; Lingle and Clark, 1985; Hindmarsh *et al.*, 1989; Alley, 1992) (Chapter 3).

In tandem with these findings has been the increasing knowledge concerning Pre-Pleistocene glaciations and the overall causative mechanisms that may lead to global glaciation (Chapter 2). In pondering how global glaciations began and ended, our knowledge of past global climates, carbon dioxide levels, biomass productivity and other habitat environmental indicators has dramatically improved (cf. Kasting, 1984), thereby permitting predictions of the possible effects of global greenhouse warming to be better understood if not anticipated (Meier *et al.*, 1985; Raynaud *et al.*, 1993).

What has often been ignored by glacial geo(morpho)logists in focusing on landforms and sediments is an attendant and critical understanding of ice physics. A tacit recognition of this subdiscipline has always existed, but attempts to directly merge the findings of glaciological research into glacial geo(morpho)logy has been latently

resisted. The result has been the divergence at times between glaciology and glacial geo(morpho)logy. Today this past separation is diminishing and the value of combining both disciplines in a concerted effort to understand glaciers and their environments is gradually succeeding (cf. Drewry, 1986). The emergence of glaciology, especially after the 1940s as witnessed, for example, in the pages of the Journal of Glaciology and Zeitschrift für Gletscherkunde und Glazialgeologie, has resulted in an immense effort by many scientists to comprehend all aspects of ice mechanics, ice movement and hydrology (Chapters 4, 5 and 6). It is no longer possible, for example, to try to understand the deposition of subglacial sediments without first considering the thermal and mechanical parameters of subglacial environments (Menzies, 1989a). What has emerged from the chaos of ideas and mis-developed theories concerning glacial sedimentological processes is that glacial environments are much more complex than perhaps previously considered. Yet from a sound glaciological base, new ideas and with them paradigms (Boulton, 1986a; Menzies, 1989b) may emerge that will allow improved insights into glacial environments, their dynamics, processes and sediments.

Just as glaciology has undergone profound changes and evolution, the wider field of glacial geo(morpho)logy has undergone maturation. When the Geikie's, Esmark, Agassiz, Bernhardi, Lyell, Torell and Buckland in Europe and Leverett, Chamberlin, Gilbert and Shaler in North America, first began studies of glaciated terrains they brought with them a profound geological appreciation of sedimentary stratigraphy as well as a spatial awareness of topography (physiography). Descriptions of sections in the field were often so detailed and accurate that even today when visiting these same sites, the accuracy of the precise details commented upon by these early 'glacialists' is evident. Somewhere in the early twentieth century, the science became diverted into a dominance of form over sediment. Landforms and physiographic studies became de rigueur and increasingly detailed studies of sediments and depositional processes were often relegated to a subordinate place or in many studies virtually ignored. From this developed what might be termed a 'spatio-morphological school' of thought that placed emphasis on spatial interrelationships and supposed geographical locational cause and effect. Likewise, an intense period of glacial chronological studies evolved. Glacial sediments and interrelationships were only lightly considered in pursuit of the establishment of temporal–spatial relationships between landforms and landform assemblages in order to develop scenarios of glacier fluctuations at local, regional and continental scales (Nilsson, 1983). This paradigm achieved enormous successes in explaining and developing global and regional glacial chronologies. However, in detailed studies of local glacial variations this paradigm failed to provide answers if only for want of sound sedimentological and glaciological inputs.

Although many workers held to a strong sedimentological methodology, only since the 1970s has a general swing toward a 'glacio-sedimentological school' of thought begun to evolve. Emphasis was placed on all aspects of glacial processes and glacial ice dynamics in an attempt to establish glacial process-patterns within a time frame constrained not by potentially perilous spatial–topographical relationships but a broader, more complex and stochastic, appreciation of sediment processes, ice mechanics and transient environmental conditions. At the same time a concomitant shift away from a unique appreciation of landforms has emerged that places landforms as part of suites of associated forms developing in similar environmental conditions (cf. Aario, 1977). The land-system approach advocated by Boulton and Eyles (1979) is illustrative of an early attempt to place landforms and related sediments within a contextual environment rather than as isolated separate forms that demand equally isolated, and stultifying, unique explanations of origin. Research into the origin of drumlins is a particularly illustrative case of this maturation of ideas (cf. Menzies, 1984, 1987). What has developed, and is illustrated in the various chapters of this book, is the emergence of the twinning of glaciological research with a sedimentological and geomorphological recognition and perception of the conditions found within glacial environments.

1.5. SYNOPSIS OF CHAPTERS

Each chapter in this book tries to introduce the reader to new ideas and concepts often not collectively brought together before or synthesised within the broad field of glacial studies. Literature is reviewed and placed within the context of present knowledge always with the intent or aim of developing a greater understanding of glacial environments. In many chapters new data are introduced for the first time and novel ways of considering older ideas and concepts proposed.

In Chapter 2, Calkin and Young bring together the newest ideas on the causes of global glaciation as interpreted from the recent GRIP ice core data and from the geological record. The chronology of global glacial events are also synthesised for both the Pleistocene and Pre-Pleistocene glaciations. Evidence supportive of the Milankovitch theory of orbital forcing continues to be viewed as the most likely theory for global glacial initiation but the recent results from Devils' Hole in Nevada are discussed and the significance of these results contrasted with the marine and isotope records from both deep sea drilling and ice core projects. Gaps in the geological record indicate that our understanding of the reasons for warm and cold phases during glacials and interglacials remains less well understood than previously imagined.

The basic thesis behind Chapter 3, as stated by Hughes, is to make ice sheet modelling as simple and clear as possible for the non-mathematical reader. This chapter displays the two schools of thought that have developed in ice sheet modelling and the ongoing developmental stage where aspects of both schools can be combined to satisfy both geological and glaciological parameters and constraints. It is also readily apparent that in successful modelling, the combined disciplines of ice physics and glacial geo(morpho)logy are prerequisites.

Chapters 4 and 5 introduce glacial physics from a glacial geo(morpho)logist's perception. I have attempted to distil glaciology to its relevance in relation to glacial processes and sediments. To many practicing glacial geo(morpho)logists the physics of ice remains an obscure, almost forbidden, territory that can only be entered hurriedly with considerable trepidation. It is the intention of these chapters to dispel this notion and to present ice physics in as palatable a form as possible without losing its scientific rigour. As in Chapter 3, these chapters combine facets of both glaciology and geo(morpho)logy, relating various concepts such as the deformable bed within the wider context of glacial environments.

In the last few years it has become increasingly apparent that one of the critical parameters in understanding many glacial processes is the hydrology of glacial environments. Chapter 6 explores the many, still poorly understood, aspects of glacial hydrology with an emphasis on subglacial environments. Many of the remaining problems in glacial physics and subglacial sedimentology would appear to hinge on the hydraulics of subglacial bed conditions. For example, the phenomena of glacier surging and associated jökulhlaups would appear to be allied to changes in glacier bed hydraulics, possibly the result of subglacial drainage 'short-circuits' under certain topographic and ice dynamic constraints. Likewise, subglacial deposition and erosional processes are intrinsically linked to subglacial hydraulics in more critical relationships than have been assumed in the past.

Considering the vast outpouring of literature on glacial processes, especially in the past several decades, glacial erosion has often been neglected. Iverson, in Chapter 7, introduces new developments and strives to rectify this virtual research vacuum. Before discussions of sediment transport and deposition, or bedform evolution can be considered, it is imperative that the fundamental processes of erosion be understood. Few researchers have ventured to formulate sound physical models of glacial erosion. This chapter synthesises this novel research and places emphasis on the mechanics of erosion processes. Iverson demonstrates that a new approach to erosional processes can be established by a detailed understanding of the wear processes occurring at the many interfaces that abound in glacial environments.

Both Kirkbride and Whiteman in Chapters 8 and 9 review modern ideas and concepts on transport and deposition in glacial environments. Kirkbride concentrates on evidence from modern glaciers in

New Zealand and Europe emphasising the several septa of transport pathways that exist at times independent of one another while often uniting. The separation of these septa on the basis of provenance and particle size, for example, remains a major research concern. Whiteman introduces the wide range of terrestrial depositional environments and the process that occur within them. In the mid-1970s a plateau seemed to be reached where our understanding of depositional mechanics became well established. However, since then the field of glacial deposition has entered into a new stage of flux with many more research questions arising than being answered. A prime example has been the discovery of deformable bed conditions beneath Ice Steam B in Antarctica (Alley *et al.*, 1986) and many other valley glaciers (Chapter 5). This discovery has resulted in renewed interest in understanding the depositional processes occurring beneath glaciers and ice sheets and, in turn, will necessitate a re-examination of sediments and stratigraphies throughout the glaciated parts of the world.

In Chapter 10, van der Wateren utilises existing knowledge in structural geology in concert with data gleaned from modern glacier margins to demonstrate the mechanics of glacio-tectonism. By considering, for example, the development of accretionary wedge mechanics, the evolution and subsequent sedimentary sequences developed in push moraines can be understood. This chapter focuses on the need to fully understand the deformational mechanics at all levels of scale within glacial environments. Deformational processes occur prior to, during or following on from deposition. It is critical that our understanding of these processes, their chronologies and stress conditions be fully comprehended if the process-patterns of each sedimentological environment within the glacial system be fully understood (cf. Preston and Jones, 1987; Maltman, 1994).

Lawson, in Chapter 11, brings a wealth of new data from modern glacier margins to introduce the topics of subglacial and supraglacial sedimentary environments. As major sources of information, modern glaciers can act as important, but cautionary, analogues for past glacial conditions in terms of rates and mechanisms of processes. Nowhere else can the vicissitudes of climate and related glacier mass balance be better examined than at the margins of these glaciers where variations in the rate of sediment output, meltwater geochemistry, for example, allude to changing conditions within the whole ice mass.

Anyone acquainted with glaciers has trudged across what, at first, seems the barren stretch of the proglacial zone; yet this area is particularly germane to and testimony of the many complex processes that have occurred supra- and subglacially to a specific glacier in recent times (Chapter 12). In this area exposed subglacial and dumped supraglacial sediments in close association with proglacial sediments suffer the first stages of diagenesis, mass movement and other subaerial processes. It is in this zone that the character and structure of these sediments can be altered or affected to the degree observed in past glacial sediments. The proglacial environment is an especially dynamic geomorphic zone that is also a reflection of ice activity, glacial hydraulics and subglacial conditions. Maizels reviews the sedimentology of this glacial subenvironment, discussing in detail some of the processes occurring under braided stream conditions and across proglacial sandur following jökulhlaups.

Modern glaciolacustrine environments (Chapter 13) remain an area of intense interest especially since our understanding of past glaciolacustrine sediments and sedimentary processes are so uneven. Ashley draws on a wealth of data to review the complexity of glacier-fed lake processes and dynamics. The limnological factors that control sediment dispersal and deposition with these lakes is discussed as are the distinctive sedimentary records that are produced in allied subenvironments in ice-contact and distal lakes.

In the past two decades a plethora of data on modern glaciomarine environments has accrued (Chapter 14) especially from work in the oceans surrounding Antarctica, the Arctic Ocean, the Barents, Beaufort and Weddell Seas, the North Atlantic and Pacific Oceans and the Gulf of Alaska. The output of data has been so great that texts such as those by Molnia (1983), Dowdeswell and Scourse (1991) and Anderson and Ashley (1992) can hardly

keep pace with new findings. The chapter by Domack and Powell synthesises this modern data within the context of previous findings. The areal extent of past glaciomarine sedimentation has been unquestionably underestimated, thus only with a clearer understanding of present glaciomarine environments can recognition and knowledge of past glacial sediments be achieved.

One of the novel techniques of the past 25 years has been the advent of scanning electron microscopy (SEM) in studies of clastic particles, especially those from glacial environments. Mahaney, in Chapter 15, offers startling examples of the use of this tool in analysing particles from modern, recent and past glacial environments. SEM tends to reveal more particle surface detail as the magnification increases. Hitherto hidden, unexplained structures and morphological architecture of particle surfaces are discovered that have generated intense debates and speculation as to their meaning and significance. A glimpse of the effects of erosional processes in all glacial environments can be garnered from an understanding of the fracture processes reflected in SEM photoimages of individual particles. Only in modern glacial environments can guaranteed, uniquely-derived, particles from specific glacial subenvironments be obtained for comparison with particles that have undergone multi-environment overprinting as is found in ancient glacial sediments.

As the complex details of the dynamics, related processes and resultant forms in glacial environments increases, many new questions arise to be added to existing questions and problems. Studies in modern glacial environments are approaching a new and exciting level of research endeavour. Study of these fascinating, but often perceived exotic environments, are no longer in any sense academically whimsical but are, of necessity, fundamental to present and future human and global habitats.

When any topic shifts from a peripheral to a more central focus in human concerns, great advances and discoveries are likely to accrue — it can be anticipated that this will be the case with glacial environments.

Chapter 2

GLOBAL GLACIAL CHRONOLOGIES AND CAUSES OF GLACIATION

P. E. Calkin
(With Contributions by G. M. Young)

2.1. INTRODUCTION

It was only 150 years ago that Louis Agassiz announced his theory of a 'great ice period' at a meeting of the Helvetic Society (Chorley *et al.*, 1964). In his publication, Etudes sur les Glaciers, Agassiz (1840) suggested that a vast sheet of ice had once extended from the North Pole to the Alps. He also presented a model involving a succession of ice ages driven by long-term trends of radiational cooling (Imbrie, 1985); however, evidence was lacking for more than one episode of glacial period. In its original form, Agassiz's (1840) whole theory had several serious flaws, including some catastrophic concepts (Chorley *et al.*, 1964), but his ideas were a remarkably significant and far-sighted beginning.

Field observations and indirect probing via deep sea and ice sheet coring have now confirmed a long and complicated record of glaciations, or intervals of global cooling, and major ice sheet advances which extend over at least the last 2.4 million years (Ma) of Late Cenozoic time. Furthermore, geological studies on every continent have helped to document a record of at least four major intervals of geologic time preceding this last 'ice age' when fluctuating ice

sheets blanketed a major portion of the Earth's crust. The earliest of these pre-Late Cenozoic ice ages (Wright and Moseley, 1975) began as early as 3200 million years ago (Ma) and each spanned tens to hundreds of millions of years. They were separated by intervals when little or no terrestrial ice occurred and temperatures were often higher than those of today (Crowell and Frakes, 1970).

This chapter will provide a brief review of the glacial stratigraphic record with some consideration of causal factors and mechanisms responsible for driving climatic changes and the spectrum of glacial advances and retreats.

Agassiz was not the first to discern the evidence of former extensive continental glaciations, the idea of continental glaciation had already been articulated by others before 1837 (Nilsson, 1983). Agassiz had been convinced of the idea by observations at the Swiss glaciers of Diablerets and Chamonix while in the company of Jean de Charpentier, Director of Mines for the Swiss canton of Valais (Chorley *et al.*, 1964). Charpentier (1841) had read a paper in 1834 strongly supporting a concept of a former much-enlarged Alpine glaciation. Similar innovative, but unaccepted, arguments had been presented even earlier, in 1821, to the Helvetic Society by the Swiss engineer, Venetz-Sitten. Both Charpentier (1841) and Venetz had acquired the idea of extensive glacier advances from the mountaineer Jean-Pierre Perrudin of Val de Bagnes. Venetz had also supported the idea presented by Hutton (1795) who suggested that erratic boulders were transported from the Alps to the Jura Mountains by a great mass of ice. Elsewhere, Esmarck (1824) proposed comparable erratic transport by glaciers in Norway as had Bernhardi (1832) for Scandinavia and north Germany (Nilsson, 1983).

These ideas were disseminated rapidly by Buckland and Lyell in Britain and by Edward Hitchcock (1841) and eventually Agassiz in North America. Lyell had already modified his own views toward those of Agassiz by 1840, despite the fact that only a few years earlier, in his 1830 edition of Principles of Geology, he had presented the 'actualistic principle' of uniformitarianism as espoused by Hutton much earlier. However, by 1835 Lyell realised that many of the surface sediments in Britain contained marine shells and a novel theory to explain these sediments was required. Lyell's 'Drift Theory' which was intended to combat the prevailing ideas of catastrophic flooding linked these 'diluvium' deposits to transport by icebergs (Chorley *et al.*, 1964). The term 'drift' deposits persists today especially where sediments of imprecise glacial origin are initially mapped.

The early glacialists accepted the concept of a single integrated glacial event forming all glacial deposits. However, this mono-glacial hypothesis was soon dispelled as two glaciations were inferred from distinct drift layers in the Vosges Mountains by Collomb (1847), in Wales by Ramsay (1852) and in Scotland by Chambers (1853).

Interdrift strata were referred to as interglacial and the interval of time between two successive glaciations (glacials or glacial stages) were interglaciations (interglacials). The glacial stages were subsequently subdivided into stades (stadials), marking times of climatic deterioration and secondary glacial readvance, and interstades (interstadials), when climate was incompatible with advance and ice marginal stands or minor retreats were recorded by the ice margins. Interglacial times

have been variously defined as times when the climate was essentially as warm as present or, in recent times, they imply intervals of world-wide reduction in ice cover to near present levels as measured by sea level or marine oxygen-isotope data (Section 2.4). Summaries of the history of the glacial theory are given by North (1943), Chorley *et al.* (1964) and Carrozzi (1967); and integrated with Quaternary geology by Flint (1971), Bowen (1978) and Nilsson (1983) among others.

Long before the extent of this last ice age had been delimited, the search for ice ages earlier in the geologic record had begun. In Britain, Ramsay (1855) suggested that the presence of angular, polished and striated rock fragments of boulders in Permian breccias of Shropshire and Worcestershire indicated the existence of glacier ice during the Permian. Although these were discounted as glacigenic (glacial-formed), this find stimulated the search for evidence of a Permian glaciation elsewhere and led to the verified report of the glaciogenic 'Talchir' Boulder Beds in India (Blanford *et al.*, 1859; Coleman, 1926; Hambrey and Harland, 1981). Agassiz's work was barely published when Adhemar (1842), a French mathematician, suggested that the primary instigator of ice ages might be the Earth's orbital path around the sun. Adhemar's astronomical ideas were based on the processional wobble of the Earth's axis and that the relative length of the warm and cold seasons might cause an ice age (Imbrie and Imbrie, 1979). Croll (1864) reinvestigated these ideas and added the effect of orbital eccentricity or change in Earth's orbital shape to build on Adhemar's work. Croll suggested that long cold winters and short hot summers of high latitudes would favour glaciation and, in addition, the precession effect caused ice ages to occur alternately in northern and southern hemispheres (Lamb, 1977).

Milankovitch (1930), a Yugoslavian mathematician, suggested that the climatic effect of change in radiation due to precession, eccentricity and the axial tilt (obliquity) would be sufficient to cause the glaciations and interglaciations that comprise ice ages. The basis of the astronomical theory of paleoclimates and, in particular, Milankovitch's (1941) theory of variable solar insolation for climatic change has been intensely studied and has, since the late 1970s, been considered the major basis for pacing of Late Cenozoic glacials and interglacials (Berger, 1978, 1981b; Imbrie and Imbrie, 1979; Berger *et al.*, 1984; Ruddiman and Wright, 1987; Ludwig *et al.*, 1992; Winograd *et al.*, 1992).

2.2. PRE-CENOZOIC GLACIATION

As noted above, with the realisation that the glaciated regions of the world are but remnants of a formerly much more extensive ice cover, many ancient glaciogenic sedimentary successions have been recognised (Fig. 2.1) (Hambrey and Harland, 1981). Study of ancient glaciations has been spurred primarily by scientific curiosity and, today, concerns about possible anthropogenic modification of Earth's climate have placed such studies in a different light. There has been a tendency for environmental concerns to be directed almost exclusively at the present and the future (van Andel, 1989), yet detection and interpretation of climatic change can

only succeed if carried out against a background of previous records, both short- and long-term (Young, 1991). The only long-term record of climatic change is the geologic record. Glaciations represent significant perturbations in the Earth's climatic regime. The evidence of such glaciations is commonly preserved and relatively easy to recognize in the geologic record.

2.2.1. Astronomical Background

There is a widely held belief (Sagan and Mullen, 1972; Gough, 1981; Kasting, 1987, 1989; Kasting and Toon, 1989; Gilliland, 1989) that stars have a history involving a gradual increase in luminosity followed by a decrease. According to this theory, the radiative power of the sun, during the early part of Earth's history, was probably about 70% of its present value. According to Kasting *et al.* (1984), the average temperature on Earth in Early Archean times (~3–4 billion years ago (Ga)) would have been about 30°C

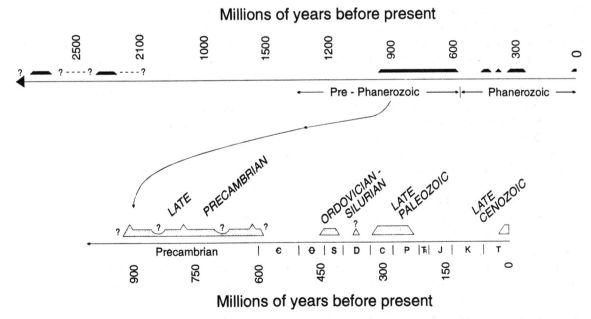

FIG. 2.1. Distribution in time of ice ages plotted linearly, at two scales. ε, Cambrian; θ, Ordovician; S, Silurian; D, Devonian; C, Carboniferous; P, Permian; TSR, Triassic; J, Jurassic; K, Cretaceous; T, Tertiary (from Crowell, 1982; reprinted with permission from *Climate in Earth History*. Copyright 1982 by the National Academy of Sciences. Courtesy of the National Academy Press, Washington D.C.).

lower than at present, given the present day concentrations of atmospheric greenhouse gases. The geological record is strongly at odds with such an interpretation, for the oldest rocks on Earth (~4 Ga) contain clear evidence of the presence of liquid water (not frozen). As discussed below, there is also a near-complete absence of evidence of glaciation in the stratigraphic record for the ancient (Archean) part of geologic history. These contradictions led to what Kasting (1987) called the "faint young Sun paradox". A solution to the problem is that the composition of Earth's ancient atmosphere differed significantly from that of the present (Sagan and Mullen, 1972). This hypothesis invoked an enhanced greenhouse effect due to the presence of ammonia. Subsequently, Kasting (1984) showed that ammonia is an unlikely candidate for a greenhouse gas due to its susceptibility to photochemical dissociation. It was then proposed that the early atmosphere was rich in CO_2 (Hart, 1978). Carbon dioxide is a small but climatically important component of today's atmosphere. There is, however, a vast reservoir of CO_2 (equivalent to about 60 bar) currently trapped in carbonate rocks at or near the surface of the Earth's crust. Thus, in spite of the "faint young Sun", geological evidence suggests that the climate on the early Earth was warm enough to permit flowing water and there is a general absence in Archean rocks (up to about 2.5 Ga) of glaciogenic sedimentary deposits.

2.2.2. Possible Relevance of Plate Tectonics to Ancient Glaciations

2.2.2.1. The Paleomagnetic record

Since the development of plate tectonic theory in the early 1960s there have been many attempts to apply these theories to ancient glacial deposits. Most have involved attempts to correlate glaciation on a particular continent with periods when the continent was carried into high paleolatitudes. The analogy is simple. Glaciated areas, at present, are mostly in high latitudes. Therefore, it seems reasonable that ancient glaciations occurred in a similar paleogeographic setting. This line of reasoning has been successfully applied to glacial deposits of the combined southern

continents (Gondwanaland) during the Paleozoic. According to Crowell (1978) and Caputo and Crowell (1985), successive passage of different parts of Gondwanaland across the south polar region led to glaciations ranging from the Devonian to Permo-Carboniferous times.

Attempts to explain the very extensive Late Proterozoic glacial deposits by a similar mechanism have been frustrated by conflicting results from paleomagnetic studies. For example, early work by Tarling (1974) on Late Proterozoic glacial deposits in Scotland suggested deposition in relatively low paleolatitudes. Stupavsky et al. (1982) restudied the same rocks and suggested that the low paleolatitudes obtained by earlier workers were the result of subsequent metamorphic remagnetization. On the other hand, other evidence has been provided favouring widespread glaciation in low paleolatitudes during the Late Proterozoic (900 to 600 Ma) (Embleton and Williams, 1986; Chumakov and Elston, 1989). Attempts to obtain meaningful paleolatitudes for ancient glacial deposits become more difficult with age. Thus, the most reliable data are from Phanerozoic rocks, where there is a common relationship between glaciation and high paleolatitudes. In the Late Proterozoic, many paleomagnetic results remain equivocal, but in recent years the idea of low latitude glaciation appears to be gaining ground. In the Early Proterozoic of North America, paleomagnetic evidence suggests that glaciation took place in middle latitudes (Morris, 1977).

2.2.2.2. The Supercontinental cycle

It has been hypothesized that the Phanerozoic history of North America was characterised by a series of relative rises and falls in sea level leading to periods of flooding and periods of withdrawal of the sea (Sloss, 1963). Based on this concept and ideas published by Anderson (1982), Fischer (1984) proposed that Phanerozoic climatic history was largely controlled by a 'supercycle', involving flooding and exposure of the continents to produce what was termed alternating 'greenhouse' and 'icehouse' conditions, on a scale of about 400 Ma. Worsley et al. (1984), Nance et al. (1988), Worsley and Nance (1989) and Young (1991) extended the concept of the supercycle back into the Precambrian, where it had earlier been espoused by Sutton (1963) in his chelogenic cycle concept. The supercycle involves the periodic amalgamation of the continental crust into a supercontinent and subsequent fragmentation. During periods of 'super-continentality' the amalgamated mass of continental lithosphere has a blanketing effect on the release of thermal energy from the Earth's interior (Nance *et al.*, 1988; Worsley and Nance, 1989), resulting in its elevation (Anderson, 1982) and consequent relative lowering of sea level. Subaerial exposure of such large areas of continental crust leads to enhanced weathering which, in turn, causes drawdown of large amounts of atmospheric CO_2. Up to 80% of the present withdrawal of CO_2 from the atmosphere is due to this mechanism (Houghton and Woodwell, 1989). Thus, there is a possible connection between periods of supercontinentality (emergence of the continental crust) and reduction of atmospheric CO_2. The reduced greenhouse effect could result in significant cooling, leading to the onset of a glacial period (Section 2.9).

2.2.3. The Archean Record

One of the most striking aspects of the geological record is the near complete absence of evidence of glaciation during the first half of geologic history (Fig. 2.1). The Witwatersrand Succession of South Africa contains some possible evidence of glaciation in Archean times (Harland, 1981). Page (1981) reported the occurrence of possible dropstones in Archean sedimentary rocks associated with the Stillwater complex in Montana. Most older reports of Archean glacigenic rocks have been shown to be spurious so that the record of Archean glaciation is very sparse. One possible interpretation is that, in spite of the inferred low solar luminosity, large amounts of CO_2 in the ancient atmosphere helped to maintain relatively high surface temperatures.

2.2.4. The Early Proterozoic Ice Age

The earliest evidence of glaciation is recorded by terrestrial and marine tillites and proglacial deposits.

The oldest substantiated evidence of widespread glaciation is found in rocks of Early Proterozoic age (Harland, 1981). Glacigenic deposits are reported from North America, Finland, South Africa, Australia and India. In South Africa these sediments are represented by units of Griquatown West and Transvaal Basins, suggesting the occurrence of a vast continental ice sheet, possibly in a high polar position (Visser, 1981).

Early Proterozoic deposits of North America include the Huronian Supergroup of Ontario (Coleman, 1926) and a near-identical succession in southeastern Wyoming, the Snowy Pass Supergroup (Houston *et al.*, 1981). Less extensive or less well-documented successions include those of the Hurwitz Group on the west side of Hudson Bay (Bell, 1970; Young, 1973; Young and McLennan, 1981), the Chibougamau Formation of northern Quebec (Long, 1981), tillites in Michigan's Upper Peninsula possibly indicative of later mountain glaciation and various locally developed diamictite-rich successions around the west end of Lake Superior. Finally, continental glaciation is reported in a narrow, 90 km long band of unmetamorphosed rock in central India.

In all of these areas it has been suggested that the glacial sediments were preserved by virtue of an episode of continental rifting (Zolnai *et al.*, 1984; Young and Nesbitt, 1985). In areas where a thick stratigraphic succession is preserved there is evidence of several glacial episodes (Fig. 2.2) separated by rocks containing both geological and chemical evidence of warm climate and relatively intense weathering (Chandler *et al.*, 1969; Young, 1973; Nesbitt and Young, 1982).

The time framework of these climatic fluctuations is not known, but deposition of the entire Huronian Supergroup is constrained between about 2.5 and 2.2 Ga. If a significant part of this time interval were involved in Huronian deposition, then the timescale of the inferred climatic oscillations would be much greater than those attributed to the Milankovitch Effect.

The alternation of cold and warm climatic conditions is attributed here to a negative feedback mechanism. It has been inferred (Taylor and McLennan, 1985) that there was a period of significant addition to the continental crust at the end

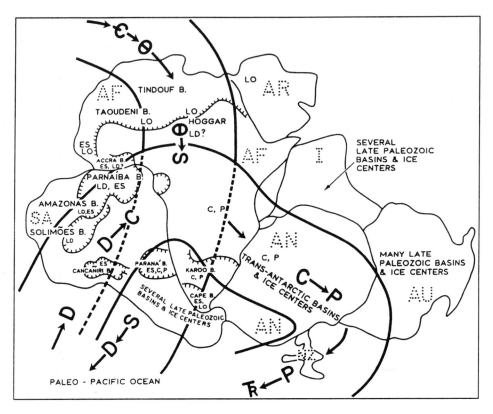

FIG. 2.2. Path of glacial centres during migration across western Gondwanaland during the Paleozoic (from Caputo and Crowell, 1985; reprinted with permission from the authors).

of the Archean. This process would have resulted in uplift and enhanced weathering of the continental crust, with concomitant decrease in atmospheric CO_2. Once continental glaciation was initiated, weathering would be inhibited in ice-covered regions by lowered temperatures. Reduced weathering rates would have led eventually to a gradual build-up of atmospheric CO_2, subsequent destruction of the ice sheets and establishment of a warm climatic regime (Fig. 2.3). This long-term alternation of warm and cold climatic regimes would continue until it was interrupted by some other factor such as fragmentation of the 'supercontinent'. The tectonic history of the Huronian has been interpreted in terms of continental break-up (Young, 1983; Zolnai *et al.*, 1984). Continental break-up is accompanied by a rise in sea level (Worsley *et al.*, 1984) which would interrupt the previous cycle and usher in a period of warm climate. If the CO_2 content of the early atmosphere were much

higher than today (Hart, 1978; Kasting, 1989), and if the 'faint young sun' theory is valid, then the onset of global glacial conditions may have been possible at much higher partial pressures of CO_2.

2.2.5. The Mid-Proterozoic

The Early Proterozoic ice age was followed by an interval of 1300 Ma when evidence is lacking for glaciation. The Middle Proterozoic has been inferred by several authors (Windley, 1977; Piper, 1978; Hoffman, 1989) to have been characterized by the existence of one or more supercontinents and, according to the ideas outlined above, should have been ideal for initiation of glaciation. Hoffman (1989) proposed that the Mid-Proterozoic supercontinent underwent abortive attempts at fragmentation. Evidence for this includes widespread mafic and acid magmatism from ~1.8 to 1.3 Ga, including a rare

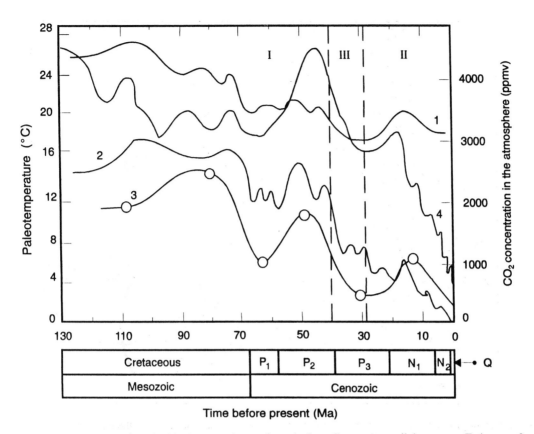

FIG. 2.3. Temperature and carbon dioxide change estimates from the Late Cretaceous until the present. Estimates of ocean surface temperatures in low latitudes (curve 1) and high latitudes (curve 2) of atmospheric CO_2 concentration (curve 3) and of surface air temperature over the Russian plains (curve 4). Time periods are divided into I, non-glacial regime; II, glacial regime; III, transition regime; P, Paleogene: respectively Paleocene (P_1), Eocene (P_2), Oligocene (P_3); N, Neogene: respectively, Miocene (N_1), Pliocene (N_2), and Q, Quaternary (from MacCracken *et al.*, 1990; reprinted with permission from *Prospects for Future Climate*. Copyright 1990 Lewis Publishers, a subsidiary of CRC Press, Boca Raton, Florida).

period of anorthosite and rapakivi granite intrusion. These events may indicate a unique stage in the Earth's thermal evolution, involving interaction between the first extensive supercontinent and the internal heat of the Earth. Outgassing of CO_2, related to widespread continental magnetism, may have been sufficient to balance the predicted large drawdown caused by weathering of the highstanding supercontinent.

2.2.6. The Late Proterozoic Ice Age

There is evidence of Late Proterozoic glaciation from all continents except perhaps Antarctica. This widespread and long-ranging glacial episode may have had three peaks at about 940, 770 and 615 Ma (Williams, 1975; Crowell, 1981). Two of these are recorded in the same stratigraphic sequence around the North Atlantic Basin and in central and southwestern Africa, Brazil, western North America and Australia. In North America, the sediments are distributed between the Alaska–Yukon border, Appalachia and eastern and northern Greenland. Long stratigraphic sequences in some areas (e.g. Africa) suggest multiple glaciations over a period of 'several hundred million years' (Hambrey and Harland, 1981, p. 944).

Many Late Proterozoic glaciogenic successions display evidence of contemporaneous rift activity (Schermerhorn, 1974; Coats, 1981; Preiss, 1987; Young and Gostin, 1989a, b). In some places, associated iron-formations and manganese-rich sedimentary rocks have been interpreted as products of hydrothermal activity related to rift-related volcanism (Yeo, 1981; Breitkopf, 1988).

Interpretation of the widespread glacigenic rocks of the Late Proterozoic remains enigmatic. Harland (1965) proposed that they might signify a worldwide glacial episode. Earlier paleomagnetic results (Harland and Bidgood, 1959; Bidgood and Harland, 1961) suggested that some Late Proterozoic glacigenic rocks might have formed in tropical latitudes (McWilliams and McElhinny, 1980). Many Late Proterozoic rocks are associated with dolostones, red beds and other rock types generally considered to indicate warm climatic conditions. Spencer (1971) supported the notion of a vast ice cap in equatorial latitudes to explain the Late Proterozoic Port Askaig Tillite of western Scotland and supposedly correlative glacial facies in Greenland and elsewhere in the North Atlantic region. This interpretation was favoured by Fairchild and Hambrey (1984), who suggested that in the Late Proterozoic of Svalbard, glacial conditions were developed in an otherwise warm climatic regime.

G.E. Williams (1975), in an attempt to resolve these enigmatic associations of rock types, proposed that there had been significant variations in the tilt of the Earth's spin axis relative to the rotational plane of the solar system (variations in the obliquity of the ecliptic). According to this theory, during periods of greatly increased obliquity (>54°), annual insolation in polar regions would be greater than that in low latitudes. In the event of a glaciation, the equatorial belt would have been preferentially glaciated. By itself, this theory does not, however, explain the widespread nature of the glacial deposits of this period.

Crowell (1983), using the analogy with the Permo-Carboniferous glaciation of Gondwanaland, suggested that the Late Proterozoic was a period of rapid plate tectonic movement. He proposed that the continents could have moved into high latitudes where they were glaciated, but that a magnetic signature was not imprinted on the sediments until much later when they were located in low paleolatitudes. Stupavsky *et al.* (1982) were of the opinion that some paleomagnetic results, such as those of Tarling (1974) from the Scottish Late Proterozoic diamictites, were due to later magnetic overprinting. Recently, Embleton and Williams (1986) and Chumakov and Elston (1989) have, however, contributed to the idea that during much of the Middle and Late Proterozoic, most of the continental mass was situated in low paleolatitudes. Both paleomagnetic results and geological arguments (Windley, 1977; Hoffman, 1989) support the concept of supercontinentality throughout this period.

If the concept of a supercontinent in near-equatorial latitudes is adopted, then active chemical weathering and reduction in atmospheric CO_2 could have led to glaciation. The first effects of such a cooling episode would have been freezing of the polar seas but eventually the continental region would have been glaciated (Fig. 2.4). Glaciation would have been initiated in regions situated at high altitudes (Schermerhorn, 1983), but ultimately would have extended to sea level. With the development of widespread glaciation and lowering of temperatures, negative feedback would have led to a build up of atmospheric CO_2 again and introduction of a warm climatic period. Such a mechanism would be expected to produce a sedimentological record displaying evidence of alternation of warm and cold climatic episodes. The Late Proterozoic record in many parts of the world displays such an alternation. Smaller scale oscillations such as the numerous advance retreat cycles reported by Spencer (1971) from the Late Proterozoic Port Askaig Tillite in Scotland are more likely due to phenomena acting on a much smaller timescale, such as Milankovitch forcing (Hambrey, 1983).

2.2.7. The Phanerozoic Ice Age Record

There is evidence of glaciation in the Ordovician, Devonian and Permo-Carboniferous. Evidence is mostly found on the continents of the southern hemisphere that were then amalgamated into the Gondwana supercontinent. There appears to be a relationship between the sequential glaciations of the

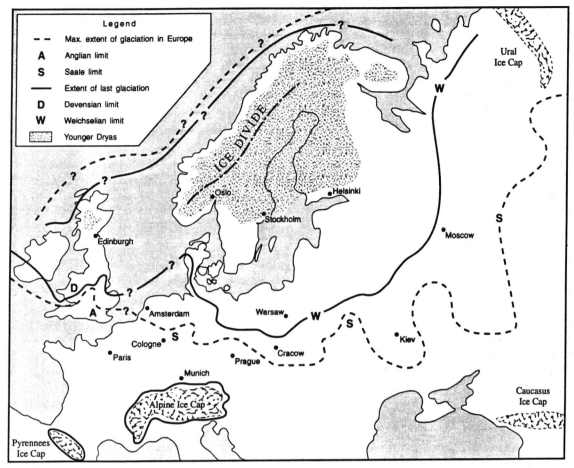

FIG. 2.4. Extent of Pleistocene glaciation in Europe. Compiled from data in Nilsson (1983) (Reprinted with permission of Kluwer Academic Publishers).

Permo-Carboniferous and the passage of different parts of the supercontinent across the south polar region (Crowell, 1978). Two main factors may have contributed to the development of Late Paleozoic glaciation. The emergent state of the Gondwana supercontinent could have led to CO_2 depletion (anti-greenhouse effect) of the atmosphere. Positioning of part of the supercontinent at high latitudes would also have contributed to lower average summer temperatures leading to the build up of snow and the development of continental glaciers.

2.2.7.1. The Ordovician–Silurian ice age

The early Paleozoic paleogeography, like that of

the Precambrian, was unlike that of the present and involved wide dispersion of cratons (Scotese *et al.*, 1979). Most of the North American, European and Siberian cratons were located in the tropics or subtropics. However, the Gondwanaland supercontinent, which formed in late Precambrian time and encompassed South America, Africa, Arabia, Madagascar, India, Australia and Antarctica, as well as Florida, southern and central Europe, Turkey, Iran, Afghanistan, Tibet and New Zealand, extended between the equator and the South Pole throughout the Paleozoic. The Paleozoic glacial centres followed roughly the drift of the South Pole (Fig. 2.2), with apparent paths of the pole crossing glacial sites in chronological order (Morel and Irving,

1978; Caputo and Crowell, 1985; Crowley et al., 1987). Continental reconstructions for the Late Ordovician (Smith et al., 1981) suggest that the northern continents were scattered but the Gondwana continents were joined together. North Africa, where the best evidence of Late Ordovician glaciation is preserved, was in a position close to the South Pole.

An Ordovician–Silurian ice age is recorded locally by Early Ordovician terrestrial and glaciomarine deposits in Britain. It culminated during the Late Ordovician with extensive continental glaciation in north and western Africa and contemporaneous glaciomarine deposition to the north in southwestern Europe (Robardet and Doré, 1988) and in the south. The Late Ordovician phase is well represented in North Africa where there is spectacular preservation of evidence of continental glacial phenomena (Deynoux and Trompette, 1981; Biju-Duval et al., 1981). Less extensive ice masses occurred in South Africa and eastern North America. The African ice sheets may have been only somewhat smaller than that of Antarctica today (Biju-Duval et al., 1981). As the African ice sheet was retreating in Early Silurian time, the glacial centre moved into Andean South America (Crowell et al., 1980). Ice apparently disappeared during the early part of Late Silurian Period, about 450 Ma, after an interval of ~50 Ma.

2.2.7.2. The Devonian period

Glaciation during the Devonian and Permo-Carboniferous periods would appear to have developed in a manner similar to that during the Ordovician–Silurian periods (Crowell, 1978; Caputo and Crowell, 1985; Veevers and Powell, 1987), spanning a period of almost 100 Ma from about 360 to 255 Ma when Gondwanaland began to drift away from the south pole (Smith and Briden, 1977; Smith et al., 1981). Indisputable tillites, which may be Devonian in age, occur in West Africa and Antarctica (Hambrey and Harland, 1981). Devonian rocks of possible glacial origin (Crowell, 1982) have also been reported from the Amazon basin, Brazil (Caputo and Crowell, 1985) but by far the most widespread glaciation of the Phanerozoic is that of the Permo-Carboniferous.

2.2.7.3. Permo-Carboniferous ice age and later periods

Evidence for this pre-Late Cenozoic ice age was first discovered in Africa and Australia in 1859. Subsequently, it has been reported at deeply eroded shield margins and within local basins on all Gondwanaland continents, including the first Antarctic encounters in the Transantarctic Mountains (Long, 1959). It is the best known of the early ice ages with the most abundant evidence. Paleolatitudes range generally from 48 to 80°S. However, Permian glaciogenic units are also known from northeastern Siberia, which lay near the North Pole at that time.

Beginning 330 Ma in South America and south west Africa, large scale, lowland, ice sheet glaciation covered much of Gondwanaland (Fig. 2.2), including India, but particularly in Africa where, at ~280 Ma, a subpolar, marine-based ice sheet has been described in the Dwyka Formation (Du Toit, 1921; Visser, 1989). Glaciation ended after a 90 Ma interval at 240 Ma in Antarctica and Australia (Crowell, 1981). The glaciogenic Tolchir Boulder Beds of Peninsular India were deposited in Early Permian time. The termination of glaciation in Australia coincided with the formation of hot deserts in eastern South America and northern Africa, which were in the tropical latitudes at this time.

The evidence for glaciation is extremely detailed, particularly, for example, from ice tongues that reached into Brazil and Uruguay from Africa and deposited the Itarare strata (Rocha-Campos and Dos Santos, 1981). In some areas of Gondwanaland, ten or more glacial episodes have been recorded with glacial units as thick as 1000 m, and multiple flow centres, many of which may have never merged. However, widespread, synchronous glaciation may have been limited, as evidence suggests that sea level was not drastically lowered over the 90 Ma interval (Crowell, 1982). This was a time of fusing of the Gondwanaland and Laurasian supercontinents to form Pangaea (Scotese et al., 1979) and the creation of large outpourings of lava and construction of mountain chains. The glacial evidence on diverse continents provided a major basis for early hypothesis of continental drift (Wegener, 1966) and plate tectonics.

Glaciation terminated in Late Permian times as Gondwanaland moved away from the polar zone. The likelihood of glaciation was greatly reduced when, in the Early Jurassic, about 180 Ma, the supercontinent began to break up, causing elevated sea levels and ushering in a period of global warming.

In the Late Cretaceous, about 80 Ma, Antarctica and Australia drifted back into polar latitudes. In addition, there was a lowering of sea level due to the increase in average age of the maturing Atlantic Ocean (Worsley et al., 1984) which would have caused withdrawal of sea water from the continents. The combined effect of these factors probably introduced a cooling phase that culminated in glaciation in Antarctica about 36 Ma (Barrett, 1981).

In this section of the chapter an attempt has been made to explain Proterozoic and Paleozoic glaciations in terms of variations in atmospheric composition, controlled largely by relative sea level changes which, in turn, can be traced to plate tectonic causes. Although this set of causes has been emphasized, it is not suggested that it is the sole cause of all glaciations. The model outlined above does not provide an explanation for the complex advance–retreat cycles displayed by many glacigenic successions. Other factors such as the Milankovitch effect and other poorly understood, still smaller scale cycles, manifested particularly in Cenozoic glacial successions probably also played a role in generation of the ancient glacial deposits.

2.2.8. The Pattern of Ice Ages and Events Leading to Late Cenozoic Glaciation

Poor preservation, uncertainty in dating and limited knowledge of cratonic positions during Precambrian times have restricted our knowledge of pre-Cenozoic ice ages (Fig. 2.1). However, glaciation was apparently extensive and prolonged in the Precambrian relative to that of the Phanerozoic.

It is unlikely that all glaciations have a single cause, rather, like most natural phenomena, they are probably the result of a complex interplay of conditions deep within the earth and at its surface. The near-complete absence of Archean glacial deposits in spite of the inferred 'faint sun', is attributed to high CO_2 levels and a relatively small area of exposed continental crust. In the Early Proterozoic the first widespread glaciations occurred.

These are interpreted as a response to the increase in continental crust (Taylor and McLennan, 1985), which would have caused a reduction in CO_2. Possibly due to the still faint nature of solar radiation, global temperatures could have been significantly lowered by reducing atmospheric CO_2. Alternation of glacial and warm climatic conditions, on a long timescale, is evident in some Late Proterozoic successions, lending support to the idea of a negative feedback mechanism, causing rapid re-establishment of relatively high CO_2 levels between the glaciations.

There is an absence of glaciation during the long period from about 2.0 to 1.0 Ga, despite evidence for the existence of a supercontinental configuration. This is attributed to a protracted period of unusually widespread magmatic activity, postulated to have maintained high atmospheric CO_2 levels and therefore precluded development of glaciation.

The Late Proterozoic record reveals the greatest proliferation of ice sheets the world has known. Available paleomagnetic data suggest that many of these glacial centres developed at low paleolatitudes. Widespread evidence of rifting and some paleomagnetic data support the concept of the existence of a supercontinent or supercontinents during this period. Such a supercontinental configuration at tropical latitudes could have led to an unprecedented reduction in atmospheric CO_2 and extensive glaciation into low paleolatitudes. A negative feedback mechanism would eventually have induced a warming trend and a return to a warm climatic regime. By this means, an alternation of warm and cold climatic episodes could have been produced. In many parts of the world there is evidence, both in the Early and Late Proterozoic, of such climatic oscillations. The cycle of warm and cold climatic episodes was broken when continental fragmentation took place at or near to the end of the Precambrian, yet no simple cycle can be distinguished for these ancient ice ages (Crowell, 1982).

Data from the Phanerozoic are much more closely constrained, both paleontologically and paleomagnetically. These more recent glaciations appear to have taken place on continents in high paleolatitudes (Young, 1991; Worsley and Kidder, 1991). Given that the CO_2 content of the atmosphere was probably much reduced by this time, proximity of the continent to the pole may have been the dominant factor in controlling the climatic regime. There also appears to be some correspondence, however, between periods of glaciation and lowered sea levels, related to the existence of supercontinents (Gondwanaland). Likewise, the Cenozoic glaciation, and possibly also the Late Ordovician event, affected regions in high latitudes at periods when relative sea level was low, perhaps due to the presence of a wide, mature, Atlantic-style ocean (Worsley et al., 1984).

No Triassic glaciogenic deposits are known and there are no tillites recognized from units of Jurassic and Cretaceous age (Hambrey and Harland, 1981). The Mesozoic was an era of equitable global warmth as the progressive breakup of supercontinents, Laurasia and Gondwanaland, created high sea levels, extensive continental flooding and new ocean basins. In general, this and probably earlier non-glacial climate conditions were characterized by concentrations of the greenhouse gas CO_2 that were more than five times those of the pre-industrial world with temperatures 10 to 15°C above those of the present at high latitudes. These conditions were compatible with the north–south distribution and low emergence of Mesozoic continents that would have retarded weathering, but allowed CO_2 buildup in the atmosphere (Worsley and Kidder, 1991). Because there were no ice sheets, no cold bottom waters would have been produced in the oceans. North–south temperature gradients were less than present and deserts may have been absent (MacCracken et al., 1990).

Several major paleogeographic changes were taking place in the Mesozoic and in early Tertiary time. These were accompanied from the Late Cretaceous by a somewhat irregular decrease in global temperatures that was associated with a decrease in the atmospheric concentration of CO_2 (Barron, 1985; MacCracken et al., 1990) (Fig. 2.3). From Middle to Late Eocene time when the drop in CO_2 was most marked, temperatures fell about 10°C or more through the Pleistocene. Gradual southward retreat or elimination of temperate or subtropical vegetation followed the initiation of these temperature changes (Wolfe, 1985) as did the development of the Arctic sea ice pack (Theide et al., 1990) and North Atlantic Deep Water currents (Woodruff and Savin, 1989).

Antarctica had moved into the South Pole area by the Tertiary Period and Australia had started its northward movement from Antarctica, marking a major step toward thermal isolation of that continent (Stump and Fitzgerald, 1992). A second major phase in Antarctica's thermal isolation occurred between Late Oligocene to Miocene time (~24 Ma) (Table 2.1) with the opening and deepening of the Drake Passage between Antarctica and South America (Matthews, 1984). Simultaneously, the Arctic Basin was forming and northwest and northeast extensions of the North Atlantic were opened by the spreading of the North Atlantic Ridge. This allowed the separation of Greenland from Europe about 37 Ma (Herman et al., 1989), leading toward production of North Atlantic deep water flow (Section 2.9).

Strong cooling must have occurred in Antarctica during the Middle to Late Eocene time (Fig. 2.3) bringing winter snow cover and mountain ice buildup between 52 and 36 Ma. By at least the early Oligocene (~36 Ma), an extensive, full-scale ice sheet had formed. This is suggested by thick diamictite accumulations at widely separated locations at the periphery of Antarctica (Barker et al., 1987; Barrett et al., 1987, 1989; Harwood and Webb, 1990; Webb, 1990).

Some interpretations of the Antarctic glacial record suggest that major fluctuations of the ice sheet(s) may have occurred at intervals of as little as one to two million years over the last 36 Ma (Harwood et al., 1989; Webb 1990; Clapperton and Sugden, 1990; Scherer, 1991) and that the ice cover, over several periods, was probably more extensive than today (Robin, 1988). Significant enlargement of the East Antarctic Ice Sheet did occur in Middle Miocene time following opening of the Drake Passage. Continued cooling allowed development of ice shelves in West Antarctica and finally the grounding of the West Antarctic Ice Sheet below sea level by 10 Ma. Sediments collected beneath an ice stream of this ice sheet contain marine and non-marine microfossils that appear to attest to ice-free intervals here, both prior to, and subsequent to the ice-sheet grounding (Scherer, 1991).

Appreciable retreat of Antarctic ice occurred after several million years at this maximum stage of ice expansion ~10 Ma, but further cooling caused accumulation and mass budgets of the ice sheets to

TABLE 2.1. Timing of some Cenozoic global events related to glaciation (modified from Herman *et al.*, 1989)

Time (Millions of Years)	Tectonic Events	Climatic Events
0	0.9-0 Himalayan, Alpine, etc. orogenies underway	
0.9	Orogeny peak	0.9-0 Major glacial-interglacial cycles; onset of large-amplitude climatic fluctuations
1.6	2-1 Increased Tibetan, Himalayan, and Sierra Nevadan orogenies	
2.4-2.5		First major Northern Hemisphere ice sheets and onset of marked global cooling
2.5-2.6		Fluctuating temperature
3.5	Uplift of Panama Isthmus; opening of Bering Strait	Gradual temperature decline. Southern Hemisphere lowland glaciation
5.5	Isolation of Mediterranean Sea	
6		Strong global cooling. 6-4 Expansion of Antarctic Ice Sheet
14		14-12 major Antarctic ice buildup?
18	18-16 Subsidence of Iceland-Faeroes Ridge	
	18-14 Renewed Himalayan Tibetan orogeny	
24	Drake Passage open	24-14 Increased glaciation of Antarctica; intensification of global climatic gradients
36.6	Greenland separates from Eurasia; Tasman Seaway opens	Antarctic continental glaciation; slow cooling
37	37-35 major Himalayan and Alpine orogenies	Cooling at high and low latitudes; mountain glaciation in Antarctica

decrease (Robin, 1988). A warm Early Pliocene may have caused further deglaciation (Webb, 1990; Scherer, 1991); however, a period of rapid uplift in the Transantarctic Mountains followed that may have corresponded with the start of Late Pliocene cooling and glacier readvances in the northern hemisphere about 2.5–2.4 Ma (Harwood and Webb, 1990; Behrendt and Cooper, 1991). Elsewhere in the southern hemisphere, Miocene uplift and cooling had caused mountain glaciers or ice caps more extensive than the present to form in the mid-latitude Patagonian Andes by 4.7 Ma (Mercer, 1976, 1983; Rabassa and Clapperton, 1990) (Table 2.1).

In the northern hemisphere, direct evidence of growth of the Greenland Ice Sheet is unknown, but mountain glaciers were established by 10–2.7 Ma in mountain ranges of East and West Greenland. An ice sheet at least as big as that of the present probably formed over the continent by 2.4 Ma (Robin, 1988; Funder, 1989). Small-scale glaciation by about 3 Ma in Iceland is recorded by multiple drifts (McDougall and Wensink, 1966). Glaciation in Alaska may have developed during the same period in response to uplift that accompanied convergence of the Pacific and North American Plates 6 Ma (Eyles and Eyles, 1989).

Mountain building at plate margins was an important consequence of the rifting and plate tectonic movements separating the supercontinents and of the rearrangement of land masses. The American Cordilleran system, the Alps–Caucasus–Central Asian mountain system, as well as the Andes Mountains, reached near-present heights in Late Cenozoic time only shortly before mid-latitude ice sheets were developed (Ruddiman and Raymo, 1988; Raymo, 1991; Unruh, 1991) (Table 2.1). Explosive volcanic activity had also apparently reached peaks in Miocene to mid-Pliocene time, continuing to increase from this time into the Quaternary (Kennett and Thunell, 1975).

2.3. LATE CENOZOIC GLACIATION AND THE CLASSIC SUBDIVISIONS

2.3.1. Scope and Classification

In the previous sections, we were concerned with ice ages that occurred in the geologic past at intervals of the order of 150 to 300 Ma and apparently discontinuously over spans of up to several hundred million years. Discussion will now concentrate upon the global terrestrial and marine records of the present ice age. These records delineate glacial and interglacial intervals of 10,000 to 100,000 years, stadial and interstadial events within the glaciations that span 1000 to 10,000 years, as well as major fluctuations of mountain glaciers in postglacial and historic times involving decades to centuries to millennia.

The time of onset of the present ice age may be fixed in a number of different ways. The continental glaciation in Antarctica began ~36 Ma, whereas initiation of glaciation in middle latitude lowland areas of the northern hemisphere began around 2.5 Ma. The latter is based on evidence of major North Atlantic ice rafting considered further below (Shackleton et al., 1984; Hodell et al., 1985; Ruddiman and Wright, 1987). In any case, it is now understood that initiation of glacial conditions on a global scale apparently does not correspond with most definitions of any of the widely accepted Cenozoic chronostratigraphic terms.

Desnoyers (1829) was the first to apply the term Quaternary to post-Tertiary terrestrial strata and Reboul (1833) modified the application to include flora and fauna still living. Later, Lyell (1839) introduced Pleistocene (most recent) to marine units that contained more than 70% of mollusca still living. Forbes (1846) who, following acceptance of the Glacial Theory, modified Lyell's definition to make 'Pleistocene' equivalent to the Glacial Epoch and suggested the term Recent for post-glacial time. 'Recent' was replaced by Holocene (wholly recent) by the International Geological Congress of 1885. At present, the internationally proposed Pliocene–Pleistocene boundary stratotype is a highly fossiliferous marine strata of the Vrica section in Calabria, southern Italy (Nilsson, 1983; Aguirre and Pasini, 1984, 1985; Richmond and Fullerton, 1986a, b; Rio et al., 1991). This boundary is given an age of 1.64 Ma (Aguirre and Passini, 1985). The Pleistocene–Holocene boundary is usually taken arbitrarily at 10,000 BP (Before Present) in North America and elsewhere (Hopkins, 1975; Watson and Wright, 1980) (Menzies, 1995b, Chapter 8).

TABLE 2.2. Major glaciations and interglaciations of western Europe and North America following the classical systems. The Alpine sequence and the pre-Illinoian sequence of North America are not considered useful for stratigraphic correlations. Major glaciation began following the Cromerian temperate stage in Europe

European Alps	Northwest Europe		Britain	North America
Würm[+]	**Weichsel**		**Devensian**	**Wisconsinan**
*R/W**	*Eem*		*Ipswichian*	*Sangamon*
Riss	**Saale**	Warthe	**Wolstonian**	**Illinoian**
		Drenthe		
M/R	*Holstein*		*Hoxnian*	*Yarmouth*
Mindel	**Elster**		**Anglian**	**Kansan**
G/M	*Cromerian*		*Cromerian*	*Aftonian*
Gunz				**Nebraskan**

* Interglacials in *Italics*
+ Glacials in Bold

2.3.2. Classic Subdivisions of the Pleistocene

As discoveries of multiple Pleistocene sets of glacial deposits (drifts) became more common, following acceptance of the glacial theory, several regional schemes of classification emerged (Table 2.2). The Pleistocene glacial classifications of the Alps, northwestern Europe, and central North America, outlined below, are well-known, long-established and often still used in some form (Menzies, 1995b, Chapter 8). However, each although much modified, has been unsatisfactory and even misleading in various ways (Bowen, 1978). As new evidence of glaciation was discovered, the value of the original classification was often neutralized, especially when sites were often a continent distant from type localities. Regional classification schemes were well established before development of the marine oxygen isotope records or the long terrestrial sequences tied by numerical ages and/or the

paleomagnetic time (Section 2.4). The most well-known succession of glaciations and interglaciations of the Pleistocene was the four-part Alpine classification of Penck and Brückner (1909) based on river terraces. This attained the status of a continental to worldwide framework for glacial sequences despite its lack of stratigraphic basis (e.g. local identification, description and correlation of rock units in sequence).

2.3.2.1. Alpine terrace model of Penck and Brückner

The Alpine ice mass (Fig. 2.4) at its maximum, covered about 150,000 km² flowing as a network among mountain peaks and divides, but with coalescing valley and piedmont glaciers that reached down toward sea level on the north and south flanks (Flint, 1971; Nilsson, 1983). In north-draining valleys of the German Alpine foreland south of Munich,

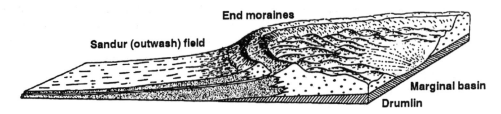

FIG. 2.5. Alpine terraces and moraines of Penck and Brückner (from Nilsson, 1983; reprinted with permission of Kluwer Academic Publishers).

Penck and Brückner (1909) delimited the Günz (G), Mindel (M), Riss (R) and Würm (W) glaciations, and the G/M, M/R and R/W interglaciations (Table 2.2). Named after the four Bavarian tributaries of the Danube (Donau) and Isar, all have their type localities at the Iller-Lech Platte; however, no true stratotypes were designated (Kukla, 1977). Subsequently, the four-part sequence was extended by the discovery of two older glacial stages: the Donau (Eberl, 1930) and Biber (Schaefer, 1953). The four principal glacial stages were represented by a sequence of terraced glaciofluvial outwash plains (schötter); each younger (and generally lower) plain was tied successively farther up slope to end moraines representing the contemporaneous glaciation (Fig. 2.5). Interglacial times were represented only by terracing of the previous outwash gravels to produce steps.

A chronosequence was established based on the depth of step cutting between one terrace and the next, the M/R interglaciation was estimated by Penck and Brückner to be about four times longer than the other interglaciations. Furthermore, based on the interval of postglacial time, the whole Pleistocene was estimated to have lasted about 650,000 years (Imbrie and Imbrie, 1979). The timing of the four glaciations corresponded well with estimates of radiation minima at high to mid-latitudes that were calculated by Milankovitch. The estimates implied short glacial and long intervening interglacial intervals (Köppen and Wegener, 1924).

Problems with this classic four-part Alpine sequence have been delineated almost since its inception (Eberl, 1930; Schaefer, 1953, 1968; Kukla, 1977; Bowen, 1978; Nilsson, 1983; Billard, 1985; Šibrava, 1986). The terrace sequence is more complicated than originally envisioned and contains both interglacial and postglacial deposits; the erosion

intervals (unconformities between terraces) are largely glacial rather than interglacial features; and deposits may represent only a few millennia of glaciation (Schaefer, 1953; Kukla, 1977). Revisions to the Penck and Brückner sequence are reviewed by Kukla (1977), Nilsson (1983) and current stratigraphic correlations are considered by several authors in Šibrava et al. (1986). The classical nomenclature have little more than morpho-stratigraphic significance in the originally designated area and must be abandoned for external correlations (Šibrava, 1986).

2.3.2.2. Northern European glacial chronology

While the more celebrated stratigraphic subdivision originated in the Alps, more important may be the scheme relating to the continental glaciation due to the Scandinavian Ice Sheet (Table 2.2). At its maximum, the Scandinavian Ice Sheet extended eastward to the Ural Mountains, southeast to beyond Kiev, south into central Germany and westward to the British Isles (Fig. 2.4). The fundamental origin of this classic subdivision is morphostratigraphic based on a progressively younging end moraine system marking glaciation limits from southern central Europe northward into Scandinavia (Nilsson, 1983; Šibrava, 1986). The classic stages (Table 2.2) were named (from oldest) Elster, Saale and Weichsel by Keilhack (1926) after rivers on the northwestern European plain. Later, the Warthe moraine stage was added, originally with the Weichsel, then the Saale (Woldstedt, 1954) and more recently singled out as a distinct glacial stage between Saale and Weichsel (Picard, 1964). All stages are now known to fall into the Brunhes paleomagnetic chron (~0.9 Ma) (Šibrava, 1986). However, some evidence of earlier Pleistocene glacial advances is reported locally (e.g. Britain and Poland) (Šibrava, 1986; Bowen *et al.*, 1988).

Deposits of the Elster stage mark the earliest clear and well-dated evidence of major glaciation in the type area of northwestern Germany and Europe. They occur within the north German lowland associated with a widespread series of deep, buried channels considered by many workers to be formed by subglacial meltwater (Grube *et al.*, 1986). While the Elster deposits display no original topographic relief, remains of the Late Saale and the Weichsel (youngest) are little altered by subaerial processes (Woldstedt, 1954). The equivalent glacial stages of the Pleistocene in Britain are generally considered to be the Anglian, Wolstonian and Devensian, respectively (West *et al.*, 1988) (Table 2.2).

The system of moraine-based units in northern Europe is supported by classical interglacial stages (from oldest): the pre-Elster, Cromer, named after the Forest Bed on the North Sea bluffs of East Anglia, England; the Holstein (Hoxnian in Britain); and the Eem (Ipswichian in Britain) (Madsen, 1928; Kukla, 1977). The interglacial units are represented by deposits of marine transgressions in the lowland areas, and also by peat bogs with pollen documenting temperate, hardwood forests in northwestern Europe.

Important problems relative to the northern European scheme have been the difficulties in correlating the end moraines with interglacial deposits, repetitive glacial and interglacial deposits, the fragmentary nature of individual Pleistocene deposits together with poor subsurface control and the scarcity of distinguishing fossils in the Pleistocene strata (Kukla, 1977; Cepek, 1986; Šibrava, 1986; Zagwijn, 1986). Furthermore, reorganization is needed to account for the deposits including those of known pre-Elster age (Bowen *et al.*, 1988).

In the British Isles, several centres of glaciation existed during the Quaternary, with frequent linkage with the Scandinavian Ice Sheet as it invaded from across the North Sea. Consequently, the moraine sequence is less clear here than in continental Europe. Subdivision of the Quaternary has been based in part on climatic change with temperate stages alternating with cold stages that are characterized as glacial when

such deposits occur (West, 1967). One of the classic areas of Quaternary study in the British Isles has been on the North Sea coast of East Anglia because it contains a remarkably complete and varied succession (West, 1967; Mitchel *et al.*, 1973; Bowen *et al.*, 1986a).

2.3.2.3. Central North American sequence

The Laurentide Ice Sheet extended from the Arctic Ocean in the Canadian Arctic Archipelago to the mid-western states of the United States in the south and from the Canadian Rocky Mountains in the west to the Grand Banks off the coast of Newfoundland (Fig. 2.6a), possibly incorporating 35% of the world's ice volume during the Late Wisconsinan maximum and 60–70% of the ice wasted at its conclusion (Fulton and Prest, 1987; Fulton, 1989; Dreimanis, 1991). The most extensive record of fluctuation at its southern margin occurs in the north central United States. Chamberlin and Leverett were the principal leaders in the development of a Pleistocene sequence of four major glaciations in this area (Table 2.2) (Chamberlin, 1894, 1895, 1896; Leverett, 1898; 1899; Shimek, 1909; Flint, 1971). Their conceptual framework has lasted for many decades with minor modification, partly because further work did not appear necessary (Ruhe, 1969; Hallberg, 1986). Named after the states where they were best characterized or easily studied, the Nebraskan, Kansan, Illinoian and Wisconsinan glaciations represented the glacial sequence of the interior lowlands between the Appalachian and the Rocky Mountains. In many areas, their drift boundaries were thought to be nearly coincident (Fig. 2.6b); however, the Kansan has been generally considered to be the most extensive of these classical glaciations in the northern mid-continent of the United States (Flint *et al.*, 1959) obscuring the older Nebraskan drift in all but a few areas.

The original basis of the earliest three glaciations were till sheets, while the Wisconsinan was once based on the little-modified constructional topography of ubiquitous end moraines. The interglaciations, Yarmouth and Sangamon, were represented by paleosols (buried soils) developed in tills or other deposits; the Aftonian interglaciation

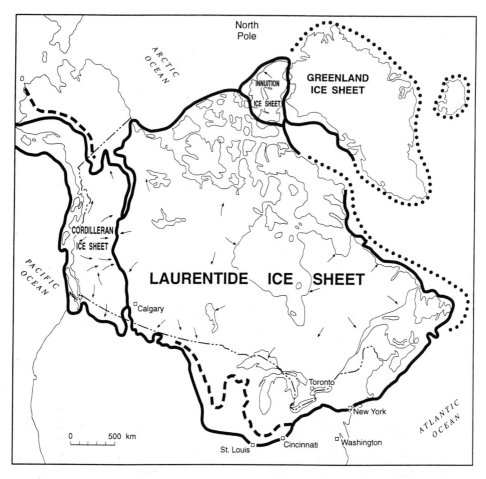

FIG. 2.6. (a) Approximate maximum extent of glaciation and main ice sheets in North America. Inner dashed lines at the southern boundary show generalized limits of Late Wisconsinan glaciation (modified from Flint, 1971 and Fulton, 1989).

was represented by Kansan outwash gravel before a major soil horizon was later identified (Frye *et al.*, 1968).

The Aftonian, Yarmouth and Sangamon interglacial paleosols were, for many years, believed to have formed during discrete soil-forming intervals that corresponded to the respective interglacial. Estimates of their duration (Leverett, 1898) inferred from soil thicknesses and development suggested that like the Penck and Brückner (1909) estimates for the Alps, the second interglaciation or Yarmouth was the longest interval during the Pleistocene. Subsequent studies have shown these estimates to be unreliable. For example, colluvial flow, rather than *in situ*

soil-forming processes, plays a major part in the thickening of some of the 'soils' (Willman and Frye, 1970). In addition, away from the type areas, the soils show extreme variation and differences in interval of formation (Follmer, 1983).

Data gathered from more precise, numerical dating and correlation methods, together with information from deep coring through the drift sheets have suggested that a complete reorganisation is necessary in the central plains region of Nebraska, Iowa and Missouri (Boellstorff, 1978; Hallberg, 1980). For example, major correlations based on an assumed Yarmouth-Kansan age of Pearlette ash with a source in the Yellowstone National Park area

Legend

Area glaciated during Wisconsin Glacial Age

Additional area glaciated during earlier glacial ages

Conspicuous end moraines of Wisconsin age

End moraines of earlier glacial ages

Glaciated area in Cordilleran region is not differentiated and is only approximate

FIG. 2.6. (b) Extent of glaciation in the northern United States and southern Canada (modified from Flint, 1971).

(Reed and Dreeszen, 1965) had to be revised when this widespread tephra was discovered to consist of three ash beds of quite different ages (Hallberg, 1986). In general, glacial and non-glacial deposits formerly assigned to the Kansan and Nebraskan glaciations, as well as Aftonian and Yarmouth interglacials, have often been miscorrelated and are of such diverse age as to make the terms almost meaningless, even in the type region (Hallberg, 1986). Furthermore, three distinct tills are now defined below the oldest 'type' Nebraskan till. Therefore, the terms Nebraskan, Aftonian, Kansan and Yarmouth have been abandoned in stratigraphic nomenclature (Hallberg, 1980, 1986; Richmond and Fullerton, 1986b, p. 183).

Stratigraphic relations are relatively well known for the Illinoian (Late Middle Pleistocene) and in considerable detail for Wisconsinan Stage deposits. These latter sediments occurred following a major, and locally well-documented, shrinkage of the Laurentide Ice Sheet during the Sangamon interglacial.

2.4. OXYGEN ISOTOPE STRATIGRAPHY AND THE MARINE RECORD OF GLACIATION

Our present knowledge of the character, magnitude and timing of Late Cenozoic glaciation has come through the development of new, detailed records throughout both glaciated and unglaciated areas, as well as from major modifications of the classical glacial–interglacial chronologies (Imbrie and Imbrie, 1979). For example, radiocarbon dating (Libby, 1955) provided an opportunity to focus on details of the latest Quaternary glacial fluctuations and therefore to infer changes in continental ice volume and climates (Menzies, 1995b, Chapter 14). At about the same time, similar global glacial and climatic inferences were made by taking advantage of the development in 1947 of deep sea piston cores. These allowed long undisturbed records of ocean bottom sediment to be obtained that were deposited at relatively constant rates over long time periods.

Perhaps the main revolution in thinking of ice ages comes from the lithostratigraphy and resulting chronologies obtained from deep sea cores that involve the acquisition of the oxygen-isotope ratios ($^{18}O/^{16}O$) of micro-organisms, principally the calcareous, planktonic and benthonic foraminifera. Cores have provided a continuous time series of the worldwide variations in continental ice volume for the Late Cenozoic that generally cannot be duplicated by the more fragmentary and shorter continental records (Mix, 1987; Ruddiman and Wright, 1987). Because the mixing time of oceans is relatively short (of the order of millennia) the trends in isotopic ratio variations expressed in the deep sea cores are, globally, nearly synchronous; in addition, they are easily correlated on the basis of unique stratigraphic horizons (Prell et al., 1986). Therefore, they have been used as a standard against which both continental and other marine chronologies are measured (Fig. 2.7).

Instead of the four classical major glaciations differentiated in central North America or Europe (Table 2.2), these deep sea isotopic records indicate that as many as 16 or more major ice advances and retreats occurred during the Pleistocene and a greater number when the record is extended into the Pliocene. The patterns of isotopic change have also provided convincing support for the astronomical theories of Croll and Milankovitch (Section 2.9) as describing the principle 'pacemaker' in global climatic changes (cf. Ludwig et al., 1992; Winograd et al., 1992). This implies that the orbital factors set the phase and frequency of the climatic changes if not also driving them. The technique of $\delta^{18}O$ with its adaptation to defining worldwide glaciations was pioneered by Emiliani (1955) on Caribbean cores and based on principles articulated by Urey (1947) of isotopic fractionation during crystallization of marine microfaunal tests.

The size of ice sheets on surrounding continents may also be estimated from deep sea cores by measurement of the input of ice-rafted continental debris, the estimation of sea surface temperature (SST) determined from cores using biological transformations and the $CaCO_3\%$ controlled by the depth of dissolution (Ruddiman and Wright, 1987; Rea and Leinen, 1989). These methods have supported the δ methods and chronologies that are considered below.

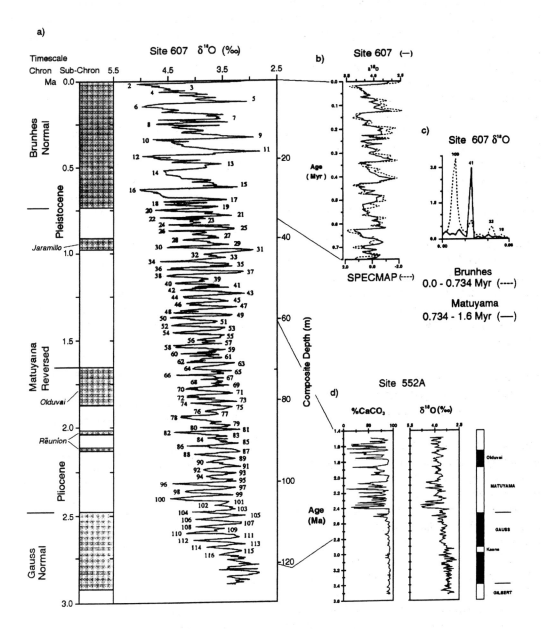

FIG. 2.7. Oxygen isotope record for the North Atlantic over the last 3.5 Ma. DSDP (Deep Sea Drilling Project) composite cores for site 607 (41° 00'N, 32° 58'W) with: (a) isotope stages against magnetic timescale (modified after Raymo *et al.*, 1989 and Ruddiman *et al.*, 1989, reprinted with permission of the American Geophysical Union); (b) Brunhes Chron portion overlaid onto the SPECMAP δ^{18}O stack (dashed line of Imbrie *et al.*, 1984) (from Ruddiman *et al.*, 1989; reprinted with permission of the American Geophysical Union). (c) Spectral analysis of Brunhes and Matuyama portions of δ^{18}O record of site 607 (from Ruddiman *et al.*, 1989; reprinted with permission of the American Geophysical Union). Note strong obliquity (41) cycle for early record compared to eccentricity (100) and precession (23 and 19) cycles for Brunhes. (d) North Atlantic DSDP site 552 records show major (about 30%) increase in northern hemisphere ice sheets at 2.4 Ma (about Stage 100) (data of Shackleton *et al.*, 1984; reproduced from Ruddiman and Wright, 1987; reprinted with permission from the authors). Abrupt decrease of $CaCO_3$% values mark initiation of deposition of ice-rafted sand.

2.4.1. Principles of Oxygen-Isotope Analysis

Oxygen occurs in three isotopic forms in nature: ^{16}O (99.763%), ^{17}O (0.037%) and ^{18}O, the heavy element (0.200%). The stratigraphic and correlation technique of the oxygen isotopes in deep sea cores is based on the fact that the proportions of these isotopes in sea water has changed through time. It also rests on the premise that the calcareous microfauna (largely foraminifera), whose tests collect on the sea floor, have incorporated the two major isotopes of oxygen (^{18}O and ^{16}O), in proportion to the isotopic abundance in the surrounding water. This isotopic abundance is dependent largely on the amount of water stored in glaciers on land; however, it is also dependent on sea water temperature and to a minor degree on the water's salinity (Duplessy *et al.*, 1970; Duplessy, 1981; Mix, 1987).

The $^{18}O/^{16}O$ ratios determined from analysis of the marine organisms are expressed as a departure ($\delta^{18}O$) from a standard as follows:

$$\delta^{18}O \equiv \frac{(^{18}O/^{16}O)-(^{18}O/^{16}O)\ REF}{(^{18}O/^{16}O)\ REF} \qquad (2.1)$$

where for calcareous fauna in the deep sea cores, REF is PDB, a belemnite from the Pee Dee Formation, and δ^{18} is expressed in per mille (‰). In the case of applications to ice core analyses (Section 2.6), the deuterium to hydrogen (D/H) ratio may be substituted for the $^{18}O/^{16}O$ ratio to determine δD, but for both determinations, REF is SMOW (Standard Mean Ocean Water). In practice, separate references of the National Bureau of Standards are used (Mix, 1987; Paterson and Hammer, 1987).

Both bottom dwelling (benthonic) and free floating (planktonic) foraminifera are used. Some work suggests that the planktonic fauna yield the best measures of continental ice volume (Shackleton and Opdyke, 1973; Prentice and Matthews, 1988). Different faunal components (e.g. calcareous coccolith vs foraminifera-dominated sediment) may yield different stratigraphic results (Paull *et al.*, 1991). Siliceous microfossils require special chemical processing and are rarely used to constrain isotopic curves.

The paleoclimatic application of deep sea cores, as well as ice cores (considered in a later section) is controlled by the rate of isotopic fractionation as water is evaporated from the oceans (Dansgaard *et al.*, 1973; Dansgaard, 1981; Mix, 1987). Because hydrogen also occurs as ^{1}H and ^{2}H (deuterium), water molecules may exist in any of nine isotopic states, of which only three are important in paleoclimatic research. These are $^{1}H_2H^{16}O$ (written HDO), $^{1}H_2^{18}O$ and $^{1}H_2^{16}O$. The light molecule $H_2^{16}O$ has a vapour pressure that is about 10% or 100‰ higher than that of HDO and 1% (10‰) higher than $H_2^{18}O$. Therefore, with respect to evaporation and condensation cycles, the heavy molecules of HDO and $H_2^{18}O$ are less able to evaporate and more readily condensed from the vapour than the light $H_2^{16}O$. At equilibrium, the water vapour in the atmosphere contains 10% less ^{18}O and 100% less deuterium than ocean water.

These relationships mean that the heavy molecules are enriched in the ocean water relative to its original condition during initial evaporation and also as the evaporate moves toward the polar regions; this occurs since cooling, resulting condensation, and rain or snowfall preferentially removes the heavier isotopes. Precipitation of snow on ice sheets leads to their enrichment in light $H_2^{16}O$ molecules, and this is the basis for isotopic analysis of ice cores considered in Section 2.6 (Paterson and Hammer, 1987). In paleoclimatic analysis, the oxygen isotope has been more commonly used than the isotopes of hydrogen, which are less abundant in nature and harder to measure in the laboratory (Bradley, 1985).

The marine oxygen-isotope record is largely a composite of ice volume and temperature. In a pioneering study, Emiliani (1955) used the foraminiferal delta (δ) values, in part, as a paleothermometer. However, Shackleton (1967) argued that ice volume dominated in the $\delta^{18}O$ signal since benthonic and planktonic fauna have nearly the same down-core pattern of $\delta^{18}O$ change in many areas. Bottom waters everywhere in the oceans today have temperatures that vary only between about 1 and 30°C (Rea and Leinen, 1989). Dansgaard and Tauber (1969) estimated that glacial age ocean water accounted for about 70% of the $\delta^{18}O$ signal in deep sea cores, and Shackleton (1987) arrived at similar figures for benthonic foraminifera of the deep Pacific Ocean. Nevertheless, the glacial–interglacial amplitude of $\delta^{18}O$ displayed in marine records

(~1.7‰) (Mix, 1987) appears to be accounted for to an even greater percentage than 70% by ice volume changes without other contributing effects (e.g. temperature) (Paterson, 1972b; Hughes *et al.*, 1981). The maximum contributor to the $\delta^{18}O$ signal in deep sea cores are the Laurentide ice sheet (~50%) and the excess Antarctic ice (beyond that at present) (Hughes *et al.*, 1981).

2.4.2. Timescale, Milankovitch Controls and Phase Lags of the Deep Sea $\delta^{18}O$ Marine Chronology

A succession of quasi-periodic fluctuations or stages corresponding to cooler (glacial) and warmer (interglacial) cycles has been demonstrated from isotopic changes noted in foraminiferal tests (Emiliani, 1955, 1966). Subsequent work on cores from all oceans has revealed the same trends. Stages in the cores are numbered from 1 for the present (Holocene) warm stage backward in time, with colder stages (glacials or stadials) given even numbers and warm stages (interglacials or interstadials) odd numbers. Boundaries are placed at the mid-point between temperature maxima and minima (Fig. 2.7). Only 16 stages were defined by Emiliani (1966) but the number has since been extended several times (Shackleton and Opdyke, 1973; van Donk, 1976; Kominz *et al.*, 1979; Berggren *et al.*, 1980; Johnson, 1982; Prell *et al.*, 1986; Raymo *et al.*, 1989).

The timescale for marine $\delta^{18}O$ records was initially adjusted to numerical time, based on dated paleomagnetic reversal boundaries and assumed constant sedimentation rates (Shackleton and Opdyke, 1973; Emiliani and Shackleton, 1974) as well as correlations with coral terraces dated with uranium-series isotopes (Broecker and van Donk, 1970). Since Hays *et al.* (1976) demonstrated the

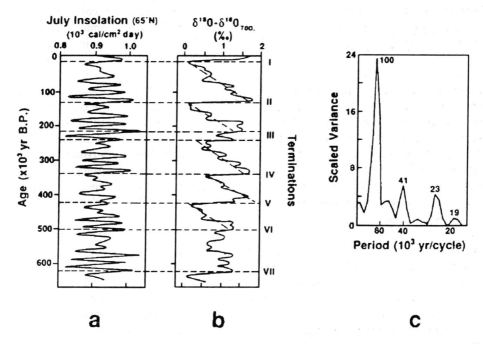

FIG. 2.8. Comparison of Milankovitch July insolation record for 65°N (a) with composite oxygen-isotope record (SPECMAP) from Imbrie *et al.* (1984) (Reprinted with permission of Kluwer Academic Publishers), (b) showing $\delta^{18}O$ spectral analysis (from Broecker and Denton, 1989) (Reprinted from *Geochimica et Cosmochimica Acta,* **53,** Broecker, W.S. and Denton, G.H., The role of ocean atmosphere reorganisations in glacial cycles, 2465–2501, 1989, with kind permission from Elsevier Science Ltd, The Boulevard, Langford Lane, Kidlington, Oxford OX5 1GB, U.K.). (c) Scaled variance per cycles. Horizontal dashed lines mark the terminations while diagonal dashed lines suggest the periods of gradual ice buildup.

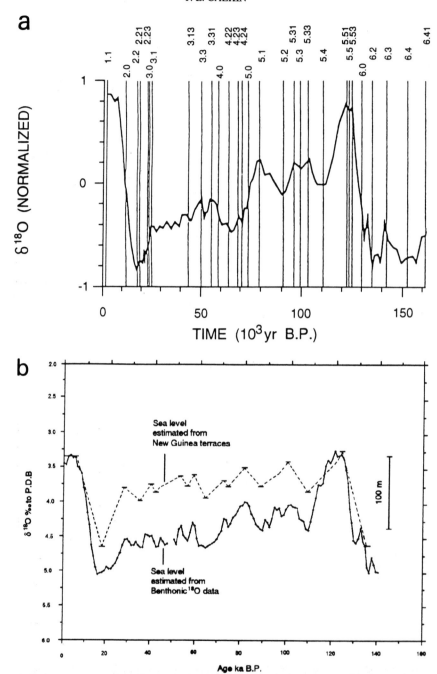

FIG. 2.9. (a) Orbitally-based chronostratigraphy for the last 150,000 years compiled from the stacked oxygen-isotope record of Pisias *et al.* (1984) (from Martinson *et al.* 1987; reprinted with permission of the authors and Academic Press). Numbered vertical lines indicate identifiable features of the record, with age estimates given by Martinson *et al.* (1987). Isotopic points 5.1 to 5.5 correspond approximately with Stages 5a to 5e, respectively. Ages for the last glacial–cycle compare well with ²³⁰Th/²³⁴U measurements on raised coral terraces from various sources. (b) Benthonic oxygen isotope record of East Pacific core V19-30 compared with sea level record derived from raised carbonate terraces of the Huon Peninsula, New Guinea (b,c,d from Shackleton, 1987).

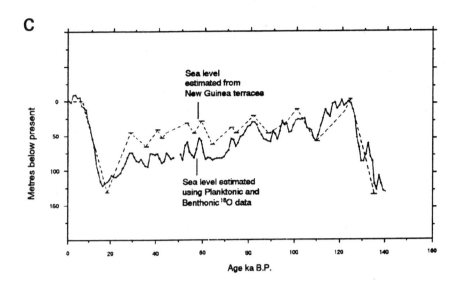

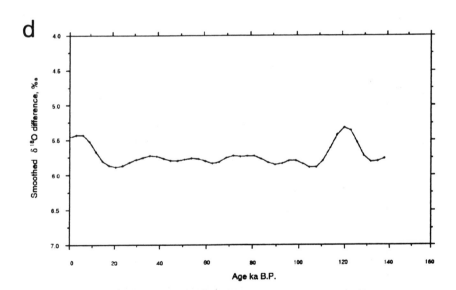

FIG. 2.9. (c) Improved oxygen isotope record of sea level (and ice volume) change obtained from records of Fig. 2.9b after subtraction of the temperature component in the isotope record (shown in lower panel d). (d) A smoothed record of temperature changes during peaks of Terminations II and I (interglacials) obtained by subtracting the planktonic (surface) oxygen isotopes record of West Pacific core RC17-177 (considered a perfect record of isotopic change formed in area of constant temperature) from core V19-30. (Reprinted from *Quaternary Science Reviews*, **6**, Shackleton, N.J., Oxygen isotopes, ice volumes and sea level, 183–190, 1987, with kind permission from Elsevier Science Ltd, The Boulevard, Langford Lane, Kidlington, Oxford OX5 1GB, U.K.)

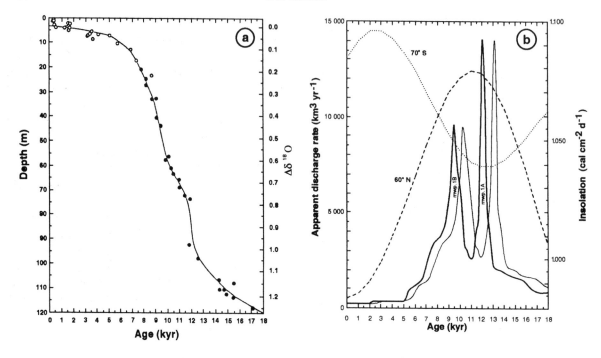

FIG. 2.10. (a) Sea level curve for deglaciation from the last Pleistocene maximum (Late Wisconsinan) when the level was 121±5 m below present level. Curve is scaled to δ^{18}O using the calibration of 0.011‰ per metre change in sea level to provide an upper limit for mean ice volume for global deglaciation (from Fairbanks, 1989; reprinted with permission from *Nature*, **342**, 637–642, 1989, Macmillan Magazines Limited). (b) Rate of glacial meltwater discharge (solid curve) calculated from the Barbados sea level curve (a) compared with summer insolation (dashed and dotted) for two latitudes. The faint curve shows correction for atmospheric ^{14}C changes (from Fairbanks, 1989; reprinted with permission from *Nature*, **342**, 637–642, 1989, Macmillan Magazines Limited). (c) AMS ^{14}C-dated oxygen isotope record for the deglacial interval starting 15,000 BP (from Fairbanks, 1989; reprinted with permission from *Nature*, **342**, 637–642, 1989, Macmillan Magazines Limited). Meltwater pulses spanning the Younger Dryas interval are indicated.

close correspondence of orbital periods in δ^{18}O data, the timescales have been adjusted by matching to orbital variations. This correspondence was achieved for particular latitudes and seasons using insolation curves (Broecker and van Donk, 1970) and later stacking (averaging) of multiple core records from the world's oceans that are matched to curves showing variations in obliquity and precession (Imbrie *et al.*, 1984; Martinson *et al.*, 1987; Rea and Leinen, 1989) (Figs 2.7b, 2.8, 2.9a). Single cores display wide variance in detail from site to site and only broad features can be extracted. It is noteworthy that not all fluctuations and amplitudes detected in core signals can be resolved by orbital parameters. Large climatic changes are commonly detected in the records and frequently occur too rapidly to be explained by orbital forcing.

Typical frequencies noted within cores (Fig. 2.8c) are: (1) a 100,000 year rhythm similar to the periodicity of orbital eccentricity which increases in importance from the Pliocene and dominates the Brunhes Magnetic Chron; (2) a 41,000 year period of obliquity (tilt) which dominates in the Late Pliocene through the Matuyama Chron; and (3) weaker 23,000 and 19,000 year signals corresponding to precessional periodicity that influences mid-latitude insolation levels. The precession signal is amplified when in phase with the earth's orbital eccentricity (Imbrie *et al.*, 1984). The δ^{18}O, δ^{13}C, CaCO$_3$ and the sea surface temperature (SST) all appear to exhibit a coherent, in-phase correlative response to these frequencies in cores. Ice sheets appear to have responded to orbital forcing and then transferred their responses to the surface regionally (Ruddiman *et al.*, 1989) (Section 2.9).

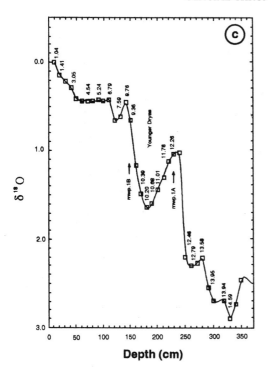

FIG. 2.10. (*continued*)

In the data correlation from deep sea cores, phase lags play an important role. The $\delta^{18}O$ signal, for example, lags behind terrestrial ice volume change by ~500 years during ice sheet shrinkage and by ~3000 years during ice sheet growth. This is a result of the anisotropic composition of ice containing the $\delta^{18}O$ isotope within ice sheets where $\delta^{18}O$ depletion is greatest in the central areas of ice sheets and least on the margins. In the marine cores the $\delta^{18}O$ signal lags insolation, forcing at the year 41,000 year obliquity by ~8000–10,000 years; while at the 19,000 and 23,000 precessional periods a lag of ~5000 to 6000 years occurs (Mix and Ruddiman, 1984; Imbrie *et al.*, 1984; Ruddiman, 1987a). This latter lag has been ascribed to the inherent slow response of ice sheets to climatic change (Weertman, 1964b).

2.4.3. Interpretation of Late Cenozoic Glaciation from the Deep Sea Records

2.4.3.1. The overall record

Initiation of continental ice sheets may have occurred in North America and Europe by ~3.0–3.1 Ma (Einarsson and Albertson, 1988; Raymo *et al.*, 1989) (Fig. 2.7). At about 2.4 Ma (~Stage 100), $\delta^{18}O$ values increased abruptly and $CaCO_3$ suddenly decreased, indicating that ice sheets may have increased from $\frac{1}{4}$ to $\frac{1}{2}$ as large as those of the Late Pleistocene. Also sufficient ice buildup had occurred to initiate the first major episode of marine ice rafting of terrigenous debris into areas south of 40°N (Shackleton *et al.*, 1984; Hoddell *et al.*, 1985; Loubere and Moss, 1986; Raymo *et al.*, 1989).

The ice sheets existing after 2.4 Ma through the Matuyama chron fluctuated at least 40 full climatic cycles with a periodicity of ~41,000 years (Berger, 1978; Ruddiman and Wright, 1987). By 0.9–0.6 Ma, signals indicate a doubling effect such that both the North American and European ice sheets had probably doubled in size (Williams *et al.*, 1988; Ruddiman *et al.*, 1989).

2.4.3.2. Details of the last glacial–interglacial cycle and comparison with data from uplifted coral reef terraces

The $\delta^{18}O$ record for the last 130,000 years (Fig. 2.9) has been divided into five stages and in turn into Substages 5e to 5a (Shackleton, 1969) or 5.5 to 5.1 (Prell *et al.*, 1986). These isotopic events, like earlier stages, can be correlated via deep sea records over widespread areas of the world's oceans. Stage 5e has been correlated with the Eemian interglacial of northern Europe (Mangerud, 1989) and the Sangamon in North America (Fulton, 1986). Terrace studies and deep sea data support a distinct temperature change at this time and indicate that at Stage 5e sea level was ~6 m below present sea level (Fig. 2.9) (Shackleton, 1969; Frenzel and Bludau, 1987).

Subsequent isotopic events (Stages 4–1) mainly represent ice volume changes on the continents. The last global glacial maximum occurred during Stage 2 and was followed by rapid deglaciation (Termination I) to the Holocene (Berger *et al.*, 1987).

Tectonically uplifted coral terraces, for example, in Barbados (Mesolella *et al.*, 1969; Fairbanks, 1989) and New Guinea (Bloom *et al.*, 1974; Chappell, 1983) (Fig. 2.9), dated by carbon-14 and

Uranium-series analyses, have been utilised as first-order indicators of sea level change and hence ice volume fluctuations. The close association of coral growth with specific sea levels validates this method. Detailed records of coral growth curves over the last 150,000 to 240,000 years compare very favourably with $\delta^{18}O$ curves. The growth curves confirm the 20,000 year $\delta^{18}O$ rhythm and provide isotopic ages for sea level maxima at 124,000, 107,000 and 84,000 years ago (Mesolella *et al.*, 1969; Chen *et al.*, 1991). However, there are several 'mismatches' between the growth and isotopic curves that may be a function of temperature effects (Chappell and Shackleton, 1986; Mix, 1987).

A detailed examination of those cores with high sedimentation rates or with stacked (averaged) sets of well-dated cores over the last 20,000 years (Mix, 1987) reveals ice volume maxima during the last glaciation from 20,000–16,000 BP followed by a major deglaciation and a rapid decrease in $\delta^{18}O$ at about 14,000 to 12,000 BP. A second termination occurred between about 12,000 and 9000 BP (Fig. 2.10). These latter periods straddle an interval of cooling in the North Atlantic (Ruddiman, 1987a; Paterson and Hammer, 1987; Fairbanks, 1989) and perhaps elsewhere known in European pollen zone terminology as the Younger Dryas, a term designated after remains of the arctic flower *Dryas octapetala* in deposits near Alleröd, Denmark.

The first, almost continuous sea level curve constructed for deglaciation has been obtained from borings in uplifted interglacial coral reef terraces off Barbados (Fig. 2.10) (Fairbanks, 1989). This curve indicates a sea level 121 ± 5 m below present during the last glacial maximum. Furthermore, the steep parts of this curve, which represent unusually rapid melting and sea level rise, appear to correspond with the $\delta^{18}O$ steps considered above and, in turn, the peak of Milankovitch summer insolation for 60°N at 11,000 BP.

2.4.4. Limitations of the Deep Sea Chronology

The resolution of the oceanic record is limited by the normal slow pelagic sedimentation rates (1–4 cm ka^{-1}), the bioturbation of the bottom sediments and the rate of ocean water circulation. It may take 1000–1500 years for complete ocean mixing (Mix, 1987;

Catt, 1988b). The lags inherent in the orbital, ice sheet and oceanic records lead to inaccuracies and, in addition, the isotopic record has its deficiencies, the most important of which is the unknown contribution of temperature change vs. ice volume to down-core $\delta^{18}O$ records. The temperature contribution is, in turn, tied in part to an understanding of the foraminifera ecology (Mix, 1987; Paull *et al.*, 1991).

2.5. LONG TERRESTRIAL RECORDS OF CLIMATE AND GLACIAL FLUCTUATIONS FOR THE QUATERNARY

Loess deposition is largely a Quaternary phenomenon (Menzies, 1995b, Chapter 6): sequences of alternating loess and dark soils that occur, for example, in central Europe are particularly well known and closely correlated with glacial fluctuations. Very extensive central Asian loess deposits, however, furnish more complete terrestrial paleoclimatic records for the Late Cenozoic. In addition, several long lacustrine sequences with their associated sediments and pollen records extend well back into the Pleistocene and show promise for the future. These types of records are dated using methods such as faunal changes, carbon-14, fission track-tephrochronology, thermoluminescence and paleomagnetic measurements; in addition, many of the records can be correlated directly with the glacial–interglacial $\delta^{18}O$ changes of deep sea records (Kukla, 1977, 1987a, b, 1989) providing details of the Middle Pleistocene. Ice cores, which possess an extremely important and detailed terrestrial record, are considered in a subsequent section.

2.5.1. Loess Sequences of the Czech Republic, Slovakia and Austria

Loess is a silt transported and deposited by wind in unconsolidated and unstratified accumulations that are usually loosely cemented by syngenetic carbonate. Of importance for many Pleistocene glacial correlations are the loess deposits that occur in the unglaciated area between the former limits of Alpine glacier lobes and the Scandinavian ice sheet in central Europe (Fig. 2.11).

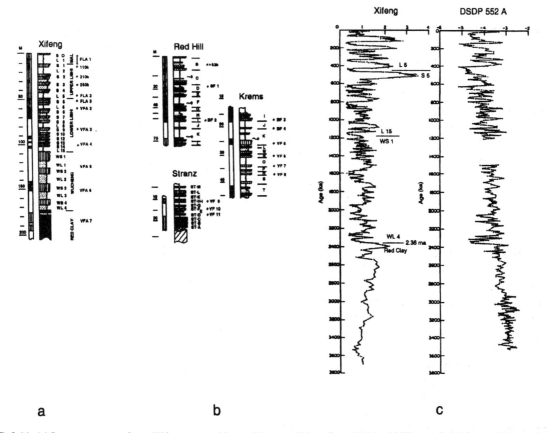

a   b   c

FIG. 2.11. (a) Loess sequence from Xifeng, central Loess Plateau, China (from Kukla, 1987b; reprinted from: *Palaeocology of Africa – and the surrounding islands* (K. Heine, ed.), Volume 18, 37–45, 1987; courtesy of A.A. Balkema, Rotterdam). (b) Combined loess record from three areas in central Europe (from Kukla, 1987b; reprinted from: *Palaeocology of Africa – and the surrounding islands* (K. Heine, ed.), Volume 18, 37–45, 1987). Lithologic columns a and b display soils and soil groups – hatched, strongly weathered rubefied soils and red clays – crosshatched, weathered loess or loess interspersed with hillwash – dotted, other loess – blank, gravels – open circles. SO – WS4 are soil units; L1 – WL4 are loess units; B . U are glacial cycles. Deep erosion shown by horizontal wave lines. VF, vertebrates of Villafranchian age; BF, of Biharian age. (c) A comparison of magnetic susceptibility record in Xifeng loess with the oxygen isotope record of DSDP 552A in central North Atlantic (from Kukla, 1987a; reprinted from *Quaternary Science Reviews*, 6, Kukla, G. Loess stratigraphy in central China, 191–219, 1987, with kind permission from Elsevier Science Ltd, The Boulevard, Langford Lane, Kidlington, Oxford OX5 1GB, U.K.).

As the respective glaciations reached toward their maxima, loess accumulated in the intervening regions under the influence of continental, periglacial tundra-type climates. Cold-resistant gastropod fauna lived in the deposits and soils exhibiting gley conditions and periglacial features and developed under these 'full glacial' conditions. During ice marginal retreats and true interglacial conditions, climates were similar to the present or warmer and wetter. These conditions persisted for several centuries, resulting in leached soils with clay-enriched B horizons (spodosols) and deciduous forest cover. Both parameters are used to help define the European, pollen-based interglacials (Turner and West, 1968; Kukla, 1977). Interstades were recorded in the loess by soils developed under more boreal (northern) coniferous forests and other faunal assemblages. The 17 or more glacial–interglacial loess cycles, defined by these sections, resemble the deep sea record and contrast strongly with the three or four cycles known from the classical studies (Kukla, 1977, 1987a).

2.5.2. Loess Sequences of Central China

The most extensive loess deposits in the world underlie the semi-arid Loess Plateau of central China (Fig. 2.11), where they reach 130–300 m in thickness (Richthofen, 1877; Liu *et al.*, 1986; Kukla, 1987a, b). Other very thick units occur in central Asia (Tadzhikistan and Uzbekistan) (Dodonov, 1981, 1984) and elsewhere, such as near Fairbanks in unglaciated, central Alaska (Westgate *et al.*, 1990; Menzies, 1995b, Chapter 6).

At sites in Luochuan, Xifeng and Xian of central China (Fig. 2.11), a record of worldwide climatic oscillations, comparable with that of deep sea cores, is displayed in a sequence of alternating soil horizons and loess totalling over 200 m. This stratigraphic sequence extends back over the last 2.5 Ma. Here, three primary loess formations unconformably overlie Late Pliocene red clay (Liu, 1987; Kukla, 1987a, b, 1989; Kukla and An, 1989) (Fig. 2.11).

During the cold or windy conditions in central China that accompanied global glaciations, vegetation was sparse and the rate of loess deposition from local sources was high. In contrast, under interglacial conditions, like those today, the Loess Plateau had a temperate, wet climate (Liu *et al.*, 1986) with dense grassland and woodland. Dust deposition, perhaps derived from more distant sources, was thin and may have been incorporated in the strong, decalcified soil horizons typical of interglacial and/or early glacial environments. Incorporated gastropod and vertebrate fauna, pollen and variably concentrated fine ferrous oxides (e.g. magnetite) provide additional climatic signatures.

The red clay underlying the loess formations is a product of fluvial, aeolian and pedologic processes developed under a warm humid climate; the transition to loess at 2.5–2.35 Ma may represent an abrupt climatic deterioration to cold and dry steppe, correlative with inception of major marine ice rafting in deep sea cores. The cooling is so pronounced at 2.4 Ma that it is recognised, in China, as the formal base of the Pleistocene (Liu, 1987). These events in China coincide with the time of uplift of the Tibetan Plateau (Liu *et al.*, 1986), a correlation important in the search for ice age initiation.

A high paleomagnetic susceptibility signal in the interglacial soil related to petrogenic concentration of detrital magnetite and/or in situ production of magnetic particles (Maher and Thompson, 1992) contrasts with the low signal in the cool-glacial loess (Kukla, 1987b; Kukla and An, 1989). These signals produce profiles and a timescale closely parallel with and directly linked to the deep sea oxygen isotope record, both in terms of individual peaks and in their timing (Burbank and Li, 1985; Kukla, 1987a, b; Hovan *et al.*, 1989) (Fig. 2.11). The paleomagnetic evidence delineates 44 major shifts in glacial–interglacial climate over the last 2.5 Ma and indicates 11 prolonged periods of loess deposition, each lasting >40,000 years. Major shifts to cooler and dryer climates were recorded at 2.4, 1.2 and 0.5 Ma (Kukla and An, 1989) (Fig. 2.11).

2.5.3. Pollen Records

Pollen from a continuous 280 m long peat core from Tenaghi-Philippan, Macedonia, Greece, reveals a cyclical vegetation and climatic changes from open steppe to closed hardwood forest over the last 900,000 years (Wijmstra, 1969; van der Wiel and Wijmstra, 1987; Kukla, 1989). Variations in the ratios or abundance of arboreal pollen to that of grass and herb pollen, as well as that in the oak to pine pollen (Fig. 2.12a), allowed differentiation of cycles comparable to the last 25 oxygen isotope stages, from dry cool intervals of open steppe to relatively warmer times under hardwood forests. This record displays, in somewhat more detail, all the major features (e.g long dry intervals corresponding with ^{18}O Stages 16, 12 etc.) of the ^{18}O deep sea records and of the Chinese loess noted above (Kukla, 1989). In addition, the Macedonian sequence shows frequent large amplitude climate oscillations lasting a few millennia throughout the core.

Particularly long lacustrine cores come from the intermontane basins of the tropical Andes and southern Japan, and from the subsiding rift basins of the Dead Sea or northern California (Figs 2.12b, c, d). These cores include sequences up to 900 m that contain multiple tephras datable by potassium-argon, fission-track and paleomagnetic methods. Such pollen cores display very high sedimentation rates that allow detailed extraction of climatic changes not available from the deep sea cores.

Hooghiemstra (1984) has defined 27 major climatic cycles spanning the last 3.5 Ma from a long core in the Andes near Bogota, Columbia (Fig. 2.12b). The core records very cold intervals at 3.5–3.4 Ma, 2.6–2.5 Ma, and again at 1.8–1.7 Ma. Improved dating suggests that the latter two events are equivalent to cooling of 2.4 and 1.7–1.6 Ma determined elsewhere for the deep sea and loess records (Kukla, 1989).

The upper 900 m of a 1400 m core from Lake Biwa, near Kyoto, Japan, has yielded a sequence of 37 pollen zones over the last ~3.0 Ma (Fig. 2.12c). These show fluctuations of subpolar to warm temperate climates associated with glacial–interglacial episodes (Fuji, 1988).

Some of the most important continental climatic records may be obtained from drilling studies initiated in 1989 in the 1600 m deep Lake Baikal, Russia (GeoTimes, 1991). Located between 52° and 58° N, and far from marine influences in unglaciated southeastern Siberia, this area is particularly sensitive to changes in solar insolation.

2.5.4. Summary of Loess and Pollen Core Sequences

The records cited above and others of a similar nature (Kukla, 1989) indicate that 40–50 glacial–interglacial fluctuations are recorded in the middle latitudes of the northern hemisphere at 40,000 and 100,000 year intervals. Ten to twelve of the 20 or so glacial intervals were distinctly longer and more intense than average. Multiple oscillations of stadial–interstadial rank are also recognizable over apparent intervals of several thousand years to roughly 20,000–40,000 years. The most pronounced local cooling and environmental change occurred at about 2.4 Ma (Kukla, 1989).

The long terrestrial records therefore reveal strong parallels with the deep sea records. The climate changes (e.g. monsoon shifts in the case of the loess in China) may be due directly to the buildup of glaciers in high latitudes. However, early climatic changes prior to considerable ice buildup at ~2.4 Ma suggest that both terrestrial climate and ice volume changes represent parallel responses to other triggering events (Kukla, 1989).

An exception to the parallelism of terrestrial climate change and ice buildup is shown in deep sea $\delta^{18}O$ oscillations occurring between 1.9 and 1.2 Ma. Two markedly intense glacial climatic episodes 200,000–300,000 years apart took place at this time, while the deep sea record reveals a monotonous 40,000 year fluctuation (Hooghiemstra, 1984; Zagwijn, 1986; Kukla and An, 1989; Kukla, 1989) (Fig. 2. 13). It has been suggested that this may reflect a cooling on land without major ice buildup (Kukla, 1989).

2.5.5. Vein Calcite Record at Devils Hole, Nevada, U.S.A.

A continuous 500,000 year $\delta^{18}O$ climate record (Fig. 2.13a) has been obtained from a 36 m long core of calcite precipitated along an open fault zone at Devils Hole, Nevada (Winograd et al., 1992). The record 'mimics' major features displayed in marine $\delta^{18}O$ ice-volume records and ice cores from Greenland and Antarctica (Fig. 2.13a, Section 2.5). Because of these similarities, the Devils Hole record is thought to reflect global climate. However, locally, the saw-toothed pattern of changes may be driven by winter–spring land surface temperatures in the southern Great Basin that yield isotopic variations in the local atmospheric precipitation (Winograd et al., 1992).

Of particular note is the minimum timing of specific climatic events, based upon 21 mass spectrometer uranium-series ages. These appear to be at odds with those observed in marine cores considered to reflect orbitally-controlled variations in solar insolation. For example, the intervals spanned by each of the last four glacial cycles recorded in the calcite core increased from 80,000 to 130,000 years, suggesting that climatic changes may have been aperiodic. Furthermore, glacial terminations II and III of the Devils Hole record occurred at least by 253,000 and 140,000 years ago, respectively, or well before those of SPECMAP $\delta^{18}O$ marine records at 244,000 and 128,000 years ago which display major peaks in the high latitude northern hemisphere summer insolation. The 20,000 year long penultimate interglacial is similar to that interpreted from the Greenland Ice Core Project (GRIP) Summit Greenland core and the Vostock ice core, Antarctica. Resolution of the Devils Hole record with Milankovitch pacing is a problem (e.g. Kerr, 1992; Broecker, 1992a, b; Imbrie et al., 1993).

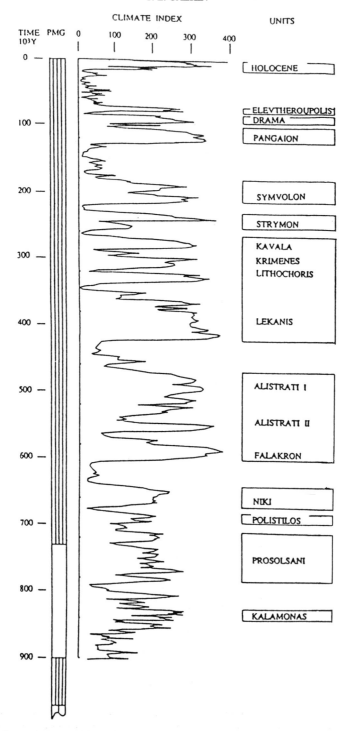

FIG. 2.12. (a) Climate changes expressed by an index based on oak, pine and total tree pollen records obtained by Wijmstra *et al.* from the Tenaghi-Phillippan peat bog in Macedonia, Greece (reproduced from Kukla, 1989, reproduced with permission of Elsevier Science Publishers BV, Amsterdam).

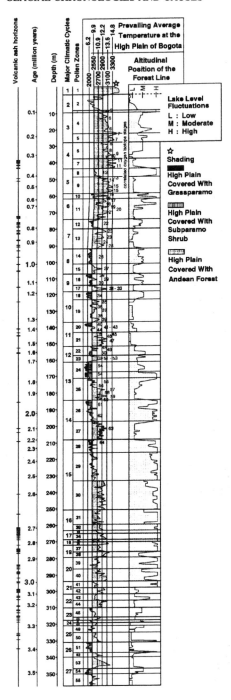

FIG. 2.12. (b) Climate changes, as well as changes in vegetation, lake level and volcanic ash deposition in the high plain of Bogota, Columbia, revealed by a 357 m pollen core through lake sediments (from Hooghiemstra, 1989). The pollen diagram on which this was based (Hooghiemstra, 1984) reveals climatic oscillations registered by altitudinal shifts of the Andean vegetation belts along the slopes of the eastern Cordillera at Funza. Twenty-seven major climatic (glacial–interglacial?) cycles are shown (reproduced with permission of Elsevier Science Publishers BV, Amsterdam).

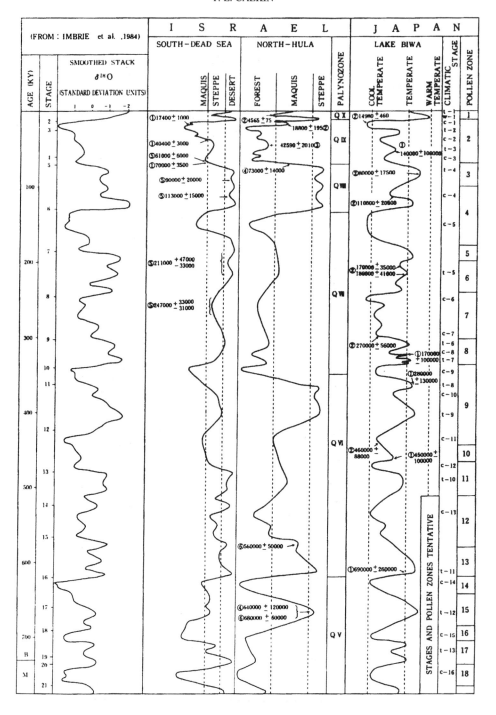

FIG. 2.12. (c) Correlation of climatic records for the Brunhes Chron along the Jordan River–Dead Sea area, Israel, and from Lake Biwa, Japan, based on continuous pollen diagrams of lake sediments (from Fuji and Horowitz, 1989, reproduced with permission of Elsevier Science Publishers BV, Amsterdam).

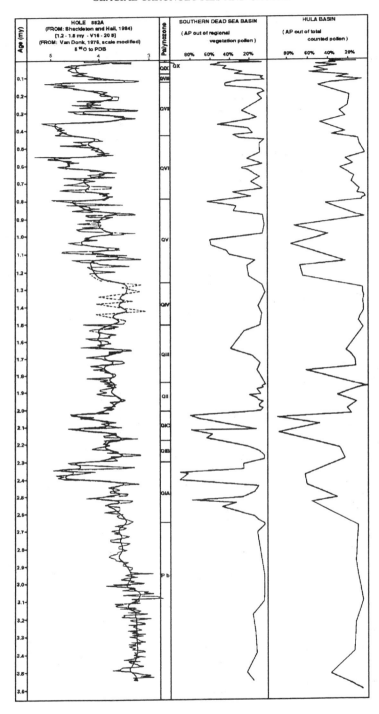

FIG. 2.12. (d) Curves based on the arboreal–non-aboreal pollen from Jordan River valley–Dead Sea area cores compared with the North Atlantic deep sea core 552A (from Horowitz, 1989, reproduced with permission of Elsevier Science Publishers BV, Amsterdam).

2.6. THE LATE PLEISTOCENE CLIMATIC RECORD REVEALED BY DEEP ICE CORES

2.6.1. General

A valuable and continuous record of Late Pleistocene paleoclimate and paleoenvironmental changes, especially through the last glacial cycle, and hints of their cause can be obtained from ice cores through isotopic and chemical analysis of ice (Dansgaard *et al.*, 1973), entrapped gases (Neftel *et al.*, 1982; Stauffer *et al.*, 1985) and particulate matter (Petit *et al.*, 1981) taken through both polar (high latitude) ice sheets and mountain glaciers (Plate 2.1).

The longest and most useful cores are obtained from the polar ice sheets where there is no surface melting and ice temperatures are usually below the pressure melting point (Figs 2.13a, b). A stratigraphic record encompassing ~150,000 years through the penultimate interglacial has also been extracted from surface samples along the outer edge of the ice sheet in central west Greenland near Pakitsaq. This is within the ablation zone where the older ice layers emerge (Reeh *et al.*, 1991). Long cores have been obtained from Devon Island Ice Cap, Canada; Camp Century, Renland, Dye 3, Greenland Ice Sheet Project 2 (GISP2) and GRIP sites in Greenland; and Dome C, Byrd, and Vostock sites in Antarctica (Figs 2.13a, b, 2.14, 2.15). All of these extend back chronologically into the past Pleistocene maximum and beyond (Paterson and Hammer, 1987; GRIP Members, 1993; Jouzel *et al.*, 1993). A larger diameter core of the GRIP2 Project at Summit (Wahlen *et al.*, 1991) reaches 1.5 m into bedrock below the ice sheet for a depth of 3054.3 m (Grootes *et al.*, pers. commun. 1993). Twenty kilometres away at the summit divide of the central Greenland Ice Sheet is the 3028.8 m core complete to the ice sheet base by the European multinational GRIP project (Fig. 2.15) (GRIP Members, 1993). Both records are believed to extend back through 250,000 years or two interglacials before the Holocene (Morrison, 1993). A 2546 m core from Vostock Station, Antarctica (core 4Γ, Fig. 2.14) succeeds a somewhat shorter one (core 3Γ) (Lorius *et al.*, 1985, 1989) and spans the last 220,000 years (Jouzel *et al.*, 1993). Future coring to bedrock at ~3700 m is possible.

The extraction of paleoclimatic data from ice cores is not without problems; for example, it is difficult to date cores to within errors ±5% (Bradley, 1985; Lorius *et al.*, 1985) and ice deformation and diffusion can destroy data in the lower portions. Nevertheless, these records provide much better resolution of short-term change over the last glacial cycle than deep sea cores (Paterson and Hammer, 1987). Ice cores are not disturbed by bioturbation as are marine cores, and environmental changes of the last glaciation are measured through annual snow or ice layers that, although gradually compressed and stretched as they are buried (Fig. 2.16), span intervals of hundreds to thousands of metres compared with a

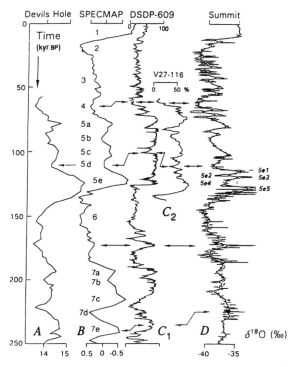

FIG. 2.13. (a) Climate records plotted to a common linear timescale (modified from Dansgaard, 1993). (a)δ^{18}O variation in vein calcite, Devils Hole, Nevada (Winograd *et al.*, 1992). (b) Orbitally tuned SPECMAP δ^{18}O curve (Martinson *et al.*, 1987). (c) Part 1, grey-scale measurements along marine core at site DSDP 609; Part 2, %Ca CO₃ in core V27-116 from WGW + W of Ireland. (d) δ^{18}O along upper 2982 m of the GRIP ice core, Greenland. (e) δD record from Vostock, East Antarctica, converted to a δ^{18}O record ($\delta = 8.\delta^{18}$O + 10%). Interpretation in European terminology. (Reprinted with permission from *Nature*, **364**, 218–220, 1993, Macmillan Magazines Limited.)

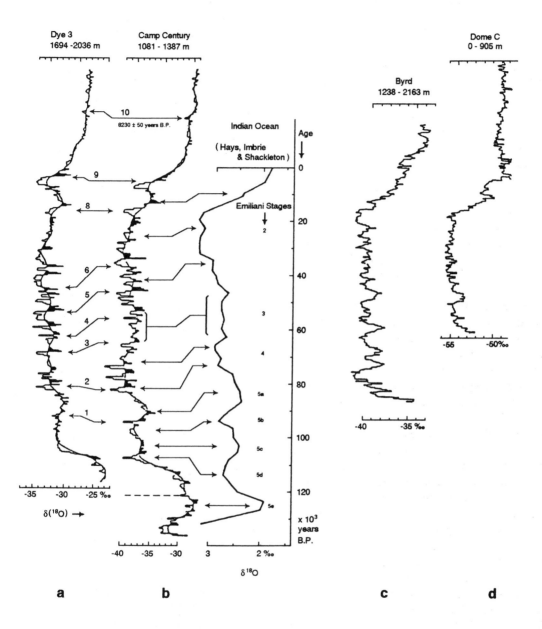

FIG. 2.13. (b) δ18O profiles along the five deep ice cores from Greenland (Dye 3 and Camp Century) and Antarctica (Byrd and Dome C) plotted on linear depth scales for intervals indicated (modified from Dansgaard, 1987; reprinted with permission of Kluwer Academic Publishers). Ages on Greenland core arrows and Byrd and Dome C, Antarctica, logs are added from Paterson and Hammer (1987). The Late Wisconsinan maximum is estimated to be at a depth of 1854 m in the Dye 3 core where the δ18O is at a minimum and the 10Be and dust reach their highest concentrations. Planktonic foraminiferal δ18O profile from Hays et al. (1976) has been adjusted to Camp Century profile (now believed to end within Stage 5e) as plotted by Dansgaard et al. (1982). (Reprinted with permission from the authors and Science, **218**, 1273–1277, 1982. Copyright 1982 by AAAS.)

few metres or more in the deep sea cores. Finally, the multiple proxy records of climate via chemical ions, gases and solid particles provided by ice cores allows the development of models that link temperature and precipitation ($\delta^{18}O$, dD), solar variability (^{10}Be), atmospheric chemistry (CH_4) or biological productivity (CO_2, CH_4, $\delta^{18}O$ in air bubbles) and atmospheric circulation (turbidity, storminess — dust) (Hecht *et al.*, 1989, p. 384). These parameters strongly reflect climatic variations of global and local interest and together help reveal the causes of climatic change.

2.6.2. Principles of Ice Core Sample Analysis

The oxygen-isotope ratio ($^{18}O/^{16}O$) or the deuterium to hydrogen ratio (D/H) provide very important primary data from which one can infer surface temperatures and precipitation rates. The δ ratios, expressed as the relative deviation from the concentrations in standard mean ocean water (SMOW) as δ ‰ (per mil) (Section 2.4.1), largely depend on the condensation temperature of the precipitation where the snow formed (Dansgaard *et al.*, 1973). As previously discussed, differences in the vapour pressure (an equilibrium effect) and in molecular diffusivity (a kinetic effect not directly temperature-related) cause isotopic fractionation of the H_2O, HDO and $H_2^{18}O$ molecules.

During cycles of evaporation and condensation from the oceans, molecules of $H_2^{18}O$ and HDO evaporate less readily than $H_2^{16}O$. Thus, as water is moved from its original site of evaporation and toward the cooler polar areas, condensation occurs, producing precipitates of progressively more negative δ_s ($\delta^{18}O$ and δD). Cooling is less important over the relatively stable sea surface than it is over the polar ice sheets where the air mass is orographically cooled toward the ice sheet centre (Paterson and Hammer, 1987).

Temperatures have usually not been computed from the δ_s since many factors other than climate can alter these, including changes related to precipitation source area, seasonal distribution of precipitation,

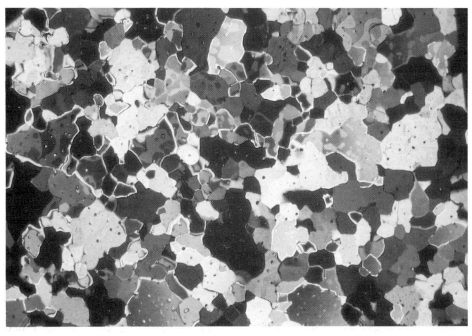

PLATE 2.1. Thin section of an ice core sample taken from 160 m below snow surface at Camp Milcent, Greenland. Section is photographed under polarized light, showing the irregular shape of individual ice crystals after nearly 300 years of transformation from snow flakes to a polycrystalline aggregate. The diameters of the large crystals are approximately 1 cm. The colours depict the different orientations of individual crystals (photo courtesy of Chester C. Langway Jr).

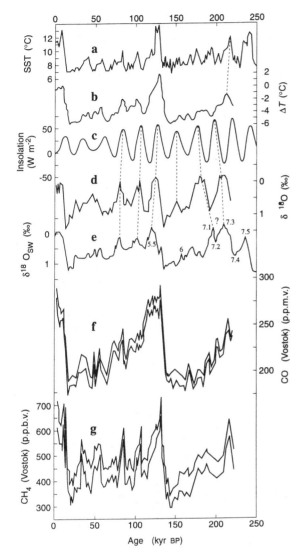

FIG. 2.14. Vostock ice core records and correlative marine and orbital variations (modified after Jouzel *et al.*, 1993, reprinted with permission from *Nature*, **364**, 407–412, 1993, Macmillan Magazines Limited). (a) Summer sea surface temperature at Indian Ocean site; (b) Atmospheric temperature change at Vostock derived from the deuterium data of cores 3Γ and 4Γ; (c) Summer insolation at 20°N; (d) Vostock δ¹⁸O profile; (e) Isotope change in sea water from marine V19-30 δ¹⁸O record; (f,g) Vostock CO₂ and CH₄ profiles, respectively.

changes in pattern of ice flow and thickness with time (Paterson and Hammer, 1987). However, in most polar areas, there is a linear relationship for modern precipitation between annual means of the δ_s and

surface temperature (Johnsen *et al.*, 1992). Deuterium (δD) profiles in the Vostock core, for example, are translated into surface temperature using a 6‰/°C slope based on this relationship (Jouzel *et al.*, 1987, 1989). The δD rather than δ¹⁸O is used in temperature approximations because the deuterium concentrations are considered to be less affected by the kinetic effects during fractionation (Fig. 2.14) (Jouzel *et al.*, 1989). Both seasonal and long-term temperature changes are recorded: however, the isotopic composition reveals weather data during the short periods of snowfall and *not* during the much longer intervals between precipitation (Dansgaard *et al.*, 1984). Past precipitation rates may be estimated using ¹⁰Be concentrations. Beryllium-10, which is produced by cosmic radiation in the atmosphere, becomes attached to atmospheric aerosols and is removed from the atmosphere during precipitation after a mean residence time of 1–2 years. In addition, in polar areas, particularly Antarctica, the present precipitation rate is strongly correlated with surface temperature (Robin, 1977). This allows precipitation rates to be estimated from the temperature data in Antarctica (Jouzel *et al.*, 1987).

2.6.3. Dating of Ice Core Profiles

The most precise method of obtaining age–depth relationships is by counting seasonal changes distinguished by δ¹⁸O, dust particles or acidity (Langway, 1970; Dansgaard *et al.*, 1973, 1993; Hammer *et al.*, 1985). This is undertaken where accumulation rates are high (e.g. >25 cm a⁻¹) and at some height above the basal layers where diffusion may make counting impossible (Fig. 2.16). The chronology of the GRIP core (Fig. 2.15) is determined by counting annual layers back 14,500 years and, beyond this date, by ice flow modelling (GRIP Members, 1993). Annual layers have been counted back almost 80,000 years in the GISP2 core (Taylor *et al.*, 1993; P. Mayewski, *pers. commun.*, 1993).

Beyond the Holocene or latest Pleistocene in the case of GISP2, less precise methods are used such as: (1) matching reference horizons of known age, for example, in the Byrd core by volcanic acidity maxima (Hammer, 1989); (2) radioactive dating of carbon-14

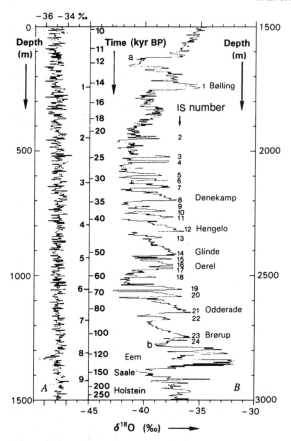

FIG. 2.15. Continuous GRIP δ¹⁸O record from Summit, Greenland, plotted with 2.2 m increments in two sections on a linear depth scale (modified after Dansgaard *et al.*, 1993). (a) From surface to 1500 m; (b) from 1500 to 300 m depth. IS, interstadials with longer ones assigned European pollen horizons. YD, Younger Dryas; OD, Older Dryas; and A, Alleröd have been added. (Reprinted with permission from *Nature*, **364**, 218–220, 1993, Macmillan Magazines Limited.)

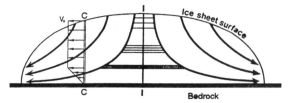

FIG. 2.16. Ice flow, compression and deep ice-core dating in an ice sheet (after Dansgaard *et al.*, 1969). Bold arrows show path of ice particles in an ice sheet to the margin where it ablates away or forms an iceberg. Annual layers (centre horizontal lines) are stretched and therefore thinned as they approach bedrock. Deeper layers encountered in core C-C originate farther inland. Ice velocity components V_x reach zero close to the bedrock if the ice is below the pressure melting point at the base. (Reprinted with permission from the authors and *Science*, **166**, 377–381, 1969. Copyright 1969 by AAAS.)

2.6.4. Summary of Major Environmental Changes Revealed by Ice Cores and Possible Causes

2.6.4.1. The overall record

The ice core δ records qualitatively show the same general features as do the deep sea δ records over the last glacial cycle and some important differences during earlier intervals (Fig. 2.15). The ice core timescales are independent of orbital tuning used in the marine cores; thus they may provide a valuable check on changes of environmental parameters obtained from cores in adjoining seas. These environmental parameters appear to have frequently shifted rather simultaneously.

The GRIP and Vostock profiles (Figs 2.13a, 2.14), for example, display two interglacials preceding the Holocene and both show the penultimate and last glacial maxima. The Vostock atmospheric temperature profile (Fig. 2.14) indicates three similar, well-marked temperature minima during the last glaciation that appear to follow the Milankovitch tilt cycle at about 100,000, 60,000 and 20,000 years (Jouzel *et al.*, 1993); the stadials are separated by two interstadials. These are similar to fluctuations shown in the marine SPECMAP profile. The strong relationship between local annual insolation, which largely reflects obliquity, is absent between 200,000

from CO_2 bubbles; (3) by $^{36}Cl/^{10}Be$ ratios, or for a shorter term, ^{32}Si or ^{210}Pb; (4) matching of the δ_s record with another dated climatic record (principally deep sea records (Dansgaard and Oeschger, 1989)); and (5) by the less precise method of ice-flow models (Chapter 3). Ice-flow models are used in the lower regions of most polar ice cores and for the entire length of the two last Vostock cores, where accumulation rates are too low for annual counting (Lorius *et al.*, 1985, 1989; Jouzel *et al.*, 1993).

and 40,000 years ago. This penultimate glaciation was particularly long and cold in East Antarctica compared with the last glacial interval. The Vostock interstadials and others recognized on the more precise chronology of GRIP (Figs 2.14, 2.15) can be reconciled provisionally with European pollen zones.

Classification of the penultimate interglacial (Eem, Sangamon) record has been a major contribution of the Vostock and GRIP core interpretations (GRIP Members, 1993; Jouzel *et al.*, 1993). These interpretations suggest that the Eemian (marine isotope stage 5e) began ~10,000 and 5000 years, respectively, before that age shown by marine ^{18}O data (e.g. SPECMAP) and lasted about 20,000 years (135,000–115,000 years ago). The Devils Hole record is similar to the GRIP interval but distinct from the 11,000 year Eemian interval obtained from SPECMAP. The GRIP temperature changes are believed to precede sea level changes by ~4000 years (GRIP Members, 1993).

Both GRIP and Vostock records also show that warm intervals of the Eemian were ~2°C higher than present. However, unlike the less refined Vostock records, early modelling of the GRIP core (Fig. 2.15) suggests that temperatures plunged from 2°C warmer than today to –5°C colder in as little as a few decades during two major intervals within the Eem (GRIP Members, 1993). These mode switches, which also occur in the greenhouse gases and other chemical and physical properties of the ice core, are considered to mark large changes in atmospheric and ocean circulation patterns over the North Atlantic. This preliminary GRIP interglacial record, therefore, contrasts markedly with all the data showing climatic stability through the Holocene. GISP2 data appear to indicate a somewhat different history before the last glaciation (P. Mayewski, *pers. commun.*, 1993).

Greenland ice cores, including GRIP (Johnsen *et al.*, 1992; Dansgaard *et al.*, 1993) and GISP2 (Taylor *et al.*, 1993; Alley *et al.*, 1993) and the previous long Greenland records (Dansgaard *et al.*, 1984; Dansgaard and Oeschger, 1989) show abrupt mode switches in the isotope as well as dust and gas records (Figs 2.13a, b). The mode switches are only weakly shown in Antarctic cores (Jouzel *et al.*, 1987). The switches, sometimes called 'interstadials' (Johnsen *et al.*, 1992) or 'Dansgaard–Oeschger' cycles (Kerr, 1992), take about 50 years and represent $\delta^{18}O$

changes of 2‰–3‰ or temperature shifts averaging 5°–7°C. These changes occur every 500 to 2000 years, at least between 40,000 and 20,000 years ago. The Younger Dryas cooling beginning ~13,000 years ago represents the most recent of these major cycles (Dansgaard, 1987; Alley *et al.*, 1993). Furthermore, electrical conductivity studies of the GISP2 records show fluctuations (called 'flickers', Taylor *et al.*, 1993) of dust on scales of less than 5–20 years within and between these longer term, warm–cold Dansgaard–Oeschger cycles. All these events seem to depend on rapid reorganization in atmospheric circulation and may be, in turn, related to modulations in the strength of the North Atlantic Deep Water circulation pattern.

A total temperature amplitude of about 11°–12°C is suggested over the last 100,000 year cycle for both Greenland and Antarctic data with maxima of 14°C. This cycle is compatible with the 14°C sea surface temperature anomaly estimated by CLIMAP (1976) for the North Atlantic from deep sea core records.

The estimated precipitation rates from d_s and ^{10}Be in the Vostock core (Lorius *et al.*, 1989) follow the same pattern as the temperature, suggesting that the precipitation rate was about 50% lower during the glacial conditions of the Pleistocene than through the Holocene. Mean precipitation rate measurements on Greenland cores indicate that during the Late Wisconsinan the rate was about one-third of the Holocene (Herron and Langway, 1985; Paterson and Hammer, 1987; Reeh, 1990).

The abundant presence of impurities (dust and ionic constituents) in the Greenland and Antarctica ice have a strong relationship with temperature, showing high concentrations during the coldest Late Pleistocene Stages relative to the Holocene. Concentrations of dust in Greenland's Dye 3, for example, core build up irregularly from the Sangamon to the last glacial (Wisconsinan) maximum, reaching levels of 3–70 times that found in the Holocene ice. These increases may reflect more extensive areas of aridity, greater exposure of continental shelves due to low sea level, and more efficient meridional transport linked to high wind velocities during times of large ice sheets (Paterson and Hammer, 1987).

Carbon dioxide and methane records in Greenland

(Dansgaard and Oeschger, 1989) and CO_2 in Antarctica (Genthon *et al.*, 1987; Stauffer *et al.*, 1988; Jouzel *et al.*, 1993; Peel, 1993; Raynaud *et al.*, 1993) display a similar high, in-phase correlation with the climatic change (Figs 2.14f, g, h). Both are infrared active gases that contribute to the greenhouse effect; however, CH_4 is 21 times as powerful a greenhouse gas as CO_2. Both gases have also been increased markedly by anthropogenic emissions. The ice core records are characterized by a CO_2 and CH_4 long-term decrease during cold times, reaching a minimum at 20,000 BP during the last glaciation, for example, before concentrations increased during the shift to warmer times (Dansgaard and Oeschger, 1989). The correlation of CO_2 and temperature records differ in some details and CO_2 changes also differ between some core intervals in Antarctica and Greenland. For example, Greenland CO_2 values show major fluctuations of lowered CO_2 that appear to persist for periods of millennia. These correspond with parallel highs in $\delta^{18}O$, dust and ^{10}Be concentrations (Dansgaard and Oeschger, 1989).

In terms of leads and lags between the atmospheric gas abundances and temperature changes shown by the ice cores, there are some contradictions in the details of the generally in-phase relationship. For example, Antarctic ice core records suggest that CO_2 levels had risen by 35 ppmv before continental ice began to be noticeably reduced during the penultimate interglacial. Similar phase relationships are recorded by the deep sea cores (Raynaud *et al.*, 1993). During the onset of glaciation, the fall of CO_2 levels in the Vostock core lag behind the temperature drops by several thousand years (Jouzel *et al.*, 1993). However, the CH_4 changes are roughly in-phase with the cooling. Furthermore, the GRIP data from Greenland show that CH_4 levels closely parallel the inferred temperature changes, including those during periods of extremely unstable climate late in the last glacial cycle and even during the Little Ice Age (Peel, 1993).

Such strong linear relationships of CO_2 and especially CH_4 with temperature and other climatic parameters through the Greenland and Antarctic ice cores suggests that these gases are important factors that amplify the relatively weak orbital forcing of temperature change. There is a strong relationship

between orbital and solar radiation spectra in the Vostock ice core (Genthon *et al.*, 1987). The most direct drivers of atmospheric CO_2 and CH_4 changes are ocean current mechanisms and continental wetland changes, respectively (Raynaud *et al.*, 1993).

2.6.4.2. The last deglaciation

The $\delta^{18}O$ minimum (Late Wisconsinan maximum), dated in Greenland ice cores at approximately 18,000 BP and in Antarctica at about 20,000 BP (Fig. 2.13) was followed by a general warming trend until about 17,000 BP when Greenland cores show major temperature drops. These represent the last Dansgaard–Oeschger δ_s cycles and intervals that are closely correlative with northern European pollen zones. The Older Dryas, the first of these coolings, ended with a sudden increase δ_s and temperature at about 14,700 BP that correlates with the Bølling–Allerød warming (Fig. 2.15). The Younger Dryas interrupted this warming about 13,000 years ago with a temperature drop to full glacial conditions that lasted until 11,500 years ago according to annual layer counting in the GISP2 core (Alley *et al.*, 1993; Taylor *et al.*, 1993). The shifts to warm intervals following the Older and Younger Dryas took place in about 50 years according to isotope data, but as fast as 3–5 years based upon the dust profiles. The end of the Younger Dryas was marked by a temperature rise of about 7°C and a doubling of snow accumulation in Greenland (Alley *et al.*, 1993). Antarctic cores show more muted reversals during these and earlier intervals (Jouzel *et al.*, 1987).

The interpretation, extent and causes of the Younger Dryas or last major Dansgaard–Oeschger cycle have been extensively debated (Berger and Labeyrie, 1987; Broecker *et al.*, 1988; Dansgaard *et al.*, 1989; Fairbanks, 1989; Berger, 1990; Bond *et al.*, 1992; Alley *et al.*, 1993). While most of the various hypotheses have centred around changes in North Atlantic currents, ice sheet changes may also play a part in the temperature cycles. The Younger Dryas and certainly other coolings extending back through at least several hundred thousand years appear to correspond with one of many intervals of prolific iceberg sweeps across the North Atlantic Ocean. Called 'Heinrich events' after their discoverer

(Heinrich, 1988), these occurred at intervals of about 12,000 years over at least the last 250,000 years.

Whether the Pleistocene–Holocene transition occurred as a result of rapid climatic jumps (Taylor et al., 1993), or more gradually and uniformly as some data suggest (Ruddiman, 1987b), typical Holocene temperatures were established by about 9000 BP simultaneously in Greenland and Antarctica (Paterson and Hammer, 1987). The total shift in δ_s during the transition in Greenland cores varied from 7‰ to 11‰ representing a local warming of 10.5–16.4°C (Dansgaard and Oeschger, 1989), somewhat higher than in Antarctica. The Greenland cores, however, display a greater amplitude of climatic change than is recorded at lower latitudes during such global events (Lamb, 1977).

Coincident with the drop in $\delta^{18}O$ was a general increase in CO_2 from Pleistocene values of 18°–280 ppmv. By itself this fluctuation in CO_2 would have caused a significant change in the global radiation balance and temperature rise (Oeschger et al., 1984). With a positive feedback system established through such mechanisms as a reduction in sea ice cover and increasing vegetation cover, overall lowered percentage of reflected insolation and ocean current changes, the CO_2 may have been a major factor in rapid deglaciation.

Shorter-term climatic fluctuations are more difficult to extract from $\delta^{18}O$ ice core records than are the large amplitude fluctuations partly as a result of large variations between drill sites (Robin, 1983). Nevertheless, ice-core data from some Greenland cores suggest that Holocene ice surface temperatures reached a long-term maximum 1°C to 2°C above present temperatures between about 6000 and 3000 BP (Robin, 1983; Dansgaard and Oeschger, 1989). However, elevated snow accumulation shown in the GISP2 core between 9200 and 7300 years ago suggest that this may have been the Holocene 'climatic optimum' or Hypsithermal warm interval (Meese et al., unpublished results). Antarctic cores indicate relatively high δ_s and high temperatures about 7500–5000 BP. These ages correspond with a range of estimates of the Early–Middle Holocene temperature decline (Hypsithermal) recorded by other proxy data (Robin, 1983; Bradley, 1985; MacCracken et al., 1990).

The δ_s quasi oscillation of approximately 2500 years recorded in the Greenland cores, although less pronounced during the Holocene, correlates crudely with mountain glacier oscillations that took place during the post-Pleistocene interval (Denton and Karlén, 1973; Dansgaard et al., 1984; Denton et al., 1986). Nevertheless, general cooling since the Hypsithermal (Fig. 2.15) culminated between about 1200 and 1900 AD in the most pronounced Holocene advance of mountain glaciers. This is referred to as the 'Little Ice Age' (Lamb, 1977; Porter, 1986; Grove, 1988).

Both the Camp Century and Dye 3 core appear to display an oscillation of local surface temperatures of 1–3°C lower than average over the period of the Little Ice Age. Accumulation records from GISP2 increases and decreases that correspond with more warming as well as cooling episodes within the Little Ice Age (Meese et al., unpublished results). Another representative index curve for Greenland was assembled from four $\delta^{18}O$ profiles and compared with northern hemisphere temperatures and volcanic acidity data (Fig. 2.17). Shorter ice cores from the tropical Quelccaya Ice Cap suggest that the Little Ice Age occurred in South America between 1490 and 1880 A.D. with onset and termination within a few decades (Thompson et al., 1986).

2.7. CORRELATION OF LATE CENOZOIC GLACIATIONS IN THE NORTHERN AND SOUTHERN HEMISPHERES

Figure 2.18a from Bowen et al. (1986b) represents an attempted correlation and dating of Quaternary ice sheet and mountain glaciations across Eurasia, Europe and North America. The chart and the regional compilations from which this was extracted (Šibrava et al., 1986) is an outgrowth of the International Geological Correlation Program (IGCP) begun in 1971. The primary bases for the correlation are the even-numbered stages of the oxygen isotope signal of deep sea cores to which the glaciations have been ascribed.

Figure 2.18b from Clapperton (1990a, b) is constructed around the same chronological framework as was Fig. 2.18a. However, detailed records are limited because, except for the Antarctic

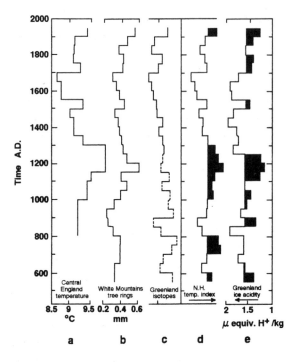

FIG. 2.17. Correlation of northern hemisphere temperature variations (curve d integrated from a, b and c), with volcanic activity expressed from the Crete ice core, central Greenland, acidity profile with 50 year average values (from Hammer *et al.*, 1980). Greenland isotopes (c) represents combination of normalized δ¹⁸O curves back to 1250 A.D. and to 550 A.D. Blackened portion at 1200 A.D. represents medieval warm interval prior to Little Ice Age cooling (1250–1850 A.D.). (Reprinted with permission from *Nature*, **288**, 230–235, 1980 Macmillan Magazines Limited.)

ice sheet, only two percent of the land area lies in cool temperate latitudes similar to glaciated areas of the northern hemisphere (Clapperton, 1990b). The discussion of basic assumptions below is taken largely from Bowen *et al.* (1986b) and Richmond and Fullerton (1986a, b).

2.7.1. Basic Assumptions

The primary indicators used for actual glacial advance and retreat in northern hemisphere correlations (Fig. 2.18a) were units such as tills or proglacial deposits. Proxy indicators such as loess, or 'cold' periods marked by pollen in non-glacial deposits were considered secondary and are not shown on the chart. However, glacial advances in the

European Alps were based on the river terraces since they are tied to glacial deposits (Bowen *et al.*, 1986b). Each glaciation is indicated by a wedge for which the bottom line indicates advance and the top line retreat. Wedges, generally, are not proportional in size to glaciation magnitude except for latest Pleistocene with Holocene maxima shown for Canada.

Chronological control from dating of lava and tephra by potassium–argon and fission-track methods allowed bracketing or limiting ages and reliable independent chronometric control in parts of the western and mid-continent United States. The latest 50,000 years of the Pleistocene chronology may be the most reliable, especially where carbon-14 ages are applicable in the central Great Lakes area. In Europe, some chronometric methods such as carbon-14 and particularly thermal luminescence have been applied in correlation; however, lithological, paleontological and archaeological techniques, and to a lesser extent, paleomagnetic data form the major basis of what must be even more provisional correlations.

As a basis for correlation, synchroneity of glacial events is assumed, although this is problematic (Richmond and Fullerton, 1986a, b). These deposits are separated from one another based upon the presence of non-glacial units, tephras, major unconformities, or paleomagnetic reversals. Furthermore, the glacial deposits, or glaciations they represent, are grouped, where possible, into arbitrary time divisions of the even-numbered marine oxygen isotope stages (Shackleton and Opdyke, 1973; van Donk, 1976; Richmond and Fullerton, 1986a). Other timescales have been used (e.g. SPECMAP of Imbrie *et al.*, 1984) which yield a similar classification.

2.7.2. The North American Record

In North America, several pre-Illinoian glaciations of the Laurentide and Cordilleran Ice Sheets, as well as of mountain glaciers, have been inferred (Hallberg, 1986; Johnson, 1986; Richmond and Fullerton, 1986b). Major mountain glacier advances occurred by Miocene time in Alaska (Hamilton, 1986) (Section 2.2.5) and probable Late Pliocene mountain glacier fluctuations are recorded in the Rocky Mountains of Wyoming and the Cascades of Washington, U.S.A. (Fig. 2.18a, glaciations L, J and/or K). The Cordilleran continental Ice Sheet, a mountain ice complex that spread over the north–south trending mountain ranges west of the Canadian plains (Prest, 1984), reached into Washington State to form the Puget Sound lobe by Late Pliocene or Early Pleistocene time. This advance is interpreted from paleomagnetic signatures, stratigraphy and weathering characteristics of tills (Fig. 2.18a) (Easterbrook, 1986).

The evidence to indicate that an advance of the Laurentide Ice Sheet occurred in the Late Pliocene is based upon a specific glacial diamicton in the plains of Iowa. This till displays a paleomagnetic reversal that can be assigned to Glaciation K (Fig. 2.18a). A possible correlative till occurs in western Wisconsin (Boellstorff, 1978; Hallberg, 1986; Matsch and Schneider, 1986; Richmond and Fullerton, 1986b). Further conclusive evidence confirming this event is required since other advances of the Laurentide or Scandinavian Ice Sheets are unknown until ~0.9 Ma (Nilsson, 1983; Kukla, 1987b).

Deposits of possible glaciofluvial origin in the Canadian prairies are also inferred to be of Late Pliocene or Early Pleistocene age (Fig. 2.18a) (Fulton *et al.*, 1986). The occurrence of continental glaciation at about 2.2 Ma is consistent with the $\delta^{18}O$ maximum recorded by major ice rafting and other evidence of sudden climate change (Section 2.4.3.1).

During the interval 1.5–0.9 Ma, no persuasive evidence exists of glaciation in the United States (van Donk, 1976; Fullerton and Richmond, 1986b). Nevertheless, mountain glaciers in the southern hemisphere may have reached their greatest extent during this period (Fig. 2.18b).

Beginning at Isotope Stage 22, glaciations have been inferred to have occurred in the mountainous areas of western North America and in areas influenced by the Laurentide and Cordilleran Ice Sheets. These glaciations occurred at intervals of 40,000 to 140,000 BP, corresponding to the $\delta^{18}O$ stages.

Late Middle Pleistocene (Illinoian) glaciation is recorded by an early, as well as a two-pronged, late glaciation; a relatively well-developed soil marks the non-glacial interval in the type area of Illinois (Johnson, 1986). The Sangamon paleosol, which is markedly dichronous, is developed locally on these Illinoian units in the mid-western United States.

A particularly extensive terrestrial sediment record illustrating the Sangamon interglacial is found at sites along 1600 km of the southwest margin of Hudson Bay, Canada (Dredge and Cowan, 1989, p. 232). These sediments may represent an early glacial lake formed by the receding Laurentide Ice Sheet margin, a high level sea that later regressed to levels similar or below present, a subaerial interval of deposition and a possible succeeding proglacial lake that preceded ice margin readvance over the area.

Early Wisconsinan advances are generally considered to be less extensive than those of the Late Wisconsinan, but a considerable increase in global ice volume did occur during isotope Stage 4 (Fig. 2.9). Advances along the southern margin of the Laurentide Ice Sheet are recorded from Michigan eastward into New England (Richmond and Fullerton, 1986b; Dreimanis, 1991); however, some of these advances may be of older or younger age because dating is insecure. During the Middle Wisconsinan interval, the Cordilleran Ice Sheet may have disappeared (Fulton *et al.*, 1986); the southern margin of the Laurentide Ice Sheet was restricted generally to Quebec, eastern Ontario and possibly north central New York State (Fullerton, 1986a; Andrews, 1987c; Dreimanis, 1991). At least partial deglaciation and marine invasion of Hudson Strait and Hudson Bay had also occurred (Laymon, 1991).

The Wisconsinan maximum of the Laurentide Ice Sheet at its southern margin, in the United States, occurred between 24,000 to 14,000 BP during oxygen isotope Stage 2 (Fig. 2.9); retreat from the northern and eastern margins in Canada occurred between 12,000 to 8000 BP (Porter, 1983; Andrews, 1987c) (Fig. 2.6). The mountain glacier maximum in the United States may have occurred generally before ca. 22,000 BP and before its record in deep sea cores by ca. 18,000 BP (CLIMAP, 1976; Fullerton and Richmond, 1986). The maximum extension of the Cordilleran Ice Sheet occurred 15,000–14,000 BP (Booth, 1987).

2.7.3. The European Record

Pliocene continental glaciation, is generally unknown in Europe or Eurasia and although glaciation has been frequently assigned for the Alps (Billard and Orombelli, 1986), it is yet unsubstantiated until about 800 Ma in the Early

Chart 1. Correlation of Quaternary Glaciations in the Northern Hemisphere

Bowen, Richmond, Fullerton, Šibrava, Fulton and Velichko — Correlation

TIME DIVISIONS, TIME SCALE, AND SELECTED CHRONOMETRIC CONTROLS OF GLACIATION IN THE UNITED STATES OF AMERICA   CORRELATION OF GLACIAL DEPOSITS IN CANADA

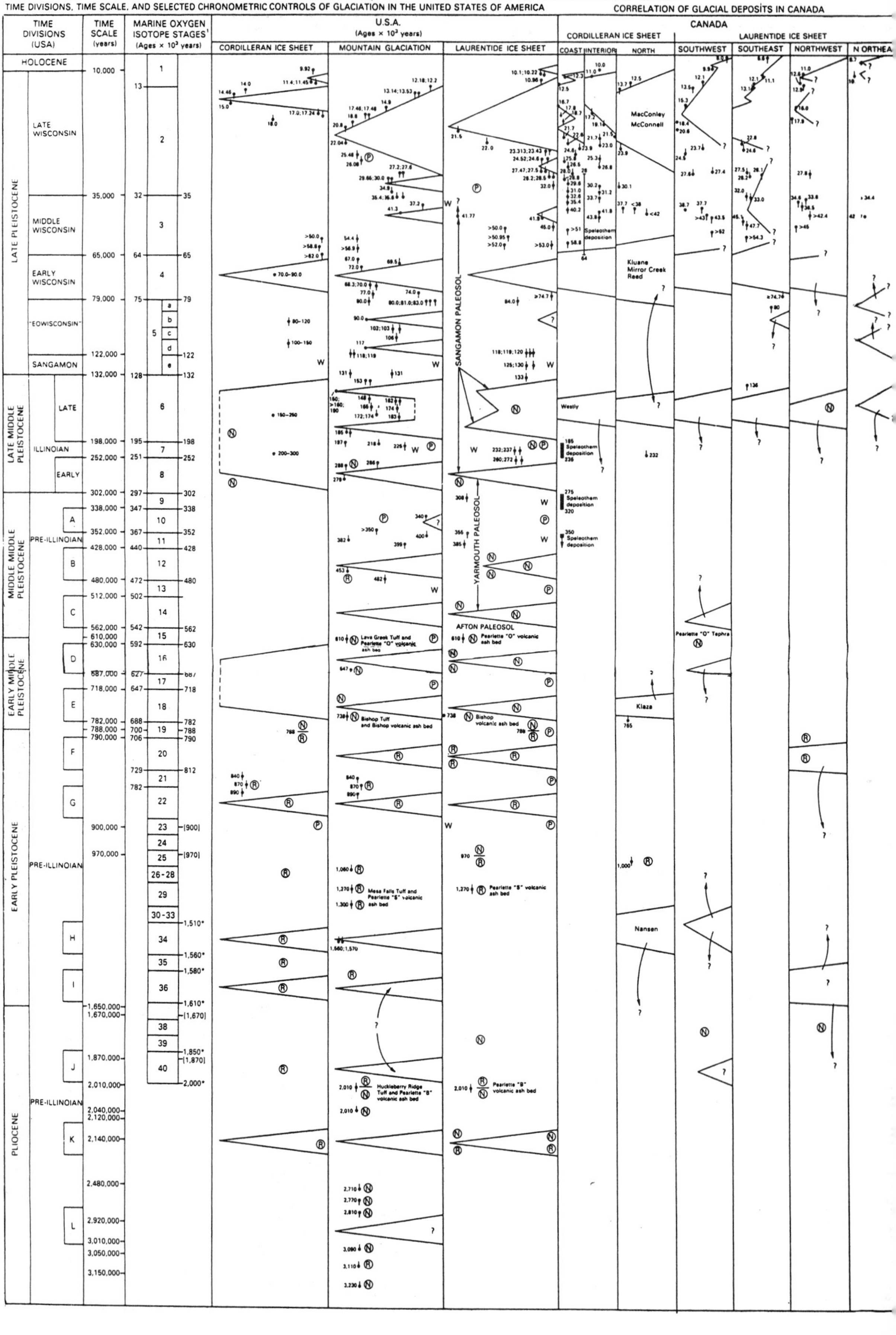

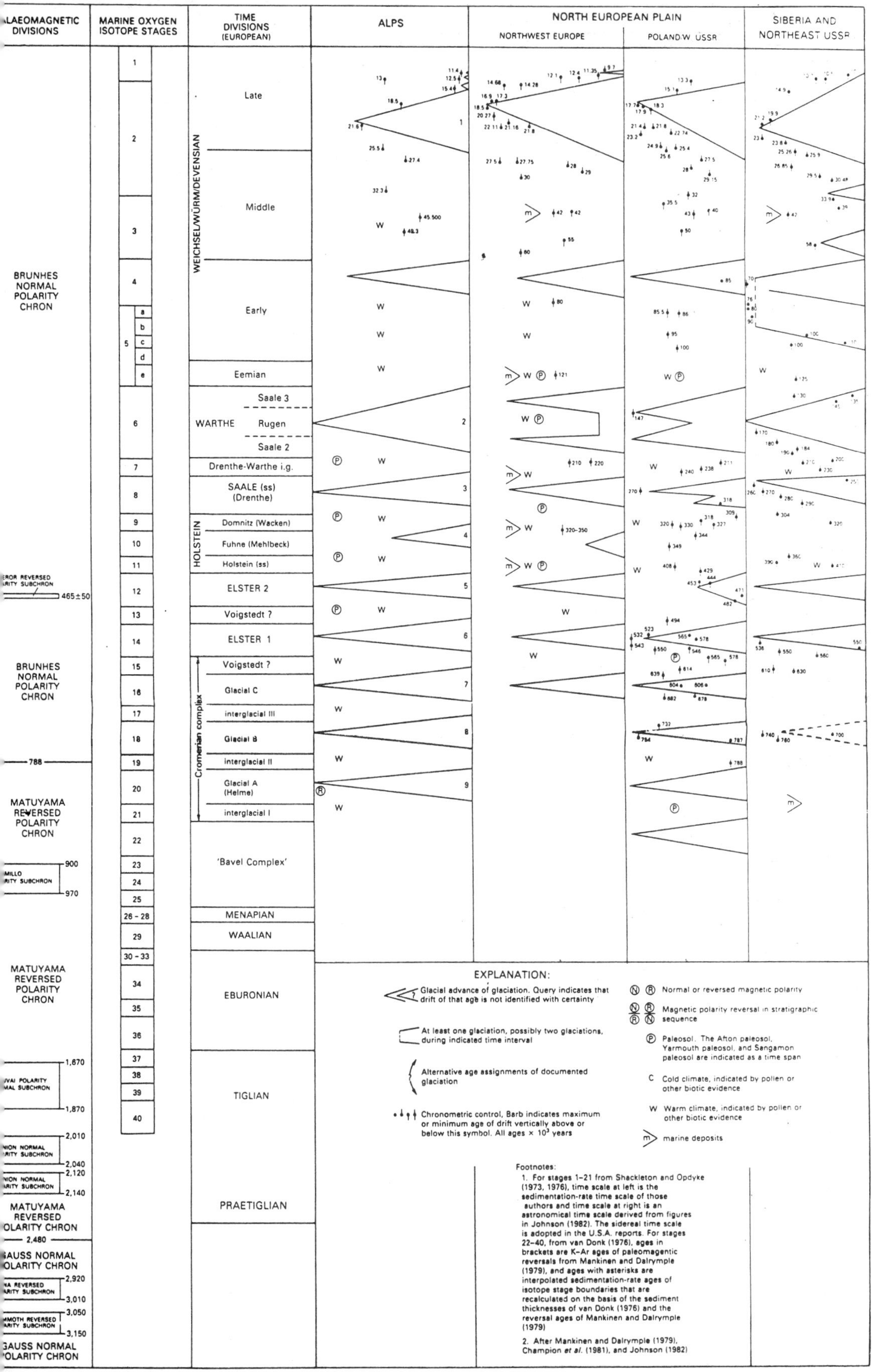

FIG. 2.18. (a) Correlation of Late Cenozoic glaciation in the northern hemisphere (from Bowen et al., 1986; reprinted from *Quaternary Science Reviews*, **5**, Bowen, D.Q. *et al.*, Correlation of Quaternary Glaciation, Chart 1, 1986, with kind permission from Elsevier Science Ltd, The Boulevard, Langford Lane, Kidlington, Oxford OX5 1GB, U.K.).

TIME DIVISIONS (USA)	TIME SCALE (years)	MARINE OXYGEN ISOTOPE STAGES (Ages x 10³ years)	SOUTH AMERICA			NEW ZEALAND	TASMANI
			NORTHERN ANDES	CENTRAL ANDES	SOUTHERN ANDES		

HOLOCENE

10,000

13

1

LATE WISCONSIN   2

35,000   32   35

MIDDLE WISCONSIN   3

65,000   64   65

EARLY WISCONSIN   4

79,000   75   79

a
b
EOWISCONSIN   5   c

122,000   d
SANGAMON   e
132,000   128   132

LATE   6

198,000   195   198
7
252,000   251   252

ILLINOIAN   EARLY   8

302,000   297   302
9
338,000   347   338
10
PRE-ILLINOIAN   A   352,000   367   352
11
428,000   440   428
12
B   480,000   472   480
13
512,000   502
14
C   562,000   542   562
15
610,000
630,000   592   630
16
D   687,000   627   687
17
718,000   647   718
18
E   782,000   688   782
788,000   700   788
790,000   706   790

F

G

900,000

970,000
PRE-ILLINOIAN

H

I

1,650,000
1,670,000

J   1,870,000

2,010,000
PRE-ILLINOIAN
2,040,000
2,120,000
K   2,140,000

2,480,000

L   2,920,000

3,010,000
3,050,000

3,150,000

LATE PLEISTOCENE
LATE MIDDLE PLEISTOCENE
MIDDLE MIDDLE PLEISTOCENE
EARLY MIDDLE PLEISTOCENE
EARLY PLEISTOCENE
PLIOCENE

NORTHERN ANDES labels: 9.3, 10.6, 11.4, 13.0, 14.7, 19, 22, 25, 33, P, 34, 43, 90, 110, RW

CENTRAL ANDES labels: 9.9, 12.2, 13.9, 14.5, 23.9, 33.5, 33.6, 42, RW, I, 2.2 Ma, 2.48 Ma, 3.27 Ma

SOUTHERN ANDES labels: 9.7, 9.9, 12.2, 10.5, Llanquihue III, 13.8, 14.2, 16.2, 18.1, 18.9, 20.1, 23.3, Llanquihue II, 30.7, 32.8, 42.4, 45, 57.9, RW, I, Llanquihue I?, 20?, 250-300, Greatest Patagonian Glaciation, 1.2, 1.8, 5 tills in Patagonia, 2.4 Ma, 8 tills in Patagonia, 3.3 Ma

NEW ZEALAND labels: 9.4, 8.6, 10.5, 13.5, 13.4, 11.9, 15.1, 14, 19.2, 22.2, 34.1, 3, 50, C, I, w, m, Uplift and erosion, Porika Glacn, Ross Glacn

TASMANI labels: 11.4, 13, Late Ma Glacn, 18.8, 21.1, C, 26, 48.8, RW, Early Glacn, A, W, RW, Henty Glacn, W, A, RW, Moore Glac, Linda Glacn, Porika Glacn, Ross Glacn

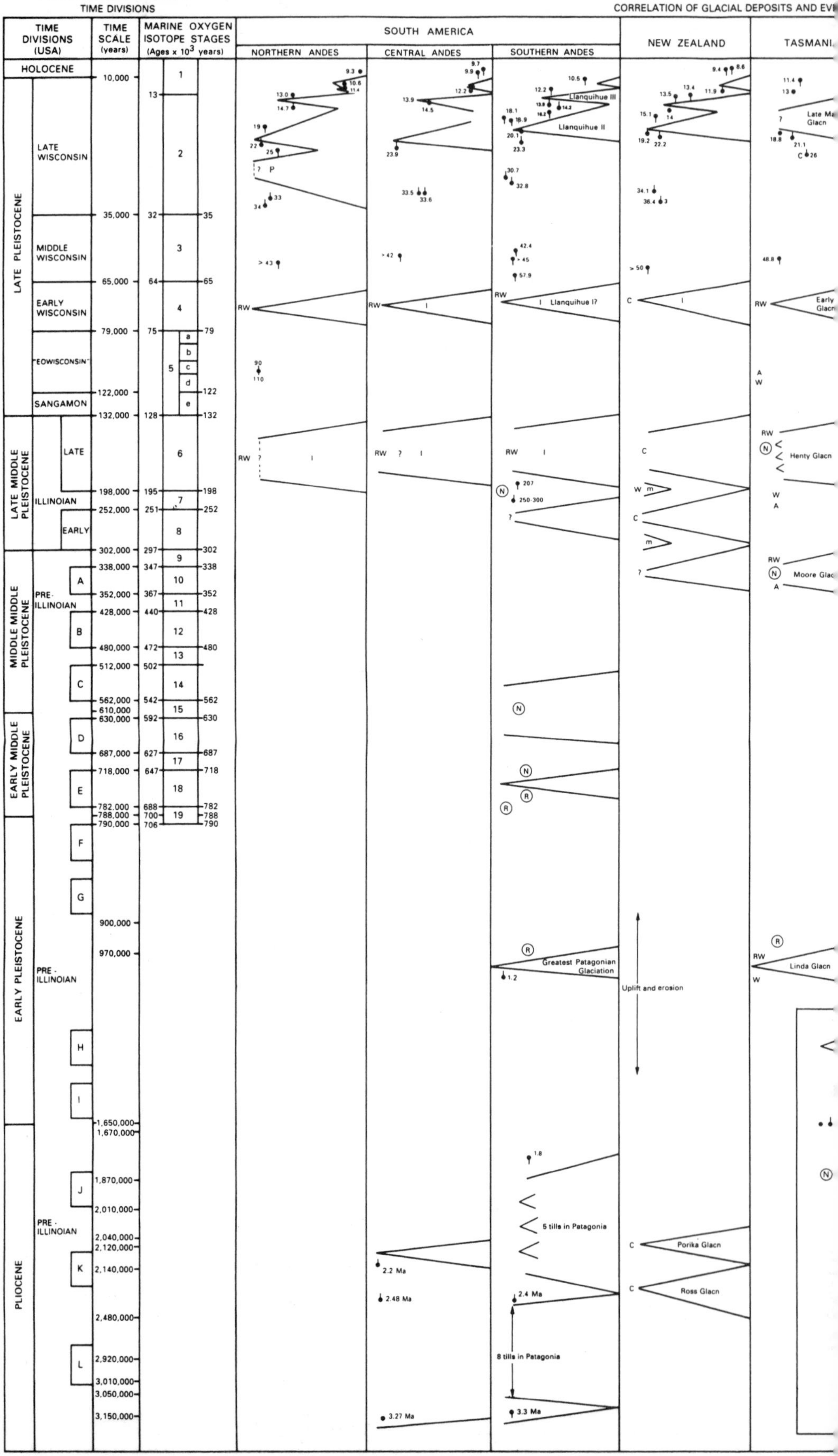

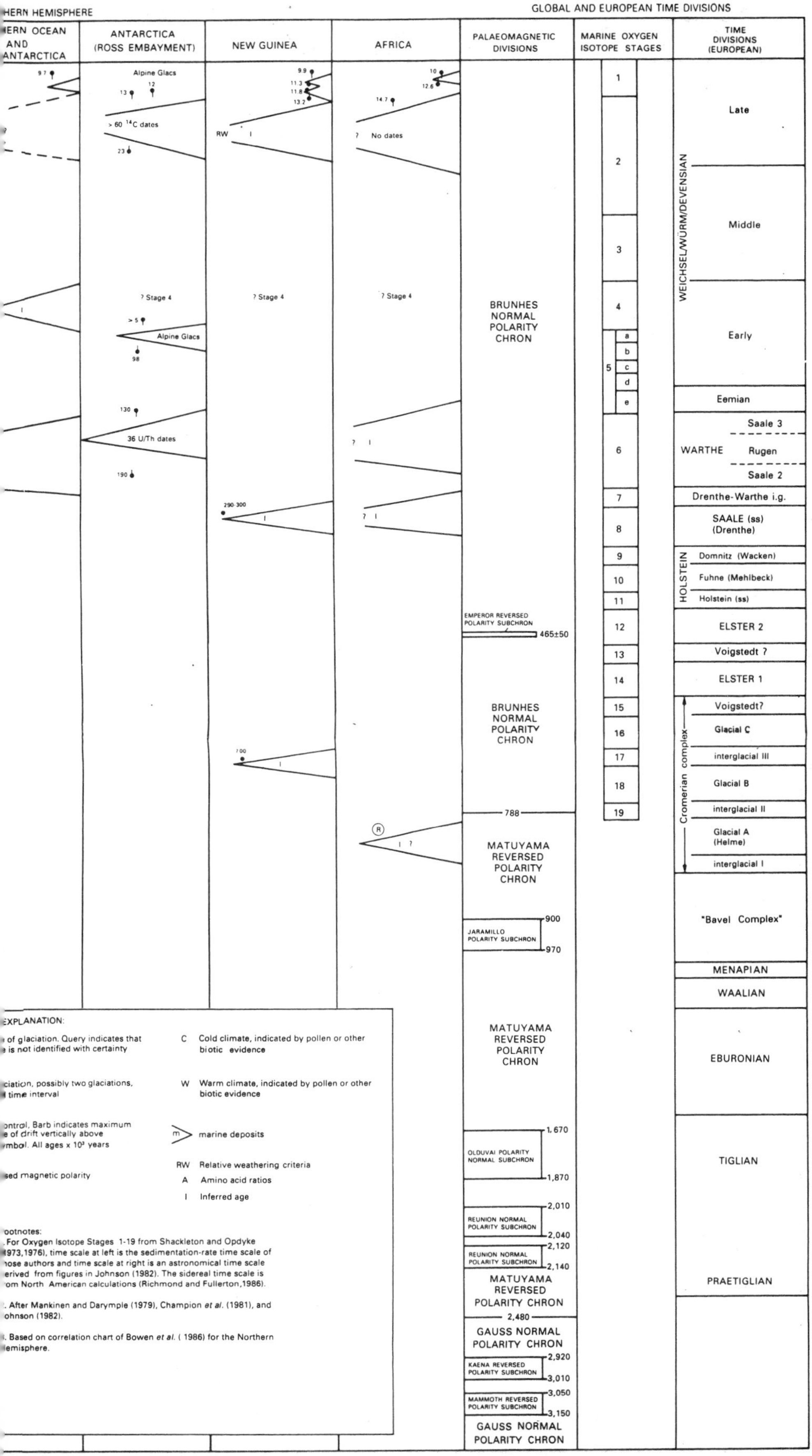

HERN OCEAN AND ANTARCTICA	ANTARCTICA (ROSS EMBAYMENT)	NEW GUINEA	AFRICA	PALAEOMAGNETIC DIVISIONS	MARINE OXYGEN ISOTOPE STAGES	TIME DIVISIONS (EUROPEAN)

Antarctica (Ross Embayment): Alpine Glacs 12, 13, 13.2; > 60 ¹⁴C dates; 23; ? Stage 4; > 5; Alpine Glacs 98; 130; 36 U/Th dates; 190

New Guinea: 9.9, 11.3, 11.8, 13.2; RW; ? Stage 4; 290-300; ? 00

Africa: 10, 12.6; 14.7; ? No dates; ? Stage 4; ?; ?; ®

Palaeomagnetic Divisions:
BRUNHES NORMAL POLARITY CHRON
EMPEROR REVERSED POLARITY SUBCHRON — 465±50
BRUNHES NORMAL POLARITY CHRON
— 788 —
MATUYAMA REVERSED POLARITY CHRON
JARAMILLO POLARITY SUBCHRON — 900 / — 970
MATUYAMA REVERSED POLARITY CHRON
OLDUVAI POLARITY NORMAL SUBCHRON — 1,670 / — 1,870
— 2,010 —
REUNION NORMAL POLARITY SUBCHRON — 2,040
REUNION NORMAL POLARITY SUBCHRON — 2,120 / — 2,140
MATUYAMA REVERSED POLARITY CHRON
— 2,480 —
GAUSS NORMAL POLARITY CHRON
KAENA REVERSED POLARITY SUBCHRON — 2,920 / — 3,010
MAMMOTH REVERSED POLARITY SUBCHRON — 3,050 / — 3,150
GAUSS NORMAL POLARITY CHRON

Marine Oxygen Isotope Stages: 1, 2, 3, 4, 5 (a, b, c, d, e), 6, 7, 8, 9, 10, 11, 12, 13, 14, 15, 16, 17, 18, 19

Time Divisions (European):
WEICHSEL/WÜRM/DEVENSIAN — Late, Middle, Early
Eemian
Saale 3
WARTHE — Rugen
Saale 2
Drenthe-Warthe i.g.
SAALE (ss) (Drenthe)
HOLSTEIN — Domnitz (Wacken), Fuhne (Mehlbeck), Holstein (ss)
ELSTER 2
Voigstedt ?
ELSTER 1
Cromerian complex — Voigstedt?, Glacial C, interglacial III, Glacial B, interglacial II, Glacial A (Helme), interglacial I
"Bavel Complex"
MENAPIAN
WAALIAN
EBURONIAN
TIGLIAN
PRAETIGLIAN

EXPLANATION:
  ...n of glaciation. Query indicates that ...is not identified with certainty
  C Cold climate, indicated by pollen or other biotic evidence
  ...ciation, possibly two glaciations, ...l time interval
  W Warm climate, indicated by pollen or other biotic evidence
  ...ontrol. Barb indicates maximum ...e of drift vertically above
  ...ymbol. All ages x 10³ years
  m marine deposits
  ...sed magnetic polarity
  RW Relative weathering criteria
  A Amino acid ratios
  I Inferred age

Footnotes:
  . For Oxygen Isotope Stages 1-19 from Shackleton and Opdyke
  1973,1976), time scale at left is the sedimentation-rate time scale of
  hose authors and time scale at right is an astronomical time scale
  erived from figures in Johnson (1982). The sidereal time scale is
  om North American calculations (Richmond and Fullerton, 1986).

  . After Mankinen and Darymple (1979), Champion et al. (1981), and
  ohnson (1982).

  i. Based on correlation chart of Bowen et al. (1986) for the Northern
  hemisphere.

FIG. 2.18. (b) Correlation of Late Cenozoic glaciation in the southern hemisphere (from Clapperton, 1990b; reprinted from *Quaternary Science Reviews*, **9**, Clapperton, C.M. *et al.*, Quaternary glaciations in the southern hemisphere, Chart 1, 1990, with kind permission from Elsevier Science Ltd, The Boulevard, Langford Lane, Kidlington, Oxford OX5 1GB, U.K.). See text for explanation.

Pleistocene. Some Scandinavian erratics found in Pliocene deposits of northwestern Germany, however, require explanation (Ehlers *et al.*, 1984). Furthermore, pollen data record the distinctive cooling at or near the Gauss–Matuyama paleomagnetic boundary in many areas of Europe (Zagwijn, 1959, 1986; Hooghiemstra, 1984; Velichko and Faustova, 1986). The oldest loess sheets of the Alpine foreland have ages up to 2.5 Ma (Šibrava, 1986). These events correspond generally with the evidence of major northern hemisphere ice rafting and glaciation indicated by deep sea cores.

Evidence of an Early Pleistocene advance by the Scandinavian Ice Sheet is also sparse and poorly dated. However, deposits of two advances may be recorded in west central Poland (Rzechowski, 1986) and adjacent C.I.S. (Velichko and Faustova, 1986) (Fig. 2.18a). Somewhat more recent (Early Middle Pleistocene) glaciations are reported from Britain as well as in Poland and the C.I.S. A lobe of the Scandinavian Ice Sheet is reported to have advanced into the Don River basin in the Central Russian Plain to at least 52°N at about this time (Arkhipov *et al.*, 1986).

The primary advances, which left widespread evidence across northern Europe and Eurasia, may not have occurred until Elster time (Fig. 2.18). This is correlated with $\delta^{18}O$ Stages 14 (Fig. 2.7) and/or 12 by Šibrava (1986). During the Pleistocene, maximum glacial activity shifted westward in Europe (Arkhipov *et al.*, 1986).

The Middle and Late Pleistocene glaciations show strong correlations, in general, with respective events in North America. For example, they display (1) weak mid-Holstein, Stage 10, advances; (2) a tri-partite Saale as in the Illinoian of North America; (3) a weak early Weichsel as in the Wisconsinan glaciation; and (4) a strong Late Weichsel glacial maxima (Fig. 2.18a) at about 20,000 BP, post-dating a Weichselian interstadial between 29,000 to 26,000 BP (Šibrava, 1986).

2.7.4. The Southern Hemisphere Record

Large glaciers have existed in Antarctica since the Oligocene as confirmed by the diamicts of the poorly-dated Sirius Formation. The interpretation of this formation has important consequences for the stability of the Antarctic Ice Sheet through the Cenozoic (Clapperton and Sugden, 1990). According to one hypothesis, the Sirius Formation is associated with a very large Mid-Miocene ice sheet. Since that time there has been a reduction in size to its present form (Prentice *et al.*, 1986; Denton *et al.*, 1989). A further hypothesis of the Sirius Formation suggests that warm marine waters penetrated beneath the ice sheet with subsequent partial collapse during a Late Pliocene warm interval (Webb *et al.*, 1984; Webb, 1990; Scherer, 1991).

The Southern (Patagonian) Andes of Argentina and Chile have undergone repeated glaciation during the Late Cenozoic and may provide one of the best preserved and well-dated terrestrial glacial records in the world (Rabassa and Clapperton, 1990; Clapperton, 1993). Glaciers, which formed here as early as in Late Miocene time and before 4.7 Ma, left multiple drifts on the mountain flanks through the Pliocene that may correlate with less well-dated glacier deposits in the Peruvian–Bolivian Andes and New Zealand (Fig. 2.18b). The most extensive Patagonian glaciation occurred during the Early Pleistocene when evidence of northern hemisphere glaciation is limited. At this time, 1.2–1.0 Ma, southern Patagonian glaciers reached 200 km beyond (east of) the mountains to the Atlantic Continental Shelf (Rabassa and Clapperton, 1990).

Glacial advances are inferred to have occurred throughout the southern hemisphere mountain region during the Late Middle Pleistocene (Fig. 2.18b), but are only well-dated in Antarctica. Here, grounding of the ice sheet in the Ross Embayment allowed eastward advance and penetration into the seaward ends of ice-free valleys of the Transantarctic Mountains. This penetration is dated between about 190,000 to 130,000 BP according to U/Th series analyses or during Isotope Stage 6 (Denton *et al.*, 1989). Subsequent retreat of the ice sheet from the Ross Sea in Antarctica provided moisture for advances of the local mountain glaciers and ice sheet outlet glaciers in the adjoining Transantarctic Mountains. These advances occurred during Stage 5a, out of phase with global advances elsewhere (Hendy *et al.*, 1979; Clapperton and Sugden, 1990).

Drifts and morainic limits of the late Quaternary have been assigned to Isotope Stage 4 in South

America, New Zealand, Tasmania and in the sub-Antarctic. However, these occurrences are dated largely on the basis of stratigraphic relations or relative degree of dissection and placed in Stage 4 because the $\delta^{18}O$ record indicates that this is the time of greatest global ice build up during the last glaciation. The last glacial maximum probably peaked in mountainous areas of the southern hemisphere between about 24,000 and 18,000 BP, being well-dated in all but the African, New Guinea and Southern Ocean–Sub-Antarctic region (Clapperton, 1990a, 1993) (Figs 2.18b and 2.19).

2.8. THE LAST 25,000 YEARS AND SYNCHRONEITY OF NORTHERN AND SOUTHERN HEMISPHERE GLACIAL CYCLES

A variety of data from deep sea and ice cores, as well as mountain glacier records, attest to the severity and near-synchroneity of most climatic changes in both hemispheres during the last Pleistocene glaciation and Holocene when reasonably precise dating is possible (Nilsson, 1983; Mercer, 1984; Nelson et al., 1985; Broecker and Denton, 1989; Clapperton, 1990a, b). The problems of causes for such near interhemispheric synchroneity are discussed in Section 2.9.

2.8.1. Last Glacial Maximum

2.8.1.1. Greenland and Antarctica data

That glacial cycles of both northern and southern hemisphere ice sheets are in phase has been well documented by the $\delta^{18}O$ or δD records of the Greenland and Antarctic ice cores (Figs 2.13, 2.14, 2.15 and 2.19). Vostock and Byrd cores, as well as a shorter core from Dome Circe in Antarctica, suggest that the last glacial maximum occurred between about 25,000 and 14,000 BP with 50% less annual accumulation and 6–9°C colder temperatures than now (Jouzel et al., 1989). Similar results are obtained from Greenland (Paterson and Hammer, 1987).

Data from the Ross Sea region and adjoining Transantarctic Mountains of Antarctica (Denton et al., 1989) indicate ice thickening with subsequent grounding and advance of ice through the Ross Sea, coincident with the maxima along the southern margins of the northern hemisphere ice sheets. A similar, but earlier, series of ice advances in the inner Ross Sea correspond with the penultimate (Illinoian) glacial cycle.

Fluctuations of the Antarctic Ice Sheet are directly influenced by the growth and disintegration of the largely terrestrial ice sheets of the northern hemisphere. This direct effect occurs due to the interplay of fluctuating global sea level change and continental ice sheet growth and diminution. Since much of the Antarctic Ice Sheet is grounded below sea level, sea level recession (northern hemisphere ice growth) results in Antarctic Ice Sheet expansion and, conversely, Antarctic ice retreat ensues with increased calving and disintegration that accompanies sea level rise (northern hemisphere ice wastage) (Hollin, 1962; Stuiver et al., 1981; Denton et al., 1986, 1989). Southern hemisphere mountain glaciers that were not controlled by global sea level may thus have reached their maximum positions several hundred years or more earlier than tidewater glaciers. The Antarctic Ice Sheet should have culminated its advance slightly after northern hemisphere glaciers reached their maximum (Clapperton, 1990b).

2.8.1.2. The mountain snowline

One of the clearest demonstrations of interhemispheric coupling of climate change is shown by the uniform depression of mountain snowlines during the last glacial maximum (Broecker and Denton, 1989). The transect in Fig. 2.20 from along the western Cordillera of North and South America indicates that a maximum depression of 900–950 m occurred in tropical and subtropical areas. The depression has been related mostly to cooling (~4–6°C) (Rind and Peteet, 1985; Broecker and Denton, 1989) in high mountain areas rather than to increased precipitation. These findings are supported by recent data that indicate lowland terrestrial temperatures as well as sea surface temperatures were uniformly ~3–4°C lower in glacials than during interglacials in the tropics (van Campo et al., 1990). However, these temperatures conflict with the

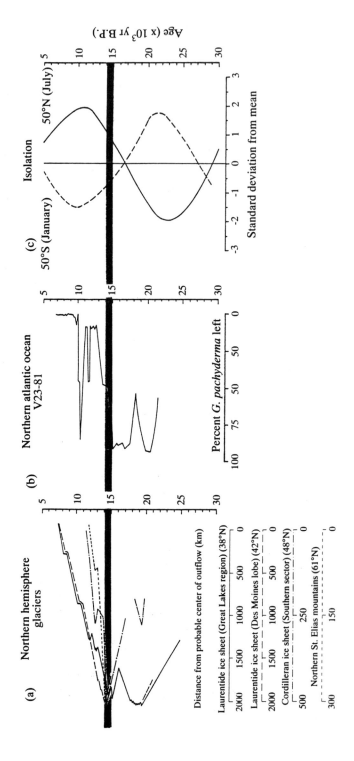

FIG. 2.19. (a–c) Comparison of Late Pleistocene proxy climate records of ice sheet and mountain glacier advances, deep sea δ¹⁸O record V23-81 and Dome C (Cire) ice core profiles with insolation curves from northern and southern hemispheres (from Broecker and Denton, 1989) (Reprinted from *Geochimica et Cosmochimica Acta*, **53**, Broecker, W.S. and Denton, G.H., The role of ocean–atmosphere reorganisations in glacial cycles, 2465–2501, 1989, with kind permission from Elsevier Science Ltd, The Boulevard, Langford Lane, Kidlington, Oxford OX5 1GB, U.K.)

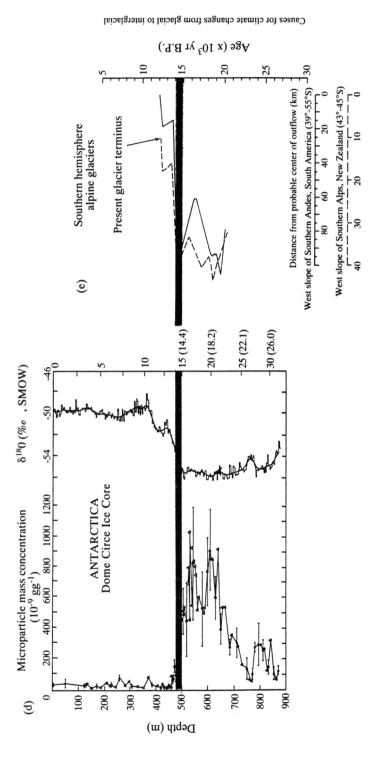

FIG. 2.19. (d, e) Comparison of Late Pleistocene proxy climate records of ice sheet and mountain glacier advances, deep sea δ¹⁸O record V23-81 and Dome C (Circe) ice core profiles with insolation curves from northern and southern hemispheres (from Broecker and Denton, 1989) (Reprinted from *Geochimica et Cosmochimica Acta*, **53**, Broecker, W.S. and Denton, G.H., The role of ocean–atmosphere reorganisations in glacial cycles, 2465–2501, 1989, with kind permission from Elsevier Science Ltd, The Boulevard, Langford Lane, Kidlington, Oxford OX5 1GB, U.K.)

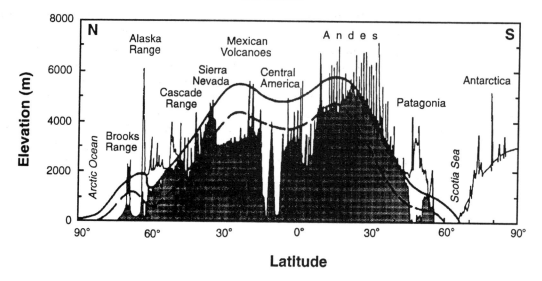

FIG. 2.20. Average mountain snowlines of present day (heavy solid line) and last glacial maximum (heavy dashed line) on a meridional topographic profile along the cordillera of North and South America (from Broecker and Denton, 1989) (Reprinted from *Geochimica et Cosmochimica Acta*, **53**, Broecker, W.S. and Denton, G.H., The role of ocean–atmosphere reorganizations in glacial cycles, 2465–2501, 1989, with kind permission from Elsevier Science Ltd, The Boulevard, Langford Lane, Kidlington, Oxford OX5 1GB, U.K.)

findings of CLIMAP (1981) and some deep sea $\delta^{18}O$ data that indicate tropical ocean surface temperatures remained within ±2°C of their present values (Broecker, 1986).

2.8.1.3. Terrestrial ice advances in middle to high latitudes

Figure 2.19 suggests a correlation of climatic data for both hemispheres across the last Pleistocene glacial maximum (Broecker and Denton, 1989). The time–distance diagrams for the northern hemisphere (Fig. 2.19a) represent the southern margins of the Laurentide and Cordilleran Ice Sheets and mountain glaciers of the St. Elias Mountains of the Alaskan–Yukon border area (Fig. 2.18a). Similar data are available elsewhere (e.g. for northern Europe — Nillson, 1983; Lundqvist, 1986). The southern hemisphere diagrams of Fig. 2.19e show generalized indications of mountain glacier advance in the Southern Andes and the Southern Alps of New Zealand.

Two major maxima of the latest Pleistocene glaciation are prominent in both hemispheres; one occurs at ~20,000 BP and the other at ~15,000 BP; the latter more sharply defined, although of slightly lesser extent in many areas (Mercer, 1984). Evidence in the Huron and Erie basins of the Great Lakes, for example, indicates that the Laurentide ice margin retreated 200–300 km between 20,000 and 15,000 BP (Mickelson *et al.*, 1983; Fullerton, 1986a). General mid-latitude ice recession, with minor readvances, began in both hemispheres by about 14,000 BP.

2.8.1.4. Sudden glacial readvances of the Younger Dryas in Northwestern Europe

The glacial recession that began 14,000 years ago was interrupted in northwestern Europe at ~13,000 years ago by a rather sudden cooling back to almost glacial conditions between warm Bølling-Allerød and cool Younger Dryas European chronozones that lasted until ~11,500 years ago (Nilsson, 1983; Alley *et al.*, 1993) (Fig. 2.15). The Younger Dryas termination is dated ~10,000–10,200 BP in Europe; this age has been calibrated to within a few centuries of 11,500 BP (Alley *et al.*, 1993). Ice core data indicate that this probably represents only one perturbation of a quasi-cycle during the Late Quaternary representing a 7°C temperature change within a few centuries. During this cooling interval, the Scandinavian Ice Sheet paused and built ice margin deposits nearly continuously around Scandinavia (Fig. 2.4). At the same time, small mountain glaciers reformed in the marginal highlands of northwestern Europe (Sissons, 1979; Mangerud, 1987) or readvanced to distinct moraines on the coast

of Norway where net balance reinforcement occurred due to higher precipitation.

There are widespread climatic events through both hemispheres that correspond in time with the Younger Dryas (e.g. Broecker *et al.*, 1985, 1988; Berger and Labeyrie, 1987; Andrews, 1987c; Heusser, 1989; Berger, 1990; Clapperton, 1990b; La Salle and Shilts, 1993). However, glacial advances and climatic change are best documented and of greatest magnitude in northern Europe and the North Atlantic region in general (Fig. 2.10).

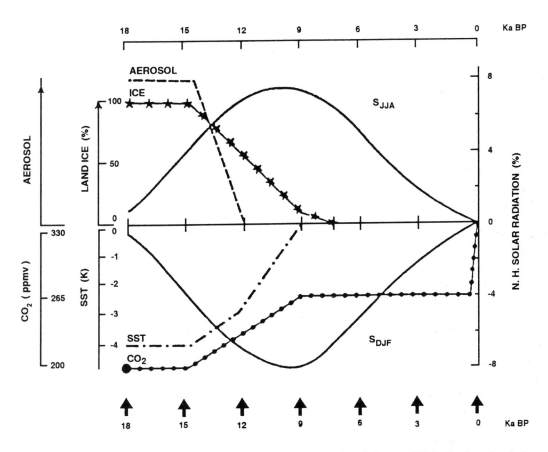

FIG. 2.21. Changes in external climate-forcing elements (from Kutzbach and Guetter, 1986). Northern hemisphere solar radiation in June through August (S_{JJA}) and December through February (S_{DJF}) as percent difference from the radiation at present; land ice (ICE) as percent of 18,000 BP ice volume; global mean annual sea surface temperature (SST), including calculated surface temperature over sea ice, as departure from present, K; excess glacial-age dust (AEROSOL); and atmospheric CO_2 concentration. (Reproduced with permission of the American Meteorological Society.)

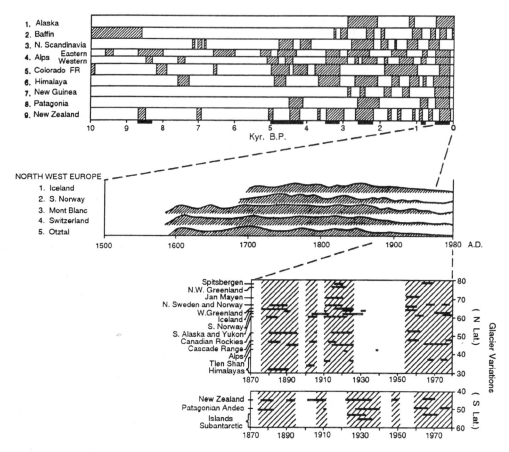

FIG. 2.22. Holocene phases of glacial advance with indications of recent advances (enlargement) that may be globally synchronous (from Grove, 1988) (Reprinted with permission from Grove, J.M., The Little Ice Age. Copyright 1988 Routledge Publishers).

2.8.2. Holocene Glaciation and the Little Ice Age

Late Pleistocene retreat of ice sheets and major global warming led by ~9000 BP to average Holocene climates being similar to today. Increased summer insolation brought on by a solar radiation peak after ~10,000 BP caused temperatures (Fig. 2.21) in Eurasia and western North America to reach 2–4°C higher than present. However, south of the remaining ice sheets, in continental interiors, temperatures did not reach these Holocene maxima until ~6000 BP (COHMAP Members, 1988; ~9200 to 7300 years ago according to GISP2 accumulation record) (Section 2.6.4.2). This period is the so-called 'Hypsithermal' (Altithermal or Climatic Optimum). Subsequent deterioration of temperatures due to

decreased summer insolation is usually considered to have culminated in the Little Ice Age of recent centuries, following a brief medieval warm phase (Lamb, 1977; Kutzbach and Guetter, 1986; MacCracken et al., 1990).

Superimposed on these broad Holocene climatic changes have been shorter-term temperature variations with periods of mountain glacier advance and retreat (Fig. 2.22). Dating is as yet too imprecise at these timescales to certify that Holocene glacier expansions have been synchronous worldwide (Röthlisberger, 1986; Grove, 1988; Davis and Osborn, 1988).

The Little Ice Age, the last and most ubiquitously recorded of the Holocene climatic events, was a global phenomenon of temperature variations, at its maximum, of the order of 1 to 2°C in mid-latitudes. It appears to have begun with minor glacial expansions in the thirteenth and fourteenth centuries (Figs 2.17 and 2.22). Following a brief interval of warmer weather, it reached a major phase of repeated glacial expansions and retreats at ~1600 A.D., culminating with global warming and retreat of temperate and subpolar mountain glaciers in the mid nineteenth to early twentieth centuries (Denton and Karlén, 1973; Porter, 1986; Grove, 1988; Mayewski *et al.*, 1993).

The availability of precise dating by tree rings and particularly, the many historical observations in Europe, suggests that glacial fluctuations were nearly synchronous within separate regions spanning several hundred kilometres (Gribbin and Lamb, 1978; Flohn and Fantechi, 1984; Grove, 1988). Ice core records from polar ice sheets and low latitude ice caps substantiate the Little Ice Age record and indicate, at least locally, that onset and termination occurred abruptly within a few years to decades (Thompson and Mosley-Thompson, 1987). The data illustrate the sensitivity of glaciers as an indicator of short-term climatic change (Grove, 1988; Broecker and Denton, 1989).

2.9. CAUSES OF CLIMATIC CHANGE FOR LATE CENOZOIC GLACIATION

Suggested causes of glaciation are much more numerous than the glaciations themselves. Both extra-terrestrial and terrestrial causes have been invoked. Extra-terrestrial mechanisms include proposed changes in solar radiation, dust clouds interposed between Earth and Sun, changes related to the galactic year (Steiner and Grillmair, 1973), shading of Earth's equatorial zone by an icy ring similar to that of Saturn and changes in Earth's orbital parameters (Milankovitch effect). Terrestrial causes include explosive vulcanism, producing a globe-encircling dust cloud (offset by increased CO_2 levels?), anti-greenhouse effect related to decreases in CO_2 content caused by precipitation of large amounts of carbonate rock (Roberts, 1976; Schermerhorn, 1983) or decreased volcanic activity (Schermerhorn, 1983; Frakes, 1986). Plate tectonic positioning of the continents can have a major effect on oceanic circulation and heat distribution across the surface of the Earth (Ewing and Donn, 1956). Broecker and

Denton (1989) also explained climatic fluctuations in terms of changes in oceanic circulation.

2.9.1. A Range of Causes

There is some general agreement as to the most important factors affecting climatic change over the Late Cenozoic as well as distant geologic past (Berger, 1981a, b; Nilsson, 1983; Grove, 1988; MacCracken *et al.*, 1990). These causes are listed below with a few of their interactions. All may have played some part in triggering, forcing, sustaining, or otherwise controlling Late Cenozoic glaciation:

(1) Changes in solar radiation related to solar output.
(2) Changes in land–ocean distribution associated with plate tectonic movements with consequent changes in mountain elevation, ocean circulation, sea level, and atmospheric composition.
(3) Changes in atmospheric composition, particularly of greenhouse gases carbon dioxide and methane (CO_2 and CH_4). These gases may be linked to major reorganisation of the ocean–atmospheric circulation system as well as to tectonic events.
(4) Changes in the albedo of the Earth's surface.
(5) Changes in the Earth's orbital parameters (Milankovitch hypothesis).
(6) Other changes including catastrophic events such as extended periods of volcanic eruption.

2.9.2. Initiation of Late Cenozoic Glaciation

Since the Permo-Carboniferous ice age (~250 Ma), the solar flux increased by ~1% (MacCracken *et al.*, 1990), the continents joined and split apart again and moved to their present positions, and the heights of plateaux and mountains have increased. Changes in ocean geometry and of atmospheric vapour transport and current pathways must have had profound effects on ocean temperatures and salinity, on rainfall and temperature patterns and atmospheric composition (e.g. CO_2). Geological evidence suggest that CO_2 concentrations were 5–10 times greater during the

Mesozoic compared with present values. The apparent decrease in atmospheric CO_2 from the Late Cretaceous through the Pleistocene (Fig. 2.3) and the consequent decreased greenhouse warming (Barron, 1985; MacCracken et al., 1990), help explain the general but stepped cooling over this interval. Nevertheless, the causes of carbon dioxide changes and rapid cooling are uncertain. A critical point in the search for causes of the Cenozoic cooling trend is that it culminated in development of the Antarctic and mid-latitude ice sheets by 36 and 2.4 Ma, respectively. The decline in atmospheric CO_2 levels over this interval has been ascribed by climate modellers to decreases in global sea-floor spreading rates and volcanic outgassing. Decreased global chemical weathering (which removes CO_2 from the atmosphere) related to elevated sea levels and hence a reduction in land area, may also have been a major cause of climatic change (Raymo, 1991).

The movement of land masses into polar latitudes has often been considered critical in bringing on continental glaciation (Crowell and Frakes, 1970; Donn and Shaw, 1977; Crowley et al., 1987). Paleogeographic reconstructions indicate, however, that Antarctica and the northern continents have been near their present latitudes for the past 100 Ma, long before the glaciations developed in either location.

Many local geographic factors were once also considered to be critical to the growth of Late Cenozoic glaciation at high latitudes. These included: the uplift of ice-sheet nucleation centres in arctic Canada (Flint, 1943); the opening of the Drake Passage isolating Antarctica from South America (Kennett, 1977); and the formation of the Isthmus of Panama and the possible diversion of vital ocean currents to more northerly paths (Keigwin, 1982). According to climatic models, these geographic changes may not provide clear explanations for long-term, global mean annual cooling (Barron, 1985). In addition, increased Cenozoic volcanism and solar shielding (Kennett and Thunell, 1975) has not proven to have provided adequate climatic forcing at appropriate times for Antarctic and mid-latitude northern hemisphere glaciations (Ruddiman and Raymo, 1988; Raymo, 1991).

Two complementary hypotheses for intensification of cooling and for triggering of Late Cenozoic

glaciation are related to the significant, post-Eocene uplift of the Himalayan Mountains, Tibetan Plateau, Andes Mountains and plateaux and mountains of western North America. Uplift appears to have culminated by ~5 Ma and in the Himalayan Mountains and Tibet between 3 to 1 Ma (Ruddiman et al., 1986; Ruddiman and Raymo, 1988; Raymo, 1991; Unruh, 1991) (Table 2.1). Models of uplift in the northern hemisphere indicate that the modification of the trajectory of the upper atmosphere jet stream would have altered the Rossby wave pattern to enhance cold polar air mass penetration over more southerly areas of the northern hemisphere continents (Birchfield and Weertman, 1983; Ruddiman and Kutzbach, 1989). However, the amount of cooling distal to the uplifted areas would have been modest, especially during the critical summer ablation period; thus a more powerful hemispheric-scale explanation for cooling such as a decrease in CO_2 is demanded (Ruddiman, 1991).

In a related hypothesis it has been argued that uplift of the Tibetan Plateau, the Himalayas and probably the Andes took place during the Late Neogene resulting in a dramatic increase in subaerial weathering and erosion which would have significantly decreased atmospheric CO_2 concentration, thus resulting in global cooling (Raymo et al., 1988; Raymo, 1991). Chamberlin (1899) advanced a similar argument, and recent records of $^{87}Sr/^{86}Sr$ preserved in marine biogenic sediment cores may substantiate this hypothesis. The strontium ratios indicate that a 50% increase in global average weathering rate has occurred since the Miocene, despite declining global temperatures (Raymo, 1991).

2.9.3. The Milankovitch Astronomical Theory

2.9.3.1. General principles

Strong and continuous global cooling since the Mid-Pliocene, and the increase in intensity of glacial/interglacial oscillations over the past 1 to 2 Ma, have apparently been paced and to some extent partly driven by changes in the Earth's orbital parameters. However, the long-term pattern of this forcing is not known to have changed sufficiently to

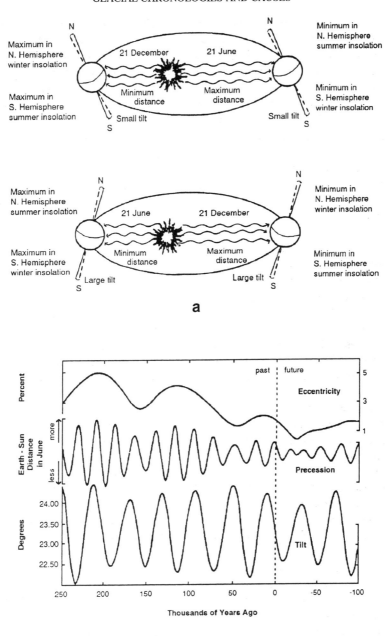

FIG. 2.23. (a) Diagrammatic representation of the Earth's orbital elements (eccentricity, tilt, and precession) (from Ruddiman and Wright, 1987; reprinted with permission from the authors). Eccentricity is the distance from the ellipse centre to one focus divided by the length of the semimajor axis (i.e. halfway between perihelion and aphelion). It varies from a maximum where the receipt of radiation outside the atmosphere varied by 30% between perihelion and aphelion to a minimum when the orbit is nearly circular and precession has little climatic effect. (b) Changes in eccentricity, tilt and precession of the Earth's orbit (from Imbrie and Imbrie, 1989). The addition of these curves yield insolation curves (e.g. Fig. 2.8a). (Reproduced with permission of the authors and Enslow Publishers.)

explain selective and sudden glacial initiation or intensification.

The Milankovitch (1941) theory holds that these orbital changes, resulting from gravitational perturbations by other planets, cause cyclic changes in latitudinal and seasonal distribution of solar radiation which, in turn, induce climatic fluctuations (Imbrie and Imbrie, 1979; Berger, 1981b, 1988; Berger et al., 1984). Summer insolation received in middle latitudes is considered to control the volume of ice stored on continents. The seasonal insolation cycle arises mainly from the effects of: (1) changes in eccentricity (e) of the Earth's orbit around the Sun; (2) change in tilt (obliquity of the ecliptic — E) of the Earth's (spin) axis relative to the orbital plane (ecliptic); and (3) changes in the relative position of the equinoxes due to precession of the Earth's axis of rotation about the axis of the orbital ellipse (e sin w), where w is the orbital between vernal equinox and perihelion (Fig. 2.23). Orbital mechanics indicate that e, E and e sin w vary quasi-harmonically with periods of 100,000, 41,000 and ~23,000 and 19,000 years, respectively (Figs 2.7c and 2.8c). These cyclical variations, whose geometric changes have been calculated for past and future (Vernekar, 1972; Berger, 1978), affect the Earth's distance from the Sun, changes in the latitude of the tropics, the position of the polar circles by a few degrees, and the angle of incidence of the solar beam at different seasons (Berger, 1981a, b).

Local variations in monthly mean insolation of ±12% from present day values can occur, although the insolation affects the eccentricity term by only ±0.1% (Berger, 1981b). Nevertheless, eccentricity is important as it modulates the precessional cycle. Furthermore, non-linear response characteristics of ice sheet growth and decay (see below) may permit the eccentricity alone to affect climate (Imbrie and Imbrie, 1979; Ruddiman and Wright, 1987).

Orbital solar forcing is virtually constant on a global-mean annually averaged basis; therefore a feedback mechanism is needed to convert seasonal and latitude-specific atmospheric changes to a global climatic response. The Milankovitch orbital theory maintains that northern high latitude summers receiving minimum radiation may be cooled sufficiently to allow winter snows and sea ice to persist, thereby permitting a positive annual accumulation budget to develop. This effect may eventually initiate a positive feedback cooling response over the Earth leading to a further latitudinal extension of snow and sea ice and a consequent increase in surface albedo (Berger, 1981b; Oerlemans, 1991). Recent evidence, however, contradicts the effectiveness of this mechanism in initiating and terminating glaciations (Broecker and Denton, 1989). It may be concluded that the importance of the Milankovitch orbital mechanism in driving or pacing global climatic change and glaciation will be continually evaluated as long as problems of such feedback mechanisms are debated, or as controversies such as the Devils Hole vein calcite record (Section 2.5.5) are raised.

2.9.3.2. Problems of synchronous glaciation and feedback mechanisms for Milankovitch forcing

Synchronous behaviour of glaciers and climate fluctuations between the two polar hemispheres are not compatible with orbital forcing. Under conditions of relatively high eccentricity, the effects of precession, considered by some to be the driving force of glaciation in the Brunhes Chron, are in antiphase between northern and southern hemispheres (Fig. 2.23). Counter arguments (Berger, 1981b) suggest that the minimum winter insolation in the southern hemisphere, accompanying minimum summer irradiation in the northern hemisphere, would favour major ice shelf buildup and hence induce similar albedo feedback effects as those of cool summers in the north. Moreover, it has been contended that sea level changes initiated by phases of terrestrial ice sheet extension and contraction play a critical interactive role in the stability–instability of marine ice sheets or those ice masses at least partially grounded below sea level (Denton and Hughes, 1983; Denton et al., 1986) (Chapters 4 and 5). Thus, the Antarctic Ice Sheet (the West Antarctic Ice Sheet being particularly sensitive) expanded after a lag period following the growth of the Laurentide Ice Sheet (Clapperton and Sugden, 1990) further strongly increasing the albedo effects in the polar regions of the southern hemisphere.

Some modelling experiments (Manabe and

Broccoli, 1985; Broccoli and Manabe, 1987) indicate that the terrestrial ice albedo–temperature feedback, by itself, is not enough to explain amplification of the Milankovitch orbital effect and production of global ice buildup. However, alternate climatic models appear to dispel some of these arguments (Berger *et al.*, 1990). Nevertheless, the Manabe and Broccoli studies suggest that the linkage, both within or between hemispheres, was not through the ice-sheet–albedo effects even with simultaneous ice-sheet growth at both poles. It has also been speculated that the Antarctic Ice Sheet may have made no significant change in climatic conditions to the rest of the southern hemisphere. However, model studies suggest that other mechanisms such as global CO_2 variations may provide the most likely link (Broccoli and Manabe, 1987).

The recently reported isotope dates from a calcite vein within Devils Hole, Nevada, adds increasing doubt to the validity of the Milankovitch theory (Ludwig *et al.*, 1992; Winograd *et al.*, 1992; Broecker, 1992). If the isotope date of 145,000 years, indicative of the close of the penultimate ice age (Termination II), is correct then an asynchroneity between this date and the later accepted date of 128,000 years occurs. If, in fact, the transit time for groundwater to enter the fault is added (15,000 years) an even greater discrepancy exists that cannot be reconciled with the normal Milankovitch cycle.

2.9.4. Ocean–Atmosphere Interchanges and CO_2

Orbital insolation variations in the northern hemisphere may be the principal trigger for changes in ocean circulation or biology and ensuing changes in the absorption and release of CO_2 to the atmosphere. The atmospheric CO_2 reservoir is small compared with that of the oceans (or carbonaceous rocks) and holds about 60 times less CO_2 than the ocean (MacCracken *et al.*, 1990). However, the gas diffuses easily between ocean surface and atmosphere and its concentration in surface waters provides a regulator for the atmospheric concentrations. Mixing of CO_2 between hemispheres occurs on a timescale of about one year

and would provide a mechanism for climate feedback amplification through variations in the greenhouse effect (Broccoli and Manabe, 1987; Broecker and Peng, 1989; Broecker *et al.*, 1990). By itself, CO_2, for example, may have had a small effect on climate, while ice cores display an 80 ppmv decrease in CO_2 with glaciation (Figs 2.15 and 2.21), model studies show only a 1.2°C decrease in temperature (Broccoli and Manabe, 1987). Fluctuations in high-latitude evaporation, surface salinity, sea ice extent and overall pulsed meltwater input could possibly be the direct root of ocean current changes (Broecker *et al.*, 1985; Bryan, 1986; Berger *et al.*, 1987; MacCracken *et al.*, 1990).

2.9.5. Changes in North Atlantic Circulation

An important hypothesis for rapid global cooling and glaciation revolves around the thermohaline circulation (conveyor belt) of the North Atlantic ocean (Broecker and Denton, 1989, 1990; Broecker *et al.*, 1990). Under present conditions, warm waters of relatively high salinity in the equatorial areas of the Atlantic move northward at surface and intermediate depths. In transit north, the waters evaporate and release considerable heat into the Greenland–Norwegian Sea and Labrador Sea areas. Here, the resulting cooler, denser water sinks to the ocean bottom forming the North Atlantic Deep Water (NADW) current(s). An annual release of about 5×10^{21} calories of heat into the atmosphere occurs over the North Atlantic as a result of this process. The cool NADW, thus formed, exports cool salty water (excess salt) southward, eventually reaching around the tip of Africa into the southern oceans (Broecker *et al.*, 1990).

If some mechanism should 'shut off' or simply decrease the present flow of warm salty waters to the north and hence salt-laden North Atlantic Deep Water (NADW) to the southern oceans, the result would be cooling of North Atlantic surface waters, a decrease in salinity, and possible growth of ice sheets in the surrounding area. In the reverse process, deglaciation may have been brought on by the 'turning on' or increase of NADW. Such major changes must have had pronounced effects on atmospheric CO_2 compositions to reinforce climatic changes (Broecker *et al.*, 1990). The rapidity of the CO_2 exchange between ocean and atmosphere may be a major uncertainty (Sarnthein *et al.*, 1987).

There appear to be strong isotope patterns in deep water sediments that show that NADW varied on a glacial to interglacial scale and that its production was checked during glaciations (Berger, 1990). Strong fluctuations in the intensity of NADW production appear to have been associated with up to four climatic oscillations during major deglaciations (such as the Younger Dryas cooling) over the period 15,000 to 8000 BP, as recorded in the Greenland ice cores (Berger *et al.*, 1987, p. 279; Fairbanks, 1989; Broecker *et al.*, 1990; Lehman *et al.*, 1991). Some of these fluctuations may have been orbitally induced by seasonal changes. For example, NADW may have been altered by decreases in the level of water vapour transport from the North Atlantic to the Pacific (Broecker *et al.*, 1990). Other fluctuations were probably related to changes in meltwater output from ice sheets and consequent CO_2 changes in the atmosphere (Section 2.9.4). Broecker *et al.* (1990) have speculated that a switching between the 'on' and 'off' states of thermohaline circulation (NADW current system) is driven by natural oscillations in the salt budget of the Atlantic Ocean that occur during times of major ice cover on adjacent continents. When the Atlantic's conveyor system is 'on', the normal salt export, plus a greatly enlarged meltwater input, causes a decrease in the salinity and density of Atlantic water. When water density is sufficiently low, the NADW conveyor stops. Consequent exchange of CO_2 between atmosphere and ocean would then be impeded, lowering atmospheric CO_2. The conveyor is turned on when, as result of greatly reduced rates of salt export and glacial meltwater input, the salt content of the Atlantic increases to a new threshold. Substantial feedback effects dependent on water vapour content, cloud cover and changes in ocean temperature or surface albedo allow additional changes of the order of 1°C to occur within decades (Mitchell *et al.*, 1989; MacCracken *et al.*, 1990).

2.9.6. Causes of Shorter-Term Climate Change and Holocene Glacier Fluctuations

Short-term changes in climate such as those recorded during the Pleistocene–Holocene transition

or during the Holocene, itself, cannot rely for explanation on the longer-term temporal cycles of Milankovitch-type orbital perturbations. However, while Pleistocene ice sheets exerted a strong influence on northern mid-latitude climates, model studies (Kutzbach and Guetter, 1986) indicate that as late as 6000 BP seasonal radiation produced by perihelion in the northern summer and Earth axis tilt of almost 1° more than today (Fig. 2.21) were exerting overall control on global climate (Section 2.8.2). That ocean–atmosphere or CO_2 mechanisms could have played a part, at least in the early Holocene, is suggested in the discussions above. Two mechanisms are considered below for climatic change that may occur at any time, but are most evident in the late Holocene mountain glaciations.

2.9.6.1. Fluctuations in solar radiation

Climatic imprints left by factors external to the climate system, such as variations in solar radiation or changes due to volcanic blocking of incident radiation, are difficult to distinguish from internal climatic variability. Had it been otherwise, the relationship of climate and solar output might have been clarified long ago (Eddy, 1977; Stuiver *et al.*, 1991).

It has been recognised that dark sunspots on the visible disk, which have cycles of 11 and 22 years as well as longer ones, correlate directly with intervals of increased internal activity of the sun, increased solar brightness, surface activity and with a slight increase in the solar radiation constant (Pecker and Runcorn, 1990). An analysis of historic records of sunspot activity has revealed, for example, a remarkable correlation of sunspot minima with some of the historic and proxy data of cooling over the last millennia. This includes a solar maximum about 1200 A.D. (recognised as the medieval warm interval) followed by three sunspot minima, the Wolf (1280–1340 A.D.), Sporer (1420–1530 A.D.) and the well known Maunder Minimum (1645–1715 A.D.) with no recorded sunspots (Eddy, 1977). The latter two minima correspond with what may have been the stormiest and coldest three centuries in the mid-latitudes of the northern hemisphere (Little Ice Age) (Willett, 1951, 1987; Eddy, 1977; Lamb,

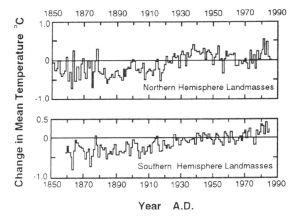

FIG. 2.24. Recent northern and southern hemisphere landmass temperatures (from Broecker and Denton, 1989) (Reprinted from *Geochimica et Cosmochimica Acta,* **53**, Broecker, W.S. and Denton, G.H., The role of ocean–atmosphere reorganisations in glacial cycles, 2465–2501, 1989, with kind permission from Elsevier Science Ltd, The Boulevard, Langford Lane, Kidlington, Oxford, OX5 1GB, U.K.).

1977). Furthermore, the data support the existence of a 200 year solar cycle. However, comparison of recent solar constant changes and sunspot activity suggest that predicted global temperature fluctuations for the Maunder Minimum, the most marked, may only be of the order of ± 0.1–0.3°C (Gérard and Pecker, 1990). This is inconsistent with the ± 1–2°C global temperature change estimated for the Little Ice Age fluctuations. Therefore some mechanism of solar amplification is required if this were the sole cause of these climatic changes (Stuiver *et al.*, 1991).

A correlation of high radiocarbon production, as recorded in tree rings, with times of low solar activity occurs because cosmic rays, which produce ^{14}C, are correspondingly less deflected away from the Earth by the low solar wind at these times (Stuiver, 1961; Suess, 1973). This antiphase relationship of carbon-14 and solar intensity has allowed the recognition of solar minima in prehistoric times, including three significant minima with periods in the Little Ice Age (Stuiver and Quay, 1980). This correlation has also clearly defined a dominant solar cycle of ~200 years and one of ~2300 years (Pecker and Runcorn, 1990).

Isotope [10]Be data from ice cores also confirm the 200 year cycle since both [10]Be and ^{14}C isotopes are regulated by solar output but the 2300 year cycle is absent in the [10]Be data. Runcorn (1990) has suggested that major control of ^{14}C is derived from ocean–atmosphere changes that have an approximate 2000 year cycle.

Attempts to correlate climatic events recorded by a variety of paleoclimatic proxy data, including glaciers with ^{14}C solar proxy data have yielded mixed results (cf. Magny, 1993). Some of the Little Ice Age solar minima can be correlated with periods of decreased CO_2 in ice cores (Stuiver and Braziunas, 1989); however, in general, the CO_2 concentration is rather constant in ice cores. A statistically significant correlation of changes in ^{14}C concentration in the atmosphere with a worldwide record of Holocene mountain glacier advances (Röthlisberger, 1986) was found by Wigley and Kelley (1990), but a study by Stuiver *et al.* (1991) using the same data suggested no measurable past influence of the sun on climate. It is likely that the worldwide Holocene glacial correlations are not sufficiently well-dated to portray details of global change. Furthermore, the global increase in temperature over the past 150 years (Fig. 2.24) coincides with mountain glacier retreats, a ^{14}C decrease and apparently with increased solar activity. Therefore, some component of the observed temperature increase may be due to elevated solar energy output (Pecker and Runcorn, 1990).

2.9.6.2. Changes in atmospheric transparency and volcanism

Changes in atmospheric transparency and incident radiation caused by volcanic eruptions emitting micro-particles and gases may compliment changes in solar activity (Landscheidt, 1987). Convincing correlations of volcanic eruptions with historic climatic change are available, but often offer conflicting evidence (Lamb, 1977; Bryson and Goodman, 1980; Rampino and Self, 1982; Broecker, 1992). Estimates of the effects of volcanism on surface temperatures over the last century suggest that cooling may have been limited to a few tenths of a degree for a one- to two-year period after each event (Mass and Portman, 1989; MacCracken *et al.*,

1990). Other studies of modulations in atmospheric transparency fluctuations indicate that major interdecadal trends of mean surface air temperatures do occur (MacCracken *et al.*, 1990). Furthermore, the long-term correlations seem to substantiate volcanism as a major, if not primary, factor in climate change.

The most important climatic effect produced by volcanic eruptions is considered to be the injection of sulphur dioxide into the atmosphere. This gas often reaches the stratosphere where it is quickly converted to sulphuric acid aerosols that remain aloft for a year or more and reflect sunlight away from the Earth. As a consequence of this chemical reaction, acidity measurement of the annual ice layers together with indications of tephra in polar ice sheets may provide the most complete record of such eruptions (Hammer *et al.*, 1980; Grove, 1988; Zielinski *et al.*, 1993). Fifty-year acidity averages of a 404 m long core from Crete, central Greenland (Hammer *et al.*, 1980) show a significant correlation of temperature events combined and normalised from proxy data for central England, California and the Greenland Ice Sheet since 600 A.D. (Fig. 2.17). Warm periods, generally, match intervals of reduced volcanic activity. The Crete acidity profile as well as the SO_4 concentrations in the GISP2 core show that episodes of cooling and northern hemisphere mountain glacier advances are probably associated with the intervals of high volcanic aerosol production (Porter, 1981a, 1986; Zielinski *et al.*, 1993).

Little Ice Age glacier advance, for example, may reflect temperature depressions of 0.5–1.2°C, which are of the same magnitude as some of those attributable to long-lasting volcanic eruptions (Grove, 1988). Both mountain glaciers and some ice core studies suggest that sulphur-rich aerosols derived from volcanic eruptions may be an important and possible primary-forcing mechanism of climate change on decadal or even century timescales. Nevertheless, further detailed ice core studies are needed to confirm the Crete and GISP2 data because some ice cores (e.g. Dye 3) show somewhat conflicting correlations of acidity and global climatic events. Moreover, Little Ice Age mountain glaciation is rarely dated to the decadal scale necessary for confirmatory tests of the volcanic hypothesis.

2.9.7. Which Cause?

The search for specific causes of long- or short-term climatic fluctuations within glacial or interglacial intervals, or for the triggers of longer-term, orbitally-regulated coolings or warmings may remain an enigma for some time to come. It has been suggested, when considering the Younger Dryas cooling phase, that at the incipient point of major orbital changes, small disturbances may produce large climatic changes by positive feedback amplification (Berger, 1990). However, the ability to distinguish between cause and effect in such feedback loops remains problematic. Furthermore, causes external to the climate system may greatly complicate the search for mechanisms to explain glaciations. For example, in either the Younger Dryas or the Little Ice Age, evidence for both solar and volcanic causes of cooling are viable. Nevertheless, it may be possible, eventually, to identify the one cause with the most impact.

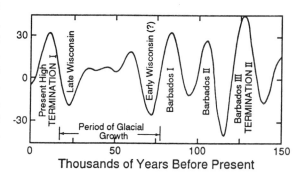

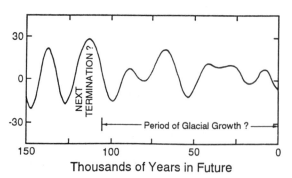

FIG. 2.25. Insolation curves for 45°N covering the past 150,000 and future 150,000 years calculated from orbital data of Vernekar (1968) (after Broecker and van Donk, 1970; reprinted with permission of the American Geophysical Union).

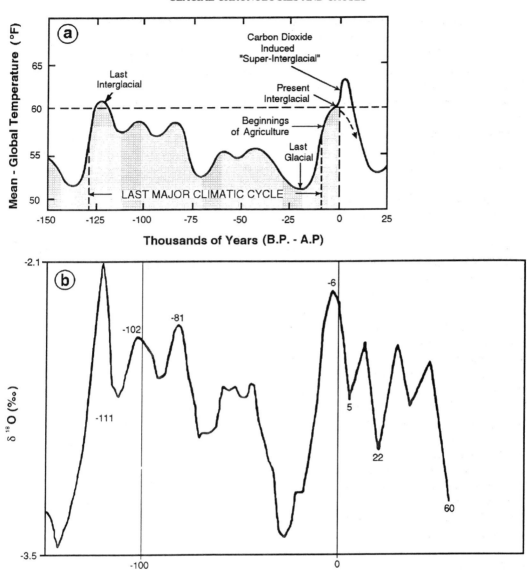

FIG. 2.26. Temperature curves based on deep sea $\delta^{18}O$ data and projection into the future. (a) Solid curve suggests rise in temperature due to greenhouse effect before temperatures eventually fall again to join cooling trend projected by astronomical forcing (from Imbrie and Imbrie, 1979). (b) Future climate simulated by auto-regressive insolation model in terms of $\delta^{18}O$ (modified from Berger, 1981b; reprinted with permission of Kluwer Academic Publishers).

2.10. FUTURE CLIMATE CHANGE AND GLACIATION

As understanding of past global climates and mechanisms of climatic change improves, the need to be able to predict future climatic change and the growth of ice sheets becomes increasingly pertinent. Such predictions must be couched in terms of the geological, climatic and astronomical factors and the impact of human civilizations on past and present global climates. From the review of Late Cenozoic climatic change and glaciation, the Earth appears to be experiencing an interglacial phase. This interglacial began between 13,000 and 10,000 BP (Section 2.4.3.2).

In Europe, interglacials have been classically defined as the point in time when deciduous forest succession over coniferous forest has taken place. If Pleistocene interglacials are defined similar to Isotope Substage 5e in the deep sea sediment record (Figs 2.7–2.9a), then these non-glacial intervals lasted ≤12,000 years. Therefore, the Earth, based upon this marine geological record, may be nearing a cooling phase of more glacial conditions. On the other hand, the Eemian Interglacial of Isotope Stage 5e, as defined in a similar way by the vein calcite record of Devils Hole, Nevada, or by the preliminary interpretation of the GRIP core from Summit, Greenland (Fig. 2.13a), lasted about 20,000 years. Furthermore, the interpretations of the GRIP core in particular, suggests that in contrast to climatic stability of the Holocene, the penultimate or Eemian Interglacial and to a lesser extent, the preceding Holstein Interglacial, were times of large climatic oscillations. The drops in temperature occurred in one to two decades and lasted from 70 to 5000 years. Is this record of high-speed reversals and major atmospheric and ocean circulation changes in the North Atlantic a proxy for our future? Or, are these changes typical only of high-latitude areas (cf. Lehman, 1993)?

Short-term climatic trends give only mixed clues to any impending glaciation. For example, mean global annual surface air temperatures have been rising since the end of the Little Ice Age approximately 100 years ago (Fig. 2.24), although there was a distinct cooling phase during the period from the 1940s to the 1960s (Folland et al., 1990; MacCracken et al., 1990). Nevertheless, paleoclimate data, particularly from pollen, do allow estimates demonstrating that mean annual temperatures of the northern hemisphere have fallen ~1°C from the Mid-Holocene Climatic Optimum to present values. Some model-based estimates, however, suggest this temperature depression to be near zero (Folland et al., 1990; MacCracken et al., 1990).

Since an apparent correlation of past climatic change with astronomical control of peak summer radiation for the northern hemisphere can be shown (perhaps excluding the Devils Hole records), it can be argued that long-term orbital curves may provide the single best available long-term climatic extrapolations for the future (Imbrie and Imbrie, 1979). Taken alone, the solar insolation curve calculated for 45°N over the next 150,000 years (Fig. 2.25) suggests that glaciers may start to build up rapidly over the next 50,000 years, reaching a maximum stage of development before the next interglacial some 100,000 years after present (AP). The marked seasonal contrast that characterized the post-Sangamon interval will not occur in the up-coming cycle because there will be a relatively low orbital eccentricity and antiphasing between the tilt and precession components (Fig. 2.23).

A more realistic projection of the orbital data comes from models that use the insolation curve of Fig. 2.25 or similar data as input, but take into account the history of climate response to changes in the solar radiation pattern (e.g. from deep sea sediment curves of $\delta^{18}O$) (Imbrie and Imbrie, 1979, 1980; Berger, 1981a, b). These irradiation simulation models indicate that temperatures will continue the general cooling trend begun in the Mid-Holocene, (~6000 BP) leading toward global cooling and probably glaciation by ~23,000 AP (Fig. 2.26b). However, an advance more comparable with the last glacial cycle (Late Wisconsinan), is proposed by ~60,000 AP (Berger, 1981b). These orbital effects must be superimposed over and above climate fluctuations that have frequencies higher than the 19,000 year precession cycle and also over climate changes forced by anthropogenic effects (Broecker, 1975; Imbrie and Imbrie, 1979, 1980). For example, climatic frequencies between a few centuries and 2–3 millennia (Thompson, 1989; Stuiver et al., 1991) that may encompass the Younger Dryas and Little Ice Age events, will undoubtedly modify the astronomically driven cycles. Magny (1993) as well as others before (cf. Lamb, 1985) have noted that during the next few centuries, the solar activity may be at a high between two, 2300 year cyclic low points. Nevertheless, a model by Wigley and Kelly (1990) suggests that if the Little Ice Age were repeated in the up-coming centuries, it would compensate in part for anthropogenic warming (Fig. 2.26a)

It is well known that the human-induced climate change largely results from rapid increases in the concentrations of greenhouse gases in the atmosphere (Mitchell, 1989; Houghton *et al.*, 1990). These gases include carbon dioxide, methane, nitrous oxide (N_2O), carbon monoxide (CO) and other atmospheric gases that absorb infrared (long-wave) radiation. Atmospheric carbon dioxide, which may now be at levels above any registered in the last 160,000 years, is by far the most abundant of these gases in the atmosphere. It has increased by 25% over the last 150 years and from 315 ppmv to about 351 ppmv during the last 30 years alone, as a result of rapid industrial development and agricultural expansion (Houghton *et al.*, 1990; MacCracken *et al.*, 1990).

Projection of atmospheric gases, based on fossil fuel use, suggest that CO_2 will double its pre-industrial level value by the middle to late decades of the twenty first century. Model and empirical analyses further indicate that the increases in greenhouse gases have already led to global temperature increases, although it is difficult to separate these from the possible increases in solar irradiation. By the middle of the twenty first century, mean global surface air temperatures will have reached 1°C above pre-industrial levels and perhaps a few degrees higher in middle to high latitudes. This increase may occur at a rate of 0.75°C every 25–50 years (MacCracken *et al.*, 1990).

Extrapolation of the greenhouse effect beyond the next century is very speculative; however, unrestricted burning of fossil fuels without counteracting climatic fluctuations may involve a 'super-interglacial' unlike any interval experienced by earth during the past million years (Imbrie and Imbrie, 1979; Manabe and Stouffer, 1993) (Fig. 2.26a). Furthermore, because of the slow absorption of CO_2 by the oceans, the effects of this carbon dioxide may endure for a thousand years after fossil fuels are exhausted (Imbrie and Imbrie, 1979). A greater understanding of past natural climatic cycles and of atmosphere–ocean CO_2 interactions may help to predict whether temperatures may peak in decades or millennia before there is a return to the cooling of the orbitally or otherwise driven cooling trends (Broecker, 1975; Mitchell, J.M., 1977; Mitchell, J.F.B., 1989; Imbrie and Imbrie, 1979).

What are the possibilities of ice sheets and major mountain glaciation occurring in the immediate future? Miller and Vernal (1991) have demonstrated that initial ice sheet growth at the start of the last glacial cycle (~120,000 BP) occurred in high northern latitudes (65–80°N) sustaining climatic conditions quite similar to those of today. The geologic record indicates that these optimal ice-growth conditions include a warm northern ocean, strong meridional atmospheric–oceanic circulation, and low summer temperatures as well as depressed winter temperatures. Since greenhouse warming is expected to be most pronounced in the Arctic and during winter months (Berger, 1978), together with decreasing summer insolation (Fig. 2.21), this anthropogenic warming could lead to snowline depression and ice sheet growth in the high northern latitudes. Both geological and historic evidence exists demonstrating glacier growth and expansion in the Arctic during periods of global warming (Miller and Vernal, 1991). Increased precipitation during the Holocene in Antarctica is countered by increased calving at the ice margin contemporaneous with sea level rise (Denton *et al.*, 1989; Clapperton and Sugden, 1990).

Chapter 3

ICE SHEET MODELLING AND THE RECONSTRUCTION OF FORMER ICE SHEETS FROM GLACIAL GEO(MORPHO)LOGICAL FIELD DATA

T. J. Hughes

3.1. INTRODUCTION

It has recently been argued that glaciology has made the transition from being a descriptive to a quantitative natural science with the development of ice sheet models (Lliboutry, 1987b; Waddington and Walder, 1987). Any model is a complex approximation of reality and the success of an individual model is dependent upon how close it comes to replicating that reality (Fig. 3.1). However, in the field of glaciology, verification of a model remains a complex and almost insurmountable problem (Paterson, 1980; Waddington, 1987; Meier, 1987; Payne and Sugden, 1987). Unlike other sciences where experimentation is possible and necessary, due to the timescale of geological events and processes, ice sheet models act as forms of surrogate experimentation.

Ice sheet models are developed for two principal purposes: (1) to determine (predict) marginal fluctuations of an ice sheet over time with inputs of net balance, internal temperature distribution, ice flow parameters and basal ice/bed conditions; and (2)

to interpret the past distribution of former glaciations based upon sediment and landform associations and, in conjunction, draw inferences about past climates and global atmospheric/oceanic circulation conditions and interrelationships.

These models, although designed as predictive, essentially attempt to 'explain' phenomena. The term *explain*, however, is subjective and needs some interpretation, itself, since in natural science this term implies only a 'best possible' approximation of reality given the constraints of data and general knowledge at any particular moment in time. In general, ice sheet models do not explain specific problems but rather usefully highlight inconsistencies and discrepancies either in the quality and type of input data to the model or in the assumptions made by the model or field data used in verification. Payne and Sugden (1987), using the Scottish Late Devensian Ice Sheet as an example, illustrate this divergence in agreement between the model and reality, and the intrinsic utility that can be garnered from model/field data inconsistency. A danger that must be guarded against is the false assumption that because a particular model appears to be physically reasonable and accurate and appears to explain observed phenomena, that an explanation has indeed been achieved (Waddington, 1987). In the use of any model there always remains the possibility of coincidental agreement which must not be confused with actual scientific agreement and verification (Meier, 1987). In 'tuning' a model, the predictive power of a model in approaching reality is enhanced but some input data must always be retained to test the model to avoid the pitfall of circular reasoning (Fig. 3.1). The question of what constitutes verification is, in itself, fraught with problems and remains a major enigma in any predictive ice sheet model's validation.

3.1.1. Uses of Ice Sheet Models

With the development of complex models, the uses to which these models can be applied have become increasingly pertinent at this stage in human history. Ice sheet models are of value in both paleoclimatic and paleoenvironmental reconstructions, in interpreting and dating ice core chronologies (Chapter 2), in understanding past ice sheet spatial distributions and marginal fluctuations due to both climatic and glaciologic change, and in understanding mantle rheology and sea level changes due to ice sheet growth and decay (Meier, 1987). The predictive value of these models, for example, in anticipating the possible effects of 'greenhouse' warming on marine ice sheet stability and sea level change is of enormous global concern and significance.

3.1.2. The Link Between Glacial Geology and Ice Sheet Models

Ice sheet models make use of glacial geologic information in two ways: geologic data are used as critical input to geomorphic models, and as an independent check on output from mechanical models (Hughes, T.J. 1987). The use of glacial geo(morpho)logy as an input emphasizes the relationship between the pattern of flowlines at the surface and the distribution of meltwater at the bed of an ice sheet, recognizing the fact that the local glacial geo(morpho)logy must be compatible with the

subglacial hydrology associated with the meltwater distribution. This link was central in the glacial geo(morpho)logy/ice sheet model scheme employed in reconstructing ice sheets worldwide at the last glacial maximum for CLIMAP (Climate: Long-range Investigation, Mapping and Prediction, a project of the International Decade of Ocean Exploration, 1970–1980). Details of this scheme were presented by Hughes (1981), and the scheme was then used to reconstruct of former ice sheets (Hughes *et al.*, 1981). Alternative glacial geo(morpho)logy/ice sheet model schemes have been developed by Boulton *et al.* (1985) and Fisher *et al.* (1985). All these schemes reconstruct models based upon the assumption of a steady-state ice sheet. A physical model for steady-state ice sheets that ignores glacial geo(morpho)logy but which predicts the distribution of basal meltwater, compatible with surface flowlines, was developed by Budd *et al.* (1971) for Antarctica and applied by Sugden (1977) for the reconstruction of the Laurentide Ice Sheet during its Late Wisconsinan maximum (~18,000 BP). A time-dependent ice sheet model, not reliant on glacial geo(morpho)logy, was developed by Budd and Smith (1981) to simulate the last glaciation cycle of North America. Many newer models have extended these pioneering efforts.

3.2. MODELLING STRATEGIES

Broadly speaking, two strategies for ice sheet modelling have emerged in the past two decades. These strategies are now converging, but were originally quite distinct.

3.2.1. The Melbourne School

The first strategy disregards glacial geo(morpho)logy as input to the model, and uses it only as a control on model output, primarily for comparison of the maximum extent of glaciation produced by the model against reality. This strategy was pioneered at the University of Melbourne (Australia) in the early 1970s, and will be referred to as the 'Melbourne school'. The Melbourne school, in its original theoretical work, held the view that if all the essential physics of ice dynamics and boundary conditions were specified correctly, the model would compute the basal thermal conditions (whether the bed is thawed or frozen and, if thawed, the rates of freezing and melting) and their distribution over the subglacial landscape. This information would then permit an understanding to be achieved of the glacial processes that modify the landscape. Hence, glacial geo(morpho)logy is relegated to an ancillary role at the tail-end of model output (Fig. 3.1). Moreover, glacial geo(morpho)logy acquires significance only insofar as model output can account for it.

In principle, the Melbourne school is correct in producing a model so sophisticated in its internal ice dynamics and so precise in accounting for all external boundary conditions that the topographical expressions of glacial geo(morpho)logy have only one explanation, the one provided by the mechanical model. It is a twentieth century expression of the nineteenth century mechanical universe that was rooted in Newtonian physics. An example of this approach was provided by Sugden (1977), who took the original mechanical model produced by the Melbourne school for determining the physical characteristics of the present-day Antarctic Ice Sheet (Budd et al., 1971), adapted it to compute the physical characteristics of the Laurentide Ice Sheet at its Late Wisconsinan maximum extent, and used the model output to explain the macro-scale aspects of North American glacial landscapes (Sugden, 1978). The primary assumption made was that, at the glacial maximum of the Late Wisconsinan Laurentide Ice Sheet, a steady state was achieved, compared to advance and retreat stages, and thus steady-state processes of glacial erosion and deposition made the dominant imprint on the subglacial landscape.

The Melbourne school has concentrated on developing models based upon the surface interaction between ice sheets and the atmosphere, particularly the surface mass balance between ice accumulation and ablation. The original Melbourne steady-state, one-dimensional mechanical model, developed for the Antarctic Ice Sheet by Budd et al. (1971), was converted into a two-dimensional model and applied as a time-dependent model to simulate a Quaternary cycle of North American glaciation (Budd and Smith, 1981). It was further applied as a steady-state model to compute physical characteristics of the present-day Greenland Ice Sheet by Radok et al. (1982), and applied as a time-dependent model to study the effect of climatic warming on the West Antarctic Ice Sheet (Budd et al., 1987). In all cases great care was devoted to the ice-atmosphere boundary. However, the ice-water and ice-bed boundaries were ignored entirely by Budd and Smith (1981) in their simulation of a North American glaciation cycle since only flow over a frozen bed was permitted and glacial geo(morpho)logy was not used. This eliminated changing sea level along marine margins, ice calving along marine and lacustrine margins, conversion of sheet flow to stream flow toward all margins, and glacial erosion and deposition at the bed as contributing to ice dynamics and boundary conditions.

3.2.2. The Maine School

The second strategy for model construction uses glacial sediments and landforms as primary data input, based on the premise that they are direct records of the existence of former ice sheets, and indirect reflections of ice sheet internal dynamics and the external boundary conditions that constrained

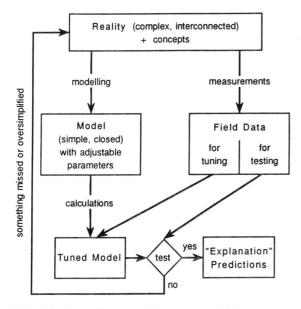

FIG. 3.1. Flow diagram of ice mass modelling process (reproduced with permission of the IAHS).

them. This strategy was pioneered at the University of Maine (U.S.A.) in the late 1970s, hereafter referred to as the 'Maine school'. Its guiding philosophy is that both internal ice dynamics and external boundary conditions for former ice sheets must have been compatible with processes of glacial erosion and deposition that produced the glacial landscapes. Glacial sediments and landforms then become tools for understanding ice dynamics and boundary conditions of former ice sheets. An innovative application of this strategy is to assist in understanding the ice dynamics of ice streams and their subglacial and marine grounding line boundary conditions, especially during deglaciation. About 90% of the ice within the Antarctic Ice Sheet today is discharged via ice streams, and most of the West Antarctic Ice Sheet is grounded far below sea level on the Antarctic continental shelf (Hughes *et al.*, 1985). Thus, if the glacial geo(morpho)logy of North America provides evidence for ice streams and marine ice sheets, any ice sheet model must include ice streams as a mechanism of ice dynamics and the marine grounding line as a boundary constraint. Rather than relying on the purely mechanical model produced by the Melbourne school, the Maine school produced a model that incorporated ice dynamics and boundary constraints imposed by glacial geo(morpho)logy (Hughes, T.J., 1981, 1985, 1987; Hughes *et al.*, 1981).

In modelling former ice sheets, both schools conceived a distinction between first-order and second-order glacial landscapes (Sugden, 1977, 1978; Hughes, T.J., 1981, 1985, 1987). It is apparent that any glaciated landscape is a complex mosaic of both macro-scale and meso-/micro-scale temporal and spatial components (Sissons, 1974; Thorp, 1986). The latter group tends to indicate the last glacial advance and retreat from an area, whereas the former reflects forms and terrains that have survived possibly repeated glacial movements. The meso-/micro-scale forms include moraines, striations, grooves, kames, eskers, drumlins and associated transitional landforms. Most of these are time-transgressive features that show the motion of the ice margin over time during retreat or close to maximum ice sheet extension and are small and delicate enough to be obliterated or over-printed by a subsequent glacial

advance. Because of their transitory and small-scale nature, these features constitute a second-order of glacial landscapes. The first-order or macro-scale forms include fjords, glaciated troughs and basins, large roche moutonnée and scoured and fluted terrains. Most of these are products of long-term and repeated glaciation and thus make up first-order glacial landscape features.

In the Maine school, glacial sediments, landforms and landscapes were employed in the primary development of ice sheet models. These models attempt to simulate both advance and retreat of ice sheets, during which time steady-state processes of glacial erosion and deposition make a dominant imprint that remains as a first-order glacial landscape. This landscape consists of large-scale glacial landforms that are not altered by the more transitory small-scale glacial erosion and deposition processes that produce a second-order glacial landscape (Hughes, T.J., 1987). In considering Laurentide Ice Sheet glaciation, first-order glacial landscapes consist of such features as: (1) the continuing glacio-isostatically depressed Hudson Bay and its surrounding lowlands, where the ice load was greatest and most prolonged; (2) the exposed Precambrian Shield surrounding Hudson Bay, which locates the region of sustained sheet-flow of ice sliding on a thawed bed and the regions of basal freezing and thawing; (3) the linear troughs along the outer edge of the Precambrian shield that generally radiate from Hudson Bay, which locate the region where sheet-flow characteristic of ice sheet interiors began to break up into stream-flow characteristic of ice sheet margins and therefore identifies the outer limit of the relatively steady-state core of the Laurentide Ice Sheet; and (4) the outermost terminal moraines on land and marine sills or sediment fans at the seaward ends of inter-island channels, which locate the maximum ice sheet extent and were therefore not glacially altered during retreat (cf. Aylsworth and Shilts, 1989; Bouchard, 1989; Fulton, 1989).

A two-dimensional, time-dependent version of the Maine school model was developed to study the role of ice streams in ice sheet dynamics and the role of changing sea level on the stability of marine ice sheets (Fastook, 1985, 1987; Fastook and Chapman, 1989). Most recently, it has been used to simulate a full glaciation cycle for ice sheets in North America, Eurasia and Antarctica (Fastook and Hughes, 1990). Glacial geo(morpho)logy is used to specify whether the bed is frozen or thawed, with thawed beds ranging from rough bedrock providing high traction to soft till providing low traction. Ice stream flow exists as channelized low-traction basal sliding. The distribution of melting, freezing and frozen conditions beneath an ice sheet is determined in large part by the distribution of accumulation, ablation and temperature conditions on the ice sheet surface. Fisher *et al.* (1985) used this glaciated terrain–ice sheet linkage to postulate surface conditions that would be compatible with basal conditions deduced from glacial geo(morpho)logy in their two-dimensional reconstruction of the Laurentide Ice Sheet at the last glacial maximum. Time-dependent glaciated terrain–ice sheet models (hereafter referred to as the GT–IS model) use second-order glacial geo(morpho)logy to control ice sheet dynamics during retreat. Boulton *et al.* (1985) have shown that computer-reconstructed ice sheets can be quite different, depending on whether primary emphasis is placed on first-order or second-order glacial geo(morpho)logy. T.J. Hughes (1987) dealt with this problem by emphasizing first-order glacial geo(morpho)logy in a shrinking inner region of a retreating ice sheet within which nearly steady-state conditions could be maintained, and by emphasizing second-order glacial geo(morpho)logy in an outer region within which transitory conditions prevailed.

3.2.3. The Merger of Both Schools

Ultimately, a merger of the mechanical and geomorphic approaches to modelling ice sheets during a glaciation cycle will produce a glaciological synthesis comparable to that in hardrock geology produced by plate tectonics. This merger will require incorporating the equations of heat transfer with the equations of mass transfer in the ice dynamics so that surface temperature and mass balance can be quantitatively related to the basal thermal conditions that produce glacial terrains, and in relating changes in sea level caused by ice-volume changes to glacio-isostatic adjustments along marine ice sheet margins. The first steps in this direction have been taken by

MacAyeal (1989b), who integrated the momentum balance equations in the vertical direction for a West Antarctic ice stream sliding on a deformable bed, and by Lindstrom and MacAyeal (1989) and Lindstrom (1990), who simulated a full cycle of Eurasian glaciation in which a climatic cooling and warming cycle produced surface temperature and mass balance changes that produced changing frozen and thawed conditions at the bed. Lingle (1985) has made a rigorous attempt to include glacio-isostatic adjustments in assessing the stability of marine ice sheets. His model applied to an ice stream of the West Antarctic Ice Sheet and included the buttressing capacity of the Ross Ice Shelf.

3.2.4. The Glaciated Terrain–Ice Sheet Model (GT–IS)

At the opposite extreme of the complexity spectrum for reconstructing former ice sheets is the simple GT–IS model. The model relies on glacial geo(morpho)logy to reveal the changing pattern of flowlines for an ice sheet, to determine whether ice sheet flow was over a melting, freezing, or frozen bed, and to show how the vitality of ice stream flow is related to a thawed bed that can change, over time, from soft sediment to rough bedrock due to glacial erosion and deposition. All of these boundary condition aspects of the model cannot be presented here, but they all cause bed traction to change in space and time; however, the method for reconstructing accurate flowline surface profiles from these variations in basal traction will be presented. In this method, knowledge of the surface mass balance is not necessary, but surface accumulation and ablation rates can be inferred from basal traction, since these rates produce changes in ice velocity and traction changes in response to velocity changes.

Glacial geo(morpho)logical input to the GT–IS model is illustrated in Figs 3.2 and 3.3. Figure 3.2 shows boundary conditions for a generalized ice sheet, and these conditions are related to first-order glacial geo(morpho)logy in Fig. 3.3. First-order glacial terrains, being created when an ice sheet most closely approaches true steady-state conditions, get reinforced by each successive glaciation. First-order features are found in: (1) the central regions of

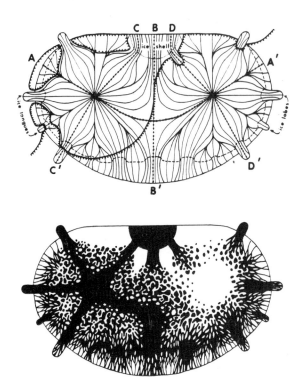

FIG. 3.2. Idealized ice-sheet flow regime. Flow in plan view for surface ice (top) and basal water (bottom). Shown at top are surface flowlines (solid lines) radiating from a terrestrial dome (inside hachured line) and a marine dome (outside hachured line), a surface equilibrium line (dashed line) that is highest on the equatorward flank of the ice sheet and an ice-shelf grounding line (dotted line) on the poleward flank of the ice sheet. Shown at the bottom are thawed patches (isolated black areas) where quarrying creates lakes, frozen patches (isolated white areas) where lodgement till creates drumlins, the arc of exhumation (black areas beneath the surface equilibrium line) where quarrying and regelation occur along a basal equilibrium line that separates an inner melting regime from an outer freezing regime, the arc of deposition (between the arc of exhumation and the ice margin) where regelation ice melts, selective linear erosion in ice-stream channels (broad black bands radiating from ice domes) and selective linear deposition in eskers (narrow black lines between broad black bands).

postglacial isostatic rebound, whether this region was frozen continental highlands or a thawed marine basin; (2) the heavily pitted erosion zones of basal melting and freezing that surround this central region, and where eroded debris is incorporated into a basal layer of regelation ice; (3) an outer zone of deposition

where the regelation ice melts, either during steady-state conditions at the glacial maximum or during transient conditions that accompany glacial retreat; and (4) in fore-deepened channels that cut across the outer zone and were occupied by either terrestrial or marine ice streams (Aylsworth and Shilts, 1989; Bouchard, 1989; Menzies, 1995b, Chapter 2). Ice streams tend to form in sedimentary marine troughs or in river valleys and lowland depressions, where soft sediments become mobilized by porewater pressure that dilates the sediments during shear deformation (Alley, 1989a, b). The resulting deformable bed persists until glacial erosion has removed the sediments or ice-bed interface conditions alter (Shoemaker, 1986a; Menzies, 1989a, b). Ice sheet surface profiles can be constructed along flowlines that include all these basal conditions. Moreover, reliable profiles can be constructed by knowing only the topography and glacial geo(morpho)logy of the deglaciated landscape (Thorp, 1986, 1991); no knowledge is required of mean annual temperatures and rates of accumulation and ablation over the former ice sheet surface.

3.3. SPECIFYING BOUNDARY CONDITIONS

3.3.1. Glacial Geo(morpho)logy as Model Input

Boundary conditions imposed by glacial geo(morpho)logy on ice sheet reconstructions arise from the relationship between the pattern of surface flowlines and the distribution of basal meltwater for a generalized ice sheet (Fig. 3.2). Each boundary condition makes a unique contribution to the GT–IS model in reconstructing steady-state ice sheets. The ice margin marks the outer limit of glaciated terrain and is, therefore, used to reconstruct ice sheet flowline profiles from the ice margin to the ice divide. The ice divide is where glacial flow originates, and is located where flowline profiles reconstructed from opposite margins have the same elevation. Divergence and convergence of flow from the ice divide to the ice margin is measured by the changing width of flowbands between adjacent flowlines. The ice bed creates the glaciated terrain that is used to draw flowbands of glacial motion along which the ice thickness profiles are computed. The ice surface

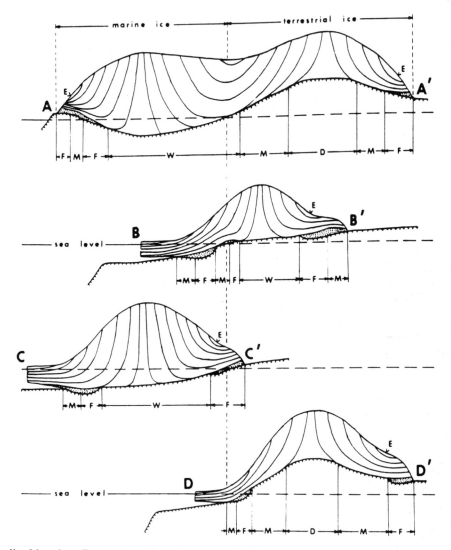

FIG. 3.3. Idealized ice-sheet flow regime. Flow of ice in vertical cross-section is represented by thin lines along dotted surface flowlines A–A' through D–D' shown in Fig. 3.2. Shown above are longitudinal flowline profiles along the crest of the ice divide (section A–A'), along opposite flanks of a saddle on the ice divide (section B–B'), along opposite flanks of the marine ice dome(section C–C'), and along opposite flanks of the terrestrial ice dome (section D–D'). At the base (hachured line), a dry bed (D) is frozen everywhere, a wet bed (W) is thawed everywhere, melting beds (M) and freezing beds (F) have a mix of frozen and thawed areas, and regelation ice (dotted areas) forms at ice-stream headwalls, and over the thawed parts of a freezing bed

experiences major changes in mass balance and temperature along a flowband, and these changes ultimately determine the basal hydrology that produces glacial terrain features and controls the ice thickness profile over a given subglacial topography. Therefore, glacial terrains can be used to reconstruct both former flowlines and ice thicknesses. These boundary conditions revealed by glacial geo(morpho)logy are input to the GT–IS model for reconstructing ice sheets and specifying conditions at these boundaries is the first step in using the model. Figures 3.2 and 3.3 illustrate the variety of conditions at each boundary and should be consulted in the following discussion of these ice sheet boundaries.

3.3.2. Ice Margins

Ice sheets can have terrestrial margins where ice terminates in continental interiors or marine margins in the tidewater zone along continental coastlines, or where ice spreads over continental shelves to form ice shelves. Owing to these various terminating environments, margins are different at midcontinent and tidewater positions and at the calving fronts and the grounding lines of ice shelves. Ice cliffs lie along tidewater margins and float along ice shelf calving fronts (Hughes, 1989, 1992b).

Ice streams are fast-moving segments of ice within an ice sheet; they form toward both terrestrial and marine margins, provided that subglacial topography is able to channel the flow of grounded ice (Chapter 5). Ice streams are more common along marine margins because coastal fjords and channels and inter-islands facilitate flow better than the river valleys of interior plains. This is seen by comparing the profusion of ice streams along marine margins with the poverty of ice streams along terrestrial margins of the present-day Greenland Ice Sheet, in relation to relative linearity of topography along these margins. A terrestrial ice stream forms an ice lobe. A marine ice stream forms an ice tongue, which may float freely or be imbedded in a floating ice shelf. To remain stable, ice shelves must lie in confined embayments or be pinned to islands or shoals. Shoaling creates ice rises or ice rumples on the ice shelf, depending on how firmly the ice is pinned. Shelf flow moves around an ice rise (strong pinning) and over ice rumples (weak pinning) (Chapter 5).

3.3.3. The Ice Divide

Flow toward the terrestrial and marine margins of an ice sheet begins at the ice divide. An alternating sequence of domes and saddles commonly exists along the ice divide. Marine domes may form when sea-ice thickens and becomes grounded over broad areas of shallow seas surrounded by islands (archipelagoes) or in shallow embayments on polar continental shelves. Terrestrial domes form on polar plateaux when climatic cooling lowers the snowline enough so that snowfields can expand and thicken to become ice caps. Saddles form when domes merge or

when a major ice stream surges. Ice downdrawn into a surging ice stream can convert a dome into a saddle on the ice divide.

Owing to the much smaller surface slope and much greater divergence, flow from dome to saddle along an ice divide crest is very slow compared to flow down the flanks of the ice divide. Flow diverges downslope from domes and converges downslope from saddles. Domes and saddles are different boundary conditions at the ice divide.

3.3.4. The Ice Bed

The ice bed has frozen, thawed, freezing and melting boundary conditions. Freezing and melting beds consist of a mosaic of frozen and thawed patches or of sticky and slippery spots that determine if the bed is rough or smooth and hard or soft by controlling the types of glacial erosion and deposition processes (Menzies, 1995b, Chapter 2). A great variety of complex boundary conditions is therefore possible at the bed (Hughes, 1981; Menzies, 1987). A smooth, soft bed exists in depositional environments, especially marine sedimentary basins and river valleys, where porewater is abundant. As these sediments are glacially eroded, bedrock 'sticky spots' eventually project above or within the sediments and become frozen patches (Baranowski, 1970; Habbe, 1989, 1990; Menzies, 1989a). These regions grow in size and number until isolated thawed patches of sediment remain as 'slippery spots' that are lower than surrounding terrain. This transformation through time becomes a transformation through space from the inner erosional zone to the outer depositional zone of an ice sheet, and produces a pitted or hummocky subglacial landscape (Eyles and Menzies, 1983).

Ice creeping over the bed can also slide if the bed is thawed. Basal heat is supplied by the geothermal flux and the friction in ice shearing over the bed (Chapter 4). Although ice moves too slowly beneath ice divides to generate much frictional heat, basal melting may still occur if surface ice accumulation rates are low, so that cold surface ice moves downward slowly, and if the ice is thick enough to inhibit conduction of geothermal heat to the surface and to depress the pressure melting point to the basal ice temperature. A marine dome formed in a polar embayment is more likely to have a thawed bed than a terrestrial dome that formed on a polar plateau since the terrestrial dome would form on colder ground and would be thinner, see Figs 3.2 and 3.3.

Frictional heat is greatest where ice velocity is highest and ice is still strongly coupled to the bed. These conditions exist at the bed beneath the surface equilibrium line, where the mass-balance velocity is greatest and at the heads of ice streams where converging flow is greatest. Ice progressively slows in the ablation zone and becomes progressively uncoupled from the bed in an ice stream. A thawed bed under the equilibrium line is likely, but a freezing bed may exist beneath the ablation zone where deceleration reduces frictional heat. The freezing bed may become frozen if the basal temperature gradient in the ablation zone is sharp enough to conduct latent heat of freezing to the surface. A thawed bed at the head of an ice stream should remain thawed in the ice stream because ice streams accelerate as they become progressively uncoupled from the bed. However, a layer of regulation basal ice may form in the ice stream because uncoupling also reduces basal frictional heat.

Basal frictional heat increases with ice velocity, which is slowest beneath the ice divide and fastest beneath the equilibrium line or in ice streams. Hence, if the bed is frozen beneath the ice divide and thawed beneath the equilibrium line and in ice streams, the intervening bed must be transitional from fully frozen to fully thawed. By definition, this is a melting bed, and latent heat of melting is supplied by basal frictional heat. Therefore, if the bed beneath the ice divide, the equilibrium line and ice streams is thawed, the intervening bed must also be thawed. However, when burial of cold ice accumulated at the ice divide is considered, frictional basal heat is conducted upward more rapidly because of the progressive movement of cold surface temperatures toward the bed as ice moves away from the ice divide (Robin, 1955). This can cause basal ice to move over a bed that is thawed, freezing, frozen, melting and thawed, or thawed, freezing, melting and thawed, on the way to the equilibrium line or to ice streams (Hughes, 1981). Glacial geo(morpho)logy can specify which of these possibilities was actually the case. Freezing and melting zones contain a mosaic of frozen and thawed

patches. A melting zone consists of isolated thawed patches toward the ice divide and isolated frozen patches toward the ice margin, with a transition between. A freezing zone consists of isolated frozen patches toward the ice divide, isolated thawed patches toward the ice margin, a transition zone in between and a basal layer of regelation ice that forms in the thawed portions. The basal regelation ice would pass over a frozen zone that might lie between freezing and melting zones, and regelation ice would be melted away in the melting zone. Each of these sequences produces a distinctive glaciated terrain.

3.3.5. The Ice Surface

An ice sheet has a drainage pattern revealed by surface flowlines (Boulton and Clark, 1990a, b; Clark, 1993). Surface curvature changes along a surface flowline are caused by major changes in boundary conditions especially basal traction and bed topography. Transitions from ice sheet flow to ice stream flow to ice shelf flow cause transitions in the surface curvature from convex to concave to almost flat due to progressively less bed tractional basal water increases. Sheet flow predominates in the interior where thick ice overrides all but the largest variations in bed topography. Stream flow predominates along the margins where thin ice can be strongly channelled by seemingly minor variations in bed topography. Shelf flow predominates where the ice sheet becomes afloat in deep water beyond the grounded margins. Shelf flow is channelled only by variations in bed topography that create embayments and ice rises where a floating ice shelf is confined laterally and pinned to the bed.

A second boundary condition manifested at the surface is the division of the ice sheet into a number of drainage basins separated by ice divides, with secondary ice divides branching from the primary interior ice divide to form ice ridges between adjacent ice streams. Major ice streams have drainage basins that include a portion of the major ice divide, whereas the basins of minor ice streams develop closer to the ice sheet margin. Downdraw in ice stream drainage basins can create ice ridges or even local ice domes between ice streams. Ice flow diverges from these ridges and domes, which are therefore centres of spreading along the grounded margin of an ice sheet. In Antarctica, a whole hierarchy of ice streams and spreading centres exists in the grounded ice sheet (Drewry, 1983b; Hughes *et al.*, 1985).

A third surface boundary condition is the surface mass balance. The surface equilibrium line that separates the interior zone of net ice accumulation from the peripheral zone of net ice ablation is the most prominent manifestation of this boundary condition. All ice sheet models must include the surface mass balance because ice sheet dynamics arise from the surface mass balance, and advance or retreat of an ice sheet is an accommodation to changes in the surface mass balance.

3.4. FIRST-ORDER GLACIAL TERRAINS

3.4.1. Interpreting Glacial Terrains

Perhaps the most controversial aspect of this scheme for reconstructing former ice sheets is making a distinction between first-order and second-order glacial terrains (Hughes, T.J., 1981, 1987; Hughes *et al.*, 1981). Glacial processes producing the first-order glacial terrains used as input for this GT–IS model are illustrated in Figs 3.2 and 3.3. First-order glacial geo(morpho)logy is linked directly to basal traction in this model. The interpretation of first-order glacial geo(morpho)logy is similar to the interpretation by Sugden (1977, 1978) in that it emphasizes large-scale glacial landscapes as described by Sugden and John (1976). The GT–IS model, presented here, depends upon the assumption that a nearly steady-state flow regime exists in the core area of an ice sheet during most of each glaciation cycle. Glacial isostasy, erosion and deposition create stable zones in the core. Diagnostic glaciated terrain is created in each zone during most of a glaciation and is enhanced during subsequent glaciations. Eventually these diagnostic zones of stable glacial activity impart the dominant imprint on the subglacial landscape, and this imprint constitutes first-order glaciated terrain. The imprint is legitimate input for a geomorphic model that reconstructs former ice sheets at a glacial maximum when steady-state flow was approximated most closely.

The assumption that steady-state glacial activity produces broad zones of first-order glacial terrains implies that second-order glacial terrains are less permanent and smaller in scale. Glacial geo(morpho)logy has traditionally been concerned with these second-order features, which range in size from drumlins and eskers to erratics and striations. Their predominant orientation, distribution or direction reflects ice flow closer to the margin of an ice sheet. As directional indicators, therefore, they are most useful in determining ice flow near the margin during the last retreat and are, therefore, time-transgressive. However, a distinction between first-order and second-order glacial terrains cannot be made if the ice sheets were so unstable during a glaciation cycle that ice margins advanced and retreated repeatedly over even the core area. Geomorphic reconstructions of former ice sheets must then depend on theories for producing drumlins (Boulton, 1979; Menzies, 1981), eskers (Shreve, 1985), erratic fans (Shilts, 1980), striations (Hallet, 1979) and other glacial landforms of similar scale. Boulton *et al.* (1985) and Fisher *et al.* (1985) have employed ice sheet models using glacial geo(morpho)logy on this second-order scale as model input. These features are often found within the first-order glacial terrains discussed below and illustrated in Fig. 3.3.

3.4.2. Weathered Rebounding Subglacial Highlands

Subglacial highlands located in the interior of a region covered by a former ice sheet will be weathered and have little or no record of glacial erosion or deposition if a major ice dome were located there and the bed beneath the dome were frozen. Evidence for the dome will be in the form of raised marine beaches surrounding the highlands and highland lakes having old shorelines that tilt up toward the dome. The Queen Elizabeth Islands in the Canadian Arctic Archipelago are an example where an ice dome appears to have existed. Blake (1970) has postulated the existence of a Late Wisconsinan Innuitian Ice Sheet based on the dated pattern of raised marine beaches in these islands. Glacial erosion was limited because there was no, or limited, ice motion at the interface with a frozen bed, and glacial deposition was restricted since no, or very limited, debris enters the basal ice region under these conditions (cf. Koerner and Fisher, 1979) (Fig. 3.4a). This leaves only isostatic rebound, as recorded by raised beaches and negative gravity anomalies, to document the existence of a former ice sheet.

3.4.3. Scoured Rebounding Subglacial Basin

A subglacial basin in the interior of a region covered by a former ice sheet will be an areally scoured erosion zone if it were thawed and lay beneath an ice dome on the ice-divide. Ice flow lines beginning at the dome and those nearest the dome would have intersected the bed if it were thawed (Fig. 3.4b). Glacial erosion, rather than deposition, would occur because ice can slide over a thawed bed, and the ice sliding at the ice–bed interface would be clean because flow lines for this basal ice originated at the ice surface. The only possible glacial deposition is of debris eroded at the ice–bed interface by the sliding action (Menzies, 1995b, Chapter 2). In the case of an ice dome formed over a marine, lacustrine or previously glaciated depositional basin, ice flow may deform subglacial sediments if they are thawed and have high pore water pressure. Deformation will thin the sediments and, given sufficient time, may eventually remove them.

A smooth landscape of areal scouring in the centre of a region occupied by a former ice sheet, with only occasional glacial sediments deposited, is proof that the bed was thawed and is a strong indication that the ice divide was located there, especially if bedrock striations show ice motion radiating out of the region (Aylsworth and Shilts, 1989). If this central region includes a marine embayment, such as Hudson Bay in the Canadian shield or the Gulf of Bothnia in the Baltic shield, raised beaches will be highest and gravity anomalies will be most negative around the embayment, assuming that here a prolonged heavy mantle of ice caused maximum crustal depression. An ice dome over an interior subglacial basin could collapse during deglaciation, in which case the final pattern of bedrock striations would show ice moving into the basin, especially from flanking highlands such as those in Keewatin and the Ungava peninsula west and east of Hudson Bay (Bouchard, 1989).

3.4.4. Pitted Zone of Net Basal Melting

A pitted erosion zone should exist in lowlands surrounding highlands where a major dome of a former ice sheet rested on a frozen bed. The lowlands act as a zone of net basal melting, across which isolated thawed patches nearest the highlands become bigger and more numerous with increasing distance from the highlands, until they coalesce, leaving isolated frozen patches that become smaller and less numerous toward the outer limit of the melting zone. Erosion should occur in the thawed patches because basal sliding is possible there. Under steady-state conditions, basal water volume is constant in each thawed patch, so basal melting in the upstream half of a patch reverts to basal freezing in the downstream half. Bands of regelation ice therefore extend downstream from each thawed patch, until they encounter a larger thawed patch where the regelation ice melts. Debris eroded and entrained in the regelation ice from one thawed patch is then deposited in a thawed patch downstream, where it is either comminuted for further transport or becomes deposited as a diamicton. An increasingly thick basal layer of debris-charged regelation ice therefore forms as ice moves across a melting erosion zone, as shown in Fig. 3.4c, and the entrained rock particles act as cutting tools that erode the bed as the regelation ice melts in the downstream thawed patch (Chapter 7). All regelation ice is melted after thawed patches coalesce. Each thawed patch may be deeply eroded, creating a pitted subglacial landscape.

A melting erosion zone can be identified by a deeply pitted landscape in lowlands surrounding a weathered and rebounding highland zone. The postglacial landscape will consist of lakes that become larger and more numerous with distance from the highlands, with each lake being the site of a formerly thawed patch. Even farther from the highlands, hills will be the predominant feature as sites of formerly frozen patches. These hills will not be striated if the patches remained frozen during deglaciation. If the patches become thawed, the hills may be streamlined. A patchy diamicton may be found across the melting zone, where sediment was deposited when the debris-laden layer of regelation ice melted during deglaciation.

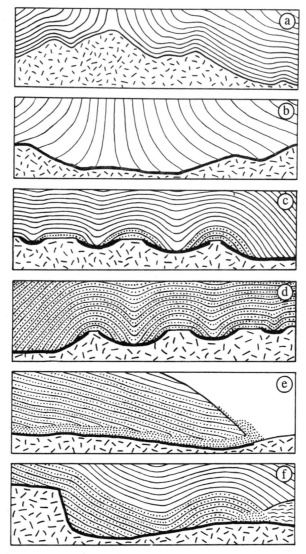

FIG. 3.4. Erosion and deposition beneath an ice sheet. Thin lines are flowline trajectories in vertical longitudinal cross-section. Dotted lines show eroded debris entrained in basal regelation ice. Dotted areas show steady-state deposition of eroded debris as regelation ice melts. Stippled areas denote bedrock. Light line between bedrock and ice denotes a frozen bed. Heavy line between bedrock and ice denotes a thawed bed. (a) Flow over frozen highlands beneath the central ice divide. Ice flowlines descend to the bed and diverge parallel to the bed, with no motion at the ice–bed interface. (b) Flow over a thawed basin beneath the central ice divide. Ice flowlines descend and intersect the bed, with sliding at the ice–bed interface. (c) Flow at the inner part of a basal melting zone. Ice melts in the upstream end of isolated thawed patches and the meltwater freezes in the downstream end, creating bands of regelation ice that eventually melt in the outer part of the basal melting zone. (d) Flow at the outer part of a basal freezing zone. Regelation ice from the inner part of the basal freezing zone melts in the upstream end of isolated thawed patches and the meltwater freezes in the downstream end, creating new bands of regelation ice that lie beneath the older regelation ice. Active steady-state quarrying takes place in the thawed patches shown in (c) and (d), creating pits that become the sites of lakes after deglaciation. (e) Flow in the outer deposition zone of basal melting near the ice-sheet margin. Debris entrained in regelation ice melts out to create a ground moraine across the zone and a terminal moraine along the ice margin. (f) Flow in a fjord occupied by an ice stream. Regelation ice melts above the headwall and the meltwater refreezes on the headwall to create new regelation ice that melts, refreezes, and melts on the fjord floor, leaving an end moraine as a sill.

3.4.5. Pitted Zone of Net Basal Freezing

A pitted erosion zone should also develop in the lowlands surrounding an interior marine embayment over which a major dome of the former ice sheet rested on a thawed bed. Isolated frozen patches may exist nearest the embayment and these patches will become larger and more numerous with increasing distance from the embayment. Eventually they coalesce and leave isolated thawed patches that become smaller and less numerous toward the outer limit of the zone. Hence, net basal freezing occurs across the zone and erosion occurs where the bed has thawed. The decrease in ice overburden pressure with distance from the ice dome drives basal meltwater from the thawed subglacial basin into the freezing zone where it refreezes to form a basal layer of debris-charged regelation ice. Steady-state flow requires that meltwater produced in the upstream half of isolated thawed patches be refrozen in the downstream half; thus, bands of debris-charged regelation ice form downstream from these patches. This situation is illustrated in Fig. 3.4d.

A freezing erosion zone is identified by the lake-spattered landscape associated with isolated thawed patches toward its outer perimeter, and the hilly landscape associated with isolated frozen patches toward its inner perimeter. After deglaciation, the freezing zone will be covered by a till sheet that melted out from debris-charged regelation ice formed in the zone. This till sheet will become thicker and more continuous toward the outer limit of the zone, because the layer of regelation ice thickened in that direction.

3.4.6. Outer Zones of Net Deposition

A thick layer of regelation ice exists at the outer limit of a zone of net basal freezing, but not a zone of net basal melting. When a basal melting zone surrounds a frozen bed beneath the ice divide, only clean ice would continue onward across the subsequent basal freezing zone where dirty regelation ice would form and be transported across a second basal melting zone along the equatorward margin of an ice sheet and a basal frozen zone along its poleward margin. When a basal freezing zone surrounds a thawed bed beneath the ice divide, dirty regelation ice formed in the freezing zone would cross a basal melting zone along the equatorward margin of the ice sheet and a basal frozen zone along the poleward margin. In both cases, the basal layer of regelation ice would be melted in the basal melting zone along the equatorward margin, depositing a thick basal sediment sheet across the zone. Any regelation ice remaining unmelted at the base would appear in the surface ablation zone where its entrained debris would form a terminal moraine at the steady-state glacial maximum and recessional moraines during retreat. The outer zone at basal melting along the equatorward margin would therefore be a zone of net deposition, as illustrated in Fig. 3.4e.

The outer frozen zone along the poleward margin is not a deposition zone during the steady-state glacial maximum, but a debris-charged layer of regelation ice passes over it so that it can become a deposition zone as the regelation ice melts during glacial retreat. The sheet of sediment deposited during retreat would be thin if retreat were rapid. During the steady-state glacial maximum, debris entrained in the regelation ice would either be deposited as a terminal moraine if the poleward margin of the ice sheet ended on land and had a surface ablation zone, or the entrained debris would be funnelled into marine ice streams where the poleward margin advanced into deep water on the continental shelf.

3.4.7. Erosion and Deposition in Ice Streams

Debris entrained in basal regelation ice can be melted out and deposited in the region of strongly converging flow at the head of ice streams because considerable basal frictional heat is generated there. This heat keeps the bed thawed locally, even in an overall frozen basal zone, and it can also melt basal regelation ice being funnelled into the region of converging flow. Much debris deposited in the region of converging flow will be heavily comminuted, but in that process it provides cutting tools for wear processes. Fore-deepened troughs are characteristic of ice stream channels whether produced by intensive wear and down-cutting or strong hydraulic erosion by high-pressure subglacial meltwater stream convergence or a combination of both processes in concert. These troughs are characteristic, for example, of the inter-island channels of the Canadian Arctic Archipelago, large parts of the western coasts of British Columbia, Patagonia, Norway and New Zealand, and around Greenland and Antartica.

Erosion might be replaced by deposition in the outermost part of an ice stream advancing into deep water, because buoyancy eventually uncouples ice from the bed. Basal uncoupling reduces basal frictional heat so that bedrock eroded in the fore-deepened end of an inter-island channel may be frozen into a new layer of regelation ice. If that regelation ice is melted by heat arising from tidal pumping or estuarine circulation where the ice stream becomes afloat, the entrained debris will be deposited as a terminal moraine that forms a basal sill along the grounding line of floating ice.

Channels extending to the edge of the continental shelf often begin at the headwalls of fjords through coastal highlands, such as those in Labrador and Baffin Island. Ice streams occupying these fjords begin in the region where ice converges on the headwall. Basal meltwater produced in the region of converging flow possibly pours over the headwall and refreezes because frictional heat diminishes. The refreezing rate controls the rate at which latent heat of freezing is converted into mechanical energy needed for quarrying. The regelation ice frozen onto the headwall as the meltwater refreezes is pulled forward by the overlying ice, thereby producing a strong quarrying mechanism at the headwall (Chapter 7). Blocks plucked from the headwall by the regelation ice cause the headwall to retreat. These blocks may become powerful cutting tools that sculpt the fjord sidewalls into nearly vertical rock cliffs and fore-deepen the bedrock floor of the fjord. Frictional heat generated by ice shearing along the sidewalls and base of the fjord can melt regelation ice created at the headwall and, above it, regelation ice surviving from the region of converging flow upglacier from the headwall. Entrained debris in both layers of regelation ice may be deposited at the mouth of the fjord as ice begins to float and regelation ice melts. This debris forms a sill at the entrance to the fjord. Figure 3.4f illustrates these mechanisms of erosion and deposition in fjords occupied by ice streams.

Ice streams occupying inter-island channels and fjords are marine ice streams. Terrestrial ice streams can develop in river valleys at the equatorward margins of an ice sheet. The Des Moines and James lobes of the Laurentide Ice Sheet occupied valleys of the Des Moines River and the James River, and may have been the termini of ice streams that originated in Lake Winnipeg and Lake Manitoba. Other Laurentide ice lobes formed beyond the long axis of each of the Great Lakes and may have been extensions of Laurentide terrestrial ice streams that developed in these lakes and were responsible for eroding the lake floors. Basal diamictons and recessional moraines delineating these lobes include rock debris entrained in regelation ice formed in the lake troughs.

3.5. COMPUTING FLOWLINE PROFILES

3.5.1. The Basal Shear Stress

The fundamental formula in glaciology is the expression for basal shear stress τ_o, which depends upon the ice thickness h_i, ice surface slope α, ice density ρ_i and gravitational acceleration g, as follows:

$$\tau_o = \rho_i \, g \, h_i \, \alpha. \qquad (3.1)$$

Equation (3.1) is a statement of Newton's law that a force acting on a body is the product of the body's mass and acceleration. All bodies on the Earth's surface, including ice sheets, experience the acceleration of gravity. It is the force of gravity that causes an ice sheet to spread under its own weight and τ_o is a measure of the bed traction that resists spreading. The smoother and more lubricated by meltwater is the bed, the smaller is τ_o, and τ_o vanishes altogether when the ice sheet floats.

3.5.2. Recursive Ice Elevation Computations

Figure 3.5 demonstrates how flowlines of length L from the ice divide to the ice margin of an ice sheet can be divided into a number of steps having equal length Δx. The ice elevation changes a variable amount Δh at each fixed distance Δx along the flowline. If each Δx step is identified by an integer i, numbered consecutively from the ice margin to the

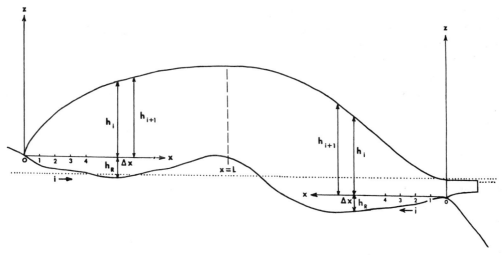

FIG. 3.5. A scheme for reconstructing ice elevation profiles along ice-sheet flowlines from the ice margin to the ice divide. Flowline length L is divided into i = L/Δx steps having equal length Δx, with ice elevation h computed from an initial elevation at one step, and known values of bedrock elevation h_R, basal shear stress τ_o, and degree of isostatic equilibrium r at all steps. Flowline surface and bed elevations are measured with respect to ice margins.

ice divide, then ice surface slope at step i is α_i and is defined by the expression: $\alpha_i = \Delta h_i / \Delta x$, where $\Delta h_i = h_{i+1} - h_i$ to an acceptable approximation, h_i and h_{i+1} being respectively ice elevations at steps i and i + 1. An alternative way to write Eqn. (3.1) is, therefore:

$$h_{h+i} = h_i + \left(\frac{\tau_o}{h_i}\right)_i \frac{\Delta x}{\rho_i \; g}. \qquad (3.2)$$

Equation (3.2) is an initial-value, finite-difference, recursive formula and it allows ice elevation h to be computed at step i + 1 if ice elevation h, ice thickness h_i and basal shear stress τ_o are all known at step i. Initial values of h must therefore be specified at one step before h can be computed at other steps. Also, values of h_i and τ_o must be known at each Δx, where Δx is the finite difference $x_{i+1} - x_i$. Recurring applications of Eqn. (3.2) allows sequential calculations of ice elevation along the flowline.

3.5.3. Basal Topography and Isostasy

The flowline passes over a bed which also has elevation variations, and the changing ice thickness along the flowline causes variations in isostatic depression of the bed. These effects are included in the following modified version of Eqn. (3.2), in which $h_i = h - h_R$ is obtained by setting $r_{sr} = 0$:

$$h_{i+1} = h_i + \left(\frac{\tau_o}{(1 + r_{sr}) \, h - (1 + r_{sr})^{1/2} \, h_R}\right)_i \frac{\Delta x}{\rho_i \; g} \qquad (3.3)$$

Here, h_R is present-day elevation of bedrock above (positive h_R) or below (negative h_R) the margin of the ice sheet for a given flowline and r_{sr} is a ratio such that r = 0 for no isostatic depression and $r_{sr} = \rho_i/(\rho_R - \rho_i) \approx 1/3$ for complete isostatic depression, taking $\rho_i \approx 900$ kg m^{-3} for glacial ice density and $\rho_R \approx 3600$ kg m^{-3} for mantle rock density. Intermediate isostatic conditions are accommodated by using intermediate values of r_{sr}.

3.5.4. Ice Margin Elevations

In Eqn. (3.3), values of τ_o, h_R and r must be specified at each step i along the flowline, but h needs to be specified at only the initial step. It is best to take the initial ice elevation one Δx step in from the terminal moraine along a terrestrial ice margin, and at the grounding line of an ice shelf along a marine ice margin. In the absence of specific information, the marine grounding line can be taken as the edge of the continental shelf.

For terrestrial ice margins, if we specify that τ_o is constant and the bed is horizontal along the first Δx step in from the terminal moraine, Eqn. (3.1) can be integrated for $\alpha = dh/dx \simeq \Delta h/\Delta x$ over distance x to give:

$$h = h_i = (\frac{2\tau_o}{\rho_i} \frac{\Delta x}{g})^{1/2} \qquad (3.4)$$

at $x = \Delta x$, where $h = 0$ is ice elevation at the terminal moraine. Taking $i = 0$ at the moraine, h from Eqn. (3.4) can be entered for $i = 1$ in Eqn. (3.3) as the initial ice elevation that allows computation to begin.

For marine ice margins, the buoyancy condition for floating ice requires that:

$$h = h_i = h_w (\frac{\rho_{sw}}{\rho_i}) \qquad (3.5)$$

at $x = 0$, where $h = h_i$ is ice elevation above the grounding line of floating ice, h_w is water depth at the grounding line and $\rho_{sw} \approx 1020$ kg m^{-3} is the density of sea water. This value of h can be entered for $i = 0$ in Eqn. (3.3) as the initial ice elevation above the bed.

The length Δx of steps along an ice sheet flowline should be 20 km and 50 km. Assumptions made in deriving Eqn. (3.1) break down when $x < 20$ km and important topographic features can be stepped over when $\Delta x > 50$ km. A good compromise is to let $\Delta x = 50$ km or $\Delta x = 20$ km over a rugged bed and $\Delta x = 100$ km over a smooth bed.

3.6. BASAL SHEAR STRESS VARIATIONS

3.6.1. Steady-State Flow

An ice sheet advances and retreats when the ice discharge flux across any transverse vertical section of a flowband is balanced by the upstream rates of net ice accumulation and ice thickening (advance) or thinning (retreat). During steady-state, net upslope accumulation alone balances the discharge flux, thus with no ice thickening or thinning, the ice margin remains in place. In any case, basal shear stress τ_o for sheet flow adjusts naturally to provide basal traction that rises when ice velocity is high and falls when ice velocity is low. Equation (3.1) was derived for ice sheet flow and shows that τ_o varies with the product of ice thickness h_i and surface slope α. For steady-

state sheet flow, h_i decreases as α increases and this gives the flowband a convex surface if the bed is smooth. For steady-state stream flow, on the other hand, net upslope accumulation is balanced by a discharge flux controlled by the pulling power of ice uncoupled from its bed, as seen most dramatically in the pulling force acting on an unconfined floating ice shelf (Weertman, 1957a). Since the stream-flow 'pulling force' was ignored in the sheet flow derivation of Eqn. (3.1), it must be accounted for by allowing the traction force to decrease as the 'pulling force' increases. A decrease of τ_o in the downslope direction is possible in Eqn. (3.1) only if τ_o and α decrease, which then gives the ice stream a concave surface for a smooth bed.

In reconstructing former ice sheets, mechanical models must specify rates of ice accumulation and ablation over the surface to determine whether the mass balance is positive, zero, or negative so the ice sheet can advance, halt or retreat (Chapter 4). Ideally, the surface mass balance should also be used to compute the temperature and velocity field through the ice so that thermal conditions at the bed are obtained as model output (Hooke et al., 1979). Moreover, mechanical models should include the pulling power of ice streams because ice streams can drain up to 90% of the ice. These are nearly impossible tasks, even for the most sophisticated mechanical models, because ice accumulation and ablation rates are unknown, thus they must be parameterized to give temporal and spatial mass balance changes for an ice sheet; and the dynamics of ice streams are not understood. We can avoid specifying these unknown quantities by letting glaciated terrain features indicate what variations in basal traction had to occur in order to produce the first-order glaciated landscape. This is the justification for using GT–IS models to reconstruct former ice sheets. Basal shear stress variations deduced from first-order glacial terrains that were created by sustained steady-state glacial flow are presented below. Fluctuations in rates of surface accumulation and ablation that are compatible with these basal shear stress variations can be computed readily (Fastook, 1984; Hughes, 1985), and are as reliable as those used to parameterize mass balance in mechanical models.

3.6.2. Frozen and Thawed Beds

Basal shear stress beneath a spreading ice sheet is reduced when a frozen bed becomes thawed, since basal meltwater reduces basal traction by lubricating the bed. When the bed is frozen, no large-scale sliding motion exists at the ice–bed interface (cf. Echelmeyer and Wang, 1987), and all significant glacial motion is by internal shear within the ice column where spreading velocity u varies through vertical ice thickness z according to the flow law of ice when laminar flow predominates in the ice column (Glen, 1955):

$$\frac{\delta_u}{\delta_z} = 2(\frac{\tau_{xz}}{A})^n \qquad (3.6)$$

where $\tau_{xz} = \rho_i\, g\, (h - z)\, \alpha$ is the internal shear stress at height z in an ice column of height h above the ice margin, $n \approx 3$ is the viscoplastic exponent for creep deformation in ice and A is an ice hardness coefficient that becomes smaller as ice becomes warmer and develops an 'easy glide' polycrystal line fabric (Hooke and Hudleston, 1980). An ice sheet can slide over its bed at places where the bed is thawed. In these places, the basal sliding velocity u_b is given by the sliding law of ice (Weertman, 1957b):

$$u_b = (\frac{\tau_b}{B})^m \qquad (3.7)$$

where $\tau_o = \tau_b = \rho_i\, g\, h_i\, \alpha$ is the basal shear stress, m is a sliding exponent and B is a sliding coefficient. In simple sliding theory, m = 2 and B gets smaller as the bed becomes smoother. In advanced sliding theories, larger values of m are possible and B varies with hydrostatic pressure in the basal meltwater (Paterson, 1981). Hughes et al. (1981) obtained plausible fits to present day flowlines in Antarctica and Greenland by allowing A to range between 0.5 and 4.0 bar year$^{1/3}$ and B to range between 0.01 to 0.04 bar year$^{1/2}$.

Equation (3.6) can be integrated to give an average velocity $\bar{u}$ in the ice column that can be related to τ_o, which is τ_{xz} at $z = h_R$. At a given distance $x = i\Delta x$ from the ice margin, $u_x = \bar{u} + u_o$ is the velocity of an ice column of height h and width w transverse to the flowline. Under steady-state flow, the ice flux h, w, u_x at position x must equal the net accumulation of ice

over upslope distance (L –x) from x to x = L at the ice divide. Equating the input ice flux upslope from χ to the discharge ice flux at x allows curves of τ_o vs. x to be computed from both the flow law and the sliding law for steady-state ice sheets.

The simplest expressions for the variation of τ_o along the flowline are computed when the flowline is the centreline of a flowband having a constant width and a constant accumulation rate along its entire length. Figure 3.6 gives the τ_o curves for this case where the flow law leads to $\tau_o = \tau_D$ for a dry bed and the sliding law leads to $\tau_o = \tau_W$ for a wet bed. The τ_D curve lies above the τ_W curve because a frozen bed provides more traction than a thawed bed. For the special case of a flat horizontal bed, such that $h_R = r = 0$ in Eqn. (3.3), Fig. 3.6 also shows ice surface profiles that are obtained along the flowline by entering τ_o values at each Δx step in Eqn. (3.3) from either the τ_D or τ_W curve. These are steady-state flowline profiles, subject to the condition that accumulation over distance L–Δx is balanced by ablation in the first Δx step. Ablation is primarily melting along terrestrial margins and calving along marine margins.

In Fig. 3.6, the value of τ_o at a given value of x is specified by values assigned to n, m, A and B in Eqns (3.6) and (3.7). There is enough uncertainty in these quantities to allow $\tau_o = 100 \pm 50$ kPa. The τ_o scale in Fig. 3.6 can therefore be adjusted to reflect this variation from flowline to flowline, just as L can be adjusted in the x scale to accommodate flowlines of different lengths. In subsequent figures, τ_o and x values will not be specified, in order to emphasize this scaling flexibility. However, the ratio of $\tau_o : \chi/L$ along all these curves is precise, and should be respected.

3.6.3. Melting and Freezing Beds

The basal shear stress for ice passing over melting and freezing beds is simply:

$$\tau_b = f_t\, \tau_W + (1 - f_t)\tau_D \qquad (3.8)$$

where f_t is the fraction of the bed that is thawed at a given position x along the flowline. The τ_o curves for the cases described in Section 3.3 are plotted in Fig. 3.7. Figure 3.7a shows the τ_o variation along a

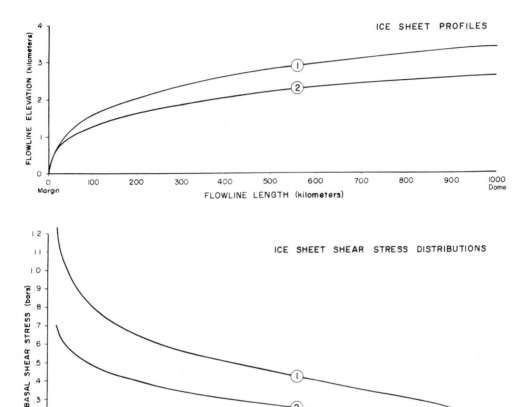

FIG. 3.6. Traction for ice sheets on frozen and thawed horizontal beds. Surface-elevation profiles (top) and basal shear stresses (bottom) are shown along a flowline for a frozen bed, curves 1, and a thawed bed, curves 2. The flowline is for a flowband of constant width over which the ice accumulation rate is constant. A basal shear stress of 1.0 bar can be as low as 0.5 bar along warm equatorward margins as high as 1.5 bar along cold poleward margins, owing to the strong temperature dependence of creep rates.

flowline that moves equatorward from a frozen central region, across a melting zone, a freezing zone and another melting zone, in succession. Figure 3.7b shows the τ_o variation along a flowline that moves poleward from a frozen central region, across a melting zone, a freezing zone and a frozen zone, in succession. Figure 3.7c presents the τ_o variation along a flowline that moves equatorward from a thawed central region, across a freezing zone and a melting zone, in succession. Figure 3.7d shows the τ_o variation along a flowline that moves poleward from a thawed central region, across a freezing zone and a

frozen zone, in succession. All these curves are for constant flowband width and constant accumulation rates along length $L-\Delta x$.

3.6.4. Ice Streams

In simple basal sliding theory, the basal water layer is a thin film that coats all bedrock projections equally and the basal sliding velocity is controlled by projections having a critical size (Weertman, 1957b) (Chapter 5). An ice stream should develop when the basal water layer thickens progressively downstream

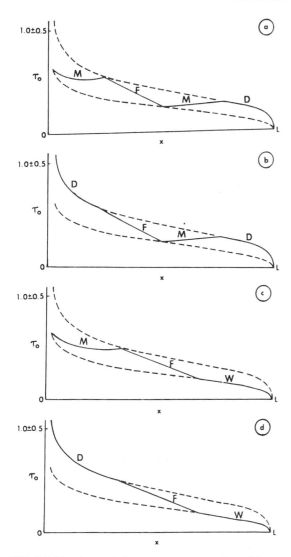

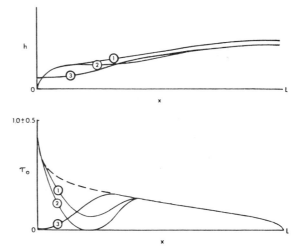

FIG. 3.8. Traction for ice sheets on thawed horizontal beds when stream flow develops toward the margin. Surface-elevation profiles (top) and basal shear stresses (bottom) are shown for terrestrial ice streams in which the minimum surface slope occurs where buoyancy is partial (curve 1) and total (curve 2), with buoyancy decreasing as a cosine curve both upslope and downslope; and for a marine ice stream in which the minimum surface slope occurs at total buoyancy, where ice becomes afloat (curve 3), with buoyancy decreasing upslope as a cosine curve. Flowband width and accumulation rate are constant, and units are metres for h, kilometres for x and bars for τ_0.

FIG. 3.7. Traction variations across a succession of frozen (D), freezing (F), melting (M) and thawed (W) beds beneath an ice-sheet flowline. From the ice divide to the ice margin, the bed conditions are (a) frozen, melting, freezing, melting, (b) frozen, melting, freezing, frozen, (c) thawed, freezing, melting, and (d) thawed, freezing, frozen. Flowband width and accumulation rate are constant, and units are kilometres for x and bars for τ_0.

the pulling power of buoyant ice (Hughes, 1992a). In Fig. 3.8, equivalent cosine variations of τ_0 with x along an ice stream are shown for three cases of the 'pulling force'. For a marine ice stream, the 'pulling force' decreases as a half cosine cycle upslope from full buoyancy ($\tau_0 = 0$) at the grounding line of its floating ice tongue. For a terrestrial ice stream, the pulling force varies as a full cosine cycle upslope from the terminus of its grounded ice lobe, with two maximum values at the point of minimum surface slope: for full buoyancy ($\tau_0 = 0$) and for partial buoyancy ($\tau_0 > 0$), respectively. Constant flowband widths and accumulation rates are assumed in plotting the τ_0 curves in Fig. 3.8.

3.6.5. Variable Flowband Widths

Four kinds of ice sheet flowbands are possible: (1) flowbands will have nearly constant width downslope from an ice divide having a nearly constant elevation; (2) flowbands will diverge strongly downslope from

along a flowline, and progressively drowns the controlling bedrock projections (Hughes, 1981). However, recent studies indicate that ice streams can also lie on a deforming debris layer (Rooney et al., 1986; Alley et al., 1987a, b) or a frozen bed (Scofield, 1986). Their concave surface profile can be related to a prominent dome on the ice divide; (3) flowbands may diverge from a dome and then converge into an ice stream; and (4) flowbands will converge strongly downslope from a prominent saddle on the ice divide. Figure 3.9 illustrates these four kinds of flowbands, the variations in basal shear stress they produce and their flowline profiles for ice spreading over a flat horizontal bed.

The ice elevation profile for a flowband that first diverges and then converges is little different from one for a flowband of constant width. However, the ice elevation profile is lowered over 5% for diverging flow and is raised over 5% for converging flow. This means that the flowline from a saddle must be substantially shorter than the flowline from a dome in order for the dome to be higher than the saddle. This requirement favours a dome over Hudson Bay, rather than a saddle, if the Laurentide Ice Sheet had attained steady-state equilibrium at its glacial maximum.

3.6.6. Variable Accumulation and Ablation Rates

Variations of τ_0 in Figs 3.6, 3.7, 3.8 and 3.9 were for constant rates of ice accumulation along the flowline, with all ablation restricted to the first $\Delta\chi$ step in from the ice-sheet margin. Actually, ablation can occur over up to $\frac{1}{3}$ the distance to the ice divide for a flowline terminating at an equatorward ice margin. Also, a peak in the accumulation rate may occur at an ice divide low enough to be crossed by convective storm systems, but peak accumulation rates will occur between the equilibrium line and an ice divide above these storm systems. Figure 3.10 shows variations of τ_0, along with variations of h above a horizontal bed, for four conditions of mass balance equilibrium. The first condition has a constant rate of ice accumulation along the entire flowline, with ice ablation restricted to calving along a marine margin (Hughes, 1989, 1992b). The second condition has constant ice accumulation over $\frac{2}{3}$ of the flowband nearest the ice divide and constant ice ablation over the $\frac{1}{3}$ of the flowband nearest a terrestrial ice margin. The third condition has accumulation decreasing from a maximum at the ice divide to zero at the equilibrium line and ablation increasing from zero at the equilibrium line to a maximum at the terrestrial ice

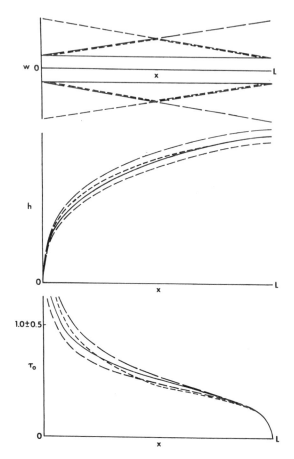

FIG. 3.9. Traction for ice sheets with variable flowband widths. Lines with dashes of different lengths are used to show the effect of parallel (solid), converging (long dashes), diverging (intermediate dashes), and initially diverging followed by subsequent converging (short dashes) conditions of flowbands (top) on ice elevation profiles above a horizontal bed (centre) and the basal shear stress (bottom). Units are kilometres for x, bars for τ_0 and τ_0 curves are for a frozen bed. For a thawed bed, 1.0 ± 0.5 bar should be replaced by 0.5 ± 0.3 bar for these τ_0 curves.

margin. The final condition has accumulation increasing from zero at the ice divide and the equilibrium line to a maximum in between, and ablation increasing from zero at the equilibrium line to a maximum at the terrestrial ice margin. In these cases, the equilibrium line is $\frac{1}{3}$ the flowline distance from the ice margin to the ice divide and rates of accumulation and ablation change linearly with distance along a flowband of constant width.

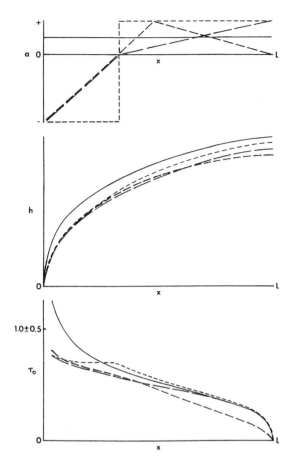

FIG. 3.10. Traction for ice sheets with variable accumulation and ablation rates. Lines with dashes of different lengths are used to show the effect of constant and linear accumulation rates and the equilibrium-line position (top) on ice elevation profiles above a horizontal bed (centre) and the basal shear stress (bottom). Mass balance equilibrium exists in all cases, with the equilibrium line at either the ice margin or one-third the flowline length in from the ice margin. Units are kilometres for x, bars for τ_o and τ_o curves are for a frozen bed. For a thawed bed, 1.0 ± 0.5 bar should be replaced by 0.5 ± 0.3 bar for these τ_o curves.

The mass balance variations show that ice elevation climbs much more rapidly near the margin when the equilibrium line is at the margin (marine iceberg calving) than when it is $\frac{1}{3}$ the flowline length in from the margin (terrestrial surface melting). Since these extremes are end members of a continuum, they can be used to scale basal shear stresses for equilibrium lines at intermediate distances in from an ice margin. It is also noteworthy that the ice elevation

climbs more slowly near the ice divide when the peak accumulation rate lies between the equilibrium line and the ice divide than when it is at the ice divide. However, the elevation difference at the ice divide is <5% of the total elevation for these two cases, whereas both ice divides are ~10% lower than the ice divide when the accumulation rate is constant to the ice margin. This implies that a flowline to a terrestrial margin must be longer than a flowline to a marine margin, if both flowlines are to have the same elevation at the ice divide. Stream flow at the marine margin and sheet flow at the terrestrial margin can overcome this elevation differential because an ice stream has a concave surface profile.

3.7. APPLICATIONS

Equation (3.3) can be used to compute ice elevation profiles quickly and easily along the flowlines of former ice sheets. The flowline trajectories first must be drawn on a contoured topographic map of the glaciated region using the professional judgement of a glacial geo(morpho)logist, based on field observations. Then the flowlines must be divided into equal steps of length Δx, with $\Delta x = 20$ km appropriate for short flowlines over rugged topography and $\Delta x = 50$ km acceptable for long flowlines over smooth topography. Ice elevation for the initial step at the ice margin must then be specified, using Eqn. (3.4) for terrestrial margins and Eqn. (3.5) for marine margins. The contoured topographic map is used to determine bed elevation h_R at each Δx step along the flowline, with h_R positive above and negative below the ice margin. A value of r_{sr} between 0 and $\frac{1}{3}$ is chosen, depending on an assessment of the degree of isostatic equilibrium that exists now and at the time when the ice sheet existed. This value of r_{sr} can be constant for all steps, or can vary from step to step. In general, r = 0 is improved along the former ice margin, while $r_{sr} = \frac{1}{3}$ is better beneath the former ice divide at the glacial maximum. Finally, an interpretation of the glacial geology is made along the flowline in order to decide: where the bed was frozen, thawed, freezing, or melting; where surface flow was parallel, converging, or diverging; and where ice stream flow replaced sheet flow, using criteria presented in Sections 3.3

and 3.4. The equilibrium line should be placed along poleward marine margins and up to $\frac{1}{3}$ the flowline length in from equatorward terrestrial margins. Values of τ_o at each δx step are then specified from the curves in Figs 3.6–3.10. Ice elevations h at each δx step can now be computed from Eqn. (3.3) using the specified value of h at the initial step and specified values of h_R, r_{sr} and τ_o at all subsequent steps. Allowing the maximum τ_o for sheet flow to be scaled between limits of 50–150 kPa for each flowline, which is equivalent to varying A and B in the flow and sliding laws for ice, allows all flowlines to have compatible elevations at the ice divide. Any remaining mismatch should be eliminated by shifting the ice divide toward the higher flowline.

To illustrate the utility of Eqn. (3.3) when units for h and h_R are metres above (also below, for h_R) the ice margin, δx are kilometres, and τ_o are kilopascals (100kPa = 1bar), Eqn. (3.3) reduces to:

$$h_{i+1} = h_i + \left(\frac{\tau_o}{(1 + r_{sr}) h - (1 + r_{sr})^{1/2} h_R} \right)_i \times \frac{\Delta x}{(9 \times 10^{-3} \text{ kPa km m}^{-2})} \quad (3.9)$$

The Cordilleran Ice Sheet and the Laurentide Ice Sheet at their Late Wisconsinan glacial maximum were close to the two end-member applications of Eqn. (3.9). At the Cordilleran maximum extension, a relatively thin ice sheet covered a relatively rugged bed for a relatively short time; thus, isostatic depression was minor and bed topographic variations changed rapidly along flowlines. These requirements are satisfied approximately by taking $r_{sr} = 0$ and $\Delta \chi = 20$ km, so that Eqn. (3.9) becomes:

$$h_{i+1} = h_i + \frac{(2.22 \times 10^3 \text{ m}^2 \text{ kPa}^{-1}) (\tau_o)_i}{(h - h_R)_i} \quad (3.10)$$

At the Laurentide maximum, a relatively thick ice sheet covered a relatively smooth bed for a relatively long time; therefore, major isostatic depression occurred and bed topographic variations changed slowly along flowlines. These requirements are satisfied approximately by taking $r_{sr} = \frac{1}{3}$ and $\Delta \chi = 100$ km, so that Eqn. (3.9) becomes:

$$h_{i+1} = h_i + \frac{(1.11 \times 10^4 \text{ m}^2 \text{ kPa}^{-1}) (\tau_o)_i}{(1.33 h - 1.54 h_R)_i} \quad (3.11)$$

It is important to remember that h and h_R in Eqn. (3.3) are vertical distances above the ice margin, so that elevation above sea level for terrestrial margins and depth below sea level for marine margins must be added and subtracted, respectively, from h and h_R to obtain ice-surface elevations referenced to modern sea level. However, h_W in Eqn. (3.5) is referenced to ancient sea level at the time when the ice sheet existed.

The graphical method for reconstructing former ice sheets that is presented here gives reliable reconstructions if the constraints of the method are recognized. These constraints are the boundary conditions discussed in Section 3.3. In practical terms, these constraints amount to interpreting glacial geology correctly and then using the appropriate basal shear stress curves in Figs 3.6–3.10 to plot Eqn. (3.9). In these curves, the upper limit of basal shear stress at the ice margin can range from 50–150 kPa for ice sheet flow. The lower value is appropriate toward equatorward margins, where ice is warm and surface melting rates are high. The upper value is appropriate toward poleward margins, where ice is cold and surface melting rates are low. A value of 100 kPa is appropriate for intermediate margins. Surface temperature is more important than surface melting rates in making this judgement, as ice thickness varies with the melting rate raised to the $\frac{1}{8}$ power for n = 3 in the flow law and to the $\frac{1}{5}$ power for m = 2 in the sliding law (Hughes, 1981, p. 249). The choice of r_{sr} in Eqn. (3.9) is also important. In general, r will tend to approach zero toward the ice margin and to approach $\frac{1}{3}$ toward the ice divide of a former ice sheet during its glacial maximum. Intermediate values of r_{sr} can be entered in Eqn. (3.9) for intermediate positions along reconstructed flowlines. For ice stream flow, $\tau_o = 0$ should be used at the ice shelf basal grounding line of marine ice streams and $0 < \tau_o < 200$ kPa should be used at the minimum-slope surface-inflection line of terrestrial ice streams, as shown in Fig. 3.8.

3.8. SUMMARY

In summary, the GT–IS model method for reconstructing former ice sheets, their dimensions and spatial and temporal distribution provides a useful technique for combining glacial geomorphic data as initial input into ice sheet parameters prior to the development of a model. An example of the use of a GT–IS style-model is exemplified by Thorp's work (1986, 1991) on the spatial and temporal distribution of a Late Glacial mountain ice field in western Scotland. Other examples of the GT–IS model approach can be found in the work by Aarseth and Mangerud (1974) in Norway, Porter (1975a) in New Zealand and Porter (1976) and Thorson (1980) in Washington State, U.S.A.

Finally, the use of these models in combining both glaciated terrain field evidence with ice sheet dynamics and glaciological parameters permits the possibility of a greater understanding of glaciogenic processes. Conversely, it is essential that explanations of glaciogenic processes be developed within the boundary conditions that such models impose. However, the possibility of new models and paradigms evolving must always be anticipated and exploited where inadequate explanations are uncovered. In the following chapters, explanations of glaciogenic processes and bedform and landform development are couched within the boundary conditions that present glaciology dictates. Where departures occur, obvious areas of new research will be highlighted.

Chapter 4

GLACIERS AND ICE SHEETS

J. Menzies

4.1. INTRODUCTION

In studying any glacial environment a fundamental appreciation of the glaciological factors that influence the glacial processes and sediments must be acquired. It is the purpose of Chapters 4, 5 and 6 to consider those aspects of glacial physics pertinent to the understanding of sedimentological processes within glacial environments. For an exhaustive study of the wider field of glacial physics and glaciology, see Paterson (1981), Colbeck (1980), Hutter (1983) or Lliboutry (1987b).

The mechanics of ice movement, basal ice stresses, pressure melting, basal ice thermal conditions, glacial hydrology and the response of ice masses to climatic changes are a few examples of the influence exerted by the glaciological component upon the glacial system (Chapter 3). When periodic increases of snow within the accumulation area are revealed in the development of kinematic waves, increased but localized basal slip velocity may occur. In time, the kinematic wave may induce an ice advance, perhaps localized deposition of subglacial debris melt-out beneath the area of the wave or increased mobilization of a deforming basal debris layer. Likewise, reduced accumulation or increasing ablation may lead to ice retreat and increased debris melt-out in the near marginal supraglacial areas of an ice mass. Both examples depict complex and interrelated facets of the glacier system. The impact such glaciodynamic variations may have upon sedimentological processes and sediments is

exhibited in the sediment/landform associations formed within specific glacial environments (Menzies, 1995b, Chapter 9). Recently, Powell (1991) has illustrated the myriad of complex relationships that exist between glacial system changes and the movement of a floating ice front (Fig. 4.1). Field observations can only be fully elucidated where glacial mechanics are understood. In understanding the mechanics of glacial erosion, debris entrainment, transportation and subsequent deposition, the relationships between ice mechanics, underlying topography and sediment source and transport pathways must be considered. Each landform or bedform produced within a glacial environment is a product of, or a reaction to, glaciological factors that were either of ephemeral or long-term effect.

Our past understanding of glaciers, of how they move over their beds, erode and transport debris, has been partly hampered by their general inaccessibility. Early pioneering work, for example, by Forbes, Tyndall and Agassiz was largely based upon an understanding of alpine valley glaciers, a trend that dominated at least until the 1950s (Boulton, 1987b). However, with the exploration of Greenland and Antarctica, an awareness of the importance of continental ice sheet glaciation emerged. As knowledge of the Pleistocene Laurentide and Fenno-Scandinavian Ice Sheets unfolded, the necessity to understand ice sheet glaciodynamics steadily increased (n.b. Boulton, 1987b; Clarke, 1987a; Robin and Swithinbank, 1987). It became apparent that the landscapes of glaciated northern and southern hemispheres could only be understood if ice sheet glaciodynamics were fully understood.

By 1957, the International Geophysical Year, studies of the Antarctic and Greenland Ice Sheets had begun and have since continued apace (Weertman, 1987b; Oescheger and Langway, 1989). Since then studies of glaciers can be roughly partitioned into: (a) theoretical and applied glaciological studies; and (b) glacial geo(morpho)logical studies of Pre-Quaternary sediments, Quaternary sediments and landforms and present-day glaciated areas. Both branches of enquiry have, at times, diverged but each has added to the other.

4.2. ICE MASS TYPES

Any classification of ice masses is fraught with difficulties since no taxonomy can satisfactorily encompass all the differing aspects of ice mass types. Classifications based upon morphology, thermal

structure, geographical location and velocity have all been used to various degrees of success (Armstrong *et al.*, 1973; Sugden and John, 1976; Paterson, 1981). As new information has been acquired, ice mass classifications have been revised or abandoned. At present, no universally accepted classification exists. Ahlmann's early thermal classification (1948) had been revised (Avsyuk, 1955; Court, 1957; Miller, 1973) and later altered to accommodate increasing knowledge of ice mass thermal structures and, in particular, subglacial thermal states (Hutter, 1983). Recently, newly developed classifications of the subglacial environment with a perspective on sedimentology have been proposed (e.g. Shoemaker, 1986a; Menzies, 1987). As new information accrues new taxa or sub-taxa will be introduced. For example, tidewater and marine ice masses can be distinguished on the basis of being grounded or floating, or above or below present sea-level (Dowdeswell and Scourse, 1990; Anderson and Ashley, 1991).

The value of any ice mass classification is in its

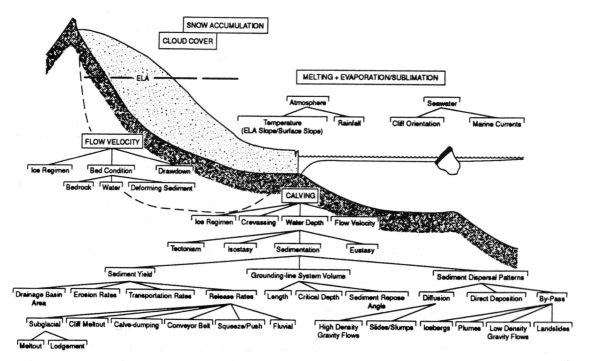

FIG. 4.1. Factors controlling mass balance of ice masses, including sedimentological controls for tidewater termini. Many factors are dependent variables, feedback loops are not illustrated for simplicity (after Powell, 1991; reproduced with permission of the author).

ability to provide some clues as to likely glaciologic conditions and sedimentologic processes and environments. Therefore, where an ice mass is expected to have been a grounded tidewater-type glacier, as was the case in certain instances along the edges of the Great Lakes during the Late Wisconsinan, then specific environments are likely to be encountered when unravelling the stratigraphic record from such a site (Clayton and Attig, 1990). Where the record does not seem to correspond with the expected lithofacies associations, either the classification is faulty or inappropriate, or a greater understanding of a particular environment is necessary such that the classification, if it is to be effective and functional, needs to be modified and probably expanded.

4.3. FORMATION OF GLACIER ICE

As snow accumulates on the surface of a glacier in a loose and highly porous state, it begins a series of transformations termed diagenesis or metamorphosis (Kingery, 1963; Gow, 1975; Michel, 1978; Schwander, 1989; Lock, 1990). Following burial by the new year's snowfall, a process of densification or consolidation ensues (Herron and Langway, 1980). The transformation of snow to ice occurs, under dry conditions, when increased grain packing, rounding (sintering), reducing porosity and permeability, and recrystallization takes place (Plate 2.1). Under these conditions, densification can be expressed by Sorge's Law where density becomes an exponential function of depth below the original accumulation surface (Bader, 1963). Under these dry conditions, the process of transformation of snow to firn can take over 100 years. Where wet conditions of surface melting and percolation prevail, the above process occurs more rapidly since meltwater from surface snow melt percolates into the snow and refreezes at depth thus reducing porosity, reducing snow grain size and accelerating transformation (e.g. within 3–5 years on the Seward Glacier (Paterson, 1981) (Fig. 4.2).

Glacier ice is, by definition, a polycrystalline substance impermeable to air, at the macroscopic level, but containing air bubbles and inclusions of other chemical species. Polycrystalline ice is strongly anisotropic due to the variety of crystal shapes, sizes and orientations (Plate 2.1). These properties account for the remarkable ductility and strength exhibited by glacier ice.

4.3.1. Glacier Ice Density

The process of snow transformation passes through two phases before becoming glacier ice viz: firn — a compacted snow usually of the previous year's accumulation (density — 0.4–0.8 Mg m^{-3}) and névé — densified snow of reduced air permeability (density — 0.7–0.8 Mg m^{-3}) (Paterson, 1981; Lock, 1990). Glacier ice, normally, has a density of $\approx$0.9 Mg m^{-3}; however, the input of impurities and particles can alter this figure slightly. Air bubbles are often found within glacier ice due to the relative rapidity of the densification process. Where surface melting and refreezing occurs fewer air bubbles are present (Koerner and Fisher, 1979). Highly attenuated, sheared sets of air bubbles are commonly encountered in the lower basal zones of temperate ice masses where repeated regelation processes have resulted in the incorporation of air.

4.3.2. Glacier Ice Crystals

Sugden and John (1976) have noted that glacial ice polycrystals have two unique qualities. First, ice crystals are relatively weak along the basal plane, allowing comparatively easy adjacent crystalline slip (dislocation climb) (Michel, 1978) (Chapter 5). Internal deformation of glacier ice thus occurs under very small stress applications resulting in glacier movement (Plate 4.1). Second, glacier ice is less dense than water and thus is buoyant. The implications of this factor will be observed later in discussing basal water films and the characteristics of floating ice shelves.

Ice crystal size, generally, increases from the surface downwards in a step-like fashion until, at $\approx$100m below the surface, a relative constant size (typically $\approx$10–30 mm $\sqrt{a}$) persists until close to the glacier sole where very large crystals ($\approx$5–10 cm $\sqrt{a}$) are commonly found (Fig. 4.3) (Lliboutry, 1965; Hobbs, 1974; Vallon et al., 1976; Michel, 1978; Herron and Langway, 1979; Hutter, 1983). Crystals

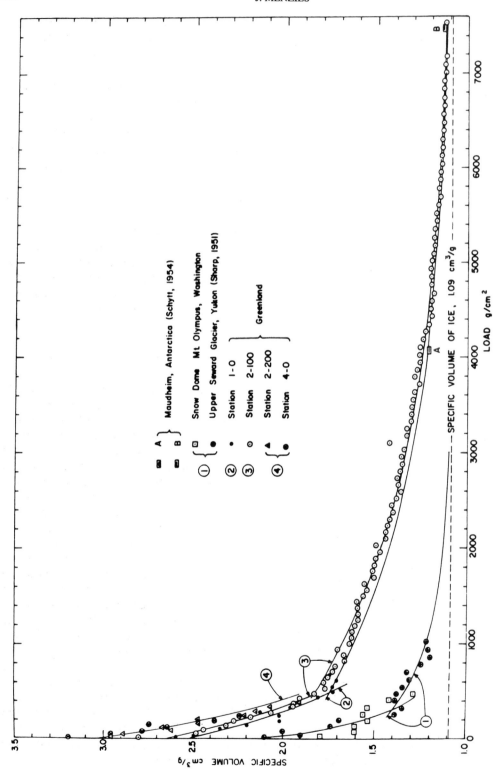

FIG. 4.2. Specific volume vs. snow accumulation load. Curve 1 illustrates the influence of complete soaking, whereas Curves 2, 3 and 4 exhibit limited to negligible soaking (Kingery, 1963; reprinted with permission from Kingery, W.D., *Snow and Ice; Properties, Processes and Applications*, MIT Press).

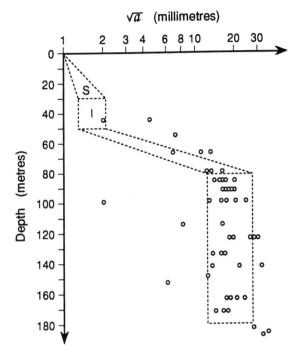

FIG. 4.3. Average crystal size vs. depth in the Massif du Mont Blanc where (S) is in firn and (I) is glacier ice (after Michel, 1978; reproduced with permission of Laval Université Presse).

may vary from fine to coarse size depending upon the age and debris content of the ice and past changing climatic temperatures (Holdsworth, 1974; Vivian and Bocquet, 1973; Michel, 1978; Koerner and Fisher, 1979; Lock, 1990). Crystals are typically round within firn and evolve into more complex forms with continued burial and the development of enclosing stress fields. Small crystal sizes are likely due to consolidation and/or crystallization around microparticles. With increasing stress application, ice crystals develop a stress-anisotropic fabric the result of syntectonic recrystallization (Duval, 1976; Gow and Williamson, 1976; Hudleston, 1976, 1977; Russell-Head and Budd, 1979; Herron and Langway, 1982; Herron et al., 1985; Alley, 1988; Lipenkov et al., 1989; Lock, 1990).

4.4. MASS BALANCE AND GLACIER SENSITIVITY

The surface morphology, overall dimensions and geographical location of all ice masses is largely a function of the factors controlling mass balance (Fig. 4.4). Where low snowfall increments accumulate in high latitude, polar dry continental sites such as central Greenland, low surface temperatures and low levels of solar radiation combine to preserve almost the complete annual snowfall. In contrast, in sub-arctic and mountainous areas in middle latitudes, snowfall increments are often extremely large such that higher summer solar radiation values and high ablation rates result in large meltwater production. The preservation of snow is restricted to the higher parts of the glacier (accumulation area) where the influence of lower surface temperatures with increasing elevation is effective. Finally, in regions of high altitude, low latitude, high solar radiation and summer temperatures, snow accumulation may only survive if sufficiently large accumulations occur. (For an extensive review of mass balance data see *Geografiska Annaler*, **66a**, 1984; Østrem and Brugman, 1991.)

4.4.1. Factors Influencing Mass Balance

The flow diagram in Fig. 4.5 shows, in simple form, the linkages between regional climate and ice mass response via the impact of mass balance change (Meier, 1965; Furbish and Andrews, 1984; Andrews, 1975). Considerable research on these linkages continues without substantive agreement on the ability to predict or explain ice mass response to both short- and long-term climatic change (Kuhn, 1981, 1984, 1989; Smith and Budd, 1981; Mayo, 1984; Reynaud et al., 1984; Weidick, 1984; Greuell and Oerlemans, 1986; Letréguilly, 1988; Powell, 1991; Sturm et al., 1991). This lack of universal correlation of mass balance to specific influencing factors can be illustrated over relatively short distances where comparable climatic conditions prevail (Fig. 4.6). As an example, the mass balance relationships in the Canadian Rockies reveal that the Peyto Glacier, Alberta is dependant upon summer temperatures, the Sentinel Glacier, British Columbia upon winter precipitation, while the Place Glacier, British Columbia is influenced by both elements (Letréguilly, 1988; Sturm et al., 1991). Powell (1991) has illustrated the complex relationships that exist within any tidewater/floating glacier system and mass balance controls (Fig. 4.1).

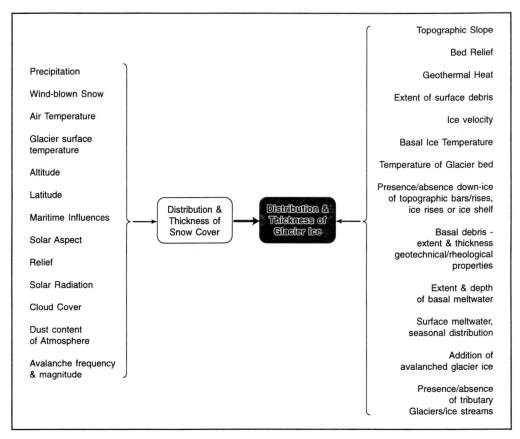

FIG. 4.4. A generalized linkage diagram exhibiting the relationships between external geographical and atmospheric variables, glacier mass balance and ice mass thickness and extent.

The influence and effect of mass balance upon any ice mass is difficult to 'isolate' but often is illustrated by the position and movement of the equilibrium line (equilibrium-line altitude — ELA). The equilibrium line, the boundary line between the upper accumulation and lower ablation sections of an ice mass, fluctuates, over time and space, as a function of climatic variables over time (Sutherland, 1984; Oerlemans, 1989). Regression analyses have been employed to relate mass balance to climatic variables using, for example, annual specific mass balance, ELA, summer temperatures and precipitation, and annual temperatures and precipitation. The gradient of mass balance with altitude termed the mass balance gradient has become a common comparative value for mass balance and ice mass sensitivity analyses

(Lliboutry, 1974; Mayo, 1984; Boulton et al., 1984; Oerlemans, 1989; Oerlemans and Hoogendoorn, 1989; Sturm et al., 1991). It is difficult to generalize about mass balance gradients since they exhibit considerable variation (Kuhn, 1984). As shown in Fig. 4.7, gradients may exhibit a steep ablation section and a shallower accumulation zone and vice versa. Oerlemans and Hoogendoorn (1989) point out that two assumptions are implicitly accepted when considering mass balance change and ice mass response viz. (a) that mass balance perturbations are independent of altitude; and (b) that the mass balance profile is constant, it only shifts up or down. Neither of these assumptions can be generally substantiated either from theory or actual observation. Mass balance gradient is a function of several contributing factors viz:

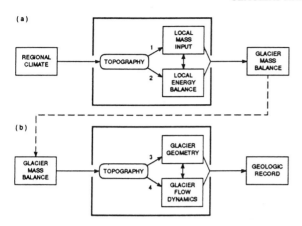

FIG. 4.5. (a) Simplified linkages between regional climate and glacier mass balance where local topographic control exerts a local specific influence. (b) Reverse linkage diagram showing mass balance and glacier response as reflected in the geologic record (after Furbish and Andrews, 1984; reproduced by courtesy of the *International Glaciological Society from Journal of Glaciology*, **30**(105), 1984, p. 200, Fig. 1).

(a) accumulation from precipitation is influenced orographically;

(b) wind drift snow accumulation tends to be higher at high altitudes;

(c) surface albedo increases with altitude as a function of short-wave absorption;

(d) air temperature lapse rate varies as a function of altitude;

(e) the fraction of precipitation falling as snow increases with altitude; and

(f) snow cover increases with altitude thus heat gain from side-wall areas in valley glaciers decreases.

4.4.2. Equilibrium Line: Sensitivity to Change

In periods of higher snowfall and/or reduced ablation, the equilibrium line will move down-ice as the accumulation area enlarges depending upon factors constraining flow and glacier dimensions. Glacier retreat, after a time lag, may follow periods of higher surface temperatures and/or reduced snowfall accumulation causing enlargement of the ablation zone. The time lag (or relaxation or response time) following a change in input or output components to the glacier is the time delay necessary for a system change to pass through the ice mass and thus be

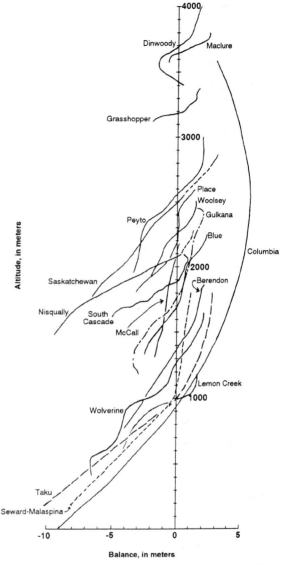

FIG. 4.6. Annual mass balance gradients measured from western North American glaciers (after Mayo, 1984; reprinted from Glacier mass balance and runoff research in the U.S.A., by L.R. Mayo, *Geografiska Annaler*, 1984, 66A, 215–227, by permission of the Scandinavian University Press).

evident as a visible expression of that change at the terminus. This input/output response relationship, however, is too simple and rarely can be detected in reality. An increase in surface mass flux may often lead to readjustments in local ice thickness, surface gradient and flow rate (thus basal shear stress field).

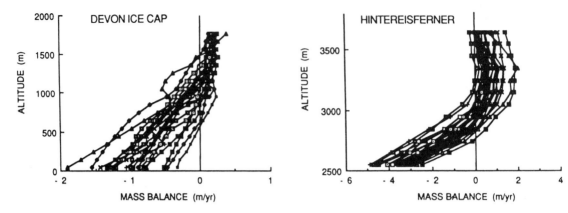

FIG. 4.7. Actual annual mass balance profiles from Devon Ice Cap, Canada (lat. 75° 25' N., long. 83° 15' W.) and Hintereisferner, Austria (lat. 46° 48' N., long. 10° 46' E.). Data from Kasser (1967, 1973) and Haeberli (1985) (after Oerlemanns and Hoogendoorn, 1989; reproduced by courtesy of the International Glaciological Society from *Journal of Glaciology*, **35**(121), 1989, p. 400, Fig. 1).

As these variables alter, a feedback response in terms of ice mass geometry and flow dynamics usually occurs. Major constraints affecting ice mass response to input/output fluctuations are extra-glacial factors, such as topographic constraint and the relation of any part of the glacier system to major and minor tributary ice masses. For instance, a long, narrowly constrained valley glacier entering into a large dendritic glacier system may experience a very slow response: so slow that such a response may be difficult, if not impossible, to detect. In contrast, a small, cirque glacier will respond very rapidly with a change at the ice terminus being readily noticeable.

The position of the ELA is a reflection of climatic variability as it relates to altitude, latitude, continentality, maritime influences, and hemispheric atmospheric circulation. The Global ELA, today, varies from being very high in the equatorial regions of the world where the line may be at 6000 m (as on Mount Kenya) to 3000 m in the Sierra Nevada of California, to 200 m in coastal northern British Columbia, to sea level at around 66 degrees of latitude in Glacier Bay, Alaska. In comparison to the present-day, during the Pleistocene, the ELA at sea-level was at least 30 degrees closer to the Equator.

4.4.3. Net Mass Balance

A glacier's net mass balance is said to be in a steady state when the inputs of fresh snow, blown snow and avalanched snow are in equilibrium with outputs due to ablation, and loss by blowing wind and calving (Paterson, 1981; Østrem and Brugman, 1991). In mathematical terms, the mass balance (b) can be expressed as follows (after Paterson, 1981):

$$b = c + a = \int_{t1}^{t} (\dot{c} + \dot{a}) dt \qquad (4.1)$$

$$b_n = b_w + b_s = c_t + a_t \qquad (4.2)$$

$$b_n = c_w + a_w + c_s + a_s = \int_{t1}^{t_m} (\dot{c} + \dot{a}) dt + \int_{tm}^{t_2} (\dot{c} + \dot{a}) dt \qquad (4.3)$$

where a is ablation (as a_s — summer ablation, a_w — winter ablation), $\dot{a}$ is ablation rate, b is balance or mass balance, c is accumulation, $\dot{c}$ is accumulation rate, a_t is total ablation, c_t is total accumulation, b_n is net balance, b_s is summer balance, b_w is winter balance, c_s is summer accumulation, and c_w is winter accumulation within time frames for the winter (t_l to t_m) and summer (t_m to t_2) seasons.

Thus the glacier's accumulation area has a $b_n > 0$, and the ablation area a value of $b_n < 0$, while at the equilibrium line $b_n = 0$. When and if $b_n = 0$, for the whole ice mass, a steady state is reached. In reality, this state is rarely, if ever, attained due to the changing spatial geometry of an ice mass over time and the impact of past variations in quantities b_w and b_s. This latter problem increases when mass balance calculations are made for large ice sheets (Chapter 3).

Values obtained of total accumulation for the Antarctic and Greenland Ice Sheets are, within a reasonable margin of error, accurate, but total ablation values remain elusive (Hollin, 1970; Oerlemans, 1989; Østrem and Brugman, 1991). The question, therefore, remains as to whether these major ice sheets are stable or whether they are gaining or losing mass at present (Hughes, 1973, 1975; Drewry, 1983b).

4.4.4. Mass Balance Gradients

The calculation of mass balance can be made by considering an energy balance equation (Oerlemans and Hoogendoorn, 1989; Østrem and Brugman, 1991) that is altitude dependent. This approach utilizing the entire balance for a year (over a time step-wise function of 30 min) involves the atmospheric temperature, snowfall and atmospheric transmissivity for solar radiation related to altitude allows a balance gradient to be calculated. The basic mass balance model equation is:

$$M = \int [\min(0; -B_e/L_m) + P]dt \qquad (4.4)$$

where M is the annual mass balance, L_m is the latent heat of melting, B_e is the energy balance at the surface of the ice mass and P is solid precipitation. Surface ablation occurs when the energy balance becomes positive. The response of mass balance gradients to several climatic parameters is shown for Sönnblick, Hintereisferner and Kesselwandferner (Austrian Alps) (Fig. 4.8), perhaps indicating a certain consensus for mid-latitude ice masses. However, for

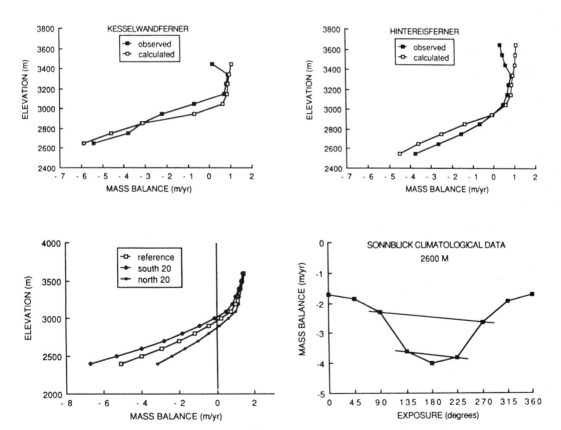

FIG. 4.8. Austrian glacier mass balance data both calculated and observed with account taken of exposure, surface slope and shading. Note balance gradient difference in Sonnblick data between accumulation and ablation areas (after Oerlemanns and Hoogendoorn, 1989; reproduced by courtesy of the International Glaciological Society from *Journal of Glaciology*, **35**(121), 1989, p. 403, Figs 4 and 5).

ice masses at high latitudes with a slower turnover rate ((sub)polar ice masses) the mass balance model, above, may not be as applicable (Østrem and Brugman, 1991).

The response of mass balance gradients of differing glacier types to variations in climate and thus mass balance parameters is a complex one (Nye, 1960, 1963; Paterson, 1981; Hutter, 1983; Furbish and Andrews, 1984; Kuhn, 1984). The variations may consist of one or a combination of all of the following effects (Hutter, 1983; Kuhn, 1984):

(1) a change in accumulation to the ice mass at the surface or base;

(2) a change in the energy flux from the atmosphere to the surface of the ice mass;

(3) a change in the geothermal heat flux;

(4) a change in the length of the ablation season.

Mass balance gradients reflect the contrasting characteristics of ice masses located in varying geographical areas under differing thermal and glaciodynamic regimes (Kuhn, 1984; Boulton et al., 1984; Hindmarsh et al., 1989). Kuhn (1984) has attempted to categorise glacier types with certain balance gradient characteristics based upon imbalances between positive and negative periods viz:

(a) Polar–Extrapolar: exhibit greatest variation due to the length of the ablation period and have the lowest balance gradients;

(b) Dry Continental: are affected by albedo changes close to the ELA;

(c) Maritime: are most affected by changes in the accumulation area and have steep balance gradients;

(d) Alpine: shows least variation from the linear balance model (Lliboutry, 1974) and imbalances appear independent of altitude.

The above classification has considerable merit. This ice mass taxonomy introduces a process-related relationship between the factors that affect mass balance, ice mass growth and decay and glaciodynamics. No other classification has attempted to relate changing ice mass geometry due to climatic and other responses to the effect such changes have upon subglacial conditions of temperature, basal shear stress and sliding velocity (Chapter 3). This approach to ice mass differentiation introduces a process-oriented logic that has wide implications for glacial sedimentology integrated to climatic change and individual (ice mass) bed geometry over the long-term period of ice sheet growth and decay.

4.4.5. Ice Sheet Growth Models and Mass Balance

In attempting to reconstruct past ice sheet dimensions and processes of growth and decay, Boulton et al. (1984) have suggested the following scenario: as the large Pleistocene ice sheets of the northern hemisphere developed, in the early stages of expansion, high accumulation rates could be expected to occur with steep mass balance gradients, high ice flux rates and high basal shear stress values similar to maritime conditions. However, as ocean surface temperatures and total ocean surface area decreased, an associated diminution of precipitation onto the developing ice sheets must have occurred leading to a reduction in mass balance and a more dry continental state being established with lower balance gradients and basal shear stress values. With ice sheet outward expansion, moisture-laden air masses along the southern edge of the ice sheets were prevented from penetrating the dry elevated continental centres of the ice sheets resulting in most precipitation being deposited within the ablation areas which, with southern expansion, became areas of very high ablation. The effect of these two competing processes was to balance each other out. This latter effect, at near maximum ice sheet spread, was to 'flatten-out' the balance gradient even to the point of producing a reverse gradient for a short interval of time (Fig. 4.9) (Chapter 3).

An additional effect to be considered is the impact of isostatic depression upon the geometry of the overall ice sheet (Peltier, 1982; Boulton et al., 1984; Boulton, 1990). It has been suggested that where the rate of ELA adjustment to ice sheet advance is in disequilibrium with the rate of crustal depression, an unstable state will develop in which ice retreat or at least an ice front standstill will take place. Whether such an unstable state could lead to catastrophic collapse of an ice sheet is unlikely since in retreat a new equilibrium between the ELA and the rate of crustal adjustment should re-establish and a new advance could then occur.

It has been noted that both modern and Pleistocene ice sheets exhibit(ed) asymmetry across their ice divides. The possibility that this form was an effect of differential mass balance seems unlikely. The major cause would appear to be the result of differences in basal topography, thermal regimens and margin sensitivity to glaciodynamic and extra-glacial influences such as floating ice fronts, grounding lines and proglacial terrains (Chapter 3).

In discussing the general flow of ice, it will be demonstrated that the influence of thermal variations impels both surface and basal ice velocity and the

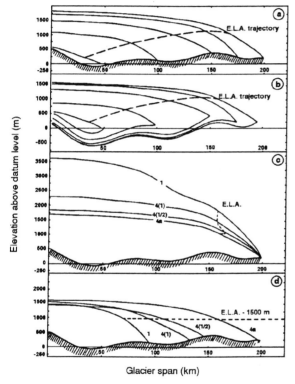

FIG. 4.9. Relationship between a series of equilibrium profiles and ELA trajectories under differing constraints: (a) the relationship between the ELA trajectory for an ice sheet of constant net balance gradient advancing form southwest Scotland to the Vale of York, England with zero isostatic compensation; (b) as for (a) but with full isostatic compensation; (c) a comparative series of equilibrium profiles for an ice sheet terminating in the Vale of York; (d) different spans of ice sheet extent for different balance patterns in northern England with an ELA of 1500 m (after Boulton et al., 1984; reproduced by courtesy of the International Glaciological Society from Journal of Glaciology, 30(105), 1984, p. 150, Fig. 13).

strain rate of the deforming ice mass to a profound degree (Hutter, 1982a, b; Boulton et al., 1984; Hindmarsh et al., 1989). The impact of such changes is considerable in terms of the response of ice mass glaciodynamics upon glacial processes and sedimentological environments. It can also be expected that responses, either negative or positive, will be of a wide temporal range, depending on the distance from the major snowfield accumulation areas to the glacier terminus and ablation area, the velocity of the glacier, the impact of glacier bed and valley side-wall topographic constraints and the addition or otherwise of tributary glaciers. A specific example of change in mass balance in response to climatic change is illustrated by the work of Müller (1976) on the White Glacier in the Canadian Arctic (Fig. 4.10).

It can be expected that glaciers in the same mountain region, even neighbouring glaciers, will exhibit differing responses at least over a period of tens of years (Jóhannesson et al., 1989). The response time (T_m) over which a valley glacier reacts to prior climatic changes manifesting as mass balance changes can be quantified. Paterson (1981) and Hutter (1983) have demonstrated using a linearized theory of kinematic wave propagation travelling down a glacier that the following semi-quantitative equation adequately describes the response time viz:

$$T_m \sim fl/u_L \qquad (4.5)$$

where l is glacier length, u_L is terminus velocity and f is a constant of approximately $\frac{1}{2}$ (Paterson, 1981). This equation predicates responses of 10^2–10^3 years. However, modern observations appear to indicate shorter responses. Jóhannesson et al. (1989) have rewritten this semi-quantitative equation:

$$T_m \sim h/[-b_L] \qquad (4.6)$$

where h is the thickness scale for the glacier and b_L is the balance rate at the terminus. In the case of major ice sheets, response variations, unless of global magnitude, may be of the order of 1000s of years (Chapter 2). For example, the present Antarctic Ice Sheet exhibits wide variations in glacier activity between those glaciers flowing into the Ross Shelf in

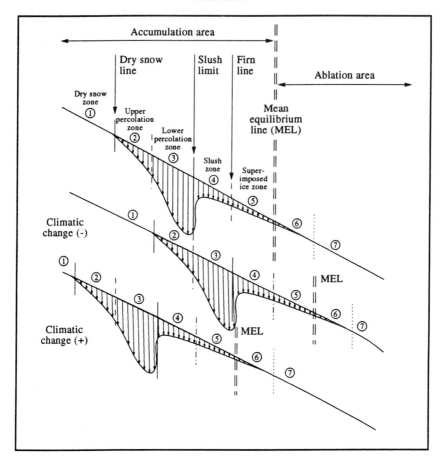

FIG. 4.10. Graphical presentation of the thermal influence on spatial pattern of accumulation zones (Zones 6: transitional superimposed ice; Zone 7: ablation) (Müller, 1974; reproduced by courtesy of the International Glaciological Society from *Journal of Glaciology*, **16**(74), 1976, p. 129, Fig. 8).

West Antarctica and those outlet glaciers flowing to the coast in the Enderby Land and Kemp Coast areas of East Antarctica (Morgan *et al.*, 1982; Drewry, 1983). The response of the Laurentide Ice Sheet, during the Pleistocene, was not synchronous with the Fennoscandian Ice Sheet. Within the same massive ice sheet variations in response can be expected along its edges. The response of the Laurentide Ice Sheet along its front in southern Ontario and New York State, for example, can be expected to have been widely different and asynchronous with the ice termini in Labrador-Ungava and the western portions of the Beaufort Sea. Along the edges of any ice sheet, separate ice streams and lobes will react at differing rates, frequencies and magnitudes. Unless synchroneity is considered in millennia, smaller asynchronous temporal variations can be expected. If marginal areas of past ice sheets have been characterized by thin ice thicknesses, rates of advance and retreat over considerable distances of terrain may have been on a scale of centuries rather than millennia. Chronological correlation, therefore, along the margin of any ice sheet becomes extremely complicated.

Interglacial and interstadial phases of the Laurentide Ice Sheet during the Quaternary, when higher temperatures followed on ice retreat or when short ice-free periods of tundra-like conditions prevailed, should not necessarily correlate with similar periods in Europe. For example, the Alleröd of Northwest Europe need not be expected to be synchronous with a similar phase in North America if such a phase exists at all in the latter instance (cf. Mathewes *et al.*, 1993).

The response of the present Eastern Antarctic to rapid increases in accumulation may lead to the possible triggering of basal creep instability leading to continental-scale surging. It has been suggested that this feedback mechanism may have led to major global glaciation and ice age initiation in the past (Schubert and Yuen, 1982) (Chapter 2).

4.4.6. Kinematic Waves

A further response to positive changes in mass balance is the development of the phenomena of surface bulges, termed kinematic waves, that travel down-ice several times (3–5) faster than the average velocity of the valley glacier, outlet glacier or ice stream (see Eqns (4.5) and (4.6)) (Plate 4.2) (Nye, 1958a; Weertman, 1958; Lick, 1970; Lliboutry and Reynaud, 1981). These waves occur as local increases in the balance gradient and have been closely linked to the triggering of glacial surges (Palmer, 1972; Hutter, 1983; Clarke *et al.*, 1984; Lliboutry, 1987b; Jóhannesson *et al.*, 1989). Waves are typically 10 m high and may be approximately 1 km in length. Lliboutry (1987b) suggests that kinematic waves form down-ice of seracs and other fast moving sections of ice masses. Since the velocity gradient is high through such fast moving regions, ice rapidly accumulates in the downstream area where increases in accumulation are quickly manifested as surface waves. It is possible that such localized ice thickening may have an influence upon subglacial conditions of shear stress and temperature, and subsequently upon processes of erosion, transport and deposition.

4.4.7. Ice Mass Flux or 'Activity'

Since ice flow is caused by unbalanced stresses within the ice mass, mass balance (accumulation/ablation) is a major component in understanding ice mass dynamics at all scales (Hindmarsh *et al.*, 1989). As will be discussed later, ice flow is essentially a thermomechanical process involving gravity, mass balance, surface and basal temperatures, and the geothermal heat flux. The contribution of mass balance changes to total ice mass varies by several magnitudes in comparison to the relative constancy of all other factors and is therefore the driving force in ice movement. What Meier (1961) termed the **activity index** is a measure of the net mass balance gradient. This gradient is calculated as a function of the net accumulation with altitude (mm^{-1}) up-ice. The greater the gradient up-ice, the faster glacier flow will tend to be. In considering the activity index on a geographical basis, it can be observed that the 'gradient' tends to decrease with increasing latitude poleward and also with increasing continentality (Sugden and John, 1976).

In reviewing indices of activity, Andrews (1972a, 1975) introduced the concept of 'energy of glacierization'. By relating mass balance changes on an ice mass to changes in mass transfer (ice flux) and thus to ice velocity, Andrews points out that significant relationships exist between mass balance changes and the processes of glacial erosion, entrainment and deposition. Such a relationship is exceedingly complex and far from fully understood.

4.5. STRUCTURE AND THERMAL CHARACTERISTICS OF ICE MASSES

4.5.1. Thermal Structure of Ice Masses

The distribution of temperature within an ice mass is a function of: (a) the rate of geothermal heat flux at the base of the ice mass, (b) the near-surface and atmospheric temperatures, (c) the vertical and horizontal components of velocity at any point within the ice mass as a function of shear heating from internal deformation, and (d) basal ice friction at the glacier bed (Hooke, 1977; Yuen and Schubert, 1979; Paterson, 1981; Hutter, 1983). Thermal changes within ice masses affect ice mass dynamics and the possible sedimentological environments within an ice mass system viz:

(a) paleo-surface temperatures are reflected at depth within an ice mass, thus influencing strain rate and therefore rate of internal deformation (n.b. concept of strain heating) (n.b. Glen's Law) (Robin, 1983) (Fig. 4.11);

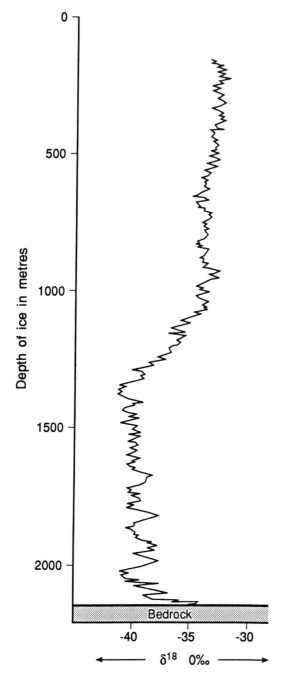

(b) basal thermal regimes directly influence the form and type of basal ice glaciodynamic flow;

(c) the rate and form of basal ice deformation;

(d) the character of the ice–bed interface(s);

(e) the presence or absence of meltwater and its discharge flux; and

(f) the nature of sedimentological processes at the ice–bed interface.

Close to the surface of a glacier within the snow/firn/névé zone temperature fluctuations occur in response to seasonal and diurnal changes in air temperature. Temperature changes reduce in amplitude with depth such that below 7–16 m fluctuations become negligible (Mellor, 1963; Paterson, 1969, 1980, 1981). Typically, the 10 m temperature is taken as a standard criterion by which comparison between ice masses can be made.

Meltwater infiltration into the percolation zone influences englacial temperatures at depth, introducing higher temperatures into the upper layers and reducing temperatures due to the latent heat of freezing as meltwater refreezes at depth (Benson,

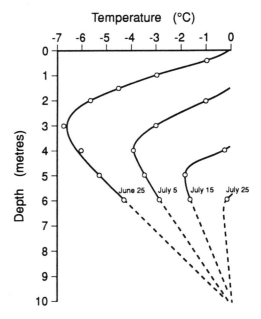

FIG. 4.11. Paleoclimate as indicated from measurements of oxygen isotope ratios on the ice core at Byrd Station, Antarctica. Ice at 1000 m depth was deposited ~11,000 BP while the oldest ice is ~10^5 BP, (data from S.J. Johnsen) (after Robin, 1983; reprinted with permission of the author from *Climate Record in Polar Ice Sheets*, edited by G. deQ Robin, 1983, Cambridge University Press).

FIG. 4.12. Removal of winter 'cold wave' from firn (adapted from Sverdrup, 1935) (after Paterson, 1981; reprinted from Paterson, W.S.B., *Physics of Glaciers*, 1981, p. 189, Fig. 10.1, with kind permission of Elsevier Science Ltd, The Boulevard, Langford Lane, Kidlington, Oxford OX5 1GB, U.K.).

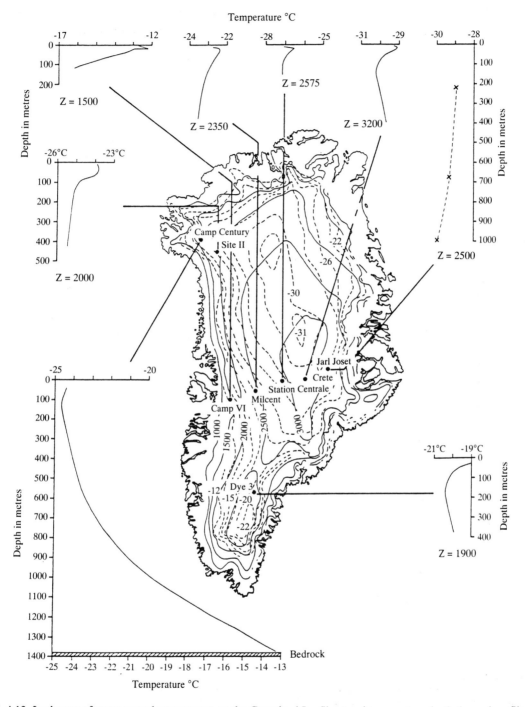

FIG. 4.13. Isotherms of mean annual temperature on the Greenland Ice Sheet and temperature depth thermal profiles for selected sites on the ice sheet. Ice thickness (Z) shown at each site; Camp Century profile to bedrock (after Robin, 1983; reprinted with permission of the author from *Climate Record in Polar Ice Sheets*, edited by G. deQ Robin, 1983, Cambridge University Press).

1961; Müller, 1962, 1976). In winter, ice mass upper layer temperatures are typically higher than the ambient surface air temperature resulting in heat conduction toward the ice surface. The effect of these processes is to create or propagate a cold wave into the upper 15 m of an ice mass (Fig. 4.12) (Paterson, 1981). Examples of cold waves are illustrated in Fig. 4.13.

Thermal anomalies such as cold waves and percolation-zone layers are preserved within the ice resulting in the development of distinct paleo-thermal layers. In time these layers fade as the ice mass equilibrates to steady-state temperatures. The time (t_h) required (in years) for an ice mass (based upon a simple slab model) to equilibrate with surface and bottom thermal boundary conditions (thermal relaxation time) can be defined as:

$$t_h = \frac{h^2}{4\kappa} \qquad (4.7)$$

where κ is the thermal diffusivity of ice = 1.16 x 10^{-6} m^{2}s^{-1} = 36.6 m^2 year^{-1}, and where h is the half width of the thermal anomaly which in this special case can be equated with the thickness of the ice mass. As an approximation $t_h = \frac{h^2}{150}$ can be used as a reasonable estimate (Clarke and Collins, 1984). In terms of glaciodynamics, this thermal anomaly relationship accounts for the existence and persistence of thermal anomalies such as occurs when surge events occur and surface meltwater penetrates crevasse systems to depths within an ice mass, or when basal ice thermal phase changes take place in association with basal glaciodynamic response alterations and bed 'irregularities'. In both instances where the time period of anomalous thermal conditions incidence (δt) is <t_h, these thermal disturbances will have had insufficient time to dissipate, thus enabling thermal anomaly persistence and down-flow line anomalous thermal superimposition within the ice mass (Boulton, 1983; Boulton and Spring, 1986).

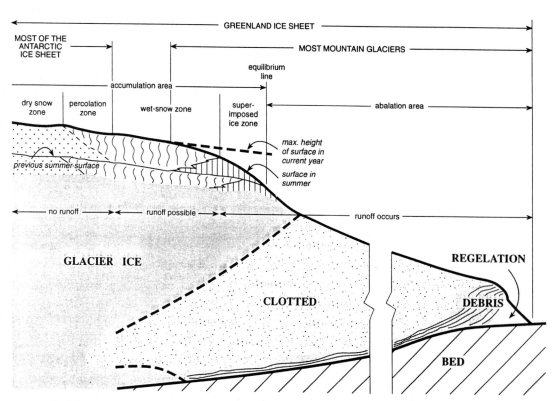

FIG. 4.14. Ice facies zonation on an idealised ice mass (modified after Benson, 1962 and Knight, 1992).

The temperature distribution through an ice mass at a specific location is influenced by several interacting parameters viz: atmospheric surface temperatures; the rate of snow to ice transformation; the lateral and vertical extent, and rate of surface meltwater percolation; the rate and form of 'new' ice incorporation (a function of the angle of descent and velocity along the flow line); internal ice crystal growth; internal creep deformation associated with grain to grain edge friction, melting and refreezing; the depth and extent of crevasse propagation; the basal ice frictional melting, refreezing, and meltwater production; and the basal geothermal heat flux. The consequence of these associated components is to produce no standard or typical vertical temperature profile within ice masses (Fig. 4.13).

The distribution of temperatures within ice shelves has received limited study (Bender and Gow, 1961; Paterson, 1980, 1981; Thomas et al., 1984). The influential parameters affecting ice shelf temperatures are the temperature of circulating water, its salinity and annual temperature fluctuations and gradients, and the addition of up-flow grounded ice (Powell, 1991) (n.b. Fig. 4.1).

4.5.1.1. Ice Facies

In an early classification of the surface facies zones of ice masses, Benson (1961) and Müller (1962) demonstrated that the accumulation area can be subdivided into separate zones as shown in Fig. 4.14 (Williams et al., 1991). Below approximately 10 m, a further grouping on the basis of crystalline structures, ice fabrics, and debris content ice facies has been used to separate glacier ice into a further seven facies types (Sugden et al., 1987; Hubbard, 1991).

4.5.2. Basal Temperatures and Thermal Conditions

The thermal boundary condition at the ice–bed interface is a critical factor in the processes of erosion, entrainment and deposition at the base of an ice mass. In temperate, wet-based glaciers basal temperatures are found at $\approx-1- -3°C$. In polar, cold-based glaciers the ice mass is frozen to its bed with basal temperatures of $\approx-13- -18°C$ (Paterson, 1981; Drewry, 1986; Huang, 1990).

Thermal conditions at the interface between the glacier ice and its bed are much more complex than implied by these two thermal states. Basal ice temperatures may vary both temporally and spatially (Oswald and Robin, 1973; Hughes, 1973, 1981; Sugden, 1977; Hooke, 1977; Goodman et al., 1979; Hughes et al., 1981) producing, in temperate ice masses, polythermal bed conditions (Fowler, 1979b; Hutter, 1983).

4.5.3. Polythermal Bed Models

In investigating thermal regimes beneath active temperate ice masses, it is apparent that local, short-term fluctuations of temperature are very likely to occur at the ice ice–bed interface. Since temperatures are so close to the zero degree isotherm and are closely coupled to short-term pressure variations and widely fluctuating discharges of meltwater, spatially and temporally transient patches of frozen and melting sections develop across the bed interface (Hughes, 1971; Eyles and Menzies, 1983) (Chapter 3). The generation of polythermal bed conditions appears to be the result of the thermo-viscous nature of ice flow across its bed (Robin, 1976; Goodman et al., 1979; Hutter, 1983; Hutter and Engelhardt, 1988; Hindmarsh et al., 1989). The presence of 'cold patches' may explain the 'stick-slip' motion of ice observed in subglacial tunnels (Fig. 4.15). However, the significance of polythermal conditions is one of the cornerstones of modern subglacial sedimentology (Eyles and Menzies, 1983). Where, in the past, erosional and depositional processes were spatially separate, polythermal conditions allow these processes to occur penecontemporaneously in local juxtaposition, and to overlap and affect the same bed area at different times. Thus the fact that subglacial diamictons are of dominantly local origin undergoing only short-distance transport can be explained.

When considering the temperature fluctuations at an ice–bed interface the following parameters converge to establish the specific thermal conditions: (a) rate of snow accumulation (densification/transformation) and snow temperature on deposition; (b) geothermal heat flux; (c) mean annual surface air temperature; (d) ice surface velocity and

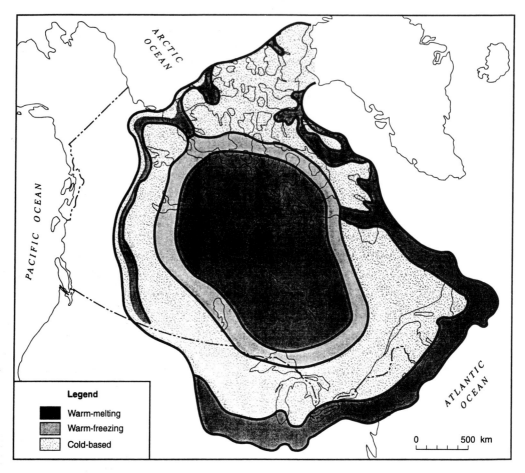

FIG. 4.15. Spatial zonation of basal ice thermal regime beneath the Laurentide Ice Sheet at its maximum (~18,000 BP) (after Sugden, 1977; reproduced by permission of the University of Colorado, Institute of Arctic and Alpine Research, 1977. Courtesy of the Regents of the University of Colorado).

vectors; (e) basal ice velocity and vectors; (f) subglacial meltwater discharge flux and temperature; and (g) the imprinted 'memory' of previous thermal conditions.

In appraising polythermal bed models, it is assumed that steady-state conditions of dynamic equilibrium (net mass balance zero) prevail. In transposing these concepts to the Laurentide Ice Sheet, for example, it is assumed that the ice sheet was in a steady-state phase during its maximum expansion ~18,000 BP. Utilizing several existing models of polythermal conditions, Sugden developed a graphic example of how bed conditions are thought to alter across an ice sheet from centre to margin (Fig. 4.15) (Sugden, 1977; Fisher et al.,

1985). The question of steady state, however, needs to be redefined. In the past, it has been assumed that steady state implies a static ice sheet centre that directly responds to external climatic forcing without significant dynamic feedback loops developing. However, recent evidence based upon patterns of cross subglacial lineations suggests that, in the case of the Laurentide Ice Sheet, the ice sheet centre shifted over time probably reflecting a coupled ice sheet/global climatic temporal response in terms of spatial fluctuating accumulation and ablation areas on the ice sheet surface and feedback effects causing fluctuating atmospheric conditions (Kuhn, 1984; Ambach and Kuhn, 1985; Boulton and Clark, 1990a, b) (Chapter 3).

4.5.4. Subglacial Thermal Conditions of Existing Ice Masses

Investigations within subglacial tunnels and from ice cliffs and boreholes have supplied much needed information on the basal thermal regimes of valley glaciers and ice sheets. The information can be subdivided into data on: (a) ice temperatures and subglacial cavity air temperatures; (b) rates of net loss from the glacier sole by melting (liquid and/or sublimate); (c) rates of net gain to the glacier sole by freezing (liquid and/or sublimate); and (d) basal, englacial debris content and concentrations. The data can be further summarized under: (a) wet-based ice conditions; (b) cold-based ice conditions; (c) polythermal bed conditions; and (d) theoretical and observed rates of basal melting and freezing.

The evidence that wet-based and cold-based glaciers exist seems fairly well established but whether such a simple two-fold classification accurately portrays conditions beneath most ice masses today or in the past remains questionable. The relationship between such mono-thermal basal conditions and polythermal conditions is complex and as yet little understood. In Fig. 4.16, a suggested relationship is drawn between these markedly different basal thermal conditions. It seems certain that such a relationship must exist and it can be suggested that the areal extent, frequency of fluctuations and inherent thermal stability/instability of bed conditions are the correlative factors.

4.5.4.1. Wet-based Ice Conditions

Air temperatures recorded beneath Østerdalisen, Norway, within a subglacial cavity some 65 m from the ice front and beneath 22 m of ice, indicate an approximate constant value of –1°C in summer (Theakstone, 1979). Winter temperatures are rather lower and unless a connection exists to the ice surface or ice margin, the air temperatures remain relatively constant often passing below zero Celsius, thus causing the glacier to freeze to its bed (Anderson *et al.*, 1982). Under wet-based or temperate ice conditions, ice is at or above pressure melting point throughout the ice mass, water is observed occasionally being squeezed from ice in the ceilings of cavities and immediately refreezing on the ice

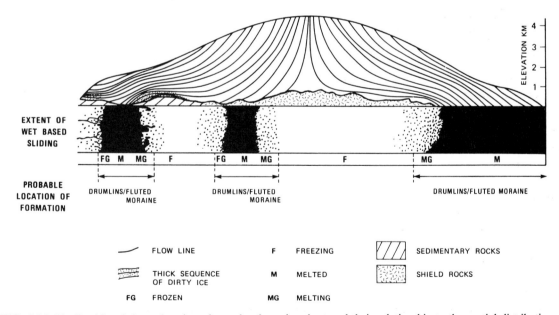

FIG. 4.16. Idealized basal thermal regimes beneath a large ice sheet and their relationship to the spatial distribution of zones of erosion and deposition (modified from Denton and Hughes, 1981, after Eyles and Menzies, 1983). Reprinted from Eyles, N. (ed.), *Glacial Geology*, 1983, p. 20, Fig. 2.1, with kind permission of Elsevier Science Ltd, The Boulevard, Langford Lane, Kidlington, Oxford, OX5 1GB, U.K.

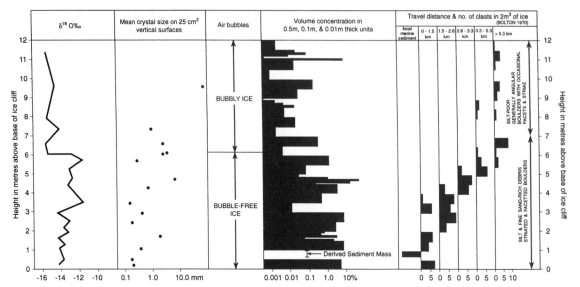

FIG. 4.17. Debris content and other characteristics of basal ice exposed in an ice cliff at the margin of Aavatsmarkbreen, Spitsbergen. Note the basal 6 m of ice are thought to be regelation ice (after Boulton, 1983; reprinted with permission of the author from *Climate Record in Polar Ice Sheets*, edited by G. deQ Robin, 1983, Cambridge University Press).

surface as fragile ice crystals or spicules (La Chapelle, 1968; Souchez *et al.*, 1973; Vivian and Bocquet, 1973; Theakstone, 1979). The base of most cavities has a refrozen meltwater ice glaze and patches of wet or trickling meltwater (Plates 4.3 and 4.4). Debris (Ice Facies 4) is found in basal ice as striped or strata-form englacial inclusions. The debris exists as discreet bands typically 0.2–0.9 m in thickness (Plate 4.5). The ice encasing the debris is termed amber ice having a fine textured crystal size and a low air or gas content (0.2 cm³/100g for Byrd Station, Antarctica). Debris-rich ice rarely exceeds 0.4 m in thickness at the base of the ice. The relative thinness of basal debris-laden ice under these thermal conditions poses several intriguing questions when compared with the much thicker debris–ice zones observed beneath cold-based ice masses. The influence of basal ice tectonisation, in particular, fold structures, complicates basal debris sequences (Boulton and Spring, 1986). The debris provenance usually indicates increasing exotic sources with elevation above the bed (Fig. 4.17) (Chapter 11).

The lowest basal ice and debris bands appear to have been formed and entrained by two possible processes. First, a process of 'Weertman regelation' of meltwater in the lee of bed obstacles to form

'laminated ice' (Weertman, 1966; Theakstone, 1967, 1979; Boulton, 1970b, 1983; Hallet *et al.*, 1978; Gow *et al.*, 1979; Anderson *et al.*, 1982; Souchez and Lorrain, 1991) or by the formation of dispersed or 'clotted' ice typical of subpolar ice masses (Lawson, 1979b; Sugden *et al.*, 1987; Souchez *et al.*, 1988; Hubbard, 1991); and second, by 'net basal adfreezing' (Tison and Lorrain, 1987; Hubbard, 1991) in which meltwater penetrating into the surface of saturated sediment of rock materials refreezes at the freezing front and thus envelops both debris and ice as part of the mobile basal ice. Under wet-based conditions, for example, beneath Østerdalisen, Norway (Theakstone, 1979); Grinnel glacier, Montana (Anderson *et al.*, 1982), Glacier d'Argentiére, France (Vivian and Bocquet, 1973) and the Casement (Peterson, 1970) and Matanuska glaciers (Lawson, 1981a, b), Alaska, evidence was found to support the 'freezing-on' hypothesis. What is apparent from any study of wet-based conditions is the delicate balance between melting and freezing conditions coupled to the pressure melting point of ice. Where deviations in stress conditions occur due to either local increases or decreases in stress, ice temperatures fluctuate across the freezing line. In a phase of freezing conditions, meltwater and debris mixtures will become frozen to the sole of the glacier and entrained within the ice. Once this process is repeated a series of layers or laminae of debris-rich ice accretes to the base of the ice. The thickness of these layers must, however, remain relatively thin since net melting of the ice occurs over time. The concentration of debris within basal ice exhibits marked variations from layer to layer and place to place in the same ice mass and between differing ice masses. Values of debris to ice concentrations range from 4 to almost 50% debris content by volume (Chapter 8).

A major consequence of debris inclusion is the alteration of rheological properties of the ice (Alkire and Andersland, 1973; Nickling and Bennett, 1984). With increasing debris content the value of cohesion and the yield strength (in the Mohr–Coulomb Failure Law) of an overall debris/ice sample increases exponentially. This experimental observation is particularly significant when considering the development of debris-rich englacial ice structures such as shear zones, shear planes and ogive bands.

4.5.4.2. Cold-Based Ice Conditions

Glaciers that exhibit cold-based conditions are of two types: (a) those completely frozen to their bed commonly termed polar, exhibiting ice conditions below pressure melting point throughout, and (b) those with frozen terminal zones but temperate wet-based interiors termed subpolar, in this case frozen conditions tend to be climatically controlled. This latter group will be discussed under the heading of polythermal-based ice conditions.

Typically, under cold-based ice conditions a sharp ice-frozen bed contact exists without cavities forming. Basal interface temperatures are well below the pressure melting point (Holdsworth, 1974; Koerner and Fisher, 1979). The ice at the base is often white, debris free and composed of fine textured ice crystals. The ice crystal size may be the result of either basal regelation in the past or from névé field snow burial down the ice flow line. Debris that is detected is thought to result from the burial of atmospheric dust particles (Koerner, 1977; Koerner and Fisher, 1979; Thompson et al., 1975; Thompson and Mosley-Thompson, 1982). In a core taken from

Devon Island Ice Cap, an amber ice zone containing clay and pebbles was found to exist for 10–60 cm above the bed. The emplacement of this debris, often as pockets or 'clots' within a polar glacier (Ice facies 5 and 6), appears to give clues to past temperate-based thermal conditions and to the probability of subglacial 'freeze-on' as the dominant means of debris incorporation.

A related phenomenon is the inversion observed between younger and older ice at depth within the Greenland and Antarctic Ice Sheets. Ice stratigraphic inversion has been noted from an analyses of oxygen isotope profiles in several glaciers (Paterson et al., 1977; Lawson and Kulla, 1978; Gow et al., 1979; Dansgaard et al., 1982; Fisher et al., 1983). A convection theory was an early explanation for this phenomenon (Hughes, 1976); however, later work has revealed that movement does occur at the basal interface of Polar glaciers (Koerner and Fisher, 1979; Shreve, 1984; Echelmeyer and Wang, 1987). Boulton and Spring (1986) have demonstrated that isotopic fractionation of meltwater at the bed of polar and subpolar ice masses that have undergone subsequent tectonism has led to massive overfolding, recumbent and thrust folding resulting in enormous attenuation of differing ice layers and related debris content (Fig. 4.17). This basal ice tectonism explains the development of an inverted ice stratigraphy and the presence of debris-rich ice at high levels above the bed within polar ice masses.

4.5.4.3. Polythermal-Based Ice Conditions

A growing body of data reveals that many ice masses have or have had varying thermal conditions at and across their base. The Barnes Ice Cap on Baffin Island, Canada is an example of such a type, exhibiting, today, a wet-based interior and a frozen cold-based snout or frontal zone. Ice masses with past or presently fluctuating thermal basal conditions can be classified into three major subgroups under the general heading of polythermal-based ice masses (Fig. 4.18) viz:

(a) Ice masses that are now polar cold-based but at some time in the past had warm-based bed conditions. (e.g. Camp Century, Greenland Ice Sheet (Herron and Langway, 1979)).

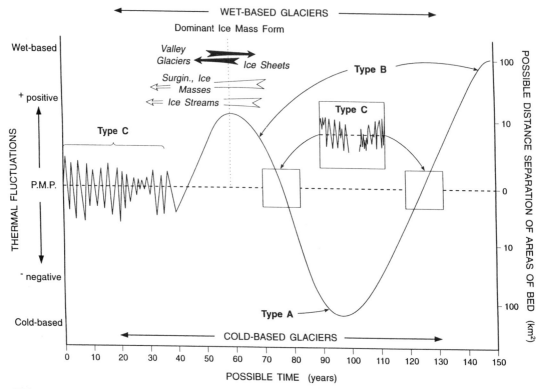

FIG. 4.18. Classification of polythermal subglacial bed types and their relationship to differing ice mass types.

(b) Ice masses that are now temperate wet-based but in the past had cold-based bed conditions. (e.g. Byrd Station, West Antarctica (Gow *et al.*, 1979)).

(c) Ice masses that have at present polythermal-based bed conditions. (e.g. Glacier d'Argentiére, French Alps (Goodman *et al.*, 1979)).

This classification helps explain the position of debris within ice masses due to past or present thermal regimens and the subsequent transport pathways such debris will follow. Ultimately, these critical transformations in thermal regimen are ubiquitous in affecting glacial sedimentological processes and glacio-lithofacies environments. In any of these polythermal-based ice mass types the presence of deformable beds is viable: n.b. the 'A' type — Urumqi Glacier, China (Echelmeyer and Wang, 1987); the 'B' type — Ice Stream B, Antarctica (Alley *et al.*, 1986); and the 'C' type — Breiðamerkurjökull, Iceland (Boulton *et al.*, 1974).

Polythermal Type A: From ice cores, retrieved from beneath the polar-based Greenland Ice Sheet at

Camp Century, evidence reveals debris encased within basal ice to a height of 15.7 m above the bed (Fig. 4.19) (Herron and Langway, 1979). The debris appears to have been glacially abraded and has a distribution of pitted angular to smoothed surrounded particles (Whalley and Langway, 1980). The average volume concentration of debris was 11.9%. To explain this debris incorporation, it is thought likely that wet-based temperate ice conditions prevailed in the past. Kumai (1977) and Herron and Langway (1979) conclude that the debris is of basal origin and was 'frozen on' to the ice mass during the Pleistocene (Menzies, 1995b, Chapter 13). As a consequence these ice masses have considerable volumes of englacial debris found at high levels within the ice.

Polythermal Type B: Some ice masses that exhibit wet-based conditions today appear to have been frozen to their bed in the past. On penetrating the West Antarctic Ice Sheet at Byrd Station, Gow *et al.* (1968, 1979) found temperate wet-based conditions prevailing. Debris was observed to exist up to 4.83 m

(b)

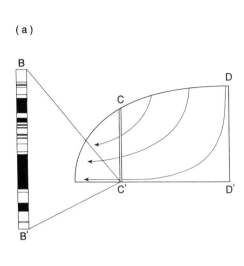

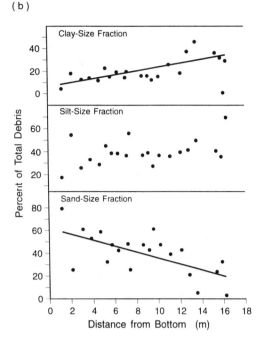

FIG. 4.19. (a) Graphical representation of Camp Century ice core, Greenland. D–D' represents the ice sheet divide; C–C' represents the Camp Century core; and B–B' represents the debris-regelation ice facies. (b) Debris particle size distribution in section B–B' (after Herron and Langway, 1979; reproduced by courtesy of the International Glaciological Society from *Journal of Glaciology*, **23**(89), 1979, pp. 194, 198, Figs 1 and 3).

above the bed (Plate 4.6) in volumetric concentrations of <7% and was described as being well stratified, ranging from silt to cobble sized particles (Fig. 4.20). Typically, the finer debris occurred as discrete 'mud clots' often of a subangular shape but not exhibiting boudinage as might be expected if shearing had been involved either in the entrainment mechanism or englacial deformation. Gow *et al.* (1979) concluded that, due to the lack of entrapped air within the lower 4.83 m of basal ice, this debris-filled ice was a consequence of basal meltwater 'freezing-on' to the glacier sole. The layered nature of the debris may indicate past episodic periods of freezing and melting or phases of freezing of debris rich mixtures from meltwater and basal debris, alternating with freezing of meltwater devoid of basal debris.

In Type B beds debris is being gradually released through melt-out from the glacier bed while in contrast with Type A beds, debris incorporation, over wide areas of the glacier bed, is virtually zero and englacial debris at height is rare.

Polythermal Type C: As the number of observations and our understanding of subglacial environments increase, it seems likely that almost all ice masses are of the Type C form with Types A and B being only end-members of a fluctuating thermal system. Type C polythermal beds exist where spatial and/or temporal fluctuations at the bed occur either over adjacent areas of less than 1–10 km² and/or over periods of less than a decade. Conversely, where thermal fluctuations range over tens of square kilometres of bed and/or persist for periods of decades polythermal beds of Type A or B may be found (Fig. 4.18).

The attributes of Type C basal ice conditions conform to our present understanding of wet- and cold-based ice masses and thus have been sufficiently discussed under these two headings. However, in a polythermal Type C ice mass, it can be expected that debris content should be within a few centimetres of the glacier sole, debris is likely to be dominated by locally derived materials and have an englacial

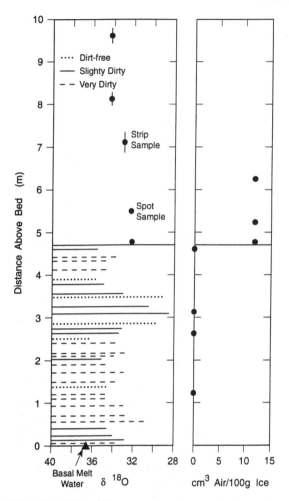

FIG. 4.20. $\delta^{18}O$ and entrapped-gas data values in the bottom 10 m of ice at Byrd Station, Antarctica. Change from debris ice facies to glacier ice facies at ~4.83 m above bed. ▲ indicates presence of basal meltwater (after Gow, 1979; reproduced by courtesy of the International Glaciological Society from *Journal of Glaciology*, **23**(89), 1979, p. 190, Fig. 5).

transport 'duration' of a few years rather than decades as in the case of Types A and B. The presence of deformable beds and conditions equivalent to 'Q'-beds as defined by Menzies (1989a) can be anticipated.

Present evidence suggests that rapid basal thermal fluctuations are likely to occur either in thin or short or steep valley glaciers, or in surging ice masses of either valley glacier or ice sheet forms and ice streams of major ice sheets. In any of these forms of ice,

polythermal Type C bed conditions are likely to predominate. The implications for understanding glacial sediments and sediment associations of polythermal Type C bed conditions are considerable and are discussed in more detail in Chapter 5.

4.6. ICE STRUCTURES

4.6.1. Primary

In wide and relatively flat accumulation areas the surface transformation, remelting and refreezing of snow and ice produce a distinctive stratification. This form of quasi-horizontal banding can be distinguished by layers of coarse bubbly ice (winter snow) alternating with coarse clear ice (refrozen meltwater) (Plate 4.7). The influence of deformation due to ice flow is virtually absent from such structures.

4.6.2. Secondary

Where internal ice deformation occurs glacier ice exhibits several characteristic structures. Deformation may be further induced by changes in bottom topography, valley wall constrictions, ice stream bifurcation or tributary addition. The major secondary structures are surface and bottom crevasses, folds, faults and ogives.

4.6.2.1. Crevasses

The development and disposition of surface crevasses is an example of stress-induced phenomena. The formation of crevasses are due to tensile stresses developing within the ice, overcoming the internal yield strength of the ice which in essence is the result of ice moving at a velocity in excess of its normal strain rate (Nye, 1952; Hambrey and Müller, 1978; Paterson, 1981). Crevasses may range from minute cracks a few millimetres across to major fractures several metres wide (Plate 4.8). It can be shown from Glen's Flow Law (Chapter 5, Eqn. (5.13)) and assuming dry (no meltwater) conditions that crevasse depth is limited to between 25 and 35 m; below 35 m the pressure of internal deformation creep acts to seal

PLATE 4.1. Landsat 1 MSS digitally enhanced false-colour composite image of part of Byrd Glacier and surrounding glaciers and the Ross Ice Shelf, Antarctica (photograph courtesy of B.K. Lucchitta and the U.S.G.S. Flagstaff, Arizona; annotated version of landsat photo from Williams, R.J., Jr and Ferrigno, J.G. in USGS Prof. Pap. 1386-B, 1988, p. B27).

PLATE 4.2. A large wave-like bulge growing at the boundary between warm-based and cold-based ice on the Trapridge Glacier, Yukon Territory prior to the next surge. The bulge is approximately 40 m high and the glacier width at the bulge is 0.75 km. Photograph was taken June 1980 (photograph courtesy of Garry Clarke).

PLATE 4.3. Portal of subglacial meltwater tunnel at the front of the Canwell Glacier, Alaska. Note debris banding and fissuring within the glacier ice. Person for scale standing on supraglacial debris (photograph courtesy of Ed Evenson).

PLATE 4.4. Subglacial cavity beneath Nigardsbreen, Norway. Note the debris lying in the leeside of a bedrock rise. Ice moving from left to right.

PLATE 4.5. Ice cliff of the southern lobe of the Barnes Ice Cap, Baffin Island revealing debris bands and folds (far left centre) as it terminates in Generator Lake (photograph taken 1969) (photograph courtsey of Gerry Holdsworth).

PLATE 4.6. Surface debris bands revealed along the edge of a large crevasse on Omsbreen, Norway.

PLATE 4.7. Example of regelation banding within an englacial tunnel (Saskatchewan Glacier, Alberta/British Columbia). Wooden plank on floor of tunnel is approximately 15 cm.

PLATE 4.8. Crevasses on an outlet glacier from Highland Ice Field, east-central Baffin Island (photograph courtesy of Gerry Holdsworth).

PLATE 4.9. Seracs and wave ogives on an outlet glacier from the Highland Ice Field approximately 50 km northeast of the Barnes Ice Cap, Baffin Island (photograph courtesy of Gerry Holdsworth).

(a)

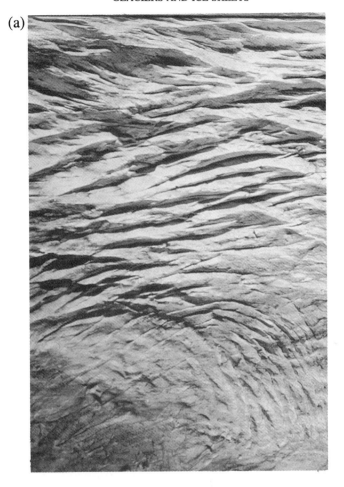

(b)

PLATE 4.10. (a) Folds on the surface of an unknown glacier Görner region of Switzerland. (b) Surface foliations on an alpine glacier east central Baffin Island (photographs courtesy of Gerry Holdsworth).

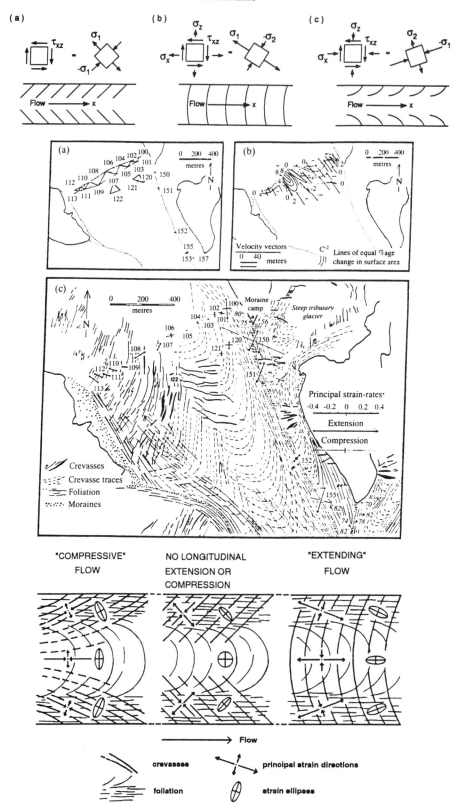

FIG. 4.21. Types of crevasses and crevasse patterns.

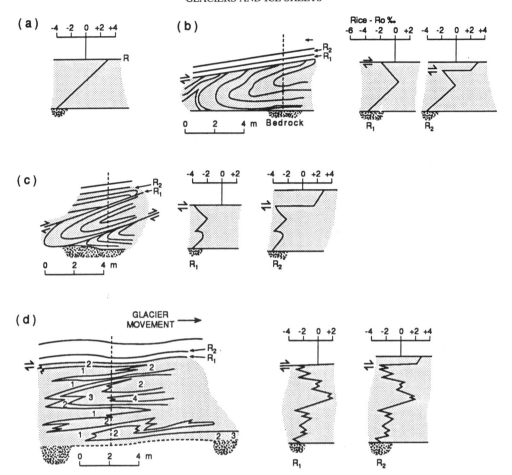

FIG. 4.22. Complex patterns of basal ice deformation, folding and faulting as illustrated by isotopic profiles (b)–(d). These patterns often occur beneath a planar décollement surface. (a) Isotopic pattern of simple laminar flow in basal regelation ice. (b) Basal ice folding beneath a sharp décollement in Sørbreen, Spitsbergen. (c) Basal folding and faulting in Aavatsmarkbreen, Spitsbergen. (d) Intense folding in Lower Wright Glacier, Antarctica. The direction of displacement of the isotopic profile at the top of the regelation ice depends on whether the upper surface coincides with the décollement (R1) or whether it lies above the décollement (R2). Vertical dashed lines illustrate the positions of hypothetical isotopic profiles in the deformed ice (after Boulton and Spring, 1986; reproduced by courtesy of the International Glaciological Society from *Journal of Glaciology*, **32**(112), 1986, p. 481, Fig. 6).

the fracture. Where meltwater is present and flows into a crevasse (moulin) the influence of water turbidity, meltwater heat, frictional heat and erosive effects causes the crevasse to remain open to considerable depths (Stenborg, 1968) (see Chapter 6). In winter, without meltwater, these crevasses may seal up by *in situ* freezing of meltwater or internal deformation on becoming dry. The duration of any crevasse is relatively short, rarely extending beyond one melt-season.

Several types of surface crevasses are found (Fig. 4.21). These crevasses reflect tensional fractures due to the effects of valley wall frictional drag, ice stream edge drag, extending flow, and sudden changes in the longitudinal surface velocity of a glacier as, for example, at an ice fall (serac) (Plate 4.9).

Theories of subglacial or bottom crevasses have appeared in the past to explain localized debris accumulations (Alden, 1911). It has been a common assumption that such subglacial crevasses would

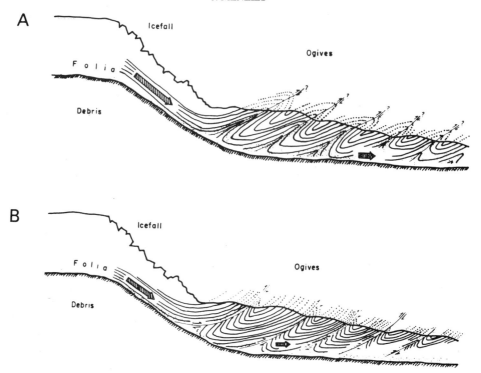

FIG. 4.23. Models of ogive formation according to reverse faulting and drag folding (after Posamentier, 1978; reproduced by courtesy of the International Glaciological Society from *Journal of Glaciology*, **20**(82), 1978, p. 219, Figs 1 and 2).

most likely form beneath inactive stagnant ice. Under normal active ice conditions, it is assumed that basal ice deformation and flow would prevent dry crevasses forming for anything but a very brief period. However, recent work by Shaw (1983b) and colleagues has hypothesized that subglacial crevasses and cavities, associated with high meltwater discharges, may help explain drumlin and associated bedform initiation. That subglacial cavities and crevasses can be maintained or enlarged by subglacial water seems reasonable (Walder, 1986), but whether the scale of discharge envisioned for drumlin development is possible remains controversial (Menzies, 1990b; Shoemaker, 1992). In the case of ice shelves, however, where the effect of buoyancy must be considered, the formation and persistence of bottom crevasses has been documented (e.g. observations by Clough, 1974; Swithinbank, 1978; for theoretical discussions see Weertman, 1973; 1980).

4.6.3. Non-Laminar Flow in Basal Ice

4.6.3.1. Folds, Fractures and Thrust Planes

Basal ice flow is often neither parallel to its bed nor 'laminar'. Due to inhomogenieties and rheological anisotropy within basal ice due to debris content, ice crystal fabric and temperatures, and bottom topographic irregularity, basal ice flow often exhibits non-laminar flow. Observations made in subglacial tunnels or in natural ice cliffs show that glacier ice is often folded (Plate 4.10). Folds develop, in general, due to local stress concentrations or due to inhomogenieties within the ice. Folding in ice is similar in style to that exhibited by rocks but has received only limited study (Hudleston, 1976, 1977; Hudleston and Hooke, 1980; Paterson, 1981; Boulton and Spring, 1986) (Fig. 4.22). As in rock folding where, in compressive strain, a variation in viscosity or permeability occurs, folding is likely to take place along the boundary contact between the two variants

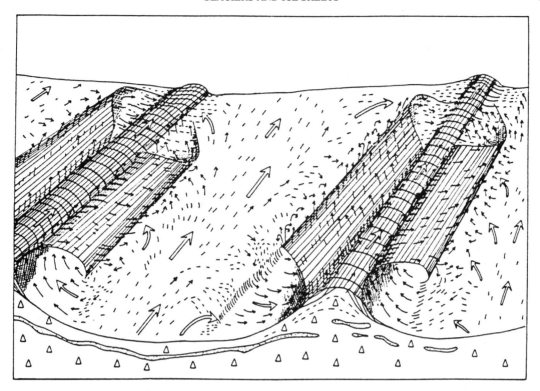

FIG. 4.24. Model of helicoidal basal ice motion (after Aario, 1977; reprinted from classification and terminology of morainic landforms in Finland, by R. Aario, *Boreas*, 1977, **6**, 87–100, by permission of the Scandinavian University Press).

of the same material (Biot, 1961; Ramsay and Huber, 1983, 1987; Lliboutry, 1987b). In ice such internal variants are most likely to be found where regelation has caused ice crystal size variations (Lawson and Kulla, 1978; Jouzel and Souchez, 1982) and/or debris concentrations (Hooke *et al.*, 1972) to reduce viscosity and/or permeability at or near the base of an ice mass (Fig. 4.22). Boulton (1983) suggests that the frequency of such basal folding is a function of: (a) the rate of stress buildup and decay; (b) the magnitude and direction of ice velocity; and (c) the effective ice shear stress. The locus of the folding is partly a function of: (a) ice-bed topographic contra-relationships; and (b) internal ice inhomogenieties and anisotropic behaviour; the latter function leading to differential shear and/or internal ice dislocation along shear planes (Boulton and Spring, 1986).

Shear fractures or faults have been observed in crevasse walls and subglacial tunnels. Most faults are of the strike-step type. Many faults are found to be

generated immediately down-ice of ice falls (seracs), in the submarginal areas of valley glaciers, along the edges of ice streams and in the proximity of ice rises, grounding-lines and fringing land-based ice on ice shelves. For example, where high-angle shear lines occur thrust planes may develop (Boulton, 1983). Such features are of two types: (a) steady, and (b) transient.

(a) Considerable evidence exists of planes of décollement developing approximately parallel to the glacier bed (Boulton, 1970a, 1972a, b; Holdsworth, 1974; Colbeck and Gow, 1979). Such planes may take various forms viz: (1) wholly within the ice and overlying 'dead' or 'cold' ice due to bed topography; (2) planes within the ice but directed away from or toward a deformable bed; and (3) planes within either an ice–debris-rich tractive basal layer or within a deforming debris layer completely.

(b) Much past discussion has attributed high-angle transient thrust planes as the means whereby debris

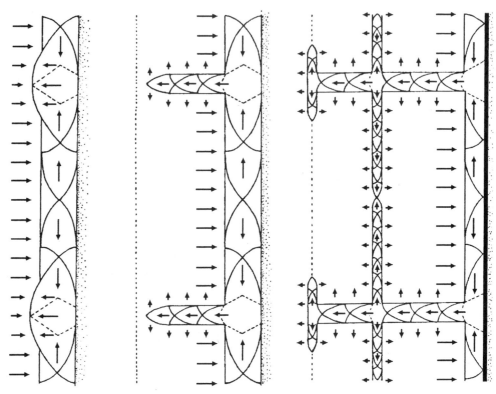

FIG. 4.25. The initiation and development of convection cells within an ice sheet. Arrows illustrate local ice flow direction and orthogonal cycloid segments show ice slip-line field in dykes and sills (after Hughes, 1976; reproduced by courtesy of the International Glaciological Society from *Journal of Glaciology*, **16**(74), 1976, p. 57, Fig. 7).

can be entrained within the ice (Harrison, 1957; Hooke, 1968; Hambrey and Müller, 1978; Hambrey, 1979; Weertman, 1961b). This mechanism would seem to have little merit since the amount of displacement along such planes is limited. Also, ice that has a debris content is rheologically 'stronger' than debris-free ice, the former being strengthened rather than weakened (Nickling and Bennet, 1984) along the internal glide planes and crystal lattice junctions. Where thrust planes do occur, principal stresses are likely to approach the horizontal; therefore, these features are not likely to be typical of basal ice. In areas at the glacier bed where ice compression occurs, as on the up-stream of obstructions, thrust planes will develop and subsequently, once out of the local stress field, attenuate with shear deformation and incline at a high angle to the glacier bed (n.b. observations of boudinage by Hambrey and Milnes, 1975).

4.6.3.2. Ogives

A distinctive feature of many glacier surfaces is the existence of transverse bands of debris-rich ice. Ogives are of two types: (1) Forbes-bands formed of alternating light and dark ice; and (2) Wave ogives formed of trough and ridge construction (Plate 4.9). Both features are relatively common but as yet without an adequate mechanism to explain their origin (Posamentier, 1978; Lliboutry and Reynaud, 1981; Paterson, 1981; Waddington, 1986). Ogives occur in many parts of the glaciated world (Waddington, 1986). The amplitude of the waves varies but is often ≈10 m from crest-to-trough. Wave-trains down-ice of a serac usually dissipate and amplitudes concomitantly reduce, but in some cases the wave-trains may persist for some considerable distance down-ice and for several years. Ogives tend to form at the rate of one per year at ice falls on some confined glaciers or on outlet glaciers of ice sheets.

Several hypotheses have been suggested viz: (a) 'seasonal' pressure waves; (b) longitudinal stretching of ice through passage via ice falls; and (c) buckling. The association of ogives with ice falls would seem to be a causal relationship (Ragan, 1969). The ablation–plastic attenuation hypothesis (b) put forward by Nye (1958a) would seem the most likely. As a result of differential velocity with passage of ice via an ice fall during summer, the ice in the fall loses more surface ablation than the ice above the fall resulting in a trough forming at the bottom of the fall. In those glaciers where either/or a combination of high ice fall velocities and large annual mass balances occur, wave ogives are likely to appear. The possible relationships between ice falls, ogives and internal glacier faulting and folding are illustrated in Fig. 4.23.

4.6.4. Macro-Scale Hypotheses of Non-Laminar Flow

In general, those ice structures described above are of a micro- to meso-scale occurrence (<50–75 m), rarely influencing large areas of the ice at any one time. Evidence for these ice structures and of non-laminar flow is well established in theory and observation. In contrast, two forms of non-laminar flow of macro-scale dimensions (>100 m), affecting large areas of an ice mass, have been postulated but with little basis in observation. These forms, convective flow and secondary flow, at present are theoretical constructs. In both cases, large zones within the basal ice are claimed to be affected by these flow regimes. The basis for these macro-flow forms stems from attempts to explain observations, for example, 'cold spikes' in oxygen-isotope profiles within the Antarctic Ice Sheet (convective flow); and 'herring-bone' till fabrics within glacially streamlined landforms (secondary flow).

4.6.4.1. Secondary Flow

The concept of secondary flow (sometimes termed helicoidal flow) has its origins in earlier work by Demorest (1938) and more especially by Folk (1971). The concept has been applied by Shaw and Freschauf (1973), Shaw (1975a) and Aario (1977a, b) among

others. It is argued that, under diverging basal ice flow and fluctuating transverse stress gradients, the components u_{bz} and u_{by} (transverse and vertical flow vectors, respectively) achieve an above average magnitude and velocity component. The effect of these two secondary flow directions is to produce a 'cork-screw' or helicoidal cell within the ice of relatively short duration travelling down-ice (Fig. 4.24). Thompson (1979), in a major theoretical discussion on flow perturbations, suggests that such instabilities are unlikely. It might also be argued that where subglacial debris clasts are encased within a sufficiently mobile matrix to be deformed into an overall 'herring-bone' fabric, it is more likely that any flow cells are more apt to develop within the deforming sediment than in the basal ice. This concept and its application to landforming processes therefore remains controversial.

4.6.4.2. Convective Flow

In a series of papers, Hughes (1976) (and others referred therein) suggested that due to thermal irregularities at the base of the Antarctic Ice Sheet convective flow might occur. This concept, which has varied with further refinement by Hughes, assumes basal ice moving vertically in a 'dyke-sill' flow pattern (Fig. 4.25). It is suggested that convection may develop in a static ice sheet with limited shear due to changes in ice viscosity and density resulting from basal thermal fluctuations. These temperature changes influence the relationship as proposed in the Flow Law of Ice. It is therefore envisaged that 'warm' ice becomes slowly injected upward while 'cold' ice sinks within the central portions of large ice sheets. Most of the evidence used by Hughes has since been refuted (Paterson, 1981, pp. 170–171; Boulton and Spring, 1986) and the concept found untenable and questionable.

On reflection, the reader might query the inclusion of these two disputable hypotheses. What they do reflect and represent are paradigm departures from our present but limited understanding of basal flow mechanics. These concepts indicate where future glaciological research might concentrate in seeking explanations for anomalous features and phenomena found in glacier ice and glacial sediments.

4.7. CONCLUDING REMARKS

It is imperative that the theoretical and field observations of glaciology be tied closely to glacial sedimentological processes if a more accurate understanding of the latter is to be finally achieved. In this chapter, discussion centres upon understanding the principles of snow/glacier ice transformation, the concept of mass balance and the thermal characteristics of ice masses. Until relatively recently these glaciological parameters and processes have been largely left separate from any attempts to comprehend the basic sedimentological processes of the glacial environmental system. This chapter strives to interrelate many of these aspects of glaciers within a unifying rationale that interlinks, for example, mass balance gradient to basal ice temperatures to their combined effects upon glacial sedimentological processes. It is evident, for example, that the response of an ice mass to a series of kinematic waves and their rate of translation through an ice mass, down the flowline, has fundamental effects upon the whole glacial system from meltwater activity to basal ice stresses to the mobility or otherwise of a potentially mobile bed and ultimately to the resultant subglacial bedforms produced.

Chapter 5 will introduce the mechanics of ice flow within ice sheets, valley glaciers and ice shelves, and relate specific flow mechanisms to particular glacier bed conditions and ultimately to glacial sedimentological processes.

Chapter 5

THE DYNAMICS OF ICE FLOW

J. Menzies

5.1. INTRODUCTION

Anyone who has visited glaciers, explored and observed them is patently aware not only of their magnificent grandeur but also that here is a moving, dynamic natural phenomenon. Movement is invariably imperceptible yet repeated visits to the same glacier indicate that an almost relentless process of motion occurs. Motion is apparent in how the glacier surface changes, crevasses appear while others close-up, new debris arrives at the surface, moulins enlarge and fresh moraines are built-up. Without question it is an environment of enormous change driven by the dynamics of ice flow (Plate 5.1).

This chapter will review ice dynamics and rheology, assess how and why ice flows and the various forms such flow takes both at the surface and glacier bed.

5.2. ICE MECHANICS AND THE THERMO-MECHANICAL PROBLEM

In an endeavour to integrate the field of ice rheology into ice physics, Hutter (1983) pointed out the inadequacy of past approaches to the constitutive 'oneness' of ice mechanics and the thermo-mechanical nature of ice flow. Previously, attention had focused upon either stress/velocity or temperature fluctuations and their interrelationships. However, it is apparent that ice flow and temperature distribution problems require an integrated thermo-mechanical solution. Although two-dimensional thermo-mechanical solutions are available for polar or temperate ice masses, no adequate solution(s) exist(s) yet for polythermal ice masses (Fowler and Larson, 1978; Hutter et al., 1988; Blatter and Hutter, 1991). Complex three-dimensional stress/strain conditions within ice masses are symptomatic of ice conditions. However, progress with three-dimensional problems of ice mechanics and how they relate to ice flow conditions are exacting and remain unresolved (Dahl-Jensen, 1985; MacAyeal, 1987; van der Veen and Whillans, 1989). These questions remain enduring unsolved glacial thermo-mechanic problems.

5.3. RHEOLOGICAL BEHAVIOUR OF ICE

Natural glacier ice is crystalline and may be either mono- or polycrystalline, usually the latter. The major difference between the two types is that polycrystalline ice is composed of a range of differently oriented single crystals (Plate 2.1) and therefore any rheological understanding of polycrystalline is an average computed value based upon the relative amounts of single crystal types. Single crystal ice is best described as exhibiting a viscoelastic response to stress application (Michel, 1978; Paterson, 1981; Hutter, 1983; Drewry, 1986; Lliboutry, 1987a). The stress–strain curve of ice exhibits three creep 'states' (Fig. 5.1) and has been described, in the first instance, by the equation known as Glen's Flow Law (Glen, 1955) (Chapter 3). Polycrystalline ice under stress exhibits linear viscoelastic behaviour (Section 5.5.3). Generally, polycrystalline ice is regarded as isotropic but under

PLATE 5.1. Photograph of the retreat stages of the Athabasca Glacier, Alberta taken in 1985 with marker stone in foreground of 1965 and 1970 in middle ground (photograph courtesy of Hugh Gayler).

the effects of recrystallization, especially at the base of ice masses, becomes strongly anisotropic (Budd and Jacka, 1989). Observations of polycrystalline ice under stress (Rigsby, 1960; Wakahama, 1964; Michel, 1978) reveal that as the ice deforms several interrelated processes occur viz: dislocation climb where both elastic and plastic deformation takes place; grain boundary slip; cavity formation at grain boundaries, formation of low-angle boundaries; polygonization and recrystallization. The various creep processes are illustrated in Fig. 5.2 (after Michel, 1978) and are the result of crystal lattice

defects within their atomic structure. These defects are referred to as dislocations and may occur singly or collectively as point, edge or stacked lattice defects (Michel, 1978). Inter-crystalline movement occurs when, as a consequence of dislocations, crystalline units glide past each other in a slip or climb process (Budd and Jacka, 1989).

5.3.1. Stress Conditions within Ice Masses

Stress is transmitted within any ice mass through intergranular ice crystal contacts, and through bottom and side drag effects that translate as longitudinal 'push–pull' transient influences (van der Veen and Whillans, 1989, Eqn. (1)). In the vertical plane, normal stress (σ_i) at a point within the ice is a function of ice density and ice thickness above that point (Fig. 5.3). Where water filled intercrystalline or basal cavities, joints or veins exist an effective stress $(\sigma_i - \rho_w)$ can be envisaged where ρ_w is porewater pressure.

In the central vertical plane of an ice sheet it is assumed that there is no shear stress (τ). Shear stresses develop as the ice deforms toward its margins, in simple shear, under its own weight. In a two-dimensional model, where consideration is given only to a parallel-sided ice slab, it has been shown (Nye, 1952) that simple shear stresses vary with depth and ice velocity from a value of zero at the upper surface to a value at base computed as follows:

$$\tau_{xy} = \rho_i\, gh_i\alpha \qquad (5.1)$$

For simple shear the Flow Law can then be reduced to:

$$\frac{du}{dy} = 2A[\rho_i\, gh_i\alpha]^n \qquad (5.2)$$

In this model, flow is assumed to be laminar and therefore parallel to the boundary surfaces of the ice

slab. Increasingly, observation and research reveals this latter assumption to be untenable, evidence being put forward of 'convective', 'turbulent' and 'tectonic' style flow states in, at least, basal areas of major ice sheets (Hughes, 1970, 1976; Smith and Morland, 1981; Boulton, 1983; Hutter, 1983; Boulton and Spring, 1986) (see Chapter 4).

However simplistic this two-dimensional model is, several important implications for ice flow are revealed (Bindschadler *et al.*, 1977; Paterson, 1981, p. 86; Hutter, 1983). First, it is assumed that the value τ can be calculated from known values of α and h, provided ice exhibits perfect plasticity and a 'standard' yield strength of approximately 100 kPa (100 kPa = 1bar). Secondly, it can be seen from Eqn. (5.1) that basal shear stresses (where $\tau_{xy} = \tau_b$) are assumed to be a function of the surface profile of an ice mass and not bottom topography. The value of α is generally taken to be very small and approximately constant over large areas of an ice mass. Paterson (1981) investigated the relationship of τ_b to α and concluded that several scalar relationships exist. For example, at the large scale, over distances >20 h, it can be assumed that τ_b is relatively constant; whereas

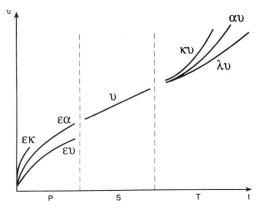

FIG. 5.2. Creep curves in polycrystalline ice from primary, secondary, tertiary creep. Various forms of deformation are shown where ε is pure elastic deformation, υ is pure plastic deformation, α is deformation caused by an increase in the number of dislocations in the ice and a corresponding decrease in ice viscosity (strain softening), κ is deformation accompanied by micro-cracking activity in the ice which also decreases the apparent viscosity of the ice and δ is deformation associated with decreased viscosity as a result of syn-tectonic recrystallization (after Michel, 1978; reproduced with permission of Les Presses de L'Université Laval).

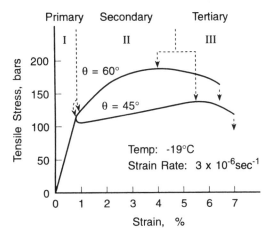

FIG. 5.1. Stress–strain curves for non-basal glide of single ice crystals with orientations of 45° and 60°. The tensile axes lies in the basal plane (after Hutter, 1983; reprinted by permission of Kluwer Academic Publishers).

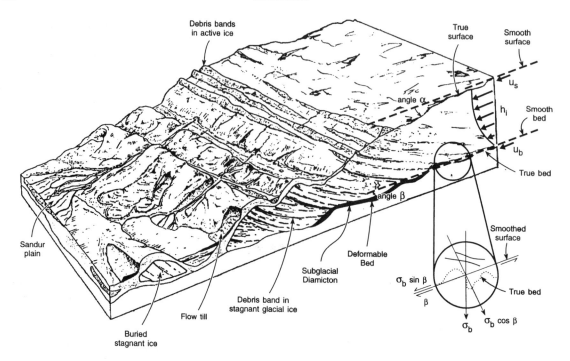

FIG. 5.3. Model of ideal ice sheet/glacier. Dashed lines illustrate real top and bottom interface surfaces (adapted from Boulton, 1972 and Drewry, 1986).

at intermediate horizontal distances of 1–4 h, the influence of bottom topography and gradient becomes significant and the relationships shown in Eqn. (5.1) breaks down. At the local scale, with horizontal distances of ≤h, a derivative equation of Eqn. (5.1) is necessary (Paterson, 1981, Eqn. (4.5)) where the bottom topographic gradient is integrated into the equation. Since many glaciological problems are of a 'local' nature, a more sophisticated three-dimensional approach is demanded (Hutter, 1983). Thirdly, in the idealized case, deformation within the ice is assumed to occur in only its lowest layers. In reality, ice flows three-dimensionally, thus shear stress conditions are not as simply defined. Finally, ice flow within confined valleys is inhibited due to side-wall friction and the shape of the conduit. Basal shear stress calculations in these conditions are incomplete. Assuming continuity of flow discharge, it can be expected that over short distances of a few kilometres where valleys become constricted and narrow, ice velocities and basal and side-wall stresses will increase, and vice versa (Raymond, 1980;

Frolich *et al.*, 1987; Harbor, 1992). A similar state will also occur as a boundary effect in ice streams.

Stresses within ice shelves at basal boundary surfaces are zero. Only where the ice shelf impinges on the coast, ice rises and up-ice of the grounding-line are stress fields of significance (Thomas, 1979; MacAyeal, 1987; Budd and Jacka, 1989).

5.3.2. Glacier Bed Stress Conditions

Our understanding of stress level fluctuations and stress fields at glacier beds remains imprecise (Robin and Swithinbank, 1987). Glacier beds are extremely complex and exhibit transient temporal and spatial fluctuations of thermal, sedimentological and glaciological conditions. Over an area of bed the impact of meltwater, varying thermal regimens, ice pressures, bed topography and debris all collectively, singly or in variable association alter the local stress fields. The development of local, site-specific stress fields, in turn, affect ever-radiating stress conditions across the glacier bed. Nevertheless, normal stress levels acting in the vertical plane can be described by the following expression:

$$\sigma_i = \rho_i \, gh_i \qquad (5.3)$$

Significant stress levels are only likely to develop where no water is present. In general, this is a rare state beneath an ice mass except where ice is frozen directly in contact with minute asperities on the bed (Chapter 7). Under these circumstances very high local stress levels capable of overcoming intact yield strengths of bedrock can be achieved. Typically, normal stresses are effective stresses defined as:

$$\sigma_\varepsilon = (\rho_i \, gh_i - \rho_w) \qquad (5.4)$$

Normal effective stress levels develop as a function of the presence of free meltwater at the glacier bed, or porewater within subglacial debris or a combination of both. Values of normal effective stress lie within a broad range from <20 kPa to >10^3 kPa. Values of σ_ε beneath Ice Stream B, West Antarctica have been reported in the range 50±40 kPa (Blankenship et al., 1987); while beneath the Blue Glacier values of ≈1100 kPa have been described (Englehardt et al., 1978).

When the bed becomes locally 'drowned' (i.e. no contact points exist between the ice and its bed due to the presence of an incompressible water layer or film) effective stresses approach zero. Under these circumstances a state of potential ice mass instability is reached (Shoemaker, 1992). With the ice mass locally decoupled from its bed, shear stresses will also disappear and basal ice velocity should rapidly increase. However, as Weertman (1966, 1972, 1979) has demonstrated this unstable condition is self-limiting if the source of water is totally derived from geothermal and frictional melting at the base. As the water layer increases in thickness, production of continued meltwater as a result of melting due to ice-bed frictional energy will reduce. In the longer term to the consequent thinning of the water layer will occur and a return to stable conditions. Clearly, where water is obtained from other sources, especially the supraglacial environment or up-ice impounded meltwater, a state of instability due to bed asperity 'drowning' may lead to some form of catastrophic ice advance (Robin and Weertman, 1973), enormous subglacial flood discharges (Shaw, 1988a;

Shoemaker, 1992) or vast and sudden short-lived subglacial cavity enlargement and development (Walder, 1986). Weertman and Birchfield (1983a, b) have argued that any thick water film is unstable and may, over a very short time period, develop into a system of R-channels before collapsing back into a water film (Weertman, 1986).

The simple basal shear stress developed at the base of an ice mass can be defined as follows:

$$\tau_b = \rho_i \, gh_i \alpha \qquad (5.5)$$

Basal shear stresses are found to range from 50–150 kPa (Paterson, 1981; Andrews, 1987a, c; Robin and Swithinbank, 1987) and often an average value of 100 kPa is assumed in calculations. However, under fast moving ice streams, outlet glaciers and probably basal zones of ice masses where deformable debris layers exist, values as low as 4–20 kPa have been suggested (Mathews, 1974; Boulton and Jones, 1979; Fisher et al., 1985; Shreve, 1985b; Beget, 1986; Blankenship et al., 1987; Boulton and Hindmarsh, 1987; Ridky and Bindschadler, 1990).

The impact of shear stresses, over a small area of a few square metres or across minute asperities on the bed, can be considerable. Generally, both tensile and compressive stresses much higher than the average are involved with one or other dominating in any one given instant or location. The effect of these stresses on bed topographic hummocks, steps, joints and fracture plane geometry largely determines the resultant glaciated landscapes (Addison, 1981; Röthlisberger, 1981; Sugden and John, 1981). Where the shear strength of stressed materials is overcome, material failure ensues and erosion of the material results. Material may, as a consequence, become either entrained within the ice (englacial) or as a tractive load (subglacial), or squeezed (intruded or extruded), or pushed into cavities, depressions, cracks of underlying materials.

For example, as ice moves across a bed protuberance normal effective pressure fluctuations occur. A higher than average normal stress occurs up-ice and a reduced stress in the immediate lee of the obstacle (Fig. 5.4) (Boulton, 1974, 1982; Morland and Boulton, 1975). It can be shown that the stress

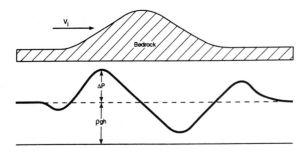

FIG. 5.4. Idealized normal pressure distribution over a glacier bed as ice flows over a bed hummock (after Boulton, 1974).

fluctuation ($\Delta\sigma_p$) across a sinusoidal protuberance is as follows:

$$\Delta\sigma_p = ((a_p/\lambda)^2 \, 2\pi^2)^{1/2} \, \eta u_b K_* \qquad (5.6)$$

where a_p is sinusoidal bed amplitude, λ is wavelength, η is ice viscosity of $\approx 100^{-1}$ kPa and K_* is a material constant of ~ 0.1 cm^{-1}. As the ice moves across the protuberance a cavity may or may not develop as a function of ice velocity, viscosity and the dimensions of the protuberance.

(a) No Cavitation

The normal pressure fluctuation can therefore be defined as:

$$\Delta\sigma_p \approx \frac{10\eta u_b}{\lambda} \qquad (5.7)$$

assuming a wavelength: amplitude ratio of 4:1 and wavelength >10 m. Compressive and tensile stresses acting upon the protuberance facilitate erosion.

(b) Cavitation

The relationship between basal ice velocity and ice overburden normal stress changes with increasing velocity or decreasing ice thickness, or where reduction in obstacle size (low bed roughness) cavitation occurs (Fig. 5.4). When cavitation occurs

$$\frac{\Delta\sigma_p}{\rho_i g h_i} > 1 \qquad (5.8)$$

and the normal pressure fluctuation will become:

$$\Delta\sigma_p = (\Delta\sigma_p + \rho_i g h_i) \qquad (5.9)$$

As a result, tensile stresses, due to ice adhesion, approach virtually zero, only affecting a small area of

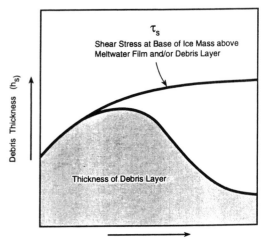

FIG. 5.5. The effect of bed decoupling on the thickness of a deforming layer as both meltwater production and shear stress at the Upper interface of the deforming layer increases (after Menzies, 1989; reproduced with permission from Elsevier Science Publishers BV).

the protuberance at the initial point of ice–bed separation. It is at that point that maximum erosive energy may operate (Chapter 7).

This 'idealized' example demonstrates the complex stress history that exists at the ice–bed interface. In reality, the influence of subglacial water, patches of unconsolidated materials of varying rheology, local thermal fluctuations, discontinuities in bedrock and deposited materials, adjacent stress fields and anisotropies within the ice such as debris bands and extraneous stress fluctuations must be considered in totality if local ice–bed interface conditions are to be understood.

5.3.2.1. Stress Conditions and Bed Interfaces

In examining the ice/bed interface, the nature of the interface boundary must be evaluated. One or several interfaces may exist simultaneously at the bed of an ice mass (Chapter 6) (Shoemaker, 1986; Menzies, 1987, 1989a). The stress fields at these various interface boundaries are complicated since stress fluctuations across boundaries are both temporally and spatially transient (Shoemaker, 1986;

PLATE 5.2. (a) Basal interface between glacier ice and bed, note shear planes within debris-rich ice at base of glacier inclined at approximately 45° to ice–bed interface, ice moving from left to right. Large unit of ice-rich debris containing tight chevron-folds detached from moving glacier to the right of cavity (unnamed glacier Alberta/British Columbia border, Rocky Mountains). Scale shown by embedded clast in the ice at top left-centre measuring approximately 18 cm. (b) Basal dirt bands in Meserve Glacier, Wright Valley, Antarctica. Glacier bed at the base of hammer shaft (photograph courtesy of Gerry Holdsworth).

Boulton and Hindmarsh, 1987; Brown *et al.*, 1987; Menzies, 1987; 1989a; Ridky and Bindschadler, 1990). An example of the complex potential relationship between stress and interface boundaries is shown in Fig. 5.5.

5.4. THE FLOW OF ICE

Ice flows as a viscoplastic material under the influence of gravity, and is opposed by boundary interface friction. Although the gravitational effects are well understood, the impact of longitudinally transmitted frictional influences such as bed drag and side-wall friction remain inadequately understood (Budd, 1970a; Hutter, 1983; van der Veen and Whillans, 1989). Ice flow can be studied from mechanical or thermo-mechanical perspectives. At present, the former approach dominates glaciological research but greater advancement would seem possible from a thermo-mechanical understanding of ice flow processes (Hutter, 1983; Lliboutry, 1987b).

Ice flow differs between ice sheets, valley glaciers and ice shelves, viz: ice sheet flow is unconfined by topography or competing ice mass units; valley glaciers are confined and side-wall friction is of major importance in understanding rates of flow. Ice shelves are confined within coastal embayments but overall have zero basal friction. There are exceptions to the above separation; for example, within ice sheets, fast moving ice streams are influenced by side-wall friction due to slower moving adjacent ice. During surges of an ice sheet or valley glacier basal

friction values, at least locally, may approach zero, and where piedmont glaciers emanate from confining mountainous terrain, a valley glacier may have little side-wall friction. Further distinctions that can discriminate all three types of ice mass are variations in mass balance, response time, and the location and possible spatial fluctuations of the equilibrium line. The physical principles of ice flow for all three are identical; only the overall behaviour, growth, surface morphological expression, hydrological characteristics, sediment entrainment mechanisms, subglacial processes and proglacial processes and forms vary in frequency and magnitude depending upon type and location of ice mass.

5.4.1. Extending and Compressing Flow

The terms extending and compressive flow were applied by Demorest (1943) to describe the longitudinal flow of the Greenland Ice Sheet. As glacier ice moves under the influence of gravity and the internal yield strength of the ice is overcome, a longitudinal strain rate develops. The strain rate gradually increases through the accumulation area to a maximum at the equilibrium line beyond which, in the ablation area, the effects of ablation cause the overall ice body to lose mass and the strain rate to decelerate toward zero at the glacier snout. This characteristic longitudinal attenuation and reduction is termed extending and compressing flow, respectively (Hutter, 1983).

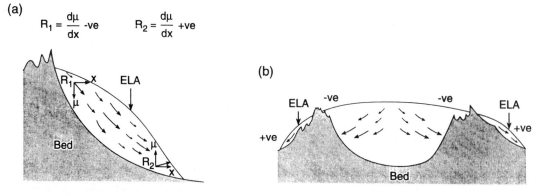

FIG. 5.6. Models of a valley glacier (a) and ice sheet (b) showing flowlines and longitudinal strain both positive — extending flow — in accumulation areas, and negative — compressive flow — in ablation areas. (R_1 and R_2 longitudinal strain in accumulation and ablation areas, respectively; du and dx flow vectors in vertical and horizontal directions, respectively; and ELA — the equilibrium line altitude.

The appearance at the margins of valley glaciers of the corpses of mountain goats, and, infrequently, ill-fated mountaineers and, recently, a lost medieval traveller all of whom have gone astray and/or fallen by accident into crevasses high in the accumulation areas of Alpine glaciers in Europe atest to the slow efficacy of ice flow pathways (Ambach *et al.*, 1991).

In the accumulation area, this net downward ice movement, relative to the surface, results in tensile stresses developing within the ice mass and the occurrence of extending flow (Nye, 1957; Budd, 1970b; Raymond, 1971, 1980; Paterson, 1981). This flow condition manifests a positive strain rate increasing to a maximum at the equilibrium line (McMeeking and Johnson, 1985; Reeh, 1988) (Fig. 5.6).

In the ablation area where net movement is upward, compressive stresses develop and compressive flow occurs. Strain rate becomes negative decreasing from a maximum at the equilibrium line to zero at the glacier terminus (Fig. 5.6).

Typical longitudinal strain rates are of the order 10^{-1} year^{-1} in temperate valley glaciers to 10^{-5} year^{-1} for the central parts of Antarctica (Paterson, 1981). However, Müller and Ritter (1988) illustrate the enormous variation in strain rate that occurs in the Ekström Ice Shelf, Antarctica based upon geodetic measurements.

One of the striking features of surging glaciers is the sudden increase in longitudinal strain rates immediately prior to surge initiation (McMeeking and Johnson, 1985; Raymond, 1987).

The importance of these flow types and stress states is two-fold. First, under extending flow conditions, ice movement is generally toward the glacier bed while under compressive conditions ice flow is away from the bed. At the macroscale these directions of ice movement have important ramifications for subglacial depositional and erosional processes. Where ice is moving toward the bed, conditions suitable for melting and debris release from the ice occur. When ice flow is away from the bed, processes of regelation and debris entrainment can potentially result. These flow processes, when considered in a local context, allow for sediment distribution and re-distribution at the ice–bed

interface. Secondly, these flow conditions permit the differentiation of long distance transport pathways of glacially-entrained sediment (Boulton, 1978; Aylsworth and Shilts, 1989). This fact has important ramifications for clast provenance recognition and drift prospecting (Menzies, 1995b, Chapters 15 and 16), and, in part, for the reconstruction of past ice sheet dimensions and the positions of ice divides and termini (Fisher *et al.*, 1985; Fulton and Andrews, 1987) (Chapter 2).

5.5. THE FLOW OF ICE SHEETS

Ice rheological behaviour can be subdivided into thermal and mechanical components. At present this artificial division is utilised for simplicity sake but, as our understanding of ice masses and their polythermal nature increases, there is a growing need for the development of an overall thermo-mechanical approach (Hutter, 1982b; 1983; Hutter *et al.*, 1988; Blatter and Hutter, 1991). Ice masses flow across the Earth's surface as a result of three processes all of which may occur independently of each other or in concert. These mechanisms are: (a) internal deformation; (b) basal slip or sliding; and (c) ice mass movement due to a deforming bed. The first two processes are well established and discussed in the literature; the third process is of recent derivation and its pervasiveness or otherwise remains to be established. In any one ice mass all types of flow are possible either in different parts of the ice mass or at different times depending upon basal thermo-mechanical conditions, topographic constraints and seasonal variations (Hughes, 1977; Hutter and Engelhardt, 1988). These variations of flow mechanisms and regimes have critical ramifications for glacial sedimentological processes. Considerable research work into these interrelationships between ice flow and sedimentology remains to be done.

5.5.1. Ice Sheets and Bed States

As will be discussed in detail when considering subglacial sedimentological processes (Chapter 11; Menzies, 1995b, Chapter 2), bed states are fundamental not only to the subglacial processes likely to take place at any specific locality beneath an ice mass but also in the probable mode of glacier movement. Essentially, ice masses can be subdivided into those which have Hard or Rigid Beds and those with Soft, Mobile or Deformable Beds (Hutter and Engelhardt, 1988; Menzies, 1989a, b). In the former case, it is assumed that the glacier bed beneath the ice is rigid, smooth with zero permeability and is undeformable. This case is the one most often modelled in the past. These beds are likely to be composed of unfractured bedrock of high intact shear strength or, if sediment-based, to be frozen and again impenetrable to meltwater (H-beds). Under these conditions, if the ice is temperate, meltwater will move along the single basal interface and thus motion of the ice mass is likely to be by basal slip. In the case of polar ice masses, the ice will adhere to the bed and internal deformation is likely to account for glacier motion. Where soft Mobile beds occur (M-beds), meltwater can penetrate the underlying sediment to a lesser or greater degree and the debris may be mobilised such that ice motion is largely accounted for by this underlying mobile bed. In this case both temperate and polar ice (Echelmeyer and Wang, 1987) conditions seem possible with limited upper interface meltwater present. Unlike the H-bed situation, several interfaces may temporally exist beneath the ice mass and act as shear zones to carry the pervasive/or non-pervasive tangential shear stress imparted to the mobile sediment by the overlying ice mass (Menzies and Maltman, 1992). It is also likely that beneath any ice mass both bed types may exist at differing times but in the same location. Under these, perhaps more common conditions, a Quasi-Hard/Mobile Bed state prevails (Q-beds) where all three forms of glacier motion may periodically occur.

5.5.2. Ice Sheet Models

In the past twenty years several ice sheet models predicting growth rates (Chapter 3), sliding velocities and surface morphological development have emerged (Paterson, 1980, 1981; Raymond, 1980). Each model has tried by the inclusion or rearrangement of specific controlling parameters to improve our understanding of ice sheet 'evolution' (see Raymond, 1980; Paterson, 1981; Hutter and Engelhardt, 1988; Maohuan et al., 1989). Hindmarsh

et al. (1989) have proposed a numerical model that considers the temperature field (Robin, 1955) by attempting to introduce thermal advective, diffusive and conductive effects over a long time period to model ice sheet growth and decay (Hutter, 1983; Morland, 1984; Morland et al., 1984). Several important implications emerge from this model viz: (a) ice sheet growth and decay; (b) sedimentological implications; and (c) general indications of ice sheet dynamics.

(a) Ice sheets will expand until the area within the ablation is equal to that of the accumulation area. Ice velocity in the ablation area tends to be much higher than in the accumulation area by an order of magnitude; therefore, it is concluded that the areal extent of the accumulation area is probably a magnitude in area greater than that of the ablation area. As a consequence the ablation area is sensitive to the geometry of marginal areas where the greatest ablation occurs. There is considerable evidence that basal movement varies between the inner zones of an ice sheet and the outer margin where indications point to the likelihood of deformable beds (M-beds). One problem with this analysis is that high dissipation of warm ice should result in an unfrozen marginal area. The possibility of proglacial marginal lakes acting as insulators preventing freezing may partly explain frontal ice sheet dynamics. Evidence of deforming sediments over-running frozen ground may give credence to this concept; however, other researchers find evidence which points to a frozen submarginal area (Habbe, 1989; Menzies, 1990a, b; Menzies and Habbe, 1992). Clearly, the kinematical sensitivity of ice sheet margins cannot be explained by reliance upon equations governing ice sheet development since any perturbation within the ice sheet, especially within the accumulation area, may be greatly amplified at the margin (Kuhn, 1984; Morland et al., 1984; Oerlemans and van der Veen, 1984; Oerlemans and Hoogendoorn, 1989). At present, there is no verifiable universal sliding law that can be used successfully to define an ice sheet profile and associated basal temperature spatial mosaic (Morland et al., 1984). The implication of this statement suggests that great care must be taken, depending upon which sliding relationship is employed, in developing general assertions regarding ice sheet development, and their climatic and marginal sensitivity (Chapter 4).

(b) As our understanding of glaciodynamics increases, it is evident that the sensitivity of ice sheet margins to minor perturbations in the accumulation area is of great significance. The sedimentological evidence from marginal areas should reveal considerable clues as to overall ice sheet dynamics. The sediment structures and glaciotectonics of ice sheet margins contain elements indicative of thermal conditions at the margin that carry information regarding ice sheet profiles, thermal and stress fields.

(c) Hindmarsh et al. (1989) suggest that continental ice sheets have temperature and conductive heat flow of ≈105 a. Dissipation is important in marginal areas, while advective processes occur dominantly in the upper portion of the divide area and conduction in basal divide areas. While shear heating is an important, if not a pervasive, aspect of all ice sheets since heat conduction is very slow, mechanical adjustments by the ice mass are relatively insensitive to this effect. The implication of this shear-heat phenomenon is that this effect is not likely to act as a source of potential instability to explain ice sheet/valley glacier catastrophic collapse and/or surging (Clarke et al., 1977; Yuen and Schubert, 1979; Fowler, 1980; Schubert and Yuen, 1982; Yuen et al., 1986). An opposing view advocates the feasibility of ice sheet instability being induced by sudden shear-heating. Shear-heating occurs as heat generated by internal deformation of ice causes frictional energy to be produced. This heat energy is expended as sensible heat within the ice mass, thereby affecting the strain rate of ice, as demonstrated by Glen's Flow Law. As the ice continues to deform, the quantity of frictional heat produced creates a positive feedback mechanism allowing ice to internally deform at an ever faster rate, thus producing more frictional heat. Finally, a point of explosive (catastrophic) 'overheating' is reached leading to massive internal deformation, thermodynamic instability and short-term ice mass collapse. In Yuen et al.'s analysis (1986), it is assumed that no horizontal cold ice advection or fluctuations in basal shear stress due to local thermal effects occur. Both effects would tend to act as counterbalances to shear-heating (to an unknown

extent) and maintain stability, as noted above. By combining equations for temperature evolution within a one-dimensional ice sheet model (Eqn. 5.10) with the down-slope velocity of an ice sheet as derived from an analysis of the creep law of ice (Eqn. 5.11), the relationship between thermal change and ice flow can be noted:

$$\frac{dT}{dt} = \kappa \frac{d^2T}{dy^2} - u(y)\frac{dT}{dy} + \frac{2A}{k_{tc}}[\rho_i \; g \; (h_i - y)\alpha]^4 \; \exp\{-\frac{E^*}{RT}\}$$
(5.10)

$$\frac{du}{dy} = 2A[\rho_i g(y - h_i)\alpha]^3 \; \exp\{\frac{-E^*}{RT}\} \quad (5.11)$$

where **dT** is the change in temperature; k_{cc} the thermal conductivity; κ is the thermal diffusivity, **E*** the activation energy for the creep of ice and R is the universal gas constant. Yuen et al. (1986) suggest that a sudden explosive shear-heating instability is possible if an ice sheet, such as the Eastern Antarctica, experiences a rapid increase in accumulation due to sudden climatic deterioration. Rapid shear–heating instabilities appear to be favoured where low activation energies and large exponential factors in the flow law prevail; thus instability is possible where high levels of ice ductility occur. The validity of this analysis in realistic ice sheet dimensions seems debateable but, as a contributing factor in fluctuations of ice sheet flow, this mechanism requires consideration and further research.

5.5.3. Internal Deformation

As snow transforms to glacier ice, the ice begins to internally deform under the influence of gravity. The movement due to gravity within an ice mass is described typically, as a downward concave flowline within the accumulation area and upward concave beyond the equilibrium line, in the ablation area (Fig. 5.6). In order for ice to move, stresses must be generated that first overcome the shear strength of its own internal crystalline structure (yield strength) (Fig. 5.2). From that point, beyond the yield strength of the ice, the ice deforms under stress as a viscoplastic material (Michel, 1978; Paterson, 1981; Drewry, 1986; Lliboutry, 1987b; Lock, 1990). The process of deformation within glacier ice has been observed to consist of a series of intercrystalline dislocations (Webb and Hayes, 1967; Glen, 1975; Lliboutry, 1987b). Since glacier ice crystals are anisotropic, irregular, with lattice defects, slippage of planes of atoms over each other is possible at even very low applied stresses. This process of slippage is termed creep. Creep occurs in two forms: (a) creep relatively insensitive to confining pressure, and (b) creep that is prevented by high confining pressures. The former type involves microscopic inter- and intracrystal slippage and is termed dislocation creep or climb (discussed above); the latter concerns microcrack growth, propagation and healing. Where stress levels are too high or confining pressures are too low, healing of the cracks either does not occur or a disequilibrium between new crack growth and healed cracks occurs and brittle creep takes place (Fig. 5.7). Initially, dislocation movements appear to be few but gradually increase by a positive feedback mechanism leading eventually to a 'piling-up' of dislocations possibly at 'point/edge defects' and a hardening of the ice (Paterson, 1981). Some further mechanisms, as yet not fully understood, allow dispersal of these dislocations and continued deformation. Since glacier ice is polycrystalline allied to deformation due to dislocations, movement also results from crystal boundary migration, recrystallization at micro-stress points within the ice structure, crystal to crystal rearrangement and neo-crystal growth. It is likely, however, that dislocation remains the most critical of the causative deforming mechanisms.

Within secondary deformation creep (Fig. 5.2), it can be shown that glacier ice deforms approximately to the following mathematical expression now commonly termed Glen's Flow Law such that:

$$\dot{\varepsilon} = A\tau^n \tag{5.12}$$

where $\dot{\varepsilon}$ is the average uniaxial strain or creep rate, τ is the internal shear strength of polycrystalline ice, n is a constant and A a constant depending upon temperature, ice crystal size, orientation and debris/impurity content, and confining pressure. Considerable debate exists as to the values of A and n, the latter parameter varies from 1.5 to 4.2 (Weertman, 1973) to a mean or around 3 (Paterson, 1980, 1981; Raymond, 1980; Hutter, 1983; Budd and Jacka, 1989). An illustration of this variability of parameters in the flow law is shown in Fig. 5.8. In attempts to find a more exact solution of ice flow

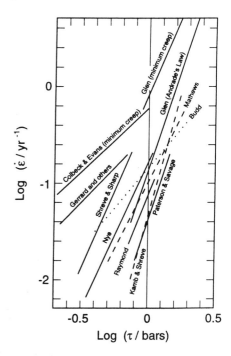

FIG. 5.8. Relationship of shear strain rate and effective shear stress for ice from laboratory experiments (Glen, 1955; Colbeck and Evans, 1973); analysis of borehole deformation (Mathews, 1959; Paterson and Savage, 1963; Shreve and Sharp, 1970; Kamb and Shreve, 1966; Kamb, 1970; Raymond, 1973); surface velocity (Budd and Jenssen, 1975) and tunnel closure (Nye, 1953) (after Raymond, 1981; reproduced with permission of the author and Academic Press).

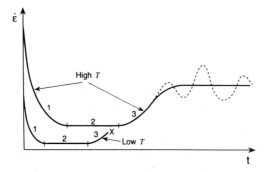

FIG. 5.7. Creep of ice under a constant load where T is temperature (after Lliboutry, 1987; reprinted by permission of Kluwer Academic Publishers).

within a temperate ice mass (i.e. at pressure melting point throughout) several differing 'best-fit' polynomial equations have been developed (Duval, 1978; Raymond, 1980, Fig. 4; Hutter, 1983; Lliboutry, 1987b).

In illustrating the influence that temperature has upon the flow of ice, Mellor and Testa (1969) have shown that ice between –10 and –60°C is subject to secondary creep, whereas between 0 and –10°C ice deformation is much more temperature dependant with grain boundary melting and regelation (Nye and Frank, 1973). At around 0°C, ice deformation is influenced more by grain growth and crystal re-orientation. Mellor and Testa observed that ice subjected to low confining pressures and thus low stress levels akin to the state of stress at the base of an ice mass, may exhibit a near linear stress/strain rate relationship. Duval (1977) has also shown that the strain rate of temperate ice is dependant upon the water content of the ice. The influence of water is thought to be a function of regelation processes occurring at ice grain boundaries.

The flow law of ice has two major relevant and interrelated thermo-mechanical aspects related to internal deformation. First, as indicated, ice movement is temperature dependant (Wolff and Doake, 1986). Thus, at the base of an ice mass where most vigorous internal flow occurs, the lower the temperature the slower deformation is likely to be and vice versa. As a consequence the impact of ice movement on subglacial processes is, in part, thermally controlled, thus affecting the efficiency and effectiveness of subglacial processes. However, there are two boundary conditions that operate in relation to ice movement and temperature viz: (1) at temperatures below pressure melting point it is generally assumed that the ice becomes cold-based and frozen to its bed, thus differential motion at the ice–bed interface ceases; and (2) if subglacial temperatures rise to an extent that basal melting rates produce large volumes of meltwater, basal topographic obstructions may become 'drowned' and frictional contact between the ice and the bed approaches zero.

Secondly, internal ice movement is influenced by confining tangential stresses. Where tangential shear stresses increase due to increasing ice thickness and/or flow constriction, an exponential increase in ice mobility occurs with the lower zones of any ice mass becoming increasingly mobile. This basal elevated mobility may translate into increased basal friction leading to high rates of melting at the basal interface. The process may lead to higher rates of sedimentation due to pressure melting at the base of the ice causing either increased debris melt-out or porewater production and decreasing basal mobile debris viscosity. However, this rudimentary single basal interface boundary zone may be over-simplified.

5.5.4. Basal Slip

In ice masses where ice is at pressure melting (temperate, warm-based), ice not only moves by internal deformation creep but by basal slip. The term basal slip characterizes the process whereby an ice mass slips over a very thin lubricating layer or sheet of meltwater, or series of interconnected water-filled cavities at the ice–bed interface (Lliboutry, 1979; Weertman, 1957b, 1964a, 1966, 1979, 1986; Morland, 1976a, b; Walder, 1986). Since the ice is at pressure melting, basal deformation results in basal melting and the production of lubricating films or water-filled channels. The ice mass therefore changes its boundary conditions from non-slip to perfect sliding (Hutter, 1983). In considering basal slip, two distinct models can be examined viz: (1) sliding without separation (no cavities), and (2) sliding with

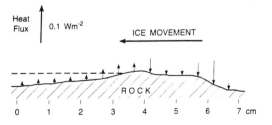

FIG. 5.9. Heat flux at the ice–bed boundary. It is assumed that ice motion is parallel to the dashed line indicating the approximate upper limit of regelation ice, therefore heat fluxes correspond to pressure melting on the up-stream face and regelation down-stream of the high point of the bed. Numerical values of flux are proportional to the length of the arrows. Heat flux toward the bed indicates melting and vice versa (after Robin, 1979; reproduced by courtesy of the International Glaciological Society from *Journal of Glaciology*, **16**(74), 1976, p. 192, Fig. 3).

separation (cavities). Basal slip is not a continuous movement but consists of a 'stick- slip' or jerky motion. This form of motion is attributable to several causes. First, obstacles occurring at the ice–bed interface may cause localized reductions in the pressure melting point and subsequent freezing and immobility. Secondly, if the thin film of water (only a few microns thick) over which the ice mass slips is removed either by freezing or drainage, localized 'grounding' of the ice mass results with subsequent increased longitudinal and transverse stresses and boundary interface friction (Goldthwait, 1963; Vivian and Bocquet, 1973). Thirdly, Robin (1976) and Goodman *et al.* (1979) have observed that as ice moves against the up-ice side of an obstacle a 'heat pump' effect is produced, and a 'cold patch' is formed at the ice–bed interface (Fig. 5.9). These 'patches' may account for the 'stick-slip' motion of temperate ice masses (Hindmarsh *et al.*, 1989).

Finally, Lliboutry (1968) and Walder (1986), among others, have pointed out that the formation or otherwise of basal cavities leads to marked changes in the value of basal friction, falling locally to zero where cavities form. Paterson (1980) summarizes the main input factors in basal sliding boundary conditions as follows: basal shear stress, bed roughness, normal pressure due to overlying ice, interstitial porewater pressures and perhaps other unknown factors, the result of the lack of accessibility in observing the process. Basal sliding also seems to be seasonally controlled, being often higher in the spring, summer and fall due to greater meltwater production (Meier, 1969; Hodge, 1974; Sugden and John, 1976; Iken, 1978). However, our understanding of what Weertman (1979) terms, "the unsolved general glacier sliding problem" remains (Hutter,

1982a, b, 1983).

The problem can be distilled into attempts to explain how ice, when not in intimate contact with a non-deformable bed, can slip when in the presence of cavities and a water film of finite thickness (Frank, 1976; Lliboutry, 1965, 1968, 1979; Robin, 1976; Hutter, 1983; Morland *et al.*, 1984; Hindmarsh *et al.*, 1989). This problem is compounded when attempting to develop theoretical solutions due to the complexity of the basal water-flow problem and the presence of debris in basal ice layers and/or bed protuberances.

5.5.4.1. Basal Sliding with no Separation (no Cavities)

When a temperate ice mass, in intimate contact with a non-deformable bed (Model #3, Table 5.1) of small bed slope variation, slips over that bed and moves past bed obstacles, two mechanisms appear to operate viz: pressure melting and enhanced plastic flow (Weertman, 1957, 1979; Nye, 1969b, 1970; Kamb, 1970; Lliboutry, 1976). As ice moves against the upstream side of an obstacle, high pressures develop leading to basal ice pressure melting and meltwater production. This meltwater flows around the obstacle, freezing on the downstream side where ice pressure is reduced (Fig. 5.10). This regelation process continues provided up-ice conduction of latent heat from the downstream (lee) to the upstream (stoss) side of the obstacle persists. Since ice deforms plastically, it can be further observed that on moving around and over an obstacle on the bed there is an increase in above average longitudinal stress within the ice that leads to an increase in the strain rate. Since the strain rate is directly related to ice velocity over distance, ice velocity will also locally increase. It follows that as the obstacle increases in size and the

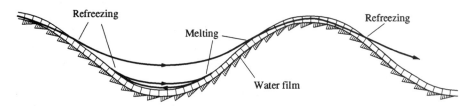

FIG. 5.10. Idealized model of flow lines in regelation ice and in the water film across a hummocky bed (after Lliboutry, 1979; reproduced by courtesy of the International Glaciological Society from *Journal of Glaciology*, **23**(89), 1979, p. 73, Fig. 1).

TABLE 5.1. Classification of various ice mass types related to thermal bed conditions, and the presence or absence of cavities and/or deformable bed conditions and the likelty bed motion of the ice mass

MODEL	ICE MASS TYPE	BED STATE	BED MOTION
1	Cold	Frozen to non-deformable	no basal slip
2	Cold	Frozen to deformable	no basal slip
3	Warm	P.M.P.[+] no cavities non-deformable	basal slip
4	Warm	P.M.P. no cavities deformable	basal slip
5	Warm	P.M.P. cavities non-deformable	basal slip
6	Warm	P.M.P. cavities deformable	basal slip

[+] **P.M.P. - Pressure melting point.**

thermodynamic system of melting and freezing deteriorates, enhanced plastic flow becomes more effective (Kamb, 1970; Duval, 1977, 1979; Lliboutry, 1986). Kamb has demonstrated that the characteristic wavelength of a bed is related to basal shear stress and ice viscosity in the form:

$$\lambda_{wr} = 2\pi (\frac{\eta}{\Gamma})^{1/2} \qquad (5.13)$$

where λ_{wr} is the characteristic bed wavelength, sometimes termed white roughness, or the transition scale where ice moves past bumps on the bed which are $<\lambda_{wr}$ by regelation and past bumps $>\lambda_{wr}$ by plastic deformation. The approximate value of λ_{wr} is $1/2$ m, and Γ, is a constant related to the regelation process (Weertman, 1957b; Kamb, 1986; Kamb and Echelmeyer, 1986a, b; Echelmeyer and Kamb, 1986). Work by Lliboutry (1979) and Iken (1981), however,

casts some doubt on the above equation. It has been shown that enhanced plastic flow would appear to be more quantitatively important than regelation (Hutter, 1983).

Under conditions of a bed in perfect contact with the ice (water-tight), basal sliding velocity can be related to basal shear stress as follows:

$$\tau_b = u_b (\Gamma \eta)^{1/2} \, m_*^2 \qquad (5.14)$$

where u_b is the basal sliding velocity and the parameter m_* relates to the spectral density of the bed in terms of the mean quadratic slope in the direction of bed slip.

Weertman (1957b) developed a simplified relationship of bed slip and basal shear stress in which the non-linear creep behaviour of ice was modeled. Using a stylised glacier bed with obstructions of known dimensions (**d**) at known distances apart and a Newtonian-fluid approximation, he developed an approximate equation as follows:

$$u_b = B_1 \tau_b^2 [1 + 10\frac{d}{d_c}], \quad d < 0 \ (d_c) \quad (5.15)$$

or

$$u_b \approx B \tau_b^{(n+1)/2} d, \quad d \geq 0(d_c) \quad (5.16)$$

where n is the power component in the Glen's Flow Law, d_c is the controlling obstacle size on an idealized glacier bed (Weertman, 1964) and B and B_1 are constants of bed 'sliding-softness' (Chapter 3). In Weertman's original hypothesis of bed sliding (1964) obstacles contribute to sliding resistance up to a maximum obstacle size (controlling size). The controlling size is usually taken as $d_c = 0$ (1–10 mm) and $B \approx 10B_1/d_c$.

Considerable heated debate surrounds the appropriate measure of bed roughness and the magnitude of B. (Alley (1989a) reports values of $B \approx 2 \times 10^{-6} Pa^{-2}m^{-1}$ for Ice Stream B, West Antarctica.) Perhaps the greatest flaw in these theoretical equations is the fact that bed roughness, basal sliding velocity and shear stress have not been measured in any one locality beneath an active ice mass. Kamb (1970), in an attempt to test these equations, used estimated values and a deglaciated section of bed. It was found that where basal velocities were > 20 m year⁻¹, bed roughness was underestimated by a factor of 2–4 times (Morland, 1976a, b; Weertman, 1979; Kamb, 1987; Alley, 1989a, b). In all of these non-separation sliding equations friction at the bed and the non-linear response of the ice is ignored (Morland et al., 1984). However, in reality, friction plays an important role and will reduce sliding locally and thereby alter the configuration of basal shear stress geometry and thermal conditions across the bed. Both of these effects have major impacts upon the effectiveness of erosional, entrainment and depositional processes at the bed by influencing and developing, for example, local stress concentrations and meltwater passageways

Fowler (1979a, b, 1980) considered an alternate approach to the sliding problem by assuming two levels of bed roughness: the bed asperity (α_{sp}) roughness defined as the wavelength {χ} relative to

ice depth (h_i), and the bed roughness slope (s_r), defined as the roughness amplitude {y} relative to wavelength {χ}. Ice flow was viewed at two scales: (1) an upper flow over a smooth bed, and (2) a lower flow within a boundary layer defined by the roughness wavelength. Ice flow is then equated as the sum of both flow conditions where $h_i >> \{\chi\}$ and the equality $h_i \to 0$ cannot be satisfied. The following equation assumes that ice satisfies Glen's Flow Law and perfect slip is achieved:

$$u_b = [a_{sp} s_r^{-(n+1)}(tauHxUBb /R_u)^n] \quad (5.17)$$

where R_u is an order-unity roughness parameter independent of α_{sp}. This equation can be re-written to allow variations in ice thickness from ice mass centres to margins (after Morland et al., 1984) as follows:

$$\tau_b = \lambda_f(h_i)B(u_b) \quad (5.18)$$

or, following Morland et al.'s analyses of a steady state sliding law (Boulton et al., 1984; Hindmarsh et al., 1989), the above sliding law takes the form:

$$\tau_b = \lambda_f(\sigma_b)u_b^{1/m} \quad (5.19)$$

where σ_b the ice overburden pressure and λ_f is a 'frictional' parameter dependant upon ice thickness, overburden pressures and bed form and conditions.

Based upon data from the Greenland Ice Sheet and Devon Island Ice Cap, Morland et al. (1984) have found that the friction parameter exhibits wide variations indicating, perhaps, that bed sliding may not be solely accounted for by direct bed slip. The essential derivative of this above equation is that in the long-term ice behaves as a rheologically incompressible non-linearly viscous material not as a perfect viscoplastic material in which ice flow perfectly adjusts to changes in net balance without a change in basal shear stress (Weertman, 1979).

In the Weertman model a highly coupled ice-bed system acts in a decoupled manner — a condition that observation of present-day ice masses repudiates. In a model of non-linear viscous ice, in contrast, it can be demonstrated that as net balance gradient increases, unless there is a commensurate increase in net balance, a greater basal shear stress is necessary to force a larger discharge through the system (Boulton et al., 1984; Oerlemans, 1989; Oerlemans and Hoogendoorn, 1989).

5.5.4.2. Basal Slip with Separation (Cavities)

In studying glacier velocities and from subglacial interface observations, it has become apparent that many ice masses slip along their basal interface as a consequence of the presence of meltwater flowing either in sheets or within cavities (Lliboutry, 1968, 1976, 1979; Iken, 1981; Bindschadler, 1983; Walder, 1986; Ridky and Bindschadler, 1990). However, it is not the cavities themselves that cause enhanced basal sliding but the effect of meltwater in these cavities, cavity systems or other forms of meltwater passage. Debate surrounds the question of meltwater: is it the volume of meltwater or the hydraulic pressure exerted by the meltwater that causes the movement? How does meltwater evacuate from beneath an ice mass — as an ultra-thin sheet or film; in interconnected conduits and cavity systems which may be partly ephemeral, partly permanent; or totally or partially via a complex boundary interface in which a permeable substrate may act as a major conduit (Shoemaker, 1986, 1992; Willis et al., 1990)?

Bed separation can only occur, no matter what the circumstance, when the normal stress distribution (P_0) across the bed drops below a minimum level. Such situations occur down-ice of bumps or obstacles on the bed where stress distributions are, in general, below the mean stress level defined as:

$$\bar{\sigma}_b = (\rho_i g h_i)\cos\alpha + \rho_a \qquad (5.20)$$

where ρ_a is the average compressive stress.

Discussion of basal slip, under active ice conditions, with separation from the bed by moving basal ice must consider several distinct aspects of bed morphology viz: (a) movement where cavities of various types are present; (b) movement as a consequence of sheet flow; or (c) sheet flow in concert with cavities.

Cavities exist beneath ice masses in two forms: either as discrete (autonomous) cavities at the base of active ice or within the glacier bed; or as elongated cavities that are best described as conduits or channels. In the latter case the channels may extend from some distance beneath the ice mass to the terminus or lateral margin or they may be pinched out after only a short distance beneath the ice. The channels incised up into the ice are termed R channels

(Röthlisberger, 1972) and those cut down into the glacier bed N channels (Nye, 1976).

If one considers a discrete cavity at the ice–glacier bed interface (Fig. 5.11), it can be observed that a marked stress distribution exists across the cavity from the up-ice point of contact (A) where high tensile stresses occur to point B in the ceiling of the cavity where zero stresses exist to the point of closure at C where high compressive stresses occur as the ice re-joins the glacier bed. Since cavities commonly tend to form in the lee of obstacles on the glacier bed, water frequently is present. Some cavities may be partially or completely filled by water while others are empty or contain varying degrees of regelation ice (Lliboutry, 1976; Vivian and Bocquet, 1973; Theakstone, 1979; Weertman and Birchfield, 1983a; Kamb et al., 1985). Any cavity beneath an active ice mass is subject to closure by the overlying ice and only the presence of meltwater, the velocity of the ice or infilling by saturated basal debris and other restraints upon cavity closure will maintain cavity size. In general, it is assumed that all cavities vary in overall dimensions continuously and on a distinctly seasonal basis.

In considering basal cavities that are infilled with meltwater, two extreme or limiting conditions can be examined for Hard bed states (Model 5, Table 5.1) (Lliboutry, 1976): (1) Type A — Cavities with an autonomous hydraulic regime; (2) Type B — Cavities with a connected hydraulic regime.

In those cavities (Type A) that have an autonomous regime, water circulation does not occur with other parts of the subglacial hydraulic network. The cavities are isolated (confined/autonomous) and the water adjusts to hydraulic pressure values in equilibrium with basal ice stresses and motion. The importance of Type A cavities is difficult to assess but, in general, it seems likely their impact on overall glacier motion and ultimately upon the glacier bed and subglacial processes is limited.

In Type B cavities, water pressures will vary considerably since interconnection exists with all or some parts of the subglacial hydraulic system(s) and glacier bed. These cavities, as can be seen from deglaciated areas, have an enormous impact upon glacier bed morphology and subglacial processes (Walder and Hallet, 1979; Hallet and Anderson,

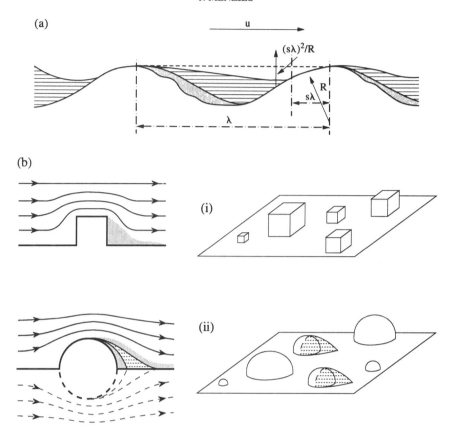

FIG. 5.11. (a) Sliding with cavitation at large basal ice velocities. (b) Bed interface models of basal ice sliding motion: (i) Weertman, 1957; and (ii) Lliboutry, 1978 (after Lliboutry, 1987; reprinted by permission of Kluwer Academic Publishers).

1980). The growth or closure of the cavity is influenced therefore by melting rates, hydraulic discharge and pressure variations (Eqn. 5.21). A cyclic pattern of evolution may exist between both cases. In the connected cavities slow infill by regelation ice or saturated debris may occur to a point where the cavity ceases to exist and becomes refrozen into the moving ice. Subsequently a new cavity (autonomous) will form in the lee of the obstacle and slowly this new cavity will develop, first in isolation and finally as a part of the larger subglacial hydraulic regime (connected). The transition from isolated Type A to connected Type B may have a 'short-circuit effect' triggering a sudden and rapid ice motion. Such a glaciodynamic 'jump' may be instrumental in surging (Kamb et al., 1985; Walder, 1986) (Section 5.10). The reverse transition may also explain the termination of a surging state.

The growth and shrinkage of cavities coupled with subglacial water pressures influences basal ice motion in a variable (non-steady-state) manner (Lliboutry, 1979; Fowler, 1987). Walder (1986) has demonstrated that cavity closure is a differential function of the normal stress applied by the moving ice to close the cavity and the opposing effect of melting of the ice attempting to keep the cavity open. This function can be expressed as:

$$\dot{u}_e = h_{bc} \left(\frac{\sigma_e}{nA} \right)^n - (1-\gamma) \frac{QGv_i}{P_c L_h} \qquad (5.21)$$

where $\dot{u}_e$ is the effective rate of cavity closure, h_{bc} is the height of the backwall of the cavity, σ_e is the effective stress due to overlying active ice, the factor

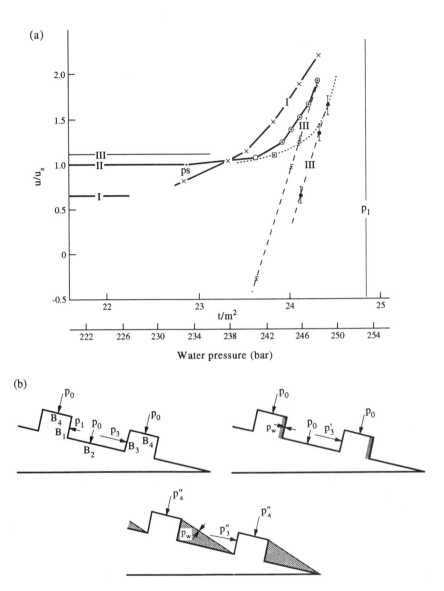

FIG. 5.12. (a) Relationship of basal ice sliding velocity against water pressure. u/u_s is the x-component of sliding velocity, the index s refers to (steady state) sliding on a sinusoidal bed at the separation pressure, u/u_s is the x-component of sliding velocity expressed as a fraction of the sliding velocity on the sinusoidal bed at the separation pressure. p_1 is the limiting basal water pressure at which sliding becomes unstable. Curves I,IV show initial, transient sliding velocity at the instant when water pressure is applied (except for the horizontal parts and two parts on lines III and IV which show steady state sliding). The dotted line marks steady-state sliding, below this line basal cavities are shrinking (after Iken, 1981; reproduced by courtesy of the International Glaciological Society from *Journal of Glaciology*, **27**(97), 1981, p. 417, Fig. 8). (b) Idealized basal interface stress distributions (basal ice motion from left to right). (i) Stress distribution before water at a given pressure (p_w) has access to lee faces; (ii) where water pressure is greater than normal ice overburden pressure (p_i) [$p_w > p_i$]; and (iii) cavity formation (after Iken, 1981; reproduced by courtesy of the International Glaciological Society from *Journal of Glaciology*, **27**(97), 1981, p. 418, Fig. 9).

$(1 - \gamma) \approx 0.68$ is a correction for the energy needed to maintain meltwater at the pressure melting point, Q is the volumetric flux (discharge) of meltwater through a cavity, G is the hydraulic gradient, v_i is the specific volume of ice, P_c is the cavity roof perimeter and L_h is the latent heat of fusion of ice per unit mass.

Iken (1981) has shown that sliding velocities are larger when cavities are growing than when a temporary steady-state size is reached for a given water pressure (Fig. 5.12). As cavity size shrinks, decreasing sliding velocities were observed to occur. In contrast, increasing sliding velocities were observed to occur and be manifested by the ice mass as a whole 'rising off' its bed due either to increased basal film thickness or to high basal water pressures associated with substantial reduction in cavitation over large areas of the bed (Iken, 1981). This unstable higher velocity condition appears to occur as basal shear stress tends toward zero and basal friction becomes extremely low (Raymond, 1980; Paterson, 1981; Lliboutry, 1987b).

5.5.4.3. Basal Slip Models and Polythermal Bed Conditions

Modelling the relationship of glacier bed hydraulics to aspects of basal ice velocity, basal shear stress and bed roughness is complicated further by the introduction of the influence of polythermal bed conditions (Blatter, 1987; Blatter and Kappenberger, 1988; Blatter and Hutter, 1991). Monothermal Models 1, 2, 4 and 6 (Table 5.1) are complex to model mathematically. They can be viewed as variants of either sliding without cavitation (Model 3) or with cavitation (Model 5) depending upon the competency or otherwise of the glacier bed, bed topography and basal hydraulic conditions. To these models must be added the possibility of polythermal bed conditions (Robin, 1976; Fowler and Larson, 1978; Goodman et al., 1979; Hughes, 1981; Hutter and Engelhardt, 1988).

In introducing more realistic bed conditions, the complexity of basal ice–bed rheological influences increase. The possibility that thermal, geotechnical (deformable, non-deformable) and hydraulic conditions may vary markedly across the glacier bed introduces geological conditions hitherto unconsidered. In the past there has been an understandable but unfortunate separation of mechanical and thermal models of ice behaviour

(Hutter, 1982a, b). With the development of thermo-mechanical models an approximation of reality can, perhaps, be achieved. Linked with this new development, a better understanding of polythermal ice masses may be obtained (Fowler, 1979b; Hutter, 1983; Hutter and Engelhardt, 1988). Our knowledge of polythermal ice masses is scant as yet, such that testing of theoretical models must lag behind for want of relevant field data (Lliboutry, 1987b; Blatter and Hutter, 1991).

5.5.5. Glacier Movement and Bed Deformation

In 1974, Boulton et al. reported a simple experiment, carried out beneath an Icelandic glacier, in which differential movement within subglacial sediment was observed. Building upon this work, Boulton and Jones (1979) suggested that some proportion of the forward motion of ice sheets might be accounted for by the existence of a deforming subglacial debris layer (a soft or deformable bed) (Engelhardt et al., 1978; Jones, 1979, Haerbli, 1981). Motion may be entirely by advective deformation within the subglacial debris layer or by 'ploughing', a process that is intermediate between basal slip on a lubricated interface and debris deformation (Brown et al., 1987; Alley, 1989a,b). Beneath Breið-amerkurjökull, Iceland, Boulton et al. noted that 88% of the basal movement was due to deformation and only 12% could be accounted for by slip at the ice–bed interface and internal ice deformation. Since then considerable evidence in the form of radar images from beneath Ice Stream B, West Antarctica where movement by the ice stream may be 100% accountable to bed deformation (Alley et al., 1986, 1987a,b; Blankenship et al., 1986, 1987), and observations beneath present-day ice masses at Storglaciären, Sweden (Brand et al., 1987); Urumqui Glacier, China (Echelmeyer and Wang 1987); the Variegated Glacier, the Matanuska Glacier, the Columbia Glacier, Alaska (Harrison et al., 1975; Engelhardt et al., 1978; Kamb et al., 1985; Fahnestock and Humphrey, 1988; Meier, 1989; Lawson, pers. commun. 1990), the Blue Glacier, the South Cascade Glacier, Washington State, United States (Englehardt et al., 1978; Hodge, 1979); and the Trapridge Glacier, Yukon Territory, Canada have noted or inferred the presence of deforming

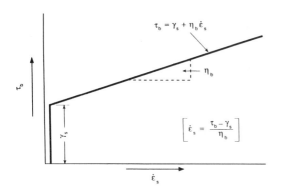

FIG. 5.13. Relationship between shear strength and strain rate in a Bingham viscous slurry (after Menzies, 1989; reproduced with permission from Elsevier Science Publishers BV).

subglacial debris layers. Discussion of the mechanics of deformation (Boulton and Hindmarsh, 1987; Boulton, 1987a; Menzies, 1989a; Alley, 1989a, b; Alley et al., 1989; Hart et al., 1990; Murray and Dowdeswell, 1992) indicates that the controlling variables influencing the thickness and mobility of a deforming layer are: porewater content and pressure; sediment yield strength, viscosity, porosity and permeability; ice mass thickness, basal sliding velocity; and basal ice interface thermal regimen (Clarke, 1987b; Hutter and Engelhardt, 1988).

If it is assumed that this debris layer acts as a Bingham viscoplastic or non-linearly viscous material, then, once yield strength of the wet debris is surpassed, the critical variable controlling debris layer thickness and deformation rate is viscosity (Fig. 5.13). Boulton and Jones (1979) calculated empirically that as a subglacial debris layer deforms, the overlying ice sheet will flatten out due to forward motion of the basal deforming layer rather than internal ice deformation or basal slip of the ice mass such that the ice sheet thickness (h_i) will be related to subglacial debris deformation as follows:

$$h_i^2 = 2 \int_x^{x_f} \left(\frac{C}{\rho_i g} + \tan\phi \ [h_i - \frac{p_w}{\rho_i g}] \right) dx \qquad (5.22)$$

where X_f is the distance from a point beneath an ice mass to the terminus in a two-dimensional ice mass; ø is the angle of internal friction of the debris layer sediment; and C is the cohesion of the sediment.

Where porewater pressures within the debris layer are less than the overlying ice pressures ($\rho_i > \rho_w$), it is thought that regelation ice will invade the debris leading to immobilization (Engelhardt, 1979). Boulton and Hindmarsh (1987) suggest that when $\rho_w \geq \rho_i$, the upper interface contact between the ice and underlying mobile sediment should be a smooth shear zone (M-bed) with a limited volume of free meltwater at the junction, if any (Shoemaker, 1986; Menzies, 1989a; Alley, 1989a, b, 1991; Murray and Dowdeswell, 1992). Depending upon the interrelationship between the internal porewater pressure of the underlying sediment and the overburden ice pressures, a steady state or an unsteady state may prevail at the ice–bed interface.

In the former instance, two possibilities arise either due to high confining ice pressures or the development of a thin skin of dilatant sediment at the upper interface between the bed and ice mass or deforming bed and bedrock in which zero deformation but positive effective pressures exist within the underlying sediment. Under an unsteady state, zero or negative effective pressures lead to sediment deformation. Boulton and Hindmarsh (1987) suggest that such a state may be the norm for ice masses overlying unlithified beds. If this latter scenario is typical then the vast percentage (70–80%) of land surfaces covered by the Pleistocene ice sheets were underlain by potentially deformable sediments. The relative importance of glacier movement upon deformable beds as compared to hard unyielding bedrock beds therefore needs to be addressed. It should be noted, unlike Shoemaker (1986) who suggests that an unsteady state, as predicted above, will develop into a 'runaway' condition, that there is strong evidence to indicate a negative feedback state will act as a balancing force on the ice/sediment system (Boulton and Hindmarsh, 1987; Menzies, 1989a). As the ice thins out and effective stress and basal shear stress reduce, so frictional drag at the upper ice–bed interface will diminish as will meltwater production thereby causing the mobile bed to reduce in thickness and eventually cease to deform, returning the glacier system back to a steady state once again.

5.5.5.1. Basal Sliding and Subglacial Sediment Deformation

Using seismic reflection techniques, a water saturated debris layer was inferred to underlie Ice Stream B, Antarctica. The following discussion will consider those aspects of deforming beds applicable to ice mass sliding (see Chapter 11; Menzies, 1995b, Chapter 2): (a) the mechanism(s) and rate of sediment (pervasive/non-pervasive) deformation; (b) the nature of clast 'ploughing' and discrete (non-pervasive) deformation; and (c) the interrelationship(s) between deformable basal sediment and meltwater at the ice–bed interface(s).

Before the discussion, it must be pointed out that only a limited amount of data from modern ice masses exist at present, that detailed subglacial observations remain meagre and the precise rheology of unconsolidated glacial debris under conditions typical of subglacial environments is poorly understood. As such 'best-fit' assumptions need to be employed, at present, in a heuristic approach to a new and major field of research (for detailed derivation of the following equations see Boulton and Hindmarsh, 1987; Clarke, 1987; Alley, 1989a, b).

(A) Mechanism(s) and rate of sediment (pervasive/non-pervasive) deformation: In a multiple layered glacier bed system where active ice overlies unconsolidated sediment and bedrock, a series of complex interface boundaries are likely to develop over a wide range of temporal and spatial scales (Menzies, 1987). Where temperate ice overlies a saturated layer of sediment, it can be shown that when the yield strength of the sediment is exceeded by the basal shear stresses imparted by the overlying ice, pervasive/non-pervasive deformation of some portion of the sediment may occur leading to forward motion of the overlying ice mass.

Deformation will cease at some depth within the sediment where the applied shear stress falls below the sediment yield strength. (It has been observed that under cold-based conditions deformation of underlying debris is also possible (Echelmeyer and Wang, 1987); however, the equations employed below all assume temperate conditions.) The rheology of the saturated sediment is modelled after a viscoplastic Bingham fluid or a non-linear viscous

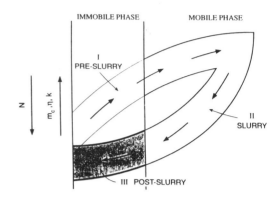

FIG. 5.14. Model of increasing and decreasing basal slurry viscosity illustrating hysteresis effects. Changes in viscosity shown as a result of variations in particle concentration within the slurry in relation to decreasing effective normal pressure, increasing moisture content and permeability (after Menzies, 1989; reproduced with permission from Elsevier Science Publishers BV).

material. As a consequence the viscosity of the saturated sediment is of great significance in controlling the flux rate and thickness of deforming sediment, and the rate of ice movement. As a subglacial sediment becomes increasingly soaked until it eventually is saturated, the individual particles within the sediment are gradually separated from each other by water-filled voids such that a Critical Volume Concentration (c.v.c.) is reached (Cheng and Richmond, 1978; Echelmeyer and Wang, 1987). At a c.v.c. of ≈ 0.6, the saturated sediment begins to act as slurry-like fluid of finite viscosity (Fig. 5.14). It is unlikely that sediment viscosities will reduce to values typical of subaerial slurries (Rodine and Johnson, 1976; Alley, 1989a; Murray and Dowdeswell, 1992). Such a highly fluid state would lead to unsteady conditions possibly typical of extrusion flow; unless the affected sediment is confined to areally narrow channels or limited mobile zones (Boulton and Hindmarsh, 1987). Evidence of viscosities measured and calculated from below ice masses, to date, indicate much higher values than subaerial slurries (Boulton and Dent, 1974; Alley *et al.*, 1987b; Fahnestock and Humphrey, 1988).

Hindmarsh and Boulton (1987) demonstrate that a subglacial 'soft' sediment should deform (using co-ordinate axes as defined in Fig. 5.15) as follows:

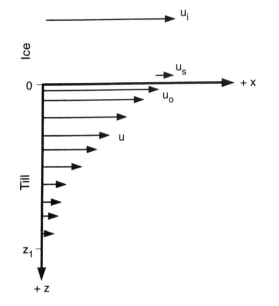

FIG. 5.15. Coordinates for till deformation equations (after Alley, 1989; reproduced by courtesy of the International Glaciological Society from *Journal of Glaciology*, 35(119), 1989, p. 122, Fig. 1).

$$\frac{du}{dz_{cs}} = \dot{\varepsilon}_d = K_b \frac{(\tau_b - \tau_c)^a}{\sigma_e^b}, \ \tau_b > \tau_c \quad (5.23)$$

$$\frac{du}{dz_{cs}} = 0, \tau_b \leq \tau_c \quad (5.24)$$

where $\dot{\varepsilon}_d$ is the strain rate of the deforming sediment, τ_c is the yield strength of the sediment where, $\tau_c = \sigma(\tan\phi + C)$, and a,b, and K_b are constants, and z_{cs} is the thickness of deforming sediment in the down-ice (χ) direction.

As basal shear stress diminishes with depth due to increasing particle to particle contact, the sediment is able to deform, from the upper interface, to a limiting depth where $\tau_c \geq \tau_b$ (Alley, 1989b; Menzies, 1989a). As the thickness of sediment increases, the effective stress will increase as must the yield strength. However, the increase in yield strength occurs much more rapidly than does effective stress. In a thick sediment layer of several tens of metres this effect will limit deformation whereas in a shallow sediment layer the effect is scarcely perceptible (Alley's 'thin-till approximation'). Alley (1989b) shows that the basal shear stress increases as a function of depth (z_{cs}) as follows:

$$\tau_b \ (z_{cs}) = \rho_i \ g \ (h_i + z_{cs})\alpha_i + (\rho_b - \rho_i) g \ z_{cs} \ \alpha_b \quad (5.25)$$

where α_i and α_b are the down-flow slope angles of the ice surface and ice/sediment interface, respectively. With an ice surface slope of $\alpha = 0.002$ and a horizontal ice/sediment interface, τ_b should increase by ≈ 20 Pa m^{-1}, while the yield strength increases by $\approx 2 \times 10^3$ Pa m^{-1}. If the sediment is saturated the effective stress $\sigma_{(z_{cs})}$ can be defined as:

$$\sigma_{(z_s)} = \sigma_e + \Delta\rho_{den} \ g \ z_{cs} \quad (5.26)$$

where σ_e is the effective pressure at $z_{cs} = 0$ (the upper interface between the sediment and basal ice) and $\Delta\rho_{den} = \kappa_b - \kappa_s$ where κ_b is sediment bulk density and κ_s is the fluid density. If it is assumed, as seems reasonable, that τ_b is constant over short distances (Alley's thin-till approximation) then the depth ($z_{cs \ (crit)}$) at which basal shear stress equals sediment yield strength (and deformation ceases) can be defined as:

$$z_{cs(crit)} = \frac{(\tau_b - C)}{\Delta\rho_{den} \ g \ \tan\phi} - \frac{\sigma_e}{\Delta\rho_{den} \ g} \quad (5.27)$$

In computing data from Upstream B and those obtained from Boulton *et al.* at Breiðamerkurjökull into the above equations, Alley (1989b) suggests that in order to maintain pervasive deformation within a subglacial layer a minimum strain rate of ≈ 5 a^{-1} is necessary. It can also be shown that the criterion for pervasive deformation within a sediment takes the form $\tau_b \geq \sigma_e \tan\phi$.

The maintenance of deformation, however, requires a much lower strain rate. As the sediment suffers stress application, it will become dilatant, reducing particle to particle contact thereby decreasing yield strength. Based upon the relationship:

$$\dot{\varepsilon}_d = \frac{du}{dz_{cs}} \quad (5.28)$$

A range of computed and measured values for z_{cs}, u and $\dot{\varepsilon}_d$ can be seen in Table 5.2 (cf. Hart *et al.*, 1990, Fig. 5). It is apparent from these values that ice masses underlain by unconsolidated sediments are likely to be affected by deformable bed conditions and therefore a considerable proportion of their forward sliding motion can be accounted for by this process.

TABLE 5.2. The relationship, both observed and calculated, between the velocity of mobile debris (u_s), debris thickness (h_s), and average simple strain rate ($\dot{\varepsilon}_\sigma$) at the upper ice/bed interface (after Menzies, 1989).

Glacier	u_s (myr^{-1})	h_s (m)	$\dot{\varepsilon}_s$ (yr^{-1})	Reference
Urumqui #1, China	~3.0	~0.35	≈8.57 ≈8.39*	Echelmeyer & Wang, 1987
Blue Glacier, Washington, U.S.A.	~4.0	~0.10	≈40.0	Engelhardt et al., 1978
Breiðamerkurjökull, Iceland	~16.0	~0.50	≈32.0	Boulton, 1979
Ice Stream B (upstream), W. Antarctica	~450.0	~6.00	≈75.0	Alley et al., 1987a
Pleistocene Puget Lobe, Washington, U.S.A.	~500.0	~15[+] ~22.5[†]	≈33.33[+] ≈22.22[†]	Brown et al., 1987

* calculation by Echelmeyer and Wang, 1987

+ undilated

† dilated

The significance of this fact is a fundamental departure from past understanding of ice mass sliding. Not only do models of ice mass sliding have to be reconsidered in light of deformable beds but, for the first time, an integration of glaciological conditions and sedimentological processes at the ice–bed interface is achieved.

(B) The nature of clast 'ploughing' and discrete (non-pervasive) deformation: An effect noticed by Boulton et al. (1974) beneath Breiðamerkurjökull is the tendency for large clasts, at the upper interface between locally deforming subglacial sediment and the ice base, to be dragged along and to 'plough' through the subjacent sediment (Krüger, 1979, 1984; Brown et al., 1987). Ploughing is a transitional state at the bed of an ice mass between pervasive deformation (mobile phase) and zero deformation (immobile phase). The process occurs when basal ice exerts a shear stress against clasts which simultaneously exceeds the local yield strength of the surrounding sediment. The consequence is that clasts plough through stable sediment (Alley, 1989b). The process depends upon local basal shear fields, and the yield strength, porewater content and clast size, shape and distribution within the sediment. The importance of this process in till fabric development is unknown. Most of these parameters have been discussed above, only clast size distribution needs to be addressed. As basal shear stresses increase (all other parameters remaining constant), the size of clasts subject to ploughing will decrease until a limit is reached where all clasts are capable of motion. At that point a transitional threshold is crossed and pervasive deformation begins. Although theoretically this scenario is correct, under certain circumstances of material anisotrophy (such as linear clay domains), a form of non-pervasive deformation may be established and only at higher confining stresses and levels of strain will pervasive deformation wholly exist. Both Brown et al. (1987) and Alley (1989b) have modelled the ploughing process and show that it can occur when the Ploughing Index is:

$$I_p^* = \frac{a_g \tau_j}{\sigma_{cj} S_j} \qquad I_p^* \geq 1 \qquad (5.29)$$

where σ_{cj} is the local effective stress upon clasts of jth size class and can be defined as:

$$\sigma_{cj} = \sigma_e \left(1 + \frac{f\tau_j}{\tau_b S_j}\right) \qquad (5.30)$$

and where α_g is a geometric factor defined as:

$$a_g = \frac{\langle 2\tan(45 - \phi/2)\rangle}{\tan\phi} \qquad (5.31)$$

and τ_j and S_j are defined in terms of an arbitrary area of the bed where the basal shear force upon a clast of the jth size of specific area S_j. Assuming that the initiation of ploughing begins when Ip=1, the average basal shear stress at initiation is:

$$\tau_b \geq \sigma_e \frac{S_j + \xi_j f}{a_g \xi_j} \qquad (5.32)$$

where $\xi_j = \tau_j/\tau_b$. Using typical values for diamicton (Brown et al., 1987; Alley, 1989b) for example, with ξ_j=0.33 0.5 for the controlling obstacle factor (Weertman, 1964); S_j=0.1; f=0.3 for water in pore spaces only; and ϕ=25°; then ploughing begins when $\tau_b \geq 0.2\sigma_e$. This last equation can be rewritten for the condition that all clasts on the upper interface are

ploughing where: $\tau_b \geq \sigma_e/\alpha_g$. Brown et al. (1987) predict, using the same parameter values, that ploughing of the entire deformable bed will occur when $\tau_b \geq 0.4\sigma_e$. With this massive degree of ploughing, a zone of shearing develops at the upper interface of the sediment. Unlike pervasive deformation, this ploughing process only involves a discretely narrow band of sediment.

It becomes apparent from the above computations that a restricted range of stress levels separate the onset of ploughing ($\approx 0.4\sigma_e$) from pervasive deformation ($\approx 0.5\sigma_e$). The precise point of transition will vary as conditions and sediment rheology fluctuate but at values of tan$\phi \geq 0.75$ only pervasive deformation takes place. It must be pointed out that these values are only pertinent to the 'onset' of both processes of ploughing and pervasive deformation in non-dilated sediment.

In general, when sediment is under stress some form of dilatancy must prevail. Such a state will lead to a reduction in yield strength and angle of internal friction, thus reducing the necessary stress levels required to maintain or begin deformation (Terzaghi and Peck, 1967). Values of tan$\phi \approx 0.2$ have been suggested for dilated diamicton, thus pervasive deformation could begin when $\tau_b \geq 0.2\sigma_e$. As clasts begin to move, with the onset of ploughing a dilatant aureole of finer sediment will develop around, and especially, in front of each moving clast (Plates 5.3 and 9.4). The ease of movement of the clasts through the sediment will require a lower stress level to maintain movement. A secondary aspect of this effect is that as each clast transits a zone of sediment, a lag period will occur before the sediment returns to a non-dilatant state. The length of time of this lag period is unknown at present (Attewell and Farmer, 1975). However, if another clast passes through this dilated zone before recovery to a non-dilatant state, passage will, in effect, be easier. It is therefore possible that as ploughing rates increase the sediment will reach a state of pervasive deformation at lower stress levels than predicted above. A converse circumstance may occur as stress levels drop, pervasive deformation ceases and only ploughing continues until a lower stress threshold is crossed. It is possible that ploughing may occur not only at the upper interface of the once deforming sediment but also at its lower interface.

PLATE 5.3. Ploughed boulder *in situ* in diamicton, Rose Township, western New York State, showing long axis orientation and surface striae. Boulder is approximately 25 cm across b-axis.

The contribution ploughing makes to basal sliding remains an unknown. Brown *et al.* (1987) suggested, from examination of Pleistocene sediments of the Vashon Ice Lobe, Washington State, that clast ploughing may have accounted for ≈10% of the basal stress where no pervasive deformation had occurred. In order for this condition to have existed, the subglacial sediment would have had to be of extremely low viscosity. Observations and inferences of Bingham viscosities of sediment from Breiðamerkurjökull (3000 Pa a), Upstream B, Ice Stream B (120 Pa a), and Columbia Glacier (30 Pa a) (Boulton *et al.*, 1974; Alley, 1989a; Fahnestock and Humphrey, 1988, respectively) indicate in the two former cases that viscosities were too high to allow ploughing as calculated to take place. It seems unlikely that discrete ploughing could occur without concomitant pervasive deformation at such viscosities. These deductions assume that a Bingham-type rheology best models the process of ploughing with potentially deforming sediment. Much remains unknown about subglacial sediment rheology with reference to the effects of porewater and fluctuating thermal conditions and other allied parameters, thus

meriting considerable new research (Alley, 1989b; Menzies, 1989a; Hart *et al.*, 1990; Murray and Dowdeswell, 1992). However, if ploughing occurred exclusively, as suggested beneath the Vashon Ice Lobe, the sediment must have been of low viscosity possibly due to a very high porewater pressure, a consequence of either excess supraglacial meltwater being supplied to the bed or very high basal velocities of the type typical of surging ice masses or very fast ice masses such as observed at Jakobshavn Isbrœ, West Greenland (Echelmeyer and Harrison, 1989, 1990). Where both pervasive deformation and discrete shearing due to ploughing occur together, ploughing may contribute as much as 30% of the total deformational velocity (Alley, 1989b). This process of discrete shearing is of considerable importance and requires further investigation (Menzies and Maltman, 1992). How widespread the phenomenon is remains to be established and when occurring without associated pervasive deformation may be indicative of exceptional glaciodynamic conditions. A final implication of ploughing is its impact upon clast fabric strength and range of dip angles (Dowdeswell and Sharp, 1986; Menzies, 1990c).

(C) The interrelationship(s) between deformable basal sediment and meltwater at the ice–bed interface(s): The influence and impact of meltwater at the base of ice masses undergoing some form of bed deformation remains enigmatic (Chapter 6). If meltwater at the bed only occupies a small fraction of the total bed area then its influence on siding velocity will be minimal (Nye, 1969; Weertman, 1979; Kamb, 1970, 1987; Bindschadler, 1983; Lliboutry, 1987a, b). In order that the proportion of the bed covered by meltwater expands it is necessary for either: (1) the rate of meltwater production to increase and/or (2) the basal hydraulic pressure to rise (Iken, 1981; Walder, 1986; Weertman, 1986; Kamb, 1987; Lingle and Brown, 1987; Shoemaker, 1992). In either instance, provided meltwater pressure exceeds normal ice stress, a larger fraction of the ice bed will decouple and the ice mass, with reduced basal friction, will exhibit higher sliding velocities.

5.6. THE FLOW OF VALLEY GLACIERS

Valley glaciers, in contrast to Ice Sheets, are confined by valley walls (Wright, 1937). The walls act to influence and retard valley glacier flow. This contrast is not as great when one considers their similarities to Ice Streams where an element of side-wall drag also occurs. The physics of ice flow and basal slippage, as described for ice sheets, are essentially similar for valley glaciers (Raymond, 1980). The major difference is in the component of side-wall frictional drag. Raymond's pioneering work on the cross-sectional transverse velocity fields of the Athabasca Glacier, Alberta illustrate this effect (Fig. 5.16). Frictional drag may reduce edge velocities by

$\approx 80\%$ compared to central maximum surface velocity (Raymond, 1978, 1980). In order to account for this frictional effect the basal sliding equation alters to:

$$u_s - u_b = [\frac{2A}{(n+1)}](\tau_b)^n h_i \qquad (5.33)$$

where, in the centre line of the glacier, the basal shear stress can be equated as

$$\tau_b \approx \tau_{xy}|bed \text{ and } \tau_{xy}|bed \approx -\rho_i g(f_{sh} h_i \alpha): \qquad (5.34)$$

where the shape factor (f_{sh}) is derived as follows:

$$f_{sh} = \frac{S}{W_b \ h_i} \qquad (5.35)$$

where S is the cross-sectional area of the channel, and W_b is the width of the channel bed (Nye, 1952; Meier et al., 1974; Raymond, 1980; Harbor, 1992). Typical shape factor values for parabolic cross-sections vary from 0.5–0.6; V-shaped have smaller values and flat-bottomed U-shaped channels larger values. The effect of the shape factor (f_{sh}), alone, on the sliding velocity equation is to reduce the overall velocity by as much as 15% compared to ice sheet flow with similar exponent values. However, sliding velocities of valley glaciers tend not to be in broad agreement with these theoretical computations. As discussed above, the mechanics of glacier basal movement is a complex thermo-mechanical process. Unlike ice sheets where basal parameter fluctuations, both lateral (transverse) and parallel with the flow direction, can and do occur, major fluctuations beneath valley glaciers are largely parallel with ice flow direction. Likewise, meltwater channel development is often restricted to a single main

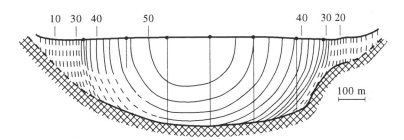

FIG. 5.16. Isolines of equal down-stream ice flow velocity observed in the Athabasca Glacier, Alberta, Canada. Isolines are metres per year (after Raymond, 1978, 1981; reproduced with permission of the author and Academic Press).

channel system somewhere close to the centre of the valley glacier, whereas under ice sheets the possibility of lateral channel migration, channel system capture and multiple systems are much more likely.

Within the more restricted system of valley glaciers it is therefore not surprising that fluctuations in meltwater have a rapid and dramatic effect upon ice morphology and most noticeably, in the short-term, glacier velocity (Hodge, 1974; Iken, 1978, 1981; Hooke *et al.*, 1983; Raymond and Harrison, 1988; Echelmeyer and Harrison, 1989; Harrison *et al.*, 1989; Hindmarsh *et al.*, 1989).

The effect of deformable beds upon valley glacier movement and stability has engendered considerable ongoing debate (Alean *et al.*, 1986; Clarke, 1987a, b). Haeberli (1981) states that wide areas of present-day European Alpine valley glaciers are underlain by unconsolidated sediments (Süsstrunk, 1951). Whether these sediments all act in the manner of mobile soft beds remains to be investigated but seems unlikely (Souchez and Lorrain, 1987). The evidence, for example, from the Trapridge Glacier, Yukon Territory, indicates the presence of deformable bed conditions, possibly associated with surging instabilities (Clarke *et al.*, 1984, 1986b). Other valley glaciers in Peru (Lliboutry *et al.*, 1977), Alaska (Engelhardt *et al.*, 1978; Kamb *et al.*, 1985), China (Echelmeyer and Wang, 1987), Sweden (Brand *et al.*, 1987), and Ellesmere Island (Koerner and Fisher, 1979) all reportedly exhibit evidence of deformable bed conditions. Over the next decade increasing research into soft bed conditions should provide fascinating clues to subglacial bed conditions and, as yet, many unknown and enigmatic facets of glaciodynamics.

5.7. THE FLOW OF ICE SHELVES

Until relatively recently, ice shelves, as components of the glacial system, were regarded as minor peripheral curiosities found only in high latitudes. However, there is increasing evidence that ice shelves may have occurred along the fringes of the Laurentide and Fennoscandian Ice Sheets during the Pleistocene and may have acted as major constraints upon ice sheet growth and ultimate decay. In North America, for example, ice shelves may have been formed in the glacial Great Lakes system and along the eastern seaboard from the Maritime Provinces north to beyond the Labrador coast of Canada (Denton and Hughes, 1981b; Eyles and Eyles, 1983a; Fulton and Andrews, 1987). The significance of ice shelves as sedimentological models in studies of both Pre-Quaternary and Quaternary glaciations has been highlighted recently (Molnia, 1983; Visser, 1983; Eyles *et al.*, 1985; Drewry, 1986; Dowdeswell and Scourse, 1990; Anderson and Ashley, 1991). To understand the processes and sedimentological conditions within glacio-lacustrine and glacio-marine environments, the physics of ice shelves is fundamentally significant. Of prime importance in understanding ice shelf growth and diminution, as well as ice shelf impact upon sedimentological processes, is the position and spatial transience of the grounding-line, the rate of bottom melting or accretion, the impact of mass balance changes, the influence of grounded ice streams entering shelves, and the effect on ice shelf dimensions of eustatic variations.

An ice shelf is a large body of glacial ice attached to the land but floating on the sea or a large lake. The ice deforms under its own weight and, in theory, bottom and surface shear stresses approach zero, if bottom and lateral drag is ignored, and it is assumed that the ice shelf is a perfect horizontal slab with top and bottom boundaries parallel. In reality, however, ice shelves exhibit variable thicknesses with gradients in the down-flow line direction. In this case shear stresses are no longer zero (Sanderson and Doake, 1979) (Fig. 5.17). Since most ice shelves have been in existence for thousands of years it has been generally assumed that if climatic and glaciodynamic controlling parameters have not significantly varied over time that those shelves may be in a state of dynamic equilibrium (Barkov, 1970). The converse that many ice shelves are, instead, in a state of considerable instability and non-equilibrium is a view that many researchers now advocate (Budd, 1966; Hughes, 1977). The latter view has profound implications concerning Global Warming and the stability or otherwise of the Antarctic Ice Sheet.

Any ice shelf, as it extends out to sea, can be regarded as the product of a complex balance of processes that control its extent, thickness and flow conditions, and thus dynamic equilibrium (Morland and Shoemaker, 1982). These processes involve the glaciodynamics of ice creep and tensile strength, edge drag, surface snow accumulation and ablation, bottom melting and freezing, the position of grounding-lines, ice input from up-ice and associated effects of glaciers ploughing into the shelf, the presence of ice rises and/or seaward islands, and the impact of ocean currents, storm tracks, waves and tidal fluctuations (Powell, 1991). Typically, ice shelves are located in a variety of topographic situations viz: confined or unconfined.

5.7.1. Confined — Unconfined Ice Shelves

In confined shelves edge-support is provided by the land normally on three sides (Fig. 5.18). Today major ice shelves such as the Ross, Ronne and Filchner Ice Shelves in Antarctica are fed by major outlet glaciers from the Antarctic Ice Sheet (McIntyre, 1985). These outlet glaciers, flowing at higher velocities than the ice shelves, plough into the shelf often becoming floating ice streams flowing for some considerable distance within the shelf (e.g., the Byrd Glacier) (Fig. 5.19). Ice rises, where the shelf locally grounds, occur where bedrock rises up to the base of the shelf or the shelf, itself, locally thickens and touches the sea bed. Ice rises and rumples cause local impounding effects leading to back-pressure up-ice within the ice shelf (Thomas, 1973; 1979; Doake, 1987).

Unconfined ice shelves are relatively rare since their stability does not hinge upon a topographic embayment but rather is dependant upon the vagaries of tidal movement, currents and wave action. Such an ice shelf extends out into a body of water at a sufficient velocity from feeder ice sources to effectively spread out and remain intact as a floating ice mass. For detailed discussion on unconfined ice

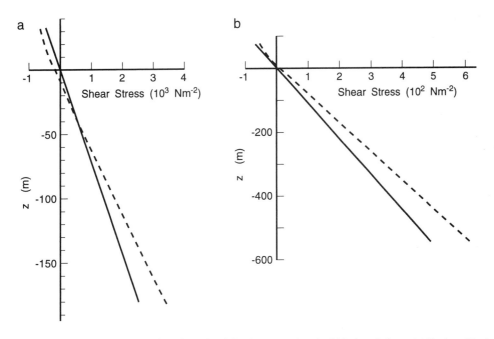

FIG. 5.17. Examples of shear stress as a function of height above sea-level within ice shelves: (a) Erebus Glacier Tongue (unconfined); (b) Amery Ice Shelf (confined). Solid lines assume density and temperature constant with depth, while dashed lines shows realistic values for density and temperature (after Sanderson and Doake, 1979; reproduced by courtesy of the International Glaciological Society from *Journal of Glaciology*, **22**(87), 1979, p. 289, Fig. 3).

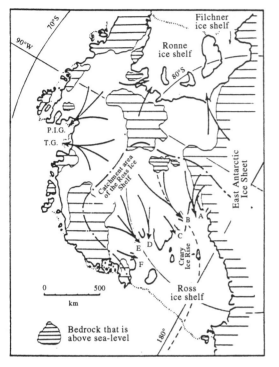

FIG. 5.18. West Antarctica showing areas of bedrock above sea-level. Arrows indicate ice streams (after Thomas, 1979; reproduced by courtesy of the International Glaciological Society from *Journal of Glaciology*, **24**(90), 1979, p. 168, Fig. 1).

shelves see the papers by Morland, Morland and Zainuddin and MacAyeal in van der Veen and Oerlemans (1987).

5.7.2. Ice Shelf Stability

At the back of an ice shelf is the grounding-line where the ice shelf ceases to float. This line demarcates the point where hydrostatic equilibrium of the ice and sea begins. At this critical junction several major thermal and glaciodynamic phase changes occur at the base and within the ice mass. In contrast to a grounded ice mass, vertical stresses in an ice shelf are almost at a constant from the upper surface to the base of the shelf (Sanderson and Doake, 1979; Sanderson, 1979). The transformation from a grounded to floating ice mass results in large-scale velocity field changes which are manifest in the thickness of the ice shelf for several kilometres down-

ice of the hinge point (Sanderson, 1979). The position of the grounding-line is especially susceptible to eustatic fluctuations (and isostatic rebound effects) resulting in potentially large increases or decreases in the amount of ice floating and thus the frontal position of the ice shelf cliff (Thomas, 1976, 1979). The sensitivity of grounding-line positions has led several researchers to hypothesize on the potential stability/instability of the marine sections of the Antarctic and northern hemisphere Pleistocene ice sheets as result of the possibility of ice shelf collapse (Hoppe, 1948; Mercer, 1968; Paterson, 1972b; Hughes, 1973, 1975, 1981; Thomas and Bentley, 1978a, b; Thomas *et al.*, 1979; Denton and Hughes, 1981a). Bentley (1984) has suggested that the susceptibility of a marine ice sheet (at present the West Antarctic Ice Sheet is the only one in the world) to climatic fluctuations is a direct function of the sensitivity of its grounding-lines. Whereas ice sheets might be expected to exhibit response times in thousands of years, in this instance, response is in hundreds of years. As sea-level rises, it is suggested that the grounding-line should begin to extend up-ice resulting in an increase in the total area of the ice shelf and therefore increasing forward (down-ice) spreading. This spreading will cause the overall ice shelf to thin and be a less effective buttress for the grounded ice mass behind it. A critical point, it is suggested, will be reached when this buttressing effect will founder and a vast outpouring of ice from the ice sheet, itself, will occur causing possible rapid ice sheet decay (Paterson, 1980; Jezek, 1984; Herterich, 1987; Lindstrom and MacAyeal, 1987; MacAyeal, 1987).

Two contrasting views exist as to the effectiveness or otherwise of the suspected buttressing action of ice shelves on ice sheets and in particular ice streams draining into the shelves. The view, as expressed above, holds that ice shelves act to restrain ice stream velocities resulting in increased ice thicknesses and the development of concave-up profiles within ice streams (Hughes, 1977, 1983). However, an opposite controversial view has been expressed that ice shelf back-pressure has little effect upon ice stream velocity fields (Vornberger and Whillans, 1986). The collapse of an ice shelf, as discussed above, and ultimately the down-draw of the rearward ice sheet

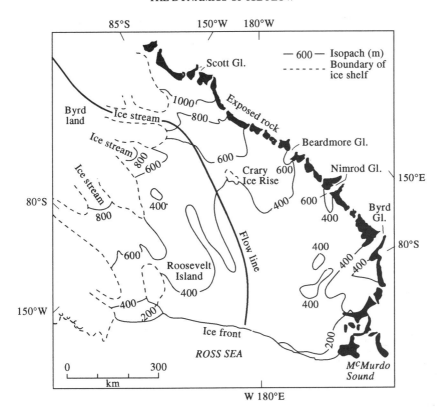

FIG. 5.19. Isopachs of ice shelf thickness for the Ross Ice Shelf. Dashed lines indicate the shelf boundaries (after Robin, 1975; reprinted with permission from *Nature*, **253**, 168–172, 1975, Macmillan Magazines Limited).

should not be confused with surges. The latter are cyclical, whereas the above form of instability is thought to be temporally random. However, it has been suggested that on a vast scale, ice sheets may surge and such an occurrence in the sedimentary record would, perhaps, be difficult to distinguish from this non-cyclic catastrophe (Wilson, 1964; Hollin, 1964, 1969).

5.7.3. Ice Shelf Stress/Strain Conditions

In theory, an unconfined ice shelf should have a longitudinal strain rate defined as (Thomas, 1979; Paterson, 1980; Morland, 1987; Muszynski and Birchfield, 1987):

$$\dot{\varepsilon}_{sh} \approx K_f(\rho_i g \Delta\rho \ h_i)^n \qquad (5.36)$$

where $\Delta\rho = (1 - \rho_i/\rho_{sw})$, K_f is a function of ice fabric, temperature and the ratio of the components of the strain rate tensor. In general, a value of n=3 for the Flow law exponent is assumed with field measurements indicating stress levels of ≤ 40 kPa. At such low stresses there is some debate that polycrystalline ice may exhibit primary or transient creep with an expected value of n close to unity (Paterson, 1980). A further concern with the above equation is the likelihood that ice density is not constant within an oceanic ice shelf (Thomas, 1973; Sanderson, 1979). An assumption of constant density can lead to an overestimation of τ by as much as 2 and in consequence an exaggeration of $\dot{\varepsilon}_{sh}$ by approximately 8.

In the case of confined ice shelves whether the shelf is pinned by land along its edges or by scattered islands acting as ice rises, a different stress field exists. In confined models (Fig. 5.20) whether of parallel, diverging or converging edges a considerable proportion of the driving force of the ice

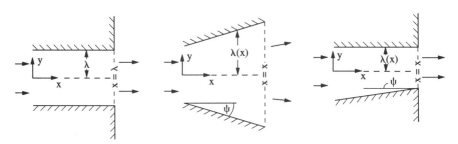

FIG. 5.20. Idealized plan views of three ice shelf configurations: (a) parallel-sided bay, (b) diverging-sided bay, (c) converging-sided bay (after Sanderson, 1979; reproduced by courtesy of the International Glaciological Society from *Journal of Glaciology*, **22**(88), 1979, p. 444, Fig. 4).

mass will be utilized in overcoming edge friction (Budd, 1966; Crary, 1966). Unlike the above equation based upon Weertman's analyses where longitudinal strain rate will increase proportionally with increased ice thickness, in the confined state this relationship breaks down. From analyses by Sanderson (1979) and Thomas (1979) the strain rate in a confined ice shelf can be defined as follows:

$$\dot{\varepsilon}_{sh} \approx K_f (\rho_i \ g \ \Delta\rho \ h_i - 2B_p)^n \qquad (5.37)$$

where B_p is the back-pressure exerted by frictional drag along ice shelf edges or behind ice rises. Back-pressure or the resistance to flow experienced by an ice shelf can be redefined in terms of geometric shape (form drag) and frictional resistance (dynamic drag) (MacAyeal, 1987). In general, the longitudinal strain rate in ice shelves, in the direction of maximum ice flow, is either positive or close to zero such that strain rate and back-pressure are coupled where:

$$\rho_i \ g \ \Delta\rho \ h_i \geq 2B_p \qquad (5.38)$$

Back-pressure manifests itself within an ice shelf as a compressive stress causing ice shelf thickness to increase and strain rate to be reduced (Budd, 1966; Jezek, 1984; MacAyeal, 1987; Bindschadler *et al.*, 1989). In an analysis of bottom crevasses beneath the Ross Ice Shelf, Jezek noted that the previously predicted penetration height of the crevasses, based on existing theory, was much less in reality, the reason being that back-pressure had not been taken into account. This field evidence indicates the considerable buttressing effect ice shelves have upon inland ice sheets. Where the back-pressure is a consequence of an ice rise a zone of highly compressed ice will extend in a zone up-ice from the

pinning point. To the leeward side of the obstruction where compressive stress has been removed, the ice shelf will rapidly thin, and where tensile stresses rise above ice yield strength fracturing of the shelf may occur. At the shelf's frontal ice cliff such fracturing results in calving and often large tabular iceberg formation (Hughes, 1982, 1989; Fastook, 1984; Jacobs *et al.*, 1986; Lange, 1987; Dowdeswell, 1989; Hughes and Nakagawa, 1989). In those areas unaffected by ice rises and in the central zones of confined ice shelves, ice shelves tend to accelerate toward the sea and in the process decrease in thickness, thus the shelf will grow at the expense of a feeding ice mass (Fig. 5.21). A confined, but diverging ice shelf will only maintain stability if it is sufficiently thick and with a low enough longitudinal strain rate that it can creep transversely and rapidly, thus avoiding longitudinal crevassing and rift development.

The location and sedimentological significance of grounding-line positions is of paramount importance in understanding glacio-marine sedimentology (see Chapter 14; Menzies, 1995b, Chapter 5). The main factors that favour changing grounding-line positions are illustrated in Table 5.3. First, it is at the grounding-line that maximum flushing-out of subglacial and meltwater transported sediments occur (Powell, 1981, 1991; Molnia, 1983; Drewry, 1986; Dowdeswell and Scourse, 1990; Anderson and Ashley, 1991). Secondly, at the grounding-line, a profound change in ice stresses, thermo-mechanics and velocity occurs (Sanderson, 1979). Thirdly, the grounding-line is the locus of rheological changes in sediments exiting the subglacial environment resulting in particular lithofacies and facies

(a)

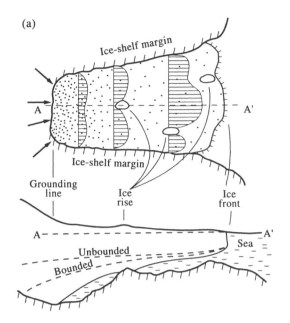

(b)

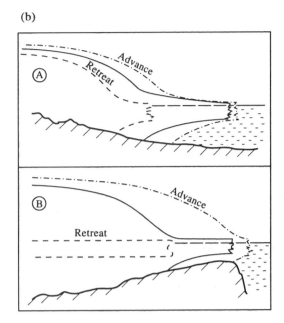

FIG. 5.21. (a) Illustration of bounded ice shelf with ice rises. Velocity cross-sections at three locations across the ice shelf. Stippling indicates areas of back-pressure exerted by margins and rises (density of stippling indicative of magnitude of back-pressure). Three bottom profile illustrate the influence of: (i) a profile with ice rises, (ii) and (iii) show profiles of bounded and unbounded shelves, respectively. Note relative position of grounding-line is strongly dependent upon type of ice shelf formed (after Thomas, 1979; reproduced by courtesy of the International Glaciological Society from *Journal of Glaciology*, **24**(90), 1979, p. 170, Fig. 2). (b) Depiction of grounding-line migration in relation to ice sheet thickness variation and sea-level fluctuation. In (A) sea bed slopes seaward thus changes of ice sheet dimensions or sea-level cause only limited fluctuation of grounding-line position. In (B) with sea bed sloping seaward the ice sheet is no longer by a bounded or partially grounded ice shelf. There is therefore only two stable modes viz. an 'advance' and a 'retreat' profile. Grounding-line thus varies in response to both ice sheet, and ice shelf dimensions as well as sea-level changes (after Thomas, 1979; reproduced by courtesy of the International Glaciological Society from *Journal of Glaciology*, **24**(90), 1979, p. 172, Fig. 3).

associations developing (Eyles and Miall, 1984). Finally, in the short term of a few hours and days, grounding-line positions are liable to change as a result of tidal or other forces causing partial floating and re-grounding of the ice shelf on already deposited sediment. This transient stress application and relief leads to local deformation of these sediments (Woodworth-Lynas and Guigné, 1990) (see Chapter 10; Menzies, 1995b, Chapter 4).

5.8. VELOCITIES OF ICE SHEETS AND GLACIERS

In considering velocity as a characteristic of a particular ice mass it must be clearly stated what timeframe is being considered (e.g. a few days, one or two months or years, several decades or centuries).

All ice masses exhibit variations in velocity; some are extrinsic to the effect of local and short duration causes or the result of long-term continental or global impacts, while others are intrinsic to a particular ice mass and are a function of mass balance change and 'internal adjustments' within the glacial system to climate, bed topography, meltwater discharge and pressure fluctuations.

5.8.1. Surface and Balance Velocities

Variations in mean surface velocities (u_s) among valley glaciers, ice sheets, ice streams and ice shelves are considerable. Values range from 10 to 200m year[-1] for many valley glaciers, to 250 to 1400m year[-1] for parts of the Antarctic Ice Sheet (Paterson, 1981; Morgan *et al.*, 1982; Clarke, 1987b; Meier and

TABLE 5.3. Response of ice shelf grounding-lines to internal and external factors affecting advance and retreat of an ice shelf (after Thomas, 1979).

Grounding-line Advance:

1. Increased size of ice shelf causing ice flowing across the grounding-line to push a larger mass of ice past its margins.

2. Increased ice shelf thickness.

3. Eustatic or Isostatic rise.

4. Overall cooling of the ice shelf, ice increasingly less elastic.

Grounding-line Retreat:

1. Decreased ice shelf size.

2. Decreased ice shelf thickness.

3. Eustatic or isostatic depression.

4. Overall warming of ice shelf, ice increasingly elastic.

5. Reduced shear stress along ice shelf margins due to ice sheet collapse and greatly increased ice velocity.

Post, 1987; Hooke *et al.*, 1989). Jakobshavns Isbrœ, West Greenland, is considered the world's fastest glacier with velocities of ≈ 8.4km year^{-1}, while several ice streams in Antarctica exhibit velocities of 400 m year^{-1} (Rutford Ice Stream) to over 800m year^{-1} (Ice Stream B). Ice Stream C, however, has a present velocity of 5m year^{-1}, but in the past 2000 years possibly moved at velocities similar to those of Ice Stream B. Typical sliding velocities are shown in Table 5.4. Short-term velocities recorded during the fast phase of a surging glacier may be exceptionally fast, for example, >3.8 km year^{-1} for the Medvezhiy Glacier, Tadzhikistan in 1973 (Dolgushin and Osipova, 1975) and >2.3 km year^{-1} for the Variegated

Glacier, Alaska 1982–83 (Kamb *et al.*, 1985); both were peak velocities recorded over a few hours only (Raymond, 1987, Fig. 1).

The distinction between glaciers as to whether they are fast or slow moving, intrinsically unstable or not is a major concern in glacial sedimentological processes. In the past, velocity variations have tended to be regarded as a function of basal ice conditions and have resulted in a search for increasingly complex and sophisticated mathematical abstractions of the basic sliding law (Hooke *et al.*, 1983 and references therein). However, it is now apparent that fast velocities are a function of fast sliding and not fast creep (Boulton, 1986a; Clarke, 1987b); the

TABLE 5.4. Sliding Velocities* (modified after Paterson, 1981).

Glacier	Ice Thickness (m)	Velocity (mmd⁻¹)	Reference
Argentière	100	600-1200	Vivian & Bocquet, 1973
Blue	26	16	Kamb & LaChapelle, 1964
Blue	65	350	Kamb & LaChapelle, 1968
Blue	65	10	Kamb & LaChapelle, 1968
Casement	20	24	McKenzie & Peterson, 1975
Grindelwald	40	250-370	Carol, 1947
Mer de Glace	100	30-80	Vivian, 1977
Mont Collon	65	30	Haefeli, 1951
Østerdalsisen	40	29-97	Theakstone, 1967
Variegated	356	250	Engelhardt et al., 1979
Vesl-Skautbreen	55	8	McCall, 1952
Susitna	520	119	Clarke, T., 1991
East Fork	410	236	Clarke, T., 1991
Storglaciären	125	180-280	Hooke et al., 1987
Urumqui No.1	12	11	Echelmeyer & Wang, 1987
Urumqui No.1	23	14	Echelmeyer & Wang, 1987
Urumqui No.1	33	16	Echelmeyer & Wang, 1987
Ice Stream B (50 km)⁺	1000	1986	Lingle & Brown, 1987
Ice Stream B (300 km)#	2000	137	Lingle & Brown, 1987

* Velocities measured may vary by at least a magnitude of 100 depending upon location, season of the year and state of bed conditions (e.g., prior to a surge event or immediately following such an event). During surge events velocities at least 10 times higher may occur.

+ Velocity taken 50 km up ice of grounding line.
# Velocity taken 300 km up ice from grounding line.

former, the effect of transient subglacial states and boundary interfaces in interaction with glaciological conditions.

Depending upon glacier type and ice velocities observed over several years, it is possible to classify glaciers on the basis of their velocities in relation to basal shear stress. As a consequence the 'state' of an individual ice mass can be crudely predicted and is of

significance both from the geological, climatological and engineering viewpoints. Major variations in ice surface and balance velocities may occur on a single ice mass due to several factors depending on where and when measurements were taken:

(a) Position in relation to the equilibrium line. In general, surface ice velocities are highest close to or at the equilibrium line and in the case of valley glaciers and ice streams in the centre of the ice mass where least side-wall drag occurs (Drewry, 1983b);

(b) Proximity to valley walls. The major influence in this instance is frictional drag; however, due to increasing surface heating from the valley walls ice flow has a slight transverse compensating component, further reducing velocity;

(c) At ice falls (seracs) where due to sharply increased bottom gradients over a short distance ice velocities may increase to several kilometres per year;

(d) In the vicinity of ice rise on ice shelves, up-ice of the ice rise typically velocities due to a 'backing-up' effect are much reduced whereas down-ice of the rise velocities are higher than mean velocities;

(e) On ice sheets where ice streams occur the influence of frictional drag along ice stream sides reduces their velocity;

(f) Where major up-ice facing bedrock rises or escarpments occur of a sufficient elevation to influence basal ice flow over a short distance, then the ice mass will reduce in velocity;

(g) The localized impact of the passage of a kinematic wave tends to accelerate ice on the down-ice side of the wave, and decelerate ice on the up-ice side. Kinematic waves average about 3–5 times the mean annual velocity (Lliboutry and Reynaud, 1981; Weertman and Birchfield, 1983a);

(h) Where an outlet glacier begins floating, velocities increase down-ice of the grounding-line due to virtually zero frictional basal drag (Meier and Post, 1987).

Other influencing factors causing velocity changes may be sudden increases in the thickness of subglacial water films, basal materials becoming unstable due to saturation or fracturing, localized surface accumulations due to valley wall avalanches or rock-ice falls, or the increase in velocity observed down-ice of a confluent glacier where the side-drag effect is removed. Hooke *et al.* (1989) discussed the problems and uncertainty of obtaining velocity measurements.

In general, most glaciers react stably to external factors that influence net mass balance, basal movement and therefore balance velocity. However, certain ice masses appear capable of switching from slow to fast mode (Fig. 5.22) both spatially along the

length of the ice mass or temporally (Ice Stream C, West Antarctica, Fastook, 1987; Shabtaie and Bentley, 1987). Both ice streams and tidewater glaciers can exhibit marked velocity changes. In the latter case, drastic retreat and rapid upstream migration of the grounding line can ensue, for example, where externally induced secular changes in tidal or eustatic levels cause grounding lines to migrate, or lower sections of the glacier to flexure and bend, or floating ice tongues or near submarginal balance gradients to alter (Krimmel and Vaughn, 1987; Meier and Post, 1987).

A similar style of short-term instability may have occurred, for example, in some regions of the Great Lakes Basin and central Prairies at the close of the Late Wisconsinan. In these regions, rapidly rising proglacial lake levels may have caused ice lobes to act in the manner of tidewater glaciers exhibiting exceptionally rapid rates of retreat (Teller and Clayton 1983; Karrow and Calkin, 1985; Fulton and Andrews, 1987). In other areas of the northern hemisphere, during the Quaternary Glaciations,

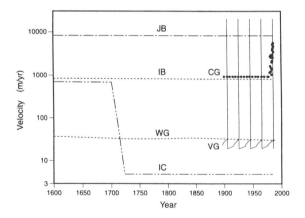

FIG. 5.22. Flow speed for various glacier types. The graphs draw upon real data but are partly extrapolated. JG — Jakobhavns Isbrœ, Greenland — a fast outlet glacier that terminates as a tidewater glacier; CG — Columbia Glacier, Alaska, U.S.A. — a tidewater glacier that in 1984 began to retreat rapidly; IB — Ice Stream B, Antarctica — a fast ice stream entering the Ross Ice Shelf; IC — Ice Stream C, Antarctica — a slow ice stream that switched from fast to slow mode ~250 years ago; WG — White Glacier, Axel Heiberg Island, Canada — a normal valley glacier; and VG — Variegated Glacier, Yukon Territory, Canada — a surging glacier (after Clarke, 1987, *Journal of Geophysical Research*, **92B**, p. 8838, copyright by the American Geophysical Union).

significant advances and retreats of large frontal sections of ice sheets have been documented (Boulton *et al.*, 1977, 1985; Denton and Hughes, 1981; Ehlers, 1983; Mickelson, 1986; Sibrava *et al.*, 1986; Dyke and Prest, 1987). These remarkable fluctuations resulted in rapid shifts and changes in velocity mode behaviour that remain important unresolved research problems. In the past, various explanations have been given including massive short-term 'snow-blitz' accumulations, surging instability, new ice stream and lobe tracks or the shifting and capture of new ice accumulation catchment areas.

A possible scenario proposed for the disintegration of the West Antarctic Ice Sheet is based upon climatic warming leading to eustatic rise and backward migration of grounding lines (Hughes, 1975; Thomas, 1979; Thomas *et al.*, 1979). Whether ice streams alter velocity mode as a consequence of external or internal factors remains unclear (Clarke, 1987a). Fastook (1987), in examining Ice Stream C, West Antarctica concluded that two explanations for the dramatic reduction in velocity of the ice stream were possible viz: (a) capture of a major part of the catchment area by another ice stream; or (b) removal of the effective deformable sliding bed. In the former case the loss of a large up-ice source of ice would result in a fundamental decrease in balance velocity, and in the latter case with the elimination of a deformable bed a much greater ice surface gradient would be necessary to drive the ice stream; in both instances a steady reduction in velocity would be manifest over the long term (ca. 2000 years).

Where velocity mode change is intrinsic to the ice mass and periods of complete quiescence almost stagnation are followed by periods of very rapid, unstable movement, a surging condition prevails.

5.8.2. Basal Velocities

Basal sliding velocities are not well documented but appear to be approximately 10–20% of the surface velocity (Paterson, 1981, Tables 5.1 and 5.2). It is difficult to accurately predict the sliding velocity from an appreciation of the influencing factors (Weertman, 1979). Where no basal sliding occurs and a polar glacier state prevails it has been suggested that basal ice velocities can be related to surface velocities as illustrated in Eqn. (5.33). Mid-summer melt season velocities are usually highest. The influence of diurnal temperature variations, rainfall, surface meltwater discharge, sky cloudiness and geothermal fluctuations all appear to reduce frictional drag at the ice base and manifest as higher than annual mean velocities (Sugden and John, 1976; Collins, 1979a, 1981; Hooke *et al.*, 1983). These higher velocities are thought to be the result of changes to the subglacial hydraulic system, increased lubrication as a result of decreased basal friction, and higher shear stresses upon those parts of the basal ice that are still in contact with the bed (Iken *et al.*, 1983). Velocities may vary between summer and winter by as much as 20–100%. Velocity mode changes may also be rather sudden especially from winter to summer (Hodge, 1974; Iken, 1978, 1981; Hooke *et al.*, 1983, 1989; Andreasen, 1985; Raymond and Harrison, 1988; Harrison *et al.*, 1989). However, not all glaciers exhibit this winter-to-summer seasonal acceleration perhaps because seasonal and related meltwater penetration is not achieved so catastrophically and effectively (e.g., Blue Glacier, Washington). In examining the velocity perturbations of the Columbia Glacier, Alaska, a tidewater glacier, Walters and Dunlap (1987) were able to distinguish three specific velocity components viz: (a) low-frequency variations on a day-to-day scale related to periods of precipitation; (b) semi-diurnal and diurnal fluctuations associated with tidal periodicity; and (c) diurnal changes due to variations in surface meltwater runoff. The greatest impact on velocity changes was observed to be due to the effects of tidal movement. Observations made upon Jakobshavns Isbræ by Echelmeyer and Harrison (1990) show that, unlike many glaciers that show delayed but direct responses to seasonal climatic changes and fluctuating surface meltwater discharge, this very fast glacier is unaffected by these transient input variables (cf. Blue Glacier, Washington, note in Echelmeyer and Harrison, 1990).

In the case of valley glaciers and ice streams the influence of conduit geometry and frictional drag due to walls and edge effects have to be included in velocity calculations (Raymond, 1980; Morgan *et al.*, 1982; Echelmeyer and Harrison, 1990). Recently, Ridky and Bindschadler (1990) have pointed out the sharp distinction that can be made between ice streams and outlet glaciers, bearing in mind that many ice streams transmute, down the flow line, into outlet glaciers (e.g. Jakobshavns Isbrœ). Ice Streams typically have very low basal shear stresses (≈10 kPa) and may exhibit high basal velocities as a result of deformable bed conditions (M-beds). Fast outlet glaciers, in contrast, appear to have basal shear stress of ≈100 kPa and lie on H-bed types with high free meltwater pressures (Echelmeyer and Harrison, 1989). The key to any explanation of this differentiation must lie in disparate mechanisms and processes of basal sliding (Kamb, 1987; Echelmeyer and Harrison, 1990).

Where ice masses enter a tidal environment, two types of glaciers can be distinguished on the basis of their termini viz: (a) floating and (b) grounded. At present no temperate floating glaciers are known to exist and typically subpolar and polar ice masses usually have floating termini (Meier and Post, 1987). Since both types of tidewater glacier tend to exhibit high basal and surface velocities those factors affecting and controlling velocity such as tidal fluctuations, calving rates and basal sliding mechanisms are of critical relevance.

5.9. ICE VELOCITIES AND BASAL ICE CONDITIONS

The velocity of an ice mass is a function of mass balance, flow from internal deformation and, if the glacier bed is at pressure melting point, additional flow due to basal slip, and where underlain by deformable sediments a large proportion of ice movement is the result of a combination of both ice and sediment motion.

5.9.1. Polar Frozen Bed Conditions

Where ice is frozen to its bed (Polar), internal deformation is the dominant cause of movement. The relationship (discussed in Chapter 3, illustrated in Eqns (3.6) and (3.7)) between basal shear stress (τ_b) and mean basal ice velocity (u_b) can be expressed as (Weertman, 1957b):

$$u_b = \left(\frac{\tau_b}{B}\right)^m \qquad (5.39)$$

where B and m are constants with m= 2. Eqn. (5.39) is difficult to test as yet due to scarcity of data and therefore the balance velocity (ζ) is calculated instead (Fig. 5.23) from mass balance and ice thickness data. Balance velocity is defined as:

$$\zeta = \frac{1}{\bar{\rho}_i\, h_i} \int_0^x M\ dx \qquad (5.40)$$

where $\bar{\rho}_i$ averaged ice density. Until relatively recently Polar ice masses were thought to move exclusively by internal deformation (i.e. they were frozen to their beds). However, studies by Shreve (1984), Fowler (1986), Echelmeyer and Wang (1987)

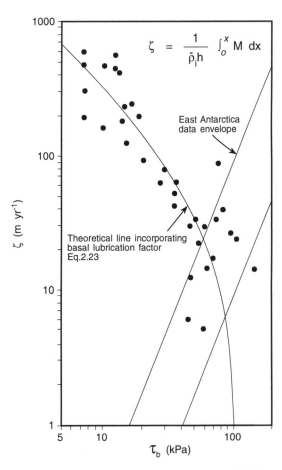

FIG. 5.23. Relationship between 'balance velocity' and basal ice shear stress for West Antarctic ice streams (circles) and for the East Antarctic ice sheet (after Drewry, 1983b, reprinted from Gardner, R. and Scoging, H. (eds) *Mega-Geomorphology*, 1983 by permission of Oxford University Press).

and Boulton and Spring (1986) illustrate that ice masses can 'slide' at their base at temperatures below pressure melting. Echlemeyer and Wang (1987) found that basal debris deformation, even under frozen conditions, can occur causing these ice masses to have higher (yet still low) than previously predicted velocities. The predicted sliding velocities of cold-based ice were formulated by Shreve (1984) based upon a model by Gilpin (1979) inter alia. Gilpin (1979, 1980) found that regelation could take place at sub-freezing temperatures due to the presence of microscopic but discrete liquid or liquid-like layers forming around foreign materials. Shreve subsequently suggested that basal sliding velocities under these 'cold' conditions could be theoretically defined as:

$$u_b = \frac{9}{4\sqrt{3}}(k_0\ a_{sr})^{-1}\ \frac{\langle\tau_b\rangle}{\eta} \qquad (5.41)$$

where k_0 is a critical wave number (Nye, 1970); a_{sr} is measure of surface roughness; and $\langle\tau_b\rangle$ is the basal drag component taken as equal to basal shear stress. While working on the Urumqui Glacier No. 1, China, Echlemeyer and Wang (1987) discovered low sliding basal velocities at sub-freezing temperatures but higher than expected from computation of Shreve's equation. Two explanations for this anomaly are thought to be due to the bed being much smoother than expected and the presence of a thin (1–2 mm) layer of low viscosity, debris-laden ice deforming at the lower boundary interface.

These findings are of importance in understanding subglacial erosion processes under ice conditions where zero interfacial boundary movement was thought not to occur. There are also important ramifications for glaciotectonics and deformed stratigraphy where, in the past, polar subglacial conditions prevailed.

5.9.2. Temperate Melting Bed Conditions

A similar general relationship exists where ice is at pressure melting point (temperate) at its bed. Basal melting occurs and ice flow is due to both basal slip and internal deformation (Eqn. (5.39)):

$$u_b \propto (\frac{\tau_b}{B})^m\ d_w \qquad (5.42)$$

where d_w is the depth of water film (Weertman, 1957b; Kamb, 1970). Improvements on the above Equation have been developed by Kamb (1970) where:

$$u_b = \tau_b\ /(c_r\ \cdot\ \eta)^{\frac{1}{2}}\ \lambda_R^2 \qquad (5.43)$$

where c_r is a constant related to regelation processes and λ_R is the bed roughness term.

Both equations for polar and temperate conditions are similar and predict increasing basal shear stress with increasing ice velocities. Data supportive of these equations are shown from East Antarctica in Fig. 5.23 (Drewry, 1983b). However, as Fig. 5.23 reveals, values of the balance velocity from West Antarctic ice streams do not fit the expected relationship.

Budd (1975) has suggested that where fast basal sliding occurs a positive feedback may exist between mechanical dissipation of heat from sliding and basal water production. Budd developed the following expression:

$$\tau_b = (\rho_i\ gh_i\ \alpha)/(1 + \Gamma\ \rho_i\ gh_i\ \alpha u_b) \qquad (5.44)$$

where Γ = a friction lubrication factor related to ice regelation where ($\Gamma \simeq \tau_b \cdot u_b^{-1}$. Budd suggested that the lubrication factor values may range from 10^{-3} to 10 kPa^{-1} m^{-1} year^{-1}. From Fig. 5.23 it can be seen that the Budd relationship provides a satisfactory explanation of the $u_b \cdot t_b$ relationship for those ice streams. However, in later work Budd et al., (1979) concluded that a simpler relationship could be shown to exist between local basal shear stress, effective normal stress and sliding velocity (Lliboutry, 1979; Bindschadler, 1983; Lingle and Brown, 1987) such that:

$$u_b = \frac{c}{\sigma_e}\tau_b^{\ m} \qquad \sigma_e > 0 \qquad (5.45)$$

where m ranges between values of 1 and 3. Where little or no debris exists beneath an ice mass or debris is present but conditions prevent deformable conditions from developing, fast ice motion as demonstrated by many outlet glaciers can still occur (Clarke, 1987b; Echelmeyer and Harrison, 1990). This fact may explain the past inconsistencies that were shown to exist between theoretical equations and actual observations.

5.9.3. Deformable Bed Conditions

In recent years it has become apparent that many glaciers and parts of ice masses flow across debris layers (1–6 m) that are themselves mobile. It is thought likely that parts of the Quaternary Ice Sheets were also underlain by deformable beds (Boulton and Jones, 1979; Hart et al., 1990; Menzies, 1990a; Alley, 1991). These soft beds act as a substantial lubricating layer causing ice masses to move at higher velocities than would be typical of ice flow solely due to basal slip and internal deformation. Boulton and Hindmarsh (1987) and Alley (1989a,b) have demonstrated that under soft bed conditions basal ice velocity (u_i) is a consequence of the merger of the component of velocity attributable to the deforming sediment (u_s) and the basal ice itself (u_b) where:

$$u_i = u_s + u_b \qquad (5.46)$$

in expanded form Eqn. 5.46 can be rewritten as:

$$u_i = \langle \ z_s \ B_f \ \frac{(\tau_o - \tau_s)^a}{\sigma_e} \rangle + \langle \ B \ \tau_b^2 \ d_w \ \rangle \qquad (5.47)$$

where a, B, and B_f are constants and τ_s is the shear strength of the deformable debris.

In general, where beds are smooth and/or soft and are spatially extensive, basal ice velocities are high. Under fast ice conditions disruption of subglacial water systems seem to occur resulting in inefficient meltwater evacuation (Kamb, 1987; Alley et al., 1989c). Smooth beds would appear to be typically undeformable, H-bed types, whereas beds under soft conditions (M-bed types) have limited free meltwater and are rough. Reality is, however, likely to be more complex with both bed types occasionally overlapping spatially and temporally (Q-bed conditions). Under Q-bed conditions, more typical ice velocities can be expected with subglacial drainage systems developing and surviving over longer periods of time. Lingle and Brown (1987) have shown that Eqn. (5.46) can be used to help explain the rapid motion of a decoupled ice stream through a much slower moving ice sheet.

The variable relationship of $u_b \cdot \tau_b$ may exhibit bifurcation in that at low values of basal shear stress there may exist high values of basal velocity and vice versa. This phenomenon and its transition from one velocity mode to another may indicate intrinsic instability in certain ice masses typical of surging behaviour. Hutter (1982b) has described these 'transition' phases and compared them, in catastrophe theory, to a 'cusp catastrophe'. However, no direct transitional relationship necessarily can be construed nor exists between different ice masses since transitory instability is an intrinsic parameter unique to each individual ice mass. Budd (1975) and Hutter (1982) recognize three distinct velocity categories or classes:

Class A: 'Ordinary glaciers' where the ice flux (discharge) for a given bed profile is of a low enough magnitude for the glacier to be in the 'slow mode', and no cavitation or any form of bed decoupling to exist. Flow can be described according to Glen's Law.

Class B: 'Fast glaciers' where the ice flux is sufficiently high to maintain the ice mass in a steady state in the 'fast mode', (i.e. where τ_b decreases with increasing u_b and some form of cavitation, as a result of high meltwater pressures, or bed decoupling, as in deformable bed development occurs) (e.g. West Antarctic Ice Streams).

Class C: 'Surging glaciers' where a sharp bifurcation 'jump' occurs cyclically between a long-term slow mode to a shorter term fast mode. This type of ice mass would appear to exhibit characteristics of both Classes A and B with an intrinsic instability that permits such a sharp velocity shift. The key to understanding these ice masses lies in understanding the mechanism that permits and cyclically maintains this inherent instability.

It is possible for glaciers to 'jump' from Class A type to Class C with no transitional phase (Budd, 1975; Budd and Young, 1979; Weertman and Birchfield, 1982; McIntyre, 1985). Significant changes in mass balance (measured in centuries), and/or bed characteristics (over tens of kilometres) and/or thermal conditions may, singly or in concert, provide the impetus for a glacier to change from one 'mode' to another. If such a change takes place, the impact upon sedimentological environments throughout the glacier system would be considerable.

It is likely, although impossible to establish clearly as yet, that each 'mode' form has a unique set of depositional facies (Sharp, 1985a, 1988; Croot, 1988b) and erosional bedform associations (Drewry, 1983b; McIntyre, 1985). Historical evidence would seem to point to such a major change in Ice Stream C (West Antarctica) and possibly parts of the Greenland Ice Sheet. Also, many outlets of the Antarctic Ice Sheet appear to exhibit similar velocity mode transitions from inland to the ice sheet margins (McIntyre, 1985).

The association of deeply incised through-valleys and fiord valleys with ice masses of outlet glacier or ice stream types (Class B) is well established (Sugden and John, 1976; Sugden, 1977, 1979; McIntyre, 1985). It has been suggested that surging glaciers (Class C) may lead to enhanced debris incorporation, increased compressive flow in the lower ablation area and rapid entrainment of debris and debris stacking near glacier snouts, thereby producing a distinctive depositional landscape (Clapperton, 1975; Croot, 1981; Sharp, 1985a). However, contradictory suggestions have been made that, under very rapid basal sliding with massive areal bed decoupling, bed debris entrainment would possibly be at a minimum -and thus a surging glacier may have little or no effect upon its bed (Sharp, 1982b). The possible permutations, however, of ice mass types, bed types, thermal fluctuations, forms of ice movement and variable debris contents all changing over time indicate that linking glacier type with sedimentological characteristics would be an arduous if not impossible task to complete, at least until more is known regarding fundamental glaciological and glacial sedimentological mechanisms.

5.10. GLACIER SURGES

Glacier surges have been described as "a brutal advance of the glacier front in a short period of time" (Michel, 1978) or as "the most impressive type of instability" (Paterson, 1980, p. 71) occurring in valley glaciers and ice sheets, ice streams and outlet glaciers (Clarke et al., 1986; Clarke, 1987a,c). There have been several conferences and review papers devoted to the 'surging' problem (Meier and Post, 1969; Dolgushin and Osipova, 1975; Budd and McInnes,

1978; Paterson, 1981; Clarke et al., 1 McMeeking and Johnson, 1986; Clarke, 198 Fowler, 1987; Raymond, 1987; Sharp, 1988; Dowdeswell et al., 1991). Ice masses subject to surging phases typically exhibit long periods of quiescent 'normal' flow behaviour that may last for several years, decades or centuries and are then interrupted periodically by short phases, lasting a few months to 2–3 years, of very rapid ice motion and advancement (Table 5.6). During the surge phase velocities may occur that are >100 times in magnitude greater than the quiescent phase. This periodicity is what is characteristic of surging ice masses but rarely can a temporal pattern be established.

Temperate and polar ice masses of all sizes, in a wide range of topographic and tectonic settings, on a variety of basement geological structures and lithologies, exhibit surging behaviour. In an extensive survey of Yukon glaciers (2356 surging and non-surging glaciers) Clarke et al. (1986) and Clarke (1991) found that there appeared to be a statistically higher probability of surging conditions occurring in long and wider glaciers with low surface slopes (Fig. 5.24). It was found that the probability of surging increased from 0.61% for short glaciers (0–1 km) to 65.1% for long glaciers (10–75 km). This result might suggest that large ice masses such as outlet lobes and ice streams of ice sheets and ice caps could have a greater tendency to surge than smaller valley glaciers. For example, it has been hypothesized that mega-surges of Antarctica may help trigger Global glaciation (Wilson, 1964, 1969). This speculation did gain some support (Hollin, 1965, 1972; Flohn, 1974), but the lack of sedimentological evidence in the Southern Ocean appears to negate the idea (Chappell and Thom, 1978). Most Antarctic surges appear to be of limited spatial impact (Lamb, 1977).

Two fundamental questions need to be answered if surging behaviour is to be understood viz. (1) what is the mechanism that enables fast ice motion to occur during a surge, and (2) what are the mechanisms whether external or internal that initiate and terminate surging? Related to these critical questions is the need to understand the pattern, if any, of the geographical distribution and topographic setting of surging ice masses, the duration and persistence of surge

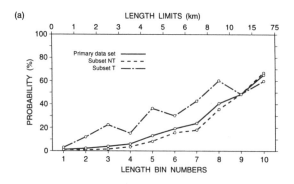

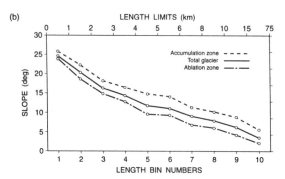

FIG. 5.24. (a) The influence of glacier length of surging. The variation of glacier length is shown for three different but related data sets: primary data set (solid line and subsets T (dot-dash line) and NT (dashed line) of the primary data set. Subset T is derived from the primary data set by selecting only glaciers that are tributaries of some larger glacier system; subset NT contains the remaining glaciers. The data set based upon Yukon glaciers has been divided into 10 BIN lengths denoted by index l. (b) The influence of slope on glacier surging. A subset of the Yukon glacier primary data was used for this analysis (after Clarke *et al.*, 1986; *Journal of Geophysical Research*, **91B**, p. 7165, copyright by the American Geophysical Union).

periodicity transformation between normal and surging flow states, the continuum of glaciological states between surging, normal and continuous fast ice flow conditions, the surging potential of present day ice sheets, and whether past ice sheets exhibited surging conditions (Raymond, 1987).

A major problem that may partly account for the proliferation of hypotheses of surging mechanisms is the lack of sound data acquired from direct observations of the surging process in progress. Until relatively recently few observations had been accurately made immediately before, during and subsequent to a surge event. New data on the 1982–83 surge of the Variegated Glacier, Alaska and work from the Trapridge Glacier, Yukon Territory, Canada and the Medvehiy Glacier, Pamirs, Tadzhikistan have provided fresh insights into surge behaviour. However, it is speculated that surging behaviour under different basal thermal regimens may be initiated by slightly disparate mechanisms (Hattersley-Smith, 1964; Raymond, 1987). If this is the case, then there are still insufficient data upon surges under differing basal thermal conditions.

5.10.1. Surface Appearance and Morphology

Glaciers that exhibit surging behaviour have a characteristic surface appearance and distinctive morphology. However, not all the characteristics are peculiar to surging ice masses nor do all such glaciers display all of these features. Surging glaciers typically display: contorted (loop-like) surface medial moraines; sheared and heavily crevassed edges; severely crevassed surfaces; a large kinematic bulge in the 'reservoir area', high in the accumulation zone, associated with a depleted ice 'receiving area', usually in the ablation zone, that is manifested as surface ice gradients higher and lower than 'normal', respectively; the presence of tributary glaciers with sheared snouts and/or ice-dammed lakes and the presence of large 'dead-ice' areas immediately in front of the glacier snout. During the surge period other features indicative of the state are: high meltwater discharges into the proglacial area; crevasses opening and closing at much higher rates than normal; the rapid appearance of up-sheared debris and clasts in the ablation area; ice gradient changes consequent on the transference of vast volumes of ice from the reservoir to the receiving areas; and the sudden draining of ice dammed lakes near the end of a surging phase (Sharp, 1988).

Based upon recent observations at the Variegated Glacier and other North-western North American glaciers, several other characteristics would appear typical of surging behaviour (Clarke and Collins, 1984; Kamb *et al.*, 1985; Brugman, 1986; Clarke *et al.*, 1986; Harrison *et al.*, 1986; Humphrey *et al.*, 1986; Fowler, 1987; Kamb, 1987; Kamb and

(a)

(b)

PLATE 5.4. (a, b) The characteristic surface moraine morphology following a surge on the Susitna Glacier, Alaska (photographs courtesy of Ed Evenson).

Engelhardt, 1987; Raymond, 1987; Sharp *et al.*, 1988; Clarke and Blake, 1991). First, it would seem well established that fast flow during a surge is a function of rapid basal sliding. Secondly, subglacial meltwater transmission across the total extent of a surging ice mass appears to take place during the surge phase but becomes restricted and/or constricted once normal flow conditions resume. Thirdly, in somewhat of a contrasting effect, it has been noted that meltwater discharge through the subglacial conduit system is much slower during the surging phase than under normal flow conditions. Fourthly, close to surge cessation or when major surge 'slow-downs' occur, vast discharges of meltwater take place. Fifthly, it is apparent that high sliding velocities are associated with high basal meltwater pressures possibly indicative of the existence of extensive basal cavitation during surging, such that on cessation of surging glacier surfaces are lowered. A sixth observation is that immediately prior to surging an episode of longitudinal shortening followed by one of elongation precedes the peak velocity passage through the glacier. This transient fluctuation in longitudinal strain rate causes differential zonation to develop within the surging ice mass such that ice above the point of velocity peak initiation undergoes continuous and cumulative elongation, while ice below the maximum velocity peak experiences continuous and cumulative shortening. Ice between these two points experiences shortening followed by elongation and thus relatively low cumulative longitudinal strain (Sharp *et al.*, 1988). Finally, at least beneath the Variegated Glacier, it has been observed that during surging basal meltwater pressures rise to within 20–50 kPa of the normal ice overburden pressure but fall, generally, to ≤15 kPa under normal flow. This correlation between high meltwater pressures and surging would appear to indicate that both aspects are closely related to each other. In any formulation of a hypothesis of surging, these above features and attributes must be satisfied.

5.10.2. The Surge/Non-Surge Cycle

As noted above, any surging glacier or ice mass manifests a two-stage flow state in which a long quiescent phase of ice flow is periodically punctuated by a short but very rapid period of ice flow (Table 5.5). As can be seen from Table 5.5 where comparisons can be made between the Variegated and the Medvezhiy Glaciers, there are many broad resemblances in behaviour across the two-stage process of surging and non-surging states. For example, similarities exist between when the surge stage began and ceased. There are, also, interesting discrepancies in ice mass displacement and daily velocity during the surge phase.

Raymond (1987) summarized the developmental stages observed on the Variegated and Medvezhiy Glaciers as manifest in changes in ice surface geometry and velocity (Fig. 5.25). As a potential surging glacier approaches the next surge, there is a buildup of ice within the reservoir area and depletion in the receiving area, resulting in increasing ice velocity within the reservoir. At this stage, in what may be termed the pre-surge phase, there develops a state in which ice within the reservoir area is increasingly active while the ice below in the receiving area may be virtually stagnant. The boundary between these two areas has been termed the dynamic balance line (DBL). It was observed on the Medvezhiy Glacier, in the pre-surge stage, that the DBL moved progressively down the flowline prior to the surge event introducing, perhaps, a critical ice mass that may act as a 'trigger' to surge initiation (Dolgushin and Osipova, 1975; Dyurgerov, 1986). In contrast, motion of the DBL on the Variegated Glacier was not readily observable since, unlike the Medvezhiy where crevasse systems became active and opened up in the reservoir area, little or no difference was noted between this glacier and other mountain glaciers in the region. However, basal sliding velocities of the Variegated Glacier in the pre-surge stage were unaccountably higher than normal during the winter period (Bindschadler *et al.*, 1977).

As a glacier approaches the succeeding surge event, occasionally sudden short pulses of rapid, almost surge-like events, have been observed. These phases were noted on the Medvezhiy where the term wavy surges was used and on the Variegated where the term mini-surges was employed (Dyurgerov, 1986). Considerable study of the mini-surges on the Variegated Glacier has occurred (Harrison *et al.*, 1986; Kamb and Engelhardt, 1987; Raymond and

TABLE 5.5. Comparison of Surging Glaciers—the Variegated and Medvezhiy Glaciers (after Raymond, 1987).

	Variegated	Medvezhiy
General		
Total length (km)	20	16
Surging length (km)	18	7-8
Surging period (yr.)	~18	9-14
Mean slope of surging length	0.094	0.11
10-m Temperature (°C)	Temperate	-0.9
Quiescent Phase		
Terminus retreat (km)	zero	<1.5
Advance of dynamic balance line	1	>4
Elevation change (m)	+60(-40)	+120(-100)
Maximum Basa shear stress (10^5Pa)	1.8	?
Maximum annual velocity (md^{-1})	0.6	1.5
Summer velocity increase (%)	80	100
Surge Phase	1982-1983 surge	1963 & 1973 surges
Number of events/duration	2, 6-8 months	1
Timing of initiation	early Winter,late Fall	late Winter ?
Timing of termination	early summer	early summer
Advance of surge front (km)	12	?
Advance of topographic peak (km)	11	?
Maximun thickness change (m)	110	120
Advance of velocity peak (km)	11	?
Maximum speed (md^{-1})	65	100
Maximum ice displacement (km)	~2	>1.6

Malone, 1987; Raymond and Harrison, 1988; Dowdeswell *et al.*, 1991). In general, these mini-surges last for only a few hours and are a manifestation of sudden peaks in basal meltwater pressure (Fig. 5.26). It would appear that these minor surge pulses were related to major structural and geometric changes taking place in the subglacial hydraulic system of the glacier and occur at the beginning of the melt season as the ice mass adjusts from winter to spring/summer flow conditions.

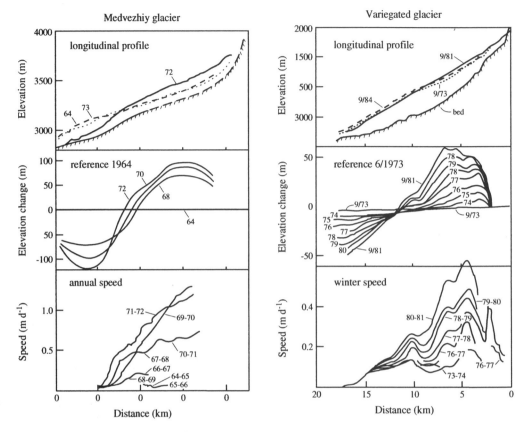

FIG. 5.25. Variation and development of longitudinal surface profiles, change in surface elevation and annual speed during quiescent periods of two surging glaciers — the Medvezhiy and Variegated Glaciers (after Raymond, 1987, *Journal of Geophysical Research*, **92B**, p. 9123, copyright by the American Geophysical Union).

However, the occurrence of mini-surges is not necessarily related to the major surge event nor indicative of surging behaviour since such events do occur on normal glaciers (Iken *et al.*, 1983).

Observations made on the Trapridge Glacier during the non-surge part of the cycle revealed a sharp spatial differentiation in basal thermal regimen with thick ice in the reservoir area underlain by a temperate bed separated by the DBL from thin ice in the receiving area underlain by a frozen bed (Clarke *et al.*, 1984). Whether this form of basal thermal development is characteristic of sub-polar surging glaciers remains unknown but elsewhere such glaciers have been observed to have warm temperate beds (Clarke and Jarvis, 1976; Harrison *et al.*, 1975).

As the ice mass enters the surge phase of the cycle there is a reversal of the geometric evolution of the

quiescent phase (Raymond, 1987; Sharp *et al.*, 1988). The typical characteristics of a surge event begin anew (see above). Any surge is perhaps best typified by the sudden rapid change in ice mass velocity usually beginning somewhere in the reservoir area and spreading down the flow line into the receiving area. This sudden acceleration of ice may occur as a single or staggered series of non-steady state events, and the surge velocity event(s) may be widespread or initially spatially restricted, but in time usually affect a major part, if not the whole, of the ice mass (Harrison, 1964; Post and LaChapelle, 1971; Kamb *et al.*, 1985; McMeeking and Johnson, 1986; Dyurgerov, 1986; Fowler, 1987; Raymond *et al.*, 1987; Raymond and Harrison, 1988; Sharp, 1988; Sharp *et al.*, 1988). As an example Raymond (1987) has illustrated the space–time evolution of the surge

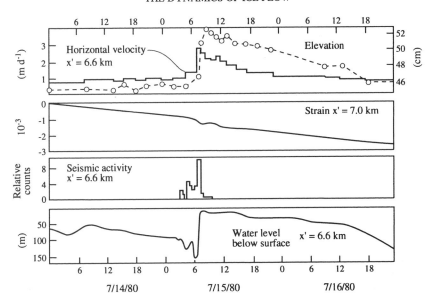

FIG. 5.26. Variation in velocity, surface elevation, longitudinal strain, seismic activity and borehole water levels on the Variegated Glacier during the 5th minisurge of 1980 (after Raymond, 1987, *Journal of Geophysical Research*, **92B**, p. 9124, copyright by the American Geophysical Union).

phase of the Variegated Glacier 1982–83 (Fig. 5.27). At present the Trapridge Glacier which is in an advance stage of pre-surge conditions, exhibits a 5-fold increase in velocity since 1969 in the upper temperate portion of the glacier; while 20% of the receiving area is at present deglaciated (Haeberli *et al.*, 1986). As the surge events approach the end, a series of decelerating stages appear to occur separated by several days; these 'slow-downs' may affect small areas in the up-ice regions of the ice mass or propagate down the whole glacier (Kamb *et al.*, 1985). As noted above, close to surge cessation, basal meltwater pressures drop and typically large meltwater discharges occur at the glacier snout.

The cycle of the surge period, based upon current limited data, is rarely rigorously constant but often exhibits irregular patterns sometimes of a single major surge event separated by differing lengths of non-surge periods or, as in the case of the Bering Glacier, Alaska, two short surge events close together followed by a longer quiet interval (Meier and Post, 1969; Post, 1972; Dowdeswell *et al.*, 1991). Surge periodicity (Table 5.6) would appear to be connected with periodicity in longitudinal strain rate change (Shoji and Langway, 1985). The timing of surge

initiation and termination have been compared for several glaciers (Raymond, 1987) but data are limited. There does, however, appear to be some consistency among the glaciers that have been studied that suggests surges often begin in the winter and terminate in the early spring/summer (Raymond and Harrison, 1988). But some glaciers such as the Muldrow in 1956 (Harrison, 1964), and Bruarjökull in 1963–64 (Thorarinsson, 1969) continued through a summer period and the Black Rapids Glacier in 1936–37 (Hance, 1937) stopped in winter. It would seem likely that there should be some type of temporal pattern to emerge as more surging glaciers are studied. However, the specific mechanism of initiation may differ from glacier to glacier and therefore the timing and pattern of initiation and termination may reveal some clue as to which particular mechanism is in operation.

5.10.3. Hypotheses of Surge Behaviour

Debate as to the specific mechanism(s) to explain surging behaviour has continued unabated for the past twenty years. In the process a profusion of hypotheses have surfaced (Table 5.7). Whether one or several

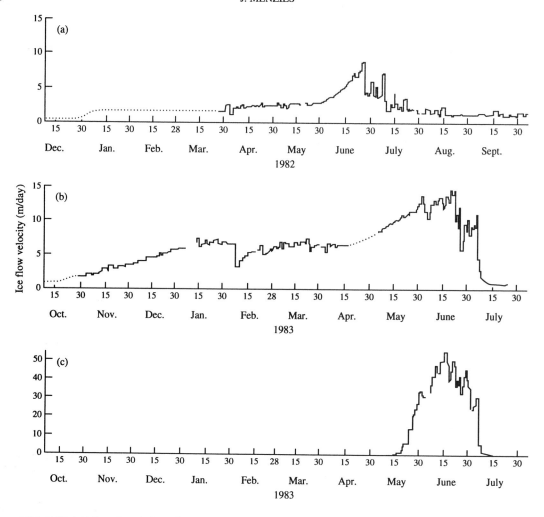

FIG. 5.27. (a) Diurnal variation of velocity on the upper (a,b) and lower (c) parts of the Variegated Glacier.

mechanisms can explain surging or whether surging is externally or internally triggered have also been sources of ongoing discussion (Clarke, 1986, p. 133). Queries as to whether surging occurs in only steep or temperate glaciers or in glaciers that are in tectonically active regions or beneath glaciers of abundant subglacial debris also continue to be raised. However, as observations on surging glaciers continue and monitoring of full surge/non-surge cycles are accomplished, the field of likely hypotheses has been significantly narrowed. There is a general consensus that the surging mechanism is associated in some manner with changes in the subglacial hydraulic system (Engelhardt, 1986, pp.

133–140; Fowler, 1987, 1989). As yet those observations on the Variegated and Trapridge Glaciers remain the most fully documented and act as the criteria for testing existing surge hypotheses.

In summary, two dominant ideas both connected with subglacial hydraulic system disruption have emerged viz: (1) Kamb's (1987) suggestion that a linked-cavity subglacial meltwater system develops to replace a hydraulic system composed of a few large drainage channels and acts to trigger the surge and, in disruption (short-circuiting), terminates the surge; and (2) Clarke et al.'s (1984) concept of subglacial permeable bed conditions ceasing due to massive basal shearing resulting in impermeable subglacial

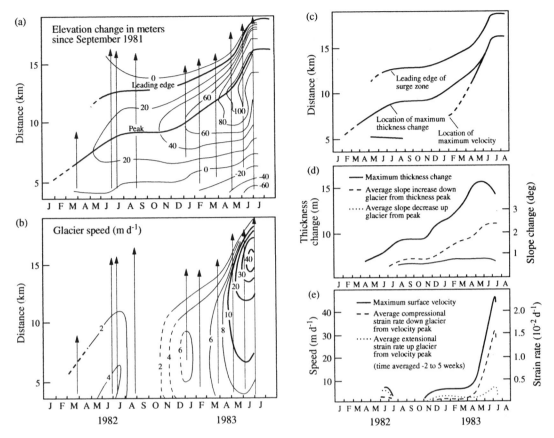

FIG. 5.27. (b) Smoothed space–time development of change in surface elevation and speed during the 1982–1983 surge of Variegated Glacier (after Raymond, 1987, *Journal of Geophysical Research*, **92B**, p. 9127, copyright by the American Geophysical Union)

bed conditions and H-bed conditions prevailing leading to high basal meltwater pressures triggering the surge which sequentially leads to basal friction reduction, drainage system deterioration and subsequent surge termination.

5.10.3.1. Linked-Cavity Mechanism

Kamb (1987) developed a surging model based upon observations at the Variegated Glacier during the 1982–83 surge. The underlying premise in Kamb's model considers how high basal meltwater pressures are maintained during a surge event and increased during spring and early summer when, according to the standard model (Röthlisberger, 1972), basal water pressure should fall as the hydraulic system enlarges and adjusts to increased meltwater flux (Chapter 6). As the hydraulic system enlarges by increasing cavity evolution, an increasing discharge of meltwater should occur; however, in contrast, Kamb found that during surges basal meltwater experiences increasing retentivity beneath the ice mass. The key to the Kamb model therefore hinges upon the maturation of the basal hydraulic system from a few major conduits to a myriad of linked cavities as the causative influence in triggering and terminating a surge event.

Under non-surging conditions the transport of basal meltwater beneath a valley glacier typically feeds into a single or a few main conduits or tunnels that are located close to the centre-line or deepest part of the glacier (Chapter 6) (Röthlisberger, 1972; Weertman, 1972; Spring and Hutter, 1981b; Lliboutry, 1983). These main tunnels are normally ≈1–5 m in diameter and are fed by lateral tributaries from both subglacial and englacial/supraglacial sources.

TABLE 5.6. Duration of the active phase of the surge cycle for glaciers in polar and mountain areas of the World* (after Dowdeswell *et al.*, 1991).

Glacier	Length (km)	Area (km²)	Duration (yr.)	Reference
Alaska, Yukon Terr. British Columbia				
Variegated, AK	20	49	2	Kamb et al., 1985
Walsh, AK	89	830	4	Post, 1960
Muldrow, AK	46	393	2	Post, 1960
Peters, AK	27	120	2	Echelmeyer et al., 1987
Tikke, BC	19	75	3	Meier & Post, 1969
Tyeen, AK	7	11	2	Field, 1969
West Fork, AK	41	311	1	Echelmeyer & Harrison, 1989
Rendu, AK	17	50	2	Field, 1969
Hazard, Yuk	8	-	2	Clarke & Collins, 1984
Childs, AK	19	20	1	Tarr & Martin, 1914
Unnamed, AK	6	-	2	Krimmel, 1988
Black Rapids, AK	45	341	1	Moffit, 1942
Carroll, AK	42	200	1	Smith, 1990
Iceland				
Sidujökull	40	350	1	Thorarinsson, 1964
Dyngjujökull	-	-	2	Thorarinsson, 1964
Tungnárjökull	25	120	1	Thorarinsson, 1964
Brúarjökull	50	1500	2	Thorarinsson, 1969
Eyjabakajökull	10	50	1	Williams, 1976
Teigadaljökull	1	1	1	Halgrímsson, 1972
Hagafellsjökull eystri	17	110	1	Sigbjarnarsson, 1976
Hagafellsjökull vestari	17	100	1	Theodórsson, 1980
Pamirs and Caucasus				
Medvezhiy	13	25	1	Dolgushin & Osipova, 1975
Byrs	2	10	2	Rototayev, 1983
Shini-bini	10	16	2	Uskov & Dil'muradov, 1983
Sugran	22	47	3	Uskov & Dil'muradov, 1983
Muzgazy	11	16	2	Desinov, 1984
Abramova	-	-	3	Glazyrin et al., 1987
Ravak	3	2	1	Dolgushin & Osipova, 1975
Tanyrnas	10	61	2	Dolgushin & Osipova, 1975
Burokurmas	7	8	2	Dolgushin & Osipova, 1975
Kolka	6	3	1	Krenke & Rototayev, 1973
Tien-shan				
Mushketov	21	70	2	Dolgushin & Osipova, 1975
Bezymyanny	6	11	2	Dolgushin & Osipova, 1975
Karakoram				
Kutiah	12	-	1	Desio, 1954
Baltbare	8	-	2	Wang et al., 1984
Andes				
Grande del Junca	10	9	1	Espizua, 1986
Grande del Nevado	6	5	1	Bruce et al., 1987

TABLE 5.7. Suggested Surge Release Mechanisms.

Release Mechanism	Instability Manifestation	References
Tectonic activity	causing increased avalanching and perhaps increased basal geothermal heating	Tarr & Martin, 1914 Post, 1965
Landsliding	increased load added to reservoir area	Gardner & Hewitt, 1990
Creep instability	(a) increased "softening" of basal ice thus increased basal ice deformation	Robin, 1955 Clarke et al., 1977 Paterson et al., 1978
	(b) change of basal thermal regime from frozen to warm bed	Yuen & Schubert, 1977
	(c) change in mass balance thus increased mass balance gradient	Clarke et al., 1977 Schubert & Yuen, 1982
Melted Core	Instability beneath reservoir area	Schytt, 1969
Lubrication	Increased basal meltwater - bed decoupling	Budd, 1975 Budd & Jenssen, 1975 Weertman, 1962 Robin & Weertman, 1973
Friction Melting	Increased basal velocity -increased meltwater	Clarke, 1976 Budd & McInnes, 1974
Basa Shear Stress Fluctuation	Increased basal slip - increased meltwater	Meier & Post, 1969 Robin, 1969 Robin & Weertman, 1973
Kinematic Wave transmission	Increased basal slip - increased meltwater	Weertman, 1969 Palmer, 1972
Super-cavitation	Increase area of bed with cavity development-decoupling	Lliboutry, 1977
Frontal instability, calving	Removal of frontal ice mass support	Hughes, 1981
Linked-cavity development	transition between localised conduit expansion and short-circuit to linked cavities-decoupling	Kamb, 1987
Deformable bed/Hydraulic instability	Decoupling of bed due to decreased viscosity within deforming subglacial sediment	Clarke et al., 1984

From the observations at the Variegated Glacier, it is apparent that a very different system of hydraulic evacuation must exist during a surge phase (Brugman, 1986). The high meltwater retentivity and high basal meltwater pressures indicate the existence of a widespread cavity system that is extensive over a large transverse region of the glacier bed (Fig. 5.28). This linked cavity model must also allow for a constricted flow (throttling) through the interconnected cavities. The model envisages cavities, averaged over beds ≈1km in width, of ≈0.2m in height distributed over ≈10–20% of the glacier bed. The large area of meltwater/bed contact may account for the high turbidity of exiting meltwater during a surge event.

The initiation of the surge lies in the 'short-circuiting' effect on the main conduit system produced as sliding velocities increase in the final pre-surge stage leading to a state in which major conduit stability in relation to water discharge and cavity geometry ceases to function and the pre-surge hydraulic system is disrupted. The disorder takes the form of an uncompensated down-ice shift leading to rapid isolation of the once fully developed drainage system. The effect of the disruption is to create transverse cavities that, in the first instance, are not connected (Fig. 5.28). The timing of the conversion from an arterial tunnel system to one of linked cavities would seem more likely during the winter when water discharge is at its lowest. Simultaneously, as drainage mutation occurs, in effect, a surge activation caused by the passage of ice mass flux both up and down glacier occurs (Fowler, 1987, Fig. 6). In time, as water pressures increase, hydraulic interconnections will evolve in the form of small connections or orifices. Kamb visualizes these

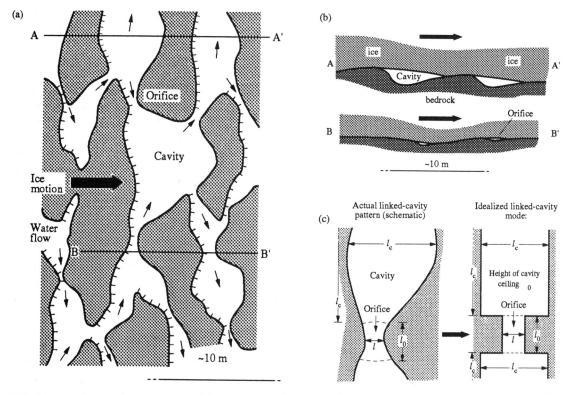

FIG. 5.28. Linked-cavity model of glacier surging. (a) linked-cavity conduit system in plan view; (b) cross-section of linked-cavities; (c) geometry of cavity system and orifice conventions (after Kamb, 1987, *Journal of Geophysical Research*, **92B**, p. 9086, copyright by the American Geophysical Union).

orifices in two forms viz: wave and step types (Fig. 5.29). The hydraulic integrity of the now linked cavity system is determined by the controlling influence of these orifices (Fig. 5.28). Water flow is now controlled by these orifices and it is envizaged that flow through the cavities would be relatively slow, but fast and turbulent through the connecting orifices where water may emerge into the next cavity as a jet-like flow. A state is therefore arrived at where basal water is under high pressure, constricted in its discharge, thus maintaining a stable linked cavity system and high basal sliding (surging). The stability of the system has been shown by Kamb, within finite perturbation limits, to be stable under conditions that differ from those under which tunnels evolve. Tunnels are essentially unstable features at the glacier bed that react immediately to either positive or negative perturbations in size and water pressure. As

considered when discussing jökulhlaups, once a tunnel begins to enlarge it does so at an ever accelerating rate. In the case of the orifices that link the cavities in Kamb's model, no such runaway development occurs provided any perturbations to the system are small. Kamb has estimated that orifice size and growth/shrinkage is a function of the process of cavitation development and the amount of cavity and orifice roof meltback caused by viscous heat dissipation as a consequence of water flow through the system. The shape and degree of roof meltback of orifices are computed in terms of a melting-stability parameter (Ξ). This parameter varies for different orifice types viz: step and wave orifices (Fig. 5.29); these orifices have quite different basal separation characteristics. Provided the melting-stability parameter values for step or wave orifices are limited to $\Xi < 1$ and $\Xi < 1.5$, respectively, the orifice will remain stable. Where these values are exceeded due to excessive viscous heat generation and dissipation, instability will develop. This delimiting size of perturbation, however, decreases as the value of Ξ approaches 1 and 1.5 for step and wave orifices, respectively. Unlike tunnel instability which is bilateral (i.e. positive and negative perturbations act to destabilize a tunnel's quasi-stability) under a linked cavity system only a unilateral increase in size or pressure will lead to a runaway instability (Fig. 5.30). It is at this stage in the surge event that if a massive and areally extensive perturbation occurs, the linked cavity system will revert to a multiple-conduit channel system which in turn will reduce to a few major tunnel conduits and the surge will quickly terminate. With termination, basal water pressures will drop and discharge will increase. At the Variegated Glacier, just prior to surge termination, a peak in water pressure was observed that may have acted as the necessary widespread perturbation to disrupt the linked cavity system and cause runaway orifice growth and eventual linked cavity system collapse.

The existence of a linked cavity system under certain ice masses leading to surge behaviour would appear to be a special conjunction of basal water pressures, hydraulic system development in relation to basal sliding conditions and a particular bed morphology. It is unlikely that under mobile

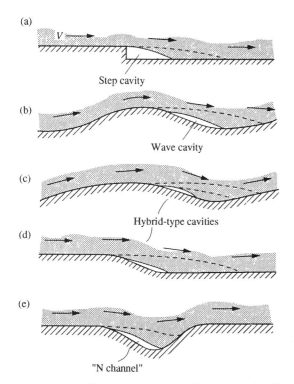

(a)

Step cavity

(b)

Wave cavity

(c)

Hybrid-type cavities

(d)

(e)

"N channel"

FIG. 5.29. Idealized models of ice–bed separation by cavitation in basal ice sliding. Scale is arbitrary but in Kamb's linked-cavity hypothesis a separation gap orifice with step or wave heights of ~0.1m (after Kamb, 1987, *Journal of Geophysical Research*, **92B**, p. 9088, copyright by the American Geophysical Union).

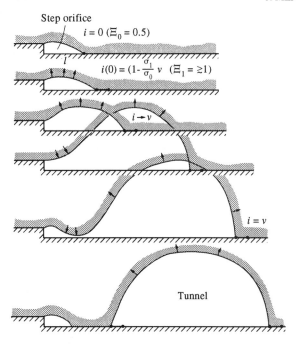

FIG. 5.30. Illustration of change in a progressively unstable step orifice with time increasing from top to bottom (after Kamb, 1987, *Journal of Geophysical Research*, **92B**, p. 9098, copyright by the American Geophysical Union).

deformable bed conditions a linked cavity system could operate. Where a deformable bed transmutates toward a Hard-bed a linked cavity system model could possibly evolve. Only as more data emerge will this model of surging behaviour be capable of detailed testing.

5.10.3.2. Subglacial Sediment/Hydraulic Sensitivity Mechanism

Based upon observations made at the Trapridge Glacier, Yukon Territory, Canada, Clarke *et al.*, (1984) have developed a preliminary model to explain surging over what would appear to be an unconsolidated sediment glacier bed. The Clarke model set out in the first instance to test the validity of shear heating and subglacial water damming as possible triggers for a surging event (Table 5.7). Both mechanisms were discounted as operating at the Trapridge Glacier. An alternate model has been proposed in which, in the pre-surge stage, a well-developed subglacial drainage system of tunnels of

the N-type variety is presumed to have been incised into undeforming unconsolidated material (i.e. H-bed conditions) within the warm-based section of the glacier and to have acted as adequate means of subglacial meltwater evacuation. It is also suggested, based upon drilling observations, that beneath the polar section of the glacier bed, meltwater continues to flow via the sediment subjacent to the frozen sediment immediately beneath the glacier. Flow in this latter position would be of a pipe-flow type controlled principally by the transient permeability of these sediments. As the surge event is approached a buildup of ice in the reservoir area occurs possibly the result of differential lateral basal thermal regimens viz: a warm bed state beneath the reservoir and a polar based area below the receiving area the latter acting to decelerate ice flow at the boundary between the basal thermal regimens. A very pronounced bulge at the junction on the Trapridge developes, and a state is reached within the unconsolidated basal debris of incipient strain-induced deformation (Plate 4.2). When the strain level attains a value sufficient to cause the unconsolidated sediments to deform, it is suggested that the N channel system is destroyed and any pipe-flow passage of meltwater is severely disrupted. The effect is to cause a build up in subglacial meltwater pressures since no adequate mechanism of evacuation now exists, thus leading to the triggering of a surge. The termination of the surge is envisaged when, due to massive volume transference of ice during the surge, the ice mass thins and strain levels diminish reducing sediment deformation and the return to an N-channel hydraulic system. Data on meltwater discharge during the surge of the Trapridge Glacier are not given, thus it is difficult to compare with the recent surge of the Variegated Glacier. However, it is intriguing to wonder if, following the destruction of the N-channel system and the triggering of the surge, a linked cavity system evolved and surge continuation and termination was similar to that described by Kamb.

Fowler (1987, 1989), in the light of the above models based upon field observations, has suggested that, provided the following 'ingredients' exist, then a particular glacier should surge. First, a sufficiently high accumulation in the reservoir area is necessary. Secondly, the glacier bed must be sufficiently rough if

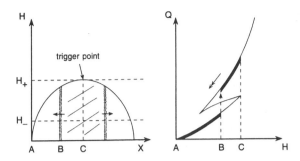

FIG. 5.31. Model of surge activation. As H reaches H_+ at point C a hydraulic transition front propagates both up and down glacier thus transferring the ice to the upper branch of the Q–H curve (after Fowler, 1987, *Journal of Geophysical Research*, **92B**, p. 9116, copyright by the American Geophysical Union).

the drainage transmutation from a tunnel to cavity system is to result in the passage of a wave both up and down the glacier from the point of initiation that triggers the surge event (Fig. 5.31). Where bed roughness is too great or small (smooth) the effect of the drainage system alteration is not transmitted through the glacier basal interface and no surge is triggered. Thirdly, even if bed roughness is within the limits necessary for a surge trigger, unless the ratio of surge to non-surge sliding velocity (u_+/u_b) is large and the value δ small, the hydraulic 'jump' from one system to the other will not occur (Fowler, 1987). Fowler defines the value δ as:

$$\delta \approx sn_K^{-1/4n} \qquad (5.48)$$

where s is area fraction of the bed that is cavity/passages free and n_K is the number of cavities/passages across the width of the glacier. As the value of δ decreases, for example as the result of increased passage tortuosity (Walder, 1986), so basal water pressures increase and a linked cavity system is more likely to be the most stable form of hydraulic evacuation.

At this stage in our understanding of surges, the model suggested by Kamb is largely restricted to hard bed conditions and that of Clarke envisages a soft bed. As yet we lack an integrated model of 'true' bed conditions just as we lack a full understanding of deformable bed conditions. Fowler (1987) hypothesizes that both hard/soft and temperate/subpolar conditions may be mathematically remarkably similar when rationalising surge activity.

5.10.4. Surging and Glacial Sedimentology

The influence of surging ice masses upon processes in proglacial, supraglacial, englacial and subglacial environments remains obscure (Sharp, 1988). However, chrono-stratigraphic evidence seems to indicate that several lobes of Quaternary Ice Sheets may have surged (Clayton *et al.*, 1985; Lagerlund, 1987; Aber *et al.*, 1989). For example, outlet lobes along the southern margin of the Laurentide Ice Sheet (Wright, 1973, 1980; Clayton and Moran, 1974; Mickelson *et al.*, 1983; Clayton *et al.*, 1985) and the eastern lobe of the British Devensian Ice Sheet (Boulton *et al.*, 1977) may have surged. The development of a model that would allow the recognition within the stratigraphic record of paleosurges, however, remains elusive (Sharp, 1988). Work in Wisconsin, Michigan and Montana indicate lobes having low surface profiles perhaps indicative of low basal stresses, basal water lubrication, deformable beds and a surge-phase (Mathews, 1974; Boulton and Jones, 1979; Clayton *et al.*, 1985; Beget, 1986).

Similarly, a relationship has been tentatively advocated between surging and the formation of flutings due to inherent flow instabilities of secondary flow patterns that may develop within the ice (Kamb and Shreve, 1963; Boulton, 1976c, 1978; Krüger, 1979; Thompson, 1979; Sharp, 1985a). The relationship of composite moraine ridges, common along the margins of several surging glaciers in Iceland and Svalbard, has also been investigated (Croot, 1981, 1988; Sharp, 1985a). Geomorphic evidence from the Fisher Glacier, Eastern Antarctica demonstrated that a series of elevated lateral moraines unique to that particular glacier are best explained by this outlet glacier surging in the past (Wellman, 1982).

Much remains to be investigated in terms of frequency, size, duration, ice mass conditions before, during and after surging and the impact of such events upon sedimentological processes and environments. The potential impact of surging involves (A) the effect of increased subglacial and proglacial meltwater activity allied with the disruption and possible destruction of all supraglacial and englacial hydraulic systems during and for some time after the surge event; and (B) the impact of stress/strain fluctuations upon subglacial and proximal proglacial sediments and bed/landforms, and the erosive effect of high basal sliding velocities upon the subglacial bed.

(A) The consequences of meltwater activity can be summarized as follows:

(i) Prior to surging most ice masses have well-developed hydraulic systems. Depending upon climatic and glacial thermal regimens an englacial hydraulic system may be in place. It is thought likely that a tunnel system of a few major conduits with smaller tributary channels will exist in equilibrium with subglacial stress conditions and bed topography. In the Clarke model, meltwater evacuation would, in the temperate areas, drain via N-channel systems and by pipe flow within subjacent unfrozen sediment within the frozen areas beneath the glacier.

(ii) On surging, the supra- and englacial hydraulic systems are probably destroyed. In the former case if the surge begins in winter, as is typical, little or no supraglacial system will have evolved, and in the latter case the englacial system may be shut down or operating at very low discharges. If Kamb's model is accepted, once the surge is triggered a completely new linked cavity system will evolve. Clarke's model predicts that meltwater N-channel systems would be destroyed and drainage via the upper interface might be similar to Kamb's linked cavity system.

(iii) With surge termination, there would be a time-lag before the pre-surge hydraulic system could re-establish. A drop in meltwater pressure would probably occur as the surge hydraulic system collapses and a jökulhlaup-style flush of turbulent and high discharge occurs just prior to the cessation of the surge or immediately afterward.

(iv) During and immediately following a surge, vast quantities of silts and clays will be transported by meltwater. The turbulence and turbidity of surge meltwater systems appear to be related to the development of a hydraulic system spreading across the glacier bed transporting sediment from areas not previously flushed since the last surge event. With meltwater under high pressure, considerable scouring erosive activity must occur during surge events.

(v) Sediment transport within supra- and englacial parts of a surging glacier systems is two-fold; first, sediment is transported into different environments. In particular, englacial debris will be transported into the subglacial system as a result of massive undermelting. Secondly, large quantities of sediment will be enclosed within sealed crevasse systems and moulins thus disrupting the 'normal' sediment transport flux in these environments.

(vi) A final effect of surging upon meltwater activity is to introduce a vast quantity of meltwater transported sediment into proximal proglacial areas.

(B) A second less precise effect of surging is the influence of transient but widely fluctuating stresses and strains that may impact upon subglacial and proglacial sediments and bedforms (Chapter 12). Raymond et al. (1987) and Sharp et al. (1988) show that both compressive and tensile stresses may impact upon various regions of the glacier bed. It can be demonstrated from the Boulton and Hallet abrasion models that as compressive stress levels increase to a critical level, increasing abrasion can be expected to take place (Chapter 7).

Where sediments underlying surging ice masses are frozen or heavily consolidated some form of brittle fracture or ploughing style deformation may occur. If the surge mobilizes a sediment bed in the manner alluded to by Clarke, sediment may suffer massive deformation with clast ploughing and intraclastic rafting occurring (Alley, 1989a, b). Deformation may be pervasive or non-pervasive and may be areally widespread or restricted to patches of the bed where localised sediment tectonisation occurs (Chapter 10). Clarke et al. (1984), Clarke (1985), Sharp (1985b) and Solheim and Pfirman (1985) inter alia report dyke-like structures that are possibly casts of basal crevasse intrusions caused by differential loading and/or deformation. The often-reported fluted diamictons beneath surging glaciers may be a consequence of rapid differential and anisotropic deformation of a mobilized slurried subglacial sediment. The influence upon deposited sediment of overriding ice during a surge may create localised compression that leads to extension within the stress field resulting in discontinuities evolving by crack tip growth extension and the re-opening of depositional or structural discontinuities, the latter acting as points of weakness for new crack development (Kazi and Knill, 1973; Banham, 1975; Derbyshire et al., 1976; Feeser, 1988).

The effect of a surging glacier advancing rapidly upon either a stagnant snout region or a proglacial zone leads to considerable lateral stress application, overriding and thus differential loading and drag effects and possibly the reactivation of debris-rich glacier ice (Mathews and MacKay, 1960; Clapperton, 1975; Moran *et al.*, 1980; Croot, 1981; Sharp, 1985a; van der Meer, 1987; Aber *et al.*, 1989). Many major push moraines (Ruegg, 1981; Oldale and O'Hara, 1984; Drozdowski, 1987; Croot, 1988b; Fernlund, 1988) and near-horizontal nappes, the latter exhibiting long distance translation of stacked sediment sequences, may be tentatively associated with surge events (Fig. 5.32) (Chapter 8).

5.11. CONCLUDING REMARKS

The rheological behaviour of any ice mass plays a fundamental role in the processes of glacial sedimentation and bed formation. To understand

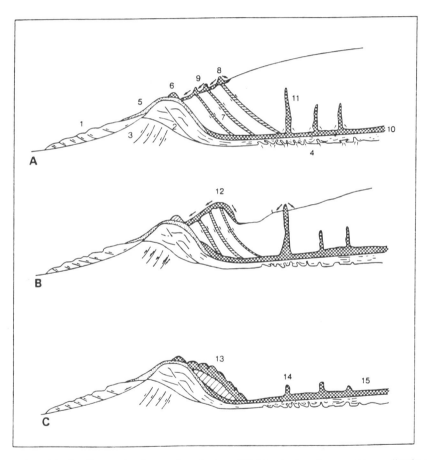

FIG. 5.32. A general model of sedimentation by surging glaciers. (A) Marginal environment immediately after a surge: (1) composite lobate gravitational slump of proglacial sediments; (2) glaciotectonically thrust ridges; (3) normal faulted sediments in ridge core; (4) water escape structures in ice-overridden sediment; (5) outwash sediments; (6) ridge composed of resedimented diamictons; (7) englacial debris bands elevated along thrust zones; (8) supraglacial debris dikes; (9) supraglacial sediment flows; (10) subglacial diamictons; (11) basal crevasse-filled dikes. (B) Marginal environment early in quiescent phase of surge cycle: (12) impact of ablation leading to sediment drapes, ridges and thermokarst relief features. (C) Marginal environment late in quiescent phase of surge cycle: (13) marginal rim of sediment isolated from the rest of the glacier characterized by ice-cored ridges and hummocky relief; (14) crevasse-fill ridges on fluted subglacial diamicton (see 11); (15) exposed subaerial surface due to glacier retreat. (after Sharp, 1985a; reproduced by permission of Academic Press).

these formative processes it is imperative that the thermomechanics of ice masses be understood. This chapter places our present understanding of ice mass rheology and thermomechanics within a context relevant to glacial sedimentology. Of particular interest are the problems associated with ice flow characteristics and how best they may be modelled and applied. The behaviour of surging ice masses and the many questions raised concerning glacial bed conditions and subglacial hydrology under rapidly changing glaciodymanic conditions emphasizes the need for continued research into the relationship between ice mass flow and glacial bed states. The potentially important role deformable bed states may have on ice motion and glacial bed conditions must be better understood. Finally, it is imperative that explanations within glacial sedimentology be generated within the context of associated rheological and glaciodynamic parameters

Chapter 6

HYDROLOGY OF GLACIERS

J. Menzies

6.1. INTRODUCTION

Water occurs as a fluid, as an interstitial liquid, as a solid and as interstitial ice within all glacial environments. As meltwater, it is active in debris erosion, entrainment and deposition. Meltwater causes channel bed erosion and scour, and the development of complex drainage networks within all glacial subenvironments. As interstitial water within debris or fractured bedrock, deformation and fracture may be facilitated and secondary sedimentological structures formed. Porewater migration and/or porewater saturation levels may lead to mineral and particle redistribution. Finally the influence of frost action and the development of cryostatic stress within most glacial subenvironments is an area of new research.

Many glaciodynamic processes are linked to various hydraulic effects whether it is the influence of porewater on the viscosity of subglacial debris or the subglacial hydraulic system fluctuations that appear to characterize the triggering of surges. Meltwater, from both subglacial, supraglacial and englacial areas of an ice mass, carries clues to other critical but unseen processes taking place within these often 'obscured' environments. Glacial hydrology, therefore, is an integral element in understanding all glacial environments and processes.

6.2. MELTWATER SOURCES

Water associated with ice masses is derived from two main sources viz: (1) atmospheric precipitation and (2) melted ice and snow. Other minor sources within the subglacial environment include water of juvenile derivation, from underlying aquifers, and from the melting of overridden permafrost sediments. Glacial water originates from melting due to surface ablation, basal ice melt due to friction, melt due to geothermal heat, basal melt due to air temperature melting in cavities and near marginal locations, melting due to the passage of meltwater over the surface of ice masses, and internal ice deformation and consequent energy dissipation (Röthlisberger and Lang, 1987).

6.3. MELTWATER SYSTEMS

Meltwater may exist either within a series of discrete and segregated hydraulic systems in each individual glacial sub-environment or as a few large systems that interconnect with all or most subenvironments. The development of glacial hydraulic systems within any ice mass is dependent upon: the type, morphology and topographic setting of the ice mass; thermal conditions at the surface, within and beneath each ice mass; the mass balance; velocity profile (surging/non-surging); basal conditions; the seasons and climatic conditions present and past; and the amount and type of debris cover and content (Fig. 6.1). Meltwater systems may be in the form of: (i) thin sheet flow; (ii) channel flow; (iii) linked cavity flow; or (iv) an integration of the above forms. A limited amount of meltwater may also seep into the ice from surface melt and runoff and through the walls of sealed or clogged crevasses (Nye and Frank, 1973; Paterson, 1981).

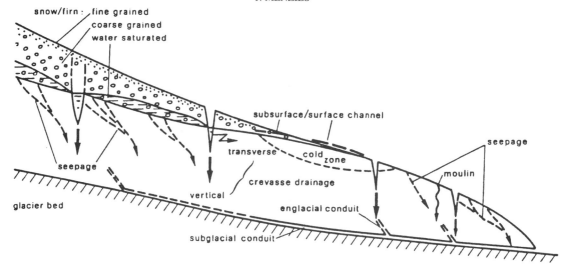

FIG. 6.1. Model of supraglacial, englacial and subglacial drainage routes (after Röthlisberger and Lang, 1987; reprinted from Gurnell, A.M. and Clark, M.J. (eds.), *Glacio-Fluvial Sediment Transfer*, by permission of John Wiley and Sons).

Meltwater systems may be seasonal or of shorter duration, and may exhibit normal or catastrophic fluctuations. The frequency and magnitude of system discharge fluctuations may vary widely from season to season, on different parts of the same ice mass and between neighbouring ice masses. The evolution of different systems may also vary enormously in time and space. Under steady-state or quasi-steady-state conditions, a hydraulic system may develop peculiar to a specific ice mass which, when interrupted, will return to that same type of system. For example, the hydraulic systems beneath surging glaciers mature during the non-surging quiescent phase but may be destroyed or altered during the surge, and then returned to the former system after a few months or years of non-surging ice flow conditions. Supraglacial and englacial hydraulic systems exhibit considerable fragility, being easily susceptible to rapid change whereas established subglacial systems are more resistant to alteration unless ice conditions are dramatically transformed as in a surge or rapid ice retreat.

6.3.1. Supraglacial Hydraulic Systems

In the supraglacial environment, hydraulic systems develop an arborescent pattern largely in accordance with subaerial fluid dynamics. As overall discharge and/or gradient increases, the hydraulic system tends to evolve from disconnected flow to an arborescent drainage pattern of connected channels (Ferguson, 1973). Supraglacial lakes may develop, especially in summer months, and form part of the drainage system over a large area of the ice surface. These lakes may create localised radial but isolated drainage systems (Björnsson, 1976) (Plate 6.1). Drainage on the surface may also be interrupted by meltwater disappearing into englacial and subglacial systems via moulins. Depending upon the ice mass activity, vertical ice thickness, state of internal stress and other local factors, connections may or may not be made between the supraglacial and englacial and/or subglacial hydraulic systems (Fig. 6.2) (Stenborg, 1969, 1970).

Drainage patterns on glacier surfaces are typically composed of several basins that may drain off laterally/frontally or into large surface lakes or major moulins (Fig. 6.2). The effect of the moulins is to frequently create, in the summer when they are best developed, a 'karstic' style surface drainage system. During massive ice downwasting and deglaciation, thermokarst drainage systems should develop to their greatest extent. Surface lakes may drain episodically during a summer season causing englacial and subglacial drainage systems to be flooded and altered accordingly (Röthlisberger, 1972; Clarke, 1982). As

(A)

(B)

PLATE 6.1. Supraglacial lakes. (A) On the surface of an unnamed glacier, Spitsbergen. (B) On the surface of the Matanuska Glacier, Alaska (photographs courtesy of Ed Evenson).

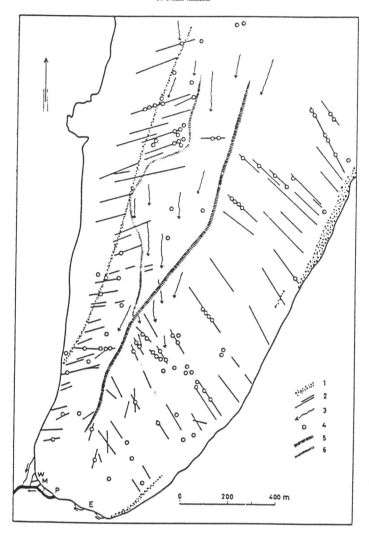

FIG. 6.2. The lower section of Mikkaglaciären, Sweden showing a generalized map of drainage collection crevasses and moulins (ca. 1958). (1) moraine covered ice, (2) crevasses, (3) direction of surface drainage, (4) moulins, (5) boundary line between crevasses striking east and west, (6) area of surface ice where drainage is characterized by predominant structure- and slope-factors (after Stenborg, 1969; reprinted from Studies on the internal drainage of glaciers, by T. Stenborg, *Geografiska Annaler*, 1970, **52A**, 1–30, by permission of the Scandinavian University Press).

winter approaches the amount of surface meltwater decreases and concomitantly moulins and small crevasses 'seal up' with snow; however, by the following summer many of these moulins re-open.

Supraglacial meltwater discharge exhibits large seasonal variations that increase from early spring to a mid-summer maximum followed by a decrease until freeze-up in the fall (Hooke *et al.*, 1989). Observations from Storglaciären, Sweden showed

that considerable surface meltwater may be stored in snow and firn during the early spring. This water is later released into the glacier hydraulic systems (Östling and Hooke, 1986; Hooke *et al.*, 1989). Superimposed upon these long-term fluctuations are diurnal fluctuations related to local weather conditions (Fig. 6.3).

The amount and form of supraglacial meltwater activity would seem to be little influenced by the

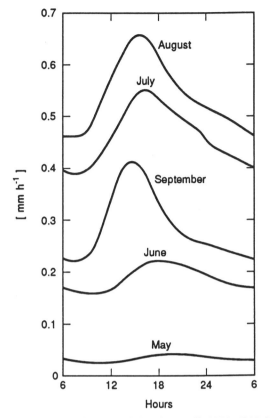

FIG. 6.3. Mean diurnal variation in runoff, 1974–1980 for Verngtbach (Austrian Alps), drainage area 11.44 km², 84% glacier covered, over the period May–September (after Röthlisberger and Lang, 1987; reprinted from Gurnell, A.M. and Clark, M.J. (eds), *Glacio-Fluvial Sediment Transfer*, by permission of John Wiley and Sons).

thermal state of the underlying ice mass (Hooke *et al.*, 1989). It might be expected that the volume of meltwater on a cold (polar) glacier might be less. However, little evidence exists to support this impression. On ice masses where the surface is heavily crevassed, the surface drainage system is adversely affected and consequently poorly developed (Sturm, 1987). In areas of ice falls (seracs), surging glaciers or along the edge of fast ice streams, crevasses intercept the natural drainage system (Sturm and Cosgrove, 1990). Crevasses act as swallow-holes or moulins (Plate 6.2).

In valley glaciers confined within a bedrock channel with high surrounding valley sides where a distinct transverse profile (the result of differential flow velocity) develops, the profile leads surface meltwater to flow towards valley sides. Lateral streams may then flow either along the edge, often disappearing within lateral moraines or sub-marginally beneath the ice (Plate 6.3). Where glaciers are extensively covered by morainal debris over large areas of the surface, supraglacial stream flow is interrupted. Moraines disrupt the supraglacial drainage system causing distinct separate drainage system patterns to develop (Plate 6.4). On debris-covered glacier surfaces, as is occasionally observed in the terminal zones of glaciers, much of the surface meltwater disappears into the debris (Plate 6.5). Excessive amounts of supraglacial debris may result from: (a) numerous coalescing medial moraines, (b) side-wall rock/debris avalanches and flows, or (c) melt-out of englacial debris.

Supraglacial meltwater influences several significant sedimentological processes viz:

(1) In the debris covered terminal zones of many glaciers where debris acts to 'soak-up' surface meltwater, debris flows, slides and subsequent flow tills may occur. Such mass transport of debris is largely a function of angle of slope, gravity and the instability caused by debris becoming saturated, reaching high internal porewater pressures, low effective stress levels and low internal shear strength.

(2) As a major source of meltwater to proglacial outwash areas, supraglacial meltwater acts to entrain large amounts of debris from the ice surface into this zone.

(3) The total hydraulic system of a glacier has been likened to an enormous 'plumbing system' in which the supraglacial system acts as one of the sources whereby debris is flushed out or through the system. Evidence of this 'flushing process' has been observed in many glaciers, especially in the early spring when melting begins or following major thunderstorms (Collins, 1979a, b, 1982).

6.3.2. Englacial Hydraulic Systems

The type of hydraulic system that will evolve within englacial and subglacial environments is much less understood. In both cases, meltwater may be under pressure due to the confining effect of the ice and the closure forces of the internally deforming ice.

PLATE 6.2. Moulin on the Athabasca Glacier, Alberta (photograph courtesy of Gerry Holdsworth).

The nature and style of development of englacial and subglacial hydraulic systems is controlled in part by the meltwater discharge, the source(s) of that discharge and the debris flux. Both systems may be linked and may also be coupled periodically to the surface supraglacial system thus deriving an extrinsic (exotic) supplementary source that is unaffected by the internal processes of either system. Neither system can be easily investigated.

The englacial drainage system, as described by several authors (Stenborg, 1969, 1973; Röthlisberger, 1972; Shreve, 1972; Krimmel et al., 1973; Nye, 1976; Sugden and John, 1976; Röthlisberger and Lang, 1987; Hooke et al., 1988; Seaberg et al., 1988; Willis et al., 1990), is usually viewed as a complex architecture of interconnected cavities and conduits within a three-dimensional gallery system. Englacial drainage systems are often compared to those found in karst terrains but, unlike such systems, the englacial system is highly dynamic and subject to the external stresses of continuous ice movement (Smart, 1990). Where ice is temperate and fast moving, englacial systems are likely to be temporary and largely autonomous. Under slow moving and temperate conditions, a well-developed englacial system that may interconnect to the subglacial system

PLATE 6.3. Lateral meltwater systems along the edges of Uversbreen, Spitzbergen (photograph courtesy of Ed Evenson).

may develop. Within polar glaciers, a limited summer near surface (< 2m) englacial system can develop if connected with those supraglacial locations where sufficient summer ablation occurs. Englacial systems within surging glaciers probably suffer total destruction during the surge phase but may resume a well-evolved system during quiescent periods if those periods are sufficiently long (Sturm and Cosgrove, 1990). However, it is very difficult to estimate the time necessary for an englacial system to become fully activated after a surge event. In stagnant ice, englacial drainage systems are very well developed and are normally fully connected to both supra- and subglacial drainage paths (n.b. englacial watertables).

From evidence gleaned through descent into moulins and by entering frontal conduit and caverns, and by the use of various types of dyes and other soluble tracers, it has been possible to establish something of the nature of englacial hydraulic systems (Carol, 1945; Behrens et al., 1971, 1982; Elliston, 1973; Stenborg, 1973; Iken, 1974; Nilsson and Sundblad, 1975; Berner et al., 1978b; Lang et al., 1979; Burkimsher, 1983; Lliboutry, 1983; Holmlund,

1986, 1988b). It seems likely that two, often interconnected, forms of englacial hydraulic systems prevail, depending upon glacier thermal regime and local rates of summer ablation viz: (i) a seepage system of tubes and veins, and (ii) a larger conduit system of channels (Raymond and Harrison, 1975).

(i) Smaller englacial systems, of connected tube and vein-like form, develop wherever seepage into the surface ice is possible (Figs 6.4 and 6.5). Thus, provided ice is at pressure melting point (i.e. temperate), such a system of small tubes and veins is operable (Nye and Mae, 1972; Shreve, 1972; Raymond and Harrison, 1975; Lliboutry, 1983; Nye, 1990). In most accumulation areas, englacial hydraulic evolution by seepage is almost negligible due to the impermeability of the ice (Hooke et al., 1983). Within temperate ice, at pressure melting point, vein development and moisture transfer is thought to be possible but limited in volume (e.g. Raymond and Harrison (1975) calculated a moisture flux of $\approx 10^{-1}$m a^{-1} from a vein system). In contrast, fine tubules only a few millimetres in diameter were observed to have a flux rate of ≈ 9 m a^{-1}. Such veins

PLATE 6.4. Separate supraglacial meltwater drainage systems on the surface of the Görnergletscher, Switzerland, August 1989. Note major drainage system on larger glaciers with smaller system developed on the tributary glacier at bottom left of photograph.

and tubes are important in transporting meltwater (Lliboutry, 1971; Berner *et al.*, 1978b; Fowler, 1984) and influencing the creep rate of polycrystalline ice (Duval, 1977). Field observations confirm that veins and tubules gradually merge into an arborescent system of drainage routes that ultimately converge into larger conduits (Shreve, 1972). Unlike conduits, veins and tubules are affected by the stress/strain behaviour of moving ice and thus tend to follow predetermined pathways (i.e. flowlines normal to equipotential planes of water pressure).

Finally, a myriad of discrete liquid veins exist within deforming ice masses, even within polar ice, that are separate from connected veins associated with the transport of meltwater (Nye, 1990). These veins of supercooled meltwater accumulate impurities from the solid ice and may act to alter pressure melting point within deforming ice (Nye, 1989) (Figs 6.4 and 6.5).

(ii) The internal geometry and evolution of englacial conduit systems is both complex and still remains, somewhat, enigmatic. Englacial conduit development may begin as a result of crevasses being exploited by either direct supraglacial streams entering them or by up-ice englacial conduits intersecting crevasses further down the flowline, or by the evolution of seepage passages into small tube-like englacial tributary systems. Crevasses become moulins or 'glacial mills' remaining open provided the supply of meltwater is sufficient to melt crevasse walls and basal tips in opposition to the self-closing effect of ice overburden pressure. The location of moulins and larger scale englacial hydraulic systems is largely dependent upon the presence of crevasses (i.e. places at or near the glacier surface where high extending strain rates occur). Therefore, in general, englacial hydraulic systems are best developed in ablation areas of ice masses and/or areas of extending longitudinal strain rates. The style, geometry and degree of development of any englacial hydraulic system is a function of: the meltwater discharge entering the system; the depth, width and number of crevasses to be exploited; the thermal distribution and rheological state of the ice mass; the conditions at the

PLATE 6.5. Meltwater penetrating morainal debris in front of an unnamed glacier, Spitsbergen (photograph courtesy of Ed Evenson).

bed of the ice mass in terms of sliding velocity and subglacial hydraulic system type and level of development; and the overall stability of the ice mass in the long-term.

In order for an englacial conduit to remain open and stable under active ice conditions after the crevasse seals, it is necessary for those forces maintaining the opening of the conduit to equal those attempting to close it. Englacial conduits remain open as a result of thermal energy generated by: (a) meltwater turbulence; (b) frictional energy due to transported debris within meltwater causing sidewall abrasion; and (c) the diffusion of direct thermal energy from the passage of warm meltwater (viscous heat dissipation) (Haefeli, 1970). The factors leading to conduit closing are those associated with the intrinsic internal deformation of the confining ice mass attempting to seal the conduit space (Nye, 1953) and, in specific instances, the freezing-on of water to conduit walls. Also, when englacial meltwater discharge wanes, for whatever reason, debris transported in these channels will settle and possibly clog the system (Saunderson and Jopling, 1980).

In explaining the englacial movement of meltwater, an important thermo-mechanical distinction needs to be made between englacial flow as seepage via veins vs. channel flow. In the former, meltwater utilises the capillaries at three-grain intersections and is dependent upon heat flow within the ice and surface energy dissipation at grain boundary interfaces (Lliboutry, 1971, 1983; Nye and Frank, 1973) (Figs. 6.4 and 6.5). The transition between these minute passages and larger conduits appears to occur at discharges of around 10^{-3}–10^{-4} m^{-3} s^{-1} (Hooke, 1984) (Fig. 6.6). In general, englacial conduits usually approximate a circular cross-section (Nye, 1965) (Plate 6.6), but variations do occur where ice stratigraphy or other anisotropic inclusions cause differential erosion of channel walls (Sugden and John, 1976). Hooke (1984) has also demonstrated that englacial and subglacial channels are often not fully filled by water and channel shape will then deviate from the circular to lesser or greater degrees. In those channels totally water filled, the hydraulic pressure of the meltwater may or may not approach the confining ice overburden pressure, whereas in partially filled channels this pressure condition does not occur and meltwater may be close to atmospheric pressure or at triple-point pressure if no open outlet exists (Lliboutry, 1983).

Englacial conduit geometry is largely unknown and is more often the subject of educated speculation based upon dye tracer experiments (Lang *et al.*, 1979;

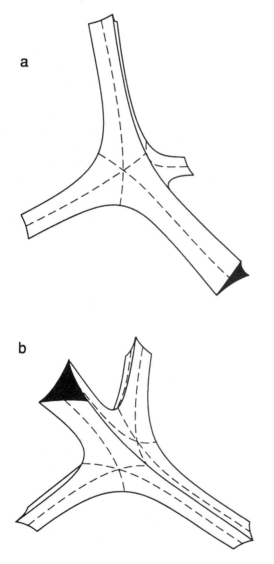

a

b

FIG. 6.4. Perspective views of a node between ice crystals (after Nye, 1989; reproduced by courtesy of the International Glaciological Society from *Journal of Glaciology*, **35**(119), 1989, p. 21, Fig. 7).

Theakstone and Knudsen, 1981; Collins, 1982; Burkimsher, 1983; Holmlund and Hooke, 1983; Iken and Bindschadler, 1986; Hooke *et al.*, 1988; Seaberg *et al.*, 1988; Sharp *et al.*, 1989; Willis *et al.*, 1990). In only a very few cases have moulins been explored and some clue as to the englacial conduit system been examined first hand (Vallot, 1898; Dewart, 1966; Reynaud, 1987; Holmlund, 1988b) (Fig. 6.7). Holmlund (1988b), working on Storglaciären, Sweden, noted the following morphological features viz: in cold ice, especially close to the ice mass surface, moulins tend to follow vertical shafts largely pre-determined by the crack geometry of the crevasse; at depth (in Storglaciären >25–30 m) the shaft tends to widen and may develop step-like ledges and small plunge pools (Reynaud, 1987) (Fig. 6.8); at these depths the shafts become more complex chambers or caverns and may have several small conduits joining the main conduit at varying depths; typically most moulins were drained by at least two channels often of short length (Fig. 6.9) followed by other vertical shafts; and finally, these moulin/englacial conduits are typically laterally interconnected, downward toward the glacier bed and upward to the surface within a system of tunnels, galleries and shafts. Cavern shape and size would appear to be a function of ice wall melting due to the effect of spray and the impact of ice mass movement. Holmlund (1988b) reports that short drainage channels connecting caverns exhibit meandering of large amplitude typical of high-speed meltwater passage (Holmlund and Hooke, 1983). The degree of evolution of any englacial system would appear to be a function of glaciodynamics and stress history; thus systems may alter or degenerate every few years.

A key to the orientation and meltwater capacity of any englacial hydraulic system is the degree of interconnectivity between parts of the galleried system and the ice pressures translated to hydraulic pressures affecting the system. Observations of moulins during the summer high melt season reveal moulins filling up at different rates, debouching into neighbouring moulins and filling them eventually to an equivalent hydraulic pressure level, thus indicative of local englacial water tables (Hantz and Lliboutry, 1983; Holmlund, 1988b). The level to which these systems can fill is, to considerable degree, controlled by ice pressure. Within any ice mass a plane of equal

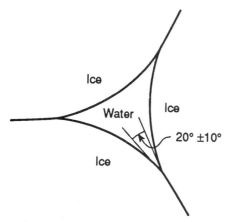

FIG. 6.5. Planar cross-section of vein-space between ice crystals (after Röthlisberger and Lang, 1987; reprinted from Gurnell, A.M. & Clark, M.J. (eds) *Glacio-Fluvial Sediment Transfer*, by permission by John Wiley and Sons).

hydraulic potential develops. The angle at which englacial conduits will descend from the ice surface will be perpendicular to the angle, $\tilde{\omega}$, made by equipotnetial lines with the ice surface. This angle is given by:

$$\tilde{\omega} \approx \tan^{-1}[\rho_i \; |\tan\alpha| \; / \; (\rho_{wt}-\rho_i)] \qquad (6.1)$$

where ρ_{wt} is the density of water.

An equipotential plane dips up-glacier such that englacial meltwater should flow normal to the plane (Röthlisberger and Lang, 1987). For example, Willis *et al.* (1990) calculate an equipotential plane on Midtdalsbreen, Norway of $\approx 58°$ and anticipate that englacial conduits should dip down-glacier at $\approx 32°$. From Eqn. (6.1), it is apparent that ice pressure and surface slope are the controlling variables delineating the equipotential plane; however, in those locations within englacial systems where hydraulic pressure is not equated to ice overburden pressure such as close to the surface where atmospheric pressures prevail, the postulated orientation of the conduit passage is not observed (Holmlund, 1988b). Röthlisberger and Lang (1987) speculated that where basal ice velocities were high or exhibited high local differential values over a small area of bed such as might occur over a bump or riegel on the bed, englacial systems might be distorted from that modelled by Shreve (1972). Englacial systems tend to develop both in the down-ice direction and toward the bed. Movement of englacial conduits in the direction of the bed would appear to be the result of excess stress convergence and crack-tip development rather

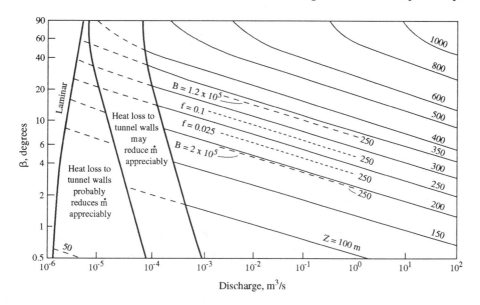

FIG. 6.6. Critical values of subglacial meltwater discharge, bed slope and ice thickness. When the discharge value in a conduit is greater than the value read on the abscissa for a given bed slope and ice thickness, the conduit is likely to be open (after Hooke, 1984; reproduced by courtesy of the International Glaciological Society from *Journal of Glaciology*, **30**(105), 1984, p. 182, Fig. 2).

(A)

(B)

PLATE 6.6. Submarginal englacial tunnels. (A) In the Matanuska Glacier, Alaska. (B) In the MacLaren Glacier, Alaska (photographs courtesy of Ed Evenson).

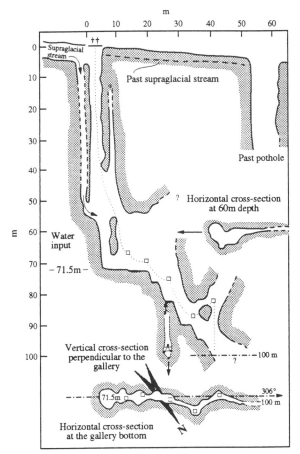

FIG. 6.7. Vertical and horizontal cross-sections of the Grand Moulin of the Mer de Glace, France from a survey of 6–7 November 1986 (after Reynaud, 1987; reproduced by courtesy of the International Glaciological Society from *Journal of Glaciology*, 33(113), 1987, p. 131, Fig. 2).

than solely differential melting (Iken, 1974; Röthlisberger and Lang, 1987). Although a crack will tend to propagate to depth and in the down-ice direction, lateral movement of conduits toward valley sides may also occur (Stenborg, 1973; Holmlund, 1986). In propagating to depth within an ice mass, the direction of the crack-tip will, in part, be influenced by differential pressures transmitted upward into the ice by subglacial pressure fluctuations developed as ice moves across the bed and around bed topographic obstacles such as bedrock riegels and overdeepenings, and subglacial channels, tunnels and cavities. Thus, there is a tendency for large englacial

conduits to intersect with and connect into the subglacial hydraulic system network. In the case of a permanent bed obstruction this may cause an englacial conduit to intersect with the bed at that point over a protracted time period (Röthlisberger and Lang, 1987). Such a preferential location of meltwater with the bed may help explain the location, size and nature of pot-holes and other P-forms in bedrock (Fig. 6.14. Likewise, the · process of overdeepening by glacial meltwater may have a similar spatial propensity. The degree to which subglacial bed conditions, in association with glaciodynamics, affect the location of englacial meltwater erosion and sediment transport remains an area of considerable research interest (Aylsworth and Shilts, 1989; Fulton, 1989).

6.3.2.1. Steady-State Channels

Several theoretical studies of englacial drainage have been made in which the existence of a steady-state channel has been assumed and in which the primary unknown has been the meltwater pressure (Röthlisberger, 1972; Shreve, 1972; Weertman, 1972) (see Röthlisberger and Lang (1987) for a detailed discussion on the derivation of the equations below). Following the analysis by Röthlisberger (1972), it can be shown, from geometric considerations, that the frictional pressure loss (δ_f) and the change in local water pressure (δ_p) within a straight conduit of length δ_s at a slope Φs are related as follows, where it is assumed that for a horizontal conduit $\delta_p = \delta_f$:

$$-\delta_p = -\delta_f + \delta_s \sin\phi_s \qquad (6.2)$$

Assuming radial closure of a circular conduit by ice pressure (Nye ,1953; Paterson, 1981) the rate of closure is given as:

$$\frac{\dot{r}}{r} = An^{-n} (p_i - p_{wt})^n \qquad (6.3)$$

where r is conduit radius and components A and n are ice deformation parameters as defined in Table 6.1. Equation (6.3) can be translated into the rate of conduit closure by volume (m_c) as a function of plastic ice flow as:

$$\dot{m}_c = 2r \pi \dot{r} \delta_s = 2\pi r^3 A n^{-n} (p_i - p_{wt})^n \delta_s \qquad (6.4)$$

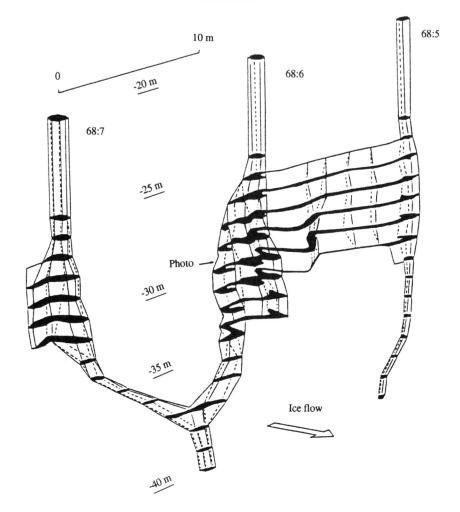

FIG. 6.8. Perspective illustration of moulins on the 2nd August 1982 on Storglaciären, Sweden (after Holmlund, 1988; reproduced by courtesy of the International Glaciological Society from *Journal of Glaciology*, **34**(117), 1988, p. 245, Fig. 5b).

Since r is an unknown value, Röthlisberger and Lang (1987, p. 240) show that Eqn. (6.4) can be re-written for a known discharge (Q) using the Gauckler–Manning–Stickler equation for turbulent flow within a conduit (Spring (1980) and Spring and Hutter (1981b) employed the Darcy–Weisbach Law, obtaining slightly different values for r (n.b. (Lliboutry, 1983)) where:

$$\dot{m}_c = C_3 \, Q^{\,q} \, (p_i - p_{wt})^n \, \left(\frac{\delta_f}{\delta_s}\right)^m \delta_s \qquad (6.5)$$

The empirical factor C_3 fluctuates dependent upon the particular mode of water flow, channel wall geometry and ice flow conditions. It is therefore difficult to generalise on rates of closure under these steady-state conditions (Röthlisberger and Lang, 1987). Röthlisberger and Lang (1987, p. 240) further note that the hydraulic gradient $(-\delta_f/\delta_s)$ within a conduit is proportional to Q^{-B}, implying that fully-filled circular englacial conduit with high discharges would exhibit low meltwater pressures and vice versa. The implications of this relationship will be discussed later when considering subglacial channels, but it indicates that large channels will tend to grow at the expense of small channels by piracy and that meltwater will be predisposed to flow in channels rather than sheets (Weertman and Birchfield, 1983b).

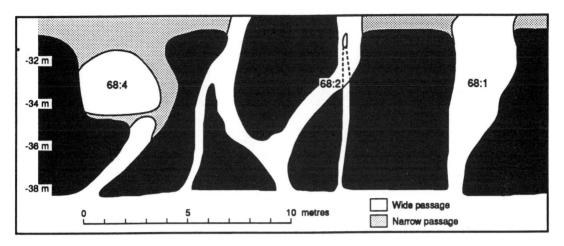

FIG. 6.9. Reconstruction of four interconnected moulins on Storglaciären drawn transverse to ice flow (after Holmlund, 1988; reproduced by courtesy of the International Glaciological Society from *Journal of Glaciology*, **34**(117), 1988, p. 247, Fig. 8).

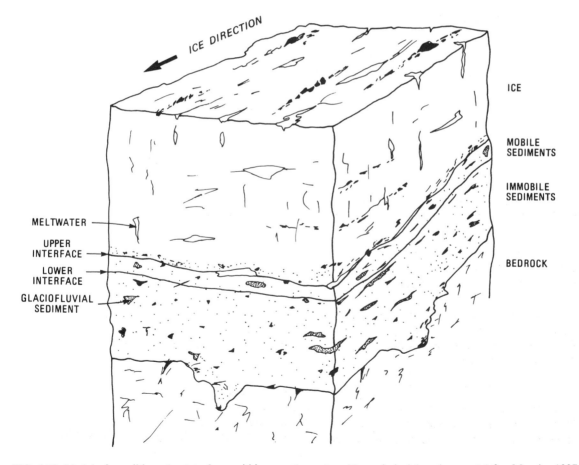

FIG. 6.10. Model of possible active interfaces within an active net-melting subglacial environment (after Menzies 1987. Reprinted from Menzies, J. and Rose, J. (eds), *Drumlin symposium* — Proceedings of the Drumlin symposium / 1st Int. Conf. Geom., Manchester, 1985, p.17, Fig. 6; courtesy of A.A. Balkema, Rotterdam).

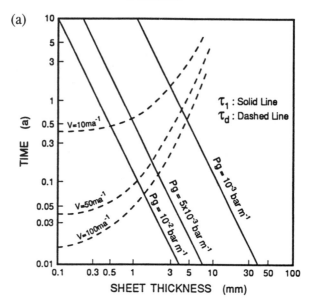

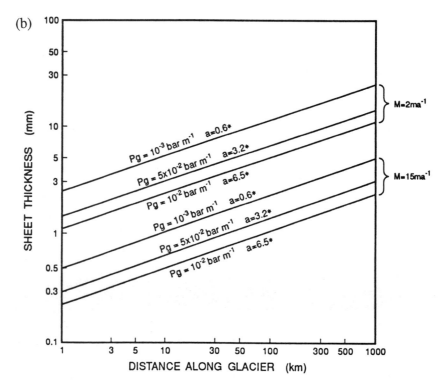

FIG. 6.11. (a) Typical development and disintegration of water-sheet perturbations at the ice–bed interface (after Walder, 1982; reproduced by courtesy of the *International Glaciological Society* from *Journal of Glaciology*, **28**(99), 1982, p. 280, Fig. 2). (b) Thickness of a water sheet as a function of distance along the ice flow direction, assuming all subglacial meltwater flows as a sheet (modified from Walder, 1982; reproduced by courtesy of the International Glaciological Society from *Journal of Glaciology*, **28**(99), 1982, p. 282, Fig. 3).

(a)

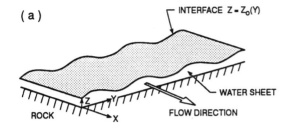

INTERFACE Z = Z₀(Y)

WATER SHEET

FLOW DIRECTION

ROCK

(b)

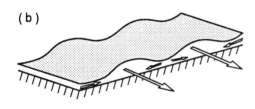

(c)

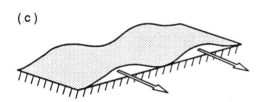

FIG. 6.12. Idealized evolution of a water film toward discrete conduit system development. (a) water film geometry; (b) incipient water film-conduit transition; (c) conduit establishment (modified from Walder, 1982; reproduced by courtesy of the International Glaciological Society from *Journal of Glaciology*, **28**(99), 1982, p. 276, Fig. 1).

6.3.2.2. Non-Steady-State Channels

Reality rarely ever matches that implied from theoretical discussions and models. This axiom is particularly germane in the case of englacial and subglacial hydraulic systems where diurnal, seasonal and intermediate variations in meltwater production and discharge occur (Nye, 1976; Spring and Hutter, 1981a; Lliboutry, 1983). As both Lliboutry (1983) and Hooke (1984), among many others, have pointed out, steady-state models of these hydraulic systems are at best 'crude approximations'. Lliboutry claims, and increasing field evidence supports his contention, that most englacial and subglacial channels, under

thin marginal ice thicknesses, are not filled for much of the time and that open-channel flow is more common than implied (Hantz and Lliboutry, 1983; Seaberg et al., 1988; Willis et al., 1990). In other locations within ice masses where ice pressures due to greater ice thicknesses cause more rapid channel closure, more channels can be expected to be filled for longer periods of time and thus approach the steady-state models discussed above. These deeper channels will also tend to be supplied from longer distances by seepage and small englacial complex hydraulic systems; therefore, discharge can be expected to fluctuate to a lesser extent and consequently approach steady-state.

The sedimentological significance of englacial meltwater systems is multifold viz:

(1) As a depositional environment especially in inactive or stagnant ice masses (n.b. englacial water tables);

(2) As a pathway via which supraglacial, subglacial and (side-wall) englacial debris can be transported to other glacial environments;

(3) As the 'connecting' segment between the supra- and subglacial environments allowing the englacial and subglacial meltwater systems to be 'flushed-out' from the ice surface.

6.3.3. Subglacial Hydraulic Systems

The imprint of subglacial hydraulic systems is unquestionably one of the most important features of any glaciated landscape. Lying as it does at the interface between the glacier and its bed, the subglacial system is capable of moving vast quantities of sediment, of cutting down into bedrock and already existing deposited sediment, and thus imparting a critical influence upon glaciodynamics that ultimately dictate the 'effect' of glaciation upon any land or near-shore ground surface (Weertman, 1962; Röthlisberger, 1972; Shreve, 1972; Nye, 1973; Iken, 1981; Bindschadler, 1983) (Chapter 3).

In those areas of the world such as the mountainous regions of the Himalayas where fresh water and/or non-polluted water is in limited or transient supply, subglacial meltwater has been found, in recent years, to be a potential and 'tappable' long-term reliable source of water. In France,

TABLE 6.1. Physical constants of ice and water at 0°C and related properties (after Drewry, 1986 and Röthlisberger & Lang, 1987).

Property	Symbol	Quantity	Units
MECHANICAL			
Density of Water	ρ_{wt}	999.8	kg m^{-3}
Density of Ice	ρ_I	920	kg m^{-3}
Young's modulus of Ice	E_I	9.10	GN m^{-2}
Yield Strength of Ice	τ_I	85.0	MN m^{-2}
Fracture Toughness of Ice	K_I	0.2	MN m$^{-3/2}$
Creep Activation Energy of Ice	E^*	6.07 x 10^4	J mol^{-1}
Flow Law Constant of Ice	A	+ 8.75 X 10^{-13}	Pa^{-3}s^{-1}
		* 5.20 x 10^{-13}	
		● 7.93 x 10^{-13}	
		▲ 7.54 x 10^{-13}	
		□ 32.5 x 10^{-13}	
Exponent in Power Law of Ice	n	3	-
Vicosity of Water	η_w	1.787 x 10^{-3}	kgs^{-1}m^{-1}
Hydraulic Roughness Parameter for Conduits	k_r	Smooth 100	m$^{\frac{1}{3}}$s^{-1}
		Medium 50	
		V. Rough 10	
THERMAL			
Melting Temperature of Ice	°K	273.1	K
Thermal Conductivity of Ice	k_{tc}	2.51	Wm^{-1}Deg^{-1}
Thermal Diffusivity of Ice	κ	1.33 x 10^{-6}	m^{2}s^{-1}
Heat Capacity of Ice	-	37.7	Jmol^{-1}Deg^{-1}
Specfic Heat of Water	c_{sw}	4.217 x 10^3	Jkg^{-1}Deg^{-1}
Change in Pressure Melting Pt. of Ice with pressure (pure water)	$c_{t,o}$	-7.4 x 10^{-8}	KPa^{-1}
Change in Pressure Melting Pt. of Ice with pressure (air-saturated water)	$c_{t,s}$	-9.8 x 10^{-8}	KPa^{-1}
Latent Heat (Fusion) of Ice	L_h	334 @ 0°C	kJkg^{-1}
		285 @ -10°C	
		241 @ -20°C	

+ Drewry, 1986
* Paterson, 1981
● Hooke, 1984
▲ Lliboutry, 1983
□ Röthisberger, 1972

Norway, Switzerland and the C.I.S., subglacial meltwater has been used for hydro-electric power generation, and in some areas such as British Columbia in Canada, meltwater has been used in mining operations. Therefore, subglacial meltwater, its mode and volume of discharge, and its long-term reliability have become important questions for a wide circle of society from terrace farmers on the hillsides of northern Pakistan to hydro-engineers beneath the French Alps.

Our understanding of both englacial and subglacial systems remains limited and often at best second-hand and derivative. However, unlike englacial systems that survive intact in a limited capacity as sedimentary expressions in the form of eskers and kames, subglacial meltwater systems, typical of late stages in glaciation, are often left as relict features in the form of channel systems and esker ridges (Mannerfelt, 1949; Charlesworth, 1957, pp. 238–245; Sissons, 1967; Walder and Hallet, 1979).

The presence of subglacial meltwater beneath ice masses is well documented not only from observations of large, apparently subglacial, meltwater portals within valley glaciers (Plate 6.7) but from borings and measurements completed some distance up-ice in the upper ablation and accumulation areas (Iken, 1972; Hodge, 1976; Engelhardt, 1978; see review by Clarke, 1987a). Until 1963 (Gow, 1963) little, even indirect, evidence existed to indicate the presence of subglacial meltwater beneath the Antarctic Ice Sheet. Since that time, with the use of seismic and radar scanning geophysical equipment and ice-boring to bedrock, the presence of meltwater, either free or as a component within a saturated debris layer beneath Antarctica, has been largely verified (Oswald and Robin, 1973; Blankenship *et al.*, 1986; Whillans, 1987). Beneath the Greenland Ice Sheet, little evidence from central areas of the ice sheet support an active meltwater presence but marginal areas appear to have subglacial water present (Paterson, 1961, 1981). The explanation for the presence or absence of meltwater reverts to an argument concerning the thermal state of the basal regions of the ice mass: where temperate conditions occur meltwater is likely to be present but under polar conditions no meltwater, unless supercooled or salt-saturated, should be found other than in an interstitial state. The conclusion to be drawn from the above statement is that subglacial hydraulic systems are only likely to be developed to any degree beneath ice masses with areas of temperate bed conditions. In those ice masses that have a subpolar thermal regimen, limited areas of the bed may have a subglacial hydraulic system; also where thermal regimens have altered from warm to cold, relict features both on the bed and within the basal ice may testify to the past existence of a subglacial hydraulic system. Where polythermal regimens occur, it is likely that interrupted, diverted and pinched-out subglacial hydraulic systems will develop, flourish and be spatially extinguished over time. This form of basal thermal conditions may typify major ice sheets over the long-term (Fisher *et al.*, 1985), whereas temperate conditions may be more characteristic of valley glacier systems as commonly observed in middle latitudes.

6.3.3.1. Basal Ice Conditions and Glaciodynamic Effects

Before considering how a subglacial hydraulic system may evolve and attain a given structural form and configuration, specific background basal ice environmental and glaciodynamic parameters need to be evaluated viz: thermal regimen, velocity, principal stress components and vectors, basal topography and terrain characteristics, the nature of the ice–bed interface, and the character and form of interconnections of the potential subglacial hydraulic system to englacial and supraglacial systems, and how all of these parameters may vary in magnitude and frequency both spatially and temporally.

It is apparent that where sudden or cyclical changes in ice velocity occur, as in a mini-surge or major surge event, a subglacial hydraulic system is likely to be drastically altered or eradicated, at least temporarily. Principal basal stress conditions vary according to position beneath the ice mass and distance from the ice front, or over significant topographic highs and lows on the bed. Thus it is possible that principal longitudinal stress vectors may exhibit compressive or extending flow behaviour that will accelerate or delay the closure rate of basal

PLATE 6.7. Subglacial meltwater channel portal, note high summer volume of glacial meltwater, Nigardsbreen, Norway.

cavities and tunnels. In addition, the influence of overburden pressures will, in part, dictate meltwater hydraulic pressures. Where no direct connection exists to the outside atmosphere, it can be expected that meltwater will be under hydrostatic pressure equal to the ice overburden pressure.

Whether bed topography is very smooth or rough, specific styles of meltwater evacuation systems may or may not be capable of development; thus, for example, Kamb's linked-cavity system is unlikely to evolve successfully across a smooth impermeable surface (Chapter 5). In those areas of the glacier bed where either highly permeable bedrock or sediment are found, it is possible that little or no meltwater will flow at the interface between the bed and the glacier sole (Shoemaker, 1986; Menzies, 1989a). It is also possible that meltwater may infiltrate into an advecting deformable bed, reducing again the presence of free meltwater at the ice–bed interface (Boulton and Hindmarsh, 1987; Menzies, 1989a). Any glacial hydraulic system will develop and/or persist to a greater or lesser degree if it can obtain exotic source(s) of water which, in the subglacial instance, is typically from englacial or supraglacial systems, both of which are unaffected by those basal ice pressure effects likely to curtail or remove subglacial water source discharges. Finally, in trying to understand how any subglacial hydraulic system develops and attains a specific form over time, it is essential to bear in mind that such a system exists within a dynamically changing environment. Thus the nature of subglacial hydraulic patterns and form will vary across the bed of a glacier, in all directions, and will alter in any one location over both the short- and long-term depending upon the nature and style of glaciodynamic changes.

6.3.3.2. Subglacial Meltwater Discharge

Perhaps no other parameter connected with the subglacial hydraulic system has been studied in such detail over the long-term as has subglacial meltwater discharge (Glen *et al.*, 1973; Glen, 1982; Young, 1985; Alean *et al.*, 1986; Bezinge, 1987; Röthlisberger and Lang, 1987). The studies, both academic and economic, have provided a wealth of information regarding the nature of meltwater removal from beneath ice masses, debris transport, the possible configuration of subglacial hydraulic systems, and the reaction time of such systems to changes in meltwater input, velocity, solar radiation and mass balance. These studies, however, almost exclusively pertain to valley glaciers and very few studies have been done on land-based ice sheets (Braithwaite, 1982; Braithwaite and Oleson, 1988).

Subglacial discharges exhibit considerable diurnal, seasonal and annual variations. These temporal scalar fluctuations decrease as the time period examined increases and/or the size of the ice mass decreases (Seaberg *et al.*, 1988). At the annual scale, a pattern of fluctuations in relation to solar energy, net mass balance and hydraulic system development can be observed (Elliston, 1973; Haefeli, 1970; Krimmel *et al.*, 1973; Müller and Iken, 1973; Röthlisberger *et al.*, 1979; Röthlisberger, 1980; Iken *et al.*, 1983; Hooke *et al.*, 1985; Röthlisberger and Lang, 1987). In general, total run-off variations from glaciers tend to smooth out on a year-to-year basis and can be observed in a wide range of glaciated regions in the world (e.g., in western North America (Meier, 1969), the European Alps (Kasser, 1959; Collins, 1985; Röthlisberger and Lang, 1987), Pakistan (Ferguson, 1985), China (Lai Zuming, 1982; Yang Zhenniang, 1982) and Greenland (Braithwaite and Olesen, 1988)).

By mid-summer, it would appear that the subglacial hydraulic system(s) (usually in association with fully developed supra- and englacial hydraulic systems) within a temperate ice mass are at the highest degree of development with high average discharge values due to high insolation and supraglacial melting. As winter approaches there is a reduction in surface melting, moulins may begin to close or seal up and, due to the relaxation time of ice movement maintaining high velocity values, subglacial cavities and channels may increase their closure rates (Hodge, 1974; Tangborn *et al.*, 1975). With further loss of surface meltwater, heat transference through the system via meltwater passage is reduced and a subglacial hydraulic system of much-reduced capacity requirements begins to develop (Sharp *et al.*, 1989). At the mid-winter stage, depending upon climatic conditions and the severity of winter temperatures, in many cases the subglacial system will seal up and cease to function. In other locales, enough basal friction and geothermal heat energies allow some subglacial meltwater to be produced but in a drastically altered and degenerated system compared with the summer. With spring, there are two significant events with associated time lags: first, crevasses and moulins have to open or re-open and an englacial hydraulic system begin to operate while the increase in snow accumulation will cause an increase in basal ice stress and marginally increase basal ice velocities leading to increasing basal meltwater production (Hodge, 1974; Hooke *et al.*, 1983; Iken *et al.*, 1983 *inter alia*). The start of the spring will therefore tend to be linked with high basal meltwater pressures that in turn lead to a new hydraulic system opening up (Seaberg *et al.*, 1988). Second, if any remnants of a subglacial hydraulic system exist it is at a much reduced and restricted level of development. This effect 'throttles' the subglacial system and obstructs meltwater flow until a new system of subglacial hydraulic passageways evolves. Once formed, this new subglacial system tends to exhibit a characteristic spring flood event or peak discharge as the dammed meltwater is released via the newly developed system (Deichmann *et al.*, 1979; Röthlisberger *et al.*, 1979; Collins, 1983; Iken *et al.*, 1983). It is during the 'spring event' that large volumes of subglacial sediment are flushed through the system where they had accumulated in cavities and on channel floors since the start of the winter period (Collins, 1979a, b, c). This sequence of subglacial hydraulic cyclical change, however, is 'ideal'. In reality, different ice masses exhibit considerable variations in terms of the timing of conduit opening and closing, and interconnection development during the ablation season as a function of glaciodynamic and environmental factors peculiar to certain ice masses and/or parts of the same ice mass (Seaberg et al., 1988; Willis et al., 1990).

6.3.3.3. The Nature of Meltwater Flow and Routing at the Ice–Bed Interface

When discussing meltwater at the base of an ice mass, two related but, for the purposes of discussion here, separate aspects must be addressed viz: the form of meltwater flow (whether sheet, channel, linked cavity or some combination), and the routing of meltwater flow viz. whether across a bedrock or sediment surface, totally or partially within a shattered bedrock or other material or by some combination of pathways.

Meltwater Routes: The development of any subglacial hydraulic system is contingent upon the nature and form of the ice–bed interface (Shoemaker, 1986; Boulton and Hindmarsh, 1987; Menzies, 1989a). There are four possible states in which meltwater may exist at the base of an ice mass viz: (1) meltwater (free) flowing in some form at the upper ice–bed interface (H-bed State); (2) meltwater (confined) flowing within rigid debris or bedrock (H-bed State); (3) meltwater (confined) flowing within a deforming sediment layer (M-bed State); and (4) meltwater (quasi-free) as interstitial films (H-bed State) (Chapter 5; Menzies, 1995b, Chapter 2).

(1) In order that a free subglacial hydraulic system fully develops, it is necessary for the largest percentage of the meltwater discharge to flow at the upper interface between the ice and bed (Fig. 6.10). There are several conditions all of which pertain to H-bed states, as discussed in the previous chapter, where this form of meltwater flow may take place viz: where ice is flowing upon bedrock of low to zero permeability (Ha) (Kamb and La Chapelle, 1964; Vivian and Bocquet, 1973), or over frozen debris of

low to zero permeability (Hb) (Freeze and Cherry, 1979), or unfrozen debris of very low permeability ($<10^{-6}$ m s^{-1}) (Hc) (Shoemaker, 1986).

(2) Where debris or shattered bedrock exists and allows the passage of meltwater through an essentially rigid (undeforming) structure, meltwater in the form of pipe flow may occur (Clarke et al., 1984). Such a condition may take place where an aquifer of sufficiently high conductivity ($>10^{-6}$ m s^{-1}) overlies an aquitard of much lower or near-zero conductivity allowing passage of meltwater (Hd), at least until fully saturated. This limiting condition may then cause the confined meltwater to flow freely at the surface. Where an aquifer overlies an aquitard that is sufficiently thick and of high permeability, meltwater may pass in a confined state down to a level defined by hydraulic pressure gradient itself related to overburden ice and sediment pressures (He).

(3) Where a deforming debris exists beneath an ice mass only a limited movement of meltwater is expected to occur at the upper ice–bed interface while the vast proportion of water will pass into deforming sediment thereby moving toward low pressure via advective means (M). The volume of meltwater likely to persist at the upper interface remains a matter of debate (Shoemaker, 1992). It can be supposed that, under typically anisotropic sediment rheological conditions, patches of sediment of much lower conductivity will occur, thereby leading to similar patches of locally H-bed conditions. This latter state is what is envisioned under Q-bed States.

(4) In cases where polar thermal bed conditions prevail or where meltwater flux is very low, only micron-thick films of water may exist at the upper interface (Nye and Frank, 1973). Such conditions may be more prevalent than considered in the past, and may be the precursor to free meltwater system evolution.

Models of Subglacial Hydraulic Systems: There are several models of subglacial hydraulic flow conditions that describe various modes of meltwater flow at the base of an ice mass, all of which have some supporting field evidence in their favour. Some of these models may be part of a continuum of hydraulic system evolution either developing from one to another over time or in space across the upper

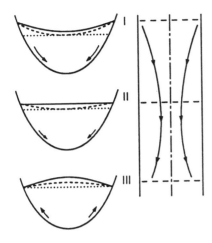

FIG. 6.13. Influence of surface ice topography of a valley glacier on the position of major subglacial meltwater conduit (after Röthlisberger and Lang, 1987; reprinted from Gurnell, A.M. and Clark, M.J. (eds), *Glacio-Fluvial Sediment Transfer*, by permission of John Wiley and Sons).

ice–bed interface. Therefore, criticism should be tempered with the realisation that certain hydraulic systems may develop only under specific glaciodynamic conditions.

Each of the models discussed below have common and limiting factors that govern their development. In order that meltwater flows freely at the ice–bed interface, it is necessary that melting (r) (cf. discussion on englacial conduits), and that the rate of meltwater supply equals or exceeds the rate of meltwater discharge. The relationship, however, between meltwater discharge and hydraulic pressure varies in accordance with the type of drainage system that evolves (i.e. under a channel dominant system an inverse relationship exists between these two variables, while in a linked-cavity system a positive relationship appears to occur) (n.b. discussion in Chapter 5 on surge mechanisms).

Meltwater flow at the ice–bed interface may occur as: (i) sheet flow in the form of a thin water film (Weertman, 1962, 1964a, 1966, 1969, 1972; Weertman and Birchfield, 1983a, b; Vivian, 1980; Walder, 1982); (ii) channel flow cut down into the glacier bed (N channels) (Nye, 1976) or cut up into the overlying active ice (R channels) (Röthlisberger, 1972), or a combination of both (*as may occur where glacier beds are composed of unconsolidated*

material) (C channels) (Clarke *et al.*, 1984); or (iii) a linked cavity system developed downstream of bedrock obstacles (Lliboutry, 1968; Walder, 1986; Kamb, 1987). In all cases, no model is exclusive of the others; thus, for example, sheet flow may co-exist adjacent to channel flow.

(i) *SHEET FLOW*: Weertman (1962) suggested that, under active ice motion, meltwater solely derived from basal ice melting would be evacuated across the ice–bed interface in the form of a thin, virtually continuous sheet of meltwater that in the process would have a positive feedback on ice motion by enhancing basal slip lubrication. It was theorised that surge behaviour might also be coupled with the subglacial water film 'drowning' obstacles at the bed thereby facilitating even greater basal ice motion. The increased basal motion would, in turn, cause increased melting due to increased basal friction leading ultimately to unstable ice motion in the form of a surge (Weertman, 1969).

Following Weertman's analysis, it is assumed that meltwater, originating solely due to basal ice melting, the result of geothermal heat and frictional heat from sliding, flows as a sheet under a pressure gradient that is largely a function of the average ice surface gradient (Weertman and Birchfield, 1983a, b; Budd and Jenssen, 1987). From Weertman (1966), the average water velocity ($\bar{v}_w$) within a film of depth d_w can be defined as:

$$\bar{v}_w \approx \frac{\rho_i \, g}{12u_b} \, \Gamma_{pg} \, d_w^2 \qquad (6.6)$$

where Γ_{pg} is the pressure gradient in the direction of mean slope defined as:

$$\Gamma_{pg} \approx \alpha + \frac{\rho_{wt} - \rho_i}{\rho_i} \, \beta \qquad (6.7)$$

where α is the ice surface slope and β is the slope of the bed.

The definition of the likely average thickness of such a water film is extremely complex considering the enormous variability of micro-topography at the ice–bed interface and the transient nature of meltwater discharge etc. (cf. Weertman and Birchfield, 1983, Appendix). Water film thickness is likely to vary both spatially and temporally across the bed and where high-pressure areas exist no water film will be found (Walder, 1982).

Assuming that the rate of basal ice melting (m_b) per unit time is uniform across the ice–bed interface, and that all the meltwater flows at that interface, the average water film thickness (after Walder, 1982, p. 283) is:

$$\bar{d} \approx (\frac{12\dot{m}_b \ x}{\Gamma})^{1/3} \qquad (6.8)$$

where x=0 is a point farthest from the glacier terminus.

Walder (1982) has shown that when typical melting rates and pressure gradients are taken into account, average film thicknesses rarely exceed a maximum of ≈2 mm and are typically ≈1 mm (Fig. 6.11). Figure 6.11 shows that where distances from the glacier snout increase to ≈1000 km, meltwater film thicknesses increase to only ≈5 mm with a basal melting rate of 15 mm year^{-1}, while for a melt rate of 2 m year^{-1} film thickness would increase to ≈20 mm. The question of the feasibility of sheet continuity over such great distances will be discussed below but with typical 'controlling obstacle sizes' with a lower size limit of ≈5 mm (Weertman, 1972, 1979), sheet continuity and integrity is unlikely to be retained except over relatively short distances of a few tens of square metres of bed area.

This lack of a widespread continuity in a meltwater film raises the query of non-steady-state vs. quasi-steady-state flow conditions. Under the parameters defined by Weertman and others where meltwater is solely derived from basal ice melting, increasing ice velocity should, in the short-term, lead to increased basal ice friction and higher melting rates. Inversely as the area of the bed in contact with the moving ice reduces due greater film thickness, the rate of basal ice melting should, in the longer term, diminish along with basal ice velocity. This state therefore verges upon apparent quasi-steady-state conditions for a meltwater film at the bed on an active ice mass. However, if ice velocity increases beyond a certain critical level and the average obstacle size at the upper interface is effectively 'drowned' (leading to an effective zero basal friction value), a state may be reached where, if incipient surge conditions are already present, a surge cycle may be maintained or precipitated (Walder, 1982) leading to non-steady-state conditions. In reality, since only a small fraction of the meltwater at the bed of a valley glacier is usually solely of subglacial derivation (<20%), transient exotic discharges of meltwater into the subglacial system will tend to cause a non-steady-state to prevail.

In ice sheets a much larger percentage of meltwater at the bed is of subglacial origin. Therefore quasi-steady-state conditions across smooth, low permeability beds where cavities are rare or infrequent may be maintained for short time periods.

Finally, it would appear that, even if quasi-steady-state water films can be maintained, they are likely to be of such limited depth to have little or no effect upon substantial alterations in basal ice velocity or bed roughness.

The concept of a thin subglacial water film raises an interrelated aspect of subglacial hydrology viz: the relationship between sheet flow and channel flow stability (Walder, 1982; Weertman and Birchfield, 1983). Walder (1982) suggested that given the nature of sheet flow, inevitably localised sheet thickenings will occur causing slight uneven, anisotropic hydraulic pressure differences to evolve. As the water flows preferentially in these 'thickenings', a local reduction in hydraulic pressure will cause meltwater to migrate increasingly to these zones leading inevitably to degeneration of the sheet flow and the development of an embryonic subglacial R-channel system (Fig. 6.12). The corollary of such sheet-to-channel evolution is that with increases in basal ice velocity, R-channels will increasingly encounter bed protuberances which if greater in size than the channels will lead to their disruption and a return perhaps to sheet flow.

Where a subglacial sheet has developed at the ice–bed interface, as basal ice velocity increases, the sheet will increasingly encounter subglacial cavities and other interface obstacles that will destroy the integrity of the sheet. In response to such occurrences a different hydraulic system will probably develop. Recent evidence (Lliboutry, 1983; Hantz and Lliboutry, 1983) suggests that where ice exceeds 100 m in thickness, basal ice exhibits a slight permeability, perhaps sufficient to absorb basal meltwater in sheet flow, at least under certain conditions. What these conditions and degrees of permeability are remains an area of continuing research.

(ii) *CHANNEL FLOW*: Meltwater may exist in sheet and channel flow side by side, or the former may evolve into the latter, or under non-steady-state conditions a hydraulic 'switch' from one form to the other may occur. Where meltwater from surface-derived sources reaches the bed, meltwater tends to preferentially flow in a channel system (Weertman, 1972; Röthlisberger, 1972; Nye, 1973; Shreve, 1972; Lliboutry, 1983; Hooke, 1984; Röthlisberger and Lang, 1987; Hooke *et al.*, 1990).

Channel Types: Channel types are of three basic forms viz: (a) R-channels, (b) N-channels, and (c) C-channels; but channel cross-section configuration and how it alters over time in response to changing meltwater discharge, temperature and debris content; bed material composition and intrinsic variability; and glaciodynamic responses remains largely unknown (Hooke *et al.*, 1990). In modelling subglacial

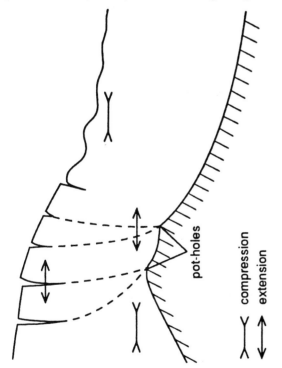

FIG. 6.14. A proposed relationship between subglacial bed pothole locations and the stress field within an ice mass (after Röthlisberger and Lang, 1987; reprinted from Gurnell, A.M. and Clark, M.J. (eds), *Glacio-Fluvial Sediment Transfer*, by permission of John Wiley and Sons).

channels, it has generally been assumed that channel cross-sections are of a symmetric semi-circular shape in response to competing processes of closure by ice pressures, opening by melting and the influences of internal ice stress fields. However, this geometry may only apply to channels oriented parallel to ice maximum flow directions and the principal longitudinal axis of stress; and/or where ice is moving at very low basal velocities, and in those channels that are totally water-filled considerable distances from the ice front. In other, more typical conditions, channels are likely to exhibit asymmetric cross-sections with considerable variation in channel width down the water-flow line, especially where channels cross at an angle to the principal ice flow direction and/or where channels are not fully filled at all times.

A further consideration of channel architecture is the influence of bed material erodibility. Since channels flow across the ice–bed interface changing bed materials and structures whether of different rock types or sediments will lead to variations in channel cross-section and long-profile geometry. Related to channel shape is the question of hydraulic roughness which may increase ten-fold between smooth englacial conduits and subglacial channels.

Which type of subglacial channel forms under a particular set of conditions depends upon the nature of the bed materials, the magnitude of meltwater discharge, ice forces and the presence or otherwise of any preferential hydraulic routes at the bed prior to glaciation or the onset of a new channel hydraulic system. In those locations where bed material is particularly resistant, R-channels will tend to form. Where prior flow routes exist or bed materials are weak, N-channels may quickly form. Clarke *et al.* (1984) have surmised that where extensive unconsolidated bed materials are found, C-channels may develop. It is likely, as examination of exhumed subglacial channels reveals, that channels may be of all types at different places and times along their length.

Channel Form: Röthlisberger's Model: The Röthlisberger model (1972), and variations on it, have been extensively used in recent years to calculate subglacial water pressures within straight subglacial R-channels (Hooke *et al.*, 1990; Willis *et al.*, 1990). Calculations of subglacial pressure at a specified distance from the ice front are based upon the following Eqn (6.9):

$$G^{11/8} - 0.316 G^{3/8} \frac{\delta P}{\delta x} =$$

$$\Omega \; D \; k_m^{-3/4} \; (nB)^{-n} \; Q^{-0.25} \; \cos^{-11/8}\beta \; \Delta P^n \quad (6.9)$$

where $\delta P/\delta x$ is the pressure gradient along a straight conduit; $G = (\delta P/\delta x + \rho_w \tan \beta)$; D is a constant that is related to ice and meltwater density and latent heat of fusion of ice; k_m is the reciprocal of the Manning roughness variable of the conduit; Q is the discharge, β is the slope of the glacier bed; ΔP is the difference between the pressure in the ice adjacent to the conduit and the water pressure in the conduit and Ω is a conduit shape factor introduced by Hooke *et al.* (1990). Hooke *et al.* (1990) demonstrate that such conduits may be typically portrayed as being broad and low which can be described by an arc (Θ) of a circle where Θ ranges from 2–36° (see Hooke *et al.*, 1990, Fig. 4 and Eqn. (2)).

In using the original equation (minus the parameter Ω) with real data, it has been found that poor agreement exists between calculated and actual observed pressure values. Typically, subglacial pressures appear to be higher than predicted. There are several possible explanations for this disparity related to ice rheology, the presence of impurities within the ice causing it to be softer than normally assumed, channel configuration in terms of sinuosity and the degree to which the channel is partially or fully filled by meltwater, and the effect of meltwater on cross-profile conduit shape (Lliboutry, 1983; Hooke, 1984; Iken and Bindschadler, 1986; Hooke *et al.*, 1990; Willis *et al.*, 1990). Setting these concerns aside, the Röthlisberger equation illustrates that discharge is inversely related to hydraulic pressure.

Hooke's Model: Where meltwater discharge in a conduit is greater than a few tens of litres per second and bed slope is more than a few degrees, the potential energy generated within the flow is sufficient to melt conduit ice walls at a rate several times faster than the rate of ice closure. Therefore internal hydraulic pressures where conduits connect externally to the margins of the ice mass are typically at atmospheric pressure and where no connection exists triple point pressure may prevail (Shreve, 1972; Hooke, 1984; Hooke, 1989). The condition m>r can be re-written for semi-circular channel discharge (Q) (6.10):

$$Q > [\frac{h_i^3}{c_4 \sin^{7/5}\beta}]^5 \quad (6.10)$$

where $c_4 = 6.55 \times 10^8 m^{12/5} s^{1/5}$. When this condition exists the melting rate will exceed the closure rate and the tunnel will be larger than necessary to carry the specific discharge available. However, it is thought that this particular condition is relatively rare in nature (*pers. commun.* Hooke, 1992). Figure 6.6 illustrates the equality relationship of Eqn. (6.10) between discharge and bed slope for various ice thicknesses. When discharge for a given bed slope and ice thickness exceeds the plotted value, the channel will or should be an 'open' channel and possibly not semi-circular but broad and low only fully filled during anomalous discharge events or over short sections of steeper slopes along the glacier bed. Although a steeper slope should lead to increased ice wall melting, some of the potential energy may be released as kinetic rather than heat energy (n.b. written comment by Shreve quoted in Hooke, 1984, p. 183). The converse effect of decreased slope may result in two consequences: (a) channels may remain open for some distance down-stream of a decrease in slope due to the delay in the ratio of closure to melting rates; and (b) back-pressure effects may lead to fully-filled channel sections in areas up-ice of the lower bed slope where discharge would allow, in theory, 'open' channel flow. These effects may have important ramifications for esker depositional processes and the siting of eskers across glaciated terrain (Shreve, 1972, 1985). The relationships in Fig. 6.6 are probably only effective under relatively thin ice masses. Below large ice sheets, as noted previously, channels will probably be fully-filled and at or close to hydrostatic pressures for distances of several tens of kilometres back from the ice margin.

Channel Maturation/Degeneration: As a subglacial channel system evolves, it may link with other channels into a drainage network. Such a network, by definition, has a hierarchical system of channel discharge which under ice pressures etc. will cause larger, trunk channels to develop at the expense of small channels (Walder, 1982, 1986). Recently, Shoemaker (1986), in modelling trunk channel spacing, found that approximately parallel channels may develop beneath large ice sheets, closest to their margins, separated by transverse distances of approximately 10^4 times the channel radius (Alley, 1989a) (n.b. major esker trains in N.W.T., Canada, Aylsworth and Shilts, 1989). Typically, valley glaciers, 2 km or less in width, are thought to have a single trunk subglacial channel (Kamb *et al.*, 1985; Humphrey *et al.*, 1986).

Channel Location: Prediction of the location of subglacial channels remains problematic. Control of channel location appears to be regulated by ice stress conditions, the presence of pre-existing channels and bed topography. Hooke (1984) argued that some channels may cross bed slopes diagonally as a result of meltwater attempting to flow directly downslope due to gravity and the influence of active ice attempting to align the channel in the direction of principal ice flow direction. The manifestation of this effect is to cause channels to 'migrate'. Fluctuations in discharge are a major controlling variable affecting channel size and ease or otherwise of ice wall melting and therefore migration.

In general, two subglacial channels types may co-exist within any one ice mass viz: 'gradient' channels and bottom channels. The former flow in relation to the gradient of the equipotential piezometric line (as defined above; also see Shreve, 1972; Röthlisberger and Lang, 1987, Fig. 10.15, p. 242) and therefore may occur at high levels along valley sides or in the deeper areas of an ice mass close to the main axis of ice flow. 'Gradient' channels may migrate downward more easily than upward due to energy considerations thus transforming from gradient to bottom channels (Röthlisberger, 1972; Hooke, 1984). In mapping glaciated terrain this transition can often be noted where subglacial channels cut across the topographic slope of hillsides and then steeply descend to the lower levels of or close to the valley floor (Fig. 6.14).

Röthlisberger and Lang (1987) have considered channel location in relation to ice mass surface slope i.e. concave, flat and convex (Fig. 6.13). Where concave and flat surfaces occur within relatively short glacier widths, typical of valley glaciers, subglacial channels might be expected to occur within the lower central parts of the glacier. Where glaciers, with convex surface relief exist, channels might preferentially migrate toward the lateral glacier margins. Whether this type of analysis can be applied to separate ice lobes or streams, within ice sheets, remains uninvestigated (Echelmeyer and Harrison, 1990).

Two further considerations of subglacial channel location remain puzzling viz: (1) how do subglacial meltwaters traverse areas of overdeepening or bed rises; and (2) how are subglacial (and englacial) channels affected by floating ice margins where tidal or ice shelf environments exist?

(1) Lliboutry (1983) suggested that where glacial overdeepening is encountered, subglacial channels would tend to bypass such areas by a marginal route. However, Röthlisberger and Lang (1987) argue that the opposite is true and that meltwater may well flow down the glacier axis and cross overdeepened areas. The problem of bed rises or riegels only becomes significant where bed slope, in the up-ice direction, exceeds the surface slope by approximately ≈ 11 times (Budd and Jenssen, 1987, p. 305). At that point, flow is either cut off or reverts to a Weertman-style film. Where up-slope topographic obstructions occur but of lesser slope, subglacial channels will often bifurcate and produce a 'horseshoe' shaped furrow (sichelwannen) on the stoss-side of the obstruction (Plate 6.8). Such furrows may range in scale from a few millimetres to tens of metres in depth. In some instances, just a short distance along the sides of the obstruction where compressive stress levels once again drop, erosion appears to have been less effective, perhaps due to the channels degenerating into sheet flow for short distances toward the lee-end before again coalescing into channel flow. The tendency for channel flow to switch to sheet flow conditions and back again will be discussed below in terms of hydraulic stability.

(2) In almost all discussions of subglacial and englacial hydraulic regimens within ice masses there is little or no discussion on the unique problems associated with the pressure fluctuations and basal stress conditions close to and at the margins of floating ice fronts. Whether the ice mass passes into an ice shelf or floats as a tidewater glacier margin, there are distinctive hydraulic conditions that must impact upon the hydraulic systems developed up-ice from the grounding-line position (Hooke, 1989; Echelmeyer and Harrison, 1990; Powell, 1990; Anderson and Ashley, 1991). First, channels, whether subglacial or englacial, entering a body of water from a floating ice mass margin must be fully-filled. Second, the hydraulic pressure developed by standing water, in a lake or sea, will be higher, for equivalent depth, than the ice just immediately up-ice of the grounding-line. Finally, this higher equipotential will cause quasi-artesian conditions to exist within the channels since the water equivalent line, at and snout-ward of the grounding-line, will be above the ice surface. This final effect explains the bubbling up of water often observed when glaciers enter lakes, and the jet-and plume-effects and related sedimentation processes observed subaquatically at many floating ice margins (Plate 6.9). Further effects of a floating margin are: (1) the tendency for climatically controlled temperature fluctuations, that influence meltwater flow and seasonal discharge, to be diminished close to the floating margin due to lake or sea water thermal effects; (2) back-pressure, from standing water up-ice of the grounding-line, will help maintain melting rates and reduce closure rates of tunnel walls, but will also critically influence longitudinal channel velocity and sediment load competency. Back-pressure effects, acting on a channel's hydraulic regime, will stretch back beneath an ice mass, where the glacier bed is essentially horizontal, for at least ≈ 1.1 times the depth of water margin-ward of the grounding-line. Where the glacier bed dips toward the ice margin and grounding-line, the inner subglacial extension of back-pressure will be greater and where the glacier bed dips up-ice the inward extension will be less (Shoemaker, 1990, 1992).

Channel Stability: Subglacial channel form is essentially an unstable form that quickly degenerates or evolves into a different hydraulic system. Where perturbations to the system occur due to fluctuations in meltwater discharge, channel dimensions and/or transient meltwater pressures, channel form quickly adjusts. The size of change required varies in magnitude dependent upon channel size, location, discharge and other glaciodynamic aspects of the particular channel's surrounding environment.

Weertman and Birchfield (1982, 1983b) have argued that over the long-term, due to ice pressures on the bed, R-channels, solely supplied with subglacially derived meltwater, would become starved of meltwater and would cease to function in a progressive up-stream to down-stream degeneration of arborescent drainage systems. In contrast, Walder (1982) has shown that sheet flow is inherently unstable leading to incipient channel development where preferential flow in deeper parts of a sheet leads to channel evolution (Fig. 6.15).

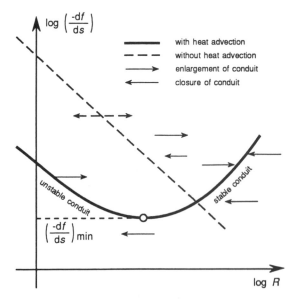

FIG. 6.15. Incipient instability within subglacial conduit draining major subglacial meltwater reservoirs (after Spring, *unpublished* and Röthlisberger and Lang, 1987; reprinted from Gurnell, A.M. and Clark, M.J. (eds), *Glacio-Fluvial Sediment Transfer*, by permission of John Wiley and Sons).

Two reasons for channel instability have been advanced viz: (1) that pressure fields along the edges of R-channels will prevent meltwater migrating into the channels thereby 'starving' them (Weertman and Birchfield, 1983b); and (2) that developing channels encountering obstacles at the ice–bed interface will be disrupted by them at a rate equal to the sliding velocity of the ice mass and as bed roughness increases (Walder, 1982; 1986). This latter proposal would cause sheet flow to be quasi-stable, provided the rate of encounters with protuberances is equal or greater than the rate of incipient channel growth. It is possible that an episodic sequence of '*sheet flow–channel flow–sheet flow*'' conditions may occur randomly at the bed of an ice mass. Walder (1982) suggests that where sheet flow occurs with depths of ≤ 4 mm, quasi-stable sheet flow may occur, otherwise channel flow will more typically develop. The only set of conditions in which sheet flow, as opposed to channel flow, may occur is where few subglacial cavities exist, bed roughness is slight and large volumes of extraneous meltwater from englacial and supraglacial sources penetrate to the subglacial bed interface (Fig. 6.15) (Spring, 1983; Röthlisberger and Lang, 1987).

A further form of channel instability occurs when, under increasing meltwater discharge typically derived from external, non-subglacial sources, channel wall melting increases at a very rapid rate in comparison to closure rate. This situation is typical of 'flood events' caused by seasonal discharge fluctuations, surges and ice-dammed lake outbursts. Under these circumstances, a limited but unstable condition is reached, at least in the short-term (a few hours or days) leading to a rapid headward and lateral expansion of the drainage system (Spring, 1983; Spring and Hutter, 1981b; Kamb, 1987). Spring (1983) considered the problem of increasing discharge in relation to hydraulic gradient and found two possible scenarios viz: (1) unlimited conduit expansion where heat advection is ignored or unimportant; and (2) limited expansion of conduits when heat advection is accounted for (Fig. 6.15). It is likely that when heat from flowing meltwater is advected along the conduit and is no longer available for continued ice wall melting, conduit closure begins

even at high discharges and thus unstable conditions resume until discharge drops to a lower volume and stable conditions once again resume (Kamb, 1987; Fig. 12). Where most of the discharge is from external sources and thus discharge at unstable high volumes is not reduced by self-regulation, the instability inherent in this condition may ultimately lead to channel degeneration and either massive sheet flow or to a linked-cavity hydraulic system (Kamb, 1987). The 'switching' of hydraulic systems under increasing discharge from sheet flow in the early stages to channel flow and, in the final stages of high discharge, to a linked cavity system and vice versa needs to be considered. Depending upon bed topography and materials, and ice basal velocity, temperature and stress; the progression of hydraulic systems may proceed from one form to another or not as the case may be. Where basal ice–bed interface conditions permit, the hydraulic system may evolve to the linked-cavity system as the ultimate manifestation of a surging glacier state, and/or alternatively a jökulhlaup event may occur.

(iii) *LINKED-CAVITY FLOW*: The development of a linked-cavity flow hydraulic system has been advocated by several workers in which increasing discharge via a series of lee-side cavities linked by narrow, short R-channels (orifices) cutting transversely across a glacier bed appears to be directly related to hydraulic pressure (Lliboutry, 1968; Humphrey *et al.*, 1986; Iken and Bindschadler, 1986; Walder, 1986; Fowler, 1987; Kamb, 1987). Figure 5.28 illustrates the nature of this cavity/orifice hydraulic system (Chapter 5). Lliboutry (1968) has suggested that within an interconnected cavity hydraulic system, cavities would be linked together as 'beads' on a string; while Humphrey *et al.* (1986) noted a constricted subglacial hydraulic system operating during a mini-surge beneath the Variegated Glacier.

Evidence for a linked-cavity hydraulic system rests upon: (a) the mapping of previously glaciated areas (Walder and Hallet, 1979; Hallet and Anderson, 1980; Sharp *et al.*, 1989); (b) modelling of time-related dispersion of injected dye tracers in present-day ice masses (Brugman, 1986; Seaberg *et al.*, 1988; Willis *et al.*, 1990); and (c) subglacial meltwater pressure observations under present-day glaciers and subsequent modelling of likely subglacial hydraulic regimens (Kamb *et al.*, 1985; Kamb and Engelhardt, 1987; Raymond, 1987).

Unlike channel systems, no intrinsic instability exists within a linked-cavity system at high meltwater discharges (Fig. 6.16). As discharges increase, the orifices act to self-regulate the through-flow of meltwater by effectively 'throttling' the flow from one cavity to another. However, a linked-cavity system would only appear capable of operation under relatively high basal meltwater discharges; once meltwater flux levels fall below the level necessary to maintain the orifice connections (due to ice wall closure) either channel or sheet flow hydraulic systems will re-form.

It would appear that the linked-cavity hydraulic system is peculiar to certain glaciers and glacier bed states and is only fully active under hyper-flow meltwater conditions probably of the type found at or immediately prior to the commencement of a surge event. However, it seems possible that such a hydraulic system may occur beneath parts of many glaciers and help explain anomalously high basal water pressures and local high basal ice velocities.

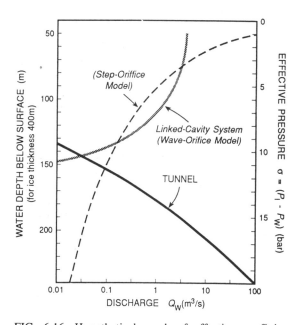

FIG. 6.16. Hypothetical graph of effective confining pressure against subglacial discharge via a conduit system types (step and wave orifices, and tunnels) (after Kamb, 1987; *Journal of Geophysical Research*, **92B**, p. 9094, copyright by the *American Geophysical Union*).

For example, Willis *et al.* (1990) have suggested that subglacial meltwater flowing in the eastern portion of the Midtdalsbreen Glacier, Norway may spend between 10–41% of overall transit time within a linked-cavity system.

6.4. HYDRAULIC SYSTEMS AND DEFORMABLE BEDS

All the discussion so far on hydraulic systems has been largely concerned with rigid beds (H-bed states). However, increasing evidence reveals that large portions of many glacier beds today and in the past were probably composed of soft, unconsolidated beds (M- or Q-bed states) termed deformable beds (Chapter 5).

Where deformable sediment, saturated by meltwater, exists beneath ice masses, deformation will occur when the applied shear stress imparted by the overlying active ice overcomes the internal shear strength of the sediment. Perhaps the greatest controlling variable influencing this process is the intrinsic viscosity of the sediment, since this parameter reflects the level of consolidation, grain size and porosity of the sediment (Boulton and Hindmarsh, 1987; Menzies, 1989b; Alley, 1989a, b). If it is assumed that the sediment at the ice–bed interface is deforming, then it is generally thought that little or no free meltwater derived from solely subglacial sources, would exist at the upper interface between the mobile sediment and glacier sole (Shoemaker, 1986; Menzies, 1989b; Murray and Dowdeswell, 1992). However, where meltwater from external sources reaches the glacier bed, a slightly different set of circumstances may occur. In the first instance, extraneous water cannot be totally absorbed by the already saturated sediment and thus must move along the upper interface. The absorbed meltwater will aid in reducing sediment viscosity, further allowing the sediment to deform at an even greater rate. The remaining unabsorbed water, however, must find a pathway along the upper interface or within the sediment itself in the form of pipe-flow (Clarke *et al.*, 1984; Kamb, 1987). Inconclusive evidence for the latter form of flow does exist within many diamictons where pipes and small intercalated channels of stratified sediment are commonly observed (Eyles *et al.*, 1982; Eyles and Menzies, 1983; Brodzikowski and Van Loon, 1991). The amount of water remaining on the upper interface surface persists as an intriguing question vis a vis subglacial hydraulic system development (Murray and Dowdeswell, 1992; Shoemaker, 1992).

Meltwater flow at the upper interface of a deformable bed cannot be directly compared to water flow under rigid bed conditions. Thus, the question as to whether hydraulic systems will develop in a manner akin to rigid bed states is predicated by the fact that sediment deformation causing channel and cavity closure must also be considered. Were channels and cavities to form, several processes need to be considered in assessing the stability of these openings (Alley, 1989a, b). If a channel is to remain open under deforming bed conditions, not only must melting rates exceed or equal ice wall closure rates but, in addition, meltwater erosion rates must equal or exceed the creep of deforming sediment entering the channel.

When the subjacent sediments fully deform, it has been suggested that all meltwater produced at the base will enter the deforming sediment and be advected via pore spaces within the sediment (Alley et al., 1986, 1987a, b; Menzies, 1989a, b). However, this strict criterion probably only applies under rare circumstances or when the sediment has a very high permeability (i.e. acts as a conducting aquifer). Under normal conditions, free meltwater will be present at the upper interface between the sediment and base of the ice mass (Shoemaker, 1986). This meltwater may be ubiquitous or not; it may exist partly as a sheet or in channels (Shoemaker, 1986). Meltwater may accumulate locally at the bed or be quickly evacuated down a hydraulic gradient. If accumulations of meltwater at the upper interface spread laterally, an interconnected drainage system may evolve. Such a system may be ephemeral or not but, in either case, the impact upon the sliding velocity of an ice mass becomes considerable even if for only a short time (Kamb, 1987; Shoemaker, 1992).

Where temperate ice overlies unconsolidated sediment capable of deformation several models of upper interface/meltwater interrelationships have been developed. These models consider the microscale conditions at the upper interface (Alley, 1989a, b); and the complex macroscale relationships between differing immobile sediment — deforming sediment — ice interface boundaries (Shoemaker, 1986; Boulton, 1987; Boulton and Hindmarsh, 1987).

In general, as meltwater is produced at the base of an ice mass and arrives at the upper interface, three routes are available: (1) solely via the upper interface; (2) via porous flow through the sediment where the sediment acts as a rigid structure; and (3) as advective flow within mobilised sediment.

(1) If meltwater has to flow at the upper interface because the permeability of underlying sediment is insufficient or the bed is frozen (or bedrock), the meltwater will initially flow as a distributed sheet or film of variable thickness (Weertman, 1972; Weertman and Birchfield, 1983b). As meltwater volume increases, water evacuation across the upper interface will adopt an interconnected channel system of conduits cut upward into the ice (R-channels), or downward into the bed (N-channels), or both up into the ice and into the bed (C-channels) or by linked-cavities (Walder, 1986; Kamb, 1987). These conduits develop as a consequence of either meltwater volumetric increase or surface irregularities or simply increased water film velocity down-gradient in the manner characteristic of subaerial overland flow/rill evolution (Walder, 1982). Unless these channels are supplied by meltwater from a source other than subglacial meltwater (i.e. englacial or supraglacial sources), these channels remain ephemeral (Weertman and Birchfield, 1983b; Weertman, 1986). The channels close due to the rapidity of ice creep closure and infilling by squeezed unconsolidated sediment.

Channel closure, within a soft bed state, is a function of the rate of sediment movement into the conduit which in turn is a consequence of the effective stress applied to the sediment, the sediment's yield strength and porewater content. Assuming a semi-circular channel of radius $\dot{r}$ with a closure rate of r and an ice driving stress of τ_b, the closure rate (Ψ_c) (due to creep closure) can be defined as (Boulton and Hindmarsh, 1987; Alley, 1989a) (see Alley, 1989a, p. 112, Eqn. (4) (n.b. Eqns (5.23) and (5.46)):

$$\Psi_c = K_b \frac{(\tau_b - \tau^*)^a}{\tau_b \, a^a} \, , \qquad \tau_b > \tau^* \qquad (6.11)$$

$$\Psi_c = 0 \, , \qquad \tau_b \leq \tau^* \qquad (6.12)$$

$$\Psi_c = \frac{\dot{r}}{r} \qquad (6.13)$$

This simplified closure rate equation takes no account of spatial variation in effective stress or sediment yield strength, and is limited to small (mm) channels. Any sediment channel (Alley's till channels) will remain open if infilling sediment can be evacuated at a rate comparable to the closure rate (Chapter 5). Alley (1989a) defines the erosion rate (Ψ_e) based upon channel geometry and discharge as:

$$\Psi_e = \frac{1}{2\pi r^2} \frac{\partial J_s}{\partial x} \qquad (6.14)$$

where x is the distance along the channel in the flow direction and J_s is an approximation of the sediment flux within the channel defined as:

$$J_s = Z_o \, \pi r^2 V_m^3$$

where V_m is the mean flow velocity (Allen, 1985) and Z_o is a constant. Since the erosion rate (Ψ_e) must equate with the closure rate (Ψ_c) to maintain an stable open channel, a further important aspect of such an equilibrium process is the discharge of water into the channel. *(Equations accounting for laminar and turbulent flow of a sediment/water mixture being eroded from a channel are derived by Alley (1989a).)* The water influx (Q_x) from across the upper interface into a channel can be approximated as:

$$Q_x \approx 10^4 r \dot{m}_b \qquad (6.15)$$

Using this approximation a channel 1 m in radius would collect meltwater from a 10 km^2 wide area of the bed beneath an ice mass. From our present understanding of subglacial hydrology this approximation seems reasonable, if perhaps an over-estimation for small channels (Alean *et al.*, 1986; Röthlisberger and Lang, 1987).

The implication of the above is that any meltwater channel system at the junction of unconsolidated deformable sediment and the base of an ice mass is likely to be short-lived and therefore of minor significance in terms of basal ice decoupling and as a proportion of ice mass sliding velocity (Fig. 6.17a,b). R-channels, if they exist, at low effective stress levels might be stable if a few millimetres in radius or less. Meltwater, if only of subglacial origin, at the upper interface is liable to be a thin film or sheet with local, ephemeral, thickenings in the lee of small obstacles. The key to sliding velocities over unconsolidated sediments is the thickness and spatial integrity of a meltwater film. Where a water film becomes continuous or micro-channels intersect and generate larger channels over a wide area of the ice bed, the ice mass will progressively decouple and high sliding velocities should occur. The spatial distribution of any film is a function of where local pressures are less than the average ice overburden pressures and therefore can laterally expand. As a result of spatial enlargement of these lower pressure areas, the fraction of the bed (f) covered by meltwater increases. This relationship can be summarized as:

$$\bar{\sigma}_b = \frac{G_{eo} \, \tau_b}{f} \qquad (6.16)$$

where G_{eo} is a geometric factor. This equation is limiting when either τ_b and σ_b approach zero (n.b. 'bed separation index'; Bindschadler, 1983). When either parameter reaches low values the effect of basal friction reduces to a self-limiting point where further meltwater production at the bed ceases and the above relationship breaks down (Iken, 1981; Humphrey, 1987; Lliboutry, 1987a, b). This state can be phrased in an alternate form such that when all low pressure points on a relatively smooth bed are infilled by meltwater and the bed contact boundary layer is 'smoothed out' it can be shown that the effective pressure (σ_e) (in the Humphrey Model) tends toward a constant (Humphrey, 1987). However, when meltwater flows across a unconsolidated sediment bed with a broad range of grain-sizes, this latter τ_b–σ_e boundary relationship does not apply.

The rate of meltwater movement at the upper interface can be related to Eqn. (6.16) since as the thickness of a water film (d_w) increases so the fraction of the bed covered by meltwater must increase in some manner related to bed upper interface geometry.

(2) Where subglacial sediment has a high hydraulic conductivity and/or is overlying an aquifer of similar or higher conductivity, a high flux rate of meltwater evacuation via porous flow may be possible assuming the sediment does not deform and acts in a 'rigid' structural manner (Shoemaker, 1986). Provided meltwater production at the upper interface does not exceed the through-flux of porewater within the sediment layer, meltwater may be removed successfully (Shoemaker, 1986, Type II). The passage of meltwater via the sediment layer can be modelled according to Darcy's Law:

$$v_w = i \, K \qquad (6.17)$$

where v_w is the specific discharge assuming that the inflow and outflow rates of water are equal through a unit area of sediment, i is the hydraulic gradient defined as the excess hydrostatic pressure in the meltwater through a specific length of sediment and K is the hydraulic conductivity. Using a range of typical values for porewater flow through a highly permeable substrate (Shoemaker, 1986; Lingle and Brown, 1987), three fundamental aspects become readily apparent. First, flow rates are so low that flux rates would be in thousands of years or through

enormous thicknesses of sediment (>200 m) (Freeze and Cherry, 1979). Second, and of critical importance in the short-term, is the rapid increase in porewater pressures that may lead to a critical hydraulic gradient developing. With a critical hydraulic gradient ($i_c \approx 1.2$), effective stresses are either zero or negative such that porewater begins to move upward and fluidisation of the sediment layer occurs (Terzaghi and Peck, 1967). Finally, it is unlikely that meltwater flow through a porous sediment can ever account exclusively for all meltwater produced at the upper interface. A substantial proportion of the meltwater must flow across the upper interface in channels and/or sheets. As a result, basal sliding conditions will be influenced and perhaps short-term ice mass stability.

(3) When subglacial sediment moves as a slurry under the applied stresses imparted from overlying active ice, meltwater enters the mobile sediment system and is advected in the direction of the basal ice pressure gradient (Alley et al., 1986, 1987a, b). At present our understanding of subglacial sediment

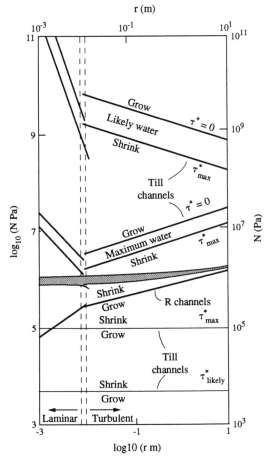

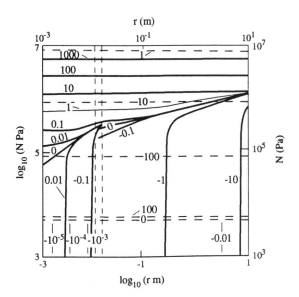

FIG. 6.17. (a) Effective confining pressure against subglacial channel radius for R and 'Till' channels. At high effective pressures 'Till' channels exhibit stability. A 'Till' channel plotted above a high effective confining pressure tends to grow and below the channel shrinks. Within the stippled zone both R and 'Till' channels shrink at equal rates (after Alley, 1989b; reproduced by courtesy of the International Glaciological Society from *Journal of Glaciology*, **35**(119), 1989, p. 113, Fig. 2).

FIG. 6.17. (b) Graph of conduit shrinkage rates for both R (solid lines) and 'Till' (dashed) channels as a function of channel radius (r) and effective confining pressure. Negative values indicate channel growth (after Alley, 1989b; reproduced by courtesy of the *International Glaciological Society* from *Journal of Glaciology*, **35**(119), 1989, p. 114, Fig. 3).

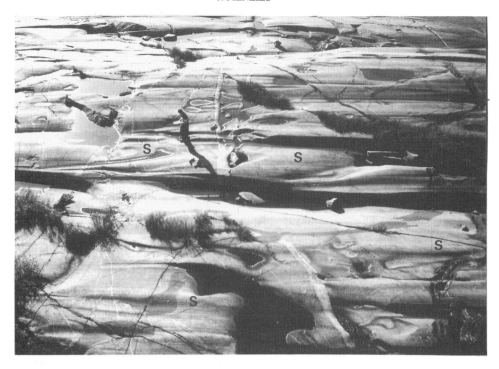

PLATE 6.8. Sichelwannen forms (s) from Georgian Bay, Ontario with differing degrees of development. Shaw has classified these forms as *comma forms*, being sichelwannen with only one arm developed or missing and are thus transitional forms of sichelwannen. Scale bar = 2 m (photograph courtesy of John Shaw).

PLATE 6.9. Jökulhlaup at Grimsvötn, Iceland in 1982 (photograph courtesy Helgi Björnsson).

rheology is restricted (Boulton and Hindmarsh, 1987; Clarke, 1987b; Menzies, 1989a; Hart et al., 1990). As the porewater content of a potential slurry increases, a critical volume concentration (c.v.c.) is reached (–0.6) beyond which inter-particle contacts are drastically reduced due to partial suspension of particles by the porewater. At this critical moisture content, the sediment begins to behave as slurry-like material of finite viscosity (Cheng and Richmond, 1978); and with increasing stress application a dilatant reaction will occur thereby increasing the sediment's intrinsic permeability (κ). (Intrinsic permeability and hydraulic conductivity are related in Darcian terms where $\kappa = K\mu_w/\rho_{wt}g$.)

A complex relationship between sediment parameters and permeability can be modelled using an empirically modified Kozeny–Carmen relation (Raffle and Greenwood, 1961):

$$k = \frac{\gamma_g}{5\eta_{sed}} \left[\frac{n_p^3}{(1-n_p)^2} \right] \frac{1}{S_0^2} \qquad (6.18)$$

where κ is permeability, n_p is porosity, η_{sed} is sediment viscosity and $S_0 \approx 6/d_0$ = the specific surface area of particles per unit volume of slurry where d_0 is mean particle size diameter. From this equation, assuming constant values of γ_g, S_0 and n_p, a magnitude increase in viscosity would lead to a 10^2 decrease in permeability. Since this simple relationship takes no account of co-related changes to the other parameters, it can be concluded that a change in sediment viscosity would lead to even greater changes in permeability. The effectiveness of this advective process of meltwater evacuation is dependent therefore upon the viscosity and porosity of the sediment both of which fluctuate as the stress applied to the sediment alters the rheology of the sediment (Chong et al., 1971). Theoretically, if basal stress increases and the thickness of deforming sediment increases correspondingly, more porewater (produced by increased basal stress leading to increased melting) will be capable of advection. However, a limiting point is reached where either the excess meltwater, unable to be advected, flows along the upper interface contact thus leading to increasing bed decoupling, or, with increasing porewater pressure, basal friction reduces and melting diminishes leading to a form of glaciodynamic equilibrium being achieved with advection remaining the dominant mode of meltwater removal.

Lingle and Brown (1987, Table 2) derived an empirical equation relating meltwater flux rates, ice mass and sediment characteristic values to the basal sliding velocity of an ice mass overlying a deforming bed. In a heuristic relationship where the basal sliding velocity (u_b) assumed to equate approximately with the sliding velocity of the upper interface of the mobile sediment (u_m) the thickness of deforming sediment is computed as:

$$z_s \approx \frac{2Q}{n_p \ W \ u_m} \qquad (6.19)$$

where Q is the subglacial meltwater flux rate and W is the width of the deforming layer transverse to maximum ice flow direction (Lingle and Brown equated W with the width of Ice Stream B, Antarctica but maybe used the transverse width of a deforming sediment zone beneath a much larger ice mass (cf. Menzies, 1989a).) Some of the implications of this equation are illustrated in a series of graphs in Fig. 6.18. The applicability of this equation to Quaternary sediments remains to be further investigated but as a beginning it draws a much needed link between glaciodynamics and sedimentology (Menzies, 1995b, Chapter 2).

6.5. SUBGLACIAL LAKES

The term subglacial lakes (basal or sub-ice lakes) is applied to large pondings of meltwater that have accumulated beneath ice sheets. These lakes appear to form under conditions of temperate basal melting but are prevented from evacuating due to the constraining effects caused by a hydraulic seal as result of high normal ice overburden pressures, thermal fluctuations in basal ice, or a topographic high. Evidence for these lakes is restricted to radio-echo data since no lake as yet has been specifically identified by boring through an ice sheet. The areal extent of these subglacial lakes is difficult to establish, but a lake near Vostok Station, East Antarctica is estimated to be 8100 km² (Oswald and Robin, 1973; Robin et al., 1977; Drewry, 1983b; Whillans and Johnsen, 1983; Cudlip and McIntyre, 1987; Whillans, 1987; Goodwin, 1988). Drewry (1986) suggests that these lakes must be deeper than 6.5 m for radio-echo attenuation to occur. Many other

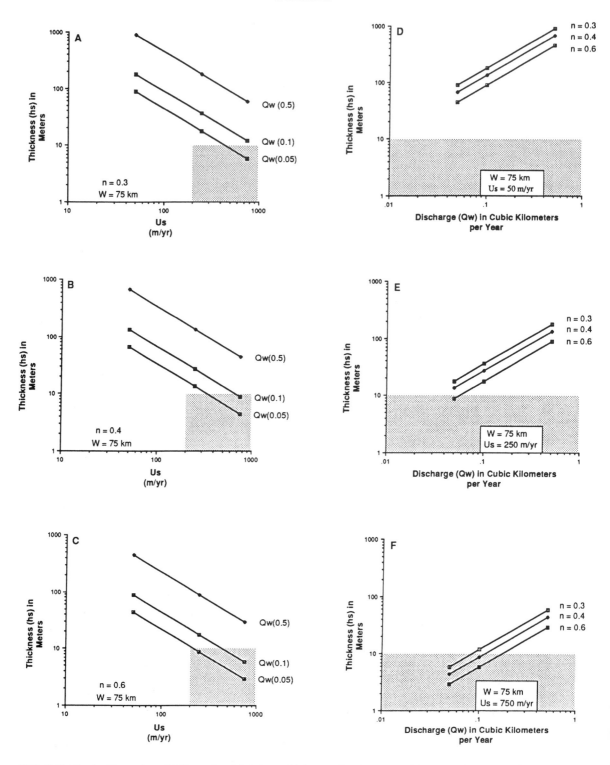

FIG. 6.18. Graphs illustrative of deformation debris layer thickness. Graphs A, B and C show debris layer thickness against upper debris layer velocity while Graphs D, E and F show debris layer thickness against subglacial meltwater discharge (after Menzies, 1989; reproduced with permission from Elsevier Science Publishers BV).

small lakes and pondings occur beneath ice masses of all sizes. However, only large subglacial lakes can appreciably affect the regional glaciodynamics of the ice sheet for some distance around the lake perimeter producing either flat areas or slight depressions within the general surface topography, the latter only close to ice divides where ice movement is slow (Whillans, 1987; Shoemaker, 1990, 1992) (Chapter 3). Since a large body of subglacial meltwater will cause the basal shear stress level of the ice mass to drop locally to near zero, resulting in an increase in the longitudinal strain rate, this effect is transmitted three-dimensionally in all directions away from the ponded area (see Shoemaker (1990) for calculations).

The geomorphological effect of these large lakes is two-fold viz: (1) as localities of fine grained sediment accumulation, and (2) to be sources of major subglacial flood events, where and when the barrier seal 'breaks' (Waitt, 1980; Mayer and Nash, 1987; Baker et al., 1988; Shaw et al., 1989; Shoemaker, 1992) (Chapter 12). The latter possibility may occur when ice thicknesses decrease during deglaciation and surging, or when thermal fluctuations at the ice base permit transmission of meltwater beyond the lake barrier. Whether a flood event would occur when a subglacial lake seal breaks depends upon other basal ice conditions, topographic constraints, and the style of barrier collapse whether catastrophic or gradual (Goodwin, 1988).

6.6. HIGH-MAGNITUDE MELTWATER DISCHARGES

There are perhaps few glacial phenomena quite as spectacular as the sudden and episodic outbursts of meltwater either from subglacial and supraglacial lakes, ice-dammed lakes or over-spilling proglacial lakes. These outbursts are due to several causes but are collectively termed 'jökulhlaups' (Chapter 12). These glacial outbursts or 'flash' floods were thought to be peculiar to temperate ice masses but recent evidence of a non-cyclic jökulhlaup has been observed in East Antarctica under polar basal ice mass conditions (Goodwin, 1988). Such catastrophic events have a considerable impact upon the entire proglacial

zone and beyond (Bretz, 1969; Liestøl, 1955a, b; Thorarinsson, 1974; Björnsson, 1976, 1992; Sissons, 1981; Clarke and Waldron, 1984; Waitt, 1980, 1985; Björnsson and Kristmannsdóttir, 1984; Maizels, 1989a, b; Maizels and Caseldine, 1991; Maizels and Russell, 1992; Mathews and Clague, 1993). The greatest effect of jökulhlaups is their devastating floods which may cause enormous sediment transport, reworking, and erosion and, in human terms, loss of human life and that of livestock, and destruction of property (Hewitt, 1982; Haeberli, 1983; Tufnell, 1984; Maizels, 1989a; Maizels and Caseldine, 1991). Consequences of jökulhlaup destruction include the washing away of bridges, roads and habitation and the breaking of pipelines (Sturm and Benson, 1985; Mathews and Clague, 1993), the latter having the added problems associated with potential air, water and land toxic pollution. It is the association of jökulhlaups with human tragedy in one form or other that has left a mark on folk histories among the peoples of the European Alps, the Pamirs and in the great saga writings of Iceland. Thorarinsson (1974) has chronicled the passage of repeated jökulhlaups for the Skiedarárjökull area of Katnajökull from 1332 to 1972 (Table 6.2). Perhaps the best studied and researched jökulhlaups are those associated with Grimsvötn in Iceland (Rist, 1955, 1973; Thorarinsson, 1974; Björnsson, 1976, 1992; Nye, 1976; Spring and Hutter, 1981; Björnsson and Kristmannsdóttir, 1984; Maizels, 1989a, b; Maizels and Russell, 1992).

Within the proximal proglacial zone large blocks of ice may be left stranded later creating 'reversed' thermokarst terrain (Knudsen and Theakstone, 1988). Substantial thicknesses of sediment are deposited and new proglacial stream networks formed and previous ones abandoned (Krigström, 1962; Bradley et al., 1972; Fahnestock and Bradley, 1973; Church and Gilbert, 1975; Maizels, 1979, 1983, 1989a, b; Price, 1980; Shakesby, 1985; Fenn and Gurnell, 1987). Annual moraines may be breached and extensive areas of bedrock scoured (Haefeli, 1963; Haeberli, 1983; McCarroll et al., 1989).

TABLE 6.2. A record of Jökulhlaups since 1332 until 1972 from Iceland (from Thorarinsson, 1974; translation by Helgi Bjornsson)

Month	Year	Interval since previous Jökulhlaup, Years	Jökulhlaup in Skeiðará	Eruption in Grímsvötn	Remarks
November	1332		§+	§	
May	1341	8½ *	§	§	
November	1598		§	§	
July	1619		§	§	
Uncertain	1629	10	#	#	
February	1638	9	#	#	
Uncertain	1659		§	§	
(December)	1684/85		#	#	
(September)	1706		§	§	
October	1716		§	§	
April	1725	9	#	#	
July	1766		§	§	
February	1774	7½	#	#	
April	1784	10	#	§?	
June	1796	12	#		
June	1816		#		
June	1838		#	§	
Uncertain	1851		#		
May	1861	10	#	§	
August	1867	6	#	§	Main eruption south, southwest of Grimsvötn
June	1873	6	#	#	
March	1883	10	#	#	Greatest production of tephra
March	1892	9	#		
January	1897	5	#		
May	1903	6½	#		Eruption at Thordarhyma
April	1913	10	#		
September	1922	9½	#	#	
March	1934	11½	#	#	
May	1938	4	#		Probable eruption north of Grimsvötn
April	1941	3	#		
September	1945	4½	#		
February	1948	2½	#		
July	1954	6½	#		
January	1960	5½	#		
September	1965	5½	#		
May	1972	6½	#		

* Only those intervals are included when it is most likely that no jökulhlaups are missing.

+ The symbol § indicates that the event is uncertain.

6.6.1. Discharge

As the Icelandic term jökulhlaup indicates, a glacial meltwater stream experiences a very rapid and enormous increase in discharge. This increased water volume usually continues for 3 to 4 days and just as quickly subsides (Fig. 6.19). In this period of increased flow, conditions are suitable for significant changes in rates of debris erosion and entrainment. Tómasson (1974) reports >3000 tons of sediment moved in the outlet streams over a seven-day period following the 1972 Grimsvötn jökulhlaup. Jökulhlaups cause de-stabilisation of proglacial (sandur) sediments leading to extensive slumping and debris flows (Sugden *et al.*, 1985; Russell *et al.*, 1990). Detailed analysis of this sediment transport and distance of particular grade size movement from the Grimsvötn events is illustrated in Fig. 6.20. In general, meltwater streams that experience jökulhlaups do so often on a cyclic basis every 1 to 5 years, or every decade (Sturm and Benson, 1985; Knudsen and Theakstone, 1988; Russell *et al.*, 1990).

Typical meltwater discharge volumes vary immensely according to the type and nature of their sources. Table 6.3 illustrates some examples of measured and estimated volumes of water from modern and Quaternary jökulhlaups. A comparison of discharge from a jökulhlaup of Strandline Lake, Alaska to the Mississippi and Missouri rivers gives some idea of the immensity of some of these catastrophic outbursts. Sturm and Benson (1985) estimated the peak discharge of Strandline Lake during the 1984 jökulhlaup was $\approx 6.1 \times 10^3$ m^3 s^{-1}

while the Mississippi River at Vicksburg has a mean discharge of $\approx 15.5 \times 10^3$ m^3 s^{-1} and the Missouri River near St.Louis has mean discharge of $\approx 2.0 \times 10^3$ m^3 s^{-1}.

Characteristically, the discharge hydrograph of a jökulhlaup event exhibits an exponential increase in discharge with gradually ascending and very steeply descending limbs (Meier, 1973; Post, 1975; Clarke and Mathews, 1981; Clarke and Waldron, 1984) (Fig. 6.19). Clague and Mathews (1973) derived an empirical relationship between peak discharge (Q_{max}) and total volume of drained meltwater (V_{max}) during a jökulhlaup event where:

$$Q_{max} \approx 75(\frac{V_{max}}{10^6})^{0.67} \qquad (6.20)$$

This equation seems to give reasonably consistent results but is not founded in theory. Clarke (1982) working on 'Hazard Lake', dammed by the Steele Glacier, Yukon Territory, Canada, found the above empirical equation to work reasonably well on the basis of measured lake volume and discharge. Clarke has developed slightly more sophisticated equations that better predict maximum discharge but also take into account lake water temperature, drainage tunnel length, and tunnel geometry, parameters not readily available (Spring and Hutter, 1981).

6.6.1.1. Sources of Meltwater

The major sources of jökulhlaup water are from either: (1) lakes impounded by active or stagnant ice, and morainic debris, or (2) lakes in natural depressions in proglacial, subglacial, or supraglacial positions (Björnsson, 1976). It is difficult to assess the contribution made by englacial water which may be supercooled and under hydrostatic pressure, or supersaturated by saline salts, or a combination of both. Except for supraglacial lakes, all other lakes form in depressions the result of glacial processes or glacial loading (isostatic depression) or in glacier-free tributary valleys (Fig. 6.21). The latter type are what are generally known as ice-dammed lakes being impounded by a barrier of glacier ice interrupting and intercepting normal drainage patterns.

Discharge from ice-dammed lakes tends to be (a) by continuous and gradual leakage subglacially, or along the ice margins, or supraglacially or by some other non-glacier ice route; or (b) by discharging on a

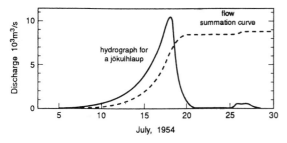

FIG. 6.19. Characteristic hydrograph of a jökulhlaup. Jökulhlaup of 1954 from Grimsvötn via Skeidará (Rist, 1955). Rist suggests accuracy of discharge estimate to be −20%. Note rapidity of flow cut off within a few hours (after Björnsson, 1974; reproduced by permission of the Iceland Glaciological Society).

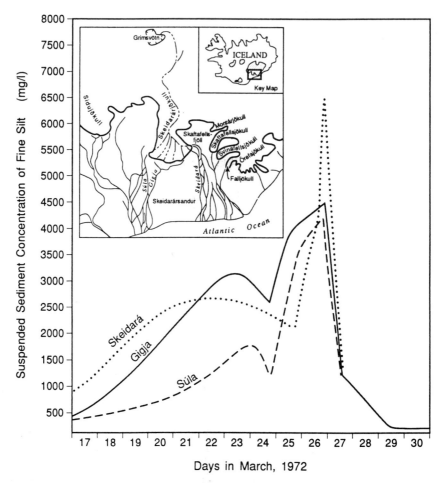

FIG. 6.20. Variation in suspended sediment load for three outlet proglacial streams of Vatnajökull due to a jökulhlaup related to Grimsvötn (Grimsvötnahlaup of 1972). Location of streams shown in inset map (after Tómasson, 1974; reproduced by permission of the Iceland Glaciological Society).

periodic basis due to overspilling; or (c) by jökulhlaup-style discharge subglacially toward a margin. In most cases reported from modern ice-dammed lakes (Gilbert, 1971; Liestøl, 1955a; Fisher, 1973; Mathews, 1973; Björnsson, 1976, 1992; Clarke, 1982) an almost continuous leakage takes place, which negates the idea of a watertight ice–bedrock seal.

6.6.1.2. Jökulhlaup Initiation Mechanisms

The process of jökulhlaup initiation is both complex and controversial. Several hypotheses have been advanced all attempting to explain specific phenomenological characteristics (Mathews, 1973; Nye, 1976; Shreve, 1972; Röthlisberger, 1972; Björnsson, 1974, 1992; Glen, 1954; Liestøl, 1955a; Spring and Hutter, 1981). These characteristics are: (1) the rapidity of the meltwater outburst; (2) the sudden cessation of flow often before a lake is emptied; (3) the periodicity of the outburst events; and (4) the lack of a watertight ice–bed seal.

At present three main hypotheses have been suggested to explain the triggering mechanism that drains ice-dammed lakes viz: (1) ice-barrier buoyancy, (2) plastic deformation of the ice barrier, and (3) englacial/subglacial hydrostatic water pressure reductions related to rates of subglacial leakage.

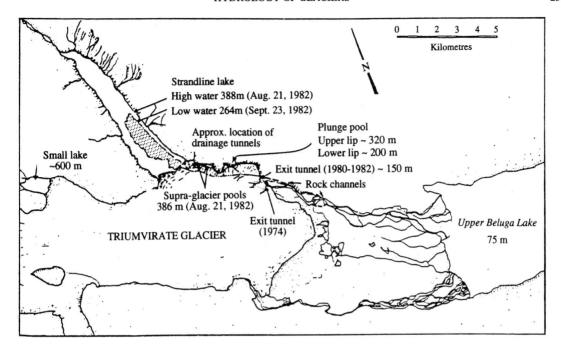

FIG. 6.21. Effects of jökulhlaup flood as shown in the Strandline Lake and Triumvirate Glacier area of Alaska. Jökulhlaup subglacial channel location shown by dot-dash line (after Sturm and Benson, 1985; reproduced by courtesy of the International Glaciological Society from *Journal of Glaciology*, 31(109), 1985, p. 273, Fig. 2).

(1) Björnsson (1974) and Liestøl (1955a) among others suggested that when the ice overburden pressure at the base of the ice barrier became less than the hydrostatic water pressure within the lake due to increasing lake water depth, the ice barrier would become buoyant.

(2) Glen (1954) considered that once the depth of an ice-dammed lake reached approximately 200m, at which time a shear stress of 100 kPa at the ice barrier would develop, plastic deformation of the ice in the barrier would result. Since jökulhlaups take place from lakes of lesser depths this explanation seems untenable. Björnsson (1974) points out that at a depth of 100 m buoyancy is possible long before plastic deformation of the ice is likely to occur.

(3) A further hypothesis relates to variations in the subglacial water pressure within a subglacial stream network (Liestøl, 1955a; Röthlisberger, 1972; Fisher, 1973). This hypothesis makes use of the known time-lag that exists between variations of water discharge and pressure within subglacial conduits and the period

of time necessary for the ice to plastically readjust to the reduced flow regime. The period of channel geometry adjustment may take up to one year following a change in flow regime (Röthlisberger, 1972). It has been noted, however, that channels with less than 50 m of overburden glacier ice tend to remain open. This hypothesis seems the least likely to explain the triggering of a jökulhlaup (Tómasson, 1974) but may be important in terms of subglacial drainage system changes once a jökulhlaup has begun.

6.6.1.3. Jökulhlaup Stages of Initiation and Development

Any jökulhlaup develops in a two-stage process, the first being a triggering mechanism, the second, the subglacial evacuation of lake water. This second phase, within temperate ice, occurs through the exploitation and enlargement of already existing subglacial drainage networks. As water penetrates these networks from the impounded lake,

TABLE 6.3. Jökulhlaups data for various examples of both modern and Pleistocene ice dammed lakes (after Clague and Matthews, 1974; Beget, 1986).

Lake	Year	H(m)	L(km)	D(m)	V_{max} $(10^6 m^3)$	Q_{max} $(m^3 s^{-1})$	Reference
Strupvatnet, Norway	1969	186	1	29[*]	2.6[†]	150	Whalley, 1971
Ekalugad Valley, Baffin Island, Canada	1967	120	2	120	4.8	200	Church, 1972
Demmevatn, Norway	1937	404	~3	79	11.6	1000	Ström, 1938
Gjánúpsvatn, Iceland	1951	167	5	20[*]	20.0	370	Arnborg, 1955
Vatnsdalur, Iceland	1898	372	10	188	120.0	3000	Thorarinsson, 1939
Tulsequah Lake, B.C., Canada	1958	150	8	73	229.0	1556	Marcus, 1960
Summit Lake, B.C., Canada	1965, 1967	620	12	200	251.0	3260	Mathews, 1965
Graenálon, Iceland	1939	535	19	230	1500.0	5000	Thorarinsson, 1939
Lake George, Alaska, U.S.A.[*]	1958	40	9	40	1730.0	10100	Stone, 1963
Lake Missoula, Montana, U.S.A.‡ (Wallula Gap)	Pleistocene	640		610		$1.87 \times 10^{4\#}$	Bretz, 1925; Pardee, 1942
Lake Missoula- Upper Spokane Valley/Rathdrum Prairie					2×10^6	$17\pm3 \times 10^6$	O'Connor & Baker, 1992
Lake Missoula- Wallula Gap						$10\pm2.5 \times 10^6$	O'Connor & Baker, 1992
Mulakvisl, Iceland	1956				3.50×10^6	50	Rist, 1967
Kaldakvisi River, Iceland	1965				6.20×10^6	260	Freysteinsson, 1972
Hazard Lake, Yukon Terr., Canada	1978				19.6×10^6	640	Clarke, 1982
Kverka, Iceland	1980				20.6×10^6	410	Rist, 1982
Katla, Iceland	1955				28.0×10^6	2500	Thorarinsson, 1957
Snow River, Alaska, U.S.A.	1967				140×10^6	780	Chapman, 1981
Flood Lake, B.C., Canada	1979				150×10^6	1200	Clarke & Waldron, 1984
Sululaup, Iceland	1978				175×10^6	3000	Rist, 1982
Skeidararsandur, Iceland	-				200×10^6	2000	Björnsson, 1977
Skaftardalur, Iceland	1970				237×10^6	1500	Björnsson, 1977
Strandline Lake, Alaska, U.S.A.	1984				710×10^6	6100	Sturm, 1984

[*] Depth to bedrock knob limits jökulhlaup magnitude =13m.
[†] Volume stored in lake = $4.6 \times 10^6 m^3$; water released = $2.6 \times 10^6 m^3$.
[#] Lowering of lake surface during jökulhlaup.

'uncontrolled' enlargement occurs due to channel-wall melting caused by higher lake water temperatures, and side-wall friction and abrasion caused by suspended debris and water turbulence (Fig. 6.15). Spring and Hutter (1981) and Clarke (1982) have suggested that the lake water temperature at the entrance portal from the lake to the subglacial channels, by releasing advected thermal energy, may be critically important in controlling and maintaining lake water evacuation. Flow continues from the lake provided channel enlargement is greater than the rate of closure. These competing forces alter as the lake level drops and ice overburden pressures increase both at the entrance portal and within the adjacent subglacial drainage network.

As is often observed jökulhlaups end abruptly. This cessation of flow may be due to subglacial channels becoming squeezed shut, to channel roof collapse and channel infill or to bedrock lips cutting off continued meltwater supply either at the entrance portal or down-ice within the subglacial network system. Where jökulhlaups exit in more than one major channel, it is thought that sharp changes in subglacial bedrock gradient have caused bifurcation (Vivian and Zumstein, 1973).

6.6.1.4. Effect of Jökulhlaups

The impact of jökulhlaups, for example, on parts of the Alaskan highway, southern Iceland and routes through the Karakoram Mountains in Pakistan are of such magnitude and potential hazard that some prediction as to peak discharges are of value to road and pipeline engineers and hydrologists. At present the only estimate available of peak discharges is the empirical Eqn. (6.20) derived by Clague and Mathews (1973). The effect of any jökulhlaup is both destructive and constructive (Costa, 1988). Evidence from modern and Quaternary glaciated areas indicate the immense erosive power of these lake outbursts (Lliboutry et al., 1977; Eisbacher and Clague, 1984; Blown and Church, 1985; Clayton and Attig, 1987; Kehew and Lord, 1987; Teller and Thorleifson, 1987). The scale of this evidence is extremely variable ranging from huge channels of the Lake Missoula Badlands (Bretz, 1969; Waitt, 1988) to small channels less than 2 m deep (Sissons, 1981). Similarly constructive evidence in the form of sediment deposition is as variable as the size of jökulhlaup (Maizels, 1989a, b; Maizels and Russell, 1992) (Chapter 12). Evidence of sediment deposition from jökulhlaups varies according to the site of deposition whether in proximal locations to the exit portal(s) or distal locations where finer laminated silts and clays are found or in flood-drowned backwaters or slackwaters of side valleys where often more rhythmite–like sediment sequences are found (Kochel, 1988; Waitt, 1988). Tómasson (1974) has shown the variability in sediment range as measures during the 1972 Grimsvötn jökulhlaup for different sites on the proglacial sandur. Sissons (1981) has suggested that large gravel beds tens of kilometres down valley from the ice-dammed lakes of Glens Roy and Spean, Scotland, may be attributable to jökulhlaups. Tómasson suggests that the volume of sediment transported and subsequently deposited from a single jökulhlaup may have more impact in the stratigraphic record than sediments from normal glacial depositional processes. Perhaps in the deliberate shift away from catastrophism towards a uniformitarian perspective, a significant sedimentological event in the sediment record has often been overlooked.

6.7. CONCLUDING REMARKS

It is increasingly apparent that glacial hydrology is a critical element in our understanding of many glaciodynamic processes. The significance of glacial meltwater in its effect on, for example, mass balance gradient, basal ice movement and surging phenomena has been perhaps underestimated in the past. At present our knowledge of subglacial and englacial hydrology continues to lag behind theoretical studies but already, in the last decade, an ever-increasing understanding of glacial hydrology has become integrated into the physics of ice masses. The role of glacial meltwater can no longer be regarded as of secondary relevance in glacial environments solely as a means of sediment transport and re-distribution within the glacial system but rather, for example, as the controlling variable in the mobility and persistence of deformable beds, in the development and evolution of certain wear phenomena across the glacier bed, in the initiation and consequent maturation of specific subglacial bedforms, and in the 'construction' and sedimentological architecture of sediment lithofacies types and associations.

Chapter 7

PROCESSES OF EROSION

N. R. Iverson

7.1. INTRODUCTION

The basic processes responsible for the spectacular landforms associated with alpine glaciation were identified over a century ago. Forbes (1846) recognized that rock clasts embedded in glacier ice abrade the underlying rock bed, and Tyndall (1864) noted that glaciers quarry rock blocks bounded by joints. Detailed observations of striation patterns and grooves made subsequently by Chamberlin (1888) established that ice behaves as a deforming fluid rather than a rigid body. These observations helped Gilbert (1903, 1906) relate ice flow at glacier beds to mechanisms of rock fracture and erosion. The most illuminating recent studies of glacial erosion have paralleled Gilbert's pioneering efforts, explicitly linking glaciological theory with the processes that erode the bed. This approach, complemented by observations in the field and laboratory, has provided a fuller understanding of erosional processes than that possible solely from inferences based upon field observations (Plate 7.1).

Three mechanisms of erosion are traditionally considered: abrasion, quarrying and the action of subglacial water. The last of these may produce conspicuous bedforms. Turbulent flow of water in zones where ice has separated from the bed erodes sinuous channels, potholes, and elongate crescentic furrows around resistant obstacles on the bed (e.g. Dahl, 1965; Gjessing, 1965; Sharpe and Shaw, 1989).

Such erosion is mainly a result of abrasion by rock particles entrained in the flow, pitting of the bed by the violent collapse of vapour bubbles in decelerating flows (cavitation) (Barnes, 1956) and dissolution. Dissolution also occurs where the ice and the bed are separated only by a meltwater film of micron thickness, and leaves a clear imprint on carbonate bedrock (Hallet, 1976). Despite the local importance of erosion by subglacial water, it is usually considered to be volumetrically subordinate to abrasion and quarrying (e.g. Drewry, 1986, p. 90). Unequivocal data in support of this viewpoint are lacking, but the ubiquity of glaciated bedrock smoothed by abrasion and roughened by quarrying that so impressed early scientists (e.g., Tyndall, 1864; Gilbert, 1903) suggests that these two processes are dominant in many areas (Plate 7.1). They are the focus of this chapter. Erosion of glacier beds consisting of sediment will not be considered, as that process, usually called entrainment, is more appropriately included in a discussion of subglacial sediment transport (Chapter 8; Menzies, 1995b, Chapter 2).

At a fundamental level, both abrasion and quarrying involve the fracture and dislodgement of rock from the bed. In the case of abrasion, fracture results from large stress differences localized beneath ice-entrained rock fragments in frictional contact with the bed. In the case of quarrying, such stresses are induced directly by ice, usually over

PLATE 7.1. Bedrock exhibiting striae and chattermarks. Glacier (Omsbreen, Norway) had flowed from left to right.

larger areas of the bed and for longer periods of time. Abrasion, therefore, produces fine debris, ranging from fine silt to coarse sand, whereas quarrying generally produces larger rock fragments. An exception would be large boulders entrained in the glacier sole that may fracture the bed on scales usually associated with quarrying. Dislodgement of quarried fragments requires that frictional forces be overcome by the combined action of water and sliding ice. Dislodgement of abraded material is usually accomplished by the abrading tool.

7.2. ABRASION

Early this century, Gilbert (1903, 1906) discussed many of the factors governing abrasion. He recognized that abrasion is: (1) a brittle process, involving fracture of the bed beneath clasts (Gilbert, 1906); (2) that it depends on the velocity of the basal ice, the quantity of subglacial debris, and the hardness contrast between the debris and the bed (Gilbert 1903, pp. 203–205); (3) that it also depends on the stress that clasts exert on the bed, in excess of that exerted adjacent to points of clast-bed contact (Gilbert, 1906); (4) that such stress concentrations beneath clasts result from the flow of ice toward the bed (Gilbert, 1906); and (5) that the rate of ice flow toward the bed depends on the bed geometry and the sliding speed of the glacier (Gilbert, 1906). As discussed hereafter, these concepts have since been incorporated in most quantitative models of glacial abrasion (Hallet, 1979, *in preparation*; Riley, 1982; Shoemaker, 1988). Virtually all studies consider the primary agents of abrasion to be clasts embedded in the basal ice of glaciers. Recently, however, Alley *et al.* (1987a, b) suggested that a subglacial deforming sediment layer that is free of ice might effectively abrade underlying bedrock, consistent with earlier conjecture by Gjessing (1965). Given the widely-proposed importance of such layers in facilitating the motion of modern and former ice masses, this bed condition will also be considered.

7.2.1. Abrasive Wear

Abrasion beneath glaciers occurs by fracture and subsequent displacement of minute particles of bedrock. Stresses sufficient to cause plastic deformation should not develop at contacts between clasts and the bed, except perhaps in a very limited zone directly adjacent to asperities on the contact. The dominance of brittle deformation is clearly indicated by close examination of glacial striae, which "present rough borders...whose edges are bruised, torn, or hackly" (Chamberlin, 1888, p. 218), and also by the main product of abrasion, rock flour, which consists of angular silt particles. Clasts indent the bed and produce such particles by indentation fracture (cf. Lawn and Wilshaw, 1975), a process well studied in materials science and rock mechanics.

Consider a point on a clast, regardless of its exact shape, impinging on the bed under a normal contact force, F_c. A good approximation is to assume the bed and point behave as linear elastic solids. In a cross-section of the bed, the most compressive (+) principal stress σ_1 is directed radially away from the contact, while the least compressive principal stress σ_3 is concentric about the contact (Fig. 7.1). This latter stress becomes tensile immediately adjacent to and also at some depth below the contact. In the extreme case of a perfect point load, this stress is tensile everywhere near the contact.

Like all rocks, the bed contains microscopic cracks of many orientations. The largest cracks that lie approximately normal to σ_3 and in zones where σ_3 is tensile will be most likely to grow, due to tensile stresses that are concentrated at the tips of such cracks (Atkinson, 1984). Consequently, if F_c is large enough, radially-oriented macroscopic cracks progressively develop beneath the contact (Fig. 7.2). The cracks eventually begin to merge and become sufficiently pervasive for the point to displace the crushed debris, and thus indent the bed (Fig. 7.2).

The cross-sectional area of the fractured zone is directly related to F_c. Crack growth is probably slow and intermittent; it is facilitated by the chemical action of water, which weakens strained bonds at the crack tips (e.g. stress corrosion, etc. Atkinson, 1984). Thus, to some extent, the process is time-dependent, suggesting that the depth of abrasion may decrease

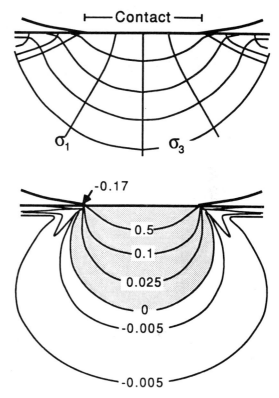

FIG. 7.1. Principal stresses beneath an elastic sphere pressed against a flat elastic bed (Hertzian contact). Cross-sectional view of trajectories of σ_1 and σ_3 (top), and contours of σ_3 normalized with respect to mean normal stress at the contact (bottom). Shaded area in bottom figure is zone where σ_3 is compressive. Crack growth should be initiated at edge of contact where stress is most tensile (−0.17) and should occur normal to σ_3 (modified from Lawn and Wilshaw, 1975; reprinted from *Journal of Materials Science*, **10**, Review of indentation fracture: principles and applications, 1049–1081, 1975, with kind permission from Elsevier Science Ltd, The Boulevard, Langford Lane, Kidlington, Oxford OX5 1GB, U.K.).

with increasing clast speed (Scholz and Engelder, 1976).

If a tangential load is applied in addition to F_c, tensile stresses adjacent to the trailing edge of the contact increase and extend deeper into the bed, and conversely beneath the leading edge (Lawn, 1967). This produces the crescentic fractures that often occupy the bottoms of glacial striae (Plate 7.2). They are concave down-glacier in plan view, and roughly demarcate the trailing edge of the former contact. Stress fields may also be asymmetric due to the

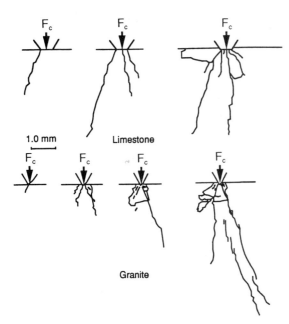

1.0 mm   Limestone

Granite

FIG. 7.2. Cross-sectional view of progressive indentation fracture of limestone and granite by a truncated and sharp wedge, respectively, under a contact force F_c, as observed with a SEM. (Modified from Lundqvist *et al.*, 1984; reprinted from *International Journal of Rock Mechanics and Mineral Sciences and Geomechanics, Abstract*, **21**(4), Indentation fracture development in rock continuously observed with a scanning electron microscope, 165–182, 1984, with kind permission from Elsevier Science Ltd, The Boulevard, Langford Lane, Kidlington, Oxford OX5 1GB, U.K.).

torque that is expected on clasts, which focuses F_c on the leading edge of the contact. This may produce tensile stresses that reach maximum values down-ice of the contact and is responsible for the primary cracks of crescentic gouges (Johnson, *unpublished*; Hill and Billings, 1985). Tensile stresses may also develop at the leading edge of a contact if the point has indented the bed, resulting in crack growth and coalescence that allows the point to plough forward (Iverson, 1991a).

A wear law relates point motion and contact force to the rate at which material is eroded. Most theoretical treatments of glacial abrasion (Boulton, 1974; Hallet, 1979, 1981; Drewry, 1986, p. 50; Shoemaker, 1988) use essentially the same relation. Hallet (1979) gives the rate of abrasion, $\dot{A}$, as:

$$\dot{A} = \alpha_{hc}\, C_r\, u_c\, F_c \qquad (7.1)$$

where α_{hc} is a constant dependent on the hardness of clasts and the bed and the geometry of the striator point, C_r is the areal concentration of clasts in contact with the bed, u_c is the clast velocity and F_c is the effective contact force between the indenting clast and the bed (cf. Drewry, 1986, p. 55). The abrasion rate, therefore, depends on the flux of abrading clasts, $C_r u_c$, and the depth to which clasts indent the bed, proportional to $\alpha_{hc} F_c$.

This relation follows from Archard's (1953) original wear model and is supported by geophysical data on frictional wear between sliding rock surfaces (e.g. Scholz, 1987). It is expected to be a reasonable approximation for average rates of abrasion after many clasts have passed over the bed, but may be a poor approximation when applied to an individual event. This is because the depth to which a clast indents the bed also depends on the shear loading at the contact. This loading depends not only on F_c, but also on the geometry of the striating element and the potential for rotation of the clast in the ice. Thus, the depths to which different clasts abrade vary significantly, even when F_c is steady (Iverson, 1991a).

7.2.2. Effective Contact Force and Clast Velocity

The glaciological parameters that control the effective contact force and the clast velocity are discussed below, initially without reference to how these variables may vary spatially on a rough glacier bed. The ice is considered to be at its pressure-melting temperature and sliding over the bed. Sliding at subfreezing temperatures occurs (Shreve, 1984; Echelmeyer and Wang, 1987), but at a speed that is probably several orders of magnitude slower than that of temperate ice. The glaciological parameters that control F_c for isolated clasts are most important because F_c influences both the depth to which clasts indent the bed and the clast velocity, as discussed hereafter. As was recognized by Gilbert (1906), McCall (1960) and Röthlisberger (1968), clasts will typically be completely embedded in basal ice, particularly along stoss surfaces where abrasion is most effective (Hallet, 1979, 1981). The deviatoric stresses in the bed that result in indentation fracture only develop if contact stresses exceed the pressure on the bed adjacent to the contact. Thus, the

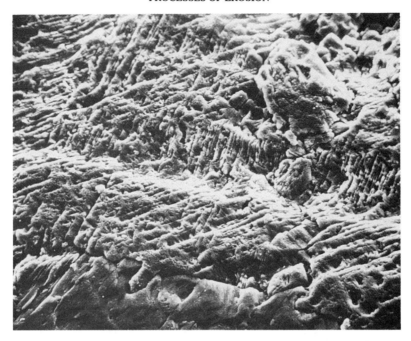

PLATE 7.2. Crescentic fractures (chatteermarks) at the bottom of a striation. SEM photomicrograph, magnifcation 700X).

hydrostatic ice pressure, which contributes equally to the contact stress and the pressure adjacent to the contact, does not contribute to F_c. Stresses, instead, are concentrated beneath clasts by motion of ice toward the bed (Gilbert, 1906; Hallet, 1979, 1981). This induces a bed-normal gradient in pressure across the clast that forces ice to regelate and deform past the clast. The resultant drag is a major component of F_c. It is represented by the term, Ψu_N, in the relation:

$$F_c = B_y + \Psi\, u_N \qquad (7.2)$$

where B_y is the buoyant weight of the clast, u_N is the component of the ice velocity normal to the bed, and Ψ is a drag coefficient that increases with the effective ice viscosity and clast size. The non-linearity of the flow law for ice can be included explicitly in the formulation of Ψ following Lliboutry and Ritz (1978). B_y is negligible, except in the case of boulder-sized clasts (Hallet, 1979).

Laboratory simulations of abrasion by sparse debris have demonstrated the direct relationship between u_N and F_c (Fig. 7.3), although theoretical and experimental results differ by as much as a factor of

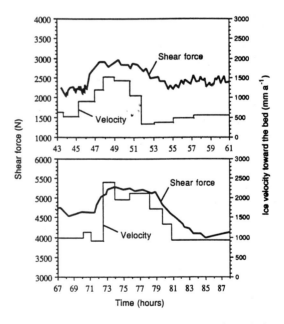

FIG. 7.3. Bed-normal ice velocity and shear force during two experiments in which a flat rock bed was slid beneath temperate ice containing sparse clasts. Shear force results from friction between clasts and bed and is approximately equal to μF_c, where μ is the coefficient of rock friction.

six (Iverson, 1990). Complications arise because the bed is a rigid boundary and a source of heat. This prohibits an exact determination of Ψ using classical regulation theory (Morris, 1979). Coupled with the possible effect of solutes on regelation (Drake and Shreve, 1973) and the effect of the bed on viscous flow past clasts (Hallet, 1981), this leads to order-of-magnitude uncertainty in theoretical estimates of Ψ.

Observations of clasts in basal ice (Boulton, 1974; 1979) and experiments (Iverson, 1990) suggest that cavities may sometimes form between clasts and the bed. Boulton (1974) argued that F_c should depend on the local effective pressure, equal to the difference between the pressure exerted by the overlying ice column and the local water pressure in such cavities. His formulation generally is not applicable, however, because the drag on fragments as ice flows toward the bed is neglected. Furthermore, if cavities are common beneath clasts, it is likely that the water pressure within them is nearly as large as the local mean ice pressure, as such cavities are likely to be isolated from the through-flowing subglacial hydraulic system. An exception would be where ice regains contact with the bed down-ice from zones of ice–bed separation.

The bed-parallel velocity of clasts will depend on the extent to which are frictionally impeded by the bed relative to the adjacent sliding ice. If μ is the coefficient of rock friction, then the tangential force on clasts is μF_c, and the velocity of clasts relative to the ice is $\mu F_c/\Psi$, where Ψ in this case is applied to bed-parallel ice flow past clasts (Boulton, 1974; Hallet, 1979). Thus, using Eqn. (7.2) and ignoring B_y, the clast velocity is given by

$$u_c = \left(u_t - \frac{\mu F_c}{\Psi}\right) = (u_t - \mu\,u_N) \qquad (7.3)$$

where u_t is the ice velocity parallel to the bed (Hallet, 1979). Thus, u_c is independent of the clast size. Strictly, this relation provides an minimum estimate of the clast velocity because clasts may rotate in the ice, resulting in a shear force at the clast-bed contact that is smaller than μF_c (Iverson, 1991a). In the extreme case of a smooth, spherical clast, $u_c = u_t$ because there would be no resistance to clast rotation due to the water film that lubricates the clast–ice interface.

Debris in the basal ice of many Alpine glaciers is generally sparse (Lawson, 1979; Wold and Østrem, 1979; Anderson et al., 1982). Thus, ice flow past clasts can be treated independently of flow past neighbouring clasts, as is done in theoretical models (Hallet, 1979, in preparation; Shoemaker, 1988). Debris may sometimes, however, occupy more than 30% of the volume of the basal ice layer (Lawson, 1979a; Echelmeyer and Wong, 1987; Ronnert and Mickelson, 1992). An important question is whether the physics embodied in Eqn. (7.2) are fundamentally different in this case.

Philip (1980) has analyzed regelation of temperate ice through a dense three-dimensional array of particles. If the particles of a debris-rich ice layer are idealized as spheres that are in mutual contact, Philip's theory can be adapted to obtain the following approximation for F_c:

$$F = B_h + \frac{4h_d\,r_p^2\,u_N}{K_p} \qquad (7.4)$$

where h_d is the thickness of the debris-rich ice layer, B_h is the buoyant weight in ice of a 'column' of particles within the debris layer, r_p is the particle radius and K_p is the apparent conductivity of the particle array to ice, proportional to the permeability of the array to ice. The permeability depends on the array porosity and on the thermal properties of the ice and particles. The expression should only apply to small particles ($r_p \ll 0.10$ m), as it is in this situation that regelation is the dominant mechanism of ice motion. The presence of the bed and the relative motion of particles add uncertainty to the result.

The form of this expression, however, should be essentially correct. It is similar to that for isolated clasts (Eqn. (7.2)). An important difference, however, is that unlike the case of isolated clasts, dense debris will affect the overall flow of ice near the bed; for example, regelation through debris will affect the temperature at the bed and thereby influence the heat flux to the bed and basal melting. A second potentially important difference is that in the case of dense debris, the rate of basal melting may exceed the rate at which ice can regelate through debris toward the bed. In that case, ice would not penetrate all the way through debris to the bed and a basal cavity would form. F_c would then be a function of the basal effective pressure and be independent of u_N (Iverson, 1993). For example, such ice–bed separation may occur as debris accumulates in basal ice and a critical debris thickness is eventually reached. Taking a basal melt rate of 100 mm year^{-1}, K_p = 2.7 x 10^{-15} m^2 Pa^{-1} s^{-1} (Philip, 1980) and an effective pressure of 0.1 MPa, and applying Philip's theory suggests that separation should occur when h_d = 85 mm. Thus, the likelihood of F_c being controlled by the effective pressure at the bed is greater for thicker layers of clast-supported basal ice.

7.2.3. Abrasion of a Rough Bed

Abrasion of a rough bed depends on the magnitude and distribution of the basal ice velocity, as this controls both the flux of particles across the bed and particle–bed contact forces. Nye's (1969b) sliding theory provides an exact solution for the distribution of pressure and velocity in linearly viscous ice sliding over a low-roughness sinusoidal bed by regelation and viscous deformation. It has provided a useful foundation for abrasion models, although it neglects subglacial cavity formation and water pressure, both of which have a strong and well-documented effect on glacier sliding (Iken and Bindschadler, 1986; Hooke et al., 1987; Schweizer and Iken, 1992). In this section, the spatial variation of each of the parameters in the wear law (Eqn. (7.1)) is evaluated within the context of Nye's model. The discussion relies heavily on the abrasion models of Hallet (1979, in preparation) and Shoemaker (1988) and the field observations by Rastas and Seppälä (1981). It applies exclusively to sparse debris in ice.

As discussed, contact forces depend primarily on u_N. The principal component of u_N results from regelation and deformation of ice around bumps on the bed during sliding. The combined effect of the resultant melting and bed-parallel extension of ice along stoss surfaces produces rates of ice convergence with the bed that are a significant fraction of the sliding velocity. An additional minor component of u_N results from the rate of basal melting due to geothermal heat and heat from sliding friction, u_0, which should be on the order of 10 to 100 mm year^{-1} (Röthlisberger, 1968). Although basal melting due to heat from sliding friction should vary along the bed, most abrasion models (Hallet, 1979; Shoemaker, 1988) approximate it as uniform because it is typically much smaller than rates of ice convergence due to regelation and ice deformation associated with sliding. Assuming u_0 is uniform and neglecting any large-scale straining of ice that might perturb the velocity field, u_N at the surface of a sinusoidal bed of wavelength λ is given by

$$u_N = u_b\, a_p\, k_w\, \Omega_w \cos kx + u_0 \qquad (7.5)$$

where u_b is the sliding velocity, a_p is the bump amplitude, k_w is the wavenumber equal to $2\pi/\lambda$ and Ω_w is a constant between 0 and 1 that depends on λ and the transition wavelength λ_*, the wavelength for which regelation and ice deformation are equally efficient (Hallet, 1979, Eqn. (17)). λ_* depends on the effective ice viscosity; thus, due to the nonlinearity of the flow law of ice, it may range from 0.05 to 1.0 m for a reasonable range of sliding velocity and bed roughness (Hallet, in preparation). Equation (7.5) suggests that contact forces should be largest halfway up stoss surfaces, where $\cos k_w x$ = 1.0. It also indicates that clasts should lose contact with the bed a short distance down-ice from the crests of bumps where $\cos k_w x$ becomes sufficiently negative so that ice flow diverges from the bed. This distance should scale with u_0/u_b.

The velocity of clasts along a rough bed depends on the normal component of the ice velocity that presses clasts against the bed and the tangential component that pushes them forward (Eqn. (7.3)) (Hallet, 1979). Observations of glacier beds suggest that the product $a_p k_w$ seldom exceeds 0.5. For a bump of short wavelength where regelation is the dominant mechanism of ice motion ($\lambda \ll \lambda_*$), Ω_w = 1 and u_N is maximized. Thus, on the steepest portions of some stoss surfaces, u_N may approach one-half of the sliding velocity. Usually, however, $a_p k_w \Omega_w$ should be less than 0.1. Therefore, a good approximation, especially for the low bed roughness required by Nye's (1969b) model, is that clasts move at the speed of the basal ice (Hallet, 1981). This was true in laboratory experiments with sparse debris (Iverson, 1990) in which u_N/u_t was quite large, 0.11, and clast velocity was 92% of the sliding velocity.

Thus, the concentration of clasts in contact with the bed, C_r, primarily controls the spatial variability of the flux of clasts. As a first approximation, Hallet (1979, 1981) treated C_r as an independent variable, and assumed it was spatially uniform on stoss surfaces. Shoemaker (1988), however, noted that as ice slides over a bed with periodic roughness elements, debris will contact only a limited portion of each stoss surface. This arises because as a clast loses contact with the bed just down-ice from the crest of a bump, its trajectory in the ice allows it to graze only the crests of stoss surfaces further down-ice (Fig. 7.4a). The length of the abraded zone on each stoss surface should be directly related to $u_0\lambda/a_pu_b$. Thus, Shoemaker treated the areally-averaged concentration of clasts in contact with the bed as a variable dependent on $u_0\lambda/a_pu_b$ and found that it may be up to several orders of magnitude less than the concentration of clasts in basal ice. Shoemaker's model, therefore, predicts spatially averaged rates of

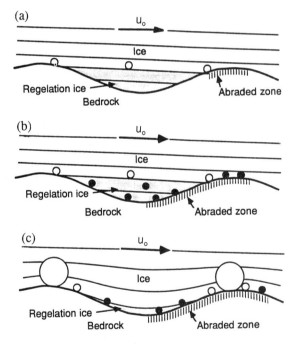

FIG. 7.4. (a) 'Shadowing effect' of Shoemaker (1988) for sliding dominated by regelation ($\lambda \ll \lambda_*$). Trajectories of clasts are inclined downglacier at a slope equal to u_0/u_b and allow abrasion of only a small portion of each stoss surface. No debris is entrained from bed. (b) Model of Hallet (*in preparation*) for sliding dominated by regelation. Clasts are quarried from lee surfaces enabling abrasion of all of stoss surfaces immediately downglacier. (c) Model of Hallet (*in preparation*) for sliding by viscous deformation with some regelation ($\lambda > \lambda_*$). Large clasts follow streamlines higher above bed, and thus, loose contact with bed further upglacier than smaller clasts (modified from Hallet, *in preparation*).

abrasion that are orders of magnitude smaller than those of Hallet's (1979) model.

Evaluating the realism of the treatments of C_r in the two models requires conjecture about how debris is supplied to the bed. u_0 is extremely important in Shoemaker's model because melting due to geothermal heat and heat from sliding friction is the only means by which clasts make contact with the bed. In the extreme case, when $u_0 = 0$, there is no contact between clasts and the bed, and $\dot{A} = 0$ (Hallet, *in preparation*). This counterintuitive result and the ubiquity of quarried surfaces on deglaciated bedrock suggest that debris entrainment from the bed must also be an important control on C_r and the rate of abrasion.

Hallet (*in preparation*) has recently modified his original model by coupling the 'shadowing effect' of Shoemaker (1988) with debris entrainment from the bed. He also includes the effect of debris comminution on C_r. He argues that in the common situation where $u_0 \ll u_b$, debris entrained from the bed should be the principal source of abrasive debris (Fig. 7.4b, c). In this case, most clasts abrade a significantly larger fraction of stoss surfaces than predicted by Shoemaker's model. Hallet's results indicate that his original model overestimated areally-averaged abrasion rates by an order-of-magnitude while Shoemaker's model underestimated them by a similar amount.

The spatial variability of F_c and C_r indicated by theoretical models provides a framework for evaluating local patterns of abrasion. Taking into consideration the shadowing effect and debris entrainment from the bed, the concentration of abrading clasts probably increases monotonically with height along stoss surfaces with a sharp increase in clast concentration at the position where clasts derived from the crest of the preceding bump regain contact with the bed (Hallet, *in preparation*). This and the knowledge that contact forces should be largest midway up sinusoidal stoss surfaces, suggest that abrasion should be focused near the tops of bumps (Fig 7.5). Thus, the result of sustained abrasion should be to smooth glacier beds in the direction of sliding by reducing the height of bedforms (Röthlisberger, 1968; Hallet, 1979). In addition, the tops of bumps that extend higher than surrounding

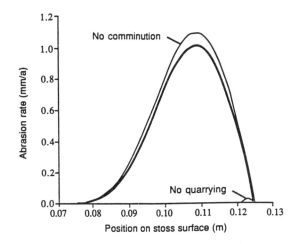

FIG. 7.5. Distribution of abrasion rate on stoss surface of sinusoidal bed extending from 0.075 to 0.125 m and centred at 0.1 m. (from Hallet, *in preparation*). Heavy line was calculated assuming some debris is lost due to comminution of clasts as they traverse stoss surface. Upper curve corresponds to no comminution; lower curve corresponds to no debris entrainment from bed.

bumps on the bed should be traversed by larger concentrations of clasts and abraded preferentially. This may explain the field observation that roche moutonées tend to be of roughly equal height in a particular region (Burke, 1969; Rastas and Seppälä, 1981).

Rastas and Seppälä (1981) reported glacial polish on gently sloping lee surfaces near the crests of roches moutonées. Such polish is presumably the result of abrasion in the absence of particles larger than silt. Abrasion models suggest that clasts should abrade the bed a short distance down-ice from the crest of bumps. The distance will depend on the particle size. Small particles will follow streamlines near the bed which should diverge only slightly from gently sloping lee surfaces, particularly if regelation is minor. Larger clasts will follow streamlines higher in the ice that should diverge more sharply from the bed (Fig. 7.4c). Thus, large clasts will lose contact with the bed further up-ice than smaller particles. Glacial polish on lee surfaces may reflect this sorting (Hallet, *in preparation*).

Hallet (*in preparation*) notes that the shadowing effect may explain the distribution of crescentic fractures that tend to be clustered near the crests of roches moutonées (Rastas and Seppälä, 1981).

Growth of such fractures requires very large contact forces (Gilbert, 1906; Johnson, *unpublished*), but rates of ice convergence with the bed are expected to be small near the crest of bumps. This suggests that the fractures were produced by large rock fragments, because both the fragment buoyant weight and viscous drag on fragments increase with size. Consistent with this expectation, the shadowing effect suggests that large rock fragments, which presumably travel long distances before being crushed, should tend to be preferentially concentrated near bump crests.

Abrasion models do not explicitly yield information about transverse variations in abrasion. However, along the sides of bumps contact forces should be small because ice convergence with the bed due to sliding is minimal there. The sides of bumps, therefore, should be more lightly striated than stoss surfaces (cf. Rastas and Seppälä, 1981).

Striae show that clasts tend to be deflected laterally around stoss surfaces. This should increase the concentration of clasts along the sides of bumps. For large bumps this deflection results from the lateral flow of ice. In the lee of such bumps, ice flow should converge and result in a lee side distribution of sediment equivalent to that up-ice from the bump. For small bumps, where regelation is the dominant mechanism of sliding, the deflection is a function of the angle between u_b and the local slope of the bed and does not result from lateral ice flow. This is an important distinction when considering the 'streaming' process advocated by Boulton (1974, Fig. 12). It implies that only in the case of small bumps should debris concentrations peak in zones down-ice from bump sides. Thus, although the convergence in the lee of large bumps may occur some distance down-ice if ice separates from the bed, it is difficult to appeal to 'streaming' to explain grooves that sometimes extend well down-ice from the sides of large hummocks and lie parallel to the overall direction of flow (Boulton, 1974; Goldthwait, 1979; Sharpe and Shaw, 1989). Similar grooves down-ice from centimetre-scale bumps (Chamberlin, 1888), however, may be the result of streaming.

Several arguments suggest that local separation between ice and the bed in the lee of bumps should increase abrasion rates by allowing clasts to abrade larger fractions of stoss surfaces than when ice contact with the bed is complete. If there is no local separation, then after traversing one stoss surface, clasts will follow streamlines that dip down-ice with a slope of u_0/u_b (relative to the mean bed slope) and thus intersect only the crests of succeeding stoss surfaces (Fig. 7.4a) (Shoemaker, 1988). If ice separates from the bed in the lee of bumps, the trajectory of the cavity ceilings should typically dip more steeply down-ice than u_0/u_b. Thus, clasts that are entrained where ice regains contact with the bed, as was observed by Vivian and Bocquet (1973), should abrade stoss surfaces further up-ice than when no cavity is present. Such clasts should be melted from the ceilings of cavities by viscous dissipation of heat associated with subglacial water flow. This process may be complemented by the ejection of particles due to the decrease in ice pressure toward the cavity ceiling (Vivian and Bocquet, 1973). As discussed hereafter, ice–bed separation should accelerate quarrying of lee surfaces, which also should increase the supply of abrasive debris on stoss surfaces.

The most important contribution of abrasion models is the clear delineation of the glaciological parameters that control the process. Hallet's conclusion that contact forces should be independent of the effective ice pressure and should depend mainly on the component of the ice velocity toward the bed is most significant. Making the reasonable assumptions that $u_c = u_b$, $u_0 \ll u_b$, and $B_y \ll \Psi u_N$ and using Eqns (7.1), (7.2) and (7.5) leads to the conclusion that the rate of abrasion is directly proportional to the square of the local sliding velocity and independent of the ice thickness. This provides a tool for quantitative analyses of the evolution of glacially eroded topography (Harbor et al., 1988; Harbor, 1992).

7.2.4. Abrasion by a Subglacial Till Layer

Some ice masses are underlain by a layer of deforming, water-saturated till. Limited field evidence suggests that such till may abrade the bed (MacClintock and Dreimanis, 1964). In this case, contact forces, in part, will depend on the effective pressure on the till. As the difference between the ice overburden pressure and the porewater pressure increases, contact forces between till particles and the bed should, on average, also increase (Chapter 5). Coarse sand and larger particles in contact with the bed should tend to be surrounded by the till matrix, which typically consists of fine sand and silt (e.g. Dreimanis and Vagners, 1971) (Chapter 15; Menzies, 1995b, Chapters 12 and 13). To some extent, the till matrix may behave as a fluid and deform around large clasts in frictional contact with underlying bedrock. Thus, on stoss surfaces, bed-parallel extension of the till matrix may induce flow toward the bed and a resultant viscous drag on clasts, similar to when clasts are surrounded by basal ice. This, together with the buoyant weight of clasts in the matrix, may result in significant contact forces beneath larger clasts, even if the effective pressure on the till is near zero, as measurements in glacier boreholes often indicate (Engelhardt et al., 1990; Hooke, 1991). Estimating the magnitude of these contact forces would require assessing the viscous drag that the matrix exerts on clasts. This is not possible because a definitive rheological model for till, or the fine fraction of a till, is not yet available (cf. Clarke, 1987b; Murray and Dowdeswell, 1992). Thus, it is not clear how contact forces would differ in this case from the more familiar situation in which clasts are surrounded by basal ice.

Mobilization of particles at the base of a till layer is a requirement for abrasion of the underlying bedrock. Field observations are few, but indicate that the lower parts of some till layers do not deform pervasively, although some movement may occur there along discrete surfaces (Boulton and Hindmarsh, 1987; Menzies and Maltman, 1992). If water escapes into permeable rock at the base of the till layer (Clarke, 1987b), deformation at depth may be suppressed by the reduction in porewater pressure and resultant increase in strength with depth. This effect may be amplified by a feedback suggested by Alley (1989a, b) in which the till loses its dilatancy below a critical strain rate; the resultant compaction strengthens the till, reduces the strain rate further, and ultimately suppresses pervasive deformation at some depth.

Perhaps the safest generalization that can be made is that as till thickness increases, the potential for abrasion of the underlying bedrock is reduced. In addition, given that mobilization of the till base may not commonly occur, abrasion by subglacial till layers should generally be less effective than abrasion by clasts in ice. This conclusion is in agreement with general observations of abrasive wear that show that when two sliding surfaces are separated by a layer of loose debris, abrasion rates are usually an order-of-magnitude smaller than when no loose debris is present (Rabinowicz, 1965).

7.3. QUARRYING

The lee sides of bumps on glacier beds are typically comprised of irregular, fractured surfaces that show little or no sign of abrasion. Such surfaces are presumably the result of quarrying, the process by which basal ice fractures the bed and dislodges rock fragments. Growth of cracks, which probably intersect joints and bedding surfaces, should detach rock fragments from the bed. Such fragments are dislodged when bed-parallel forces exerted on them by sliding ice exceed the frictional forces that hold them in place.

7.3.1. Rock Fracture

Weathering processes, particularly frost cracking, may fracture subglacial bedrock to some extent. Modern studies of frost cracking show that the expansion of water upon freezing is not principally responsible for crack growth (Walder and Hallet, 1985, 1986). Rather, slow crack growth occurs because absorbtive forces at freezing fronts draw water toward ice bodies in cracks. The process requires rock on the surface or at depth to be steadily at subfreezing temperatures. This is potentially true anywhere subfreezing air can penetrate to the bed, such as at bergschrunds (Hooke, 1991) and near glacier margins (Anderson et al., 1982). Elsewhere at the beds of temperate glaciers, however, rock should typically not be at subfreezing temperatures, and frost cracking is not expected. Subaerial weathering processes undoubtedly facilitate quarrying during a glacier advance, but this cannot be invoked to explain deep glacial valleys with quarried surfaces.

The role of joints in assisting quarrying has been cited extensively (Matthes, 1930; Crosby, 1945; Zumberge, 1955; Addison, 1981; Rastas and Seppälä,

1981). It is unlikely, however, that pre-existing joints will always be so dense and continuous that quarrying only requires that frictional forces be overcome. This is supported by highly fractured lee surfaces on some rocks that are otherwise unjointed or contain unrelated joint patterns (Boulton, 1974). An example are groups of so-called 'bullet boulders' embedded in lodgment tills. These boulders have fractured lee surfaces, but dissimilar structural characteristics as a result of their exaration and transport (Boulton, 1974; Krüger, 1979, 1984; Sharp, 1982a). Their asymmetry is strong evidence that crack growth is induced by glacier ice.

Although it is unlikely that the bed is sufficiently fragmented so that no crack growth is necessary for quarrying, it is equally unlikely that the bed contains no macroscopic cracks. This points out one difficulty in applying the Coulomb failure criterion to subglacial rock fracture, as was done in early studies (Boulton, 1974; Morland and Boulton, 1975; Morland and Morris, 1977). The criterion is empirical, based on uniaxial and triaxial compression tests on small rock specimens that generally do not contain macroscopic cracks. Such cracks are more likely to grow than microscopic voids and cracks. Therefore, the criterion overestimates the strength of larger rock bodies that have a higher probability of containing macroscopic cracks. In addition, the criterion is based on deviatoric stresses necessary to cause rapid, unstable crack growth, and, therefore, is not suitable for addressing slow, stable crack growth at lower stresses, which is more likely under long-term loading by glacier ice and pressurized water.

The approach taken here is to evaluate the subglacial geometry most likely to induce the optimum state of stress for growth of pre-existing cracks in the bed. A particularly favourable site for crack growth should be just up-ice from the point where ice separates from the bed near the crest of a bump. This, indeed, is precisely where field observations suggest quarrying is most prevalent. Ice–bed separation focuses some of the weight of the glacier on zones where the ice is in contact with the bed. Thus, normal stresses should be significantly larger on rock adjacent to cavities than elsewhere. The normal stress depends upon the extent of ice–bed separation and the effective pressure, defined in this context as the difference between the ice overburden pressure and internal water pressure within basal cavities.

Numerical models based on the finite-element method permit the study of stresses in rock steps adjacent to such cavities (Iverson, 1991b). First, the geometry of a cavity of steady size in the lee of the step and the distribution of ice pressure against the bed upstream from the cavity are calculated. For an effective pressure of 0.6 MPa and a sliding velocity of 19 m year^{-1}, pressure near the step edge is about 50% greater than the ice overburden pressure (2.7 MPa) (Fig.7.6). Although larger stress concentrations are possible when ice–bed separation is more extensive, the applied stress can never exceed the strength of the ice, which is in the order of 10 MPa in uniaxial compression (Palmer *et al.*, 1983).

All principal stresses in the bedrock near such a lee surface are compressive, with the maximum principal stress oriented approximately vertically, reflecting the bed-normal ice pressure, and the least principal stress oriented approximately horizontally, reflecting the water pressure on the lee of the step or bump. In a vertically-oriented crack near the step edge, tensile stresses should develop near the crack tips because compression parallel to the crack exceeds that normal to the crack faces (Griffith, 1924). Such tensile stresses are directly related to the crack length. In addition, the crack will normally be filled with water under pressure and this pressure also contributes to the tensile stress at the crack tips. Whether a crack actually grows depends on the rock strength, as well as stress corrosion. During this process, water reacting with silicate rock, for example, at crack tips

results in hydrolysis of strong Si—O bonds to weaker hydrogen bonded hydroxyl groups. This may reduce the tensile stress necessary to cause slow crack growth by a factor of 5 (Atkinson, 1984). Assuming that the water pressure in cracks equals the cavity water pressure, numerical calculations suggest that for the aforementioned steady cavity, stress differences in the bed would be sufficiently large to cause slow growth of centimetre-scale cracks in most sedimentary and some crystalline rocks (Iverson, 1991b).

Pre-existing cracks that are closest to being aligned parallel to the greatest principal stress are most likely to grow. As cracks extend, the orientations of principal stresses in the rock change, but final fracture surfaces should lie at a small angle to the greatest principal stress. This is indicated by so-called shear fractures produced in confined compression tests on rock, which initiate as tensile fractures and lie at less than 45° to the greatest principal stress (Jaeger and Cook, 1979, p. 92). Thus, fractures should tend to dip steeply into the bed beneath zones of ice–bed contact, because the stress that ice exerts normal to the bed, locally orients the greatest principal stress so that it is nearly normal to the bed surface.

As water discharge in glaciers varies diurnally, subglacial water pressures fluctuate. The observed range is commonly greater than 0.5 MPa (Kamb *et al.*, 1985; Iken and Bindschadler, 1986; Hooke *et al.*, 1987, 1989). Reductions in water pressure in subglacial cavities during such diurnal variations should transiently accelerate crack growth in the bed (Iverson, 1991b), as was originally recognized by Lliboutry (1962). During water-pressure reductions, some of the weight of the glacier that was formerly supported by pressurized water is shifted to the bed up-ice from cavities. Together with the decrease in water pressure against the lee surface, this increases the difference between principal stresses in the rock, which should enhance crack growth (Fig 7.7). The process is somewhat analogous to a confined compression test on rock in which the axial load is increased while the lateral confining pressure is decreased.

The effect will be largest when ice–bed separation is extensive. The mean increase in ice pressure

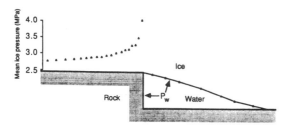

FIG. 7.6. Geometry of steady, water-filled cavity in lee of step and mean ice pressure against bed upglacier from cavity. Sliding velocity was 19 m year^{-1}; water pressure in cavity (ρ_w) was 2.1 MPa and ice overburden pressure was 2.7 MPa.

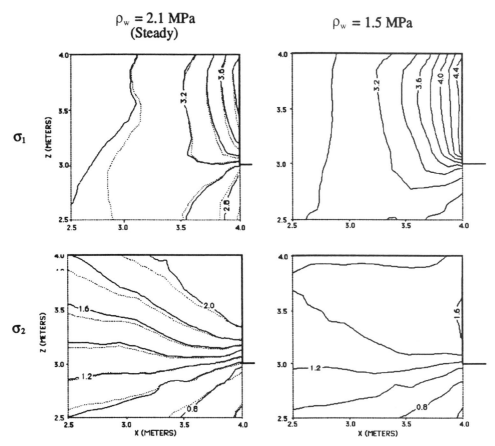

FIG. 7.7. Contours of greatest and least principal stresses in bedrock step (plane strain) for steady cavity shown in Fig. 7.6 (left) and after an instantaneous reduction in cavity water pressure of 0.6 MPa (right). Step is at x = 4.0. Contours are in MPa. Dotted lines show a second solution using larger finite-element domain. Potential for crack growth, scales with σ_1 $-\sigma_2$. This difference increases after reduction in water pressure. Orientation of σ_1 is approximately vertical throughout step and is not significantly rotated after water-pressure reduction.

against the bed is simply the product of the water pressure reduction and the ratio between the areas of ice–bed separation and contact. Thus, for a glacier that has separated from the bed over 75% of its area, which is not unrealistic for some glaciers (e.g. Anderson *et al.*, 1982), a 0.5 MPa decrease in water pressure should produce a 1.5 MPa increase in the mean normal stress between the ice and the bed. Subglacial water-pressure fluctuations during the melt season may extend cracks in fracture-resistant lithologies by this mechanism.

Hallet (pers. commun., 12/89) has suggested another bed geometry that may be particularly susceptible to crack growth. The bed consists of periodic steps, and the effective pressure is sufficiently low so that ice contacts only a small portion of each ledge (Fig. 7.8). Normal stresses between the ice and rock will be very high in these areas. Immediately up-ice, the least principal stress in the bed is tensile and parallel to the bed. There

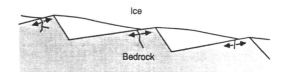

FIG. 7.8. Ice–bed configuration that may be particularly favorable for subglacial rock fracture. Least principal stress in the bed is tensile just upglacier from point where ice regains contact with bed near lee surfaces.

the potential for bed-normal crack growth exceeds that for an isolated step or bump in which all principal stresses are compressive. This situation should also be especially sensitive to fluctuations in subglacial water pressure, owing to the large fraction of the glacier sole that is supported by water under pressure.

The above discussion suggests that ice–bed separation is important, if not essential, for subglacial rock fracture. Separation requires low effective pressures, high sliding velocities and high bed roughness (e.g. Fowler, 1986) (cf. Chapter 5). Because cracks are expected to grow approximately normal to the crests of bumps, quarrying should steepen the down-ice sides of bumps. This will tend to favour continued ice–bed separation, continued crack growth and maintenance of steep lee surfaces. Growth of such cracks and their coalescence with joints and bedding planes may ultimately be responsible for maintaining bed roughness and limiting the rate of quarrying for a particular lithologic and structural setting.

7.3.2. Rock Dislodgement

Ice–bed separation and transient effects associated with subglacial water-pressure fluctuations are expected to be important elements in dislodging rock fragments that have become separated from the bed by cracks.

Röthlisberger and Iken (1981) note that once ice–bed separation occurs, the resultant lack of confining pressure on lee surfaces should allow some loosened rock fragments to spall off into the cavity. This would be particularly true for thin slabs of rock on lee surfaces (Fig. 7.9a). They also suggest a mechanism for increasing the bed-parallel force on such fragments. Expanding the earlier discussion of Robin (1976), they argue that increases in subglacial water pressure should transiently reduce the pressure on lee surfaces that are in contact with ice, but not in communication with the subglacial hydraulic system. Because some water is extruded from the basal ice under high pressure, there is not sufficient interstitial water after such reductions in ice pressure to provide the latent heat necessary to

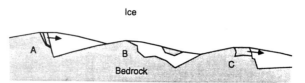

FIG. 7.9. (a) Loosened rock fragment spalling off into a water-filled cavity. (b) Rock fragment frozen to the ice as a result of pressure-release freezing (modified from Röthlisberger and Iken, 1981; reproduced by courtesy of the International Glaciological Society from *Annals of Glaciology*, Vol. 2, 1981, p. 59, Fig. 3). (c) Loosened rock fragment that must undergo frictional sliding to be dislodged.

warm the ice to the new pressure-melting temperature. Thus, the ice warms by freezing the meltwater film at the bed. The resultant tensile strength of the ice–rock bond, although it is probably small at temperatures near the melting point (< 0.2 MPa, Jellinek, 1959), may be sufficient in some cases to pull loosened rock fragments from the bed if frictional resistance is minimal (Fig 7.9b).

Other detached blocks should be pressed firmly against the adjacent rock due to the concentration in ice pressure expected near the crests of bumps (Fig 7.9c). The viscous drag exerted on such blocks by sliding ice is likely to be small due to smoothing and polishing of exposed surfaces by abrasion, and it is unlikely that shear stresses on such blocks ever exceed more than several tenths of a megapascal. Dislodgement of such blocks, therefore, should only occur when the water pressure in bounding fractures is a large fraction of the normal stress on blocks, resulting in low effective pressure and, hence, low friction on fracture surfaces.

Effective pressure and friction on fractures may be minimized during periods of rising water pressure in adjacent cavities (Iverson, 1991b). This is, in part, because sufficiently continuous fractures should be in hydraulic communication with cavities. It is mainly a result, however, of the large reduction in the normal pressure against bump crests that occurs as water pressure in adjacent cavities increases (Röthlisberger and Iken, 1981; Iverson, 1991b). In a numerical calculation presented by Röthlisberger and Iken (1981), an increase of 0.07 MPa in cavity water pressure (7 m of water column) produces a 1.26 MPa reduction in normal pressure on the crest of a

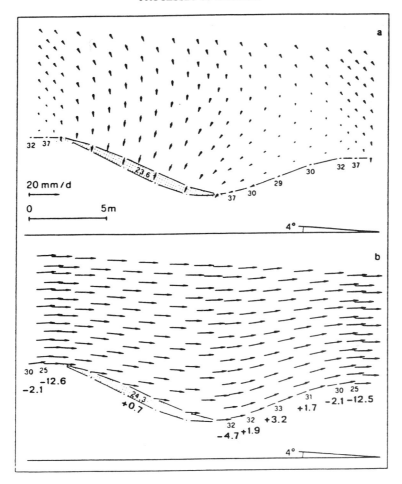

FIG. 7.10. Transient pressure distributions on sinusoidal bed with water-filled cavities. Stippled areas are cavities. Small light numbers show pressure against bed in bars (1 bar = 0.1 MPa). (a) Cavities shrinking at water pressure of 2.36 MPa. (b) Cavities growing after increasing water pressure to 2.43 MPa. Dark numbers show change in pressure against bed after water-pressure increase. Arrows are velocity vectors. Basal shear stress was 0.1 MPa and ice overburden pressure 2.7 MPa. Cavity is steady (neither shrinking or expanding) at water pressure of 2.41 MPa (from Röthlisberger and Iken, 1981; reproduced by courtesy of the *International Glaciological Society from Annals of Glaciology*, Vol. 2, 1981, p. 59, Figure 3).

sinusoidal bump (Fig. 7.10). Thus, an increase in cavity water pressure of 5% would result in a 95% reduction in effective pressure across hypothetical fractures aligned parallel to the bump crest. It should be emphasized that the effect of subglacial water-pressure fluctuations on normal stresses between ice and the bed are greatest when ice–bed separation is extensive. Thus, as was the case with rock fracture, rock fragments may be most susceptible to dislodgement when effective pressures at the bed are low and sliding velocities are large.

7.4. RATES OF EROSION

The most reliable estimates of erosion rates come from measurements of sediment discharge in meltwater streams. Such measurements, compiled by Drewry (1986, p. 87), suggest that erosion rates range from 0.07 to 30 mm year[-1]. A major assumption associated with these measurements is that over the period of measurement, water has access to most of the glacier bed so that there is negligible storage of debris. Furthermore, it is assumed that most quarried debris is sufficiently comminuted so that it can be fluvially transported. Direct measurements of abrasion were made by Boulton (1974, 1979) beneath Breiðamerkurjökull and range from 0.2 to 4 mm year^{-1} for platens of several different lithologies fixed to the bed. The experiments, however, were necessarily limited to small areas of the bed and, therefore, may not be representative of areally-averaged abrasion rates, given the spatial variability of both contact forces and the concentration of abrading clasts.

Very few data are available with which to assess the relative importance of abrasion and quarrying, but quarrying is probably dominant (Drewry, 1986, p. 90). Gilbert (1903, p. 206) observed that quarried surfaces were more abundant on eroded bedrock than abraded surfaces and stressed the importance of quarrying as an erosional mechanism. In a semi-quantitative study using the pre-existing sheet structure of roches moutonées in Massachusetts, Jahns (1943) was able to calculate that several times more rock had been quarried from lee surfaces than had been abraded from stoss surfaces. Boulton (1979) argued that if the only source of abrasive tools is quarrying, as is true for ice sheets, and if tools are comminuted at a rate comparable to the rate at which the bed is abraded, then abrasion must be subordinate to quarrying because resultant tills contain large blocks that have survived comminution. In Hallet's (*in preparation*) most recent abrasion model, quarrying is the principal source of debris because clasts derived from ice above the bed can only contact a small portion of each stoss surface due to the shadowing effect described by Shoemaker (1988). The model, therefore, provides a quantitative coupling between quarrying and abrasion, and suggests that even for very high sliding velocities, abrasion never exceeds about 40% of the total erosion rate.

7.5. LARGE-SCALE PROCESSES

In relatively few studies has current knowledge of small-scale erosional processes been used to interpret the morphology of large-scale landforms. To do so requires that large-scale patterns of glacier flow be measured or calculated and then linked with small-scale processes in order to estimate spatial variations in erosion rates. More commonly, glacier flow

models, sometimes with erosional feedbacks, have been presented without specific reference to the erosional processes involved (e.g., Nye and Martin, 1968; Hirano and Aniya, 1988; Mazo, 1989).

Hooke (1991) invoked the potential effect of subglacial water-pressure fluctuations on quarrying (Röthlisberger and Iken, 1981; Iverson, 1991b) to suggest a positive feedback that may lead to the erosion of cirques and overdeepenings in the longitudinal profiles of glacier beds. He argues that water access to the bed is generally at bergschrunds in the case of cirques and through crevasses that develop above the high points on the bed between overdeepenings (reigels). Most water, therefore, encounters the bed immediately down-ice at cirque and overdeepening headwalls. As a result, water-pressure fluctuations caused by variations in surface water input tend to be focused there. This accelerates quarrying of the headwall, which deepens the up-ice side of the overdeepening thereby increasing the potential for crevassing and further input of surface water to the headwall. The feedback is complemented by the tendency for water flow to be englacial, rather than subglacial, along the down-ice sides of overdeepenings with sufficiently steep adverse slopes. This is because as water flows up such slopes, some of it freezes to conduit walls to provide the heat necessary to keep the water at the pressure-melting temperature. The resultant constriction of the flow increases basal water pressure and forces the water into englacial conduits. This may allow a subglacial till layer to accumulate, protecting the bed from erosion, and potentially producing the characteristic asymmetry of overdeepened basins, which tend to be deeper on their up-ice ends. Measurements of water-pressure fluctuations and the distribution of basal till beneath Storglaciären, a valley glacier in northern Sweden, are consistent with the hypothesis.

Harbor (1992) recently completed a major numerical study of the progressive erosion of U-shaped valleys. He first modelled glacier flow through a V-shaped transverse cross-section in order to obtain the lateral variation in sliding velocity. In agreement with the form of Hallet's (1979, 1981) model of abrasion and Shoemaker's (1986) suggestion for a quarrying law, the rate of erosion was assumed to be proportional to the sliding velocity raised to an exponent ranging from 1 to 4. Valley evolution was iteratively simulated by calculating relative erosion rates across the profile, modifying the profile accordingly, and recalculating the distribution of sliding velocity. A sliding law was adopted in which the sliding velocity is directly proportional to the basal shear stress and inversely proportional to the effective pressure (e.g. Raymond and Harrison, 1987). Assuming a level piezometric surface across the section, the sliding velocity increases away from the glacier margins. However, near the bottom of the V-shaped valley, the constriction impedes the flow so that there is a central minimum in the basal velocity distribution. This results in erosion rates that increase away from the margins, and peak on the valley sides. This effectively broadens the valley bottom and steepens the valley walls into the U-shaped form (Fig. 7.11) (Menzies, 1995b, Chapter 2) (Plate 7.3).

The shape of the active channel eventually becomes steady with continued erosion, although the valley continues to incise. Steady valley profiles were most similar to natural examples when the exponent in the erosion law set was set equal to one or two. This is in general agreement with Hallet's abrasion model and Shoemaker's (1986) suggestion that quarrying rates vary approximately linearly with the sliding velocity.

7.6. SUMMARY

The principal glaciological variable that determines the rate of abrasion is the sliding velocity, which controls both contact forces and the flux of debris across the bed. Contact forces are largest midway up stoss surfaces of sinusoidal bumps, where the local slope of the bed dips most steeply up-ice. The flux of debris is largest near the crests of bumps, in part, because once debris has traversed one stoss surface, its trajectory in the ice prohibits its further contact with all but the crests of equal or higher stoss surfaces further down-ice. This distribution of contact force and debris flux results in abrasion rates that are largest high on stoss surfaces. Thus, the prolonged effect of abrasion should be to reduce the bed roughness. The tendency for debris to lose contact with much of the bed after abrading one stoss surface leads to another conclusion: that quarrying of debris

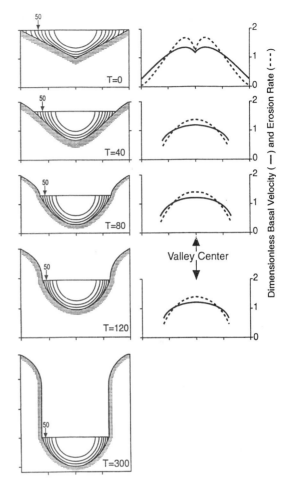

FIG. 7.11. Numerical simulation of erosion of a glacial valley. Erosion rate was scaled to local sliding velocity squared. Figure shows cross-section after different numbers of iterations (T) with model. Concentric lines are velocity contours in units of 10% of maximum velocity for section, with the most central contour being 90%. Right column shows distribution of basal velocity and erosion rate, both scaled to an average cross-sectional value of 1.0. Basal-velocity and erosion-rate distributions for T=300 (not shown) are essentially identical to that for T=120 (from Harbor, 1992; reproduced by permission of the author).

plays an important, if not dominant, role in the process by providing tools to abrade the lower portions of stoss surfaces that would not otherwise be abraded. Separation between ice and the bed probably increases abrasion rates by allowing debris greater access to stoss surfaces.

Quarrying must normally require that cracks in the bed grow and coalesce with each other and with pre-

(a)

(b)

PLATE 7.3. (a) The U-shaped valley of Lierdalen, Norway. Note kame terraces and trimline. (b) Profile of a U-shaped valley near Sogndal, Norway.

existing joints and bedding planes. The potential for crack growth is largest in bedrock immediately up-ice from the point where ice separates from the bed. In these zones, normal stresses exerted by ice on the bed are maximized, while orthogonal compressive stresses are minimized, owing to the smaller pressure exerted by water on lee surfaces. As a result, the stress differences in the bed are maximized and conditions favour the growth of cracks that lie approximately normal to the crests of bumps. Cracks grow due to tensile stresses that develop locally at crack tips. The pressure of water in cracks should contribute to such tensile stresses, while the chemical action of water at crack tips should significantly reduce the stress necessary for slow crack growth. Rapid reductions in water pressure in zones of ice–bed separation intensify normal stresses that ice exerts on the bed by shifting some of the weight of the glacier to zones of ice–bed contact. This may result in periodic crack growth as subglacial water pressure fluctuates recurrently during the melt season. The effect is largest when separation between the ice and bed is extensive, and is limited by the strength of the ice. Ice–bed separation and water-pressure fluctuations should also help dislodge rock fragments that have become loosened from the bed. As a generalization, quarrying rates are likely to be greatest beneath glaciers in which low effective pressures, large variations in water pressure, and high rates of sliding combine to produce extensive ice–bed separation and fluctuations in stress on the bed.

Chapter 8

PROCESSES OF TRANSPORTATION

M. P. Kirkbride

8.1. INTRODUCTION

The entrainment and transport of sediment by glaciers is an important link in the sediment cascade in high latitudes and at high altitudes today and, in the past, over vast areas of the mid-latitude continents. Glacial action is not only a potent erosional agent, but also provides the means by which eroded debris is removed from its source and transported to areas of deposition. Potential transport distances vary from a few hundred metres in cirque glaciers to many hundreds of kilometres in ice sheets. Without the evacuation of debris by glaciers, wear processes would not continue to attack fresh bedrock and sediments to create many of the characteristics of glaciated landscapes. Most wear and subsequent erosion is performed by rock particles (clasts) in transport at the glacier sole (Chapter 7).

Depositional landforms equally owe much to the processes of sediment delivery to ice mass margins and interfaces (Fig. 8.1). Where, for example, deposition is areally-restricted around stable margins, continuous sediment supply allows large moraines to be constructed (Boulton and Eyles, 1979); or where glaciers overrun soft sediment or easily eroded material, large basal 'layers' of glaciogenic sediment may be derived, transported and subsequently deposited (Aylsworth and Shilts, 1989). The amount, nature and distribution of sediment within, upon and beneath moving ice fundamentally influences the assemblages of landforms deposited at different stages of the mass-balance cycle (Eyles, 1983). Two

examples serve to illustrate this association. First, the large sediment loads and higher than average velocities of many warm-based valley glaciers account for their huge lateral moraines formed during the Neoglacial period. These moraines provide detailed sedimentary records of global Holocene climatic variation (Röthlisberger and Lang, 1987) (Chapter 2). Secondly, the streaming of basal debris around large obstacles on the glacier bed (Boulton, 1975; Krüger, 1979; Alley, 1989a, b) may lead to the downstream persistence of longitudinal 'ridges' in the basal transport zone that results in the deposition of fluted moraine during ice retreat (Paul and Evans, 1974; Lawson, 1976).

Glacier transport is remarkable in the diverse manner in which the sediments themselves are modified. Much basal debris can be drastically shaped and comminuted over short distances (see Chapter 15; Menzies, 1995b, Chapter 13), while part of the debris load may travel large distances with virtually no sedimentological change. In the former case, sustained mechanical crushing and fracture under the relatively low regional stresses (but large tractive forces and high local stresses) beneath ice masses produce sediment quite distinct from the products of other comminuting processes. Indeed, glacial abrasion may be responsible for most of the silt-size rock particles created over geological time. Conversely, glacial transport of large boulders over hundreds of kilometres without modification is unique in the natural sediment cascade.

It is misleading to suggest that glaciers in all parts

(a) ICE SHEET

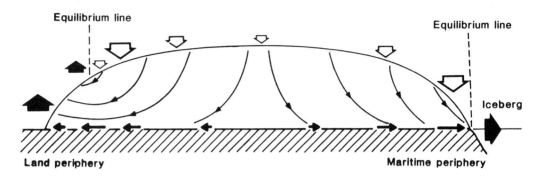

(b) ICE SHELF

(c) VALLEY GLACIER

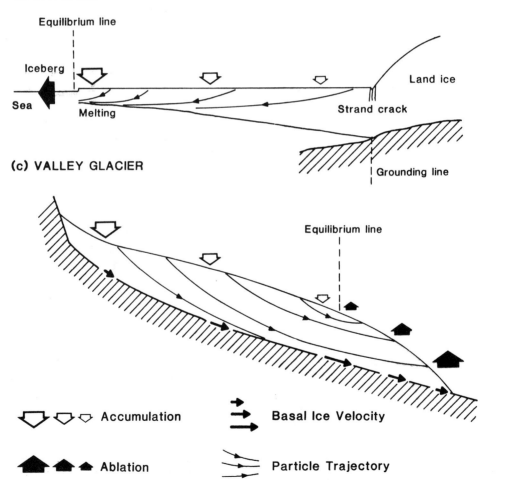

FIG. 8.1. Models of (a) ice sheet, (b) ice shelf and (c) valley glacier, showing the distribution of accumulation and ablation and related flow characteristics. Basal sliding is assumed to occur in models (a) and (c) and is at a maximum in the vicinity of the equilibrium line (from Sugden and John, 1976, *Glaciers and Landscape*, Edward Arnold).

of the world are highly erosive. Glaciers frozen to their substrates are incapable of significant geomorphic work and may even have reduced erosion rates compared to neighbouring ice-free land. Indeed, doubt exists as to whether warm-based glaciers necessarily erode and transport more material than non-glacial processes in landscapes of similar relief and precipitation (Hicks *et al.*, 1990). Nevertheless, processes of glacial transportation have played a major role in shaping landscapes and in moving vast supplies of sediment both on land and in the oceans and in the genesis and maturation of many glacial land- and bed-forms. The understanding of the principles of drift exploration (Menzies, 1995b, Chapter 15) and in mining placer deposits (Menzies, 1995b, Chapter 16) in the establishment of the geotechnical character of glacial sediments and in the decisions concerning the suitability or otherwise of landfill sites for the disposal of toxic and other waste products hinges upon the impact and influence of glacial transport processes (De Mulder and Hageman, 1989).

8.2. ICE PROPERTIES AFFECTING SEDIMENT TRANSPORT

Glaciers are commonly likened to 'rivers of ice'. Such a comparison could not be further from the truth when the mechanics of flow and the nature of sediment transport are considered! An understanding of glacial transport is helped by first examining those physical properties of ice that have a direct bearing on how rock particles are entrained and transported (Chapters 4, 5, and 6). Three properties are important in this context viz. ice temperature, density and viscosity.

8.2.1. Temperature

Within any ice mass, ice temperature varies over both time and space (Chapter 4). At glacier surfaces, melting is caused by the heat supplied from warmer superjacent atmosphere. At the glacier sole, melting is caused by: (1) heat supplied from the substrate; (2) the pressure exerted by overlying ice. A small amount of internal melting results from frictional heat generated by ice deformation. Internally, melting may

also occur as a result of the englacial passage of meltwater derived from supraglacial and subglacial sources.

Ice at the base of many ice masses is at or close to the pressure melting point. Pressure melting occurs over large areas of glacier beds where overburden pressures are high (Fig. 8.1) and in localised zones in response to pressure variations caused by substrate irregularities. In terms of sediment transport, the distribution of basal melting affects the processes and effectiveness of erosion and entrainment, and has an overall influence on the discharge of sediment through ice masses (Chapter 3). Where the ice mass bed is frozen, later transport of previously frozen debris may occur resulting in 'patches' of debris being moved long distances, in some cases, with limited comminution or evidence of deformation; or in other cases frozen debris, after transport, may melt-out and produce a distinctive lithofacies.

At glacier surfaces, the rate of melting is strongly influenced by the distribution and thickness of supraglacial debris (Paul and Eyles, 1990). Differences in ablation rate across a glacier surface are termed differential ablation. The form and development of supraglacial moraines is a response to differential ablation beneath debris.

8.2.2. Density

Ice is an unusual material in that it is less dense in the solid phase than in the liquid phase. Implications for glacial sediment transport are important where ice is in contact with water deep enough for flotation to occur, such as glaciers terminating in the sea or in deep proglacial lakes. The stress field that drives the flow of floating ice is very different from that of grounded ice, resulting in different flow paths and particle trajectories. Some ice shelves have upward particle trajectories, the reverse of those in grounded glaciers.

8.2.3. Viscosity

Newton's law of fluid friction defines viscosity as the ratio of the shear stress acting on a fluid to the rate of shear which results. Thus, highly viscous materials deform only slowly under a given applied shear stress, whereas low-viscosity materials deform very rapidly under the same applied stress.

The behaviour of ice under stress is complex. Nevertheless, ice, with a viscosity of 10^{12} to 10^{14} poises, may be likened to a viscous fluid that deforms only on attaining a critical level of stress (the yield stress). The pressure exerted by even the largest boulders resting on glacier surfaces rarely exceeds the yield stress of ice, so that glaciers transport material of about three times the density of ice without the debris sinking.

8.3. DEBRIS SOURCES

8.3.1. Entrainment at the Glacier Sole

The mechanisms by which glaciers pick up debris from their beds, and the physical controls governing these mechanisms, remain one of the most problematic, yet intriguing, areas of glaciological research. Basal debris entrainment is notoriously difficult to observe, but borehole camera observations (Kamb and La Chappelle, 1964), direct observations (Boulton, 1970a), proxy evidence such as meltwater chemistry (Souchez and Lorrain, 1987) and landform studies (Jahns, 1943; Sugden et al., 1992) now give some substance to theories of debris entrainment. Much of this research tackles the problem of how ice is added to the glacier sole, and it is through study of basal ice formation that debris entrainment can be understood (Hubbard, 1991). Most theories of basal entrainment involve some mechanism of re-freezing (regelation) of subglacial water. Regelation produces small interlayered ice crystals, containing mostly fine sediment, and so produces readily recognisable features within glacier ice. However, because of the variability in conditions controlling the location, scale, timing and rate of melting and/or regelation, and because of the variety of water sources potentially available for re-freezing, the term regelation encompasses a range of theoretically quite different processes (Hubbard and Sharp 1989).

8.3.1.1. Weertman regelation

The Weertman regelation mechanism operates under warm-based glaciers at the scale of bedrock

obstacles of <1 m in length (Weertman, 1957b, 1964a). Ice coming into contact with the upstream (stoss) side of such an obstacle experiences increased pressure which, if the ice is at the pressure-melting point, causes melting. Meltwater migrates to the downstream (lee) side of the obstacle where reduced pressure allows re-freezing and the incorporation of usually fine-grained debris into the regelation layer accreted to the glacier sole. The repeated formation and destruction of a regelation ice layer as it encounters successive obstacles limits the layer thickness to only a few centimetres. Weertman regelation has been confirmed by observation, and may be an important contributor to the sliding of warm-based glaciers.

8.3.1.2. Robin 'heat pump' effect

Under some circumstances, meltwater produced by localized pressure melting upstream of an obstacle is lost from the system, resulting in a net loss of latent heat in the lee of the obstacle and the formation of a cold patch of several square metres extent (Robin, 1976; Goodman et al., 1979). Re-freezing in the vicinity of the cold patch incorporates subglacial water not produced by pressure melting of existing ice, thus adding 'new' ice to the glacier in which thin layers of fine-grained sediment may be incorporated.

8.3.1.3. Large-scale zones of melting and re-freezing

The overburden pressure exerted on the glacier bed by glacier ice reduces where ice becomes thinner, typically around the margins or differentially across bed obstacles. Lowering of the pressure melting point commonly results in re-freezing of meltwater derived from basal melting upstream and in marginal areas from surface ablation. Ice masses experiencing such a large-scale contrast between their interiors and margins are termed polythermal glaciers. Such glaciers are capable of effective basal ice formation producing extensive sequences of heavily debris-charged ice of several metres thickness. Following Boulton (1970a, b), it is commonly believed that polythermal glaciers carry larger debris loads than fully warm-based glaciers, and may also be efficient

PLATE 2.1. Thin section of an ice core sample taken from 160 m below snow surface at Camp Milcent, Greenland. Section is photographed under polarized light, showing the irregular shape of individual ice crystals after nearly 300 years of transformation from snow flakes to a polycrystalline aggregate. The diameters of the large crystals are approximately 1 cm. The colours depict the different orientations of individual crystals (photo courtesy of Chester C. Langway Jr).

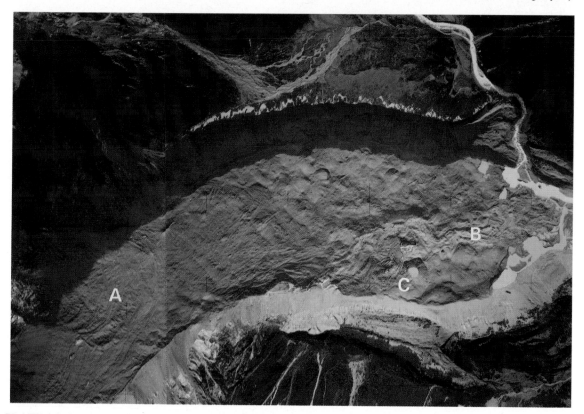

PLATE 8.2. Vertical aerial photograph of Meuller Glacier, New Zealand, showing emergence of debris at the glacier surface in zones of intense compression (A) and sinuous band of debris (B) reworked by an englacial river, now visible in collapse dolines at C and D. The river has never flowed supraglacially (photo courtesy of Lloyd Homer, Institute of Geological and Nuclear Sciences Ltd.).

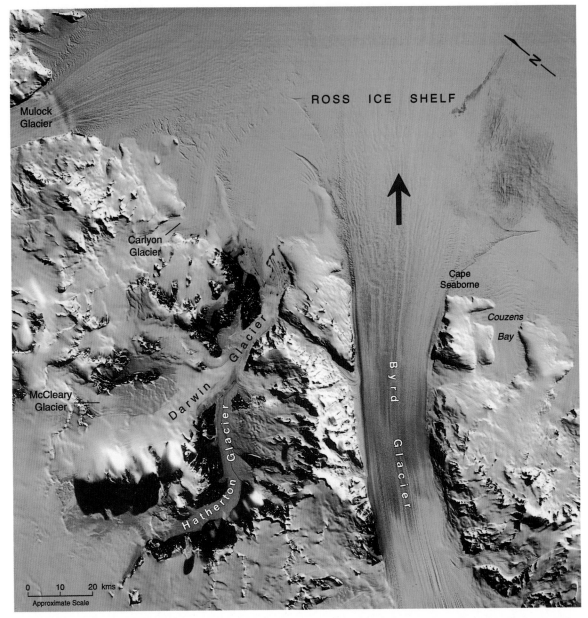

PLATE 4.1. Landsat 1 MSS digitally enhanced false-colour composite image of part of Byrd Glacier and surrounding glaciers and the Ross Ice Shelf, Antarctica (photograph courtesy of B.K. Lucchitta and the U.S.G.S. Flagstaff, Arizona; annotated version of landsat photo from Williams, R.J., Jr and Ferrigno, J.G. in USGS Prof. Pap. 1386-B, 1988, p. B27).

PLATE 4.7. Example of regelation banding within an englacial tunnel (Saskatchewan Glacier, Alberta/British Columbia). Wooden plank on floor of tunnel is approximately 15 cm.

PLATE 8.1. A large rock avalanche of 14 December 1991 which fell on Tasman Glacier, New Zealand. Such events are important debris sources for valley glaciers in tectonically-active mountain ranges (photo courtesy of Lloyd Homer, Institute of Geological and Nuclear Sciences Ltd.).

erosional agents on both local scales (Boulton, 1970a, b) and continental scales (Sugden, 1978).

Variations in subglacial water pressure may also cause widespread basal adfreezing (Chapter 6). Increased pressure in the subglacial water film reduces relative pressures on adjacent bed protrusions. A resulting rise in pressure melt point at these locations causes conduction of heat from the bed, lowering the temperature at the ice–bed interface sufficiently to cause freezing.

Variations in subglacial water pressures enhance basal entrainment. Where water pressures increase in lee-side cavities, the effective overburden pressure on the bedrock substrate is reduced and freezing encouraged. These effects in association with higher ice velocities linked with high basal water pressures increase the tractive forces which physically 'pluck' boulders away from cavity walls (Röthlisberger and Iken, 1981; Sugden et al., 1992). Zones of effective bedrock erosion may, therefore, correspond to those parts of glacier hydrological systems where water pressure variations are most marked, such as below crevassed areas (Hooke, 1991) and/or where bedrock is impermeable (Iverson, 1991b) (Chapters 6 and 7).

In addition, ice may also be incorporated onto the glacier sole by several other mechanisms. Overriding and inclusion of existing debris-rich ice by glacier surges has been recorded (Slatt, 1971; Sharp, 1985a, b), and at a smaller scale blowing snow, cold air intrusions into glaciers along crevasses and tunnels, and incorporation of frozen ground may all result in the addition of ice and debris to the glacier.

All these mechanisms, and possibly others yet to be identified, produce ice with textures and structures quite distinct from ice produced by the firnification of snow in glacier accumulation zones (Hubbard, 1991). Thus a separate zone of the ice mass, the basal ice zone, is recognized (Hubbard and Sharp, 1989). Floating ice shelves, which normally release sediment by basal melting at their grounding lines, may, under fairly rare conditions, accrete new basal ice by the adfreezing of sea water. Basal adfreezing will occur only where the ice is colder than the sea and the sea is already at its freezing point (Doake, 1976; Drewry, 1986). Very small amounts of suspended marine sediment and solutes may be entrained by this process, though occasionally larger quantities of clastic and biogenic sediment are entrained at grounding lines.

8.3.2. Supraglacial Entrainment

The extensive debris-free surfaces of ice sheet ablation zones contrast with the dirty, moraine-strewn appearance of many cirque and valley glaciers. The contrast owes much to the spatial variability of supraglacial sediment supply, which can be understood with reference to the distribution of extraglacial sediment sources and to rates of sediment supply.

The addition of debris to valley glacier surfaces is dominated by rockfall and avalanching from mountain faces, especially in temperate alpine regions (Boulton, 1978; Gordon and Birnie, 1986; Small, 1987a, b). Cliffs in alpine regions typically retreat at an average rate of between 0.5 and 3 mm year[-1] (French, 1976) providing large amounts of coarse debris. Nevertheless, many other processes acting around glacier margins can supply debris to supraglacial transport zones. Close to the margin, mass movements from hillslopes onto glacier surfaces are common in alpine regions (Birnie, 1992). Other more distant sources of debris include transport by rivers from ice-free valleys tributary to glacierized main valleys (Evenson and Clinch, 1987). Aeolian-transported fine-grained material is delivered to glaciers and ice sheets by snowfall, by direct deposition or by redistribution of unconsolidated sediment (Keys et al., 1977; Goss et al., 1985). Finally, cold-based ice sheets contain debris from extraterrestrial sources. This debris is concentrated over millenia from catchments areas of several million square kilometres into zones of flow convergence and ablation, often in the lee of nunataks. This debris includes meteorites of several varieties (chondrites, irons), smaller spherules, and cosmic dust trapped by the earth's gravitational field (Cassidy and Rancitelli, 1982; Nishiizumi et al., 1989a).

Supraglacial debris from mountain sides tends to emanate from discrete points of entrainment. Commonly, rock gullies supply rockfall, avalanche and debris-flow material repeatedly to the same point on the glacier surface. Intervening parts of glaciers receive relatively little debris (Rogerson *et al.*, 1986). By far the majority of supraglacial debris enters glacier accumulation zones, where high local relief, concave glacier slopes, an absence of marginal moraines and centripetal flow lines combine to produce a strong coupling of the mountain slope and glacier transport systems (Fig. 8.2).

In most cases, small quantities of supraglacial debris are supplied fairly constantly to the ice mass surface. However, rock avalanches may instantaneously increase the supraglacial debris load. These sudden and vast inputs of debris lead to changes in the pattern of ablation. Significant modification in net mass balance can result, over a time lag of some years, in ice margin movement. Large rock avalanches are fairly common phenomena in alpine regions, several occurring every decade in tectonically active and high-precipitation alpine ranges such as in Alaska and New Zealand (Whitehouse and Griffiths, 1983). The 1964 Alaska earthquake, for example, caused a rock avalanche of

10^6 m^3 of rock to blanket 8.25 km^2 of the Sherman Glacier (McSaveney, 1978). The largest recent example of a rock avalanche in New Zealand took place in December 1991 (Kirkbride and Sugden, 1992) and appears to have had no seismic trigger (Plate 8.1). The Tasman Glacier, onto which it fell, receives the majority of its huge supraglacial debris load from relatively large rockfalls in its accumulation zones. The 1991 avalanche almost doubled the supraglacial load in only a few minutes.

8.3.3. A Sediment Supply Model

The extreme variability in sediment supply to glaciers and ice sheets is illustrated schematically in Fig. 8.4. The amount and variety of sediment entering high-level transport depends on the nature and extent of extraglacial terrain, particularly the effectiveness of weathering and erosion. In contrast, subglacial entrainment depends on thermal regime and substrate erodibility (Fig. 8.3). The effectiveness of high-level

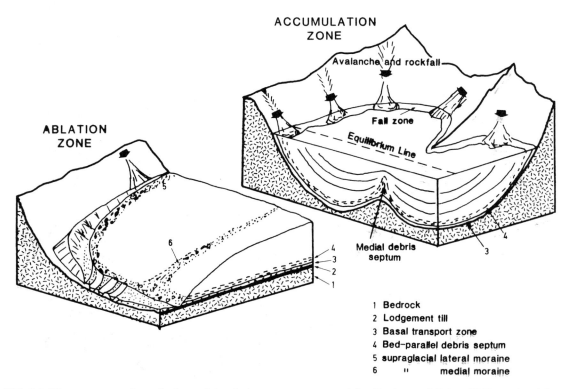

FIG. 8.2. The geometry and terminology of the glacier transport system (after Boulton and Eyles, 1979; reprinted from: *Moraines and Varves* (Ch. Schlüchter, ed.), 1979; courtesy of A.A. Balkema, Rotterdam).

PLATE 8.1. A large rock avalanche of 14 December 1991 which fell on Tasman Glacier, New Zealand. Such events are important debris sources for valley glaciers in tectonically-active mountain ranges (photo courtesy of Lloyd Homer, Institute of Geological and Nuclear Sciences Ltd.).

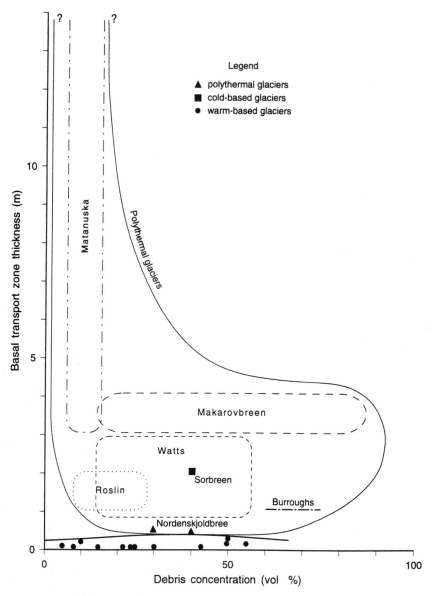

FIG. 8.3. Basal transport zone thicknesses and debris concentrations near to the termini of cold-based, polythermal and warm-based glaciers (data from Pessl and Frederick, 1981). Note that many cold glaciers have very limited basal debris loads (not shown).

and basal entrainment processes, therefore, experience quite different controls. Figure 8.4 summarizes how two of the main determinants of debris supply (namely thermal regime and degree of topographic inundation) in combination influence the amount of debris available for entrainment and transport. Two possible interpretations are shown,

following considerable debate concerning the relative erosional capabilities of warm-based and polythermal glaciers (Fig. 8.4a after Andrews, 1971, 1972a and Anderson, 1978; Fig. 8.4b after Boulton 1970b, 1971b, 1972b) (see Section 8.4.7). Both cases illustrate that overall sediment supply is less where ice cover is great and that most of this reduction is

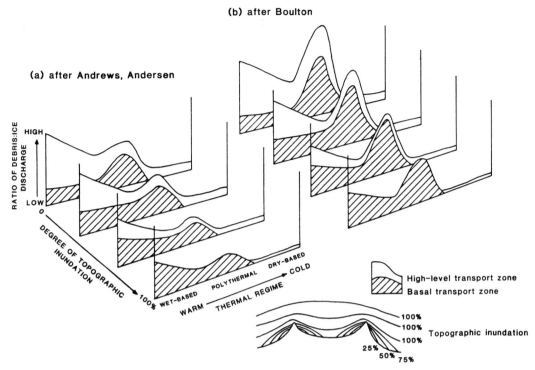

FIG. 8.4. Conjectural relationships between the relative discharge of debris to ice in glaciers, thermal regime and the degree of burial of topography by ice. Note the relative increase in basal transport zone debris as fewer extraglacial debris sources are exposed, and the extremely low debris load of cold glaciers. (a) Illustrates the case where transport by temperate glaciers exceeds that by polythermal glaciers (e.g. Andrews, 1971; Anderson, 1978). (b) Illustrates the reverse case (Boulton, 1970b). The diagram shows relative discharges: total debris discharge of large glaciers and ice sheets will be greater than that of smaller glaciers even though their thermal regime and extraglacial sources are often less favourable for entrainment.

accounted for by a relative lack of debris in high-level transport. The supply of sediment to the basal transport zone of warm-based glaciers is assumed to be a function of sliding velocity and therefore declines initially with increasingly cold thermal regime. An increase in basal sediment supply under polythermal ice reflects potent basal adfreezing in this regime: the magnitude of this increase is the fundamental difference between Figs. 8.4a and 8.4b.

In mountainous areas, the extent to which terrain is inundated by ice is in inverse proportion to the area of ice-free cliff faces from which rockfalls supply debris to the high-level transport zone. Glaciers with high supraglacial and englacial sediment loads are invariably surrounded by extensive steep cliffs and usually have 'dirty' surfaces. The degree of inundation probably plays a more important role for

warm-based than for cold-based glaciers because extraglacial climates around the former favour abundant sediment transport by water, avalanching, debris flow and rapid weathering and slope erosion; hence, supply to high-level transport is shown to decline with colder thermal regimes in all cases in Fig. 8.4. In mid-latitude glacial environments, greater burial of the landscape by ice probably causes a proportionately greater reduction in sediment availability for the high-level transport zone than in high latitudes. Around high-latitude glaciers, relatively little geomorphic work is done to supply debris to glaciers. Aeolian and, especially, subglacial processes supply debris to subpolar glaciers in quantity, but entirely cold-based glaciers depend on a very small supply of aeolian and extraterrestrial sediment.

Ice sheets, in comparison, have zero supraglacial debris in accumulation areas. In ablation areas surface debris accumulates where outlet lobes cut through marginal mountains or isolated nunataks. In ice sheet marginal areas where no such topographic relief exists, the only supraglacial debris is likely to be found very close to the margins where basally-entrained debris is elevated along upward flowlines Therefore, along the southern margins of the Laurentide Ice Sheet in Canada and the U.S.A. and the Fenno Scandian Ice Sheet across the North European Plain, supraglacial debris would have been restricted to a narrow band close to the ice sheet margins. In contrast, along the seaboards of Labrador-Ungava and western Norway, both ice sheets would have had extensive supraglacial debris cover derived from exposed ice-free mountains. A similar contrast exists today between the western and eastern edges of the Antarctic Ice Sheet.

The model provides an qualitative summary of debris supply to glaciers and ice sheets globally. Nevertheless, there are few systematic studies of glacier sediment budgets from which to draw quantitative generalizations (Fenn, 1987) (Section 8.4.7).

8.4. TRANSPORT PATHWAYS THROUGH ICE MASSES

8.4.1. Controls on Particle Trajectories

The distribution of sediment in glaciers reflects the trajectories taken by particles away from the locations of debris sources. Ice sheets, valley glaciers and ice shelves have contrasting transport pathways reflecting their different flow patterns (Fig. 8.1). The geometry of transport paths within any glacier depends on three main factors:

(1) *Locations of points of entrainment*. Debris eroded from the ice mass bed generally remains within basal ice where pressure melting prevents significant upward translocation. Debris supplied to the accumulation zone is buried by snow and enters englacial transport. Debris supplied to the ablation zone surface remains in supraglacial transport. Englacial and supraglacial transport together form the high-level transport zone.

(2) *Glacier flow lines*. A cirque or valley glacier of simple structure with no tributaries transports debris from source along flow lines that generally converge and have downward vectors in the accumulation zone, but which diverge and flow upwards in the ablation zone. Debris in transport above the equilibrium line therefore tends to move towards the centre of a glacier cross-section and, below the equilibrium line, tends to move outwards to the glacier margins. Particles in dome-shaped ice sheets with purely radial flow follow trajectories illustrated in Fig. 8.1a. The simple pattern may be disrupted by multiple centres of ice accumulation, by subglacial topography and by concentration of ice discharge into fast-flowing ice streams. Ice shelves (Fig. 8.1b) usually have downward trajectories, but this pattern is reduced or reversed where basal adfreezing predominates.

(3) *Glacier flow structure*. Many valley glaciers and topographically-constrained ice fields are composites of multiple tributaries, each with its own set of debris sources and transport paths. Ice flow is laminar, so the transport zones of each ice stream maintain their individuality after confluence with other ice streams (Stephens *et al.*, 1983).

The general model of transport path geometry in a warm-based valley glacier is described by Boulton and Eyles (1979) and Small (1987b). The model (Fig. 8.2) distinguishes between a basal transport zone and a high-level transport zone. The former includes the zone of traction, where interparticle and particle–bed collisions cause abrasion and fracture. In contrast, sediment in high-level transport is little modified during transport. Distinction between the two exists on the basis of sedimentological process-form relationships (Mills, 1977) as well as ice crystallographic characteristics (Souchez and Lorrain, 1991).

The distribution of debris in glaciers is highly inhomogeneous. Debris point sources and glacier flow structure tend to concentrate debris into planar geometric shapes termed debris septa, or into discrete bodies of variable geometry. The nature of ice deformation creates lensate and ellipsoid englacial debris bodies, though compressive flow has the potential to create more complex forms through folding. Only where debris is supplied almost continuously from a point source is a medial moraine truly continuous in form. Where inputs of debris are pulsed, such as infrequent rockfalls from the same source, medial moraines have a 'beaded' form (Rogerson *et al.*, 1986; Small, 1987a, b).

8.4.2. Basal Transport Zone

The bulk of the debris load of most ice masses is concentrated in the basal few metres, reflecting (1) the entrainment of debris at the glacier sole, and (2) concentration of debris brought from higher englacial levels to the ice–bedrock interface by basal melting. The greatest shear strain within glaciers occurs in the basal few metres, causing ice deformation and recrystallization. High strain rates yield textures and structures that reveal a history of both ice formation, debris entrainment and subsequent deformation. The basal transport zone is defined by this combination of sedimentological and crystallographic characteristics. A primary constraint on the extent and nature of the basal transport zone is the thermal regime of the glacier (Fig. 8.3). Cold-based glaciers, frozen to their substrates, entrain a very small volume of debris thus englacial debris is uncommon (Holdsworth, 1974; Koerner and Fisher, 1979; Shreve, 1984; Chinn and Dillon, 1987; Echelmeyer and Wang, 1987). What basal debris they do transport is usually entrained by overriding of marginal aprons and upward shearing to form 'inner moraines' (Chinn, 1989; Evans, 1989). Fully warm-based glaciers, on the other hand, rarely develop basal debris sequences thicker than a few tens of centimetres because basal melting is sufficiently effective to keep basal debris close to the glacier sole. Furthermore, comminution of grains during traction reduces weaker rocks to silt, much of which is flushed from the basal zone by meltwater.

Polythermal glaciers are characterized by mixed thermal regimes and are able to accumulate thick basal sequences in their marginal zones, where re-freezing adds much ice to the glacier sole (Boulton, 1979; Hutter and Olunloyo, 1981; Menzies 1981).

Perhaps the most complete picture so far of the character of the basal transport zone is provided by the Matanuska Glacier, Alaska (Lawson, 1979a) (Fig. 8.3). Here, two main facies, the englacial and basal facies, were identified on the bases of ice crystallography and debris content. Englacial facies comprise ice formed by firnification, forming by far the greater part of the glacier. Basal facies comprise ice formed at the glacier sole (Fig. 8.5). This facies is further subdivided into 'dispersed' and 'stratified' subfacies (Chapter 4). Weertman regelation during basal sliding forms the dispersed facies, which are subsequently elevated within the marginal zone by net basal adfreezing accreting the debris-rich stratified facies to the glacier sole (Souchez *et al.*, 1988; Sugden *et al.*, 1987). Thus, debris-rich bands are interstratified with relatively clean ice over several metres of vertical thickness close to glacier

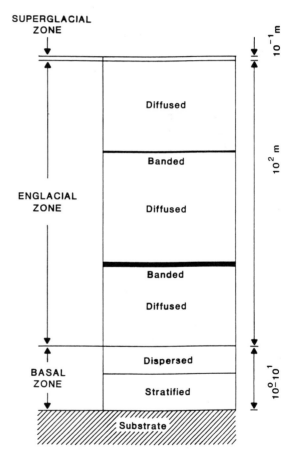

FIG. 8.5. Schematic vertical section through Matanuska Glacier, Alaska, showing the distribution of ice facies. High debris concentrations occur in the banded facies (from supraglacial sources) and in the stratified facies (from subglacial sources) (after Lawson, 1979a; reprinted by permission of the author).

margins. Such sequences are not, however, representative of the whole basal transport zone in polythermal glaciers, because thin basal transport zones associated with pressure melting and basal sliding often occur extensively beneath thicker ice upstream.

Stratified facies occupy the lower 3 to 15 m of Matanuska Glacier and comprise irregular alternating layers and bands of almost pure ice and highly debris-charged ice. A sharp contact separates stratified from dispersed facies. Intense folding and shearing characterizes the stratified facies, and abrasion, fracture and crushing of debris causes increased rounding and comminution of particles during transport. Debris may be derived from high-level transport by downward ice flow, or from glacier-bed erosion. Debris concentrations vary widely; values ranging from 5 to 55% have been reported from Matanuska Glacier.

The thickness and debris concentration of basal ice changes as ice moves across its bed by both vertical and lateral migration of debris. Vertical migration involves the slow upward dispersion of particles in the dispersed facies under vertical strain to form so-called 'amber-ice' facies (Holdsworth, 1974). Horizontal migration involves the lateral diversion of sediment-rich basal ice around bedrock obstacles (Fig. 8.6). Enhanced plastic deformation of ice due to high stresses in the vicinity of bed hummocks causes ice to accelerate around the flanks. Much less ice accelerates over the top of the obstacle, resulting in 'streaming' of sediment-charged basal ice through gaps while intervening spaces are occupied by debris-poor ice from above (Boulton, 1975). Ice moving through a field of bed hummocks thereby develops a laterally-dispersed basal sediment layer.

Recent evidence from Antarctica and elsewhere has shown the presence of a 'soft' bed of deforming mobile sediment moving as a debris traction layer between the ice and the bed (Chapter 5). This debris layer or mobile bed allows the transport of large volumes of sediment sandwiched between the overlying ice and immobile bed. Within this layer considerable comminution and deformation occurs

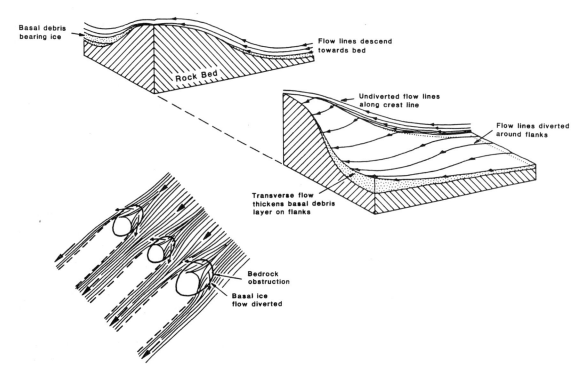

FIG. 8.6. Lateral and vertical debris migration induced by streaming of basal ice around bed undulations (from Boulton, 1974; reprinted by permission of the author).

with associated kinetic sieving, grain-size differentiation with transport and sediment block rafting. The importance of this means of glacial transport remains a major research objective. Preliminary indications suggest that large lowland areas of the Quaternary ice sheets may have been underlain by such 'soft' beds and thus this process of transport is of fundamental interest (Alley, 1991, 1992; Murray and Dowdeswell, 1992).

8.4.3. High-Level Transport Zone

The high-level transport zone includes supraglacial and englacial debris that is not subject to basal traction. Most debris in high-level transport comes from supraglacial sources, but debris is elevated from basal ice into high-level transport at glacier confluences, avalanching ice cliffs, to the lee of large bed obstacles, and near the termini of cold-based and polythermal glaciers.

Medial moraines are the most commonly recognised forms within the high-level transport zone, yet they are part of a continuum of forms comprising isolated rockfall deposits, beaded and linear medial moraines and extensive supraglacial debris mantles (Fig. 8.7). Bodies of debris manifest shapes reflecting the total strain experienced by the ice mass since entrainment. Thus, extending flow produces characteristically linear forms, elongated down glacier but compressed laterally, whereas compressing flow tends to cause crescentic and broadening forms. The shapes of debris bodies also depend on their volume, the periodicity of debris supply (for example, the magnitudes and frequencies of rockfalls from a cliff face), the duration of englacial and/or supraglacial transport, and on the depth and distribution of crevasses along the transport path.

A medial moraine forms in one of two ways: (1) at the confluence of two tributary glaciers, where lateral moraines-in-transit merge; and (2) where rockfall and avalanche debris from above cirque headwalls supply medial debris septa above the firn line (Fig. 8.8). Eyles and Rogerson (1978) classify medial moraines according to the position of emergence of the moraine relative to the firn line. 'Ablation dominant'(AD) moraines are formed by debris septa from above the firn line emerging

downstream in the ablation zone. 'Ice–stream interaction' (ISI) moraines form where two tributaries are confluent below the firn line so that two lateral moraines-in-transit join. 'Avalanche' moraines form by infrequent rockfalls leaving discrete deposits on the glacier surface. A second classification by Small *et al.* (1979) is likewise based on the position of rockfalls relative to equilibrium line position, as well as to whether or not debris enters englacial transport via crevasses or remains in supraglacial transport throughout. Small *et al.*'s (1979) four-fold classification (Fig. 8.9) relates debris supply and differential ablation to moraine morphology. 'Waxing' and 'waning' stages of medial moraines are identified by the effect of exhaustion of the englacial debris septum on moraine morphology, allowing moraine heights to decrease down glacier (Loomis, 1970).

The relative importance of ablation and ice compression in determining moraine form has been debated (Eyles, 1976; Small and Clark, 1976; Shaw, 1980; Small, 1987a, b). In general, medial moraines broaden down-glacier due to the reduction in lateral compression away from a confluence and to the aggregated effects of differential ablation and lateral movement of debris down the flanks of the moraine.

Moraines of superficially similar appearance may contain either basally-derived debris (Eyles and Rogerson, 1977; Gomez and Small, 1986) or rockfall and avalanche debris (Smiraglia, 1989; Vere and Benn, 1989), indicating fundamentally different origins. Debris from the traction zone may be elevated to the surface at confluences by the convergence of flow lines (Fig. 8.10).

Debris in basal transport is commonly elevated to high-level transport close to the termini of polythermal glaciers (Goldthwait, 1951; Swinzow, 1962; Boulton, 1967, 1968; Drozdowski, 1977; Chinn and Dillon, 1987). Often this involves upwarping of warm-based (sliding) ice upstream of an obstruction of thinner, and therefore more rigid, cold-based marginal ice (Hooke, 1968; Hudleston, 1976). The mechanisms of upwarping involve deceleration and compression leading to the formation of a shear zone between the contrasting parts of the glacier, although the extent to which debris is passively elevated, or actively affects the

(a)

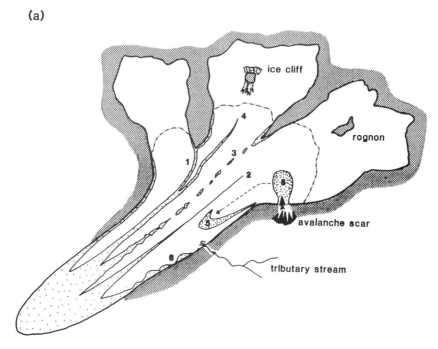

FIG. 8.7. (a) Idealized plan view of the variety of shapes of moraines in high-level transport in valley glaciers. Dashed line = equilibrium line. 1 = Ice–stream interaction medial; 2 = ablation-dominant medial; 3 = beaded (avalanche-fed) moraine; 4 = ice–cliff fed moraine (basal debris elevated to surface); 5 = rock-avalanche debris; 6 = fluvial inwash debris.

distribution of shear strain, remains uncertain (Chinn and Dillon, 1987).

8.4.4. Transport in Subglacial and Englacial Conduits

Awareness has recently increased concerning the significance of sediment transport through warm-based glaciers by water flowing on, within and under the ice (Chapter 6). Water may be supplied either by melting of glacier ice (Sharp, 1985) or from ice-marginal rivers. Valley glaciers play an important passive role by diverting and channelling the path of fluvial systems (Evenson and Clinch, 1987). This is the case where wholly or partly ice-free tributary valleys feed streams into a glacier occupying a main valley. The importance of the fluvial sediment input will increase with the area of marginal ice-free terrain and sediment and water availability. From provenance studies, Evenson and Clinch (1987) suggest that at least 90% of sediment delivered to

some Alaskan glacier termini has been routed through englacial and subglacial conduits. Such a high proportion emphasizes the much more rapid throughput of intraglacial fluvial sediments compared to the glacier sediment load, despite the constriction of transport routeways to narrow conduits.

A direct estimate of subglacial fluvial sediment discharge has been made possible by hydro tunnels beneath Bondhusbreen, Norway (Hooke et al., 1985; Bezinge, 1987). Bondhusbreen is an ice-cap outlet glacier fed by few supraglacial debris sources and flowing across resistant crystalline rocks, with a resulting low glacial sediment load. However, subglacial conduits develop each spring and transport as much as 95% of the total sediment flux of the glacier. In contrast to the Alaskan glaciers, this sediment is re-worked glacially-eroded sediment probably flushed from the basal transport zone. The transfer of sediment between englacial and intraglacial fluvial transport has also been observed in Mueller Glacier, New Zealand (Plate 8.2). Large

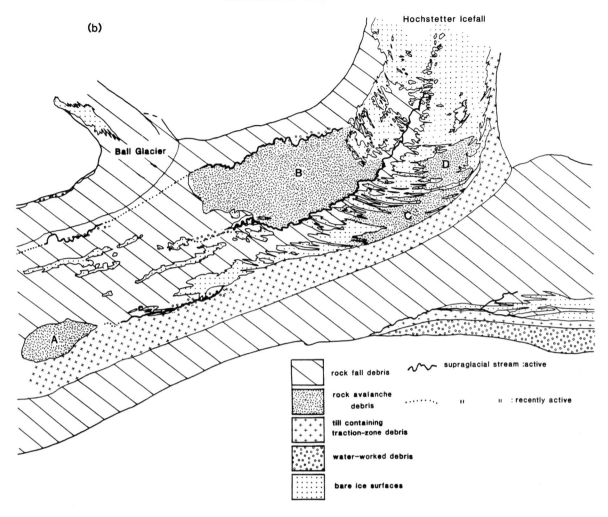

FIG. 8.7. (b) Map of moraine types in transport on the debris-covered part of Tasman Glacier, New Zealand (the glacier is 1.8 km wide). Rock avalanches A to D were all entrained in the accumulation zone above the Hochstetter icefall. This area was covered by a single rock avalanche in December 1991 (Kirkride and Sugden, 1992).

rockfall deposits which have travelled along englacial pathways have been exposed in ice caves within the ablation zone. Englacial rivers flowing within these caves repeatedly switch position, seeking to maintain their gradient towards the terminal outlet while ice motion has an upward vector: thus, lenses and ribbons of water- re-worked sediment have been successively abandoned by the river and elevated to the glacier surface. Supraglacial debris rounded by water action indicates the line of the englacial stream in Plate 8.2.

Sediment transport through glaciers by water has important implications for interpreting ice-marginal sedimentary facies. There has been an over-emphasis in the past on the glacial reworking of outwash to produce tills comprised of rounded clasts and a neglect of the effect of water-reworking of sediment *before* it leaves the glacier.

8.4.5. Transport in Floating Ice Shelves

Ice shelves are formed by the coalescence of floating glacier tongues in the sea or large lakes, or by the accumulation of snow and ice on perennial sea ice (Sugden and John, 1976). Ice shelves are attached to

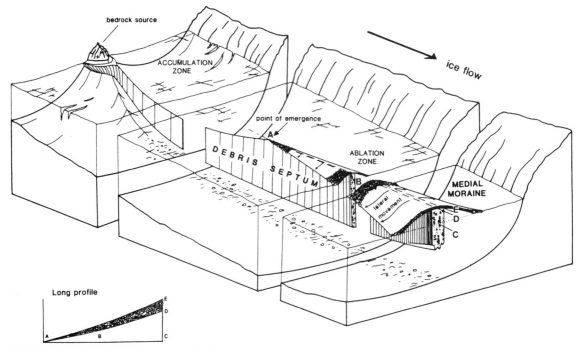

FIG. 8.8. Anatomy of a medial moraine fed by rockfalls above the equilibrium line. Vertical lines represent increments of ice movement (extension in accumulation zone, compression in ablation zone). B = ice core; AE = moraine crest; AD = ice-debris interface; AC = level of nearby glacier not affected by differential ablation. Lateral movement of debris away from the moraine crest increases the area affected by differential ablation

land-based ice at grounding-lines. Grounding-lines are of glaciological importance due to the dramatic changes in basal shear stress, temperature and flow characteristics that take place as ice becomes buoyant (Molnia, 1983; Anderson and Ashley 1991; Powell, 1991) (Chapters 4 and 5).

At the grounding-line there are major changes in debris transport paths. When the base of an ice shelf melts in contact with warmer water, sediment is released at the grounding-line from basal transport zones into the water beneath (Fig. 8.11). Conversely, in some ice shelves the majority of the debris load is entrained by basal adfreezing at the grounding-line. The thermal conditions which govern whether sediment entrainment or release prevails are complex, depending on water and ice temperatures and on salinity (Doake, 1976; Drewry and Cooper, 1981; Matsch and Ojakangas, 1991).

Particle trajectories follow flow lines that reflect the spatial distribution of accumulation and ablation (Fig. 8.12). These are often markedly different from

land-based glaciers. Ice shelves fringing Antarctica receive most accumulation close to their seaward margins, becoming more arid towards their interiors where moisture-bearing winds do not penetrate. The accumulation zone extends over the entire upper surface of most ice shelves, and ablation occurs by a combination of basal melting and/or iceberg calving from the seaward margin. The pattern of ice flow is determined by mass redistribution from the accumulation zone to the ablation zone (just as it is in land-based glaciers); therefore, particle trajectories are downward and seaward in most ice shelves. The Amery Ice Shelf (Fig. 8.12) is an exception to this pattern, having basal thermal and salinity conditions favouring adfreezing with a corresponding convergence of ice flow vectors at the distal margin.

Debris discharges in ice shelves are generally very low compared to land-based glaciers because of the lack of sediment sources in their accumulation zones. However, ice shelves nourished by land-based glaciers may receive glacially-transported debris from adjacent land masses, although basal debris commonly melts out close to the grounding-line. Morainic mounds and stripes occur supraglacially in zones of surface ablation, exceptionally in large quantities such as on McMurdo Ice Shelf, Antarctica, where basal adfreezing also entrains marine sediment (Swithinbank, 1988). Ice shelves formed from perennial sea ice have very low debris concentrations, restricted to aerosols and aeolian particles.

8.4.6. Transport Pathways and Provenance Studies

Understanding the geometry of transport routeways through glaciers has wider applications. Reconstructions of former glaciers and ice sheets commonly use erratic 'trains' and till compositions as indicators linking debris sources with zones or points of sediment deposition, thus delineating former glacier flow lines (Menzies, 1995b, Chapters 2 and 15). Such evidence is useful because flow lines in turn indicate the direction of glacier surface slope, which, especially in ice sheets, does not always mirror subglacial topography.

Glacial deposits are used as a reconnaissance tool for mineral exploration. The composition of terminal moraines of Alaskan valley glaciers has been used to locate mineral deposits in bedrock sources (Stephens *et al.*, 1983; Evenson *et al.*, 1979). Such methods are useful in alpine terrain where access is difficult and large composite valley glaciers concentrate sediment from multiple tributary valleys at a common terminus.

8.4.7. Quantitative Measurements of Debris Discharges

Direct measurement of the quantity of sediment transported by glaciers is hindered by difficulties of access to basal transport zones where most debris is usually carried. For this reason, sediment yields and denudation rates for glacierized catchments are usually estimated from outwash sediment transport rather than from glacier loads (Chapter 7). Most measurements of glacier sediment transport are based on samples from ice exposed around glacier margins, especially at termini where sediment is delivered to

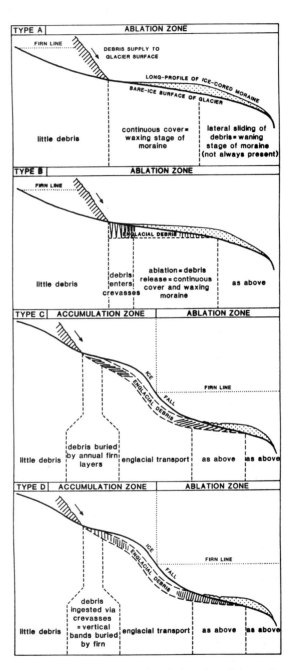

FIG. 8.9. (a) Explanation of varieties of medial moraine found on Glacier de Tsidjiore Neuve, Switzerland (after Small *et al.*, 1979; reproduced by courtesy of the International Glaciological Society from *Journal of Glaciology*, **22**(86), 1979, p. 50, Fig. 4).

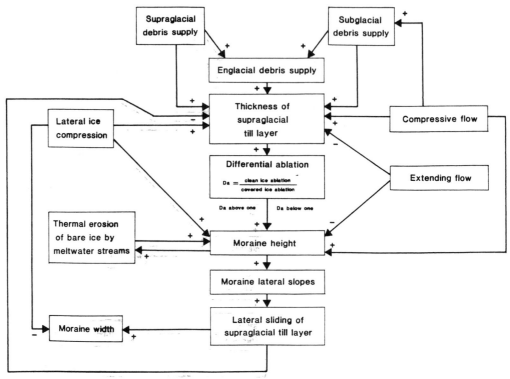

FIG. 8.9. (b) Influences on the downglacier development of medial moraines (after Small and Clark, 1976; reproduced by courtesy of the International Glaciological Society from *Journal of Glaciology*, **17**(75), 1976, p. 163, Fig. 1).

the proglacial depositional system. Measurements from the glacier sole, in more central locations, provide more representative estimates of basal transport. In a few rare cases, natural cavities and artificial tunnels have been used to observe basal debris moving over glacier beds (Boulton *et al.*, 1979; Boulton, 1978, 1979).

Sediment quantity can be expressed in several different ways (Table 8.1) so that direct comparison between different studies can only be made where similar sampling strategies have been performed: unfortunately this is almost never the case. Quantification at a hierarchy of spatial and temporal scales gives a range of information from point samples of sediment concentration to continuity equations of sediment flux through an entire glacier. There is, therefore, no simple statistic to answer the question "how much sediment does a glacier transport?"

Nevertheless, broad generalizations can be drawn from available data (Table 8.2). By far the highest

debris concentrations occur in the basal transport zone. Englacial facies are generally very low in debris content, even in glaciers where supraglacial moraines indicate apparently high debris loads. Sediment

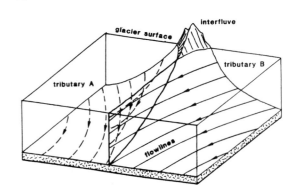

FIG. 8.10. Transfer of debris from the basal to the high-level transport zones along flowlines at the confluence of two glacier tributaries (after Boulton, 1978 and Gomez and Small, 1986). Compare with the distribution of abraded debris along the Tasman-Hochstetter confluence in Fig. 8.7(b).

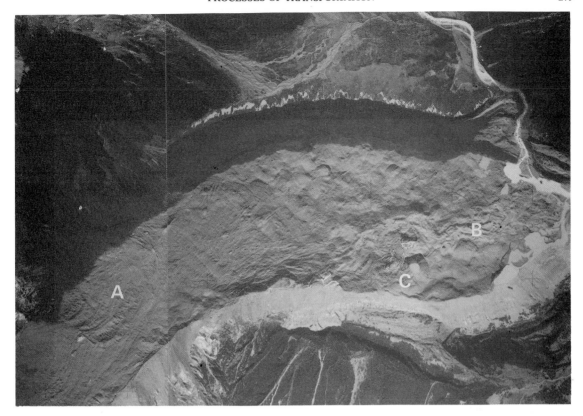

PLATE 8.2. Vertical aerial photograph of Mueller Glacier, New Zealand, showing emergence of debris at the glacier surface in zones of intense compression (A) and sinuous band of debris (B) reworked by an englacial river, now visible in collapse dolines at C and D. The river has never flowed supraglacially (photo courtesy of Lloyd Homer, Institute of Geological and Nuclear Sciences Ltd.).

concentrations vary through several orders of magnitude. Concentrations are lowest in englacial zones, where values as low as 0.0001% by weight have been recorded (Table 8.2). Ice sheets with negligible supraglacial debris supply may have even lower concentrations (Gow et al., 1979). Much higher concentrations are localized in englacial debris septa, especially those nourished by frequent rockfalls onto slow-moving ice in accumulation zones (Gomez and Small, 1986). The majority of the sediment loads of warm-based and polythermal glaciers are carried in the high debris concentrations of the basal transport zone. Concentrations vary greatly even within the basal ice zone; up to ≈80% by volume of the stratified ice facies in basal zones may consist of suspended debris, though substrate erodibility and the erosivity of ice combine to produce great variability (Table 8.2). Estimates of supraglacial and basal discharges

for six Icelandic temperate glaciers vary from 200 to 26,000 $m^3 year^{-1}$ (Eyles, 1979). Alpine outlet glaciers have similar debris loads to piedmont glaciers draining the Vatnajökull Icecap, Iceland, but the latter have more basal debris (up to 93% of total debris load) at their termini.

Sediment discharge is a measure of sediment volume passing through a unit of glacier cross-section in a given time. For valley glaciers in zero net balance (equilibrium), sediment discharges are probably at a maximum at equilibrium lines, where ice velocities are usually greatest. Most sediment discharges have been measured at glacier termini, where debris released from ice by ablation is deposited as diamicton or transported from glaciers by outwash streams. The volume of sediment released at a terminus is related to the ablation rate, ice velocity and sediment concentration (Fig. 8.13). Where a

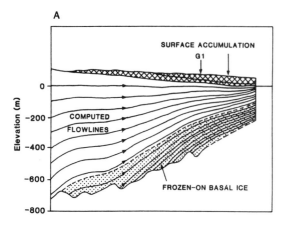

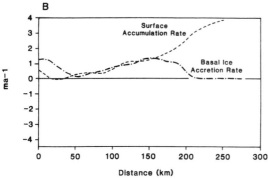

FIG. 8.11. (a) Ice thickness profile along a central flowline of the Amery Ice Shelf with calculated particle paths, zones of freezing-on at the base (hatched) and surface accumulation. (b) Mass balance components for the ice shelf. 1 = surface accumulation rate; 2= basal ice accretion rate (from Budd *et al.*, 1982; reproduced by courtesy of the International Glaciological Society from *Annals of Glaciology*, Vol. 3, 1982, p. 37 , Fig. 2).

glacier terminus becomes stagnant (during negative mass balance or the quiescent phases of surging glaciers) down-valley sediment discharge may approach zero some distance upstream, though debris is released at the terminus as long as ice ablates. Chernova (1981) summarizes the results of extensive surveys of several Asian glaciers, where sediment delivery to termini ranges over three orders of magnitude (Table 8.2).

Sediment discharges fluctuate over time in accord with mass balance changes. Debris tends to accumulate in glacier ablation zones under negative mass balance conditions due to increased compression and ablation. Such 'reservoirs' of debris can be rapidly evacuated to be deposited at glacier termini when mass balance becomes positive. Thus, the amount of sediment in high-level transport varies inversely with glacier mass balance. It is important to note that the short-term sediment discharge of a glacier is only likely to equate to the catchment denudation rate if: (a) the glacier is in equilibrium, which eliminates temporal variability; and (b) discharge is measured through the equilibrium line cross-section, which eliminates spatial variability within the glacier. This theoretical ideal is never achieved in practice and allowances must be made for disequilibrium mass balance and sediment reworking.

Published information on glacier sediment discharges is rather sketchy, so that comparisons between glaciers in different climatic and tectonic settings are difficult to make. Cirque glaciers, having logistical advantages of size and form,

TABLE 8.1. Various measures of the quantity of sediment in transport in glaciers.

Measure	Scale	Typical units
concentration	point sample to transport zone	% by volume; % by weight; g/cm^3; kg/m^3
load	transport zone to whole glacier	tonnes or m^3
discharge	glacier cross-section or debris melt-out at terminus	*gross measures:* $kg\ yr^{-1}$; $t\ yr^{-1}$; $m\ yr^{-1}$ *specific measures:* $kg\ m^{-1}\ yr^{-1}$; $t\ m^{-1}\ yr^{-1}$
sediment yield	whole catchment	tonnes or m^3
denudation rate	whole catchment	$mm\ yr^{-1}$; $kg\ m^{-2}\ yr^{-1}$

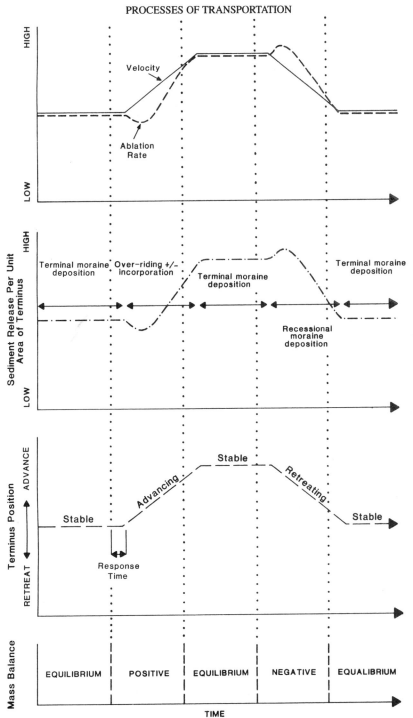

FIG. 8.12. Schematic relationships between mass balance and sediment release from a glacier terminus. As positive mass balance commences, ablation initially decreases as climate cools, then increases as the glacier advances to lower altitude. Similarly, as negative mass balance commences, ablation initially increases as climate warms, but decreases as glacier terminus retreats to higher altitude. Sediment release closely mirrors ablation, but moraine construction and preservation also depends on changes in terminus position.

TABLE 8.2. (A) Measured englacial debris concentrations.

Glacier	Debris concentration % by weight	Source
Ayutor, Tien Shan	0.02-0.33	Glazyrin (1975)
Watts, Baffin Island	0.0001-0.0097 ($\bar{x}$ = 0.0018)	Dowdeswell (1986)
Djankuat, Caucasus	0.12	Bozhinskiy et al (1986)
Tasman, New Zealand	0.028	Kirkbride (1989)
Ivory, New Zealand	0.009	Hicks et al (1990)

TABLE 8.2. (B) Measured basal debris concentrations.

Glacier	Debris concentration % by weight	% by volume	Basal Zone Thickness (m)	Source
E.Antarctic Ice Sheet (margin)	-	0-12	15	Yevteyev (1959)
E.Antarctic Ice Sheet (Byrd Core)	-	7	5	Drewry (1986)
Antarctic Ice Sheet	-	7	4.8	Gow et al (1979)
Greenland Ice Sheet	-	0.1	15.7	Herron and Langway (1979)
Nordenskjöldbreen, Spitsbergen	-	40	0.4	Boulton (1970)
Barnes Ice Cap, Baffin Island	-	8±2	8	Barnett and Holdsworth (1974)
Breiðamerkurjökull, Iceland	-	50	0.15-0.3	Boulton et al (1974)
Breiðamerkurjökull, Iceland	-	8-10	0.05-0.2	Boulton (1979)
Matanuska, Alaska: dispersed facies	-	0.04-8.4	0.2-8	Lawson (1979)
Matanuska, Alaska: stratified facies	-	0.02-74	3-15	Lawson (1979)
Glacier d'Argentiére, France	-	43	0.02-0.04	Boulton et al (1979)
Myrdalsjökull, Iceland	-	15-31	2-5	Humlum (1981)
Bondhusbreen, Norway	1	0.39	5	Hagen et al (1983)
Watts, Baffin Island	32-80	14-57	0.8-2.9	Dowdeswell (1986)

TABLE 8.2. (C) Measured debris discharges.

Glacier	Discharge	Comments	Source
Barnes Ice Cap, Baffin Island	11.5 ± 5 m^3 m^{-1} yr^{-1}	Specific discharge measured at ice cliff terminus sliding at 17myr^{-1}	Barnett and Holdsworth (1974)
Bondhusbreen, Norway	1,500 kg m^{-1} yr^{-1}	Specific discharge measured in subglacial tunnel	Hagen et al (1983)
Ivory Glacier, New Zealand	300 kg m^{-1} yr^{-1}	Specific discharge measured at ice cliff terminus with ice velocity of 30myr^{-1}	Hicks et al (1990)
Hatunraju, Peru	5,080 m^3 yr^{-1}	Gross discharge based on ice core data, measured flow rates	Lliboutry (1986)
Ayutor-3	9.6 x 10^6 kg yr^{-1}	Gross discharges measured over periods of 5 to 82 years. Correspond to erosion rates of 0.7 to 2.9 mm yr^{-1}	Chernova (1981)
Fedchenko	8,300 x 10^6 kg yr^{-1}		
Zarovshanskiy	2,200 x 10^6 kg yr^{-1}		
RGO	1,100 x 10^6 kg yr^{-1}		
IMAT	15.4 x 10^6 kg yr^{-1}		
Karabatkak (all central Asia)	13.6 x 10^6 kg yr^{-1}		

provide useful information on processes and rates of sediment transport in arid and maritime environments. For example, Reheis (1975), using sedimentological observations, estimated that 70% of debris supplied to the terminus of Arapaho Glacier (Colorado, U.S.A.) had at some time been in traction at the glacier sole. From studies of proglacial lake sedimentation, Hicks *et al.* (1990) found that of a total sediment delivery to the terminus of Ivory Glacier (New Zealand) of 840 t year^{-1}, 40% was transported in the basal transport zone, 3% in high level transport and 57% washed through, over or underneath the glacier by avalanches, water and mass movements. This latter figure reflects the potency of extraglacial processes in small, steep catchments in maritime environments.

Possibly the greatest sediment discharges in contemporary ice masses are achieved by glaciers flowing in a so-called 'fast mode', either as surging glaciers or fast ice streams draining ice sheets. Surging glaciers entrain large amounts of basal debris, particularly where they override unconsolidated sediment. Entrainment is favoured by conditions of increased cavitation and meltwater production associated with the very high velocities attained during surges (Clapperton, 1975). Ongoing research into fast ice streams may reveal very high sediment discharges from ice sheets associated with deformable subglacial sediment (Alley *et al.*, 1987a, b).

Quantitative studies of sediment transport have posed fundamental questions about the geomorphic role of glacier entrainment under different thermal regimes. Boulton (1970b, 1971b, 1972a) suggests that polythermal glaciers in general have higher sediment loads than warm-based glaciers because of the erosion and entrainment made possible by their

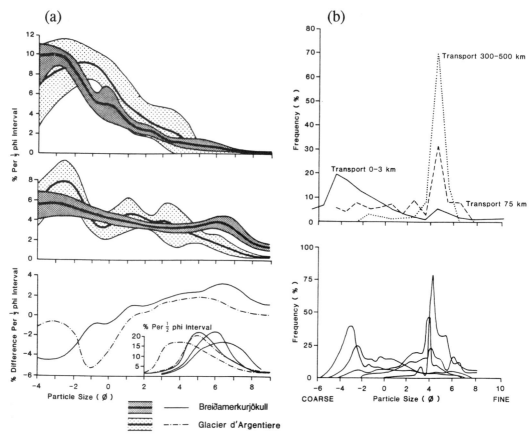

FIG. 8.13. (a) Particle size distributions from high-level transport zones (top) and basal transport zone (middle) of Glacier d'Argentiére (France) and Breiðamerkurjökull (Iceland). Bottom graph shows the difference between the curves of each glacier, indicating a deficiency in fines less than 1 phi for debris in high-level transport. Inset shows particle-size distributions of rock flour sampled at the contacts between clasts in traction and bedrock, and indicates the source of much of the fines in the basal traction zone (from Boulton, 1978; reproduced by permission of the International Association of Sedimentologists). (b) Top: Particle-size distributions of dolomite debris in tills in the Hamilton–Niagara region of southern Ontario, Canada. Comminution over increasing transport distance causes an increase in the 4 to 6 θ mode of the distribution (from Dreimanis and Vagners, 1971; reproduced by permission of Ohio State University Press). Bottom: Typical particle size distributions from basal traction debris of Matanuska Glacier (Alaska), showing the variability between samples (Lawson, 1981a; reproduced by permission of the author).

mixed thermal regime. Conversely, Andrews (1971, 1972a, b) and Anderson (1978) believe that warm-based glaciers produce between two and ten times more debris than cold-based and polythermal glaciers, based on a comparison of field studies in Colorado and Baffin Island. Aside from the questions of comparability between studies, this debate also raises questions about the distinction between erosion and entrainment capability. Warm-based glaciers appear to lose much of their potential debris load to flushing from the glacier sole by meltwater and

derived rainwater, giving the appearance of being relatively free of debris even though they are highly erosive. The argument must remain unresolved nevertheless, in the absence of sufficient direct measurements. Further important questions arise from the apparent inability of quantitative data to support the commonly-held view that glacial erosion is more rapid than non-glacial erosion (Hicks *et al.*, 1990). This debate may act as a focus for future research which may overcome the serious lack of quantitative estimates of glacial sediment transport.

8.5. MODIFICATION OF SEDIMENT DURING TRANSPORT

Sedimentological modification during glacial transport gives clues to the nature and intensity of the stresses acting on the debris load during glacier movement (Chapter 15; Menzies, 1995b, Chapter 13). Strong relationships exist between both these indicators and the trajectories of debris through glaciers. Both indicators are often employed to interpret the transport 'history' of glacial sediments in terms of comminution processes and transport distance.

8.5.1. Particle-size Distributions

Debris carried in the high-level transport zone is not usually modified significantly during transport. This is because englacial debris generally occurs at low concentrations, reducing the likelihood of inter-particle collisions. Local shear stresses at high levels within glaciers are usually very low, so that even if two particles come into contact with each other, their resistances to wear far exceed the forces potentially causing abrasion or crushing. Only occasionally does significant wear occur. Hambrey (1976) found that in a medial moraine across which significant shear occurred between ice streams, quartz and garnet grains were well preserved while softer micas were ground to powder.

Debris brought close to the glacier sole by downward ice flow and basal melting is concentrated in suspension and may even impinge upon the substrate. Here, traction occurs. Ice deformation around bed obstacles and repeated episodes of pressure melting and regelation cause numerous collisions between particles carried in basal ice (Kamb and La Chappelle, 1964; Boulton, 1978) (Chapter 15; Menzies, 1995b, Chapter 13). Particle shape is rapidly modified in the traction zone, and similar grain size distributions are created regardless of how the debris arrived at the glacier sole.

Traction-zone debris shows a high degree of granulometric variation. This reflects both the localized intensity of abrasion and crushing and the differing resistances firstly of mineral grains compared to clastic fragments and secondly of weak lithologies compared to strong. Abrasion of particles produces distributions showing a characteristic negative skew reflecting the production of fine sand to clay sizes from larger clasts. Most grain-size distributions are polymodal, though bimodal grain-size distributions have been reported (Fig. 8.14) (Dreimanis and Vagners, 1971; Boulton 1978; Gibbard, 1980; Dreimanis, 1982, 1988; Brodzikowski and van Loon, 1991). Boulton (1978) sampled debris directly at the soles of three glaciers and found that crushing produces rock fragments generally coarser than 1ø (0.5 mm), while abrasion produces material finer than 1ø. When the percentages of samples above and below the 1ø cut-off are compared, the debris modified in the traction zone is clearly discriminated. Dreimanis and Vagners (1971) conclude that the coarse mode of a bimodal distribution dominates close to the sediment source, but is progressively replaced by a fine mode with distance along a transport path as rock fragments are progressively comminuted.

Many workers recognize the concept of terminal grade of glacially-transported debris (Dreimanis and Vagners, 1971), referring to a critical particle size at which tractive forces are equalled by resistance to further wear. The concept is important because if comminution causes basal debris to attain terminal grade, particle-size distributions may eventually become independent of transport distance. However, Haldorsen (1981) and Drewry (1986) doubt the reality of terminal grades. Haldorsen indicates that for a particular mineral variable mineralogy and fracture geometry (among other factors) negate such a concept as a single terminal grade. Drewry cites experiments which suggest that further comminution would occur given sufficient time and transport distance. He concludes that bimodal particle-size distributions represent only a transient stage of the comminution process. Whether glaciers achieve such 'tertiary' comminution is uncertain, therefore it may be useful to retain the term. Both theoretical and experimental grinding studies suggest that coarse clastic fragments are crushed much more easily than fine ones. Below about 1mm grain size, the energy required to comminute debris increases exponentially with decreasing particle size and grinding of quartz grains ceases at ≈1 µm (Drewry, 1986). Terminal

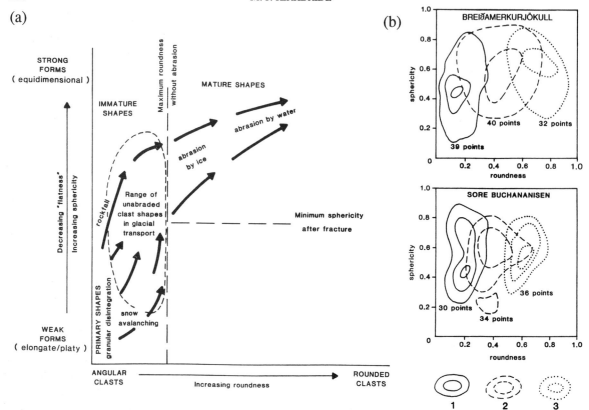

FIG. 8.14. (a) Evolution of clast shape during glacial transport, based on measurements of 88 samples of 50 clasts each on Tasman and Mueller Glacier, New Zealand. There is no time axis, because some clasts pass rapidly through the system, others never attain mature shapes depending on their exposure to high stresses in the basal transport zone or in englacial/subglacial rivers. (b) Krumbein form-roundness plots of clasts from high-level and basal transport zones of glaciers in Iceland (top) and Svalbard (bottom), showing the effect of basal traction and abrasion on particle shape. 1 = Englacial debris; 2 = basal debris; 3 = lodgement till (after Boulton, 1978; reproduced by permission of the International Association of Sedimentologists).

grades are determined as much by the lithology of source rocks, particularly the density of cracks and microcracks in mineral grains, as by the distance over which debris has been transported (Vorren, 1977; Dreimanis and Vagners, 1972; Haldorsen, 1981). Tills derived from crystalline rocks tend to be dominated by sand and coarse silt due to mechanically-resistant quartz and (to a lesser extent) feldspar in these sizes. Tills derived from fine-grained rocks tend to contain a higher proportion of silt and clay (Vorren, 1977; Aylsworth and Shilts, 1989). Terminal grades appear to be attained after fairly short transport distances: <1.6 km in New Hampshire (Drake, 1972) and only 0.5 km in Iceland (Humlum, 1985).

The terminal grade of grains of rock is determined by micro-scale processes of comminution as well as by parent lithology. Terms in common usage among glacial geologists are 'crushing' and 'grinding', but these terms do not correspond directly to the nomenclature of recognized processes of wear and redefinition for the purpose of clarification has been necessary (Sharp and Gomez, 1986). Sharp and Gomez relate crushing to the fo—

grain surf— —

to fatigue fracture in—

surfaces. Grinding, on the oth—

abrasive wear, whereby a harder pr—

groove in a softer surface by plastic —

(Chapter 7).

Th—

reproduced the granulometric characteristic of modern tills, where terminal grades reflect the comminution of clastic fragments to parent mineral grain sizes (Haldorsen, 1981, 1983). Furthermore, sand grains appear to result from the brittle fracture of gravel clasts and silt-sized grains from abrasive wear of cobbles and boulders: the granulometry of tills can be seen to be very strongly influenced by the physical processes operating within glacial transport zones. It is clearly important to understand these processes in order to explain the sedimentology of the resulting till deposits.

8.5.2. Particle Shapes

A distinction comparable to that between particle sizes in the traction zone and in high-level transport zones is revealed by the shapes of coarse clasts. Shape comprises elements of roundness (of particle faces and edges) and of form (the relative dimensions of the three principle orthogonal axes). Debris entrained by rockfall and avalanching onto glacier surfaces, and by plucking from a bedrock substrate, is typically angular (Ogilvie, 1904; Sharp, 1949; Boulton 1978; Serebryanny and Orlov, 1982; Matthews, 1987). Tills deposited at great distances from points of entrainment often have a large proportion of angular clasts. Frequent clast collisions occur in the zone of basal traction and during englacial water transport which, though very different in terms of process, produce rather similar rounded clast forms.

The trajectories of clasts in the traction zone are complex, being influenced by regional ice flow vectors, topographic irregularities on the substrate, local zones of regelation, deflection by neighbouring particles and even by the form of individual clasts. Modifications to form and roundness are correspondingly complex because clasts interact with the glacier bed and with each other.

The velocity of clasts held in suspension in sliding ice close to the glacier sole varies according to clast size, form and frictional drag. Thus, some clasts change their positions relative to surrounding ice by retardation and rotation. This is especially so when clasts at the glacier sole come into contact with the substrate. Tractional movement across the substrate is most effective on clasts in the size range -3 to -7ϕ, which move at 80–100% of the basal ice velocity. Clasts of initially equant form are more prone to being rotated, or 'rolled along' at the glacier sole with the result that edges are chipped and abraded and both roundness and, to a lesser extent, sphericity are increased. Debris from both supraglacial and subglacial sources assume similar shape properties in traction, indicating that the nature of clast contacts is the primary determinant of shape, rather than debris sources. Blade- or plate-shaped clasts ('flatirons') are common forms produced in traction. These shapes probably develop from initially platy joint-bounded fragments in much the same way that spheroidal clasts are produced where primary joint spacings are equant (Holmes, 1941, 1960). Smooth surfaces are typical of all clasts from the traction zone where sustained stresses and an abundance of wet silt and fine sand act as efficient abrasives.

In deformable bed traction zones, a complex process of interaction between advecting particles, retarded immobile substrate and superjacent mobile ice combine to produce a highly variable wear environment where many sub-wear processes act over a range of distances and areas. Comminuted sediment and individual particle shapes are produced that have characteristics similar to those found in a range of other glacial sub-environments.

A second sub-environment where boulder shapes are efficiently altered is the englacial fluvial environment. Supraglacial streams and englacial conduits commonly entrain debris from the ice walls of channels. This debris rapidly becomes rounded and platy clasts readily fracture into stronger equant forms. Deposits of water-transported debris both on and within glaciers therefore tend to contain clasts superficially resembling traction-zone debris, but with a conspicuous absence of striations and of silt and clay. Roundness and sphericity are slightly more highly developed than in traction-zone debris.

8.5.3. A Cascading System of Particle Shape Evolution

The variety of boulder shapes within a glacier reflects the persistence of some shapes over long transport distances and time periods (for example,

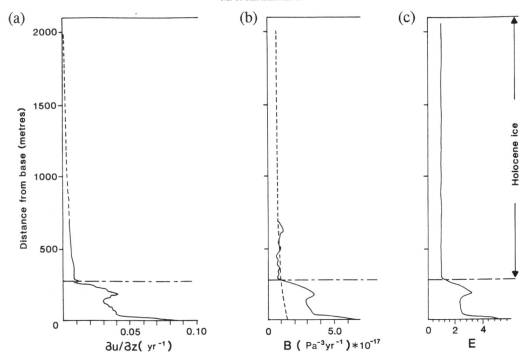

FIG. 8.15. Rheological changes with depth in a large ice sheet, measured from tilt experiments in the Dye 3 borehole, Greenland. (a) Variation of ice velocity with depth; (b) Variation of flow-law parameter B with depth; (c) Variation of flow enhancement factor E with depth. Values of all parameters increase abruptly at the Holocene–Pleistocene transition, indicating more deformable ice at depth due mostly to higher microparticle concentrations (from Dahl-Jensen, 1985; reproduced by courtesy of the International Glaciological Society from *Journal of Glaciology*, **31**(108), 1985, p. 94, Fig. 3).

unmodified medial moraine debris travelling hundreds of kilometres), while at the same time rapid and dramatic changes to boulder shapes occur almost ~ously in some parts of the transport system.
~elling between their points of
~nt of deposition follow an almost
~ort routeways, along which
~stically altered or may

                .is complexity by
           .lder shapes, rather
         .ce, as a frame of
        implies the degree to
      , are modified to rounded,
     .pes are elongate and platy
    .ractured on planes parallel to
  .nediate axes. Clasts tend to be
   , are more equant in form (i.e. their
  .xes are similar in length). Breakage

of weak clasts tends to create smaller but mechanically stronger clasts. Measurements of form and roundness of 4, 400 clasts on and around Tasman Glacier (a warm-based valley glacier in New Zealand), when plotted in a form/roundness field, reveal several groups of boulder shapes characteristic of different stages in the transport system (Fig. 8.14). Any clast may pass from one group to another either directly or via other groups depending on the combination of transport routeways followed. However, there is an overall trend from immature shapes towards the lower left corner of the graph to mature shapes towards the upper right corner. Of course, much debris is deposited before it reaches maturity, and the time period between entrainment and deposition varies greatly both within and between glaciers. It is noticeable that most subaerial processes do not greatly modify clast roundness but effectively alter weak forms into strong forms by fracture. Thus, immature debris appears more susceptible to changes in form than in roundness. Having achieved relatively equidimensional forms, subsequent shape changes in mature debris involve increasing roundness and water-transport appears more effective than subglacial traction in increasing clast roundness. The model illustrates the sedimentological distinction between the high-level and basal traction zones as well as demonstrating the complexity of shape evolution throughout the whole glacier transport system. Variations in both intensity and spatial distribution of shape-modifying processes operating during glacial transport are responsible for this complexity.

Glacial particle shape evolution demonstrates the enormous range of wear processes that impinge on any single particle over both temporal and spatial ranges and the common tendency for particle shapes from many sources and wear process combinations to exhibit equifinal shape geometries. Great caution must therefore be used in discriminating between different glacial environments solely upon particle shape and roundness.

8.6. GLACIOLOGICAL EFFECTS OF DEBRIS IN TRANSPORT

8.6.1. Rheological Effects of Englacial Debris

In recent years it has become increasingly apparent that debris is not transported entirely passively by glacier ice. Rather, particles in transport significantly alter the deformational behaviour (rheology) of ice both directly (through the inclusion of undeformable rock and mineral grains within ice) and indirectly (through affecting ice crystal growth). Glacier ice deforms along glide planes parallel to the basal crystallographic plane. The presence of undeformable sediment within an aggregate of ice crystals has been shown experimentally to reduce the creep rate of ice even at low sediment concentrations (Hooke et al., 1972; Goughnour and Andersland, 1968; Alkire and Andersland, 1973; Nickling and Bennett, 1984). In heavily sediment-charged ice, friction caused by interparticle interactions stiffens ice. In marked contrast, borehole closure measurements from ice sheets show that sediment-

charged ice deforms more easily than 'clean ice'. Experiments designed to test the influence of englacial sediment on ice rheology have given quite different results from borehole observations and have led to debate over the role and comparability of such experiments. Nickling and Bennett (1984) found that ice attained a peak strength at a sediment concentration of 25% by volume, with ice becoming more deformable at higher concentrations. They found no simple inverse relationship between debris content and creep rate. Englacial sediment indirectly affects ice rheology through its influence on ice crystal growth. Basal ice sequences usually have small crystals where sediment concentration is high. Experiments have shown that the presence of sediment actually inhibits the growth of ice crystals during recrystallisation, resulting in ice with a texture (and therefore a rheology) partly influenced by sediment content (Baker, 1978).

Understanding the rheology of debris-laden ice is important in deciphering the long-term flow of large ice sheets using deep ice cores, and therefore for establishing core chronologies for environmental reconstruction (Chapter 2). Boreholes from Greenland, Ellesmere Island, Devon Island and Antarctica have revealed a contrast in the rheologies of Holocene ice and deeper Pleistocene ice (Dahl-Jensen, 1985; Fisher and Koerner, 1986; Van der Veen and Whillans, 1990). Late Pleistocene ice has a higher concentration of fine wind-borne sediment and deforms about three times more readily under stress than 'clean' Holocene ice, a factor that must be taken into account in numerical modelling of ice sheet flow (Fig. 8.15). The reason for the different sediment concentrations is probably that low Pleistocene sea levels exposed continental shelves to subaerial wind erosion, supplying aeolian dust to the ice sheets.

8.6.2. The Effects of Supraglacial Debris on Ablation

The presence of supraglacial debris affects the transfer of heat between the atmosphere and ice surface, consequently affecting the rate of ice melt. Ablation may be increased or decreased depending on the texture, thickness and composition of supraglacial

TABLE 8.3. Estimates of the Insulating Effect of Supraglacial Debris

Location	Debris thickness (cm)	% reduction in melting of bare ice	Source
Isfallsglaciaren, Sweden	5 15 30	40 70 80	Østrem (1959)
	40	85	Østrem (1965)
Kaskawalsh Glacier, Yukon	5	42	Loomis (1970)
	10	60	
	20	82	
	30	94	
Ayutor-2, Tien Shan	4	8	Glazyrin (1975)
	8	15	
	12	50	
Sherman Glacier, Alaska	100	80	McSaveney (1975)
Djankuat Glacier, Caucasus	8	0	Bozhinskiy et al (1986)
	12	10	
Glacier de Tsidjiore Neuve, Switzerland	2-3 4	20 55	Small (1987)
Tasman Glacier, New Zealand	25	60	Kirkbride (1989)
	50	78	
	100	89	
	200	96	

debris. Small-scale morphological features commonly develop where differential ablation is caused by debris, including dirt cones and boulder tables where ablation is hindered, cryoconites and penitentes where it is enhanced (Sharp, 1949; Drewry, 1972; Gribbon, 1979).

Debris affects melting by altering the amount and rate of heat transfer by conduction and advection. Three main controls on conductive heat transfer are: (1) the thickness of the debris layer; (2) the thermal conductivity of the debris, which varies with mineralogy, porosity and moisture content; and (3)

the thermal gradient through the debris layer, which varies both daily and seasonally according to air temperature cycles.

Heat is also transferred by percolating rain- and melt-water and by air circulating within the debris, all parameters that need to be measured or estimated in order to calculate sub-debris ablation. Due to such difficulties few studies of this type have been made, yet results indicate that most melting results from downward conductive heat transfer (especially in arid environments) and that debris is a very effective insulator of ice (Østrem, 1959; Kraus, 1966;

Bozhinskiiy *et al.*, 1986). In cold environments with short ablation seasons, thin debris layers a few centimetres thick may prevent ablation from occurring, allowing long-term preservation of ice-cored landforms (Carrara, 1975; Driscoll, 1980). In warmer environments, much thicker layers are required to provide an equivalent insulating effect (Table 8.3). Recently, it has become possible to estimate sub-debris ablation indirectly from meteorological data (Nakawo and Young, 1981, 1982; Nakawo and Takahashi, 1983).

8.6.3. Supraglacial Debris and Glacier Dynamics

Many glaciers in mountainous regions such as in Alaska, the Himalaya, Karakoram, Andes and the Southern Alps of New Zealand receive so much debris into high-level transport that their ablation zones are characteristically debris-mantled. The dynamics of glaciers and their sensitivity to climatic change can be greatly altered by an extensive covering of supraglacial debris.

A fundamental effect is that debris-mantled glaciers tend to be longer than their uncovered counterparts. Such 'over-lengthening' is a consequence of reduced ablation rates being compensated by increased ablation areas to maintain equilibrium mass balance. As a result, the accumulation area ratios (AAR) of such glaciers are typically only 0.3 to 0.5, compared to an AAR of 0.5 to 0.8 for unmantled glaciers (Kirkbride, 1989).

Debris-mantled glaciers also respond differently to climate change, both during advances and retreats. Where ablation is reduced beneath debris, glacier advances tend to be more rapid and sustained (Rogerson, 1985). Furthermore, large rockfalls onto glaciers can initiate advances by reducing ablation independently of climate change. Thus, the Brenva Glacier (Italy) advanced for 20 years after a 1920 rockfall while neighbouring glaciers retreated (Porter and Orombelli, 1981). During negative mass balance, the amount of supraglacial debris tends to increase as flow velocities reduce and ablation zones become increasingly protected from melting. The terminus of the heavily-mantled (but non-surging) Tasman Glacier (New Zealand) has not retreated from its late-nineteenth century maximum position even though nearby glaciers have retreated by as much as several kilometres over the same period. Similarly, surging glaciers become heavily mantled during their quiescent phases as their abundant debris loads melt out at the surface to produce characteristic landform assemblages (Watson 1980; Sharp 1985a, b). Such protected, stagnant tongues have lead to the misleading impression that all debris-covered glaciers are necessarily stagnant: in fact, many simply have very high rates of debris supply.

The effect of supraglacial debris loads on multiple glacier fluctuations deserves further research. Revealingly, Kick (1989) suggests that the difference in timing of the last Holocene advances between Europe and the Himalaya–Karakoram may reflect the complicating effect of debris mantles. Kick observes that debris-covered glaciers in Asia retreated from their Little Ice Age maximum positions several decades later and with more variability in timing than their European counterparts (Chapter 2).

8.7. SUMMARY

The glacial transport system has two main elements: first, the entrainment of debris by the glacier sole and surface; and secondly, the differentiation of entrained debris into distinct transport zones within moving ice. The zones of basal transport and high-level transport are differentiated on the basis of ice facies and the nature and distribution of sediment within the ice, which together reflect differences in entrainment processes between the two zones.

Ice sheets, valley glaciers and floating ice shelves each have characteristic transport paths determined by basal thermal regime and ice flow trajectories. However, the natural variability of glacier size and shape, extraglacial climate, underlying topography and sediment sources combine to produce huge spatial variation in the amount, type, and trajectories of sediment in transport. Quantification of glacial sediment fluxes indicate that warm-based and polythermal glaciers can both have large sediment discharges while cold glaciers have relatively very small sediment discharges. However, comparability between different studies is poor and systematic studies of glacial sediment budgets are needed.

Glacial sediment transport is no longer regarded as a passive phenomenon. Research has shown how supraglacial debris may alter the response of glaciers to climatic variation, how fine englacial sediment increases deformation rates in ice sheets, and how saturated subglacial sediment contributes to fast ice velocities through deforming under high strain rates. Such interactions between debris in transport and glacier ice now provide a basis for understanding glacial transport processes.

Chapter 9

PROCESSES OF TERRESTRIAL DEPOSITION

C. A. Whiteman

9.1. INTRODUCTION

The study of glacial sediments and processes of glacial deposition has generated an extensive literature (cf. Goldthwait and Matsch, 1988). One of the results of this keen interest is an increasingly complex genetic classification of glacial and glaciofluvial sediments. There have been intense discussions, not only concerning the precise nature of glaciogenic depositional processes but even about what constitutes a glaciogenic sediment.

One very narrow interpretation of this debate classifies only so-called primary sedimentation by lodgement and melt-out (Lawson, 1979a). In this classification secondary processes such as flowage and deposition through water (glaciolacustrine and glaciomarine) are excluded on the grounds that particle disaggregation during flowage and under the influence of gravity removes direct evidence of glaciogenic stress (Boulton, 1980). In contrast, a more catholic view might include glacioaeolian and glaciofluvial sediments in addition to those mentioned above.

In this chapter a broader interpretation will operate: processes traditionally referred to as lodgement, melt-out and flowage, will be addressed, together with a brief discussion of glacial deformation and subglacial fluvial processes. Glaciolacustrine, glaciomarine, and glacioaeolian processes are excluded being dealt with elsewhere

(Chapters 12, 13; Menzies, 1995b, Chapters 4, 5 and 6).

This chapter is divided into three main sections dealing with (a) factors which influence glacial deposition processes, (b) mechanisms of glaciogenic deposition, and (c) examples of problems in determining depositional processes based upon sedimentary characteristics.

9.2. FACTORS CONTRIBUTING TO TERRESTRIAL GLACIOGENIC DEPOSITION

In essence, glaciogenic diamictons are deposited (Fig. 9.1) when either or both of the following conditions are fulfilled: (a) englacial debris is released from active or stagnant glacier ice in a subglacial or a supraglacial position; and/or (b) debris in motion on, or beneath, glacier ice stops moving at a location beyond the direct influence of the glacier either in a subglacial or a proglacial position. A third category of deposition involves sorted sediments (glaciofluvial) deposited in tunnels, but under strong glacial influence.

Englacial debris is released when ice, the encasing medium, melts. This process is conditioned by the thermal regime of the glacier and its ambient environment. This thermal regime is the sum of many heat sources, including atmospheric components (air, precipitation), meltwater, the geoid, friction generated by deforming ice and glacier sliding,

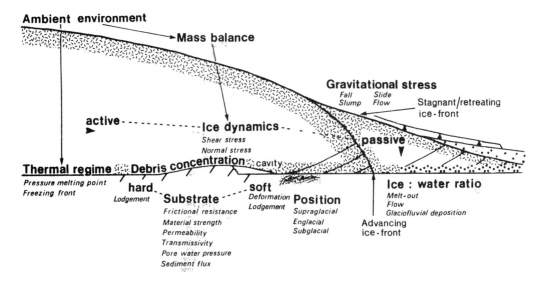

FIG. 9.1. Major factors, associated variables and processes responsible for terrestrial glaciogenic deposition.

normal pressure generated by the overburden of ice and debris, and the ice itself (cf. Paul and Eyles, 1990). A reduction in any of these heat sources may enable a freezing front to advance from the glacier into previously released debris (Mackay and Mathews, 1960; Menzies, 1981; French and Gozdzik, 1988) and return it to an englacial position, unless the porewater pressure within the deposited diamicton is sufficient to inhibit the advance of the freezing front.

Debris which is released by stagnant ice subglacially or supraglacially onto a horizontal ice surface which is gradually lowered to the ground, will remain in a stationary position and can be deemed to have been deposited. However, conversely, on release, debris may remain in motion, either beneath the sole of the glacier, or flowing across its surface, if the shear stress to which it is subjected is greater than its material strength. This condition applies whether the debris is responding to either glacier shear stress at the base of the moving glacier, or to gravitational stress on the glacier surface or within subglacial cavities.

Frictional resistance between a clast and the glacier bed is dependent upon the area of contact between them, and is proportional to the normal load (Drewry, 1986; Alley, 1989a). Movement of a particle will cease when the load exerted on the bed reaches a

critical threshold (cf. Chapter 5; Menzies, 1995b, Chapter 2). In a debate about the lodgement of till (summarized in Drewry, 1986) Boulton (1975) equated normal load to the product of clast weight and effective normal pressure (overburden pressure less porewater pressure): ice velocity was a secondary component contributing to total glacier shear stress. In contrast, Hallet (1979, 1981) considered normal load to comprise the weight of the clast together with the velocity with which it approaches the bed as basal ice melts. In Hallet's view, normal load is independent of cryostatic pressure and the downward and lateral velocity are the key components of glacial stress.

Material strength, expressed in terms of Coulomb's equation (Chapter 5), is related to the cohesion of the sediment, its angle of internal friction and its effective normal pressure. In the glacial environment the most important components are porewater pressure and material characteristics of size and sorting. These are dependent on factors such as the rate of supply of water, either from melting ice or from external hydraulic sources, the thickness and texture of the sediment, the ease with which it drains, and the lithological characteristics of the available material. These in turn are related to and determine the permeability of the substrate, either ice in the case

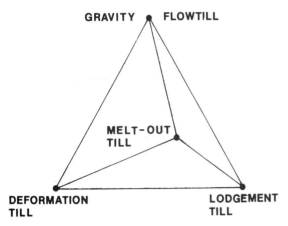

FIG. 9.2. The conceptualisation of till deposition as four principal endmembers of a tetrahedron (after Dreimanis, 1988; reprinted from Goldthwait, R.P. and Matsch, C.L. (eds), *Genetic Classification of Glacigenic Deposits*, with permission of A.A. Balkema, Rotterdam).

of supraglacial debris, or subjacent debris or bedrock in the subglacial context.

9.3. MECHANISMS OF TILL DEPOSITION

The preceding outline of factors contributing to processes of glacial deposition indicates the complexity of the glacial environment. The direct and

indirect influence of ice in the debris-release mechanism has been discussed at great length. Similarly, the roles of active and passive ice and the position of debris release have generated much debate. The outcome of these protracted discussions (Goldthwait and Matsch, 1988) is a series of genetic classifications of till. These show considerable variety in detail but most involve some or all of the following types of till: deformation, lodgement, subglacial (confined) melt-out/sublimation, supraglacial (unconfined) melt-out and flow till. The genetic interrelationships of these four basic processes have been conceptualised by Dreimanis (1988) and later Hicock (1990) as end members of a complex 'till prism' (Fig. 9.2); a view of the spatial and temporal relationships is contained in Fig. 9.3.

The processes of supraglacial melt-out and flow have been described in greatest detail (Lawson, 1979a, b, 1981a, b) possibly because, as surface phenomena, they are accessible and open to close examination. In contrast, deformation, lodgement and subglacial melt-out processes are more difficult to investigate directly, since they occur in largely inaccessible locations and are not easy to simulate because of problems of scale and time.

Debris flows under the influence of gravity when its constituent particles lose cohesion, usually due to

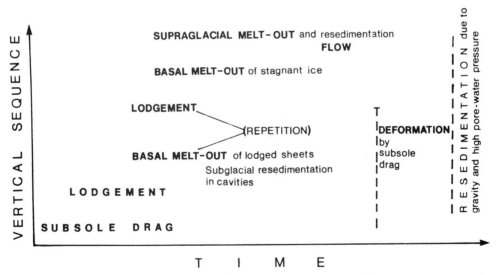

FIG. 9.3. A view of the spatial and temporal relationships between till types (after Dreimanis, 1988; reprinted from Goldthwait, R.P. and Matsch, C.L. (eds), *Genetic Classification of Glacigenic Deposits*, with permission of A.A. Balkema, Rotterdam).

PLATE 9.1. Overturned deposits (mottled Valley Farm Paleosol and sand wedges), the sheared zone and 'homogenized' zone of Hart and Boulton (1991) at Great Blakenham, Gipping Valley, Suffolk, U.K.

PLATE 9.2. Middle Pleistocene sediments at Newney Green near Chelmsford, Essex, U.K., illustrating Banham's model of glaciotectonic deformation (see Fig. 9.4).

PLATE 9.3. Vertical exposure of lodgement till, Geilston, Scotland.

PLATE 9.4. Horizontal exposure of lodgement till, Fjallsjökull foreland, Iceland. Note 'bow wave' in sediment in front of the nearest block. Scale bar is 1 m in length.

high porewater pressure. Debris melts out when ice undergoes a phase change from solid to liquid (or, in the case of sublimation, from the solid to gaseous phase). Debris lodges when resistance to movement exceeds the shear stress imparted by the glacier or gravitational force. Movement may be inhibited by adhesion between particles and a hard substrate, by the increasing resistance met by a particle ploughing into a soft bed, by localized freezing and by the gradual attenuation of stress with increasing depth in a soft, mobile bed material. Therefore, the fundamental mechanisms involved in glacial deposition are flow, melt (and sublimation) and frictional resistance, acting either separately or in concert.

9.3.1. Deformation

Deformation structures are widespread in soft, unlithified sediments (cf. Dreimanis and Hicock, 1993). Boulton and Jones (1979) demonstrated experimentally the deformation of subglacial debris beneath Breiðamerkurjökull in Iceland showing that the amount of strain increased upwards towards the base of the glacier. Earlier, Banham (1977) (Fig. 9.4)

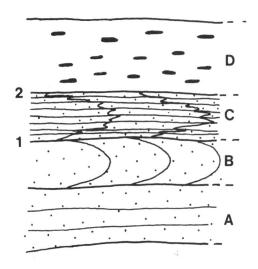

FIG. 9.4. Model of glaciotectonic deformation. (a) Undeformed bedrock; (b) weakly deformed bedrock; (c) penetratively deformed bedrock; (d) lodgement till; (1) principal structural boundary; (2) principal lithologic boundary. (After Banham, 1977; reproduced with permission of *Scandinavian University Press*.)

had proposed a model of glaciotectonic deformation using hard rock metamorphic structures as an analogue. In this model also, strain increases upwards. Banham distinguished a principal structural boundary at the base of a 'penetratively cleaved' zone in the sediments. This boundary marks the base of the zone which clearly shows the imprint of glacier stress (Chapter 10).

Boulton (1987a, Fig. 7) proposed a similar model of subglacial deformation showing zones of stability (B_2) and increasing strain (B_1 and A), below and above a plane of décollement, respectively. In zone A deformation may be so intense that the sediment becomes homogenised (Hart and Boulton, 1991) and massive in appearance, a characteristic which is often said to typify lodgement till (Menzies, 1989a). This concept of pervasive shear or homogenization has not been accepted uncritically in its application to tills of the Laurentide Ice Sheet (Clayton *et al.*, 1989) in North America and the Lowestoft Till Ice Sheet in eastern England (see below). If, however, the process of homogenization as described by Hart and Boulton (1991) does occur, then tills which have traditionally been described as lodgement tills on account of their massive appearance, overconsolidation and strong clast preferred orientations, could equally be classified as deformation tills. For this reason, Hart and Boulton (1991, p. 34) advocate the abandonment of the term non-deformed lodgement till as they envisage "most (if not all) tills have at some time been subjected to some degree of soft bed depositional processes and thus implicitly some deformation".

Less controversial, however, is the idea of deposition beneath the deforming layer. As the body of subglacial debris thickens, the strength of this substrate increases, or distance from the glacier margin changes (Boulton and Hindmarsh, 1987), strain in the lower part of the debris is attenuated until frictional resistance exceeds shear stress and the material becomes stationary below a plane of décollement (Alley *et al.*, 1986; Boulton, 1987; Boulton and Hindmarsh, 1987; Menzies, 1989a; Hart *et al.*, 1990) and can be considered to have become lodged.

9.3.2. Lodgement

In contrast to the process described above, the traditional view of the lodgement process involves the 'plastering on' of glacial debris from the sliding base of a moving glacier by pressure melting and/or other mechanical processes (Shaw, 1985; Dreimanis, 1988). Hart and Boulton (1991) have recently discussed both of these processes, viewing them as a continuum. They differentiate subglacial deposition according to the nature of the substrate, recognizing rigid and soft (i.e. deformable) beds (Fig. 9.5) (cf. Shoemaker, 1986a; Menzies, 1989a; Alley, 1989a, b). For rigid beds, lodgement is a frictional process accompanied by pressure melting during which the debris is 'plastered' against the substrate (Boulton, 1974; Hallet, 1979), while under deformable bed conditions, Hart and Boulton (1991, p. 345) suggest that "purely frictional lodgement cannot occur as there is nothing rigid for the clasts to lodge against". However, this latter statement implies that the relative velocity of ice and the deforming substrate is equal at their interface. If the velocity of the ice exceeds that of the debris beneath, some frictional resistance will be exerted by the debris which may induce lodgement in the traditional sense and allied streamlining (Menzies, 1989a).

Hart and Boulton (1991) envisage an intermediate stage between pure lodgement with no deformation and deformation showing evidence of intense strain. This process involves a large basal clast 'ploughing' into subglacial sediment and forming a bow wave of sediment which retards and eventually halts its progress (e.g. Sharp, 1982a; Alley, 1989a, b) (Chapter 5). (Whether deforming subglacial debris is ever sufficiently weak to allow large clasts to sink into it like gravitational drop-stones (Hart and Boulton (1991) is more difficult to envisage in view of the effective normal stress expected in the subglacial environment.)

9.3.3. Melt-Out

Till is also deposited by melt-out processes (Boulton 1970a; Lawson, 1979a; Shaw, 1979, 1985; Haldorsen and Shaw, 1982). Boulton (1970a) proposed the term 'melt-out till', although, as Dreimanis (1988) points out, the concept is not new (e.g. Goodchild, 1875; Carruthers, 1953). Melt-out is the slow release of debris from the lower or upper surface of stagnant ice, referred to respectively as subglacial and supraglacial melt-out. Sublimation is an associated mechanism involving the solid to gaseous phase change rather than the solid to liquid phase change as in melt-out.

During periods of marked glacier retreat melt-out may be the most productive till-forming mechanism, especially in marginal situations where warm ambient air can penetrate the glacier system, although thinning of the glacier in its marginal zone may counteract this by reducing overburden pressure. Melting may also operate preferentially over crustal hot spots where geothermal heat flux is higher than normal. Melting will also be influenced by the thermal conductivity of the debris which is controlled by factors such as its texture, composition and porosity. In the supraglacial position, slope angle and aspect influence the effect of meteorological variables such as temperature, precipitation and wind.

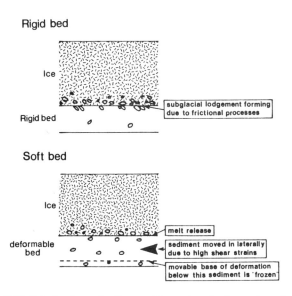

FIG. 9.5. A schematic diagram illustrating the formation of till on rigid and deformable beds (after Hart and Boulton, 1991; reprinted from *Quaternary Science Reviews*, **10**, Hart, J.K. and Boulton, G.S., The interrelation of glaciotectonic and glaciodepositional processes within the glacial environment, p. 341, 1991, with kind permission of Elsevier Science Ltd, The Boulevard, Langford Lane, Kidlington, Oxford, OX5 1GB, U.K.).

The passive nature of the ice is important to melt-out processes and contrasts with pressure melting due to sliding or internal deformation associated with the traditional active lodgement process. During melt-out subglacial debris is confined by the ice above, while supraglacial debris is confined by the accumulating debris. Such confinement should inhibit deformation and mixing (Lawson, 1979a).

Melt-out till should preserve the characteristics of the englacial debris, depending on the ratio of ice to debris. Solid layers or lenses of debris are likely to show little change of configuration on release from the ice while ice with only scattered clasts or pockets of sediment with uniform texture may lose all pre-melt structural characteristics and produce an amorphous, massive deposit. A high debris to ice ratio should allow greater preservation of englacial characteristics as fewer voids are created during ice melt. There will be less settling adjustment of the particle–particle configuration and increased particle packing. The finest debris should migrate preferentially into larger voids evacuated by melting ice between the larger grains. Only in the unusual event of there being no interstitial ice within the debris, will no modification of the englacial properties occur during melt-out. Lawson (1979b), from his observations of Matanuska Glacier, Alaska, suggests that clast preferred orientation, clast dip, debris stratification, bulk texture and deposit geometry should be well preserved (Rappol, 1985; Dowdeswell and Sharp, 1986). However, clast scatter may increase, clast dip decrease (Boulton, 1971a; Lawson, 1979b), debris stratification become more gradational and the dimensions of the deposit geometry be reduced.

Water released during melting may be available for sorting sediment especially during the melting of layers of ice with a low density of debris. The discharge of meltwater determines the degree of sorting during the melt and is dependent on the volume and rate of ice-melt, ice surface gradient, hydrostatic head and the permeability of subjacent debris. The higher the discharge the larger the calibre of material which is removed and the coarser the sorted layer which is left behind or the larger the cavity which may be developed. Shaw (1979) considers that sorted layers, and drapes of these and

interbedded diamict layers over large clasts are indicative of the basal melt-out process exemplified by what he called the Sveg Tills of Sweden (Fig. 9.6). Paul and Eyles (1990) argue that the preservation potential of melt-out till is low, especially melt-out till formed in a supraglacial position where potentially high porewater pressures, above relatively impermeable and steep ice-marginal gradients, encourage re-sedimentation by flow.

9.3.4. Flow

The concept of flow till has generated a great deal of controversy but much of the argument is terminological (cf. Dreimanis, 1988) and revolves

Stage 1

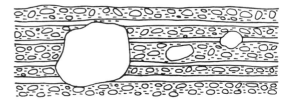

Stage 2

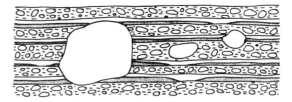

Stage 3

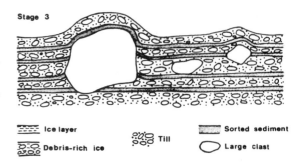

FIG. 9.6. Origin of relationships between large clasts and stratification of sorted and diamicton layers in stratified till at Overberg, Sweden, indicative of the basal melt-out process (after Shaw, 1979; reproduced with permission of Scandinavian University Press).

around the question of whether glacial debris that has flowed should be called till (Lawson, 1979b). Rather less effort has been applied to the elucidation of flow processes. Dreimanis (1988) distinguishes two subgroups of flow process, attributable to squeeze and gravity.

'Squeeze' implies active ice or at least the normal stress of stagnant ice acting on a sediment layer of low compressive strength. An upward component of movement may be present during this process. It is arguable whether such a process is more accurately described as deformation or flow.

In contrast, flow is traditionally viewed as a purely gravitational process operating down the maximum ice surface gradient (Gripp, 1929; Harrison, 1957; Hartshorn, 1958; Boulton, 1968, 1971a, 1972b). If water released during ice-melt is not easily evacuated glacial debris may become liquified and flow. Whether or not flow occurs depends on the ratio of sediment strength to shear stress, referred to in engineering terms as the factor of safety. Paul (1981) summarized a geotechnical model for the process of supraglacial deposition (melt-out) and modifications of this which induce flow. In the glacial context key parameters are porewater pressure, thickness of the sediment layer and slope angle of the buried ice. Material strength is reduced when porewater pressure is increased and this may be achieved in several ways. Meltwater may be added to the debris if not evacuated from the area; the debris may consolidate as the sediment adjusts to melting, or the debris may be loaded by the superimposition of debris flows from higher up the slope. The stress imposed on the debris is increased by the thickening of the debris pile, either by the addition of further melt-out at the base or by the superimposition of flows from upslope. An increase in the slope of the underlying ice caused by thickening or differential melt will also add to the stress on the debris.

Glacier ice typically steepens towards its margin and ice surface relief is increased by irregular melting and crevassing. This provides locations for a wide variety of mass movement processes (Varnes, 1978; Lawson, 1988) to occur, including fall, topple, slump/collapse, rotational and translational sliding, rolling and creep in addition to flow. However, the special conditions of a melting ice margin with excessive quantities of water overlying a generally impermeable glacial substrate usually ensure that flow is the dominant process, or rather the continuum of processes, of re-sedimentation.

Flow mechanics are complex. Several mechanisms of grain support and transport (grain–grain interaction, buoyancy, turbulence, fluidization, liquefaction, dispersive pressure) operate simultaneously within flowing debris (Lawson, 1981a). Relationships between texture, surface shear strength, porosity and bulk wet density all suggest that flows vary systematically with changes in water content and produce a continuum of flow morphology as flow varies from plastic to viscous in type. For discussion purposes only, Lawson (1979a) illustrated the transverse and flow-parallel cross-sections of four discrete types of sediment flow, emphasizing their transitional relationships. Flows with low water content (8–14% by weight) and high bulk density retain a measurable shear strength with shear confined to a thin zone at the base of the flow (Lawson Type I Flow). The body of material retains sufficient strength to operate as a discrete, lobate self-supported unit. As water content increases (14–19%) and strength weakens the basal shear zone increases in thickness, supporting a plug of unsheared, rafted material (Lawson Type II Flow). Multiple mechanisms of grain support and transport develop in the flows, including, within the shear zone, traction and saltation of gravel, dispersive pressures generated by grain collision, transient turbulence, mixing generated by bed irregularities and sudden bed-slope increases and localized fluidization in the shear zone, generated by increases in porewater pressure, possibly induced by the impermeability of the overlying plug. Meltwater may move over the sediment flow surface. The movement of the undeformed plug may induce sliding and rolling of clasts lying in front of the flow.

Further increases in water content (18–25%) reduce plug thickness and continuity until the whole flow is in shear and moving within an incised channel (Lawson Type III Flow). When water content exceeds saturation throughout the flow, the debris becomes liquified, fluid flow operates, grain–grain contact is lost and other grain support mechanisms largely cease (Lawson Type IV Flow).

PLATE 9.5. Supraglacial debris melting out at the margin of Solheimajökull, Iceland. Much of this material is likely to be resedimented by slumping, sliding or flowing from this steep ice-front which was advancing in 1991.

PLATE 9.6. A mixture of debris, produced by unconfined melt-out from foliated 'dirty' basal ice and the falling and sliding of coarse, angular supraglacial debris. Scale is given by the large supraglacial boulder approximately 0.5 m in length.

PLATE 9.8 and PLATE 9.9. Examples of debris flows at the margin of Fjallsjökull, southeast Iceland, showing active (Plate 9.8) and passive flows (Plate 9.9), respectively. Both features are approximately the same size as given by the bar scale in Plate 9.9.

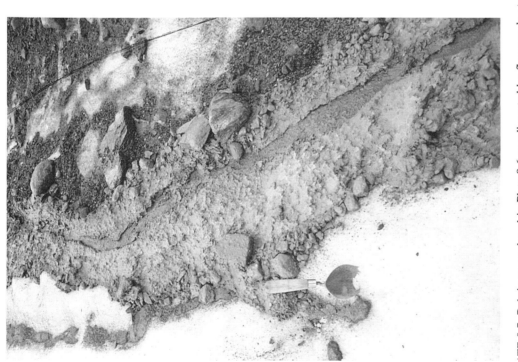

PLATE 9.7. Debris as produced in Plate 9.6 resedimented by flowage due to the addition of meltwater from supraglacial drainage. As the supply of debris decreased the centre of the flow became incised by stream flow. Scale shown by trowel (approximately 20 cm).

PLATE 9.9. Examples of debris flows at the margin of Fjallsjökull, southeast Iceland, showing active (Plate 9.8) and passive flows (Plate 9.9), respectively. Both features are approximately the same size as given by the bar scale in Plate 9.9.

PLATE 9.10. Debris flowing across the surface of Fjallsjökull, southeast Iceland, from an ice-cored ridge of debris, melting out from a shear plane in the ice. Scale bar at top right is 1 m in length.

Deposition may result from one or several of the following ambient or internal changes; a reduction in bed gradient, a decrease in flow mass thickness, the loss of interstitial fluids by drainage through permeable substrates, the blockage of the flow path, or the addition of dry sediment which increases the bulk density and reduces the relative water content of the debris.

This description of the re-sedimentation processes that can dominate some glacial margins (Lawson estimates that 95% of the deposits of the Matanuska Glacier in Alaska are 're-sedimented') shows clearly that they conform to mass wasting processes that are not exclusive to the glacial environment. Nevertheless, the debris has been accumulated by glacier activity and transported and deposited from a glacier. It is unlikely that debris flow mechanisms alone could have transported the same debris to the same place as that achieved by glacial means. On this basis it seems more reasonable to retain the term 'till' for deposits deriving from flow mechanisms as they are integral members of the assemblage of ice-marginal sediments and retain at least one property (lithology) imparted exclusively by glacial processes (Chapter 12; Menzies, 1995b, Chapter 3).

9.4. MECHANISMS OF GLACIOFLUVIAL DEPOSITION

It is clear from the preceding discussion on tills, that water is a contributory factor in most of the depositional mechanisms. In glaciofluvial contexts it becomes the predominant element especially in subaerial supraglacial and proglacial locations. Proglacial fluvial processes, especially those associated with river braiding, have been extensively discussed (e.g. Miall, 1977, 1985), and will not be discussed further here (Chapter 12).

In the subglacial context, however, 'normal' subaerial conditions may be significantly altered. In englacial and subglacial spaces confined by a glacier ice water may flow under hydrostatic pressure generated by a steep hydraulic gradient (Chapter 6). When full of water englacial and subglacial conduits function as pipes and the special conditions of 'full-pipe flow' apply (Drewry, 1986; Brodzikowski and Van Loon, 1991). Gale and Hoare (1986) discuss a useful example related to the Blakeney Esker of Eastern England.

With respect to the Blakeney Esker, the base of this sinuous, elongate ridge of sand and gravel is not conformable to the underlying topography and palaeocurrent measurements indicate upslope flow for some sections of the ridge. The following characteristics of the sediment, inversely graded beds, bimodal particle size distributions, a-axis imbrication of clasts and matrix-supported gravels, are attributed to transport under conditions of dispersive stress (Bagnold, 1954), or as a consequence of the kinetic sieving mechanism (Middleton, 1970). Textural bimodality and matrix-supported gravels are also attributed to dispersive flows and Rees (1968) is quoted as demonstrating that flow-parallel, a-axis clast preferred orientation results from repeated intergranular contact between clasts in suspension. Following Francis (1973), Gale and Hoare argue that the movement of dispersive flows will only be sustained on a slope which, in this case, greatly exceeds the longitudinal slope of the Blakeney Esker. The presence of an ice sheet supplies the requisite high hydraulic gradient.

In their investigation of flow in full glacial pipes, McDonald and Vincent (1972) found that upper flow regime dunes were succeeded not by antidunes but by full particle suspension, standing waves being inhibited by the presence of the tunnel roof. This full suspension of particles represents the final stage of the four pipe-transport regimes (a–d) distinguished by Newitt *et al.* (1955) and summarized below from Saunderson (1982).

(a) At low velocities a stationary bed with bedload and saltation, and ripples as the primary bedform.

(b) At higher velocities a moving or sliding bed, particularly for coarser particle sizes.

(c) Heterogeneous suspension, in which all of the sediment available is in suspension, the finer sizes distributed uniformly throughout the pipe cross-section and the coarser confined to the lower half of the cross-section (cf. Gale and Hoare's (1986) discussion of inversely graded sediments in the Blakeney Esker).

(d) Homogeneous suspension in which the sediment available is all in suspension but all sizes distributed uniformly throughout the cross-section.

According to Saunderson (1982), the sliding bed of the second of these stages is particularly noteworthy: this author had earlier interpreted matrix-supported gravel in the Guelph esker as sliding-bed deposits. Saunderson (1982) quotes the discussion between Acaroglu and Graf (1968) and Wilson (1965, 1969). Acaroglu and Graf (1968) equated the sliding bed of pipe flow to the antidunes of subaerial channels but were challenged by Wilson (1969) who had earlier (1965) suggested that the sliding bed mode occurred during flow decelerating from, at, or just below the limit deposit velocity, a possibility also suggested by Saunderson's (1982) tests. Whether matrix-supported sliding bed deposits are an exclusive characteristic of tunnel flow conditions, as Saunderson (1977) has claimed, has been questioned, though not conclusively disproved, by Boulton and Eyles (1979).

9.5. TILLS OF EASTERN ENGLAND — DEBATES ABOUT MECHANISMS OF TILL DEPOSITION

This discussion of the processes of glacial deposition reveals that many details associated with lodgement and deformation remain to be clarified. Eastern England provides an excellent example of the current debate concerning till-forming processes. One debate pertains to the subglacial or subaqueous origin of deformation structures in tills along the north Norfolk coast (Fig. 9.7). The other involves a model of till formation over most of the remainder of East Anglia. Both of these debates are crucial to the accurate interpretation of glacial environments in the region as well as contributing to general questions of ice sheet dynamics and till genesis.

Most authorities now recognise tills from two Pleistocene stages in the East Anglian region of Britain; the Anglian Stage (Oxygen Isotope Stage 12) and the Devensian Stage (Oxygen Isotope Stage 2). Anglian till is generally considered to be widespread (Fig. 9.7) but Devensian Till is restricted to the northwest of the region. In view of the limited distribution of the Devensian Till in the region, only the Anglian tills which occur in two lithostratigraphic formations will be discussed. In northeast Norfolk, Scandinavian ice deposited the North Sea Drifts

FIG. 9.7. Map of East Anglia, U.K., showing features and locations mentioned in the text.

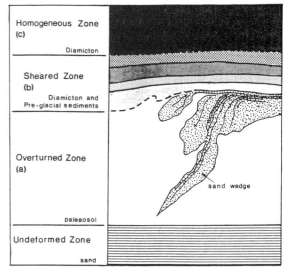

FIG. 9.8. Schematic diagram of subglacial deformation caused by the Lowestoft Till Ice Sheet, at Great Blakenham, Suffolk (after Hart and Boulton, 1991; reprinted from *Quaternary Science Reviews*, 10, Hart, J.K. and Boulton, G.S., The interrelation of glaciotectonic and glaciodepositional processes within the glacial environment, p. 346, 1991, with kind permission of Elsevier Science Ltd, The Boulevard, Langford Lane, Kidlington, Oxford, OX5 1GB, U.K.).

(Cromer Tills) (Boswell, 1916). Penecontemporaneous ice of British origin deposited the Lowestoft Till (Harmer, 1902; Baden-Powell, 1948) over most of East Anglia including some of the area previously occupied earlier by the Scandinavian ice.

Macro-scale glaciotectonic structures are found in the North Sea Drifts of the coast of Norfolk (Reid, 1882). These, and similar structures in the Lowestoft Till of the Gipping Valley in southeast Suffolk (Slater, 1926), attracted early attention but were not studied in great detail. Most research concerned provenance and relative dating (e.g. Harmer, 1928). A subglacial origin was usually assumed as the mechanism of deposition of these tills. With rare exceptions (e.g. Rose, 1974; Rose and Allen, 1977; Allen et al., 1991) the Lowestoft Till was referred to as lodgement till, while the North Sea Drifts were generally thought to have melted-out from subglacially deformed ice (e.g. Slater, 1943; Dhonau and Dhonau, 1963; Banham, 1970, 1975, 1977). More detailed analysis has been undertaken only during the last decade. A strong divergence of opinions has emerged concerning mechanisms of deposition of the North Sea Drifts (Gibbard, 1980; Banham, 1988b; Eyles et al., 1989; Hart and Boulton, 1991) and, to a lesser extent, the Lowestoft Till (Whiteman, 1987; Allen et al., 1991), and there is clearly considerable scope for further work such as that recently completed by Lunkka (1988, 1991).

The first of these debates concerns the North Sea Drifts, a complex sequence, up to 20 m thick, of diamictons (the Cromer Tills) and sorted sediments including clays, silts, sands and gravels. In the south of the region near Happisburgh (Fig. 9.7), the bedding of these sediments is subhorizontal but becomes progressively deformed towards the northwest where the deposits were referred to, in the past, as the 'Contorted Drift' (Reid, 1882). They were recently renamed the Laminated Diamicton by Hart and Boulton (1991). They are composed mostly of local lithologies but with some Scandinavian and, according to Hart and Boulton (1991), Scottish erratics.

Hart and Boulton (1991) consider the 'Contorted Drift' as a single unit, "the diamicton" (p. 338), which is deformed throughout. According to them "the most striking features of the diamicton are the laminations and the 'pods'", representing respectively, highly attenuated, conical sheath folds, and more competent masses (usually chalk and sand) with associated strain shadows, generated by simple shearing within a subglacially deforming mass.

The suggestion that the 'Laminated Diamicton/Contorted Drift' may be glaciolacustrine (Gibbard, 1980) or glaciomarine (Eyles et al., 1989) in origin is denied by Hart and Boulton (1991) on the basis of a lack of sedimentary structures and grading, and a paucity of fold noses in the laminations. The fact that deformation extends throughout the sequence and is not confined to individual beds (the laminated Cromer Tills) is also said to disprove the glacioaquatic origin. Hart and Boulton (1991) also argue against the possibility that these sediments result from the melt-out of highly deformed englacial structures generated at the margin of a cold-based ice sheet (Slater, 1926, Carruthers, 1953). They suggest that melt-out sediments are "by nature highly chaotic" (p. 339), have poor preservation potential and are unlikely to produce thick deposits (Paul and Eyles, 1990). Also, they argue that observed basal debris layers are an order of magnitude less than the 'Laminated Diamicton' (although approximating the thickness of individual Cromer Till diamict units).

An alternative view of the origin of the North Sea Drifts has been expressed by Eyles et al. (1989). Whereas Hart and Boulton (1991) concentrated their analysis largely on the area of the 'Contorted Drift' in the northwest, Eyles et al. (1989) laid considerable stress on the southeastern part of the region where the sedimentary sequence of multiple laminated diamict facies (Cromer Tills) and sorted sediments is sub-horizontally bedded and can be traced for several tens of kilometres in cliff exposures (Lunkka, 1988). Whereas Hart and Boulton (1991) treat the sequence of sediments as a single 'Laminated Diamicton' unit, Eyles et al. (1989) clearly differentiate the sequence into 'diamict facies' and 'associated facies'. The latter largely comprise sorted sediments in which they recognise an overall coarsening upwards sequence. Grading could not be discerned in the laminated and bedded diamict facies of the 'Contorted Drift'. In the flat lying and undisturbed sediments of the southeast, however, the diamicton matrix coarsens upwards into a sand-mud with large pillow structures filled with deformed sands loaded into the underlying finer diamict. Load structures are truncated by an erosion surface on which there is an accumulation of large shell fragments (<2 mm) and a pebble lag. A higher diamicton unit with similar characteristics is separated from the lower unit by a waterlain facies that shows graded laminations, wave-formed ripples, soft-sediment deformation structures and evidence of ice-rafting. In their view the intimate association of diamict with waterlain facies points to a subaqueous origin for the North Sea Drifts. They envisage plumes of fine suspended sediment emanating from an approaching ice front and ice rafted debris being deposited as a series of laminated and bedded sediments with dropstones in a proglacial marine environment. Intraformational folds are interpreted as creep and slumping of these sediments down the regional palaeoslope provided by the southeasterly dipping chalk bedrock. This is supported by the preferred direction of fold hinges determined by Banham (1975). Sand and gravels towards the top of the sedimentary sequence are interpreted as shore-face and beach sediments deposited after the ice withdrawal. These sediments are largely preserved in basins which developed by loading into the 'overpressured' diamicts. Foraminifera are being studied to determine whether the subaqueous deposits are glaciomarine or glaciolacustrine.

The second debate is concerned with the Lowestoft Till Ice Sheet in southern East Anglia. This till sheet extends over most of East Anglia and some of the adjoining area. Traditionally, the Lowestoft Till has been interpreted as a lodgement till (e.g. Rose, 1974; Perrin *et al.*, 1979), though there are references to deformation (e.g. Slater, 1926; Whiteman, 1983), melt-out (Allen, 1984; Whiteman, 1983), flow (Baker, 1977; Cheshire, 1983; Allen, 1984) and slump (Rose, 1974) in relation to a few restricted areas.

More recently, Whiteman (1987) and Allen *et al.* (1991) successfully applied Banham's (1975) three-zone model (Fig. 9.4) to a sequence of sediments deposited by the Lowestoft Till. Hart and Boulton (1990, 1991) have extended this model using evidence reported by Whiteman (1987) and Allen *et al.* (1991), to suggest that the apparently homogenous zone (their zone c (Fig. 9.8), equivalent to zone A of Boulton, 1987a) above the zone of strong shearing (zone b) actually represents a zone of extreme shear and fold attenuation culminating in a deposit which appears massive in the field, although it has yet to be demonstrated whether this degree of deformation can be recognised at the microscopic or sub-microscopic scale. Hart and Boulton (1991) interpret the whole sequence as "representing a **continuous** deposition as the ice sheet advances over the area", on the basis that "the boundaries between these different zones are **not abrupt** but each zone grades into another". In fact, this is not the case. The boundaries between the zones are usually abrupt and show very marked differences in texture, lithology and clast fabric (West and Donner, 1956; Whiteman, 1983, 1987, 1990; Ehlers *et al.*, 1992). Contrary to the Hart and Boulton model, the degree of deformation as suggested by the strength of clast fabrics decreases upwards through the till sequence. This does not itself necessarily negate the hypothesis of Hart and Boulton (1991) as it may be that the deforming sediments are responding to thresholds of change in the properties of the sediment, but it does indicate that the implications of these properties of the sedimentary sequence must be addressed before assumptions are made about the genesis of these particular British tills.

Both of these examples demonstrate the need to know far more about the processes of till deposition and till deposits.

Chapter 10

PROCESSES OF GLACIOTECTONISM

F. M. van der Wateren

In the vast range of processes operating in a glacial landscape, glaciotectonic processes produce structures and landforms which may be used to reconstruct environmental changes long after the ice has vanished. Push moraines can be reliable indicators of what went on at a glacial margin. Together, the sedimentology, paleohydrology and structural geology can give information about the mass balance condition of the glacier or ice sheet. Subglacial structures can give an insight into the conditions at the glacier bed and indicate former ice flow directions.

In this chapter current ideas on processes operating both beneath modern glaciers and at their margins are applied to explain quite distinct styles of glaciotectonic deformation that can be found in a glacial landscape. A description of the landforms alone will be rarely sufficient to identify the glacial regimes that produced them. Without a proper understanding of their internal structures the information that can be gained from a geomorphological analysis may be quite ambiguous.

Therefore, the emphasis in this chapter lies on tectonic style and strain, starting from the assumption that different glacial regimes are reflected in certain characteristic glaciotectonic styles and widely varying amounts of finite strain. The asymmetry of a glaciotectonite appears to be a reliable indicator for the reconstruction of fossil ice flow directions.

For comprehensive bibliographies of glaciotectonic structures and landforms the reader is referred to Aber (1988b) and van der Wateren (1992).

10.1. INTRODUCTION

A glacial environment offers many opportunities for deformation of unlithified and weakly lithified sediments. Deformation of the glacier bed, bulldozing of sediments at the margins, ploughing of the sea floor by calving icebergs, debris flows creeping down the glacier snout, these are but a few examples. An extensive literature list on glaciotectonics demonstrates that these deformations are analogous to structures found in many orogenic belts, however different in scale. Strain rates in hard rocks are several orders of magnitude smaller than those in unlithified sediments. While fold and thrust belts in crustal collision zones require millions of years to form, the timescale for the formation of even the largest Quaternary push moraines may vary from tens to hundreds of years. Recent studies of modern environments suggest that quite large push moraines may form even within the few years of a glacier surge. Compared to orogenic mountain chains, glaciotectonic deformations are quite shallow. They rarely exceed 200 or 300 m in depth. Yet, although push moraines are several orders of magnitude smaller than mountain chains, they may form fold and thrust belts spanning half a continent. Moreover, shear zones in sediments beneath mid-latitude ice sheets are comparable in size to crustal shear zones. The ice sheets themselves, deforming under their own weight, are continent-size metamorphic tectonites. Since glaciotectonic processes operate on very small timescales, deforming masses in a glacial environment offer interesting potentials as natural tectonic laboratories.

Since its introduction by Slater (1926) several definitions of the term glaciotectonics have been proposed, most of which differ only in detail (e.g. Viete, 1960; Moran, 1971; Woldstedt and Duphorn, 1974; Banham, 1975; Rotnicki, 1976, Ruszczyńska-Szenajch, 1980; Aber, 1985; Drewry, 1986). It is known under many names, varying from glacier tectonics and 'Eistektonik' to ice thrusting.

Glaciotectonics involves structural deformations of the upper horizon of the lithosphere caused by glacial stresses. Glaciotectonism refers to the processes leading to these deformations. Their detachment from undeformed bedrock varies from a few centimetres to a few hundred metres. This definition excludes deformations of the entire crust due to glacio-isostatic movements and reactivation of crustal faults under ice loading (Shotton, 1965). Contrary to Slater's (1926) and Drewry's (1986) usage, deformations within the ice itself are excluded as are processes of glacial erosion, such as plucking of bedrock and the particle-by-particle removal and transport of sediment by glacier ice.

Banham (1977) introduced the term glaciotectonite, meaning subglacially penetratively sheared sediments, analogous to mylonites (Lavrushin, 1971; Berthelsen, 1978; Schack Pedersen, 1988; Hart and Boulton, 1991). In this chapter a wider definition is applied in which a glaciotectonite is any body of unlithified or weakly lithified sediment that is structurally deformed by glacial stresses. These stresses include those exerted by moving glacier ice, either subglacial or at the glacier margins, and those due to loading by (static) ice masses. Sometimes relatively undeformed bodies of well-lithified rock are incorporated in glaciotectonites in which penetrative deformation is restricted to incompetent strata (Kupsch, 1962; Bluemle and Clayton, 1984; Aber et al., 1989).

Table 10.1 summarizes the structures resulting from different combinations of glacial stresses (active or passive) and position within the glacial system. Supraglacial deformations will not be discussed in this chapter. They comprise structures in kames and flow tills and these are best treated in a sedimentological context (Menzies, 1995b, Chapters 2 and 3).

10.2. SOFT SEDIMENT DEFORMATION

Deformation of unlithified sediments has been the subject of soil mechanics and engineering geology. Obviously, these sciences are mainly interested in systems experiencing quite small strains after failure. In glacial geology/geomorphology more interest centres on medium to high strains and a soil mechanical approach does not necessarily solve all problems. The most urgent problem is the present lack of knowledge of time-dependent deformation processes in unlithified sediments. Whereas these processes are quite well understood for glacier ice, the rheologies of deforming tills and glacially-pushed sediments are still largely unknown (Murray and Dowdeswell, 1992).

In striking contrast to glaciological studies, very few glacial geological studies actually report measurements of stresses and strain rates in deforming glaciotectonites. Since one of the main goals of glaciotectonic studies is to reconstruct former ice sheet conditions, a proper understanding of soft sediment deformation processes is vital. This section presents a brief introduction into soil mechanics and sediment rheology. For more detailed discussions the reader is referred to Terzaghi (1943), Terzaghi and Peck (1967), Yong and Warkentin (1975), Atkinson and Bransby (1978), Paul (1981), Allen (1985) and Jones and Preston (1987). Mase (1970) and Means (1976) are useful introductions into continuum mechanics. For a general treatment of strain in geological materials the reader is referred to Hobbs et al. (1976), Means (1976), Ramsay and Huber (1983, 1987) and Hanmer and Passchier (1991).

A sediment deforms when the shear stresses in the material exceed its shear strength. Shear strength is formulated in terms of effective stress, which is the total stress minus the (neutral) pressure of the fluid in intergranular pores. All changes in the spatial arrangement of the particles (compaction, compression, shear deformation, etc.) are produced exclusively by changes in the effective stresses (Atkinson and Bransby, 1978), which themselves reflect the interaction of the grains.

Sediment failure can be brittle or ductile. To mark off folds and shear zones from faults and other brittle

TABLE 10.1. Glaciotectonites (numbers in brackets refer to relevant sections of this chapter).

	Active	Passive
Supraglacial	-------------	collapse structures debris flows
Subglacial	shear zones streamlined bed (10.4)	diapirs & related structures water escape structures (10.6)
Marginal	compressive belts fold and thrust belts (10.5)	collapse structures diapirs & related structures water escape structures (10.6)

structures it is necessary to define ductile deformation in unlithified sediments. Contrary to metamorphic rocks, where ductile deformation pertains to crystal lattice deformation, ductile deformation of unlithified sediments is a cataclastic flow process, leaving the grains essentially intact. Ductile deformation is defined here by structures which are produced by continuous deformation on the scale in which they are viewed. Brittle deformation of soft sediments typically produces discrete undeformed blocks separated by fault planes or narrow shear zones. On a microscale, ductile structures in soft sediments are built up by numerous small sliding planes, separating aggregates or single undeformed grains (Menzies and Maltman, 1992).

The formation of fault planes is described by the Coulomb equation. Since failure is governed by the effective stresses, this law is formulated in its general form as:

$$\tau = C + \mu_s (\sigma - p_w) \qquad (10.1)$$

(Terzaghi, 1943) in which τ is the shear stress along a potential fault plane, equalling the sediment's shear strength or plastic yield strength, C is cohesion, μ_s the coefficient of friction (a material constant, $\mu_s = \tan \Phi$, where Φ is the internal friction angle of the material), σ is the normal stress on the fault plane and p_w the pressure of the pore fluid. The relationship $(\sigma - p_w) = \sigma_e$ thus represents the effective normal

stress across a fault plane (i.e. the intergranular contact pressure).

Materials such as chalk, lignite, clay and loam are cohesive sediments whereas dry sand is cohesionless. There is no permanent strain for stresses below the yield strength. When the effective stress equals the yield strength the material deforms at a strain rate that is independent of the magnitude of the applied stress. This is called perfectly plastic deformation, which is time independent. A number of investigators have attempted to model glaciotectonic processes using this principle (Mathews and Mackay, 1960; Clayton and Moran, 1974). In reality, a certain amount of irreversible strain usually occurs well below the yield strength, while the material deforms by creep (e.g. soil creep on slopes) (Allen, 1985, p. 161). In the following, an orthogonal coordinate system with the z-axis perpendicular to the shear plane xy, and the x-axis as the direction of shearing will be used.

The creep process may be modelled by linearly viscous flow of a Newtonian fluid:

$$\tau_{zx} = \eta_s \frac{du_{cs}}{dz_{cs}} \qquad (10.2)$$

where τ_{zx} is the applied shear stress, η_s the Newtonian viscosity (a material constant) and du_{cs}/dz_{cs} is the rate of change of the creep velocity with depth z_{cs} across a layer of creeping sediment where $du_{cs}/dz_{cs} = \gamma_s$, the shear strain rate in the plane of shearing. In a more general form (Eqn. (10.2)) may be written as:

$$\dot{\gamma}_s = \frac{1}{\eta_s} \tau_{zx}^{\ n} \qquad (10.3)$$

similar to Glen's power flow law (Chapters 3 and 4). Here n is a constant which, in sediments, probably depends on the internal friction angle Φ. For n = 1 flow is linearly viscous (Newton fluid), as in (Eqn. (10.2), and for other values of n flow is non-linearly viscous.

The Bingham model is a useful concept describing materials that do not deform at stress levels below their yield strength and deform by creep at higher stresses (e.g. debris flows, see Allen, 1985, p. 172). This combines plastic and viscous flow:

$$\dot{\gamma}_s = \frac{1}{\eta_s} (\tau_{zx} - \tau_c)^n , \qquad \tau_{zx} \geq \tau_c \qquad (10.4)$$

where for stresses above the yield stress deformation is either linearly or non-linearly viscous flow.

Boulton's subglacial strain experiments beneath Breiðamerkurjökull in Iceland (Boulton, 1979, 1982, 1987; Boulton and Jones, 1979; Boulton and Hindmarsh, 1987) have led to the formulation of sediment flow laws (Eqns (10.5) and (10.6)) for saturated till, relating the shear strain rate to the applied shear stress τ_{zx} (equal to the basal shear stress τ_b by the glacier) and the effective stress σ_e:

$$\dot{\gamma}_s = A(\sigma_e)^{-m} (\tau_{zx} - \tau_c)^n , \qquad \tau_{zx} \geq \tau_c \qquad (10.5)$$

in which the empirical constants A=10.62, m=1.25 and n=0.625. Stresses are in bars, strain rate in a^{-1}. Alternatively:

$$\dot{\gamma}_s = A(\sigma_e)^{-m} \tau^n \qquad (10.6)$$

where A=3.99, m=1.80 and n=1.33. Equation (10.5) models a Bingham type flow and (Eqn. (10.6)) non-linearly viscous flow. According to Boulton and Hindmarsh (1987) both flow laws fit the observed data very well. Equation (10.6) probably agrees best with the slow creep of clay-rich tills at low stress levels over long periods of time. Equations (10.5) and (10.6) illustrate the strong influence of porewater pressure on sediment viscosity. When a sediment is subjected to stress, any reduction of the effective stress by increasing pore pressure may cause slow deformation by creep. For very high porewater pressures approaching the overburden (total stress) the effective stress approaches zero. The sediment liquefies and deforms at very high strain rates.

Boulton and Hindmarsh (1987) developed a model of subglacial till deformation explaining the formation of tunnel valleys and drumlins in areas where ice sheets are underlain by soft sediments (Chapter 5; Menzies, 1995b, Chapter 3) (Boulton, 1987a; Menzies, 1989a).

Temperate ice sheets resting on deformable sediments behave in a fundamentally different way from those underlain by rigid bedrock. The latter correspond with the interior parts of Pleistocene ice sheets on the Canadian and Scandinavian shields and most present-day glaciers. The former are represented by large parts of ice sheets invading the lowland areas of North America and Northern Europe. Due to the low basal shear stresses, soft bed glaciers flow at a much higher rate and develop lower profiles than their hard bed counterparts. Growth and decay of soft bed ice sheets may be expected to be quite rapid (Boulton and Jones, 1979). Effectively, the base of the deforming layer, a cataclastic shear zone, forms the glacier sole, as was already postulated by Banham (1975).

The stability of soft bed glaciers and ice sheets is a function of the rate of deformation in the bed which in turn depends on the amount of basal meltwater draining through the subglacial sediments and upon their permeability (Murray and Dowdeswell, 1992). A positive effective stress leads to a stable state in which the bed does not deform, or deforms sufficiently slowly so as to allow steady-state ice flow. Unstable states, in which ice flow rates exceed the mass balance velocity, result from zero or negative effective stresses. The Boulton and Hindmarsh model identifies several feedback mechanisms which allow a soft-bed ice mass to return to stability. Recognition of the deformed sediments as the products of these processes will be a vital element of the reconstruction of ancient environments.

10.2.1. Frozen Sediments

It is commonly believed that frozen saturated sediments, compared to dry sediments, are rigid under glacial stresses. The argument is that sheets of sand and gravel in push moraines or large sediment 'rafts' in till can only move relatively intact if they are bound by ice (Menzies, 1990c). Fault planes and folds in sediments are regarded as evidence of frozen conditions during glaciotectonic deformation. Sometimes the argument is taken even further and the depth to décollement of glaciotectonic structures is believed to coincide with the depth of the permafrost (Richter *et al.*, 1951). Theoretical and empirical arguments challenge this assumption (van der Wateren, 1981, 1985, 1992).

Considering a stratified sediment body containing layers of contrasting rheology subjected to glacial stresses, it can be expected that these layers react differently under the same stress field. A layer of low plastic yield strength or, alternatively, one of low viscosity may deform, while more competent layers remain undeformed or deform at a much slower rate. Outcrop and map scale structures demonstrate that sand and gravel are usually more competent than clay, loam and chalk, although Croot (1988b) described a Neoglacial push moraine from Iceland in which a gravel layer forms the base of the thrust sheets, while consolidated silts appear to be more competent. Porewater pressure plays the key role since sediments of low permeability, and particularly compressible sediments such as clays, can retain high pore pressures longer than permeable sediments (Hubbert and Rubey, 1959; Paul, 1981; van der Wateren, 1985).

Soft sediments do not deform chaotically under stress. A large number of experiments confirm that faults may be generated in quite an orderly way in dry loose sand, leaving the intermediate blocks intact (Hubbert, 1951; Sandford, 1959; Horsfield, 1977; Mandl *et al.*, 1977). Slumps and landslides usually consist of a number of relatively undeformed blocks bound by fault surfaces. Submarine diapirs and compressive structures which are commonly found on large delta foreslopes further illustrate that freezing is not a precondition for faults and folds to be generated in soft sediments (Morgan, 1961; Morgan *et al.*, 1968; Aber, 1988a). Indeed, observations from Iceland (Ingólfsson, 1988) and Spitsbergen (Hart, 1987; Boulton and van der Meer, 1989) confirm that glaciotectonic deformations may occur in unfrozen saturated sediments. In both instances a glacier advanced across a submerged fjord shearing muds from the bottom and pushing them at the ice margin.

Finally, evidence from North America (Eyles and Menzies, 1983; Quigley, 1983) and Europe (van der Wateren, 1987, 1992) suggests that large lakes existed along mid-latitude ice sheets inhibiting the development of a frozen margin. In Germany, delta and lake sediments make up a large proportion of some of the largest push moraines (van der Wateren, 1992, 1994). This indicates that these lakes were present prior to the formation of these push moraines.

In situ observations of deforming frozen till indicate that ice-bound sediments may be much weaker than is commonly assumed. Strain-rate measurements in the deforming bed of a subpolar glacier in China led Echelmeyer and Wang (1987) to conclude that the effective viscosity of frozen silt is more than 100 times lower relative to ice under the same stress and temperature conditions. While in the Icelandic temperate glacier nearly 90% of the glacier motion was accomplished by deformation of the bed, comparably 60–80% of motion was accounted by deformation in the Chinese cold glacier.

Shear strain rates in the Breiðamerkurjökull experiments are between two and five times higher than those in the Chinese subglacial experiments for a comparable ice overburden. The apparent viscosities of frozen silt and water-saturated till probably differ by less than one order of magnitude. No empirically tested criteria would appear to exist to distinguish glaciotectonic structures formed under frozen and unfrozen conditions. Therefore it would appear justifiable to assume that subglacial shear zones, folds and faults may form in saturated frozen sediments just as well as in unfrozen sediments.

10.3. GLACIOTECTONIC REGIMES

Since the 1970s it is increasingly evident that deformation of the glacier bed plays a key role in the dynamics of mid-latitude ice sheets (Boulton, 1979; Boulton and Jones, 1979; Boulton and Hindmarsh, 1987; Clarke, 1987b) and ice streams (Alley *et al.*, 1986) (Chapters 5 and 11; Menzies, 1995b, Chapter

2). During glacial periods the continental ice sheets develop in the shield areas and then invade the mid-latitude sediment-covered lowlands. Large parts of these lowland areas are susceptible to subglacial shear. Conditions of the bed and the ice sheet control the production of structures and landforms by subglacial shear. It appears that the structures, which the ice left behind, can be grouped according to two main glaciotectonic regimes, reflecting the characteristics of the ice sheet and the nature of the bed (Figs 10.1a–c and 10.2a–d).

The mass balance velocity of a glacier or ice sheet terminating on land gradually increases from the accumulation area to the equilibrium line, to decrease from there to the terminus. The accumulation area is thus characterized by extending flow (strain rate

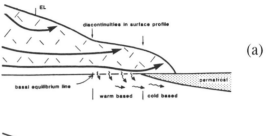

(a)

(b)

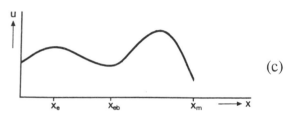

(c)

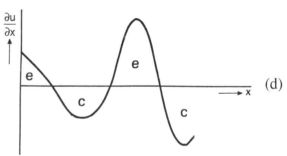

(d)

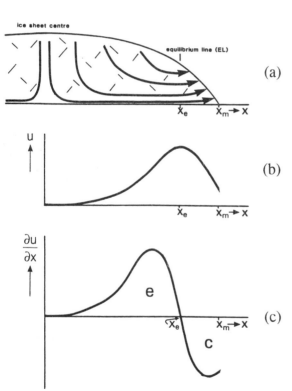

(a)

(b)

(c)

FIG. 10.1. Theoretical velocity and strain rate distributions in a continental ice sheet. (a) Flow lines in a vertical cross section (xz plane). x is distance from ice sheet centre, x_e at equilibrium line, x_m at the margin. (b) Horizontal velocity u as a function of distance x. (c) Horizontal strain rate component $\dot{\gamma}_{xx} = \partial_u/\partial_x$ is extensional (e) from the centre to the equilibrium line and compressive (c) from there to the margin.

FIG. 10.2. Theoretical velocity and strain rate distributions in a temperate continental ice sheet (after Boulton, 1972, 1987; Boulton and Jones, 1979). Discontinuities in surface profiles coincide with changes in basal shear stress. (a) Flowlines in the ablation area, related to basal thermal conditions. Ice is underlain by deformable sediments. Basal melting occurs beyond basal equilibrium line. Refreezing of basal meltwater occurs in a narrow zone near the margin. (b) Ice sheet resting on bed of low hydraulic conductivity (K) and low shearing resistance inward from the margin and high K and shearing resistance near the margin. (c) Horizontal velocity distribution for thermal and geological conditions in a and b. (d) Horizontal strain rates for a and b. High extensional strain rates near the margin may lead to formation of tunnel valleys and glacial basins. High compressive strain rates may lead to pushing at the margin.

$\partial u/\partial x > 0$) and the ablation area by compressing flow ($\partial u/\partial x < 0$), as is shown in Fig. 10.1 (Boulton, 1987a; Hart *et al.*, 1990).

Where the ice moves from a rigid bed (the shield area) into an area of deformable sediments (the lowlands), the ice can be expected to accelerate. This produces extensional strains and a low surface profile in this part of the ice sheet.

A similar distribution of strain rates will be found in deforming beds. It is therefore appropriate to distinguish two main glaciotectonic regimes, an extensional and a compressive regime (Fig. 10.2).

There are two more situations where flow in the ice sheet and within the deforming bed may change from extending to compressing flow. First, in subpolar ice sheets a zone near the margin is frozen to the bed, while further inward the ice is temperate and meltwater drains into the bed. Field experiments discussed indicate that sediments beneath a frozen margin may deform more slowly than the saturated unfrozen interior parts of the bed (Fig. 10.2a).

Secondly, if ice moves across a sediment surface of low yield strength (or viscosity) into an area of less deformable sediments, decreasing strain rates in the bed can be anticipated. This may occur where a temperate ice sheet moves from an impermeable bed to one that readily drains basal meltwater (Fig. 10.2b). Figure 10.3 gives a possible map distribution of the two glaciotectonic regimes on the scale of an ice sheet.

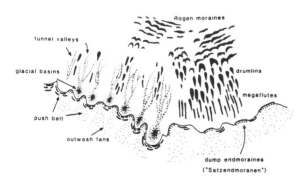

FIG. 10.3. Idealized map distribution of glaciotectonic regimes and landforms on the scale of an ice sheet. Tunnel valleys and glacial basins are landforms produced by extensional strains. Push moraines and dump end moraines (with minor pushing) mark the ice sheet margin.

10.3.1. Ice Sheets and Glaciotectonic Regimes

10.3.1.1. Subglacial Shear Zone

A zone of extension in the interior parts of glaciers and ice sheets. Under the influence of the glacier's basal shear stress subglacial sediments of low shear strength are subjected to horizontal simple shearing. Subglacial sediments are removed from this zone and transported towards the ice margin. Erosion of the substratum is strongest where $\partial u/\partial x \gg 0$. This regime is characterized by wide-spread till layers and streamlined bedforms like tunnel valleys, drumlins and megaflutes. Overdeepened glacial basins are commonly excavated on the proximal side of push moraines.

10.3.1.2. Marginal Compressive Belt

Sediments that are dumped at the terminus as subglacial, englacial and supraglacial tills and glaciofluvial deposits are subjected to horizontal compression. This zone features pushed and stacked till sheets and push moraines arranged along former glacier margins.

These regimes, subglacial shear zone and marginal compressive belt, correspond, respectively, with the streamlined and glacial thrust terrains recognized by Moran *et al.* (1980) in North America. The two regimes are obviously characterized by quite different tectonic styles. Overprinting of these styles occurs, for instance, when an ice sheet, after a period of zero mass balance, allowing a compressive belt to develop, expands and overrides the compressive structures. Steeply inclined structures will be refolded and incorporated in a subglacial shear zone (i.e. a deformation till of parautochtonous material overlying pushed structures).

Regional mapping of overprinting relations helps to identify periods of ice sheet expansion, retreat and readvance (Section 10.7). Van der Wateren (1992) presents an example of a distribution map of glaciotectonic styles in northwestern Europe.

10.4. SHEAR ZONES

Shear zones occur in several positions in a glacial environment. In the first place the ice itself principally moves by simple shear under the influence of gravity (Chapters 4 and 5). The bottom part of a glacier functions as a shear zone. When the bed deforms this becomes the lower part of the basal shear zone (Fig. 10.4b).

Shear zones also form in the bottom layers of thrust sheets and nappes which are stacked near glacier margins (Fig. 10.4a). Finally, shear zones form the base of debris flows, flow tills, landslides and slumps occurring in deposits on and around glaciers (Fig. 10.4c).

Subglacial simple shearing produces a tectonite known as deformation till (Elson 1961). Strain measurements beneath Breiðamerkurjökull (Boulton, 1979) revealed that the glacier's basal shear stress imposes a high amount of finite shear strain on the top few decimetres or metres of the bed. Strain rates in the saturated till varied from 10 to 55 a^{-1}. Thus even after a few years the finite shear strain reaches extremely high values, of the order of 10 up to a few hundred.

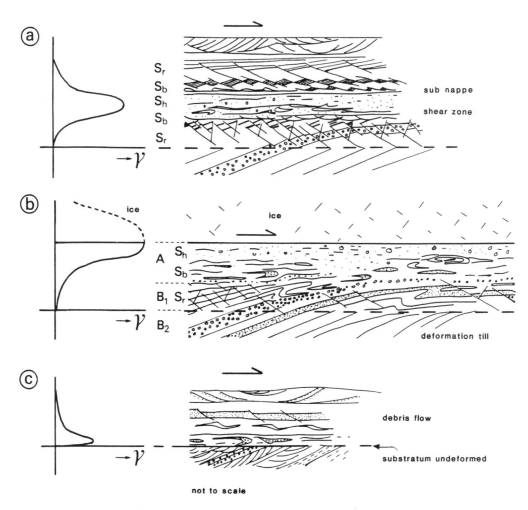

FIG. 10.4. Shear zones in different parts of a glacial environment. The graphs show finite shear strains (γ) as a function of depth. (a) Shear zone in the basal layers of a soft sediment nappe as it is found in push moraines (after van der Wateren, 1987). S_r is the minimal strain horizon containing rooted structures. S_b contains boudins, detached folds and a transposed foliation. S_h is the maximum strain, homogenized central horizon of the shear zone. Normal faulting is very common beneath nappes and thrust sheets in a push moraine. (b) Shear zone (deformation till) in subglacial sediments mirrors basal shear zone of the overlying glacier. A, B_1 and B_2 are horizons in Boulton's (1987a) deformation till model. (c) Shear zone at the base of a debris flow. Note abrupt change of shear strain from debris flow to undeformed substratum.

To illustrate this alteration in shear strain, an originally circular passive marker in the shear zone is transformed in this short time into an ellipse with a long axis more than a hundred times as long as the short axis and dipping less than 0.5°, for a shear strain $\gamma = 10$. For $g = 100$ the aspect ratio of the finite strain ellipse equals nearly 10,000 and the circle virtually becomes a horizontal line.

In view of the strong implications for the dynamic behaviour of former ice sheets, recognition of deformation tills in ancient deposits is vital. This is illustrated by seismic studies in West Antarctica revealing that the fast flow of Ice Stream B most likely is due to deformation of a subglacial till layer (Alley et al., 1987a, b; Blankenship et al., 1987). Where an ice sheet is underlain by soft sediments this may be a more general mechanism of ice stream formation (Boulton and Hindmarsh, 1987).

Surging behaviour of glaciers may also be connected with a deforming till (Boulton and Jones, 1979; Clarke et al., 1984; Sharp, 1985a; Clarke, 1987b). The large ice lobes that are characteristic of the Pleistocene ice sheet margins of Europe and North America may likewise be explained by fast ice flow or surging on a bed of very low shear strength (Boulton and Jones, 1979; Boulton et al., 1985; Lagerlund, 1987).

The next section on subglacial shear zones (or deformation tills) will be considered together with shear zones in push moraines, since they are both controlled by progressive simple shear (van der Wateren, 1987, 1992). The two are distinguished in the field by their tectonic and sedimentary relations with overlying and underlying sediment masses. A distinction on the basis of their (micro-) structures alone will be rarely possible. To ease comparison most shear zones in this chapter are presented with dextral or right lateral shear sense, when looking at the plane of the cross section.

10.4.1. Tectonic Style of Soft Shear Zones

The following ideas result from a comparative study of modern deformation tills and shear zones in push moraines of central Spitsbergen and their ancient counterparts in Germany and Norfolk (Hart, 1987, 1990; van der Wateren, 1987, 1992; Boulton and van der Meer et al., 1989; Hart et al., 1990).

Soft shear zones, or shear zones in unlithified materials, are in many ways similar to their hard rock counterparts. They share most of the macroscopic and microscopic structures known from mylonites and cataclastic shear zones, except those relating to recrystallisation of minerals at higher temperatures and pressures. (See also Maltman (1987) who discusses styles of shear zones in clay-rich sediments in relation to water content.)

Progressive simple shear is a non-coaxial deformation generating asymmetric structures, in which the principal direction of finite extension (λ_1) tends to become almost parallel to the shear zone boundaries (Choukroune et al., 1987). At medium finite shear strains, a typical soft shear zone contains folded and strongly attenuated sediment layers. At higher strains these structures disintegrate in a completely homogenized matrix.

A summary of progressive simple shear is presented in the following section (for a general discussion of progressive deformation in shear zones see Ramsay and Huber (1983, pp. 15, 217) and for a discussion of soft shear zones see van der Wateren (1992)). In an ideal shear zone, strain varies continuously from one margin to the other, being highest in the centre where the shear strain rates are at maximum, with zero shear strain at the walls (Fig. 10.5). Finite strain trajectories (λ_1) are sigmoidally shaped lines following the long axes of the finite strain ellipses (Fig. 10.6a). Heterogeneous volume change (dilation Δ) inside the shear zone, perpendicular to its walls (Fig. 10.6b) and uniform homogeneous strain affecting the shear zone together with the surrounding sediments (Fig. 10.6c) will be ignored in this discussion.

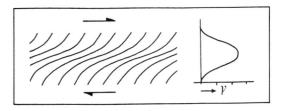

FIG. 10.5. Trajectories of finite elongation (λ_1) in a shear zone.

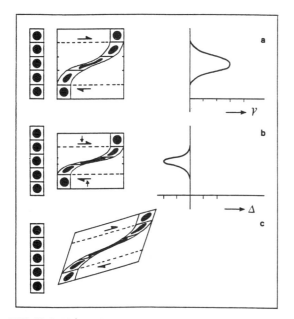

FIG. 10.6. (a) λ_1 trajectories follow long axes of finite strain ellipses. (b) Shear zone with additional dilation (Δ). (c) Homogeneous strain affecting bot the shear zone and its walls.

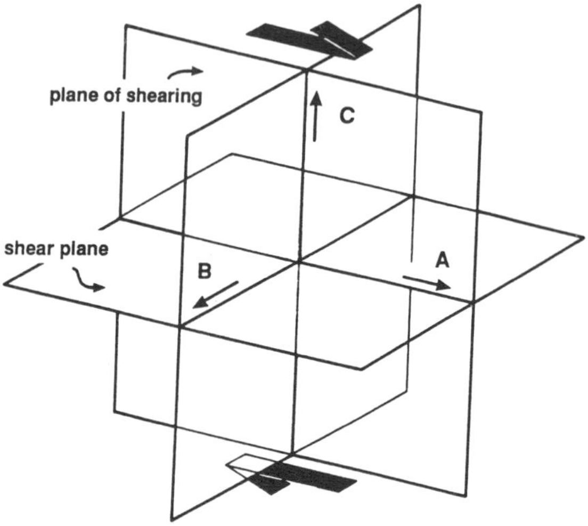

FIG. 10.7. Definition of principal directions A, B and C in a shear zone.

According to Fig. 10.7, line A is the direction of shearing, the plane AB the shear plane and the plane AC is the plane of shearing. In simple shear, all changes of length (longitudinal strain or extension, e) or changes of angles between material lines (shear strain, γ) occur in the AC plane.

An originally circular passive marker in the plane of shearing deforms to an ellipse after one increment of simple shear. If it is assumed that the strain increment is infinitesimally small, the long axis of the incremental strain ellipse will dip at 45° away from the shear direction (Fig. 10.8a). One of the lines of no incremental extension is parallel to the C axis. In progressive simple shear, the shear plane contains both the line of no incremental extension and the line of no finite extension. Finite extension is the sum of all previous incremental extensions. With each subsequent strain increment the elipticity of the finite strain ellipse increases and its long axis $(1+e_{1f})$ rotates in the direction of shearing (Fig. 10.8b).

Both the incremental and the finite strain ellipses are divided by lines of no longitudinal strain into four sectors, two in which material lines having orientations lying in this sector are shortened (e_i and e_f

negative) and two in which material lines have become longer (e_i and e_f positive). The AC plane contains the longest and the shortest principal axis of the strain ellipsoid (λ_1 and λ_3).

Superposition of the incremental strain ellipse on the finite strain ellipse yields three sectors in progressive simple shear (Fig. 10.8c) and consideration, here, will be given to the effects on a competent layer in an incompetent matrix (Figs 10.9 and 10.11):

(1) Finite longitudinal strain (e_f) and incremental longitudinal strain (e_i) are negative. Fabric elements, e.g. sediment layers having this orientation, are shortened and continue to shorten during the next increment. Shortening is accomplished by thickening and folding of competent layers.

(2) If e_f is negative and e_i is positive. Fabric elements in this sector have been initially shortened and will be stretched during the next increment. This may lead to unfolding of previously folded competent layers, but thinning and boudinage of fold limbs will be more common.

(3) Both e_f and e_i are positive. Fabric elements lying in this sector have been stretched and will

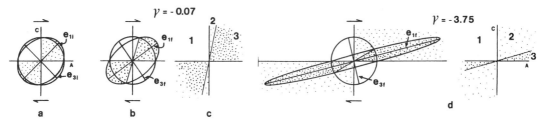

FIG. 10.8. (a) Orientation of the longest and shortest principal axes of the incremental strain ellipse. Shaded areas denote extension. (b) Orientation of the longest and shortest principal axes of the finite strain ellipse. (c) Superposition of finite and incremental strains. See text for explanation of sectors 1, 2 and 3. (d) Finite strain ellipse and compressive and extensional sectors for a finite shear strain of –3.75.

undergo stretching during all subsequent strain increments. Competent layers suffer boudinage.

Folded boudins are produced where pure shear strain overprints the simple shear fabric (i.e. when due to non-uniform flow ($\partial u/\partial x=0$) longitudinal compression alternates with longitudinal extension). This may be an indication of interference of different glaciotectonic regimes but is sometimes a product of lithological inhomogeneities.

A number of structures may be recognized which make it possible, if not to quantitatively determine the finite strain, at least to distinguish low strain and high strain fabrics and to assess the orientation of the finite strain axes and the tectonic transport direction (Platt, 1984; Chester and Logan, 1987; Malavieille, 1987; Petit, 1987).

Riedel (1929), Tchalenko (1970) and Tchalenko and Ambraseys (1970) showed that brittle shear zones of all scales share characteristic combinations of faults and fractures. These are presented below in a model of a cataclastic shear zone.

Where the material favours ductile deformation, the strain varies continuously from the shear zone walls to its centre. Materials containing inequant grains, such as clay minerals, may develop a distinct penetrative set of cleavage surfaces parallel to the λ_1 trajectories. This fabric results from preferred grain alignment in response to local strain. In soft shear zones, cleavages may be recognized, surprisingly similar to S- and C-foliations in mylonites (Berthé et al., 1979; Platt and Vissers, 1980; White et al., 1980; Malavieille, 1987; O'Brien et al., 1987).

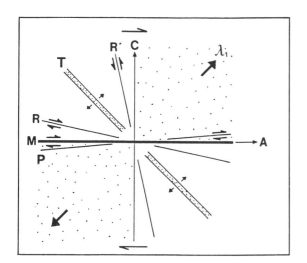

FIG. 10.9. Brittle shear zone structures.

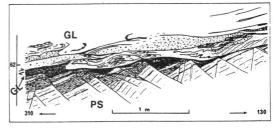

FIG. 10.10. Conjugate sets of Riedel shears beneath a nappe in the Dammer Berge push moraine (van der Wateren, 1987; reprinted from: Tills and Glaciotectonics (J.J.M. van der Meer, ed.), 1987; courtesy of A.A. Balkema, Rotterdam). Shear is clockwise.

Figure 10.9 summarizes brittle shear zone structures (after Petit, 1987). The extensional sector of the incremental strain ellipse is shaded. The inherent asymmetry is enhanced by a number of shear planes (i.e. Riedel shears (R, R') and tension fractures (T)), creating extension in the shear direction. It further contains shear planes M and P with sense of shearing compatible with the bulk shear.

The shear parallel (AC) section (containing the longest and the shortest principal axis of the strain ellipsoid) has a strong asymmetry, whereas the section normal to the shear vector (BC) shows a pattern of shear planes that is symmetric about the plane of shearing (Chester and Logan, 1987). These structures may be found on all scales. Riedel shears occur as conjugate sets of normal faults beneath tills or nappe shear zones (Fig. 10.10) or in thin sections of tills (Plate 10.1). Tension fractures (T) have been described as so-called till wedges (Åmark, 1986).

Figure 10.11 summarizes structures typical of ductile shear zones. Buckle folds, kink bands and compressive crenulations are produced by finite shortening in the compressive sector of the strain ellipse. Boudins are produced by finite extension in the direction of shearing. Folds form in lithologies with sufficient interlayer viscosity contrast. In progressive simple shear they soon become asymmetric, overturned and attenuated in the shear direction.

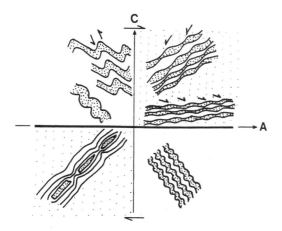

FIG. 10.11. Ductile shear zone structures. Note that at higher finite strains the boudins will be oriented almost parallel to the shear plane (AB).

Fold axes initiating at high angles to the shear vector tend to rotate towards an orientation parallel to the bulk shear direction forming tube-shaped non-cylindrical folds or sheath folds. Sheath folds typically form at shear strains of 10 or more (Cobbold and Quinquis, 1980). Although they are not diagnostic of simple shear regimes, in combination with other structures they may help to identify zones with very high shear strains. In shear parallel sections they appear as asymmetric isoclinal folds, while in the section perpendicular to the shear direction they show as eye or ring shapes (Kluiving et al., 1991).

Figure 10.12 is a block diagram summarizing the structures that may be found in association with subglacial or sub-nappe shear zones. The layering, typical of many shear zones, is a tectonic lamination produced by repeated folding and attenuation at very high finite strains. In principle, the orientation of the finite strain ellipse can be determined from the relations between compressive and extensional structures in a shear zone. The relative asymmetry of AC cross-sections and symmetry of BC sections, both on a microscale and in outcrops, indicate the direction of shearing and therefore are good indicators of former ice flow directions. Kluiving et al. (1991) applied this principle in an analysis of a section in the eastern Netherlands where three overlying tills belonging to quite different ice flow directions could be distinguished.

In progressive simple shear the direction of finite extension is almost parallel to the direction of shearing. The very high finite strains produce a transposed foliation quite similar to some highly deformed metamorphites. Transposition is the process converting a primary (e.g. sedimentary) lamination into a tectonic lamination of highly attenuated and boudinaged fold limbs and detached, isolated intrafolial folds. At first sight it may resemble sedimentary layering but, in fact, it is a tectonic layering of extremely high finite strain. Repeated folding and attenuation transforms even simple two- or three-layered sedimentary sequences into complex multi-layered transposed fabrics. Examples are given in Plates 10.2 and 10.3 showing chalk banding in a Danish clay rich till and sandy laminae in a deformation till in Germany, respectively.

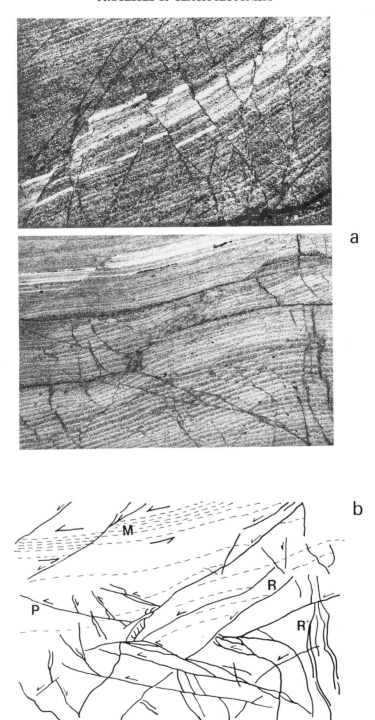

PLATE 10.1. Conjugate sets of Riedel shears in subglacially deformed rythmites, incorporated in a Saalian till in The Netherlands (van der Meer *et al.*, 1985). Thin section is viewed under crossed nicols (vertical and horizontal). Bright bands are strongly oriented clay minerals as a result of layer parallel shear (M shear). Dark cross-cutting lines are oriented clays in the extinction orientation (R and R' shears). Asymmetry indicates left lateral (anti-clockwise) shearing. Field width 6 mm from left to right.

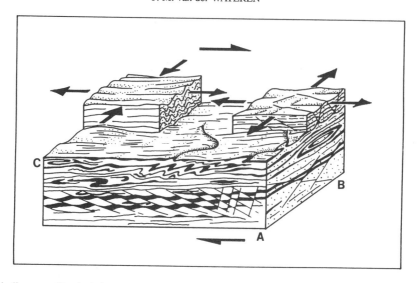

FIG. 10.12. Block diagram of typical shear zone structures. Note eye shaped structures (cross-sections of sheath folds) and symmetric fault sets in section perpendicular to shearing (from Kluiving *et al.*, 1991; reprinted from Till statigraphy and ice movements in eastern Overijessel, The Netherlands, by S.J.Kluiving, M. Rappol and F.M. van der Wateren, *Boreas*, 1991, **20**, 193–205, by permission of the Scandinavian University Press).

Continued deformation eventually produces a completely homogenized tectonite of, at first sight, deceptively low strain. Only the rare occurrence of an isolated boudin or fold nose and the typical microscopic fabric of R, R', M and P shear planes and microboudins are witness to this deformation process.

A typical soft shear zone can be divided into three units of different tectonic styles (van der Wateren, 1987). In order of increasing finite shear strain and tectonic transport these units are (Fig. 10.4a, b):

S_r—rooted recumbent structures at the lower boundary of the shear zone and low strain structures at the upper boundary;

S_b—a transposed foliation of boudins, rootless intrafolial folds and highly attenuated bands of footwall material in a more allochtonous matrix;

S_h—a completely homogenized mixture of exotic and more or less autochthonous elements, bearing witness to an extremely high finite shear strain in the central part of the shear zone.

These units may be distinguished wherever unlithified sediments have been subjected to shear stresses, whether this occurs beneath a sediment thrust sheet or beneath glacier ice. In the latter case the top half of the shear zone is in the basal ice layers and the tectonite would be called a deformation till. Units S_r and S_b correspond with the B horizon and

unit Sh with the upper, A horizon of Boulton's (1987a) deformation till. The Sb unit, of intermediate strain, may be banded if the lithological contrasts are sufficiently strong.

Boudins may reach impressive dimensions. Large, so-called rafts, floes or mega-blocks of sediment or even lithified bedrock embedded in till have been described from North America and Europe (Section 10.5.3).

Banded tills containing sediment layers and lenses of local origin have been interpreted as waterlain or meltout tills but in many cases can be proved to be the product of subglacial shearing. The chalk banding in tills from Denmark (Berthelsen, 1978) and East Anglia (Hart, 1990) is a clear illustration of this process (Plate 10.2). Rappol and Stoltenberg (1985) showed that the incorporation of footwall material had a distinct effect on the composition of the till even when banding and shearing were not immediately obvious.

10.5. MARGINAL COMPRESSIVE BELTS

Push moraines have long been known by herdsmen and farmers in the Alps. The earliest references date from the sixteenth century (cited in

PLATE 10.2. Transposed foliation of clay rich till and chalk laminae in Møns Klint, Denmark. Folding due to local compression in otherwise horizontally laminated deformation till.

Böhm, 1901). Dramatic accounts of advancing valley glaciers shoving boulders, trees and even houses entered the early glaciological literature and contributed to the glacial theories by Agassiz and A. and J. Geikie.

During the Little Ice Age many glaciers advanced, producing vast outwash fans and push moraines. Particularly those in arctic and subarctic environments have been the subject of early structural and geomorphological analyses (e.g. von Drygalski, 1897; Gilbert, 1903; Tarr and Butler, 1909; Martin, 1913; Grant and Higgins, 1913; Tarr and Martin, 1914; Slater, 1926; Gripp, 1929; Todtmann, 1936, 1960). More recent studies of Neoglacial push moraines include Croot (1988), Boulton and van der Meer (1989) and van der Wateren (1992).

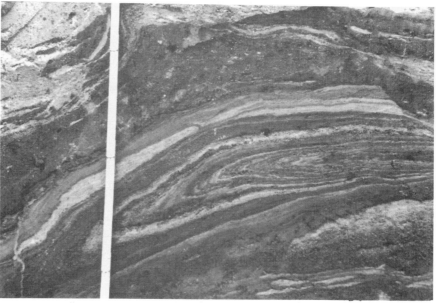

PLATE 10.3. Transposed foliation of loamy till and sand laminae from the Dammer Berge, Germany.

Only very few studies exist of modern, active push moraines. Kälin (1971) presents a geomorphological and structural analysis of the push moraine in front of the Thompson Glacier, Axel Heiberg Island, Canada, the development of which was observed over a number of years. Eybergen (1987) has done a similar study of a small push moraine in front of the Turtmann Glacier in the Swiss Alps.

From these studies it can be concluded that push moraines form under the following conditions (van der Wateren, 1987, 1992): (1) a suitable décollement close to the surface; (2) an advancing or readvancing ice sheet; and (3) a coupling of the ice mass with its bed.

Condition (1) introduces the influence of the geology of the substratum. The Rehburg line in Germany and the Netherlands coincides with a zone in which Tertiary and Pleistocene clays come close to the surface (van der Wateren, 1992). The push moraines along the escarpment of the Missouri Coteau in North America are another example of substratum controlled glaciotectonic features (Kupsch, 1962; Moran et al., 1980; Bluemle and Clayton, 1984).

Condition (2) involves glaciological conditions where an ice advance or readvance may have occurred on a continental scale due to a positive mass balance, or may be local, due to a surge or an ice stream.

Condition (3) relates to the sediment substratum over which the ice sheet advances, since the sediment mechanical properties determine whether the ice is coupled with the underlying sediments, a prerequisite for pushing. This idea will be further developed in the following sections.

10.5.1. Tectonic Style of Push Moraines

Push moraines are zones of strong horizontal compression. In Pleistocene glaciated areas, they rim vast till-covered plains dominated by strong extensional strains. On their upstream side depressions commonly occur from which the pushed masses have been removed. Typical examples are the 'hill-depression forms' in North Dakota, Alberta and Saskatchewan (Clayton and Moran, 1974; Bluemle and Clayton, 1984; Moran et al., 1980) and the push moraines of the Rehburg line in The Netherlands and Germany surrounding deep glacial basins (van der Wateren, 1987; van den Berg and Beets, 1987; de Gans et al., 1987; Meyer, 1987).

Push moraines, in the strict sense, are built of materials deposited in the foreland, before they are compressed by an advancing or re-advancing glacier or ice sheet. The sediments may consist of tills and debris accumulating at the snout during the summer that will be pushed when the glacier re-advances the next winter. The small annual push moraines in front of valley glaciers form in such a way. They may also include basal tills and glaciofluvial outwash from previous more extensive advances.

Push moraines form in a large variety of unlithified or weakly lithified sediments, such as glacial tills of varying composition, fluvial and glaciofluvial sands and gravels, fine-grained delta and lake deposits, marine clay and silt, chalk, sandstone and shale. Where the moraines are well exposed, it can usually be shown that the décollements coincide with incompetent strata.

Figures 10.13 and 10.14 are synthetic block diagrams of two general types of push moraines. Such forms are, in many ways, analogous with accretionary wedge complexes at convergent plate margins (Section 10.5.4). The models are mainly based on evidence from the Neoglacial Holmströmbreen push moraine, Spitsbergen (Boulton et al., 1989; van der Wateren, 1992) and the Pleistocene push moraine of the Dammer Berge, Germany (van der Wateren, 1987, 1992).

Folding is the dominant tectonic style in the Holmströmbreen model (Fig. 10.13). In those parts of the push moraine where thick fluvial gravel and sand sequences prevail, fold wavelengths tend to be larger than in more fine-grained and more finely laminated lithologies. Concentric style folds near the push moraine front tend to become tighter and more asymmetric in the intermediate and internal zones of the wedge, where they develop into recumbent folds, thrust and fold nappes as the strain increases. This change in tectonic style is quite similar to that in, for example, the bedrock forms of the Foothills and Front Ranges of the Canadian Rocky Mountains (Dahlstrom, 1970).

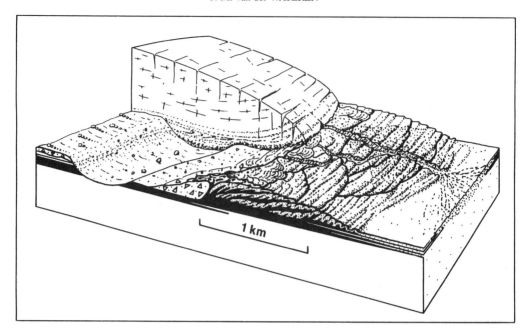

FIG. 10.13. Block diagram of the Holströmbreen push moraine, Spitsbergen (after Boulton and van der Meer *et al.*, 1989). Tectonic style changes from the foreland to the ice margin from Jura type concentric folds, through fold-thrust nappes to gravitational nappes.

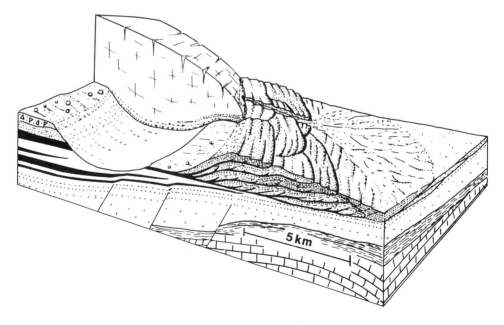

FIG. 10.14. Block diagram of the Dammer Berge push moraine (after van der Wateren, 1987). Push moraine comprises large (kilometre scale) nappes which are internally imbricated.

In the internal zone, the finite strain is at maximum and thrust nappes and gravity-driven sliding nappes predominate at the surface. Figures 10.15a, b, c are examples of structures in the external, intermediate and internal zones, respectively, of the Holmströmbreen push moraine.

The central and proximal parts of the push moraine in the Dammer Berge model (Fig. 10.14) are built of piles of large, more or less horizontal and internally-deformed nappes that have moved long distances on shear zones of Tertiary and Pleistocene clay and loam. The nappes measure up to several kilometres across and are up to 50 metres in thickness (van der Wateren, 1987). As in the

Holmströmbreen push moraine, finite shortening increases in a hinterland direction. In the distal parts, steeply inclined folds and imbricate thrusts seem to take up most of the strain. The majority of the folds and imbricate structures appear to have been initiated at an early stage of movement of the nappe and remained active throughout subsequent stages of its development.

The top layers of the proximal slopes were transformed into deformation till when the Scandinavian ice sheet advanced and finally overrode the push moraine, as occurred during the maximum advance of the Saalian glaciation in the Netherlands and Germany. It is thus bounded below and above by

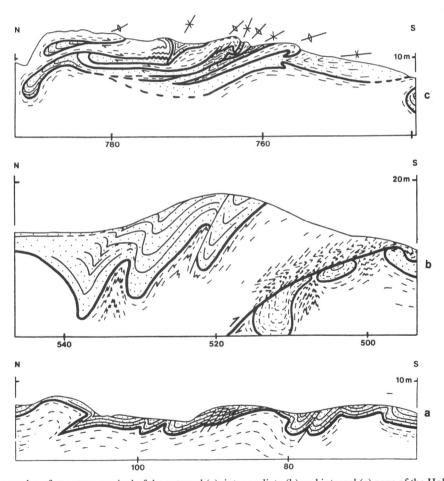

FIG. 10.15. Examples of structures typical of the external (a), intermediate (b) and internal (c) zone of the Holmströmbreen push moraine. Horizontal axis shows distance in metres from the push moraine front. (a) Concentric fold style in external zone. Low strain. (b) Fold-thrust nappe in intermediate zone. Medium strain. (c) Thrust and gravitational gliding nappes in the internal zone near the ice margin. Highest amount of finite shortening. Some of the nappes develop into mud flows.

shear zones. Debris flows (flow tills) moved down the glacier front and slumps formed at the wedge front (cf. Alley *et al.*, 1987b; King *et al.*, 1991).

When fine-grained sediments form the bulk of a push moraine the structures are dominantly ductile (i.e. folds and fold nappes). In push moraines dominated by coarse-grained sediments with a fine-grained basal slip layer, the structures tend to be brittle and folds have longer wavelengths. Large sub-horizontal nappes are the primary structural units.

Meltwater streams cutting into a push moraine while it is forming, generate syntectonic unconformity surfaces and sediment bodies. These sediments are more deformed themselves the earlier they have been deposited. Truncated anticlines with folded unconformity surfaces and growth folds (Fig. 10.16) are diagnostic of syntectonic sedimentation. These structures may help to estimate the chronosequence of tectonic events and to construct a stratigraphy of pre-, syn- and post-tectonic sediments (van der Wateren, 1987, 1992).

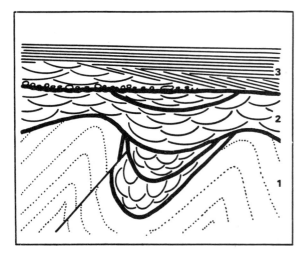

FIG. 10.16. Glaciofluvial deposition in relation to glaciotectonics (after van der Wateren, 1987). Finite strain decreases from oldest to youngest sediments. 1: pre-tectonic sediments, 2: syn-tectonic fill of synclinal valley (delta and channel deposits), 3: post-tectonic sediments (channel and alluvial fan deposits). Reprinted from: *Tills and Glaciotectonics* (J.J.M. van der Meer, ed.), 1987; courtesy of A.A. Balkema, Rotterdam.

10.5.2. Folds and Thrusts

Fold styles associated with subglacial and nappe shear zones (Section 10.4) markedly differ from macroscopic folds in push moraines. The former are produced mainly by layer-parallel simple shear and are dominated by recumbent similar and isoclinal fold styles. These folds are passive structures and the orientation of their axes bears no direct relation to the main stress field, but tends to align with the direction of maximum finite extension. The nappe shear zones are the result of horizontal compression producing steeply inclined thrusts and folds of concentric style. These active folds commonly have their axes oriented at high angles to the bulk shortening direction.

Folds, like imbricate thrusts, seem to precede nappe movement in many of the larger push moraines. In the Holmströmbreen push moraine they began as concentric or box folds, moving on a décollement over an incompetent bed. When the folded and imbricated sediments are further compressed, the structures move as a whole on top of the foreland to form folded and imbricated nappes.

In many push moraines, folds are found on a level which is too high for them to be rooted in the lowest décollement of the push moraine. This implies that they have either formed at an early stage and been transported to their present high position as part of larger nappe structures or have formed at a late stage of compression.

10.5.3. Stacked Till Sheets

Compressive zones near the margins of large ice sheets are comprised of more than just push moraines of deformed glaciofluvial deposits and other foreland materials. Compressing flow in the ablation zone produces a narrow terminal zone of thick folded and thrust till sequences. Apart from tills, steadily moving outward under basal shear, large blocks of footwall sediment or sedimentary rock may be piled up near the margin. These blocks or rafts are surrounded by till, which distinguishes them from structurally quite similar push moraines. Some of these blocks are known to have travelled very long distances. Viete (1960, p. 25, citing Moskvitin, 1938) mentions rafts of Lower Carboniferous rocks measuring 16 to 20 km² that have travelled more than 200 km from their source to the Saalian end moraines northwest of Moscow. Moran (1971) reports stacked till sheets and rafts of Cretaceous sediments from Saskatchewan and Alberta, Canada. One of these rafts measures approximately 1000 km² (Christiansen 1971, cited in Aber *et al.*, 1989, p. 94; Prange, 1978; Ringberg *et al.*, 1984; Ruszynska-Szenajch, 1987).

Stacked and folded till sheets are characteristic of the termination of the subglacial transport belt, although compressing flow may also occur on a small scale upstream from the glacier terminus, where flow is locally non-uniform. The main problem in identifying these forms is to be certain that the tills overlying and underlying the rafts are of the same age. From a process point of view they are quite similar to push moraines in that both are produced by the compressive strain regime near the ice sheet margin. Stacked till sheets are typically produced at stable margins (mass balance = 0), whereas push moraines are the products of a glacial advance or a readvance.

10.5.4. Accretionary Wedge Model of Push Moraines

It is difficult to overlook the similarities between push moraines and large mountain belts such as the Rocky Mountains or the Alps. Both are prismatic, in cross-section wedge-shaped, sediment bodies, formed by horizontal compression. Push moraines are produced wherever an advancing glacier becomes coupled to sediment masses at the margin. These are subsequently pushed to form sediment wedges, similar in shape to accretionary wedges (Stockmal, 1983).

Coupling of advancing ice and marginal sediments may occur under one or more the following conditions:

(1) When, after a period of zero mass balance, a glacier advances into glaciofluvial outwash that has accumulated during this period;

(2) When a glacier surges into outwash sediments, deposited when the margin was stationary;

(3) When a glacier advance reaches an area in which a suitable décollement layer, (e.g. a clay) comes nearer to the surface; and

(4) When a glacier advances from an area in which the surface consists of fine-grained deformable sediments into an area of more coarse-grained, resistant sediments.

Only the first of these conditions is related to mass balance changes (Boulton, 1986b). It is, therefore, important to be able to distinguish them on the basis of sedimentary sequences and glaciotectonic structures. When a glacier front is stationary over a number of years, tills, outwash fans and deltas accumulate at the terminus. If the mass balance then becomes positive, these may be pushed by the advancing glacier (Boulton, 1986b). Many push moraines, in mountain areas, are almost entirely composed of proglacial outwash material (deltas and alluvial fans).

An ice mass may be expected to be almost completely uncoupled if moving across a relatively smooth surface of fine-grained sediments, given a sufficient supply of subglacial meltwater. The bed merely deforms by simple shear, inhibiting large-scale thrusting and folding at the margin. This may also apply to frozen fine-grained sediments (Shoemaker, 1986; Menzies, 1989a, 1990a). A surface of coarse-grained sediments, since it is not susceptible to subglacial shearing, may be expected to promote pushing, provided other conditions are satisfied.

The mechanical aspects of pushing under the conditions noted above can be summarized based on an analysis by van der Wateren (1992) following similar studies of accretionary wedges (Elliott, 1976; Chapple, 1978; Davis *et al.*, 1983; Stockmal, 1983; Dahlen *et al.*, 1984; Platt, 1986). In this analogy the ice sheet performs the role of the overriding plate advancing on a surface covered by sediments of varying shear strength, scraping these off their basement and accreting them to the advancing wedge.

In longitudinal cross-section, push moraines are tapered sediment masses with their thick end abutting on the glacier. The folded and thrust masses exhibit a general spatial and temporal variation in style and intensity of deformation. Bulk finite strain increases from the thrust front towards the hinterland, with the tectonic style changing from décollement folds to large overthrusts and nappes (Fig. 10.13).

It has been observed in modern active push moraines, and confirmed by experiments in deforming wedges, that the highest longitudinal compressive strain rates occur in a rather narrow zone near the thrust front. Strain rates rapidly decrease towards the internal parts (Köster, 1958; Kälin, 1971; Cowan and Silling, 1978; Stockmal, 1983; Dahlen *et al.*, 1984; Eybergen, 1987). Accretionary wedges are mechanically analogous to the sand wedges forming in front of a bulldozer. As the bulldozer moves forward, the sand wedge continuously grows and maintains a constant surface slope which is related to the rheological properties of the sediment.

If a glacier is compared to a bulldozer a problem immediately arises. How can ice, a fairly weak material, push relatively strong and brittle sediment masses to considerable elevations? In the past (e.g. Gripp, 1975; Köster, 1958) this question has been neglected and glacier profiles have been suggested which are too low to account for the push moraines at their fronts (van der Wateren, 1985).

Since very little is known about the rheologies of unlithified sediments the deformation mechanisms of the sediment wedge must be simplified. In this approximation only bulk deformation will be considered and the details of folding and faulting ignored. Figure 10.17 shows the geometry of a deforming wedge in a horizontal orthogonal coordinate system. The approximation is valid only for variations of stress and velocity in the wedge on longitudinal length scales several times the thickness. Thus the derived equations break down for areas close to the wedge front. The theory borrows concepts from other approaches, such as 'critical state' (Davis *et al.*, 1983) and the 'stability criterion' (Platt, 1986).

The following assumptions are utilized in the light of the available field evidence:

(1) The wedge consists of two layers, a competent layer sliding on an incompetent basal layer. It is assumed to be mechanically continuous and macroscopically ductile (Price, 1973). In many push moraines the competent units are much thicker than the soft basal layers; this geometry will be adopted here;

(2) The wedge grows at the cost of the foreland by accretion of sediments to the front; and

(3) The wedge is assumed to be in a critical state, as defined in Davis *et al.* (1983): 'a critically tapered

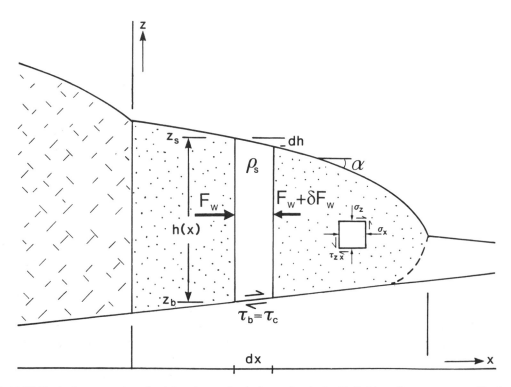

FIG. 10.17. Vertical cross-section of a deforming wedge in front of a glacier. Definition of parameters used in the text.

wedge that is not accreting fresh material is the thinnest body that can be thrust over its basal décollement without any internal deformation; it is thus on the verge of shear failure everywhere. In contrast, a critically tapered wedge that is accreting material deforms internally while sliding, in order to accommodate the influx and to maintain its critical taper.' (p. 1154). The compressive and gravitational forces are just adequate to overcome the resistance to sliding at the base (Davis *et al.*, 1983; Dahlen *et al.*, 1984).

If it is assumed that the wedge is at the point of shear failure throughout (i.e. is in a critical state), then at any one time the wedge surface strives for a condition of equilibrium between the forces propelling the wedge and those resisting its movement. This is similar to movement of a glacier where, over time, the surface profile is an expression of this equilibrium.

Budd (1970b) derived an equation for the basal shear stress in a glacier including the effect of longitudinal stress components. Compressive stresses, strains and strain rates are assumed to be negative. The force balance in a deforming wedge can therefore be derived where, neglecting the longitudinal shear force gradient, the Coulomb shear strength τ_c in the wedge just balances the gravitational and longitudinal stresses (Fig. 10.18):

$$\tau_c = \tau_g + \frac{\partial F_w}{\partial x} \qquad (10.7)$$

in which the first term on the right-hand side is the gravitational driving stress. This can be envisaged to drive shear flow mainly in the weak basal layer, similar to glacier flow:

$$\tau_g = \rho_s \, g \, h_s \, (x) \alpha \qquad (10.8)$$

This is the familiar Nye equation for small surface slopes α, where ρ_s is the bulk density of the sediments in the wedge. The second term of Equation 10.7 represents the gradient of the longitudinal normal force F_w (Fig. 10.17) driving compression and extension in the bulk of the wedge:

$$F_w = \int_{z_b}^{z_s}(\sigma_x - \sigma_z)dz = 2\int_{z_b}^{z_s}\sigma_x' dz = 2h_s\sigma_x' \qquad (10.9)$$

where $\tilde{\sigma}_x{}'$ is the longitudinal deviatoric stress averaged over the wedge thickness.

It can be shown that the following force balance (simplified from Eqn. 10.7) is a reasonable approximation for a portion ∂x of which the length is of the order of h_s, the thickness of the wedge:

$$\tau_g + 2\tilde{\sigma}_x{}' \frac{\partial h}{\partial x} - \tau_c = 0 \qquad (10.10)$$

This balance is only valid away from the leading and trailing edges of the wedge.

Assume that, at some time in segment ∂x, $\tau_g > \tau_c$. Since $\partial h_s/\partial x < 0$ it follows that $\tilde{\sigma}_x{}' > 0$ (i.e. the longitudinal deviatoric stress in this part of the wedge) is tensile. If this stress is large enough to overcome the yield strength of the wedge it will extend. Conversely, if $\tau_g < \tau_c$ then the wedge is under compression.

Translating this into terms of changing h_s, α and τ_c it can be concluded that:

(a) If either or both the surface slope α and thickness h_s attain such values that the gravitational driving stress τ_g is larger than the shear strength τ_c, the wedge will tend to extend thus reducing α and h_s until equilibrium is restored. If either α or h_s or both are too small the wedge will tend to shorten;

(b) If the shear strength τ_c of some part of the wedge (e.g. in the basal layer) is reduced the wedge will extend and if τ_c is locally increased the wedge will shorten again.

The stable wedge (Platt, 1986) is neither shortening nor extending, according to the stability criterion:

$$\tau_g - \tau_c = 0 \qquad (10.11)$$

The gravitational force just balances the shear strength, but as the wedge is in a critical state, every departure from equilibrium is followed by either extension or compression of that part of the wedge that has become unstable, which is where the F_w term enters. Horizontal compression by folding and thrusting, or horizontal extension by boudinage and normal faulting serve to maintain a stable surface profile. It will be clear that the configuration of a stable wedge is a convex parabolic surface profile. The wedge (push moraine) in a way is an extension of the glacier, albeit of a lower profile. Since the density of the sediments is about twice as high as the ice

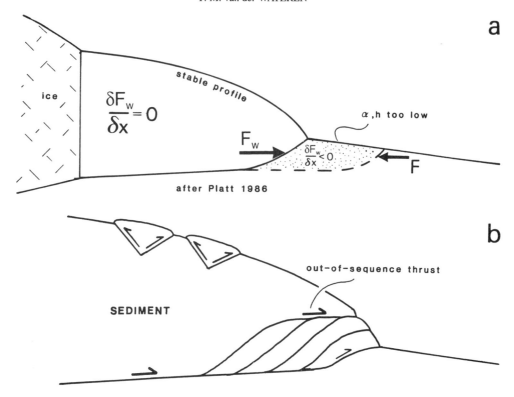

FIG. 10.18. Deforming wedge grows when the potential energy supplied by the glacier increases due to thickening of the ice under positive mass balance. The wedge develops from a stable profile (a) to an unstable profile with α and h too high in the internal part and α and h too low in the part that is to be added (b) to a stable profile again (after Platt, 1986).

density, the glacier surface near the interface with the push moraine must be twice as steep. Only if the glacier is thicker than the adjacent part of a sediment wedge will it be able to bulldoze the wedge. Thus the glacial bulldozer derives its strength from its thickness.

The two most important variables determining slope and dimensions of a push moraine are the rate of advancement of the glacier margin and deformation rate of the push moraine (van der Wateren, 1992). If the glacier advance is maintained over a sufficiently long period of time and its rate has the same order of magnitude as that of the deforming sediments, the push moraine will be large. Strong, coarse-grained sediments produce thick and rather steep push moraines, whereas relatively weak, fine-grained sediments produce extensive and flat push moraines. High advance rates produce small push moraines.

10.6. PASSIVE DEFORMATIONS

Structures produced by static loading by ice on glaciogenic sediment masses are an indication that the ice has stopped moving (Viete, 1960; Banham, 1975; Brodzikowski and van Loon, 1980, 1983; Eissmann, 1987). These structures do not normally form in the relatively narrow compressive zone near the ice sheet margin. Older load structures may be overridden and structurally modified by readvancing ice.

Two mechanisms have been proposed to explain their origin. First, they may originate beyond the margin due to loading of saturated muds by prograding sediment masses (e.g. proglacial deltas and alluvial fans). In that case they are part of an advance or readvance sequence, the latter indicating a glacial retreat (Hart, 1987; Hart et al., 1990). Secondly, they may be produced by loading of saturated tills by stationary ice masses. Surging glaciers are characterized by intensely crevassed snouts disintegrating into numerous blocks of ice, left behind after the surge has terminated. Water-saturated basal till may be squeezed up between the stagnant blocks, forming 'mud walls' as described by Gripp (1929) from several Spitsbergen glaciers (Hart, 1987; Solheim, 1988; Boulton et al., 1989; Hart et al., 1990).

10.7. CONCLUSIONS

Several generalizations can be made about sediment structural styles and their distribution in glaciated terrains. Boulton and Hindmarsh (1987) demonstrated that for undrained conditions the apparent viscosity of deforming soft sediments strongly depends of pore fluid content (Chapters 5 and 11). Since this is a non-linear relationship, small variations on the porewater pressure have strong effects on the strain rate. The abundance of meltwater under a high hydraulic head in the bed of temperate ice sheets explains the extremely high strain rates in deforming till layers. Other critical components influencing basal debris deformation are sediment porosity and permeability (Menzies, 1989a).

In nappe shear zones high porewater pressures are produced by the expulsion of water from compressible fine-grained sediments, without replacement from other sources. Since the amount of porewater present will usually be rather low, strain rates will usually be lower than in deformation tills.

These theoretical considerations are supported by the structural evidence. Shear zone geometries show great similarities, whether they are generated at the base of a nappe, or in the bed of a temperate ice sheet. They differ in the abundance of water escape and liquefaction effects and the amount of finite shear strain.

In thin section, deformation tills appear to be more intensely mixed, by alternate folding and attenuation, as compared with nappe shear zones. These tills commonly contain fragments testifying to multiple deformation episodes, such as boudins composed of a conglomerate of rounded diamict clasts (Plate 10.4). On a microscale, boudins of clay and silt are more common in tills than in nappe shear zones (van der Wateren, 1992).

The deformed substratum is the 'fingerprint' of the various strain regimes of an ice sheet. Plotting glaciotectonic styles on a map of a glaciated area may thus reliably yield locations of ice sheet margins to be used for the reconstruction of former ice sheets. The distribution of these structural styles would give the location of regions of extending and compressing flow, of ice streams and other areas of rapidly flowing ice.

A synthetic sequence of structural styles in the Holmströmbreen push moraine may serve as a basis for glaciotectonic mapping (Fig. 10.19):

(A) *Undeformed foreland*: Proglacial outwash directly related to the latest pushing event, overlying older sediments.

(B) *High-angle structures*: 'Jura-style' folding of concentric and box folds with vertical and steeply dipping axial surfaces. Thrusts are rare and steeply dipping, with small throw. Minimal horizontal tectonic shortening.

(C) *Low-angle structures*: Strongly asymmetric, overturned and recumbent folds. Low-angle thrusts. Medium shortening.

(D) *Nappes*: Extensive horizontal, relatively thin thrust sheets and internally deformed due to horizontal compression. Maximum shortening.

(E) *(Deformation) tills*: Boudinage, boudinaged folds and folded boudins, sub-horizontal shear planes, transposed foliation. Extremely high shear strain and horizontal extension. Subdivided into: Ec (compression), comprising stacked till sheets and other compressive structures; Ee (extension), with strong erosion of the substratum, comprising overdeepened (tunnel) valleys, drumlin fields and megaflutes.

B, C and D represent terrains of horizontal compression, with increasing strain from B to D, whereas E is characterized by strong horizontal extension as a result of progressive simple shear. Since the flow rate in a deforming till layer decreases towards the margin, compressive structures can be expected to occur in tills that are deposited close to the margin: hence the division of E-style terrains in a purely extensional zone and a compressive zone, Ee and Ec, respectively.

Each terrain type can be presumed to show different overprinting relationships if they are produced by either a glacial advance, or a glacial re-advance during a general retreat. Two sequences may be distinguished:

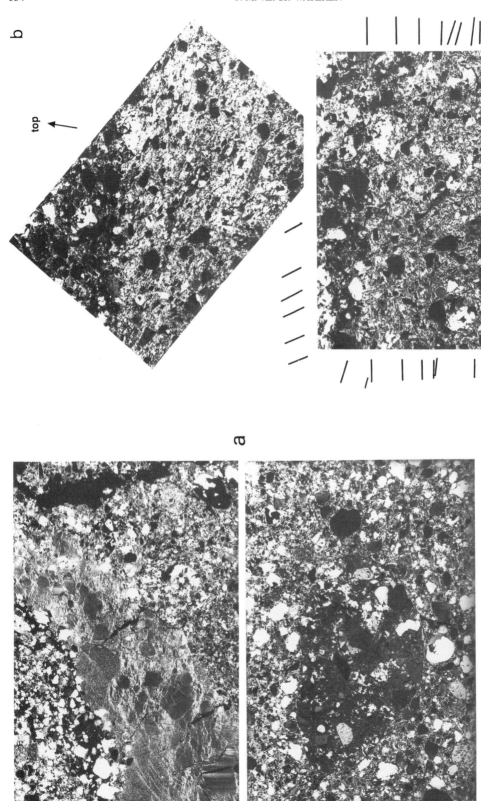

PLATE 10.4. (a) Thin section of a till from the Central Netherlands (van der Meer, 1987; van der Meer *et al.*, 1985). Crossed nicols, field width 6 mm from left to right. Rounding of fragments of a clay boudin in a sandy-silty matrix. Silt-clay lamination is a transposed foliation. Riedel shears in boudin cause it to fall apart and rotation of the small fragments produces rounded clay and till clasts testifying for progressive simple shear deformation at very high finite shear strain. (b) Two cleavage sets similar to mylonitic S–C fabric. One set parallel to the transposed foliation is cut by a steeply dipping cleavage. Cleavage asymmetry suggests right lateral (clockwise) shearing.

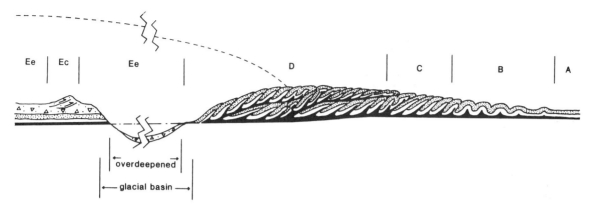

FIG. 10.19. Conceptual cross-section of a push moraine, based on the Holmströmbreen push moraine, showing glaciotectonic styles A, B, C, D and E. See text for explanation.

(1) Advance sequence (Fig. 10.20a), in which style E overprints compressive structures B, C, D and A (undeformed foreland). Boulton (1987a) argues that drumlins, cored by coarse-grained sediments, may be the result of overriding and streamlining of glacial outwash (i.e. tectonic style E overprinting A (E/A)) (Krüger, 1987). Some drumlins in northwest Germany originated as push moraines that have subsequently been overridden (Stephan, 1987). The Rehburg push moraines in western Germany (van der Wateren, 1987) and those in the northern Netherlands (van den Berg and Beets, 1987) are push moraines overlain by deformation till. These are examples of style E/(B,C,D).

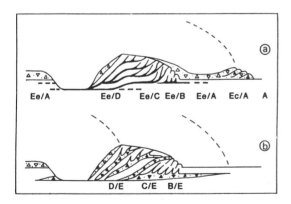

FIG. 10.20. (a) Advance sequence. Style E overprints styles A to D. Triangles denote till belonging to the glacial advance. (b) Readvance sequence. Styles B, C and D overprint style E. Black triangles denote till belonging to older advance. Open triangles denote till belonging to latest advance.

(2) Re-advance sequence (Fig. 10.20b). When a re-advance occurs during a general retreat, tills and outwash, dating from the previous glacial advance, may be incorporated into push moraines. Such moraines may be quite rare, as they will usually be eroded away by meltwater and overriding ice. Thus B, C and D overprint E style structures. The Lamstedt push moraine in Germany (van Gijssel, 1987) is an example of (C,D)/E.

In view of the different amounts of finite strain in compressive structures and subglacial shear zones, observations of interference structures E/(B,C,D) will be quite rare. The former will usually be almost completely obliterated by subglacial shear. Interference structures of successive compressive phases have a better chance of preservation. In these cases overprinting relations provide a means of relatively dating successive deformation episodes.

Mapping glaciotectonic styles is a powerful tool to reconstruct former ice sheets, if the following reservation is made. Not all types of push moraines are equally useful for a reconstruction of ice sheet margins. If the basement geology is the main factor controlling the position of a line of push moraines, individual ridges do not have to be synchronous and the use of this line for an ice sheet reconstruction is rather pointless. Stable margins are identified by stacked till sheets, dump end moraines, outwash fans and deltas, sometimes incorporated in a large push moraine.

Chapter 11

SEDIMENTARY AND HYDROLOGIC PROCESSES WITHIN MODERN TERRESTRIAL VALLEY GLACIERS

D. E. Lawson

11.1. INTRODUCTION

The sedimentary system of glaciers is defined in terms of sediment sources, processes of erosion and entrainment, modes and mediums of transport, mechanisms of primary and secondary deposition and re-sedimentation and post-depositional modifications. Glaciological and hydrological processes and factors interact with sedimentary processes to produce depositional sequences which are characteristic of the glacial environment. These deposits and sequences in turn manifest properties derived from the various processes, thereby providing a means to interpret their genesis (cf. Lawson, 1979a; Menzies, 1995b, Chapters 2, 3 and 9).

Sediments are transported directly within and on the glacier, but also as the result of the glacier's interaction with its substrate, and importantly by water flowing within and beneath the glacier. Water from several sources intimately interacts with glaciological processes, as well, thereby affecting erosion, entrainment and transport of sediment. The combined transport of sediment by ice and water results in a net flux of material to the terminus which ultimately determines the sediment yield of glacierized basins. In addition, climatic conditions that determine the rate, magnitude and timing of meltwater production and meteoric water input also determine the thermal conditions within and below the ice, conditions which affect glaciologic processes of sediment entrainment and release. Sediment discharge is therefore also linked to the nature of the englacial and subglacial drainage system, and its seasonal evolution and stability (Lawson, 1993). This chapter will focus primarily on the processes occurring beneath, within and on the surface of modern glaciers (cf. Drewry, 1986; Brodzikowski and Van Loon, 1991; Krüger, 1994).

11.2. SEDIMENT EROSION, ENTRAINMENT AND TRANSPORT

Sediment is eroded and entrained by ice from the glacier's bed and by water within the glacier's drainage system (Chapters 7 and 8). Net sediment flux of glacierized basins is therefore a combination of sediment in glacier transport and that in transport within the englacial/subglacial drainage system. Sediment sources for both ice and water include the sediments on, within and beneath the glacier, and bedrock of valley walls and floors from which new sediments are created by ice and meltwater erosion.

Streams draining tributary glacierized and non-glacierized basins may also provide significant material, while various erosional processes, including overland flow and mass wasting, may add sediment from adjacent slopes. Wind may also add a generally minor amount of sediment to the system.

11.2.1. Ice Entrainment and Transport

Sediment transport by ice is a function of the location and nature of the entrainment process and thermal and mechanical processes associated with ice mechanics (Fig. 11.1). Ice flow and sliding determine the predominant transport paths for debris derived at both the surface and the bed. In particular, sediment added to the surface of the glacier's snowpack in the accumulation area will be transported englacially, while that added to the surface in the ablation zone, without the action of processes to be described

subsequently, will be carried supraglacially (e.g. McCall, 1960; Meier, 1960; Grove, 1960).

11.2.1.1. Debris origins

Sediments in transport within or on a glacier are referred to as debris and include supraglacial, englacial and basal debris (Fig. 11.2) (e.g. Dreimanis, 1988). Subglacial sediments, which may be transported by glacier-induced deformation (e.g. Alley, 1991), form a significant erodible sediment source. Recently deposited sediments next to the ice in the terminal moraine are highly unstable, unvegetated and can be relatively rapidly eroded by any number of processes that can transport them into adjacent glacial rivers.

Numerous observations have shown that the majority of supraglacial debris, as well as englacial debris, is mainly material from valley walls that is

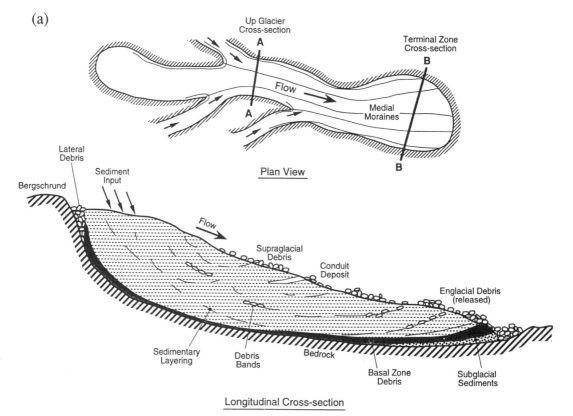

FIG. 11.1. Debris distribution within an idealized valley glacier with multiple tributary glaciers. (a) Longitudinal section and plan view.

(b)

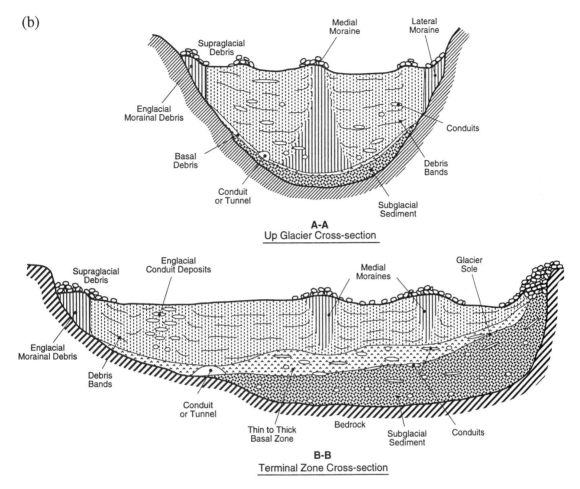

FIG. 11.1. (b) Cross sections within the central part of the glacier flow path and terminus in wide valley section.

deposited on the ice by mass movements/rockfalls, avalanches and slushflows (e.g. Tarr and Martin, 1914; Sharp, 1949; Reid, 1968). Layers of sediment deposited on snow in the accumulation area ultimately form englacial debris bands. Debris bands have a finite lateral extent and lie mostly subparallel to the ice surface. Wind may transport fine-grained material onto the glacier surface. In the accumulation area, it becomes concentrated by melt processes as it is incorporated into the perennial snowpacks, eventually forming diffuse sedimentary layers as the snow changes to ice (Meier, 1960). Streams draining adjacent valley slopes may also deposit sediments on the ice surface, which can become incorporated englacially if deposited in the accumulation area. In

the ablation zone, supraglacial streams can transport this material through moulins and conduits into englacial and subglacial locations.

Debris eroded by glacier action from the lateral confining channel walls is transported within and on the ice edge as debris septa (Sharp, 1949) to form lateral moraines that include the debris that simply falls to the ice surface from the valley walls (Fig. 11.1). Where tributary glaciers converge, lateral moraines join to form medial moraines, which are transported as longitudinal features that parallel ice flow (Plate 11.1).

Supraglacial debris is reworked and modified by weathering, especially freeze–thaw, and by rain, meltwater flow and mass movement (e.g. Sharp,

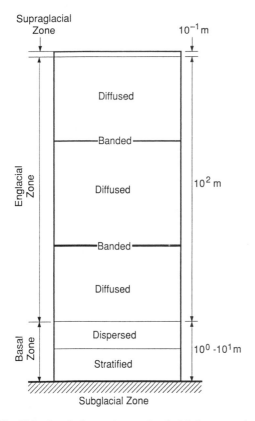

FIG. 11.2. Ice facies and associated debris zones in a vertical section (after Lawson, 1979a). Within the stratified facies, three subfacies are recognized—the suspended, solid and dispersed—each of which has different debris dispersal characteristics.

1949; Eyles and Rogerson, 1978). Supraglacial streams may rework and sort these sediments, transporting them into englacial conduits where deposition, conduit abandonment and ice creep may result in their remaining in an englacial position. Or they may be further transported into the subglacial drainage system, where basal processes will modify them. Supraglacial debris may also be incorporated into the ice by falling into crevasses that open during the course of ice flow and that subsequently close as the ice moves from zones of tensile to compressive stress.

Englacial debris derived from the accumulation area (essentially supraglacial debris) will be transported along flow lines (Fig. 11.3) and emerge at the surface of the ablation area, except if it moves to

a basal position. Transport paths from higher elevations, such as the bergschründ, may intersect the bed under proper conditions, thereby resulting in modification of the debris; here, it may be incorporated into a basal or subglacial transport mode (e.g. Boulton, 1978; Krüger, 1994). Material incorporated from lateral valley walls at a distance from the immediate ice edge is more commonly transported above the sole of the glacier and therefore out of contact with the bed (Fig. 11.3).

Sediments transported within the basal zone are primarily derived from the bedrock or sediments underlying the glacier by various processes operating at the ice/bed interface (Fig. 11.1). Particles within the ice just above the bed interact with subglacial sediments or bedrock, while effectively in tractional transport as the ice slides. In the particular case of unconsolidated bed materials, they may deform continuously, with particles being physically modified by grain-to-grain interactions (e.g. MacClintock and Dreimanis, 1964; Boulton et al., 1974). The effectiveness of subglacial deformation in causing modifications depends upon the materials' resistance to shearing, which is a function of the grain size and distribution, composition, shape and related properties, and the effective stresses generated at and below the ice/bed interface, the stress level being strongly influenced by porewater pressure (e.g. Boulton et al., 1974; Boulton, 1979; Clarke, 1987b).

Where bedrock is in contact with the base of the glacier, mechanical processes interact with meltwater to fracture and crush the rock, mainly at bed roughness features (bumps) and along existing structural features (such as joints). The strength or failure resistance of the underlying rock depends to a large degree on its composition and amount of weathering, and the presence of fractures, bedding planes, joints and other planes of weakness (e.g. Boulton, 1974, 1979; Hallet, 1974, 1979; Röthlisberger and Iken, 1981; Iverson, 1991). These particles are plucked from the bed during and following failure. The grain size of the entrained material reflects the relative importance of each mechanical process (Fig. 11.4). A detailed summary of the various mechanical and chemical processes of bedrock erosion is given by Drewry (1986).

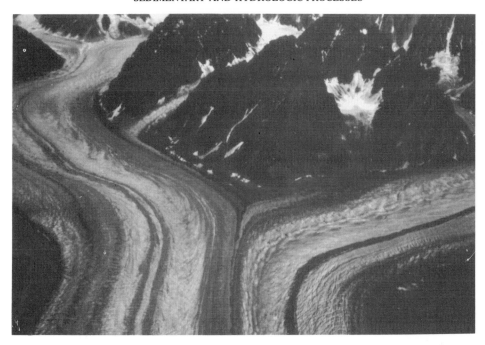

PLATE 11.1. Medial and lateral moraines developed from material of valley walls in the Alaskan Range. Medial moraines commonly separate ice from tributary valleys (photograph courtesy of Ed Evenson).

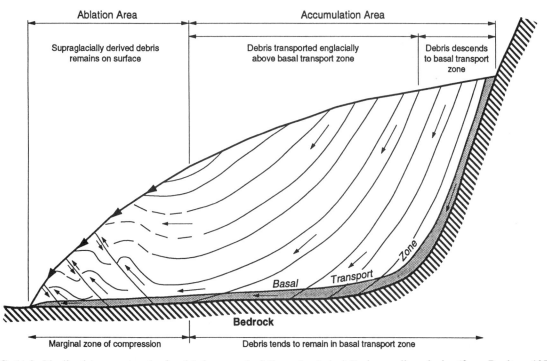

FIG. 11.3. Idealized transport paths for debris supraglacially and subglacially in a valley glacier (from Boulton, 1978; reproduced by permission of the International Association of Sedimentologists).

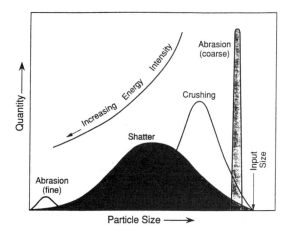

FIG. 11.4. Characteristic grain-size distributions resulting from processes of abrasion, crushing and impact (from Drewry, 1986, *Glacial Geologic Processes*, Edward Arnold).

These various mechanical processes are responsible for creating 'new' sediment, which is either entrained by the ice or by water within the subglacial drainage system. Rates of sediment production are difficult to estimate. Estimates from rock abrasion rates measured beneath glaciers and those from recently exposed bedrock substrates are of the same order of magnitude (e.g. Boulton, 1974), but clearly can differ with the rock type, degree of weathering and presence of structural weaknesses.

Sediment concentrations in meltwater discharges, which are commonly used to define a glacier's sediment yield, are also used to infer erosion rates (e.g. Thorarinsson, 1939; Østrem, 1975a; Chernova, 1981; Kjeldsen, 1981). Erosion rates approximated by this method may be reasonable where the glacier lies directly on bedrock and transports little supraglacial or englacial debris. However, such estimates become progressively less reliable as the amount of debris in transport increases, and perhaps become meaningless when the glacier substrate is predominantly sediment. In general, however, estimates based on sediment load overestimate this rate because meltwater also carries sediment entrained by thermal and mechanical erosion of debris on or within the glacier, as well as existing sediment from the glacier substrate (e.g. Gustavson and Boothroyd, 1982). Additionally, sediments can be added from adjacent glacial or glacial tributary basins

by streams entering the subglacial drainage along lateral margins and at ice tributary junctures (Evenson and Clinch, 1987), in some cases these sediments being the primary component in streams draining glaciers that lie directly on bedrock.

The method by which sediment and rock particles are incorporated into the glacier determines the thickness and debris content of the basal zone. Temperate glaciers lying on bedrock are commonly characterized by a thin (<1.0 m) debris-laden basal layer into which primarily rock fragments have been incorporated. These fragments are reasonably well-dispersed in small concentrations throughout the ice, typically at volumes of less than 5% (Lawson, 1979a; Sugden et al., 1987a; Hubbard, 1991). This material is thought to be incorporated primarily by regelation in response to pressure-melting around bedrock roughness features or obstructions to sliding (Kamb and LaChapelle, 1964; Weertman, 1964a) (Chapter 4).

Much thicker (several to 10 m or more) and more heavily debris-laden basal zones characterize glaciers where thermal conditions at the ice/bed interface permit the freezing of meltwater and the creation of new ice at the glacier sole (Plate 11.2) (e.g. Weertman, 1961b; Boulton, 1970b, 1972b; Lawson and Kulla, 1978; Lawson, 1979c; Strasser et al., 1992). During freeze-on, sediments or rock particles at or below the ice/bed interface can be included within the new ice. The resultant debris-rich basal zone is typically stratified, with layers or lenses of high and low sediment concentration (Lawson, 1979a; Sugden et al., 1987a, b). As a sediment source, limited measurements from glaciers and ice sheets suggest that this horizon commonly transports the majority of the glacier's sediment load, except in the case of an extensive supraglacial debris cover. Sediment volumes range from several percent to over 90% in individual strata, with a mean exceeding approximately 25% by volume (e.g. Gow et al., 1979; Lawson, 1979a; Hubbard, 1991).

Sediment incorporation by freeze-on results in characteristics that dominantly reflect the nature of sediments at the ice/bed interface, and hence, the processes (glacigenic, hydraulic) acting within the subglacial zone (Lawson and Kulla, 1978; Lawson, 1979a; Krüger, 1994). Their structure may be modified by subsequent ice deformation or planar

PLATE 11.2. Basal ice covered by sediment released by ablation, Matanuska Glacier, Alaska. Thick sequences of debris-laden ice develop mainly by accretion of ice and sediment to the glacier's sole. Here, the exposed sequence ranges from 5 to 7 m thick, with some overthickening by localized overthrusting of the margin to 12 m in thickness.

shearing at the new ice/bed interface, but the actual effectiveness of these processes in altering the characteristics of entrained sediments depends upon where the sediments are incorporated and the length of flow and length of time over which they are subject to basal mechanical processes (Lawson, 1988b).

Thermal conditions at the bed of glaciers are poorly characterized, either by field studies or theoretical analyses, yet appear critical to sediment entrainment mechanisms. Weertman (1961b) first theorized that, under non-steady flow conditions, changes in thermal conditions at the bed may result in large-scale entrainment of blocks or rafts of material by ice accretion. Weertman assumed that under relatively thick ice, the temperature gradient in the glacier may be insufficient to conduct viscous and geothermal heat fluxes from the bed. Dissipation of this excess heat melts basal ice, producing meltwater that flows generally outward toward the terminus in accordance with the pressure gradient. As the ice thins near the terminus margin, the temperature

gradient steepens, heat from the bed is dissipated and melting ceases. Freezing conditions develop where insufficient heat is generated to maintain the ice/bed interface at the pressure melting point. Weertman (1961b) suggested that the location of the 0°C isotherm, the boundary where temperatures closer to the terminus are below the pressure-melting point, could change in response to various factors. A shift up-glacier would depress temperatures in areas formerly at the pressure-melting point and result in freeze-on of meltwater and the water-saturated materials just below the ice/bed interface. Large-scale amounts of bed material could be included in the basal zone by this mechanism. However, because of thermal constraints, this process is probably restricted to the thinner parts of the terminus (<30 m thick) where mean annual temperatures affect bed temperatures. Thus, this mechanism may account for seasonal entrainment of sediment near the ice margin.

Boulton (1972b) suggested that the thermal regime at the base of a glacier can vary in response to certain physical parameters in accordance with Weertman (1961b). The variations result in zones where a net loss or net addition of ice and sediment can occur at the ice/bed interface. In a general sense, zones can be identified with (1) net ice loss resulting from melting by geothermal heat and the heat of sliding and deformation, (2) no net ice loss or gain, but pressure melting and refreezing (regelation) active, (3) net gain of ice by freeze-on of meltwater where the interface is slightly below the pressure-melting point, and (4) an interface where ice is frozen to the bed (Fig. 11.5). Meltwater flowing from a zone of net melting or regelation to a zone of net freeze-on would create new ice on the glacier sole and entrain sediment. The actual mechanics of this process are, however, not yet defined. The distribution of these zones beneath a glacier can theoretically vary in response to several controlling factors, including bed roughness, topography, ice flow rates and others. Complex situations, with multiple zones alternating in response to these factors, result.

Robin (1976) theorized that cold patches that are a few tenth's of a degree below the pressure-melting point and of 0.1 to 1.0 m in extent may exist, 'permanently' or intermittently at the ice/bed interface. They exist in response to heat transport with pressure and temperature variations at an interface with a thin film of water. Robin proposed that water derived from melting of ice just above the bed or from external sources can freeze to the glacier's sole in areas of either reduced stresses or reduced basal water pressures. Heat conduction by meltwater from within deforming basal ice toward the ice/bed interface (a 'heat pump' mechanism) will

temporarily cool the basal ice and thereby produce cold patches. Changes in water pressures in concert with stress variations and heat flux around bedrock roughness features could likewise produce cold patches in high stress regions. A slight but relatively rapid reduction in pressure could then temporarily result in water at temperatures below the pressure-melting point (freezing) and permit ice growth on the glacier sole. Lliboutry (1987a) similarly theorized that freeze-on over areas of several decimeters can occur on high parts of a rough bed.

More recently, researchers have hypothesized that ice accretion and sediment entrainment may result from supercooling of water within the englacial and subglacial drainage system in response to changes in pressure (Strasser et al., 1992; Lawson et al., in preparation; Alley et al., in preparation). Super-cooling in response to a rapid drop in pressure results in ice nucleation within the water, as well as ice crystal growth on conduit and cavity surfaces. Frazil ice, as the nucleating crystals are called, continues to grow and evolve within the supercooled water, forming permeable masses or flocs that attach to ice surfaces where sediments in water passing through them are trapped within the interstices. Eventually, continued ice growth forms a solid, debris-laden mass. Similarly, epitaxial ice growth on cavity walls and ceilings can both trap sediments and close conduits and cavities. Sediments in contact with the supercooled water may also be frozen to it. This new idea remains unproven, but under study.

Debris and substrate characteristics. Because of the different sources of debris, its passive or active mode of transport, and any subsequent modifications by glacigenic processes, the characteristics of debris from the englacial and supraglacial zones differ from that within the basal zone and subglacial environment (e.g. Boulton, 1970b; Lawson, 1979a, c). The primary differences are in grain size, shape and distribution. Supraglacial and englacial debris are very coarse, with few particles finer than sand-size, poorly sorted, and extremely angular (Fig. 11.6) (e.g. Tarr and Martin, 1914; Boulton, 1978; Lawson, 1979a, c; Krüger, 1994). Supraglacial meltwater flow can also remove the sand-size fraction of supraglacial debris during glacier transport, coarsening it, while processes involving freeze–thaw can break down materials into angular particles (Plate 11.3) (Sharp, 1949).

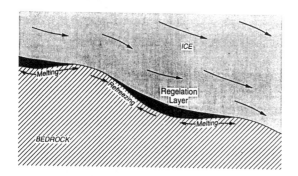

FIG. 11.5. Ice formed by regelation, with ice on the down-glacier side of bed obstacles made from refreezing of water pressure-melted from the up-glacier side of the same obstacle (after Kamb and LaChapelle, 1964).

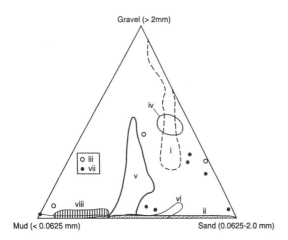

FIG. 11.6. Grain size characteristics of various ice facies from the Matanuska Glacier, Alaska. The solid subfacies has non-preferred size distribution, with extreme textural diversity and poor to very poor sorting (from Lawson, 1979a). Roman numerals refer to the following categories: (i) – supraglacial debris; (ii) – diffused facies; (iii) – banded facies; (iv) – dispersed facies; (v) – composite of stratified facies; (vi) – discontinuous subfacies; (vii) – solid subfacies; (viii) – suspended facies.

In contrast, the sediments within the basal zone exhibit a wide range of grain sizes (from clay to boulder size), with the actual grain size distribution related to the characteristics and composition of the glacier bed and the mechanisms by which they are incorporated (Lawson, 1979a, c). In addition, the amount of reworking and meltwater transport of fine-grained silt versus debris production is important in the final grain size distribution. For example, debris in basal ice of a temperate glacier lying on granitic bedrock is typically coarse, containing sand to boulder size particles, but little debris of silt or clay size (Fig. 11.7).

Debris in basal ice derived from beds of unconsolidated sediment is more complex in character (Plate 11.4). Grain size distribution ranges from clay to boulder size, exhibiting a poorly sorted, bimodal or even polymodal frequency distribution (Fig. 11.8). The actual size range, however, varies with that of the source material and the process of incorporation and therefore may vary laterally and vertically within the basal zone. Freeze-on processes produce the most variable debris characteristics

PLATE 11.3. Supraglacial debris. Meltwater erosion and topographic inversion caused by ablation rework supraglacial debris, sometimes moving it to englacial and basal positions through crevasses and moulins.

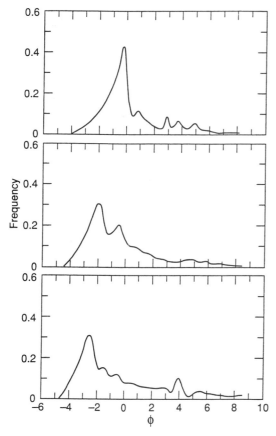

FIG. 11.7. Typical grain size distribution curves for debris from dispersed facies of the basal zone (from Lawson, 1979a); this facies is considered to be derived primarily from bedrock by the regelation process.

as meltwater (Menzies, 1995b, Chapter 15). Substrate characteristics are as varied as glaciers themselves, and will reflect the geologic history prior to glaciation and the subsequent effects of the glacier while above it. Glaciers may lie directly on bedrock stripped clean of sediment by previous glaciations; while others develop upon sediments deposited by any number of non-glacial, glacial or proglacial processes. Perhaps more commonly beneath larger glaciers and ice sheets with lengthy flow paths, materials at the ice/bed interface may be in part bedrock and in part sediment. The characteristics of the debris and ice of the basal zone of the Matanuska Glacier in south-central Alaska, for example, suggest that the bed has an up-glacier section primarily of bedrock and a down-glacier section of mainly unconsolidated sediment (Lawson, 1986, 1988b; Lawson et al., 1991). Because of the effects of bed topography, ice flow may result in localized accumulations of subglacially eroded sediment on a bed that otherwise is mostly bedrock (Hooke, 1991). With time, the interaction of the glacier with the bedrock or sediment and the erosion or deposition of these materials by subglacial meltwater will determine the overall characteristics of the bed.

Because of the inaccessibility of the subglacial environment, the nature of the ice/bed interface under active ice is poorly understood. Studies of ancient and recently exposed bedrock from beneath glaciers, some with thin, discontinuous mantles of rock particles and fragments, have improved our understanding of subglacial processes (e.g. Hallet, 1974; Walder and Hallet, 1979; Hallet and Anderson, 1980; Sharp et al., 1989; Sharpe and Shaw, 1989). Glacier interactions with beds of sediment are not as well defined by such studies because the beds are reworked by various subglacial processes during and immediately following deglaciation (e.g. Church and Ryder, 1972; Boothroyd and Ashley, 1975; Lawson, 1979a, 1982, 1988a; Sharp, 1982; Dreimanis, 1988; Krüger, 1994).

Theoretical treatments and observations of the effects of ice flow indicate that subglacial sediments are deformed (e.g. Boulton and Hindmarsh, 1987; Engelhardt et al., 1978; Clarke, 1987b). Shear stresses generated by glacier flow can mechanically abrade and comminute the coarser sedimentary

because they can incorporate sediment without regard to size. Silt from meltwater is commonly important and may dominate the distribution. Within the basal zone, debris occurs as individual lenses and layers surrounded by clear ice and can be well sorted and with a limited grain size where freeze-on incorporates material with such properties (Plate 11.5). Gravel-size clasts (granules to boulders) are less angular than supraglacial and englacial debris, often having rounded edges and spheroidal to prolate shapes, similar to those of water-worn stream gravels (e.g. Lawson, 1979a, c; Boulton, 1978; Krüger, 1994).

The characteristics of the glacier's substrate are thus important in determining basal debris characteristics and sediment transport by ice, as well

PLATE 11.4. Basal debris. Particles released by ablation of basal zone ice. Fine-grained components are being removed by meltwater flow down the sloping ice surface. Mostly subround to round shapes characterize subglacially derived gravel sized materials. Notebook for scale.

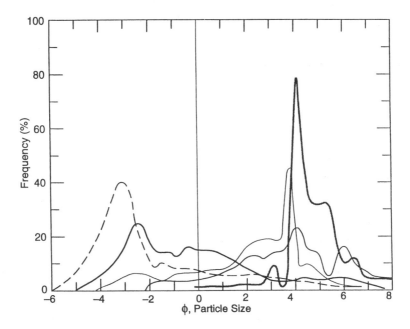

FIG. 11.8. Typical grain size distribution curves from sediment within the stratified facies of the basal zone of the Matanuska Glacier, Alaska (from Lawson, 1979a). This facies is mainly derived from net freeze-on of sediment by ice accretion to the glacier sole.

PLATE 11.5. Examples of debris in the basal zone formed by accretion processes. (a) Clear ice separates and surrounds sediment, which may occur in layers or be dispersed within it. Sediment ranges from clay to pebble size (scale in centimetres). (b) Well-stratified basal zone ice. Sediments are mainly layers of silt and sand, with occasional pebbles (scale in centimetres).

particles (e.g. Boulton *et al.*, 1974; Sharp, 1982; Boulton, 1979) and, under proper conditions, will result in transport and a net flux of sediment to the glacier terminus (e.g. Boulton *et al.*, 1974; Alley *et al.*, 1987a, b; Boulton and Hindmarsh, 1987; Clarke, 1987b; Echelmeyer and Wang, 1987; Alley, 1989b, 1991). Depths of deformation may exceed 6 to 7 m and depend on shear stress and effective pressures in the bed (Fig. 11.7) (Chapters 4 and 5).

Theory indicates that a glacier resting on unconsolidated sediment can move by discrete sliding between the ice and bed, ploughing clasts through the uppermost materials (Brown *et al.*, 1987), by pervasive deformation, or by shearing across discrete planes at various depths within the bed materials (Fig. 11.9). The type of drainage system (distributed, conduit-tunnel or linked-cavity) affects the magnitude of the sliding rate and depth of

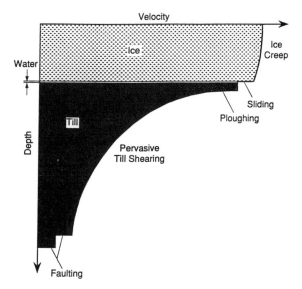

FIG. 11.9. Possible mechanisms of ice motion on a bed of unconsolidated sediment, with deformation by pervasive shearing being the principal velocity component in wet-based glaciers (after Alley, 1989b). Deformation can result in substantial flux of sediment to the terminus (Alley, 1991; reproduced by courtesy of the International Glaciological Society from *Journal of Glaciology*, 37(125), 1989, p. 69, Fig. 1).

deformation (Alley, 1989a, b) (Chapters 5 and 6). Textural properties of the bed material, particularly the number and size range of clasts, also affect basal sliding rates and the importance of ploughing of clasts at the interface. In the case of water drainage in a distributed flow system, pervasive deformation of the underlying sediments may account for up to 60 to 100% of the total velocity of the glacier's motion (Alley, 1989b).

11.2.1.2. Debris flux

The transport or flux of debris to the terminus of a glacier determines the long-term availability of sediment for entrainment by ice marginal and basin tributary streams. Sediment flux can be estimated on the basis of several parameters; however, in each case, this estimate is limited by the complex spatial and temporal variability in glacial debris, including that in a subglacially deforming layer. The total sediment budget consists of the flux of debris in transport and 'new' sediment, produced by glacier

erosion, that is entrained and transported in the subglacial drainage system.

Alley (1991) has suggested that sediment transport by bed deformation can be estimated by assuming that the depth of deformation is directly related to the basal shear stress, the sediment's yield strength and the effective pressure as it varies with depth. Velocity is determined by an inverse relationship between the change in basal shear stress and yield strength with depth as a function of the effective pressure at depth. In simplified form, the sediment flux by pervasive deformation SQ_D is

$$SQ_D = \alpha_d h_b u_b \qquad (11.1)$$

where h_b= depth of deforming layer; u_b=velocity at top of the deforming layer, approximately equal to u_i, the ice velocity; and α_d =constant dependent upon sediment layer thickness and water pressure, ranging from about 0.1 to 0.5.

The consequence of sediment reaching the glacier terminus is determined by the relationship between the sediment motion and ice sliding velocity, assuming they are equal, and the velocity of the glacier at its terminus (u_L). If $u_L \geq \alpha u_b$, the deforming layer must erode at the terminus to maintain a constant deforming thickness h_b during the advance. Without erosion, the deforming layer will progressively spread over a longer and longer distance and become thinner and thinner. If $u_L = \alpha u_b$, then there is neither erosion nor deposition and, if $u_L < \alpha u_b$, deposition occurs (Alley, 1991). These relationships clearly affect estimates of the probable long-term sediment flux and potential sediment discharge to glacier outlet streams.

Sediment flux resulting from the transport of debris within the basal ice zone can only be grossly estimated at present. Several key parameters required for sediment flux estimates, including sediment volume and its spatial and temporal distribution, are poorly defined by measurements. The principal assumptions for flux calculations are: (1) that sediment volumes are constant spatially within the basal zone and will not vary with time; (2) that the general distribution of debris-laden basal ice within the glacier is constant; (3) that a limit to the extent and thickness of the debris-laden basal ice must be based upon a presumed thermal regime at the bed;

and (4) that ice velocity and thermal regime are constant over time.

The flux of basal debris in ice per square meter width of glacier (SQ_B) can then be estimated by (Drewry, 1986)

$$SQ_B = u_b \, h_b \, c_b \qquad (11.2)$$

where h_b = thickness of basal ice layer; u_b = basal sliding velocity; and c_b = debris concentration in the ice.

Because englacial debris is widely dispersed and highly variable in concentration, it is difficult to estimate englacial transport volumes and debris flux. In most cases, englacial debris flux (SQ_E) is a small-percentage of the total sediment flux and some constant value might be assigned for long-term estimates. It might be better to estimate volumes of englacial debris from volumes of material exposed supraglacially by ablation below the equilibrium line and, as with basal debris, to make assumptions on debris dispersal and its variability with time (Chapter 5).

Supraglacial debris transport (surficial deposits, medial and lateral moraines) can be more accurately estimated if its extent and thickness can be measured. Debris concentration at depth below the moraines must be estimated as an englacial component and is usually presumed constant with depth. Thus, supraglacial and morainal band sediment flux SQ_s is

$$SQ_s = u_s \, h_{sp} \, w_{sp} + u h_m \, w_m \, c_m \qquad (11.3)$$

where u_s = surface ice velocity; h_{sp} = thickness of supraglacial debris; w_{sp} = width of debris cover; c_m = debris concentration in moraine septa; u = ice flow or creep velocity; h_m = thickness of debris septa taken equal to ice thickness; and w_m = width of debris septa.

Ice velocity u is commonly estimated from measurements of surface velocity based upon the flow law for ice (Chapter 3).

Thus, the total annual sediment flux for glacier or ice transport can be represented as the sum of each debris flux component, integrated with respect to the duration or length of flow

$$SQ_T = \int_0^t (SQ_D + SQ_E + SQ_S + SQ_B) dt) \qquad (11.4)$$

For a glacier with a substrate that is mostly bedrock, SQ_D is assumed = 0.

11.3. WATER ENTRAINMENT AND TRANSPORT

The fluvial entrainment of sediment by water moving through the englacial and subglacial drainage system, as well as its role in subglacial erosion of the bed (e.g. Sharpe and Shaw, 1989), are poorly characterized by either direct measurements or theoretical hypotheses of the thermal and mechanical processes of meltwater erosion.

Sediment concentrations in supraglacial streams and the upper reaches from englacial streams range from 0 to in excess of 60 g l^{-1}, but typically range from 50 to 400 mg l^{-1} (e.g. Dowdeswell, 1982; Hammer and Smith, 1983; Fenn, 1983; Gurnell, 1987a). This value is strongly influenced by the extent and textural character of supraglacial and englacial debris, with streams draining supraglacial moraines having the highest suspended sediment concentrations (Chapter 6).

Debris in englacial and basal ice can be released by melting of conduit and tunnel walls by water in response to viscous heat dissipation during periods of increased discharge (Röthlisberger, 1972). Similarly, expansion of a linked-cavity system can release basal debris from orifice and cavity roofs, while in a distributed flow system, pressure-melting of the glacier sole can release debris. Rates depend on the concentrations of debris within the ice and the quantity of heat available. The volume of debris released by heat dissipation in conduits and cavities has not been measured.

Similarly, there are few estimates based upon direct measurements of, or theory concerning, the rate of mechanical erosion of bedrock (abrasion, cavitation) and the volume of this material that is entrained by water flow at the ice/bed interface. Vivian (1970) estimated bedrock erosion rates of about 20 mm year^{-1} based upon analyses of pothole formation below glaciers, and correlated these rates with experiments on abrasion of quartzite. Rates of abrasion by sediment-laden meltwater depend to a large degree on the grain size, mineralogy and shape of the particles in transport, the total amount of suspended sediment, flow velocity and the composition of the bed (Bezinge and Schafer, 1968; Bogen, 1989).

Bogen (1989), based upon the work of Bouvet (1958), suggested that the abrasive capacity of sediment-laden meltwaters (s) might be represented as

$$s = Q_s r_s q_z q_s \qquad (11.5)$$

where Q_s= annual suspended sediment load; r_s= grain shape, values ranging from 1 to 5 for well-rounded to angular particles; q_z= proportion of quartz particles to all other mineral particles in the 0.063 to 0.125 mm size range; and q_s= proportion of sand-size particles (0.063 to 0.5 mm diameter) to total sediment weight.

Bogen (1989) suggested that this relationship could define the magnitude of turbine wear in hydroelectric power plants caused by annual glacial discharges. Various other theories on the processes of meltwater abrasion and cavitation are detailed by Drewry (1986) (Chapter 6).

The largest potential source of material for meltwater entrainment is unconsolidated materials in contact with the ice. Theoretical estimates of the erosive potential of conduits in unconsolidated sediment may be based upon traditional river flow theory (e.g. Allen, 1982), assuming a configuration for the subglacial passageway (R- or N-conduits and tunnels) and appropriate water and ice pressures.

Alley (1989a) estimates sediment flux through a subglacial conduit embedded within sediment by assuming that it must equal the influx of sediment by creep into the conduit. Water flow can generally be assumed to be turbulent within conduits (as well as within the 'orifice' or connecting conduit of the linked-cavity system (Kamb, 1987)) and its velocity can be defined using the Manning formula (Chapter 6). Thus, mean velocity ú is

$$\bar{u} = M^{-1} i_{1/2} r^{2/3} \qquad (11.6)$$

where M = Manning roughness coefficient; r = channel radius; and i = local hydraulic pressure gradient.

The erosion rate Ψ_e is then given by (Alley, 1989a):

$$\Psi_e = \frac{1}{2\pi r_2} \partial \frac{J_s}{\partial x} = \dot{r}/r \qquad (11.7)$$

where x is distance measured along the channel, $\dot{r}$ is the rate of change in r and J_s is sediment flux in the channel as given by:

$$J_s = J_0 \pi r_2 \bar{u}^3 \qquad (11.8)$$

where J_0 is a constant.

Alley (1989a) modifies this equation by substituting the value ú into Eqn. (11.8), differentiating with respect to x, and substituting the x derivative of J_s into Eqn. (11.7). Based upon the relationship that

$$Q = \pi r^2 \bar{u} \qquad (11.9)$$

where Q is water flux, the erosion rate for turbulent flow Ψ_{et} in terms of the water and sediment flux is

$$\Psi_{et} = \frac{3J_0 M^2 i Q_x}{4\pi r^{2/3}} . \qquad (11.10)$$

A similar expression for laminar flow Ψ_{el} is (Alley, 1989a):

$$\Psi_{el} = \frac{J_0 i^2 r^2 Q_x}{64\pi \eta_w^2} . \qquad (11.11)$$

where η_w is the viscosity of water. As stated earlier, the erosion rate must equal or exceed the creep closure rate (Ψ_c) for the conduit to remain open (see Eqns (6.14), (6.15), (6.16), Chapter 6).

The quantity of sediment eroded and transported by conduit and tunnel flow can be limited by the capacity of the meltwater discharge. This limiting condition will affect not only the sediment flux but also limit the conduit and tunnel dimensions. Theoretical estimates of sediment flux by meltwater erosion of an unconsolidated bed for various subglacial drainage types and configurations have not yet been made.

11.3.1. Sediment Flux Relationships

If we assume that the amount of sediment in transport is mainly a function of the availability of sediment at the ice/bed interface, including the debris-laden basal ice, and of the action of meltwater within the englacial–subglacial drainage network at the interfacial zone, qualitative relationships between sediment transport–sediment yield and subglacial glaciohydraulic processes can be deduced from data obtained by monitoring the waters emerging at the glacier terminus (Collins, 1979a, b). Such an assumption is required because of our lack of understanding of subglacial sediment transport processes. One of the longest multiyear records of sediment discharge in relation to water discharge and various water quality and meteorological parameters is that measured by Collins over the last 15 years at Gornergletscher and Findlngletscher near Zermatt, Switzerland, two of the glacierized basins feeding the Grande Dixence S.A. hydroelectric scheme. In addition, various hydrological and glaciological studies have been supported by Grande Dixence over the last 40 years at these and other glacierized basins in this region.

On the basis of these long-term records, Collins (1986, 1988, 1989) has concluded that the seasonal patterns in sediment flux are related directly to the seasonal development of the subglacial drainage network, and further, that short-term variations in sediment discharge are mainly a function of sediment availability for subglacial hydraulic processes. The latter relationship is also a function of the previous history of hydrological events that have removed (or deposited) bed sediments. Subglacial hydrological events may include discharge variations as the englacial–subglacial drainage system evolves each spring, sudden drainage of ice-dammed lakes, temporary blockage of subglacial passageways by ice flow, relocation or re-establishment of subglacial conduits, and flooding caused by rain. Feedbacks exist between the factors or processes, but sediment discharge may also vary irregularly (Chapter 6).

The type of subglacial drainage network will determine both the area of the bed over which water flow can withdraw sediment and the timing of active sediment withdrawal (Lawson, 1993). For example, in the case of a glacier in which subglacial drainage consists of both a linked-cavity system and a conduit-tunnel system developed on bedrock, with only a thin or discontinuous sediment layer, seasonal variations in the extent and operability of each system affect sediment discharge at the terminus (e.g. Iken et al., 1983; Iken and Bindshadler, 1986). Meltwater reaching the bed in spring progressively re-establishes and expands the linked-cavity system, resulting in increased porewater pressures, increased ice flow rates and increased sediment discharge.

Sediment transport may also vary rapidly as different parts of the bed are connected to the drainage system and sediment is progressively, but episodically, evacuated from new areas. In addition, early in the melt season, diurnal variations in meltwater input may result in episodic opening and closing of conduits connecting cavities (e.g. Iken and Bindshadler, 1986). Subsequently, discharge via the conduit-tunnel system becomes increasingly more important, relative to the linked-cavity system, as meltwater input increases into the summer season. Overall, water pressures decrease and the actual area of the bed from which sediment can be eroded decreases. Sediment discharge progressively drops through the summer, with the frequency of large variations that often characterize early summer runoff decreasing. Then, high sediment discharge results mainly from floods (e.g. Collins, 1989b).

The lower sediment flux that characterizes many glacierized basins late in the melt season apparently reflects the presence of a fully developed drainage system on a bed that is mostly rock, and an exhaustion in the supply of sediment at the bed/bed interface (e.g. Østrem et al., 1967; Fenn, 1983; Collins, 1979b, 1988). The overall seasonal trend in sediment supply reflects sediment being produced by glacier erosion during the winter period of low or absent runoff and remaining at the ice/bed interface until spring, when it is readily available for erosion and transport by meltwater runoff early in the ablation season.

As the melt season progresses, sediment is less readily available and meltwater less likely to contact it. Because meltwater erosion is more 'efficient' than ice erosion, abrasional products may be depleted during the melt season, and, thus, the volume of potentially erodible material adjacent to conduits and tunnels and across major parts of the bed may be exhausted (Collins, 1988, 1989b). Only through rapid, large volumetric increases in discharge, such as caused by precipitation or catastrophic floods, will the drainage system expand beyond primary conduits and tunnels, and thereby reach previously inaccessible parts of the bed with erodible sediment (e.g. Röthlisberger, 1980; Collins, 1986). Theoretically, this expansion is temporary and any conduits formed in either the linked-cavity or conduit/tunnel system would be unstable as the discharge decreases following passage of the flood; the drainage system would then return to its previously stable arrangement (Walder, 1986).

The nature of the bed is an integral factor in the sediment discharge–drainage system relationship. When the substrate is bedrock, it is a primary but exhaustible source of glacier erosion by-products. The bed acts as a secondary sediment source through its role as a storage medium. When the substrate is sediment it's an erodible and deformable, more or less constant, primary sediment source.

In the situation where a glacier lies on a bed of sediment, supply is apparently less important than drainage system evolution, meltwater supply or the factors determining meltwater erosion rate. The grain size distribution, consolidation history, piezometric pressure and other geotechnical properties determine the deformability and erodibility of sediment beds (Clarke, 1987b). The nature of the drainage system and the water pressure-water discharge relationship, however, determine the amount of channel or conduit erosion that can take place and the area of the bed from which sediment can actually be withdrawn (e.g. Alley, 1989a). The existence of subglacial cavities will be related to the texture and geotechnical properties of the substrate material and control whether a linked-cavity or conduit-tunnel system prevails (e.g. Fowler, 1987a; Kamb, 1987).

On bedrock substrates, similar relationships exist but other factors and potential feedback mechanisms include (1) bedrock erodibility and thus the potential rate of sediment production, (2) bed roughness, which will affect drainage system type and the amount of potential storage of sediment, (3) seasonal evolution of the drainage system, which may happen more rapidly on bedrock than on sediments owing to existing topographic controls on drainage, and (4) sediment supply, which may be exhausted rapidly. Where glaciers lie on a highly resistant bedrock substrate with a well-established conduit-tunnel system, linked-cavity system, or N-channels following bedrock structures, areas of the bed may remain inaccessible to runoff for long periods and thus the abrasional products may accumulate. Only through events that flood most if not all of the bed

would such areas be evacuated of sediment (Fowler, 1987a). Then the drainage temporarily shifts from a distributed film to conduit flow capable of sediment erosion (Alley, 1989a) and produces sudden sediment discharges. Glacier surging may represent special conditions of high water pressures that result in large, inaccessible areas of bedrock suddenly being evacuated of sediment (e.g. Humphrey, 1986, 1987; Humphrey et al., 1986).

The annual yield of sediment therefore varies with bed composition, glacier dynamics, climate and the nature of the drainage system. On substrates of bedrock, its composition and roughness are directly linked to sliding, cavitation and erosion by ice, and in a less direct way to the drainage system types and their extent across the bed. Annual variations will reflect longer-term storage of sediment in areas not drained repeatedly by meltwater flows. In a similar manner, the climatically controlled annual variations in precipitation and glacier mass balance and the frequency of extremes in precipitation (both high and low) will determine the area of the bed across which a drainage system can exist stably and erode sediment.

On substrates of sediment, supply is not as significant an issue as are the climatically controlled volumes of meltwater and precipitation, and the drainage system morphology and its annual evolution.

For both types of substrates, our lack of knowledge on glaciohydraulic processes of sediment production and transport and the glaciohydrologic processes of meltwater production, drainage and storage hamper attempts to define quantitative relationships for sediment flux (Lawson, 1993).

11.3.2. Sediment discharge characteristics

In general, rivers draining glacierized areas transport significantly more solid matter in suspension and bedload annually than non-glacial rivers under their normal flow conditions, with the suspended load increasingly important at distances downstream of the glacial source (e.g. Østrem et al., 1973; Guymon, 1974; Church and Gilbert, 1975; Bogen, 1986; Gurnell, 1987a). Guymon (1974), for example, concluded that glacial rivers in Alaska have suspended sediment concentrations of 10 to 20 times greater than those of non-glacial streams. Sediment delivery rates of glaciers and glacial streams are extremely variable, however, from glacier to glacier and from year to year (e.g. Østrem, 1975a; Guymon, 1974).

The sediment load is transported within a limited part of the year, with up to 95% transported during the melt season, the duration and timing of which vary with latitude and altitude. In south-central Alaska, for example, sediment is discharged between late May and early October, but between mid-June and late August in northern Alaska. Sediment transport during winter is severely reduced owing to the seasonal controls on the glaciohydraulic processes supplying sediment to glacial discharges.

Sediment discharge in rivers draining highly glacierized basins can be extremely variable over hours, days or seasons (Rainwater and Guy, 1961; Vivian, 1970; Vivian and Zumstein, 1973; Østrem, 1975a; Collins, 1979a; Bogen, 1980, 1989; Beecroft, 1981; Hagen et al., 1983; Gurnell and Fenn, 1984b, 1985). Volumes can change rapidly within a period of several hours or less (e.g. Østrem et al., 1967; Collins, 1977, 1979a; Tømasson et al., 1980; Beecroft, 1983; Humphrey et al., 1986), while a large proportion of the total annual yield may be released over 1 to 2 days (Fig. 11.10) (Collins, 1979a). Østrem et al. (1967), for example, measured a release of

nearly 60% of the total annual load in a single 24-hour period. Flash floods can transport large quantities of sediment (Fig. 11.11), and annual yields are highest in years with multiple floods (Bogen, 1989). Thus, depending upon the dates of measurement, sediment rating curves developed from daily totals will differ (Figs 11.12 and 11.13, Østrem, 1975a, b). Similar variabilities can characterize the differences in bed load, suspended load and sediment yield between adjacent glacierized basins (Fig. 11.14) (e.g. Gurnell et al., 1988).

Sediment discharge from surging glaciers can also be extreme, varying by an order of magnitude greater than annual sediment concentrations before the surge, yet occur with only a small increase in water discharge (Humphrey, 1986; Humphrey et al., 1986). Using the surge of Variegated Glacier (Humphrey, 1986) as an analogy, Clarke et al. (1986a) predicted that a surge in Susitna Glacier could discharge roughly 30 times the estimated annual sediment load into the proposed Watana Reservoir on the Susitna River, Alaska (Randm, 1982). Such events and the inherent variability of sediment discharge are strong evidence that aperiodic measurements are probably not representative of a basin's sediment yield over the short or long term.

The inherent variability in sediment discharge from glacierized basins reflects the characteristics of sediment in streams flowing from the ice. At the

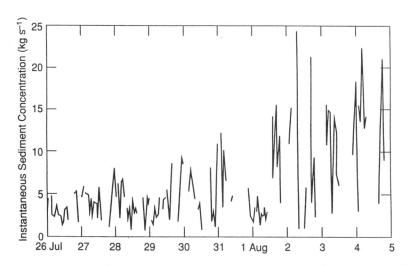

FIG. 11.10. Diurnal ranges in suspended sediment transport, Görnera stream, Görnergletscher, Switzerland (after Collins, 1979a; reproduced by courtesy of the International Glaciological Society from *Journal of Glaciology*, **23**(89), 1979, p. 254, Fig. 5). Fluctuations of over 100% occurred within 1- to 2-hour periods without corresponding large increases in water discharge.

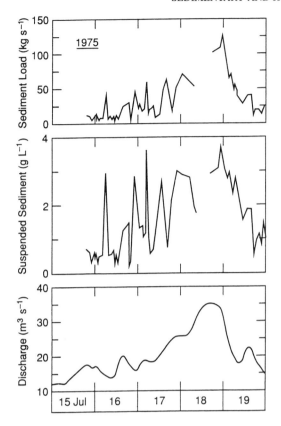

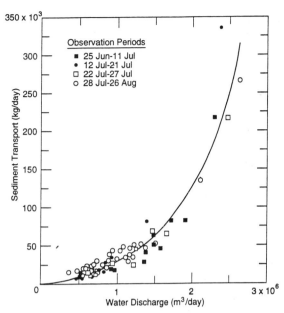

FIG. 11.12. Sediment transport rating curve obtained for a small glacier in northern Norway (Trollbergdalsbreen), based upon daily totals. The observation period is divided into four subperiods, each of them shown by different symbols (after Østrem, 1975a, reproduced by permission of *SEPM*).

FIG. 11.11. Suspended sediment discharge during the drainage of the ice-dammed lake Gornersee through the Gornergletscher, Switzerland, 15–19 July 1975. Rapid, short-term increases in suspended sediment transport characterize the rising limb of the flood hydrograph, while total sediment load peaked prior to peak flood discharge (after Collins, 1988, reproduced by permission of *Mitteilungen der Versuchsanstalt für Wassebau, Hydrologie und Glaziologie*).

glacier margin, transport measurements indicate that the proportion of suspended load to bedload varies with the nature of the sediment supply (volume, location) and the internal drainage system (Fig. 11.15). Østrem *et al.* (1973), for example, found that suspended sediment composed from 50 to ~80% of the total sediment load at the snouts of five Norwegian glaciers. Church (1972) reported that suspended load was about 20 to 80% of total load in the Lewis River, Canada, while at Nigardsbreen, Norway, Bogen (1989) found the percentage of bedload to vary between about 30 and 60% of the total load (Fig. 11.16) (Chapter 12).

Bedload may only travel several kilometers or less downstream in a year and, at distances further downstream in the basin, it may be indistinguishable from bedload derived from non-glacial sources. The proportion of total load composed of bedload therefore also decreases downstream, with suspended load exceeding bedload by 3 to 6 times within 10 km or so of the glacier, but by up to 100 times at distances of 100 to 200 km from it (Harrison, 1986). The latter relationship is illustrated by the Tanana River, at Fairbanks, Alaska, where the suspended load to bedload ratio is 99:1.

Quantitative estimates of bedload volume have been attempted by a number of researchers, including Østrem (1975a), who trapped all bedload material behind a steel fence on the river draining from Nigardsbreen Glacier in southwestern Norway. Over a 3-week period, approximately 400 metric tons of bed material was captured by this net. Bedload in multiple glacierized basins was estimated by Bezinge (1981) to range in volume from 80 to 450 m^3 km^{-2} per

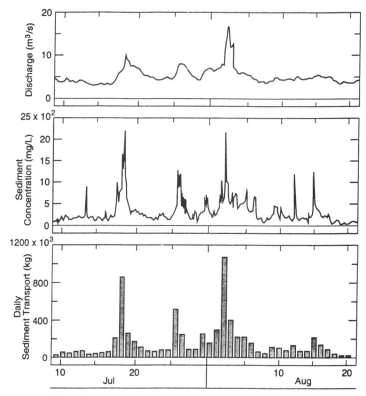

FIG. 11.13. Discharge, suspended sediment concentration and daily suspended sediment transport data for Erdalsbreen, Norway (1967). Hourly and daily variations characterize sediment concentrations (after Østrem, 1975a, reproduced by permission of *SEPM*).

year from gravel traps at the Grande Dixence hydroelectric project. More recently Bezinge *et al.* (1989) used Grande Dixence sediment trap data to estimate that bedload had an annual mean of 12 to 14 $\times 10^3$ metric tons over the period 1977–1987, composing between 25 to 46% of the total load, depending upon the analysis used. When data from multiple basins within one or more regions are compared, bedload volumes vary widely without a consistent pattern (e.g. Church, 1972; Laufer and Sommer, 1982; Hammer and Smith, 1983; Bezinge, 1981; Bezinge *et al.*, 1989; Bogen, 1986, 1989).

Variations in suspended sediment load at the glacier outlet also characterize glacially fed rivers further downstream, reflecting the sediments fine-grained nature and ease of transport. Suspended sediment concentrations in waters discharging directly from glaciers typically range from 0.5 to 5 g l^{-1} (Plate 11.6), with peak or maximum values

reported to range from 3.0 to 10 g l^{-1}, and over 20 g l^{-1} in extreme sediment discharges associated with early season floods or drainage of ice-dammed lakes (e.g. Vivian, 1970; Vivian and Zumstein, 1973; Østrem, 1975a; Bezinge, 1978; Collins, 1978; Chernova, 1981; Gustavson and Boothroyd, 1982; Gurnell, 1982; Smith *et al.*, 1982; Hagen *et al.*, 1983; Richards, 1984; Parks and Madison, 1984; Cowan *et al.*, 1988). Record values of 36 g l^{-1} on the Drance River and 150 g l^{-1} on the Isere were measured by Parde (1952) as reported by Bezinge (1978). Downstream, away from the ice sources, mean suspended sediment concentrations are typically less than 1.5 g l^{-1} (Gurnell, 1987a), but may reach 3.0 g l^{-1} in large Alaskan rivers (e.g. Guymon, 1974; Coffin and Ashton, 1986). During winter, minimal amounts of suspended sediment are in transport, with concentrations typically ranging from 0 to 0.01 g l^{-1} (Parks and Madison, 1984).

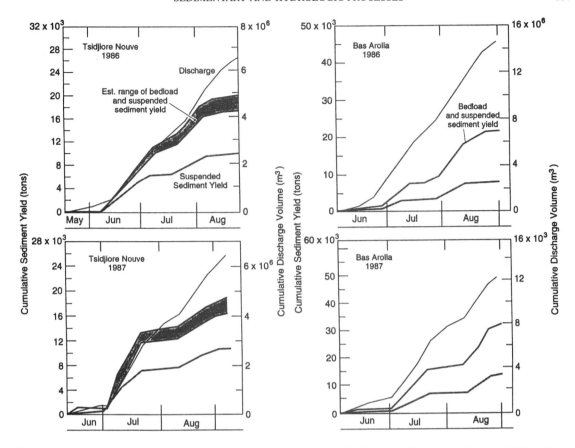

FIG. 11.14. Accumulated weekly discharge, suspended sediment load and bed load yield from the adjacent Tsidjiore Nouve and Bas Arolla basins 1986 and 1987 (after Gurnell *et al.*, 1988, reproduced with permission from the International Association of Hydrological Sciences).

Given the symbiotic nature of water and sediment in transport, studies have analyzed the possibility of a direct relationship between suspended sediment load and discharge. No simple correlation apparently exists between them, either within a basin (Fig. 11.17) or in adjacent basins (Fig. 11.18) (e.g. Østrem, 1975a; Collins, 1978, 1988; Gurnell, 1987a). Maximum suspended sediment discharge can commonly take place before maximum runoff, either daily or annually. Over the year, for example, suspended sediment discharge is generally higher early in the melt season and decreases through the season until it is proportionally lower in volume relative to water discharge in the late summer (Fig. 11.19). This seasonal trend appears to be related to the availability of sediment, which is potentially

greater after the low-flow winter when sediments accumulate below the glacier.

Over shorter periods this relationship can differ for the rising and falling limbs of the discharge hydrograph (Fig. 11.20) (Østrem, 1975; Collins, 1979a; Church and Gilbert, 1975). Plots of sequential sediment discharge versus water discharge measurements with this relationship can display a hysteresis loop to the curve (Fig. 11.21). A clockwise hysteresis means that there are greater sediment concentrations on the rising limb of the diurnal hydrograph than at equivalent discharges during the falling stage. Østrem (1975a) found hysteresis effects several times through the summer melt season for periods lasting several days or more in conjunction with abnormally high discharge events (Figs 11.21

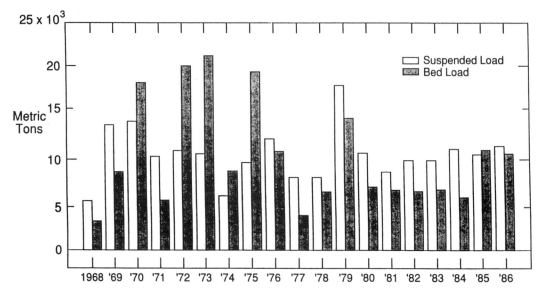

FIG. 11.15. Comparison of annual bedload and suspended load, 1968–86, in the stream draining Nigardsbreen, Norway (after Bogen, 1989, reproduced by courtesy of the International Glaciological Society from *Annals of Glaciology*, **13**, 1989, p. 7, Fig. 2).

and 11.22). Clockwise hysteresis loops that included strong involutions were also found to recur daily by Collins (1979a) during extended periods with high rates of ice ablation and snowmelt (Fig. 11.23). Involutions reflect irregular variations in sediment concentration, which Collins (1979a) interpreted as changes in sediment availability within or at the bed of the glacier. Sediment supply problems may also be responsible for hysteresis effects over longer periods

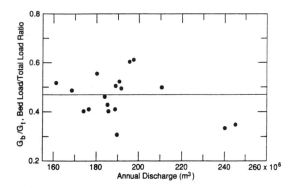

FIG. 11.16. Ratio of bedload to total load in relation to discharge for the stream draining Nigardsbreen, Norway (after Bogen, 1989, reproduced by courtesy of the International Glaciological Society from *Annals of Glaciology*, **13**, 1989, p. 7, Fig. 3).

following major meltwater discharge events (Gurnell, 1987a). Potentially, discharge vs suspended sediment relationships could be developed on the basis of different parts of such hydrographs; however, Østrem (1975) and Fenn (1989) concluded that such an analysis would only be valid during that year and could not be applied to out-years, even in the same basin.

Sporadic peaks in sediment discharge that are unrelated to flow are believed to reflect discrete inputs of sediment into the drainage system. These sudden but short and irregular increases in suspended sediment concentrations are common in streams at the glacier margin (e.g. Rudolph, 1962; Mathews, 1964a; Østrem, 1975a; Collins, 1979a; Bogen, 1980; Gurnell, 1982; Gurnell and Warburton, 1990; Johnson, 1991b). The erratic nature of these sediment bursts may result from subglacial channel migrations that capture a new supply of sediment (Collins, 1979a, 1988) or from glacier motion that also establishes new sediment sources for englacial or subglacial flows (e.g. Humphrey *et al.*, 1986). Proglacial processes tapping different sediment sources, such as unvegetated moraines (e.g. Bogen, 1980; Hammer and Smith, 1983; Richards, 1984),

PLATE 11.6. Sediment-laden meltwater in turbulent discharge from the Matanuska Glacier, Alaska, in late May 1993. Sediment concentration averaged 2.5 g l⁻¹.

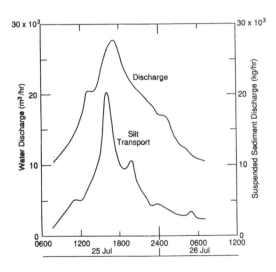

FIG. 11.17. Rapid, ten-fold increase in sediment discharge during a two-fold increase in water discharge at Storsteinfjell, Norway (after Østrem, 1975a, reproduced by permission of *SEPM*). Peak sediment discharge occurred prior to peak water discharge.

may result in sediment bursts further downstream. Sediment input may also result from reworking of sediment on moraines or moraines with ice cores (Plate 11.7). As with meltwater, sediment is effectively stored within and at the glacier bed because not all of the bed is hydrologically interconnected, nor are englacial and subglacial discharges in the smaller arterial conduits necessarily capable of entraining all available materials (e.g. Collins, 1988; Bogen, 1989).

The total sediment load or annual sediment yield of an outlet river draining a glacierized basin is apparently as variable as over a short time; quantities of 10^3 to 10^8 metric tons per year have been measured in basins of various size and percentage of glacier cover (Gurnell, 1987a). Annual sediment yields summarized by Church and Gilbert (1975) ranged from 140 metric tons km⁻² per year to over 1900 metric tons km⁻² per year, with yields from recent volcanic terrain as high as about 4000 to 8000 metric tons km⁻² per year. Østrem (1975a) presented 5 years of data from streams draining five Norwegian

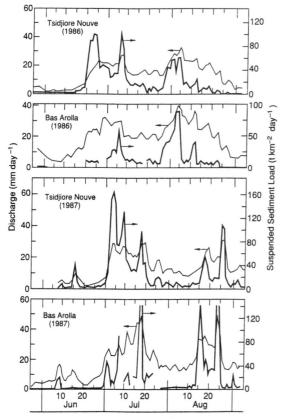

FIG. 11.18. Variations in daily suspended sediment yield and discharge through the melt season for two adjacent glacierized basins in the Swiss Alps (after Gurnell *et al.*, 1988, reproduced with permission from the International Association of Hydrological Sciences).

glaciers, where the average sediment yields ranged from 414 to 1545 metric tons km^{-2} per year, but varied by ±25 to over ±40% (Østrem, 1975a, b). Sediment data from large regional catchments with multiple glacierized basins across Alaska had sediment yields ranging from about 2 to 60 metric tons km^{-2} per day (Parks and Madison, 1985).

Bogen (1989) also reported annual sediment loads ranging from as little as 3 to 10,400 metric tons per year at 14 glacier-fed intakes for the Breheimenstryn power plant in Norway. Data from Nigardsbreen, Norway, for the period 1968 to 1986 indicate large variations in total load from year to year (10,000 to 40,000 metric tons year^{-1}) (Bogen, 1989). Mean annual sediment yield of the glaciers across Norway ranged from a highly variable 100 to 1300 metric tons

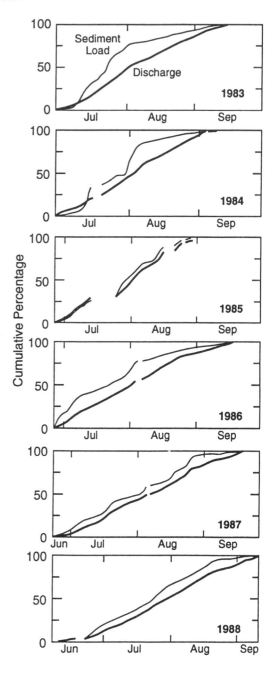

FIG. 11.19. Variability in the cumulative sediment load and water discharge for years 1983 and 1988 at Görnera, Görnergletscher, Switzerland (after Collins, 1990, reproduced with permission from the International Association of Hydrological Sciences). Large volumes of sediment may be released without increases of similar magnitude in water discharge.

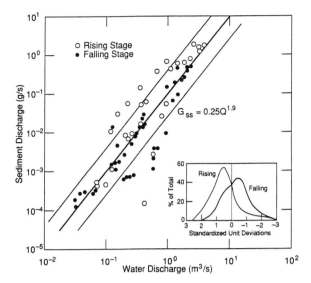

FIG. 11.20. Suspended sediment versus discharge relationship for Decade River, Baffin Island. Rising and falling stages clearly differ in terms of sediment discharge (after Church and Gilbert, 1975, reproduced by permission of *SEPM*).

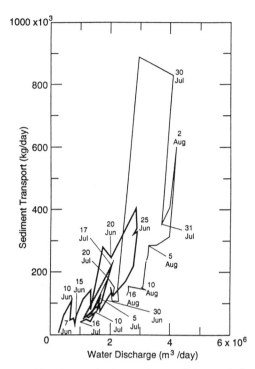

FIG. 11.22. Hysteresis loops characterizing periods of discharge over several days during the 1969 melt season at Nigardsbreen, Norway. Each loop is associated with a period of high water discharge (after Østrem, 1975a, reproduced by permission of *SEPM*).

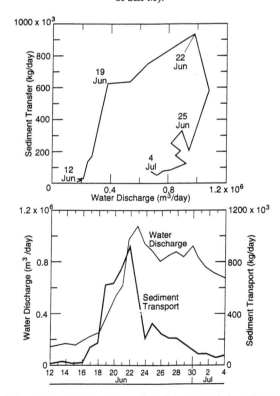

FIG. 11.21. Relationships of daily suspended sediment transport to discharge during a 1969 flood at Erdalsbreen, Norway (after Østrem, 1975a, reproduced by permission of *SEPM*). Clockwise hysteresis loop is characteristic of a peak in sediment discharge that culminates before water discharge.

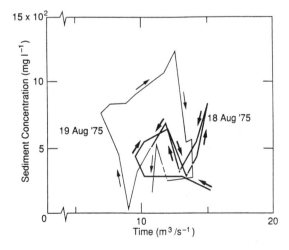

FIG. 11.23. Diurnal hysteresis loops with involutions for suspended sediment concentration as a function of discharge plotted sequentially with time for the Görnera, Görnergletscher, Switzerland (after Collins, 1979a; reproduced by courtesy of the International Glaciological Society from *Journal of Glaciology*, **23**, No. 89, 1979, p. 253, Fig. 4).

PLATE 11.7. Ice-cored moraine next to the active margin of the Matanuska Glacier, Alaska. Ice core is exposed in approximately the centre of the photograph, above the stream. Reworking moves previously deposited sediments, which originated from basal zone debris, into streams feeding the main rivers (upper left) and can cause sporadic increases in sediment concentration downstream (note people for scale).

km^{-2} per year (Bogen, 1989). In contrast, Bezinge (1978) found a much higher maximum sediment yield of about 6000 metric tons km^{-2} per year at Görnergletscher, Switzerland, a glacier of similar dimension to those in Norway. Guymon (1974) analyzed records for the Knik Glacier basin, which contains a much larger glacier and higher percentage ice cover than either Gornergletscher or Nigardsbreen; annual sediment yield here is over 20,000 metric tons km^{-2} per year.

While the basic factors affecting sediment yield have been identified, their role in determining annual trends have not been quantified and studies suggest that relationships are not straightforward (e.g. Østrem, 1975a). Sediment yield is affected by such factors as (1) basin geology and thus the erodibility of materials beneath the glacier; (2) climate and its effects on mass balance and runoff; (3) glacier dynamics and thermal regime and their effects on

erosion and sediment entrainment by ice; (4) the percentage of glacier cover and its state of change; and (5) the type and extent of the subglacial drainage system in relation to the location and quantity of sediment available for transport. Because data on sediment yield are representative of only a select few of the Earth's glacierized basins (e.g. Gurnell, 1987a), insufficient data exist to assess the role of these various factors or define the magnitude of long-term fluctuations in annual sediment yield (Church and Gilbert, 1975; Harrison, 1986).

11.4. SUMMARY

In examining the sedimentary processes and environments of modern glaciers as well as glaciers and ice sheets of the past, it is necessary to consider the environments of the glacial system, processes of erosion, entrainment and transport, and the glacio-hydrological system over the short and long term—all within the context of glaciological conditions at the glacier bed as well as within and on the ice. The mechanisms by which sediment is entrained and the modes of transport are responsible to a large degree for the material properties of deposits produced within the glacial environment. The net flux of sediment within the ice and the drainage system determines the potential volume of deposits which can be produced.

The discharge of sediment and water in the glacial environment have seasonal, annual and long-term variabilities that are unique from non-glacial rivers and watersheds. Terrestrial valley glaciers not only impinge on the lives of people in close proximity to mountainous areas but, in terms of water supply for agriculture, industrial and domestic use, to population centres in lowland areas, such as the northern plains of India and the coastal lowlands of Peru, which are far removed from glacial source areas. Applied studies have vastly increased knowledge of glaciers in general and have led to an increasing awareness of their sensitivity to global warming and climatic change. Modern valley glaciers provide a generally unparalleled opportunity to study glacigenic processes but they should not be mistakenly used, however, as unequivocal analogs for larger ice masses of the past but rather as one component within the continuum of glaciers to continental size ice sheets.

Chapter 12

SEDIMENTS AND LANDFORMS OF MODERN PROGLACIAL TERRESTRIAL ENVIRONMENTS

J. Maizels

12.1. DISTINCTIVENESS OF PROGLACIAL ENVIRONMENTS

Proglacial environments lie at and beyond the ice-margin. Sediments and landforms in proglacial environments are dominated by meltwater and sediment derived from the ice mass itself. The physical characteristics of proglacial environments are therefore largely dependent on extraneous controls, affecting input of water and sediment to the system from beyond its boundaries, and reflect the nature of those water and sediment inputs, the volume, sources and character of the inputs and the pathways and rates of transfer through the system (Fig. 12.1) (cf. Boothroyd and Ashley, 1975; Church and Gilbert, 1975; Boothroyd and Nummedal, 1978; Smith, 1985; Brodzikowski and van Loon, 1991).

12.1.1. Nature of Meltwater Inputs

Meltwater inputs to proglacial environments are characterized by large-scale regular and irregular variability in runoff magnitude. Regular variations reflect seasonal and diurnal ablation cycles and responses to periods of glacier advance (positive mass balance) or retreat (negative mass balance).

Seasonal variations in meltwater discharge are most marked, except in equatorial glacierized basins. In most climatic zones, discharge fluctuations closely follow the annual temperature curve, with little or no flow occurring during winter. Four meltwater runoff periods have been identified following the winter low flow period (Stenborg, 1970; Church, 1972; Church and Gilbert, 1975). First, spring meltwater flows occur in relation to break-up of river ice, while up to 25% of meltwater produced is retained as superimposed ice, in snow and slush areas, and in ice capillaries and channels. Runoff amounts are much lower than expected because of this storage effect. Secondly, by early to mid-summer, rising temperatures lead to a rapid rise in snow melt followed by ice melt. During this nival flood period, runoff is composed both of meltwater derived from ablation and rainfall amounts and from stored water. Meltwater flows exceed those expected from climatic conditions because of the supplementary inputs of water from storage at a time of full development of englacial water courses. Third, by late summer, runoff of stored water is expended, while the well-developed englacial channel system allows minimum delay in runoff. Runoff is directly related to ablation and rainfall, but since temperatures are decreasing

CONTROLS ON 1) Q_W AND Q_S IN GLACIER SYSTEM
2) CHANNEL FORM PATTERN AND SEDIMENTATION
IN PROGLACIAL SYSTEM, BY Q_W AND Q_S

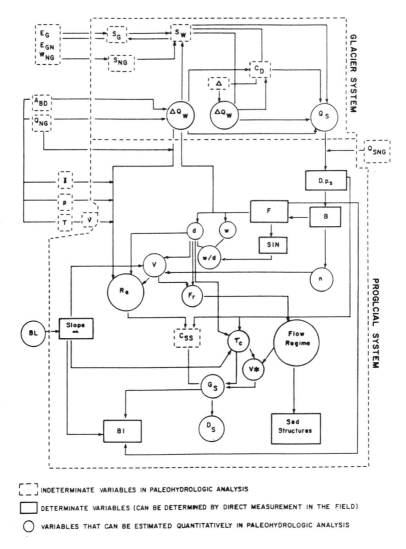

```
┌ ┐
└ ┘  INDETERMINATE VARIABLES IN PALEOHYDROLOGIC ANALYSIS

□    DETERMINATE VARIABLES (CAN BE DETERMINED BY DIRECT MEASUREMENT IN THE FIELD)

○    VARIABLES THAT CAN BE ESTIMATED QUANTITATIVELY IN PALEOHYDROLOGIC ANALYSIS
```

FIG. 12.1. A model of the main inter-relationships by which meltwater and sediment discharge from a glacier system can affect the characteristics of the proglacial drainage system. (Note: *symbols used here are defined specifically for this figure alone.*)

Key: A. GLACIER SYSTEM. E_G — Glacier erosion, E_{NG} — Non-glacial erosion, W_{NG} — Non-glacial weathering, C_D — Degree of 'channel development', Δ — Storage of water, ABL — Ablation, Q_{NG} — Water from non-glacial source, S_G — Sediment supply from glacial source, S_{NG} — Sediment supply from non-glacial source, S_W — Sediment supply to glacial meltwater system, Q_W — Discharge of meltwater, Q_S — Discharge of sediment transported in meltwaters, ΔQ_W — Variations in Q_W.

B. PROGLACIAL SYSTEM.

B — Boundary conditions, BI —'Braiding Index', C_{SS} — Suspended sediment concentration, d — Flow depth, D — Sediment size (characteristics), D_S — Amount of sediment aggradation, F — Channel form parameter, F_r — Froude No. $= (v/dg)^{\frac{1}{2}}$, g — Acceleration due to force of gravity, G_S — Rate of sediment transport, n — roughness coefficient, Q_{sNG} — Non-glacial sediment supply,

$$Re - \text{Reynolds No.} = \frac{Vd\rho}{\mu} = \frac{Vd}{\nu}$$

SIN — Channel sinuosity, T — Water temperature, τ_c — Critical tractive force $= \gamma dS$, V — Mean flow velocity, V_* — Critical shear velocity $(\tau c/\rho)^{\frac{1}{2}}$, w — Flow width, x' — Channel slope, γ — Specific weight of water, ρ — Density of water, ρ_s — Density of sediment, ν — Kinematic viscosity of water $= \mu/\rho$, μ — Dynamic viscosity of water.

towards the end of the meltseason, discharge declines in the form of a recession curve. Finally, as autumn comes to an end, the winter freeze-back begins and flows approach their seasonal minimum. Hence, the glacier acts as a natural reservoir for meltwater, storing it in winter and releasing it in summer (Chapter 6). Most glaciers also exhibit a strong diurnal cycle in runoff, reflecting diurnal temperature fluctuations. The timing of peak daily flow largely depends on the transit time for water to flow through the glacier system from different parts of the ablation zone. Transit time, in turn, varies according to the size of the ablation zone and the extent to which the englacial drainage network has developed. Hence, peak flows occur earlier in the day towards the end of the meltseason and in meltwater systems draining small glaciers. Runoff earlier in the season or draining very large glaciers and ice caps may exhibit little or no detectable diurnal cycle. However, where diurnal runoff cycles are present, the meltwater system is subject to the equivalent of a complete flood cycle every 24 hours. These daily flood flows are highly significant in terms of instigating bedload sediment transport and morphological changes in the proglacial channel system (Section 12.3).

Superimposed upon these regular seasonal and diurnal cycles of runoff are a number of irregular discharge events. The most common periods of irregular high flow are associated with summer or autumn storm events, particularly those that occur after periods of relatively warm conditions (Østrem, 1975). These rain-induced runoff events often generate the main floods of the meltseason (Fig. 12.2). Röthlisberger and Lang (1987), for example, recognized that the three highest floods in a recent 20-year runoff period (1965–1985) from the Aletschgletscher basin in Switzerland represented ablation events amplified by simultaneous rain storm activity. When plotted on a recurrence interval graph, these three events lay well above the 20-year mean recurrence interval line. Rain storm floods can be instrumental in accelerating rates of sediment transport and proglacial channel changes. Ashworth and Ferguson (1986), for example, found that a rain flood of 10 m^3 s^{-1}, compared with peak ablation flows of 5 to 8 m^3 s^{-1}, generated an order of magnitude increase in bedload sediment transport rates in a

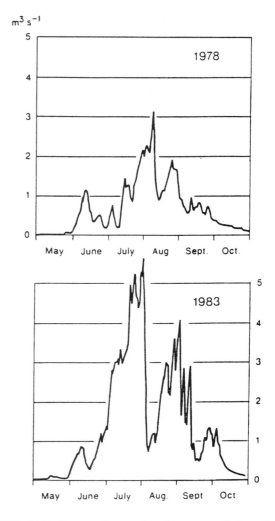

FIG. 12.2. Contrasts in annual meltwater runoff from the Vernagtgletscher, reflecting years of positive mass balance (1978) and negative mass balance (1983). Periods of summer melting were interrupted in both years by heavy snowfalls. (After Röthlisberger and Lang 1987; reprinted from Gurnell, A.M. and Clark, M.J. (eds) *Glacio-Fluvial Sediment Transfer*, by permission John Wiley and Sons.)

meltwater channel in Lyngsdalselva, northern Norway, from 0.3 to 3.0 kg m^{-1} s^{-1}.

High magnitude, periodic runoff events may also be superimposed on the regular seasonal and diurnal cycles. These are 'glacier-burst' or jökulhlaup floods (Chapter 6). Jökulhlaups may be caused by a large number of different mechanisms, the most common of which is through failure of an ice dam which has acted to pond up an ice-marginal lake. Lake drainage occurs suddenly through breaching or collapse of the ice dam and rapid enlargement of the englacial or submarginal conduit system (e.g. Clague and Mathews, 1973; Röthlisberger and Lang, 1987; Costa, 1988a; Mathews and Clague, 1993). Jökulhlaups may also be generated: by rupture of subglacial, englacial or supraglacial water bodies; by sudden breaching of a moraine or unconsolidated rock dam; as a result of rockfalls or ice avalanches on to the glacier, which terminates in a lake; during prolonged and heavy rainstorm events; glacier surges; and periods of subglacial, geothermal activity or eruptions of subglacial volcanoes (cf. Maizels and Russell, 1992).

Jökulhlaup discharges commonly exceed 'normal' ablation-related flows by up to several orders of magnitude. For example, in Baffin Island Church (1972) recorded jökulhlaup flows from sudden lake drainage of about 200 m^3 s^{-1}, compared with more normal diurnal peak flows of about 20 m^3 s^{-1} (cf. Björnsson, 1988, 1992; Russell, 1991). The Baffin Island jökulhlaup accounted for over 10% of the total meltseason runoff. By contrast, jökulhlaups from Grimsvötn, a geothermally active zone beneath the Vatnajökull Ice Cap in southern Iceland, have reached peak flows of over 40,000 m^{3}s^{-1} (early in the twentieth century, for example, in 1922), compared with more normal runoff flows of about 400 m^3 s^{-1} (Björnsson, 1988, 1992). One of the highest 'recorded' jökulhlaup flows occurred in 1918 from Katla, an active volcanic crater that underlies Myrdalsjökull ice cap in southern Iceland, and which erupts once or twice per century (peak flow was ≥1.5 x 10^6 m^3 s^{-1} compared with normal flows of only approximately 50 m^3 s^{-1}).

Jökulhlaup floods have characteristic hydrographs, reflecting the rate of development of the englacial tunnelways that accommodate the flood waters (Fig. 12.3) (Chapter 6). Hence, a jökulhlaup hydrograph is typified by a gradual rise to a peak flow associated with the progressive enlargement of tunnels, followed by a rapid recession curve as the water supply from the source reservoir becomes depleted (cf. Church, 1972; Clarke, 1982; Russell, 1991). However, complex and interrupted hydrograph forms occur where tunnels become temporarily blocked by ice and

debris, leading to short-lived shut-offs of flow (cf. Sturm et al., 1987). Jonsson (1982) noted that several major flood waves occurred during the jökulhlaups produced during the subglacial eruption of Katla in 1918. Jökulhlaups may last for less than 24 hr or, when draining larger bodies of water, may extend over several weeks (Fig. 12.3) (e.g. Björnsson, 1988, 1992). Lower magnitude, shorter-term runoff irregularities may reflect drainage of small reservoirs, collapse, blockage and breaching of ice walls and tunnel roofs, or changes in tunnel capacity. These variations may extend only over minutes or hours, and associated flow fluctuations may become rapidly dissipated downstream.

Large-scale, long-term variations in runoff also occur during periods of glacier or ice-sheet growth and decay. Reconstruction of bankfull flows at the edge of the Late Weichselian Ice Sheet in central Poland, for example, indicates a decline from about 145 m^3 s^{-1} in the Alleröd to 20 m^3 s^{-1} at the present time (cf. Rotnicki, 1991). Similarly, Teller (1990) has reconstructed meltwater flows from the Laurentide Ice Sheet during the Late Wisconsinan, demonstrating the large changes in runoff associated with repeated catastrophic drainings of Lake Agassiz through the St. Lawrence route to the North Atlantic, reaching peaks of 4000 km^3 year^{-1} between 11,000–10,000 years ago. Even in the last century or so, overall glacier retreat in the European Alps has led to declines in mean summer runoff and annual specific yields. Kasser (1973), for example, found that a 26% decrease in area of Swiss glaciers between 1876 and 1973 led to a 3.4% decrease in annual specific yield, while a 19% decline in glacier area in the French Alps between 1916 and 1968 led to a 16% decrease in mean summer runoff.

The sediments and landforms of proglacial drainage systems are therefore clearly subject to a wide range of runoff fluctuations, operating over a variety of timescales (Maizels, 1989a; Starkel, 1991) and each exhibiting a varying capacity to entrain, transport and deposit sediment, thereby modifying the proglacial landscape. However, little work has yet been directed toward identifying the role of the varying magnitude and frequency of runoff events that affect the sediments and landforms of proglacial environments. Most proglacial channel systems, for example, may be expected to reflect some degree of equilibrium with flows generated during the diurnal flood cycle; even those environments subject to less frequent, major flood events may be adjusted to these rare events, depending on the ability of the system to return to its pre-flood state. Depending on the resistance and sensitivity of the system to change, and on its relaxation time, landforms may also be

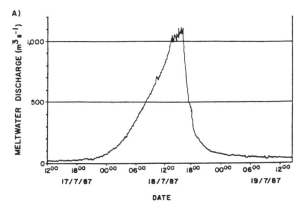

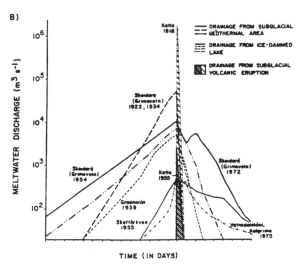

FIG. 12.3. Hydrographs typical of jökulhlaup drainage. (A) Hydrograph produced during jökulhlaup from sudden drainage of ice-dammed lake in west Greenland, July 1987. (Based on Russell, 1989.) (B) Hydrographs produced during Icelandic jökulhlaups generated from sudden drainage of subglacial reservoir in area of geothermal activity (Grimsvötn); of ice-dammed lakes (Graenalon, Skaftavkvos and Vatnsdalslon:); and from subglacial volcanic eruptions (of Katla). (Based on Thorarinsson 1957; Björnsson 1988. From Maizels and Russell, 1992.)

modified by later flows, either through lateral scour and re-working or by incision during large floods. This concept of landform response to runoff events is important in the analysis of proglacial environments, since it provides the framework for identifying (1) how glacier/ice sheet behaviour can affect the pattern of geomorphic response in proglacial environments, and (2) the proportion and character of runoff events that affect the sediments and landforms of proglacial environments, and therefore the nature of those runoff events which are represented (and not represented) in the geomorphic record.

12.1.2. The Nature of Sediment Input

Sediment inputs to proglacial environments are largely derived from the glacier source generated both by glacial and fluvioglacial transport to the ice-marginal zone. According to Hammer and Smith (1983), most suspended sediment is derived from the basal zone of an ice mass. Sediment inputs therefore reflect both the efficacy with which sediment is eroded, entrained and transported by these agents, and also the temporal and spatial variability in processes of sediment transport within the glacier system. Sediment can also be delivered to the proglacial environment from non-glacial sources, particularly in mountain areas where slope mass movements, colluvial or 'paraglacial' fan development, rain- and snow-fed valley side streams contribute sediments, and from volcanic eruptions; in coastal areas, aeolian and shoreline sediments can be made available to proglacial systems, while wind-blown sediments can be derived from both glacial and, especially, exposed fluvioglacial materials (Menzies, 1995b, Chapter 6). Sediment inputs from this wide variety of sources are hence highly variable over space and time, and are reflected in the sediments and landforms of proglacial environments (Chapter 8). Spatial variability reflects the geographic distribution of sediment pathways to the proglacial environment, while temporal variability reflects both variations in runoff and in availability of sediments (Fig. 12.4). Hammer and Smith (1983) estimated from the Hilda Glacier in Alberta that 6% of suspended loads were derived from supraglacial and englacial debris and 47% each from subglacial and bank materials; of the bedload, 46% was estimated to have been derived from supraglacial and englacial debris sources and 27% each from subglacial and bank materials.

One major control on inputs of sediment to proglacial environments is the resistance of the bedrock source to glacial and fluvioglacial erosion. In ancient crystalline shield areas, for example, suspended sediment loads of meltwater rivers have been found to be significantly lower than those of young, mountainous volcanic regions such as found in Iceland, New Zealand and parts of the Rockies–Andean chain (for otherwise similar flow and catchment conditions). Østrem et al. (1967), for example, recorded a mean suspended sediment concentration of 1060 ppm from meltwaters in Baffin

Island, compared with 39,000 ppm recorded by Klimek (1973) from meltwaters on Skeidararsandur in Southern Iceland (cf. Corbel, 1959; Church and Gilbert, 1975; Gurnell, 1987a). Nevertheless, the volumes of sediment input to almost all proglacial systems are significantly higher than for most non-glacial fluvial systems, other than for rivers flowing through thick loess sequences as in China (Walling, 1984; Lawler, 1991). Suspended sediment loads, which range in size between about 0.001 and 1 mm, generally exceed 1000 ppm (Church and Gilbert, 1975), and commonly exceed 400 metric tons km^{-2} of catchment area (Østrem, 1975; Lawler, 1991). These high values have been estimated to represent mean long-term rates of total catchment sediment yields (Table 12.1).

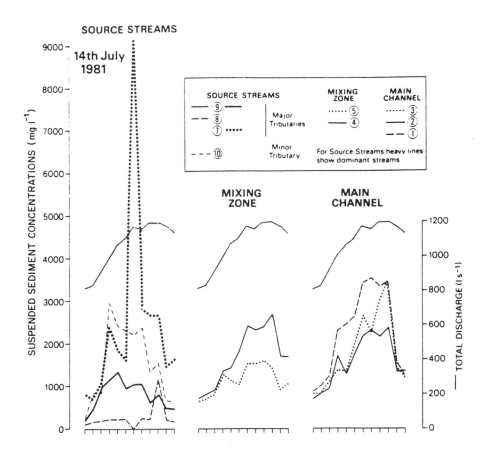

FIG. 12.4. Spatial variability in suspended sediment concentration during diurnal discharge cycles, Glacier de Tsidjiore Nouve, Swiss Alps. Time intervals shown on horizontal axis are hourly, starting at 0800. (After Gurnell 1987; reprinted from Gurnell, A.M. and Clark, M.J. (eds), *Glacio-Fluvial Sediment Transfer*, by permission of John Wiley and Sons.)

The values of glacial sediment yields in Table 12.1 can be seen to exceed the estimates of mean annual suspended sediment yields from a wide range of rivers. Hence, it can be noted that proglacial meltwaters receive very high inputs of suspended sediments in response to glacial and fluvioglacial erosion within the source catchment.

Bedload sediment yields also appear to be significantly higher than in many non-glacial streams, although there are few reliable measurements of

TABLE 12.1. Sediment of yields of selected glacial and non-glacial rivers

Location	Sediment Yield	Reference
Annual sediment yields of glacial rivers		
Hilda Glacier, Alberta, Canada	$<7.0 \times 10^2$ t yr^{-1}	Hammer and Smith 1983
Engabreen, Norway	1.7×10^4 t yr^{-1}	Østrem, 1975
Gornergletscher, Switzerland	4.8×10^4 t yr^{-1}	Collins, 1978
Solheimajökull, Iceland	1.1×10^6 t 10^{-1}	Lawler, 1991
Hunza Rv., Pakistan	$>6.3 \times 10^7$ t yr^{-1}	Ferguson, 1984

Specific annual sediment yields of glacial rivers		
Norwegian Rivers	16 -1545 t km^{-2} yr^{-1}	Østrem, 1975
Hunza Rv., Pakistan	4773 t km^{-2} yr^{-1}	Ferguson, 1984
Solheimajökull, Iceland	14482 t km^{-2} yr^{-1}	Lawler, 1991

Specific annual sediment yields of non-glaciated catchment basins		
St Lawrence Rv., Canada	5 t km^{-2} yr^{-1}	Walling, 1984
Congo Rv., Zaire	13.2 t km^{-2} yr^{-1}	Walling, 1984
Amazon Rv., Brazil	79 t km^{-2} yr^{-1}	Walling, 1984
Ganges Rv., India	537 t km^{-2} yr^{-1}	Walling, 1984
Brahmaputra Rv., India	1370 t km^{-2} yr^{-1}	Walling, 1984

bedload movements from proglacial environments because of the many difficulties in the accurate sampling of bedload (e.g. Gomez, 1987). Østrem (1975) trapped 400 t of coarse sediment in a sample net across the Nigardsbreen outlet river, Norway, over a 28-day period, before the net was destroyed by floating ice. He used these measurements to estimate that the bedload represented 30–50% of the total sediment yield for the catchment (i.e. between ca. 165–275 t km^{-2} year^{-1}). Church (1972) recorded values of 29–1114 t km^{-2} year^{-1} for the Lewis and Ekalugad meltwater rivers in Baffin Island, representing 13.5–93.6% of the total sediment yields from the catchments. For many coarse-bed rivers, however, bedload sediment transport rates are estimated, rather than measured, using a range of computational formulae (Church and Gilbert, 1975; Gomez, 1987), while measurements of bedload transport rates in proglacial rivers have tended to be restricted to short sample periods within small, minor (wadeable) channels (Ashworth and Ferguson, 1986).

The high sediment yields recorded from glacial meltwaters to some extent mask the high temporal variability that characterizes these inputs of sediment to the proglacial channel system. For example, not only do these amounts vary considerably from year to year (Østrem, 1975), but the suspended loads also vary in response to progressive 'washing out' of the supply of fines from the glacier and proglacial channel boundaries. This depletion process can occur over each meltseason, with maximum fines being made available at the start of the season, and over each periodic and diurnal flood cycle. Progressive flushing of fines leads to development of marked, clock-wise hysteresis loops during cyclic discharge fluctuations. For example, suspended sediment concentrations are higher at the start of the meltseason, and on the rising limb of cyclic events, than during similar magnitude flows at the end of the meltseason, or on the falling limb of cyclic events. Bedload only moves sporadically, once the local thresholds for entrainment have been exceeded (Fig. 12.5). The threshold conditions are not constant in a given channel, however, since they depend not only on the force of flow, but also on the resistance of the bed. Bed conditions can vary markedly through time and space, from being loose and highly mobile, to

forming a highly resistant, imbricated, armoured bed, which may only be disturbed by infrequent flood events. These events may then disrupt the armour layer and expose the wide range of sediment sizes contained in the underlying substrate. In this circumstance, almost all bed sizes become mobile and sediment transport rates increase very significantly. Østrem et al. (1967), for example, estimated that 60% of the total summer sediment load in a Baffin Island meltwater channel was removed during a 24-hour outburst flood. Similarly, Church (1972) found that between 25 and 75% of the total annual sediment transport on Baffin Island sandurs occurred during the 4 to 5 days of peak flows, while Østrem (1975) found that 50% of the load transported from Erdalsbreen, Norway, was removed in 4 to 5 days of high flows. Church (1972) estimated that jökulhlaup flows removed up to 90% of total annual sediment yield of a Baffin Island catchment, while two outburst events in an Alpine catchment were estimated to have removed about 50% of annual sediment yields (Beecroft, 1983; Gurnell, 1987b). Nummedal et al. (1987) suggest that a 2-week jökulhlaup on Skeidararsandur is able to move as much sand as would otherwise be transported in 70 to 80 years of normal meltwater flows.

The volume of material available for transport into the proglacial environment plays a crucial role in influencing the sedimentary characteristics of proglacial deposits. For example, where only small volumes of sediment are introduced, whole phases of glaciation and deglaciation may be represented in the proglacial environment (and in the sedimentary record) by only a metre or two of sediment, much of which may have been repeatedly re-worked. It becomes increasingly difficult to interpret such deposits and associated landforms in terms of the controlling processes that generated them because of their composite and interrupted history (Menzies, 1995b, Chapter 8). Hence, the smaller the volume of sediment input to the proglacial environment over a given timescale, the greater the likelihood of complex, re-worked, composite landforms evolving.

Where large volumes of sediment are repeatedly introduced to the proglacial environment, thick deposits, often forming steep outwash fans, can accumulate rapidly. These may contain sediment

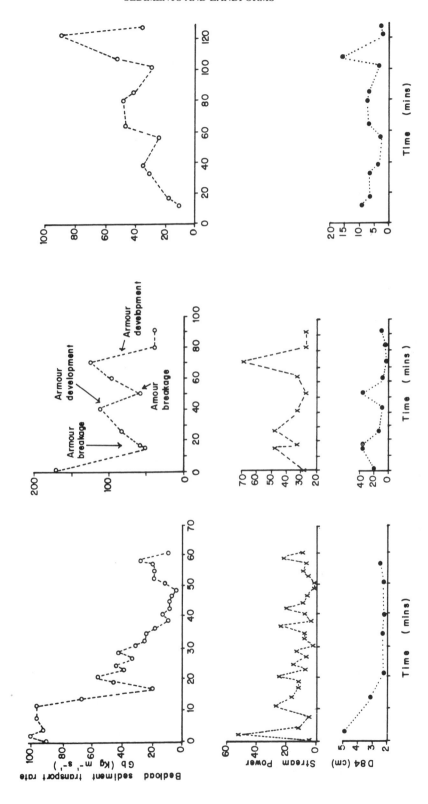

FIG. 12.5. Fluctuations in bedload sediment transport rates in a meltwater stream on a proglacial fan in west Greenland. The three examples (left to right) illustrate periods of declining, fluctuating and rising rates of sediment transport through time, respectively, independently of stream power but reflecting patterns of bed armour development.

sequences representing individual flow or flood events, or flows from successive meltseasons. With rapid steepening of fans, stream powers (W_s) in turn increase markedly $\{(W_s = \gamma_w \, QS_g)$ where γ_w is the specific weight of water, Q is stream discharge and S_g is stream gradient$\}$ and fan entrenchment and terrace formation can ensue. Terrace sequences commonly occur in proximal zones of proglacial fans, in response to fan steepening during periods of rapid accumulation (i.e. before glacier fluctuations directly affect the fan morphology), accentuated by, and possibly encouraging, entrenchment of the ice margin itself.

An important effect of these high sediment inputs lies in the response of meltwater streams to entrain, transport and deposit their loads. The balance between available stream power and that required to entrain and transport sediment through a given reach, largely controls the nature of the developing channel system. All river flows are characterized by longitudinal alterations in sediment transport capacity reflecting development of flow cells. Once deposition occurs, these alternating flow zones are accentuated by lateral flow deflections towards one or other bank. Hence, when large volumes of sediment are introduced, much of this is deposited as accretionary bar deposits in zones of flow divergence within the channel. Where lateral bank scour is possible, braided channel systems develop (Section 12.3.1). The majority of outwash plains are characterized by braided river systems, replaced by single-thread channels only where entrenchment or confinement has occurred. When little new sediment is introduced, meltwaters gain their load from the channel boundaries. Large-scale re-working of older deposits can provide the main source of sediment transported through the fluvial system and re-deposited across the outwash surface (Hammer and Smith, 1983). However, where channels are composed of mixed grain sizes, the bed can become armoured as fines become sheltered behind larger grains, filter into the bed or are winnowed away. Once the bed sediment is formed of an impacted, imbricated layer of coarse pebbles or cobbles, it becomes highly stable and resistant to entrainment. In some meltwater streams, therefore, material derived from the bed is only acquired during high magnitude flows. If the bed

cannot supply the required material, additional sediment is gained from the channel banks. This process leads to channel widening and shallowing and, in turn, reduction of bed shear stresses that could otherwise promote bed scour (see below). Deposition within the channel may ensue. Most meltwater streams gain their sediment loads from a variety of sources as fresh inputs from the glacier drainage system, re-working of older bed and bank materials and valley side slopes (Hammer and Smith, 1983). Meltwaters draining from glaciers may also transport ice blocks containing large volumes of debris.

High volumes of fine sediments transported in suspension may be deposited as a temporary veneer on channel bar deposits during waning flows, or may concentrate in proglacial pools and lakes, or be transported into the marine off-shore zone. Fines exposed on outwash surfaces are frequently removed by winds and accumulate as loessic or dune distal deposits. With the large-scale transport of fines from the fluvial system, few outwash deposits contain fines, and are commonly composed of coarse-grained sediments. The exception to this pattern occurs during meltwater flows containing hyperconcentrations of fines. Hyperconcentrated flows occur typically during sediment-rich glacier-burst floods. The effect of these flows, as will be noted later, leads to the formation of lobate, hummocky or sheet-like topography, characterized by matrix-supported, unsorted or inversely graded deposits.

12.2. MORPHOLOGY AND LANDFORMS OF PROGLACIAL ENVIRONMENTS

12.2.1. Non-Jökulhlaup Glacial Outwash

Many proglacial environments appear to be composed of extensive, featureless outwash plains. However, on inspection it is clear that proglacial environments contain a range of distinctive features, representing both active and former fluvioglacial activity. In addition, a number of landforms may be relict features of glacial or colluvial origin, modified by aeolian or permafrost activity, or affected by large-scale, long-term fluctuations in ice-limits in local or global base-level or isostatic changes.

The most distinctive characteristic of proglacial environments is the zone of outwash accumulation. The overall morphology of the outwash deposit depends on three factors: (1) the volume of sediment inputs to the outwash system, (2) the topography and accommodation of the receiving area, and (3) the nature of the processes distributing the sediment through the proglacial zone. Where the accumulation zone is confined between valley sides, outwash sediments are concentrated on the valley floor forming a 'valley-train' deposit. Valley-train deposits are found in upland areas and mountain glacier forelands (Fig. 12.6) (Plate 12.1). Infilling of glacial 'troughs' by glaciofluvial outwash leads to the development of a relatively flat valley floor, which may mask a highly irregular bedrock topography. In addition, progressive infilling may gradually bury pre-existing landforms, such as former terminal moraines, eskers, drumlins and kames. Valley-trains exhibit a wide range of down-valley slopes, depending on the three factors mentioned above. Hence, the steepest gradients are found in areas where fresh sediment inputs are greatest relative to the ability of streams to remove them, and where the controlling topography is also steep, as on mountain slopes (Menzies, 1995b, Chapter 16).

Where the proglacial accumulation zone is unconfined by valley walls and can extend laterally for a considerable distance, the zone is known as an outwash plain or sandur, the latter derived from the Icelandic term for a sand plain. Modern sandur plains are located typically in piedmont zones, commonly those which extend seawards from young glacierized mountain ranges to form coastal plain infill deposits. The coastal plains of southern Iceland, parts of south-western Alaska, and western Greenland, for example, contain some of the most extensive modern outwash plains. The largest is Skeidararsandur in southern Iceland, which is 40 km wide and 30 km from the ice-margin to the sea (Plate 12.2). Pleistocene sandur plains were more extensive, including, for example, the Canterbury Plains in South Island, New Zealand (Plate 12.3), the Patagonian gravels (of southern Argentina) (which include outwash sequences) and those bounding the margins of many segments of the Laurentide and Eurasian Ice Sheets. Where sandurs extend to a coastal margin, their distal limits are marked by steep deltaic slopes, reworked by shoreline processes to form a fan-delta or plain-delta deposit.

Outwash fans reflect relatively rapid deposition in the proximal zone, fed by a major glacial meltwater route reaching the ice-margin at a single point (Fig. 12.7) (Plate 12.4). Composite fans fed by sub-parallel, subglacial drainage routes can merge and form an 'apron' along the ice-margin. Such fans require the ice-margin to remain relatively stable, since a glacier advance leads to over-riding and burial, while retreat is likely to lead to glacier entrenchment with meltwater flows emerging at a lower level than those reaching the original fan head. Meltwaters are likely to become ponded behind the ice-contact slope of the fan and, on reaching the fan surface, will instigate rapid entrenchment of the fan itself. If the fan 'apron' deposits are banked up against a stagnating ice margin, progressive burial and differential decay of the ice leads to the formation of a hummocky, pitted outwash surface (Plate 12.5).

Where gradients exceed the threshold for incision during episodes of competent flows, bed scour may lead to development of a new channel bed at a lower level. This may happen at a local scale, around a braid bar or within a single channel reach, for example, or over the stream profile as a whole. In the former case, a channel at a lower level will promote meltwater flows to follow that routeway until progressive deposition at its head may instigate a sequence of alternate switching between adjacent braid channels. In the latter case, a new long profile may be developed at a lower level, leading to abandonment of the main outwash surface, either temporarily or permanently. If incision is repeated, a terrace sequence may develop. At present, however, little is known about the flow conditions that trigger incision and create terrace sequences in proglacial environments. In some areas, a terrace sequence may develop during a single meltwater-storm flood event, whilst in other areas ancient terrace sequences have been interpreted as forming over many thousands of years. The longitudinal profiles of proglacial rivers, terrace sequences and outwash surfaces exhibit shapes ranging from single, straight-line profiles to highly complex stepped profiles. This variety reflects the elevation of the source stream at the ice-margin, the complexity of the proglacial bedrock topography bounding the meltwater routeway, the stability and

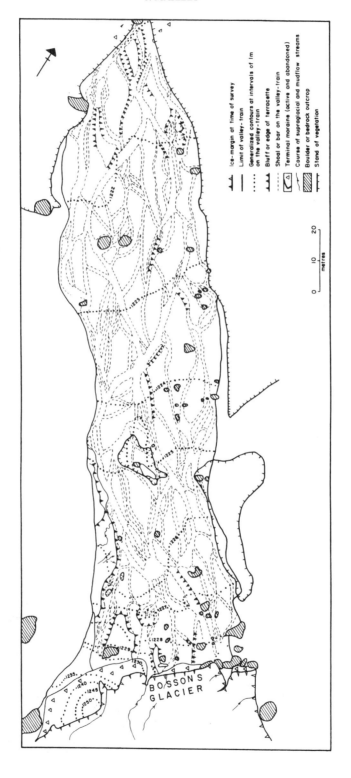

FIG. 12.6. Plane-table map of the meltwater channel system on the Bossons valley-train, April 1974.

PLATE 12.1. The ice-terminus and valley train of the Bossons glacier, French Alps, 1981. The valley-train is ca. 50 m wide and is bounded by steep Little Ice Age lateral moraines. (Photo by W. Mitchell.)

PLATE 12.2. Complex anastomosing channel systems characteristic of the Icelandic sandur plains, illustrating Sulasandur, the western edge of Skeidararsandur in southern Iceland, looking north towards Vatnajökull.

PLATE 12.3. The braided Waimakariri River extending eastwards across the Canterbury Plains, South Island, New Zealand.

relative elevation of local and regional the base-levels, and the consistency of channel processes downstream.

Distal zones exhibit significantly lower gradients than those of proximal zones and an abrupt discontinuity between the two zones has often been identified. The discontinuity may reflect statistical methods of analysis rather than physical processes. Several explanations infer an abrupt change in processes at this locality. Interestingly, this discontinuity has also been identified in many non-glacial river profiles. It may be that the boundary between proximal and distal slopes reflects a change in the size distribution of the bed material. Steeper proximal reaches may be characterized by multi-mineral or multi-grain particles, while distally the grains may have been broken down into their separate mineral or granular components to form a significantly finer load. In contrast, it has been argued that proximal channels are controlled more by bedload transport, while suspended sediment transport

dominates in distal channels, the boundary being marked by grain sizes of $\approx 2\phi$. A second explanation is that the proximal zone is characterized by the addition of sediments directly from the ice-margin and its moraines, or from valley walls, such that processes are not confined to fluvial activity and sediments are not limited to those provided solely by fluvioglacial processes. A third, and widely accepted, explanation is that the slope is adjusted to the amount of stream power required to transport inputs of sediment to the system in order for the system to achieve equilibrium. While in non-glacial, humid catchments, discharge increases downstream in response to an increased catchment area, in proglacial systems discharge remains at a similar magnitude throughout the system unless supplemented by valley-side tributaries, snow-melt inputs, or emergence of ground-water. Of the three variables of discharge, slope and sediment calibre, only the latter two variables are adjustable in the downstream direction. Therefore, large inputs of coarse sediment require steep gradients to generate

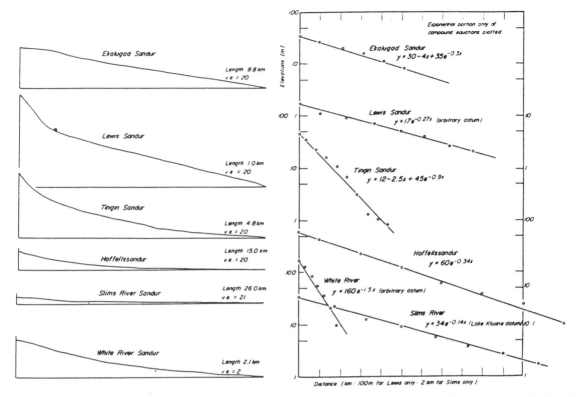

FIG. 12.7. Longitudinal profiles of sandur surfaces, showing both plotted survey data (left) and exponential fitted equations (right). (From Church, 1972; reproduced from the *Geological Survey of Canada*, Bulletin 216, p. 108, Fig. 70.)

sufficient stream power for transport, while progressive selective deposition of coarse particles creates an increasingly fine-grained load, which in turn requires lower gradients and stream powers for transport.

An additional explanation for variations in long-profile form reflects the alternating pattern of fluvial sedimentation and scour downstream. Variations in the long profiles of Baffin Island sandurs have been linked to a prominent change in channel habit (Church, 1972; Church and Gilbert, 1975; Church and Jones, 1982). Major sedimentation zones that exhibited lower gradients and were associated with wide, shallow braided channel and valley-scale flow divergence were distinguished from transportation zones that exhibited steeper gradients and were associated with movement of sediment on the steepened, downstream edge of each sedimentation zone.

In addition, outwash surfaces exhibit numerous smaller-scale relief features. Variations in relief of up to several metres occur in response to the distribution of meltwater channel networks. The relief range depends on the depth of maximum channel scour, the maximum height of flood flows, and the calibre of the bed material. Maximum bed relief occurs either in coarse-grained proximal zones where single-thread channels have formed, or in deeper channel systems of fine-grained braided distal networks. However, exceptions are common, especially where outwash is bounded downstream by rock bars or local base-levels, behind which flood waters may be temporarily ponded to form deep backwater zones.

Kettle holes are commonly found on outwash surfaces. They follow from the *in situ* melting of transported ice blocks, or from collapse of marginal ice masses buried beneath outwash deposits. The extensive development of kettle holes can lead to a

PLATE 12.4. Steeply graded outwash fan at the edge of the West Greenland ice-sheet. Note figures for scale.

pocked or pitted surface. Where the ice blocks are rich in glacial debris, kettle holes may be rimmed or filled with diamicton (Maizels, 1992). If flood flows transported the ice blocks, meltwater flows around grounded blocks can lead to longitudinal scouring. The scours may in turn be exploited by high flows, which use the scours to trigger incision of the bar surface to form chute channels, thereby promoting further braid bar division (cf. Russell, *in press*). Typically, the fossil patterns of braided or anastomosing streams are preserved on abandoned outwash surfaces. The relict paleochannel networks can exhibit surface relief of several metres. The differential extent of vegetation colonization has been widely used to evaluate the relative age and stability of various parts of proglacial braided river systems, particularly since channels are unstable and frequent migrations of the active channel zone promote repeated periods of renewed colonization by vegetation (Plate 12.6). In sub-polar and polar regions, outwash may be underlain by permafrost, and subject to pingo formation, growth of ice-wedge networks and thermokarst development.

PLATE 12.5. Terminal zone of part of Skeidararjökull, southern Iceland, illustrating complex hummocky topography related to ice stagnation (left); abandoned lateral drainage channels; a steep ice-contact slope; broad, level outwash surface, underlain by horizontally bedded sands and gravels; and zones of eskers and ponded water (right).

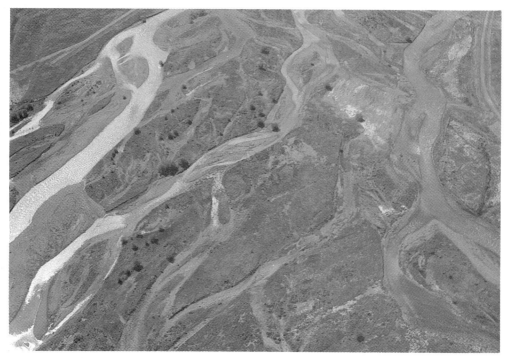

PLATE 12.6. Aerial view of braided river system subject to a variety of levels of flooding. Most stable bars are partially vegetated. Flow is towards the base of the photo. Tekapo River, South Island, New Zealand.

12.2.2. Jökulhlaup-Related Outwash

A substantial amount of literature exists on the proglacial features generated by jökulhlaups during the last glaciation (cf. Maizels and Russell, 1992). Together with modern evidence, these sources indicate that jökulhlaups can generate a distinctive suite of erosional and depositional landforms in proglacial areas. Jökulhlaups possess an enormous erosive capacity (Section 12.3.2) and can create large-scale erosional features, some of which are detectable on Landsat imagery. The channelled scablands of the Columbia Plateau, Washington State, for example, were drained by up to 70 floods from glacial Lake Missoula, to create hundreds of streamlined, residual erosional hills, longitudinal grooves and giant potholes, scours and plunge pools (Baker, 1973). All the major glacial lake basins bounding the southern edge of the Laurentide Ice Sheet are distinguished by complex systems of huge spillway channels up to 3 km wide and over 100 m deep (Kehew, 1982; Kehew and Clayton, 1983; Kehew and Lord, 1986, 1987;

Teller and Thorleifson, 1987). Kehew and Lord (1986) have suggested that a continuum of evolutionary forms can be identified in residual hills, with early stage irregular forms developing into late stage, equilibrium, lemniscate forms. On a much smaller scale, erosional forms include streamlined residual hummocks and bars, ice-block scour marks, and cavitation marks, flutes and potholes (Dahl, 1965; Maizels, 1989a, 1992; Sharpe and Shaw, 1989).

The type of landform produced during a jökulhlaup depends on the nature of the flood flow itself, its sediment load and on the topography of the flood routeway. The largest depositional landform is a jökulhlaup-sandur (e.g. Myrdalssandur) (Jonsson, 1982; Maizels, 1992). Locally, aggradation during jökulhlaups can lead to the formation of raised boulder deltas in flood routeway lakes, boulder lags, spillway expansion bars, boulder bars and pendant bars (which form downstream of bedrock obstacles) (Shakesby, 1985; Russell, 1991). Unconfined jökulhlaup depositional landforms may include streamlined boulder bars, 'washed' sandur spreads and boulder lags, pitted or kettled outwash and fields of ice-block obstacle marks, terrace sequences, lobate and hummocky fans, and a variety of distinctive sediment assemblages (Baker, 1973; Kehew, 1982; Elfström and Rossbacher, 1985; Maizels, 1989a, b, 1991, 1992). In confined proglacial routeways, the depositional impact of a jökulhlaup reflects variations in local flow dynamics and sediment supplies in different topographic positions. Where flows are constricted, depositional features may include local expansion bars, boulder berms and boulder lags (Teller and Thorleifson, 1987). Tributary bedrock channels or valleys lying at a high angle to the main flood channel commonly act as sediment traps to backwater flows to form sequences of slackwater infill deposits (Kochel and Baker, 1988).

12.3. PROGLACIAL MELTWATER CHANNEL SYSTEMS

12.3.1. Channel Pattern and Morphology

Proglacial outwash is commonly characterized by complex braided channel networks, particularly in distal zones (Plates 12.2 and 12.6). The proximal zones may have been subject to fan entrenchment, thus channels become confined between walls of older outwash or moraine and are unable to adjust their pattern and morphology freely to varying flow and sediment inputs. Proximal channels may therefore exhibit deep, single-thread courses, high in energy and transport capacity. Part of the margin of Skeidararjökull in southern Iceland, for example, is bounded by a broad zone of complex topography associated with ice retreat and stagnation (Plate 12.5) that lies up-valley from, and at a lower level than, the last major terminal moraine and its outwash surface, which date from ca. 1870 A.D. The up-glacier side of the moraine is marked by a steep, high ice-contact slope. The present river, the Gigjukvisl, drains from a series of large ice-marginal lakes, ponded up in the zone of irregular topography, and cuts through the main terminal moraine as a single-thread, sinuous channel. The moraine gap is some 50 m deep, having been incised by the present meltwater river within the past century.

In other proximal zones, where no topographic confinement exists, and where little glacial debris has contributed to form a 'blocking' terminal moraine, the proximal outwash zone may develop a braided network at the ice-margin itself. At Solheimajökull in southern Iceland, for example, the glacier has recently advanced, over-riding older outwash; it has generated only a small terminal moraine (2–3 m high), and is drained by over 100 small streams draining off the near-vertical ice-front. Here, the proximal drainage network is a multi-thread distributary system that drains downstream into the major channel which emerges from a lateral portal (Plate 12.7).

A third form of proximal channel pattern occurs where the proglacial topography is oriented parallel to the ice margin, causing lateral drainage diversion. At Skaftafellsjökull in south-eastern Iceland, for example, successive retreat stages of the glacier over the past century (since 1870 A.D.) are marked by sub-parallel terminal moraine ridges up to 3 m high, cut by a single central channel (Thompson, 1988) (Plate 12.8) (Fig. 12.8). During each recessional stage, drainage has been confined to a narrow lateral routeway, parallel to the ice margin and the confining moraine. Channels initially formed deep single-thread courses but, with progressive ice retreat and broadening of the drainage routeways, channels were able to widen, shallow, and develop complex braid bar-channel systems.

Distal drainage patterns are commonly characterized by complex braided channel networks, exhibiting highly unstable, broad, shallow channels (Plate 12.9). Width–depth ratios are widely used as a measure of channel cross-section shape. For braided proglacial rivers, width–depth values range from 5–15 in proximal Alpine streams (Maizels, 1976; Fenn and Gurnell, 1987), to 17–50 on Icelandic sandar (Churski, 1973), except during a jökulhlaup when values reached 207 (Churski, 1973). Fahnestock (1963) recorded a range of 10–71 on the White River outwash, Washington State, compared with a range of 43–154 on Lewis sandur, Baffin Island (Church, 1972), with an average of 23–81, respectively. The largest width–depth ratio of 2320 was estimated for high flows on the Slims River outwash, Alaska (Fahnestock, 1969).

PLATE 12.7. Terminal zone of the Bossons glacier illustrating large number of channels emerging from the advancing ice front and boulder moraine.

PLATE 12.8. Terminal zone of the Skaftafellsjökull glacier, southeast Iceland, illustrating braided meltwater drainage routes oriented parallel with the ice-margin (left), confiened behind a recent terminal moraine ridge (right). (Compare with Fig. 12.8.)

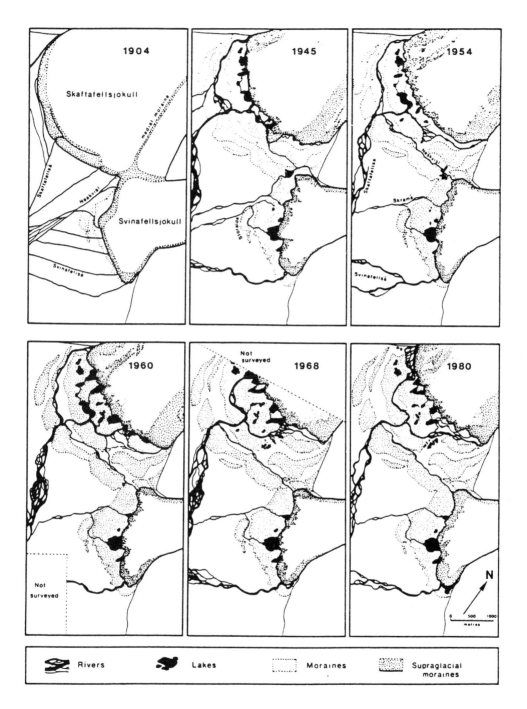

FIG. 12.8. Stages of proglacial moraine and drainage pattern development during historical glacier recession, Skaftafellsjökull, southeast Iceland. (From Thompson, 1988; reproduced by permission of the Iceland Glaciological Society.) (Compare with Fig. 12.8.)

PLATE 12.9. Broad, shallow, gravel-bed meltwater stream, west Greenland valley-sandur. Flow is towards the camera.

A number of explanations have been proposed for the high degree of instability of braided channel networks, particularly in proglacial environments. Primarily, instability is related to channel boundaries (especially the banks) being readily erodible. Bank erodibility is enhanced where bank sediments are non-cohesive containing a minimum content of silts and clays. In addition, if bank collapse occurs, flows of sufficient competence are required to entrain the sediment so that further bank faces can be exposed to erosion.

Proglacial drainage systems clearly promote the development of channel instability, according to the above criteria, since fines are normally removed in suspension, leaving only medium- and coarse-grained sediments to form bank materials. Most proglacial zones lack a full ground cover of vegetation and trees and tall shrubs are normally absent. Frequent fluctuations in meltwater flow levels promote repeated saturation of bank sediments. Finally, high magnitude diurnal flood flows allow regular removal of bank-foot material and continued attack on bank sediments.

When channel widening occurs, a complex chain reaction of feedback responses ensues. Channel widening initially leads to a shallowing of flow, which in turn reduces the bed shear stress, and hence the transport capacity of the flow. Deposition may occur within the channel cross-section or against one bank, leading to development of a mid-channel or lateral bar, respectively. According to experimental flume work by Leopold and Wolman (1957), mid-channel deposition of bar-core sediments in turn leads to concentration of flows on either side of the incipient bar. This process of braiding (Plate 12.9) is known as 'primary anastomosis'. 'Switching' of dominant flows between each braid channel contributes markedly to the instability of braided networks, particularly in rapidly aggrading, relatively fine-grained systems, where bed scour is readily achieved and not obstructed by excessive armouring.

A number of researchers have identified an alternative mechanism of braid–bar development that is believed to be common in braided river systems. Ashmore (1982), for example, described the results of

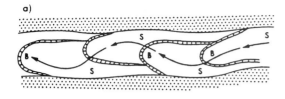

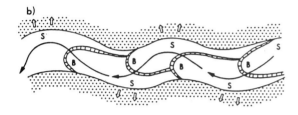

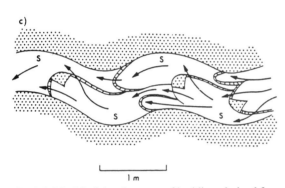

FIG. 12.9. Model of development of braiding, derived from flume experiments. (a) Shows channel pattern after one hour, showing obvious alternating bar (B) and scour (S) pattern and sinuous thalweg; (b) shows channel pattern after two hours, reflecting increase in amplitude of bends; and (c) shows channel pattern after four hours, in which alternating bar pattern has been broken down to give a braided channel. (From Ashmore, 1982; reproduced by permission from *Earth Surface Processes and Landforms*.)

a series of flume experiments in which lateral bars initially developed on alternate sides of a straight channel, leading to the division of flows into major and minor pathways, the former being diverted laterally across the bar edge, the latter continuing to flow parallel with the bank (Fig. 12.9). Lateral flow deflection directs the main flow path transversely across the channel towards the opposite bank, causing maximum bank erosion and thus enhancing the sinuosity of the flow paths and promoting the increased instability of the channel network. This mechanism highlights the close affinity between the processes operating in 'braided' and 'meandering' channels.

A third mechanism of braid–channel development has also been proposed and is known as 'secondary anastomosis'. Church and Gilbert (1975), Southard *et al.* (1984) and a number of other researchers have attributed the complexity of many braid–bar systems to dissection of outwash sheets, gravel spreads and existing bar deposits by over-bank flows. High flows spill over or 'avulse' on to bar-top and bank-top surfaces to form splays, which fan out from the first channel across the newly-inundated surface towards a second lower level channel. With repeated flooding, the distal splay points become increasingly scoured and develop into deep, narrow chute channels. They rise as abrupt 'heads' on a bar surface, remaining unconnected with the upstream channel until continued scour allows them to break through the proximal bank and form a new braid channel (Plates 12.10 and 12.11). Downstream of the 'head', small fans or lobes of sediment develop to form the core of new bars.

Many bars in proglacial river environments are highly complex in morphology and sedimentology (Section 12.4), reflecting a complicated history of aggradation (vertical), incision and accretion (especially lateral), as well as avulsion, splaying, channel abandonment, and chute and lobe development (Fig. 12.10). During high flows, bar-top sediments can become mobilized, leading to downstream migration of the bar-front to form a crudely bedded avalanche face. In gravelly environments, shallow bar-top flows lead to aggradation of a crudely horizontally bedded or sheet deposit, or an irregular spread of coarse gravels; in sandy environments, dunes associated with relatively deep, fast flows may be capped by waning stage ripples (see Section 12.4 for details). Commonly, as channels shift and switch during high flows, large areas of bar sediments may be removed, while new bar units are accreted laterally (e.g. Bluck, 1979, 1982). A braid bar may therefore contain the record of a complex history, particularly in areas with relatively low rates of vertical accretion. In zones of highly active sediment transport and widespread inundation, each bar may represent the outcome of

PLATE 12.10. Chute channel development on bars of the Sunwapta outwash, Alberta, Canada. Note that the head of the main chute channel occurs at the convergence or scour points of two shallow channels occupied during overbank flows. Flow is towards the camera.

PLATE 12.11. Complex anastomosing channel network, Myrdalssandur, south Iceland, illustrating progressive bar dissection by chute channels ('secondary anastomosis').

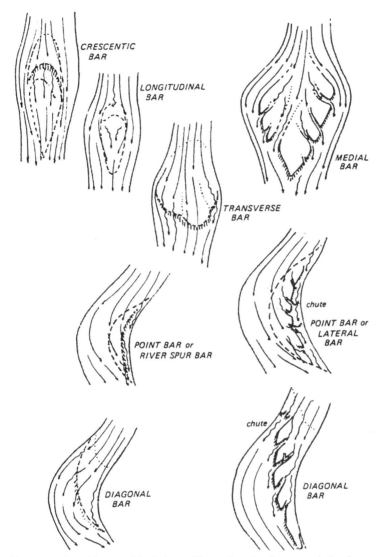

FIG. 12.10. Variety of bar types found in gravel-bed rivers, illustrating simple accumulation bars on the left ('unit bars') and erosively modified bars on the right. (From Church and Jones, 1982; reprinted from Hey, R.D. (ed.), *Gravel-Bed Rivers*, by permission of John Wiley and Sons.)

only the last single high flow event, rather than of an unknown series of multiple past events.

The complexity and transient behaviour of many bar–channel systems makes it very difficult to define or quantify the nature of the braided pattern. Quantification of the channel plan-form characteristics and degree of braiding are important inputs to studies of the physical controls on pattern development and evolution (temporal studies) and of

comparisons between channel patterns that have developed in different parts of a channel system, or wider different environmental or climatic conditions (spatial studies). Although no wholly satisfactory measure of braiding has yet been developed, a number of indices are in use at present (Brice, 1964; Williams and Rust, 1969; Rust, 1978; Maizels, 1979). Rust (1978) found, for example, that the 'braiding parameter' (the number of braids per meander

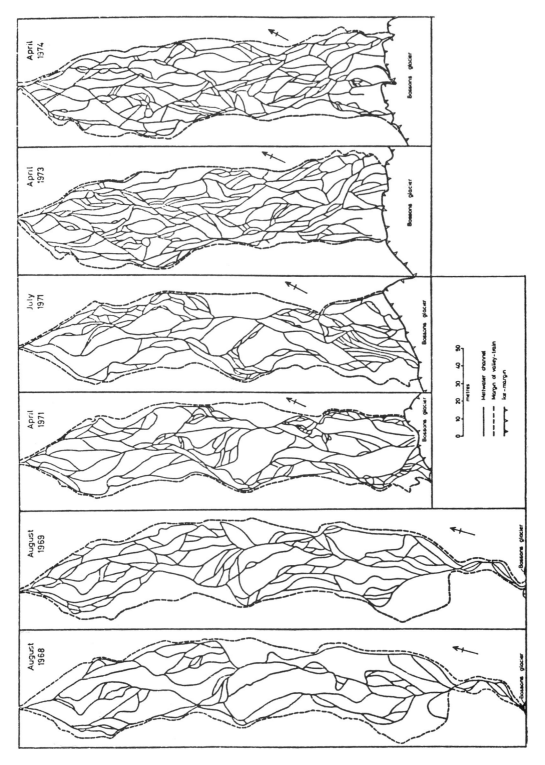

FIG. 12.11. Changes in braided channel pattern on the Bossons valley-train during a period of glacier advance, August 1968 to April 1974. (Compare with Plates 12.1, 12.7 and Fig. 12.6.)

wavelength) for first-order bars on Slims River outwash in the Yukon (a valley sandur) averaged 8.7, compared with a maximum of 32 on the middle reach of Skeidararsandur outwash plain, Iceland.

Maizels (1979) found that during an advance of the Bossons glacier, French Alps, by 114 m between 1968 and 1974, overall valley floor aggradation of 1.1 m (>9000 m² of valley train) was accompanied by a marked increase in braiding intensity. The mean 'braiding index' (number of braid channels per 100 m of outwash) was found to have increased from 5.73 to 11.11 in the six year period (Fig. 12.11), while the total number of channel segments, total channel length and number of bars increased from 344, 1890 m and 58, respectively, in 1968 to 500, 2796 and 160 m, respectively, in 1974.

Clearly, however, there is a danger that the proposed measures of braiding intensity are at least partly stage-dependent. During low flows only a single channel may be occupied, while during moderate flows a maximum number of channels is occupied, and during flood flows the full channel width may be inundated to form a single channel once more. A major problem in devising a meaningful index of braiding, or channel multiplicity, is the absence of physically-based, independent variables that conclusively control the nature of the channel pattern. Such measures need to be integrally linked to four factors: the variations of total discharge through the system as well as through individual channels; the stream power available for sediment transport; the morphology and plan-form shape of channels; and the calibre of the bar–channel sediments.

The individual channels within a braided network commonly exhibit relatively low sinuosities, reflecting bank instability. The highest sinuosity channels are found in silt- and clay-rich alluvium. In non-vegetated, non-cohesive sediments, bank collapse occurs so readily that incipient meanders cannot be sustained, and the channel expands rather than resists, resulting in formation of a straighter path. Channel sinuosity (P_{cs}) is defined by:

$$P_{cs} = \frac{Lc}{Lv} \qquad (12.1)$$

where Lc is the length of the main channel and Lv is the length of the valley axis (Leopold and Wolman, 1957).

Channel sinuosity within braided networks rarely exceeds values of 1.2. In systems that contain higher proportions of silts and/or are partially vegetated, channel sinuosity may increase substantially, such that individual channels exhibit meandering courses with sinuosity values exceeding 1.5. Meandering braided rivers are described as anastomosing. It has been demonstrated that a strong morphological link exists between the high width–depth ratios characteristic of non-cohesive sediments and the low sinuosities of their associated channels (Schumm, 1960; Ethridge and Schumm, 1978). Although using a limited data set selected only from the mid-west U.S.A. and excluding gravel-bed rivers, Schumm obtained a statistically significant relationship between channel sinuosity, P_{cs}, and width–depth ratio, F_{wd}, where :

$$P_{cs} = 3.5 F_{wd}^{-0.27} \qquad (12.2)$$

However, Leopold and Wolman's (1957) (Eqn. (12.1)) sinuosity measure does not take into account how additional energy within the system is being expended. Instead, a measure of sinuosity has been sought that allows a physically-based interpretation for both single-thread, low-sinuosity channels and multiple-thread low-sinuosity channels which characterize many proglacial braided channel systems. Richards (1982), for example, proposed an alternative index termed Total Sinuosity ($P_{cs(tot)}$) where:

$$P_{cs(tot)} = \frac{TCL}{Lv} \qquad (12.3)$$

This index attempts to take into account the total length of all braid channels (TCL) within the system. Richards found that with increasing stream power, W_s, the value of total sinuosity increases progressively from single-thread low-sinuosity channels and single-thread high-sinuosity channels, to multi-thread low sinuosity (braided) channels and, finally, to multi-thread high-sinuosity (anastomosing) channels, such that $P_{cs(tot)} = 2.64\ W_s^{0.1}$. Most proglacial channel networks exhibit patterns characteristic of high energy fluvial environments, namely multi-thread, low sinuosity channels.

12.3.2. Hydraulics

Proglacial meltwater channel systems are characterized by high energy flows, but these are concentrated in short temporal bursts. For much of the year, the braided channels and bars may remain inactive and channel switching during floods may lead either to their re-activation or more permanent abandonment and longer-term storage of sediments. Although flows in proximal channels tend to exhibit significantly higher energy conditions than in distal channels, variability in channel morphology and gradient within braided reaches can create high energy flows at any reach within the proglacial channel network. Similarly, although one might expect high energy flows in steep, alpine meltwater channels, the relatively small catchment areas of many of the glacial streams means that total flow volumes are in turn relatively low when compared with flows across the large valley-sandurs of Alaska or Baffin Island or the sandur plains of Iceland. Hence, the hydraulic conditions within each proglacial channel system are largely dependent on the magnitude of channel-forming flows, their frequency of occurrence and the nature of the constraints or boundary conditions of the system, namely, the volume and calibre of input sediments, the regional gradient and the valley floor width across which channel adjustments can occur.

During within-channel flows, measurements of a variety of hydraulic parameters indicate the generally high energy flow conditions that are commonly found in proglacial channel systems, particularly those that drain large, high relief catchments. A highly distinctive downstream change in flow conditions can also be recognized moving from proximal to distal zones. For example, the margin of part of the west Greenland Ice Sheet is drained by a downstream succession of valley sandurs, each separated by a narrow bedrock channel. Each sandur exhibits its own distinctive proximal-distal changes in hydraulic conditions and associated sediment assemblages (Russell, 1991). Jökulhlaup flows, which largely control sandur development in this ice-marginal area, exhibit a marked decline in energy and competence down-sandur. Proximal sandur zones are

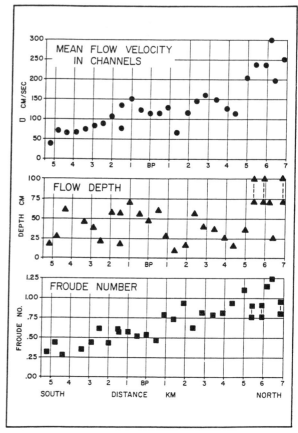

FIG. 12.12. Downstream changes in flow conditions in the Scott outwash river, Alaska. Ice-margin lies to the north. (After Boothroyd and Ashley, 1975; reproduced by permission of SEPM.)

characterized by stream powers of ca. 220 Wm^{-2} leading to development of upward coarsening, imbricated cobble and boulder sheets, dissected by chute channels and ice-block scour marks. Distal zones, with stream powers of only ca. 110 Wm^{-2}, are composed of fine-grained, upward coarsening sands and gravels and large-scale cross-beds. Measurements of stream flow on the Scott outwash in the northeastern Gulf of Alaska (Boothroyd and Ashley, 1975) also indicate marked downstream changes in a range of hydraulic parameters. Over a downstream distance of some 10 km from near the Scott glacier terminus, boundary shear stress was found to decline from about 1800 to <40 dynes cm^{-2}, stream power from about 400 to <4 ergs cm^{-2} s^{-1} x 10^3, mean velocity from about 300 to <50 cm s^{-1}, flow depth from 100 to only 25 cm, and Froude number from about 1.25 to ≈ 0.25 (Fig. 12.12). Gradient decreased from almost 18 to ≈1 m km^{-1}, while maximum grain size declined from 30 to 3.5 cm (cf. Rice, 1982).

12.3.3. Channel Pattern Adjustments to Flow

Channel characteristics represent the adjustment of its morphology and sediments to allow efficient transport of its water and sediment loads. Several studies of proglacial rivers have attempted to quantify the inter-relationships between channel shape, flow processes and sediment transport. For example, Church (1972) and Church and Gilbert (1975) examined the patterns of channel adjustment to varying inputs of water and sediment in terms of the at-a-station hydraulic geometry relations of braided and non-braided channels on Baffin Island sandurs. They adopted Leopold and Maddock's (1953) approach in which bi-variate power functions relate channel morphology (width, w and depth, d) and velocity, v, to discharge, Q, where $w = aQ^b$, $d = cQ^f$ and $v = kQ^m$, where a, c and k are coefficients distinctive to the given river and the exponents b, f and m represent the rate of change of the dependent variable in response to changing discharge. No significant difference was found between the behaviour of these two channel types (braided and non-braided) where they were both formed in non-cohesive gravels. Although most channel adjustment to rising flows occurred by bank collapse, Church found that as sediment loads increased at high flows, velocities increased most rapidly, giving mean values of b, f and m of 0.22, 0.31 and 0.48, respectively. It was argued that high sediment transport rates act to dampen turbulence, and reduce both form and boundary resistance associated with development of a mobile 'live' bed. Indeed, the flow resistance, defined by d/D_{90} (where D_{90} is the 90th percentile of the size distribution of the bedload), decreased far more rapidly than indicated by a number of predictive relations based on normal boundary particle resistance relations. The mean exponent on South River sandur, for example, was found to be 0.67, compared with predictions of between 0.16 and 0.5.

Similar patterns of channel response to rising discharges have been reported by Arnborg (1955) from the Hoffellsandur, Iceland, where values of b, f and m averaged 0.17, 0.39 and 0.44, respectively, and by Fenn and Gurnell (1987) at the Tsidjiore Nouve, Swiss Alps, with values of 0.04, 0.23 and 0.73, respectively. However, proglacial channels exhibit a wide variety of responses to changing flow conditions, depending on local boundary and hydraulic factors (e.g. Fahnestock, 1963; Maizels, 1976; Rice, 1982) and no universal model of channel response for proglacial streams can be recognized (Fig. 12.13).

As yet, few studies have focused on measuring the complex inter-relationships that exist between flow processes, sediment transport, channel form and channel pattern changes. Ashworth and Ferguson (1986) attempted to achieve this task by making simultaneous measurements of vertical velocity profiles (determining bed shear stress), bedload sediment transport and channel changes, on the Lyngsdalselva, a braided proglacial river in arctic Norway. One conclusion was that the alternating pattern of diverging, decelerating flow and converging, accelerating flow help to explain channel changes and bar formation; but as discharge increases, the expected channel response may be reversed as flows change from being only locally competent to becoming generally competent to transport high bedloads. At present, however, there is too little appropriate data from proglacial channel systems to build an integrated, process-based model of channel response to changing flows.

Proglacial outwash systems subject to extreme flood events may exhibit contrasting patterns of response. Floods can generate energies of sufficient magnitude to cause rapid exceedance of the local thresholds for channel change, disrupting any established state of dynamic equilibrium. Schumm and Lichty (1963) define this pattern of behaviour of a system as one characterized by "dynamic metastable equilibrium". Following each 'catastrophic' event, the fluvial system gradually modifies its boundaries until a new equilibrium state has been established. The extent to which post-flood flows can adjust to new conditions imposed by the flood and regain a new characteristic form, depends

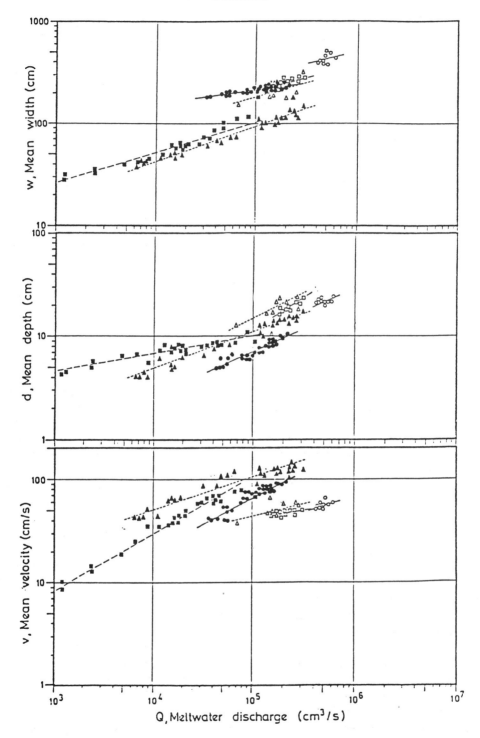

FIG. 12.13. At-a-station hydraulic geometry of meltwater channels, Bossons valley-train, 1971. Solid symbols = April flows; open symbols = July flows. Circles = distal reach, squares = western proximal reach, triangles = eastern proximal reach.

on the magnitude of the changes that have been imposed, their internal resistance to modification by subsequent lower energy, but longer duration, flows, and the time interval available in the system before a further jökulhlaup arrives, compared with the length of time required ('relaxation time') for the system to regain a state of dynamic equilibrium.

The magnitude and frequency of meltwater runoff events largely controls the scale of bar-and-channel structures. High flows generate both large-scale sediment transport and channel adjustments during the flood, and create new bar-and-channel networks that affect the rate at which lower, 'normal', ablation-relation flows can modify the channel boundaries to reach a new equilibrium state (Maizels, 1991, 1992). Because of the rarity of the event, 'normal' seasonal meltwater flows may be able to modify some of the jökulhlaup channel forms until hydraulic geometry relations match those established before the flood. By contrast, high magnitude events that occur at frequent intervals (e.g. through regular and repeated draining of an ice-dammed lake) may create a proglacial channel system that is adjusted to those flows and remains largely unmodified by intervening 'normal' flows. In these circumstances, bars are inundated only during jökulhlaup flows, so that sediment transport and bar migration can only occur during short-term (often less than 24 hours) episodic events. These circumstances have been identified on the Knik and Matanuska valley sandurs of Alaska (Fahnestock and Bradley, 1973) and on the sandurs of West Greenland (Russell, 1991). In West Greenland, channel adjustments occurred only during a 36-hour period of jökulhlaup flows, which occurred every 2 to 4 years. During these short, episodic events imbricated boulder beds were mobilized; fresh cobble and boulder sheets were deposited; new chute channels developed, particularly around stranded ice blocks; and bar fronts migrated tens of metres downstream. However, no significant change in channel courses occurred, and the overall braided channel pattern remained relatively stable. Both the Alaskan and Greenland examples represent proglacial environments in which jökulhlaups have not had a 'catastrophic' impact on the bar-channel system. Indeed, complete submergence of the sandur by jökulhlaup flows simply represents a bank-full condition.

12.4. CHARACTERISTICS OF PROGLACIAL OUTWASH SEDIMENTS

12.4.1. Characteristics of Individual Particles

12.4.1.1. Particle size and size distribution

It is widely assumed that 'all' sizes of sediment ranging from large boulders to glacially-abraded 'powder' are initially made available from the source glacier to the proglacial drainage system. Although this assumption may prove generally acceptable, there are a number of environments in which the size range of input sediments to proximal meltwater streams is curtailed or limited. First, where bedrock geology is composed of highly friable or soft materials, much of the material released at the ice margin is likely to be composed of relatively fine sediments. Secondly, where significant debris inputs to the glacier are derived from air-fall deposits such as volcanic tephra (as on many glaciers in south Iceland and in Washington State, U.S.A.) or loessic sediments, high proportions of output materials will be fine-grained. There is also some tentative evidence that suggests that the sizes of material issuing at the ice margin also partly reflect the distance of englacial transport (Maizels, 1976). In addition, coarser clasts may be found in the proximal zones of channels of larger glacierized catchments, reflecting the higher discharge and competence of these meltwater streams. Finally, there is some evidence suggesting that natural size distributions of fluvial sediments are depleted in the 2–4 mm (-1 to $-2\varnothing$) size range, possibly representing the boundary between coarser multi-mineral clasts and finer single-mineral grains, or reflecting the high entrainment mobility of these grain sizes.

The size distributions of clasts on proglacial valley-trains and sandurs have been examined in a number of studies. Mean proximal clast sizes are reported to vary between about 50 mm (Bossons in the French Alps, Maizels, 1979) and 1800 mm (Donjek River in the Yukon, Rust, 1972). Many studies have demonstrated that clast sizes decrease significantly downstream (Figs 12.14 and 12.15). The rate of downstream decrease in size is significantly related to the rate of decline in gradient (Fig. 12.16),

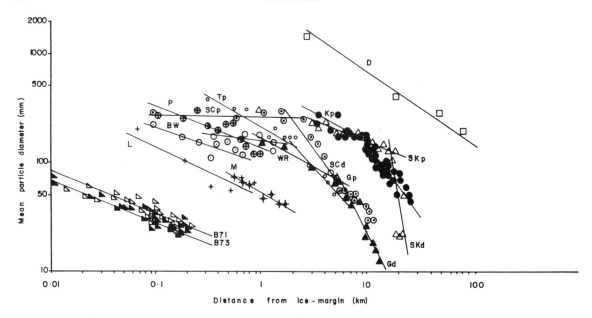

FIG. 12.14. Downstream decline in mean particle diameter in a variety of modern glacial meltwater streams. B71 = Bossons valley-train 1971; B73 = Bossons valley-train 1973; BW = Bow River, Alberta (after McDonald and Banerjee, 1971); D = Donjek River, Yukon (after Rust, 1972); G= Gigjukvisl, Skeidararsandur, Iceland (after Boothroyd and Nommedal, 1978); K = Knik River, Alaska (after Bradley *et al.*, 1972); L = Lewis River, Baffin Island (after Church, 1972); M = Mendenhall outwash, Alaska (after Ehrlich and Davies, 1968); P = Peyto outwash, Alberta (after McDonald and Banerjee, 1971), SC = Scott outwash, Alaska (after Boothroyd and Nummeddal, 1978), SK = Skeidarar River, Iceland (after Boothroyd and Nommedal, 1978), T = Tingin outwash (after Church, 1972), WR= White River, Washington State (after Fahnestock, 1963); d = distal zone, p = proximal zone.

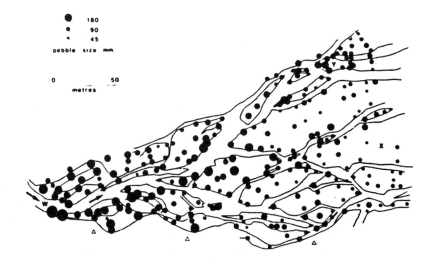

FIG. 12.15. Downstream decline in maximum surface grain size on bars over a 300 m reach in the Lyngsdalselva River, northern Norway. (From Ashworth and Ferguson, 1986; reprinted from Interrelationships of channel processes, changes and sediments in a proglacial braided river, by P.E. Ashmore and R.I. Ferguson, *Geografiska Annaler*, 1986, **68A**, 361–371, by permission of the Scandinavian University Press.)

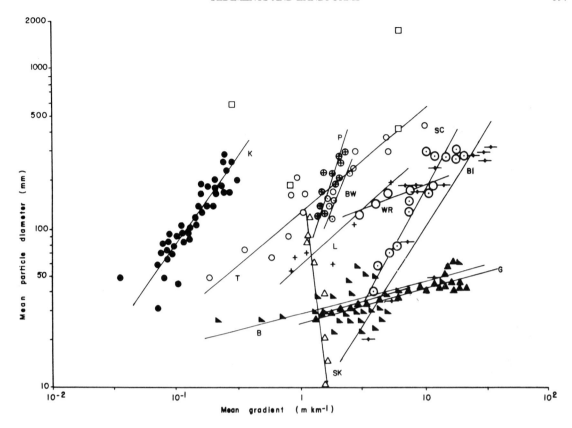

FIG. 12.16. Changes in mean particle diameter of outwash sediments in relation to sandur gradient, for a variety of modern meltwater streams. (For key see Fig. 12.14.)

as well as to the lithology of the sediment, and the initial clast size, with maximum rates of decline in mean clast diameter (b-axis) in areas with the coarsest initial sizes. Bluck (1987) has also shown that the rate of size decrease is proportional to the length of the river system itself. Bluck found that rates of size decrease are greatest on small alluvial fans, and lowest on vast alluvial plains such as that of the Indus river. The mean rates of downstream size decrease recorded in modern proglacial outwash deposits appear to conform to Bluck's model. On the 200 m long Bossons outwash (Maizels, 1976, 1979) the rate of size decrease of schist clasts was recorded as 250 mm km^{-1}, compared with 200 mm km^{-1} on the 900 m long Peyto glacier outwash (McDonald and Banerjee, 1971), while over the 16 km of the distal Knik River outwash, clast sizes decrease at a mean rate of only 8 mm km^{-1} (Bradley et al., 1972), and only 15 mm

km^{-1} on the 21.5 km of the (sampled) Skeidararsandur (Boothroyd and Nummedal, 1978).

Downstream decline in sediment size has been widely attributed to the combined effects of abrasion and selective sorting (cf. Bradley et al., 1972). Since rates of size decrease occur more rapidly than accounted for solely by abrasion, particularly for resistant lithologies, most downstream declines in sediment size are considered to reflect processes of selective deposition of coarser clasts. The processes that generate selective downstream sorting of grain sizes are, however, poorly understood and are currently the focus of research. The traditional explanation suggests that a decline in stream power downstream leads to the selective deposition of coarser clasts; thus bedload becomes progressively fine-grained. However, the fact that distal bar-heads commonly contain coarser grains than those found in proximal bar-tails (Section 12.4.3) indicates that coarse sediment can 'by-pass' zones of fine-grained sedimentation located upstream. The complexities of sediment transport pathways during sediment-transporting events (i.e. normally during floods) have not yet been resolved sufficiently to allow identification of the precise mechanisms of downstream particle sorting. Sediment size distributions of proglacial outwash deposits are highly variable, reflecting the range of source materials and depositional environments (Fig. 12.17). Hence, size distributions range from well-sorted, unimodal sands to poorly-sorted, bimodal or polymodal, coarse gravelly deposits with a variety of sand-sized matrix materials. The former may occur more commonly in distal channel zones, the latter in proximal bar and channel environments, where rapid deposition of torrential mobile-bed sediments can lead to matrix sediments becoming trapped amongst coarser clasts during deposition. Typical well-sorted,

unimodal sands may have a median grain size of between + 0.0 and + 2.0ø, a sorting index (Folk and Ward, 1957) of 0.25–0.75, a positive skewness value of 0–0.75 (i.e. with a fine tail), and a kurtosis value of 0.2–0.8 (i.e. platykurtic). By contrast, a polymodal outwash gravel is likely to exhibit equivalent values of -2 to -4ø, 2.0–3.5 (very poorly sorted), 0.3–0.8 (fine tail) and 0.2–0.4 (platykurtic), respectively. Bimodal or polymodal outwash deposits commonly exhibit a dearth of medium sands (2–4 mm), believed to result from their relatively high mobility compared with both finer (colloidal) and coarser (heavy) grains. Most outwash sediments are characterized by low concentrations of silts and clays, commonly not exceeding ~5%. The exception is where fines have selectively accumulated in sheltered pockets or pools on the outwash surface, remain as waning-stage silt-skins on bar-tops prior to removal by deflation, or where influxes of air-borne silt or dust are derived from erosion of older coastal, till, or valley side deposits.

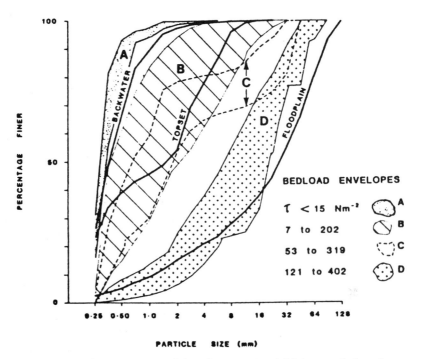

FIG. 12.17. Sediment size distribution curves (by weight) of surface (floodplain) outwash deposits, compared with those of bedload sampled in different ranges (A to D) of shear stress. Measurements of bedload were truncated at 0.25 and 76 mm. (From Ashworth and Ferguson 1986; reprinted from Interrelationships of channel processes, changes and sediments in a proglacial braided river, by P.E. Ashmore and R.I. Ferguson, *Geografiska Annaler*, 1986, **68A**, 361–371, by permission of the Scandinavian University Press.)

12.4.1.2. Particle shape

Many studies have demonstrated that clast form (especially roundness) is closely related to the agent and distance of transport and the environment of deposition, in response to the degree of abrasion (Chapter 15; Menzies, 1995b, Chapter 13). Clasts deposited by meltwaters in glaciofluvial outwash environments may therefore exhibit a wide range of roundness values depending on the distance of transport (both within the glacier and the proglacial zone), the lithology of the clast and its resistance to abrasion, and its transport history (e.g. King and Buckley, 1968). Two indices of shape have been widely adopted. First, the Cailleux–Tricart roundness index (R_I) measures the radius of curvature, r_c, of the sharpest corner on the a–b plane, in relation to the maximum particle diameter, a, such that $R_I = (2 \, r_c/a)$ x 1000. Roundness values lie between 0 and 1000, and are applied to limited size ranges only, as r_c is likely to be a function of particle size itself. Cailleux roundness values recorded for clasts deposited in modern proglacial outwash environments range from as low as 0.03 to as high as 0.40, with the majority of values lying between 0.15 and 0.25. Variability within a single deposit can also be relatively high since roundness has been widely demonstrated to increase exponentially downstream in response to progressive attrition.

A second and more widely adopted quantitative approach to clast shape was proposed by Sneed and Folk (1958). Three form ratios were derived, based on the three axial lengths of clasts, a, b and c, giving measures of symmetry (a/c), elongation ((a–b)/(a–c)), and sphericity (c^2/ab). When plotted on a triangular graph or 'form diagram', ten shape classes were distinguished. Results from proglacial outwash sediments indicate clast shape partly determines its relative ease of entrainment, with disc-shaped clasts being most mobile (Ehrlich and Davies, 1968; Bradley et al., 1972) (Fig. 12.18). However, it appears that shape is largely determined by the original structure and composition of the source rock (Maizels, 1976; Haldorsen, 1981).

Where there are influxes of debris from valley sides, tributaries, re-working of older weathered materials, or inputs directly from the ice, the spatial pattern of roundness variation can be highly complex. On the Bossons valley-train (French Alps), for example, mean roundness values of schist clasts increase significantly from 120 to 180 within 40 m of the ice-margin, but then decrease over the next 120 m because of the introduction of frost-shattered and angular debris from slopes composed of Little Ice Age lateral moraine (Fig. 12.19) (Plate 12.1). Where the additions of debris from the valley sides cease due to vegetation cover, clast roundness once more appears to increase (from values of 80 to 150, over the final 50 m of the valley-train) (Maizels, 1976).

12.4.1.3. Clast fabric

Where particles are subject to rotational stresses, such as imposed during bedload traction of clasts, deposits commonly exhibit a preferred orientation or fabric. In many fluvial environments, the more platy or discoidal clasts are typically orientated with the long axes of clasts perpendicular to the flow direction, while their a–b planes assume an upstream-dip (Harms et al., 1975). Each clast becomes banked up against other clasts or obstacles on the bed. Thus a stable imbricated or 'shingled' bed containing a strong, preferred, upstream fabric develops (Rust, 1972). A weak or random fabric is characteristic of rapidly fluctuating flows, or deposition rates that are too rapid or sediment-laden to allow clast rotation to occur freely. Where clasts are pushed along the bed, rather than rolled, their long axis becomes oriented parallel with the flow, offering minimum resistance to fluid flow.

It has been found that mean imbrication directions from 33 sites on bar surfaces of the Scott outwash, Alaska, corresponded closely to the overall valley trend, while long axis of clasts showed a strong orientation transverse to flows (Boothroyd, 1972; Boothroyd and Ashley, 1975) (Fig. 12.18). However, other workers have found that clast fabric on gravel bars can be highly variable in direction and strength, reflecting the fluctuations in magnitude and direction of flows across the bar-top at different stages of diurnal or periodic flood events (cf. Church, 1972; Bluck, 1982). The local current directions derived from gravel fabric on bars on Breiðamerkursandur, Iceland, was found to differ considerably from the

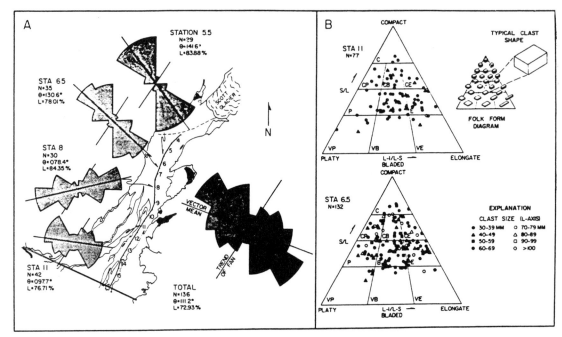

FIG. 12.18. (A) Clast long-axis orientation, Scott outwash fan, Alaska, measured within 50 x 80 cm area on bar top surfaces. (B) Clast form distributions of particles from two sites on the Scott outwash fan, following classification of Sneed and Folk (1958). (Both figures from Boothroyd and Ashley, 1975, reproduced by permission of *SEPM*.)

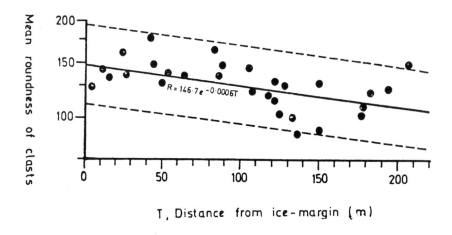

FIG. 12.19. Downstream fluctuations in mean roundness of schist clasts, Bossons valley-train, based on the Cailleux Tricart Index.

local trends of surface channels (Rust, 1975). It was argued that the contrasts reflect the different directions of flow during the period of gravel deposition, with jökulhlaup drainage creating bar-top fabric that differs from that created by later flows associated with formation of sandur channels.

12.4.2. Facies of Proglacial Outwash Sediments

Three facies types, or depositional environments, have been widely recognized from modern proglacial outwash deposits on the basis of their dominant grain size: silt, sand and gravel (Rust, 1978) (Menzies, 1995b, Chapters 8 and 9). The different facies commonly exhibit a distinct downstream distribution from gravels to sands and, finally, to silts. However, lateral variations in facies type can also occur according to the nature of the local bar-and-channel network, including its scale, stability, relative relief, and degree of inundation at different flow stages. Local variability in facies type can also occur where influxes of non-fluvioglacial sediments are deposited on the outwash. Aeolian sediments may be deposited as sheets or dunes, especially in coastal environments, while mud-flows and debris flows may be input from ice-contact slopes or valley sides (Hammer and Smith, 1983). In addition, temporal variations in facies type can occur during individual flood events, when falling-limb deposition may change from coarse- to fine-grained as flow competence declines. Hence, although large-scale downstream trends in facies types occur in most outwash deposits, local variations may also prove important.

In 1977, a coding system for annotating facies types was introduced (Miall, 1977) (Table 12.2). Although problems arise in using and applying Miall's classification scheme to a wide variety of complex sediment types, this approach has been widely adopted to describe in simple terms the sedimentary characteristics of fluvial and fluvioglacial deposits (Miall, 1983) (Menzies, 1995b, Chapter 9). Many workers have extended or modified the classification scheme (cf. Eyles and Miall, 1984; Brodzikowski and Van Loon, 1987, 1991). The use of facies codes has been widely adopted as the basis for identifying facies assemblages that characterize particular kinds of fluvial environments. These assemblages in turn form the basis of fluvial facies models. However, it must be remembered that the codings act only as descriptive summaries of the sediment characteristics that must be supplemented by field and laboratory data. They do not necessarily indicate the regional and local hydraulic and sedimentary controls that may have generated those deposits.

12.4.2.1. Fine-grained facies

The fine-grained facies include laminated (Fl) and massive (Fm) silts and clays. In proglacial environments, massive fines may be found as drapes or 'silt-skins' across bar-tops, deposited from suspension during waning flows (McDonald and Banerjee, 1971). Laminated fines accumulate in pools, abandoned channels, kettle-holes and lakes. The laminae, comprising alternating coarse and fine layers, are commonly interpreted as annual varves, in which the coarse grains represent summer influxes, with the fine grains settling out slowly during the winter. However, in many proglacial environments such sediment couplets normally reflect sedimentation from short-term 'pulsed' sediment influx events associated with turbidity currents produced during diurnal or periodic floods (Gustavson, 1975a, b; Shaw and Archer, 1978).

12.4.2.2. Sand facies

A wide range of sand facies has been recognized in proglacial meltwater environments, the most common of which are associated with the downstream migration of bars, and the formation of fluvial dunes and ripples (Plates 12.12, 12.13 and 12.14). Bars are the fundamental depositional unit within proglacial meltwater environments acting as the primary sediment storage zones. Long-term storage relates to coarse sediment which is moved only during episodic high flow events, while short-term storage relates to finer-grained sediments moved at all but the lowest flows. These storage effects may be altered where the bed is strongly armoured. A wide range of bar types has been recognized in fluvial and fluvioglacial environments and a myriad of terms has been used to describe them according to their plan-

PLATE 12.12. Large-scale trough cross-stratification formed by migrating dunes, capped by horizontally bedded sands and silts associated with shallow, waning-stage sheet flows, Myrdalssandur, south Iceland.

PLATE 12.13. Type 'A' ripple-drift cross-lamination overlain by Type 'S', sinusoidal ripple laminations, Late Devensian outwash, N.E. Scotland. Flow was from left to right.

CURRENT DIRECTION

60   10·0
Centimetres

PLATE 12.14. Type 'A' ripple-drift cross-laminations in coarse sand and gravel of Late Devensian outwash, N.E. Scotland. Note local imbrication of flatter clasts.

deposits is low, for they are likely to be swept away during subsequent high flood events. Horizontal bedding most commonly develops as upper (coarse sand) and lower (fine sand) flow regime plane beds which are confined to shallow flows across bar-tops or sandflats. Massive sands, by contrast, are more characteristic of suspension flows, in which rapid deposition occurs within deep channels (cf. Rust and Romanelli, 1975). Finally, local scours and erosion surfaces occur where a change of process or flow direction occurs, interrupting the migration or aggradation of bedforms.

12.4.2.3. Gravel facies

The most commonly occurring gravel facies within glacial outwash deposits are clast-supported, imbricated, heterogeneous, sub-rounded, bimodal or polymodal gravels exhibiting only poorly defined bedding. This facies has been described by Miall (1977) as Gm, representing massive or horizontally bedded gravel facies (Plate 12.15). However, the wide variety of textures, grading, fabric and internal structures typical of glacial outwash deposits are poorly represented by this single descriptive code. Some outwash deposits consist of well-imbricated, horizontally-bedded gravels composed of a sand matrix and a fine gravel clast fraction deposited by gradual accretion during moderate flows; while others are composed of massive, structureless, polymodal, weakly-imbricated cobble gravels, where all size fractions have been deposited rapidly during torrential flood flows. Horizontally-bedded gravels occur widely in proglacial braided river environments, although bedding planes are normally indistinct or poorly developed. Crudely-bedded imbricate gravels form the nucleus of many gravel bars and gravel sheets, where flows are shallow relative to the mean particle size of the bed material. Since many proglacial bar-channel networks exhibit a low relative relief and are subject to sheet floods across the outwash surface, this gravel facies tends to dominate many bar and channel deposits (Rust and Koster, 1984).

The most common bar type dominated by gravel facies is the longitudinal bar, which extends parallel with the flow, and is bounded by low sinuosity

form shape, location, relief and texture (Krigström, 1962; Smith, 1974; Boothroyd and Ashley, 1975; Church and Gilbert, 1975; Church and Jones, 1982).

Sandy bars typically form as point bars in meandering and anastomosing channels, and as transverse, 'linguoid' or 'spool' bars in braided environments (Fig. 12.20). While dunes develop within the major channels, rippled sands may accumulate locally during shallow flows across bar tops, and within shallow channels cut into the bar surface. However, the preservation potential of these

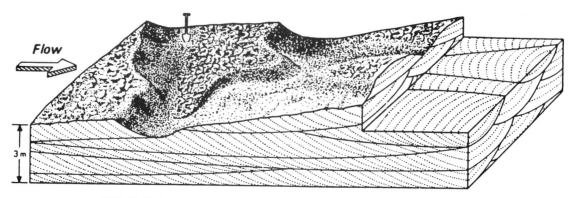

FIG. 12.20. Large-scale trough cross-stratification formed by migrating dunes.

channels (Plate 12.8) (Fig. 12.10). Longitudinal bars are composed of a core or platform of gravels on to which are accreted horizontally-bedded, imbricated gravels associated with shallow sheet flows, and exhibiting pebble clusters and other small-scale gravel bedforms (cf. McDonald and Banerjee, 1971; Boothroyd and Ashley, 1975; Koster, 1978; Bluck, 1987). Bar front sediments may exhibit crude planar cross-bedding deposited during downstream and lateral progradation of the bar (Plate 12.16).

Proglacial outwash environments may also contain, especially in proximal zones (Section 12.4.3), massive, matrix-supported gravels (described by Miall, 1977 as Gms). These deposits have a relatively high (<50%) matrix context, are poorly sorted, with no clear bedding or imbrication, and often contain large isolated, angular clasts. These deposits are widely interpreted as debris-flow or mud-flow facies, associated with viscous, sediment-rich (>70%) meltout flows from the ice-margin, solifluction deposits from adjacent slopes, or facies produced by saturation and mobilization of pre-existing sediments prior to a major flood event (Eyles and Miall, 1984). Where sediment-laden flows still

PLATE 12.15. Contact imbrication of clasts in massive gravel outwash, Late Devensian, N.E. Scotland.

PLATE 12.16. Coarse, crudely bedded gravels forming longitudinal bars, Solheimajökull outwash, south Iceland. Flow is from right to left. Note glacier snout in background.

contain sufficient fluid content to sustain low viscosity flows (i.e. ca. 40–70% sediment concentrations), hyperconcentrated grain flows occur (Plate 12.17) and an intermediate gravel facies may be deposited. Grain flow deposits are commonly composed of massive, matrix-supported gravels, which exhibit a distinct inverse grading overlying a basal, imbricated lag gravel. These deposits reflect the importance of internal dispersive stresses acting on grains in suspension, as well as basal shear stresses operating at the flow–bed contact. Debris flows commonly occur as steep, lobate deposits or 'plugs' where they are channelized, while grain flows may also be deposited as broad gravel sheets across the outwash plain (Maizels, 1989a, b, 1992, 1993).

12.4.2.4. Boulder facies

Boulder deposits are often found within proglacial meltwater environments. They may form distinctive spreads, both in proximal and distal zones, and may be composed of a wide variety of lithologies,

reflecting the geological structure of the glacierized basin, or of a single lithology, where meltwaters have been directed along selected bedrock routeways. Boulders also occur as isolated occurrences or as localized clusters, reflecting transport on ice-rafts or bergs, or by buoyant high viscosity debris or grain flows, particularly during glacier outburst floods (Maizels, 1989a, b, 1992) (Plate 12.18). Boulders accumulate particularly on the heads of bars where they can form large, imbricated speads or clusters (Russell, 1991). The clusters act as obstacles to the transport of coarse clasts from upstream, which in turn accumulate on the upstream (stoss) side of the boulder cluster, and accentuate the development of a pendant bar and flow separation into lateral channels.

12.4.3. Facies Models in Proglacial Environments

A facies model represents a general summary of the sediments that characterize a particular sedimentary environment. The summary model may take the form of an idealized facies assemblage list, a

J. MAIZELS

PLATE 12.17. Massive, hyperconcentrated grain-flow deposit, representing main flood surge during 1918 jökulhlaup from Katla, on to Myrdalssandur, south Iceland.

PLATE 12.18. Point bar in Russell Glacier outwash, west Greenland, illustrating coarse, boulder-dominated bar head (left) and sand-dominated bar tail (right) exhibiting dune forms. Note figure for scale. Flow is from left to right. (Photo by A. Russell.)

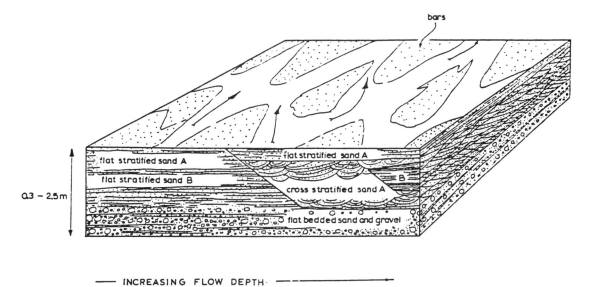

FIG. 12.21. Bluck's (1979) model of fine-grained braided stream alluvial deposits in wide, shallow channels. Flat stratified sand A deposited by bars of low relief shown on the surface of the diagram; flat stratified sand B refers to the gently inclined and flat strata. Cross-stratified sands A are deposited by dunes or megaripples; B, cross-stratified sand sheets. The left-hand sequence reflects shallower flows than that on the right. (From Bluck, 1979; reprinted by permission of the Royal Society of Edinburgh.)

vertical sequence of facies, a block diagram, a facies relationship diagram based on a transition probability matrix, a graph, or an equation or set of equations (Menzies, 1995b, Chapter 9). Facies models of proglacial meltwater environments have largely focused on the braided river sedimentary environment (Fig. 12.21). For example, important studies of braid bar and channel development, bar sedimentology, braiding intensity and spatial trends in network characteristics have been carried out in the proglacial outwash environments of Iceland (Arnborg, 1955; Hjulström, 1952; Krigström, 1962; Bluck, 1974; Boothroyd and Nummedal, 1978; Rust, 1978; Maizels, 1989, 1993), Alaska (Fahnestock and Bradley, 1973; Gustavson, 1974; Gustavson and Boothroyd, 1987), Baffin Island (Church, 1972; Church and Gilbert, 1975), the western Cordillera (Fahnestock, 1963; McDonald and Banerjee, 1971; Smith, 1974; Hein and Walker, 1977; Cant and Walker, 1978; Hammer and Smith, 1983), the European Alps (Maizels, 1979; Fenn and Gurnell, 1987), Scandinavia (Ashworth and Ferguson, 1986), Greenland (de Jong, 1992; Russell, 1991), the Himalayas (Hewitt, 1982) and New Zealand (Griffiths, 1980; Mosley, 1983; Rundle, 1985).

12.4.3.1. Bar and channel facies models for braided rivers and glacial outwash

Detailed analysis of small-scale variations in the sedimentology of braid bars and the resulting sedimentary structures have been carried out on a number of proglacial outwash deposits (Bluck, 1974, 1979, 1982). Bar-head sediments are composed of coarse imbricated gravels forming horizontally-bedded spreads, transverse ribs and local sand shadows subject to entrainment and transport only during the highest flows. The bar-tail sediments, by contrast, consist of gently inclined sheets of sand and gravel, horizontally bedded lag gravels, and crude foreset sands and gravels. These distal bar sediments are subject to re-working during all but the lowest flows (Figs 12.22 and 12.23). Bluck demonstrated that additional bar units are accreted both to the lateral and downstream edges of bars to form complex compound bar forms, with different parts of the new bar subject to modification at different stages of flow. In addition, channel facies are characterized by extensive coarse armouring and localized formation of transverse ribs in steep, shallow reaches.

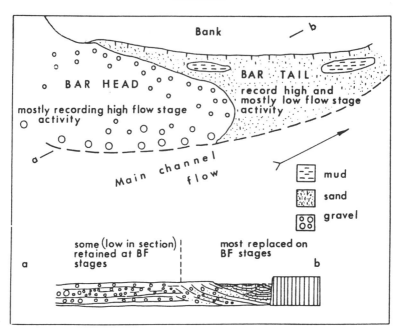

FIG. 12.22. Bluck's (1982) simplified model of lateral bar sedimentology in gravel-bed rivers. BF refers to bankfull discharge. (From Bluck, 1982; reprinted from Hey, R.D. (ed.), *Gravel-Bed Rivers*, by permission John Wiley and Sons.)

Total bar–channel relief tends to be very low, with channels typically being broad and shallow, with no clearly defined banks.

A large number of facies models have been proposed for sandy, distal braid bar (Fig. 12.24) (cf. Doeglas, 1962; Coleman, 1969; Smith, 1970; Collinson, 1970; Cant, 1978; Cant and Walker, 1978) and anastomosing channel (Smith and Smith, 1980; Smith, 1983) environments.

12.4.3.2. Proximal–distal facies models for braided rivers and glacial outwash

The most significant feature emerging from many facies studies is the widespread distinction made between the morphological and sedimentary characteristics of proximal and distal parts of the proglacial drainage system. Facies models of glacial outwash stress the importance of proximal–distal changes in sedimentology, reflecting downstream declines in gradient, stream power and maximum clast size, and a downstream succession from massive or horizontally-bedded, longitudinal gravel bars, deposited by vertical accretion in the proximal zones,

to mainly trough cross-bedded, cyclic fining-upward sequences of gravels, sands and silts in the distal zones. The proximal zones of the Scott and Yana outwash fans in Alaska, for example, are dominated by a coarse-gravel facies consisting of well-imbricated pebbles, cobbles and boulders associated with high flows across low bed relief bars (<30 cm), and longitudinal bars up to 50 m long (Boothroyd and Ashley, 1975). Towards the mid-fan zone the gravel facies exhibit a marked downstream decline in maximum clast size. The gravel is interbedded with plane- and large-scale trough cross-bedded sands resulting from 'megaripple' migration in low-stage channels. The lower fan is dominated by slip-face migration of longitudinal and linguoid braid bars, giving rise to planar cross-beds, while bar-top formation of ripples produces abundant ripple-drift cross-lamination. The meltwater channels begin to meander as the outwash sediments become increasingly fine-grained, and the most distal facies exhibit 'megaripple' migration on point-bar surfaces producing large-scale trough cross-beds and migration of the bar slip face producing planar to tangential cross-beds.

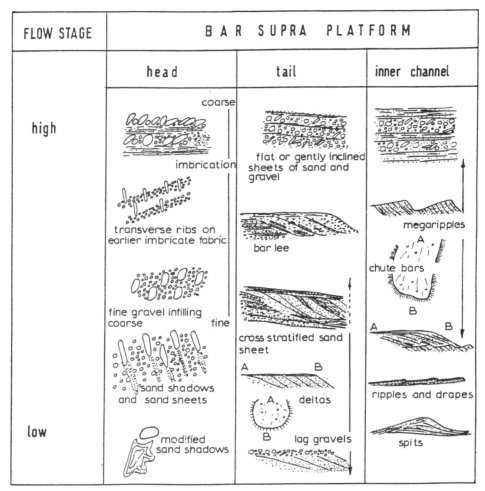

FIG. 12.23. Bluck's (1979) model of sedimentary structures of braid bars, indicating structures representative of high and low flow stages on the bar-head, bar-tail and inner (chute) channel. (From Bluck, 1979, reprinted from Hey, R.D. (ed.), *Gravel-Bed Rivers*, by permission John Wiley and Sons.)

These downstream facies trends have also been summarized in terms of vertical profile models, such as those proposed by Miall (1977, 1978) and Rust (1978). Miall recognized six principal facies assemblages in gravel- and sand-dominated braided river deposits (Fig. 12.25) (Table 12.2). Two kinds of proximal facies were identified: the Trollheim Type (equivalent to Rust's GI Lithotype (Table 12.3)), which is dominated by massive, clast-supported and matrix-supported gravels representing deposition in longitudinal bars and debris flows, respectively; and the Scott Type (Rust's GII Lithotype), which is largely composed of massive or crudely-bedded,

clast-supported gravels typical of longitudinal braid bars. Five distal facies have been recognised by these authors. A distal gravelly (gravel content of 10–90%) river facies (Miall's Donjek Type; Rust's GIII; Williams and Rust, 1969; Rust, 1972), which is dominated by fining-upward cycles of crudely-bedded coarse gravels and trough cross-bedded sands and gravels, reflecting deposition of dunes on to gravel lag during infilling of sandur channels; minor effects include the migration of linguoid bars and sand waves.

Distal braided rivers dominated by linguoid bars, dunes and ripples are summarized by Miall's South

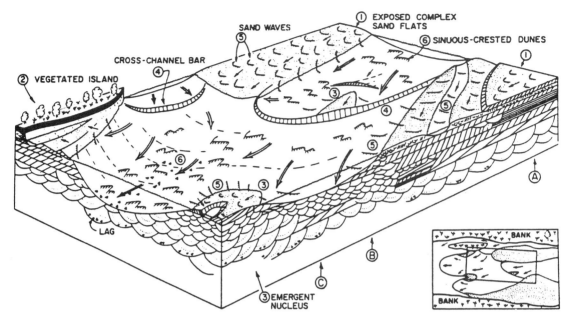

FIG. 12.24. Cant's (1978) facies model of a sandy, braided river, illustrating the sedimentary structures characteristic of both the active channel environment and of preserved deposits. The inset diagram shows a vertical view of the same area. The circled letters A, B and C shows the locations of formation of the sand flat, mixed influence and channel sequences, respectively (see text for description). (From Cant, 1978, permission of the Canadian Society of Petroleum Geologists.)

Saskatchewan Type (Rust's SII Lithotype) in which vertical sequences are mainly composed of cyclic channel fills and trough cross-bedded sands, while planar cross-bedded, horizontally-bedded, and scour fill sands, and silty laminae are of minor importance. Non-cyclic vertical sequences (Miall's Platte Type; Rust's SII Lithotype) contain higher proportions of planar cross-beds. Miall also recognizes a distal facies assemblage associated with flash floods and dominated by sheet flow deposits and horizontally-bedded sands (Miall's Bijou Creek Type; Rust's SI Lithotype). The finest-grained distal facies assemblage is that described by Rust (1978) as dominated by laminated and massive silts and clays. These are found particularly where aeolian reworking and thixotropic deformation are common, as in tidal flat areas.

12.4.3.3. Jökulhlaup facies model for glacial outwash

Most sedimentological studies of modern outwash deposits have focused on braided river sedimentation, only briefly referring to the incidental effects of large flood events. Examination of many such studies suggests, however, that sedimentation in some outwash environments may be dominated by jökulhlaup activity. In Iceland, for example, studies of sandurs during the 1970s (e.g. Klimek, 1973; Churski, 1973; Galon, 1973; Sigbjarnarson, 1973; Tomasson, 1974; Ward et al., 1976; Kaldal, 1978) emphasized the role of jökulhlaups in the development of terrace sequences, kettle holes and pitted sandur, boulder lobes and spreads, and large bar deposits. In addition, reference was made by Bluck (1979) to large-scale 'megaripple' cross-strata, some with coarsening-upward foresets, which were interpreted as flood deposits. Boothroyd and Nummedal (1978) also described planar gravel cross-strata on proximal longitudinal bars that they attributed to glacier-burst floods. Miall's (1977, 1978) Bijou Creek facies type also recognizes the impact of repeated flood events on the cyclical nature of the vertical sediment sequence, albeit in a distal, sandy environment.

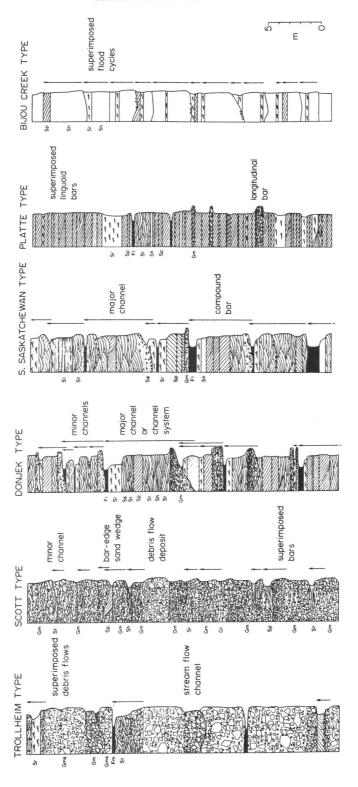

FIG. 12.25. Miall's (1978) vertical facies profile models for braided stream deposits. Facies codes to the left of each column are given in Table 1. Arrows show small-scale cyclic sequences. (From Miall, 1978, permission of the Canadian Society of Petroleum Geologists.)

TABLE 12.2. Miall's (1977, 1978) classification and coding of facies, lithofacies and sedimentary structures of modern and ancient braided stream deposits. (From Miall, 1978.)

Facies Code	Lithofacies	Sedimentary structures	Interpretation
Gms	massive, matrix supported gravel	none	debris flow deposits
Gm	massive or crudely bedded gravel	horizontal bedding, imbrication	longitudinal bars, lag deposits, sieve deposits
Gt	gravel, stratified	trough crossbeds	minor channel fills
Gp	gravel, stratified	planar crossbeds	linguoid bars or deltaic growths from older bar remnants
St	sand, medium to v. coarse, may be pebbly	solitary (theta) or grouped (pi) trough crossbeds	dunes (lower flow regime)
Sp	sand, medium to v. coarse, may be pebbly	solitary (alpha) or grouped (omikron) planar crossbeds	linguoid, transverse bars, sand waves (lower flow regime)
Sr	sand, very fine to coarse	ripple marks of all types	ripples (lower flow regime)
Sh	sand, very fine to very coarse, may be pebbly	horizontal lamination, parting or streaming lineation	planar bed flow (l. and u. flow regime)
Sl	sand, fine	low angle ($<10°$) crossbeds	scour fills, crevasse splays, antidunes
Se	erosional scours with intraclasts	crude crossbedding	scour fills
Ss	sand, fine to coarse, may be pebbly	broad, shallow scours including eta cross-stratification	scour fills
Sse, She, Spe sand		analogous to Ss, Sh, Sp	eolian deposits
Fl	sand, silt, mud	fine lamination, very small ripples	overbank or waning flood deposits
Fsc	silt, mud	laminated to massive	backswamp deposits
Fcf	mud	massive, with freshwater molluscs	backswamp pond deposits
Fm	mud, silt	massive, desiccation cracks	overbank or drape deposits
Fr	silt, mud	rootlets	seatearth
C	coal, carbonaceous mud	plants, mud films	swamp deposits
P	carbonate	pedogenic features	soil

TABLE 12.3. Miall's (1977, 1978) classification of the six main facies assemblages in gravel- and sand-dominated braided river deposits. (From Miall, 1978.)

Name	Environmental setting	Main facies	Minor facies
Trollheim type (G$_I$)	proximal rivers (predominantly alluvial fans) subject to debris flows	Gms, Gm	St, Sp, Fl, Fm
Scott type (G$_{II}$)	proximal rivers (including alluvial fans) with stream flows	Gm	Gp, Gt, Sp, St, Sr, Fl, Fm
Donjek type (G$_{III}$)	distal gravelly rivers (cyclic deposits)	Gm, Gt, St	Gp, Sh, Sr, Sp, Fl, Fm
South Saskatchewan type (S$_{II}$)	sandy braided rivers (cyclic deposits)	St	Sp, Se, Sr, Sh, Ss, Sl, Gm, Fl, Fm
Platte type (S$_{II}$)	sandy braided rivers (virtually non cyclic)	St, Sp	Sh, Sr, Ss, Gm, Fl, Fm
Bijou Creek type (S$_I$)	Ephemeral or perennial rivers subject to flash floods	Sh, Sl	Sp, Sr

In addition to large-scale gravel cross-beds, jökulhlaups generate sediment sequences associated with deposition from hyperconcentrated grain flows and debris flows, particularly in areas of abundant sediment supply. Some of the Icelandic sandurs have been found to be largely composed of thick units of black pumice, granular gravels deposited during volcanically-generated jökulhlaup events. Analysis of outwash sediments on Solheimasandur (Maizels, 1989a, b, 1992) and Myrdalssandur (Jonsson, 1982; Maizels, 1992, 1993) forms the basis of a facies model of volcano–glacial jökulhlaup sandur deposition. The sandurs are composed of a number of individual flood deposits, up to 12 m thick in proximal zones and only 1–2 m thick is distal zones (Fig. 12.26). Each flood deposit is composed of four sedimentary units, reflecting successive stages of the jökulhlaup event. The rising flood wave is marked by basal imbricated gravels with a fine pumice matrix; the main flood surge is represented by thick (<5 m), homogeneous, massive, granular pumice gravels representing grain flows; the onset of the waning stage is characterized by migration of large pumice dunes along the main flow paths leading to formation of large-scale trough cross-beds; while shallow sheet flows marking the final stages of the jökulhlaup are represented by horizontal laminations of pumice granules, sands and silts. The main flood surge unit is thickest, and the dune and sheet-flow units thinnest, in proximal zones; clast sizes exhibit little downstream decrease in the surge unit, but decline rapidly in the dune and sheet-flow units, reflecting the mass transport of sediment in the former, and bed-load traction in the latter. Debris flow deposits also occur locally, wherever older outwash, slope or soil deposits are in contact with, and entrained or mobilised by jökulhlaup flows.

Jökulhlaup outwash facies may also include a distinctive assemblage of bedform structures and morphological features. In addition to large-scale bar deposits, boulder accumulations and fields of 'megaripples', jökulhlaup outwash deposits may exhibit a range of kettle-hole forms, including: 'till-fill kettles' formed by the in situ melting of debris-rich ice blocks transported by the flood; streamlined obstacle marks, residual bars and hillocks; wash-limits on bedrock obstacles and valley sides; and thick fine-grained or laminated sequences associated

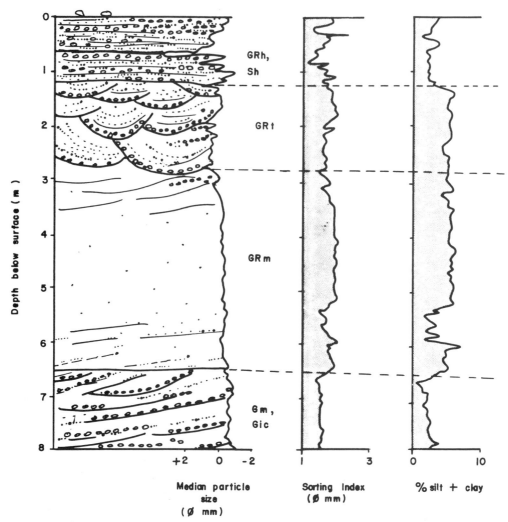

FIG. 12.26. Maizels' (1989a, b, 1991, 1992) vertical facies profile model for jökulhlaup sandur deposits where main flood surge is characterized by hyperconcentrated grain flows, to form central massive pumice, granule unit (GRm), overlying basal gravels, and overlain by waning-stage trough cross-bedded and horizontally-bedded granular gravels, sands and silts. (From Maizels, 1992; reprinted from Boulder ring structures produced during jökulhlaup flows – origin and hydraulic significance, by J.K. Maizels, *Geografiska Annaler*, 1992, **74A**, 21–33, by permission of the Scandinavian University Press.) (Compare with Plates 12.12 and 12.17.)

with temporary ponding of flood waters behind valley constrictions (cf. Maizels and Russell, 1992) (Fig. 12.27) (Plate 12.19).

12.5. CURRENT ISSUES AND FUTURE PROSPECTS

General models of the landform and sediment assemblages that characterize proglacial terrestrial environments are now relatively well-established, particularly with respect to changes in outwash morphology, channel pattern characteristics and facies variations in proximal and distal zones. There is also a substantial amount of data on seasonal and diurnal fluctuations in meltwater runoff in proglacial environments in a variety of global settings. However, there are two major areas of study that remain poorly represented. First, there are no substantive studies that

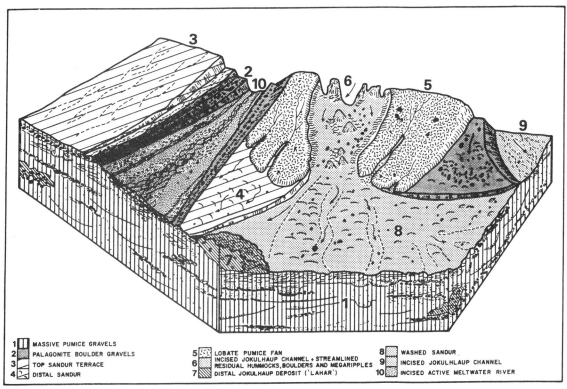

FIG. 12.27. Schematic model of jökulhlaup sandurs, illustrating range of depositional sequences and landforms, and erosional features. (From Maizels, 1991; reprinted with permission of Kluwer Academic Publishers.)

PLATE 12.19. Scattered boulders, scour marks and residual hillocks formed during jökulhlaup event on Skogasandur, south Iceland. (From Maizels, 1989a.) Flow was from right to left.

TABLE 12.4. Rust's (1978) classification of lithotypes and facies assemblages of braided alluvium. See Table 12.1 for the key to facies codes. (From Rust 1978)

Lithotype	Facies	Special characteristics
Gravel		
Fac. assemblage G_I: (alluvial fans)	*Gms, Gm, St, Sp, Fl/Fm*	Great lithological variation. Crude cycles may be present.
Fac. assemblage G_{II}: (Proximal braided rivers and alluvial plains)	*Gm, Gp, Sp, Sh*	*Gm* dominant; usually with well-developed imbrication. *Gp* more abundant in pre-U. Paleozoic deposits.
Fac. assemblage G_{III}: (Distal braided rivers and alluvial plains)	*Gt, St, Gm, Sh, Fl*	Fining-upward cycles.
Sand		
S_I: Fac. assemblages: (Proximal braided rivers and alluvial plains)	*Sh, Sp, Sr* (Bijou Ck. type) *Se, Sl, St (Ss)* (Malbaie type)	*Sh* dominant. Vertical and lateral variability.
Fac. assemblage S_{II}: (Distal braided rivers and alluvial plains)	*Se, St (Ss), Sp, Sr, Fl/Fm*	Fining-upward cycles. Transitional to deposits of meandering rivers.
Silt (Distal braided rivers and alluvial plains)	*Fl, Fm*	Lack of association with sandy channel deposits. Eolian reworking and thixotropic deformation common.

have successfully modelled channel responses to fluctuating runoff and sediment inputs to the proglacial drainage system. The complex interactions between variations in runoff, sediment transport and the processes of channel and drainage pattern development remain at a semi-quantitative stage. The largely qualitative nature of many of these studies reflects the difficulties of obtaining representative estimates of bed-load transport pathways and rates of movement and quantifying associated channel changes. Secondly, the temporal and spatial patterns of channel change and facies distributions need to be viewed within an appropriate conceptual framework. One approach that may prove fruitful is through determination of the magnitude and frequency of glacial runoff and sediment yield events that generate: (a) changes in channel morphology and pattern; (b) deposition of typical facies types; and (c) patterns of scour, sedimentation or storage that may

reflect longer-term or more permanent changes within the proglacial system. This approach should allow facies distributions and sequences to be modelled in relation not only to channel system dynamics and the controlling patterns of the runoff and sediment inputs, but also to glacier system dynamics. Seasonal and periodic fluctuations in glacier budget, for example, are likely to be reflected in fluctuations in runoff and sediment yields that may control the dominant features of the proglacial landscape. Over the longer term, the response of proglacial meltwater systems to large-scale environmental changes also needs to be modelled. While qualitative and semi-quantitative models are available at present, much still needs to be done in predicting fluvial response to deglaciation, eustatic and isostatic changes, climate change and associated changes in meltwater runoff and sediment supply regimes.

Chapter 13

GLACIOLACUSTRINE ENVIRONMENTS

G. M. Ashley

13.1. INTRODUCTION

Meltwater impounded by ice or glacial deposits, or in depressions created by glacial erosion or by isostatic loading creates a wide variety of glacier-fed lake types in terms of size, longevity and limnological characteristics. All are fed by glacial drainage and are sediment 'sinks'. Thus, typically, their deposits record important aspects of the deglacial history. The variations of sedimentation rate and organic content through time provide proxy information on climate. Interpretation of oxygen isotope records in lakes have been attempted recently in order to correlate land records with the well-established marine isotope record of climate change (Lister, 1988) (Chapter 2).

Glacier-fed water bodies are classified according to their position relative to the ice, the source of water and sediment, as either ice-contact lakes or distal lakes (physically separate from the ice) (Table 13.1, Fig. 13.1). Magnetic measurements on mineralogy have been used for provenance and sediment flux studies (Bradshaw and Thompson, 1985). More recently, high resolution seismic surveys revealed the 3-D geometry of an entire lake deposit (Mullins *et al.*, 1990). Although logistically difficult, studies of contemporary glaciolacustrine environments provide an excellent opportunity to link physical processes directly to the resulting deposits (Tables 13.2 and 13.3). Time series measurements of spatial variation in water temperature, suspended sediment

concentration and current velocities reveal insights into limnological processes. Sediment trap arrays and isotopic dating of bottom sediments (cores) yield data on sediment dispersal processes and sedimentation rates (Weirich, 1985; Leonard, 1986a, b). These new technological approaches and modern process studies have led to descriptive models which can now be used as a guide to interpret the environments of deposition of ancient sedimentary sequences.

The following sections describe the various types of glacier-fed lakes and summarize the limnological factors that control sediment dispersal and accumulation in them. Various sub-environments within ice-contact and distal lakes which produce distinctive sedimentary records are described as well as the spatial and temporal variation in deglacial sedimentation, the inevitable consequence of a continuously moving sediment source (i.e. the advancing or retreating glacier).

13.1.1. Ice-Contact Lakes

An ice-contact lake is a ponded body of meltwater with at least a portion of the lake in direct contact with either active or stagnating glacial ice (Plates 13.1a, b). The lakes may be dammed by the ice such that a change in the glacier causes complete drainage, or the lakes may be simply situated adjacent to ice. These water bodies are either subglacial, englacial, supraglacial or ice marginal (Fig. 13.1). Subglacial

TABLE 13.1. Classification of Glacier-fed Lakes (modified from Ashley, 1988).

A.   Proximal, ice-contact lakes

    1.   Subglacial and englacial lakes

    2.   Supraglacial lakes
        a) Active-ice lakes
        b) Ice-stagnation kettle lakes
        c) Ice-stagnation lake network

    3.   Ice-marginal lakes
        a) Dammed by ice
          1) River-lakes
          2) Semi-permanent lakes
        b) Dammed by topography
          1) River-lakes
          2) Semi-permanent lakes

B.   Distal lakes

lakes have been detected on airborne radar records (Drewry, 1981) and they are known to produce episodic catastrophic discharges (jökulhaups) (Thorarinsson, 1953, 1974) (Chapter 6). No modern subglacial lake has been studied in detail. There have been few documented examples of englacial lakes and although they may produce jökulhlaups they are unlikely to leave a recognizable sediment record. Supraglacial lakes develop as ablation occurs and are generally found only in the portions of active glaciers that are under compression; meltwater normally moves downward under the influence of gravity through fractures and solution conduits. Supraglacial meltwater creates isolated depressions on inactive portions of glaciers (ice-stagnation kettle lakes) (Clayton, 1964) and complex interconnected lakes (ice-stagnation lake network) (Plates 13.1c, d). Supraglacial lakes are relatively short lived and change configuration continuously as the ice melts. Essentially nothing is known about the physical limnology of these lakes, but they are likely to be shallow and thus well mixed by wind during the summer when the lake surface is generally ice free.

Ice-marginal lakes may be dammed by ice if the local slope is toward the ice or by topography if the regional slope of the land is away from the ice, but drainage is impeded locally by recent glacial deposits. Lakes dammed only by ice are likely to be ephemeral (a year to a few tens of years) and some may drain and refill more than once during the melt season (Marcus, 1960; Clement, 1984; Gilbert and Desloges, 1987a). Ice-marginal lakes that occupy pre-existing natural basins, dammed river valleys, or isostatic depressions tend to be semi-permanent (hundreds to a few thousand years) and enlarge as the glacier recedes. A river-lake, a type of ice-marginal lake, commonly develops in the poorly-drained region peripheral to the ice front where glacial discharge moves slowly but continuously away from the ice (Smith and Ashley, 1985). This body of impounded meltwater may be thought of as a low-velocity river or a high-velocity lake. These shallow, continuously moving (albeit slowly) water bodies commonly cover large areas fringing the glacier terminus.

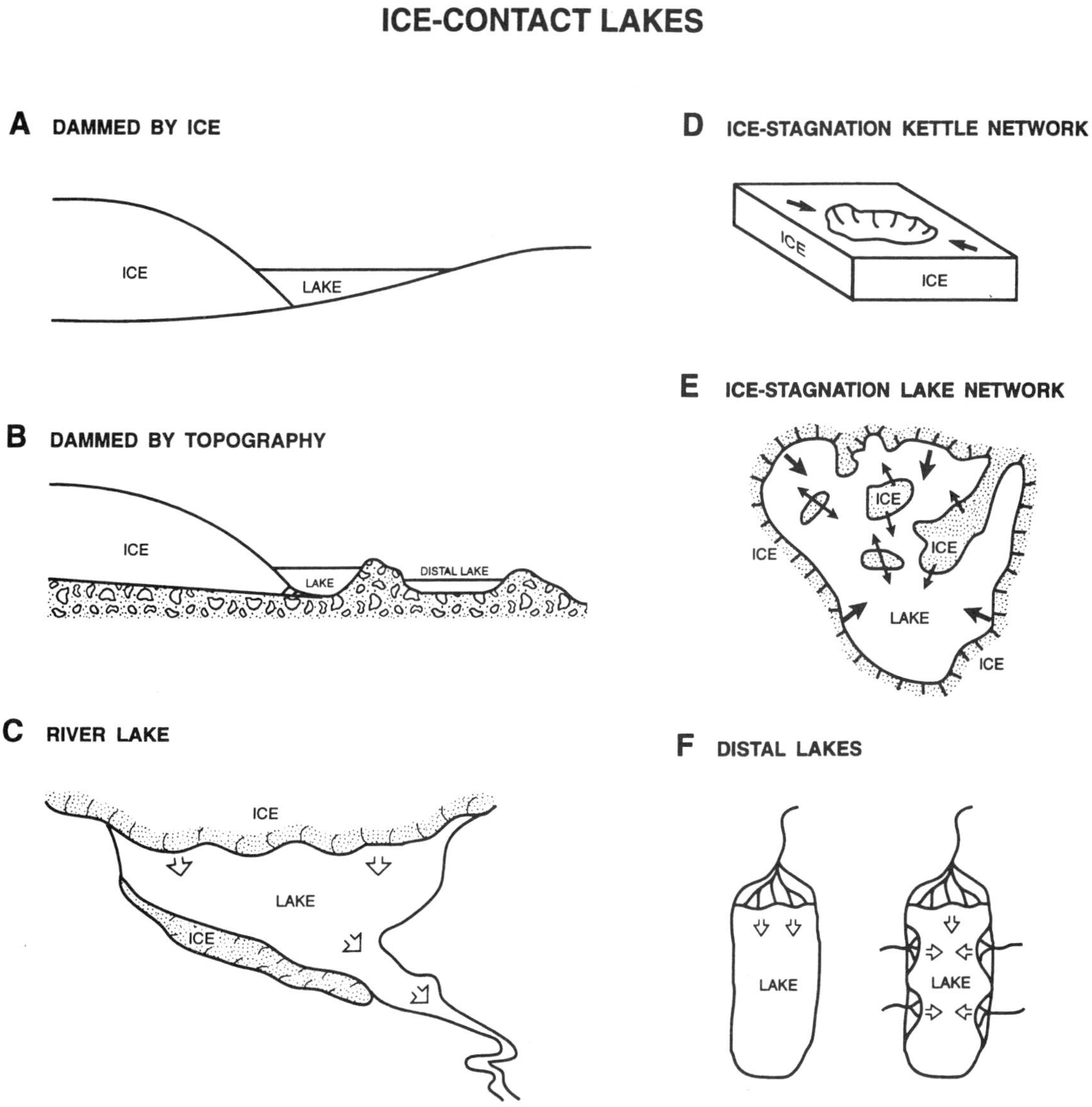

ICE-CONTACT LAKES

A DAMMED BY ICE

ICE

LAKE

B DAMMED BY TOPOGRAPHY

ICE

LAKE DISTAL LAKE

C RIVER LAKE

ICE

LAKE

ICE

D ICE-STAGNATION KETTLE NETWORK

ICE

ICE

E ICE-STAGNATION LAKE NETWORK

ICE ICE

ICE

LAKE

ICE

F DISTAL LAKES

LAKE

LAKE

FIG. 13.1. Diagrammatic sketches of the various types of glacier-fed lakes described in Table 13.1.

13.1.2. Distal Lakes

Distal lakes are physically separate from the ice but fed primarily by glacial meltwater via outwash streams (Fig. 13.1). Distal lakes may occupy pre-existing natural basins and thus have the potential to last for hundreds to thousands of years. In contrast, distal lakes impounded by sediment dams are likely to be shorter-lived (tens to hundreds of years). As the

glacier recedes from the region, the proportion of meteoric water and groundwater seepage with respect to the volume of meltwater inflow increases in distal lakes. This chapter focuses on sedimentation processes occurring during times of meltwater dominance. Meltwater effects typically include strong seasonal and weather dependent discharge variations, low salinity and temperature and high sediment loads.

TABLE 13.2. Factors Affecting Sedimentation.

Lake type (proximity to ice)

Lake stratification (thermal or sediment stratification; no stratification)

Relative densities of lake and meltwater

Position of meltwater inflow into lake water column

Sediment sources (number and direction)

Lake basin geometry

Slope stability

Ice rafting

Extent of ice cover

13.2. LIMNOLOGY OF GLACIER-FED LAKES

13.2.1. Physical Limnology

Water has a high heat capacity and thus lakes are able to absorb or give off heat in response to changes in solar radiation input with little net change in temperature. Meltwater is near 0°C as flow exits the ice and enters proximal, ice-contact lakes. Streams that flow across outwash plains before entering the lake warm slightly and may reach 10–12°C (Irwin and Pickrill, 1982). Meltwater contains varying concentrations of suspended sediment and dissolved load entrained during its passage through and under the ice. Lakes are influenced by these short-term thermal and sediment load fluctuations and generally develop some type of density stratification related to variation in temperature, chemistry or sediment concentration with depth.

13.2.1.1. Thermal stratification

Thermal stratification is produced by solar warming. Water density is a function of temperature; however, the relationship is non-linear with the maximum density ($\approx$1000 kg m^{-3}) at 4°C and lower densities at temperatures both colder and warmer than 4°C (Fig. 13.2a). On a seasonal time-scale, solar radiation absorbed near the lake surface warms the water from top down, creating a lower-density layer on the top (epilimnion) overlying a cold, higher-density layer at the bottom (hypolimnion). Daily fluctuations in solar radiation coupled with turbulent mixing by wind enhance the density contrast between the surface water and bottom so that the layers are separated by a region of rapid temperature change called the metalimnion. The thermocline is contained within the metalimnion and is located where the change of temperature with depth is the greatest (Figs

TABLE 13.3. Sedimentary Process and Lithofacies Produced in Glacier-fed Lakes (modified from Ashley, 1988).

SEDIMENTARY PROCESSES	LITHOFACIES UNITS
Mass Movement	
creep, slump, debris flow	matrix supported and clast supported massive to stratified diamicts
surge current	crossbedding, load structures, fluidization structures, massive beds, parallel and graded bedding, climbing ripples, draped lamination
Currents	
quasi-continuous	parallel bedding, graded bedding, large-scale tabular bedding, climbing ripples, multiple graded bed, load structures, fluidization structures
wind-driven	
nearshore waves, lake circulation	small-scale crossbedding, parallel lamination parallel bedding, multiple graded beds
combined	rhythmites
Ice rafting	debris blebs, dropstones
Biogenic processes	body or trace fossils, bioturbated sediments
Sedimentation	
from suspension	massive to parallel laminated silts and clays.

13.2b and 13.3a). Increased surface warming creates ever-greater density differences between the epilimnion and hypolimnion. In contrast, any wind mixing of warm epilimnial water with the water of the metalimnion tends to reduce the differences in density. On a yearly basis, the thermal profiles vary from mid-summer to winter (Fig. 13.2c). Complete mixing of the lake waters occurs when the lake is isothermal (d) which is likely in spring and autumn. Water bodies that turnover (completely mix) twice a year are termed dimictic and are common among glacier-fed lakes, particularly distal lakes. Those that circulate once per year are monomictic. Hutchinson (1957) divides monomictic lakes into 'warm' and 'cold', the former implying winter circulation above 4°C, and latter implying summer circulation below 4°C. Little limnological data exist for ice-contact lakes. They may be monomictic but most are probably polymictic, at least near the ice margin (Churski, 1973; Harris, 1976).

(a)

(b)

PLATE 13.1. Ice-contact lake environment. (a) Ice-contact lake with ice bergs (Bering Glacier, Alaska). (b) Debris flows into an ice-contact lake (Matanuska Glacier, Alaska).

(c)

(d)

PLATE 13.1. (c) Ice-stagnation kettle lakes (Malaspina Glacier, Alaska) (photo by T.C. Gustavson). (d) Ice-stagnation lake network (Malaspina Glacier, Alaska).

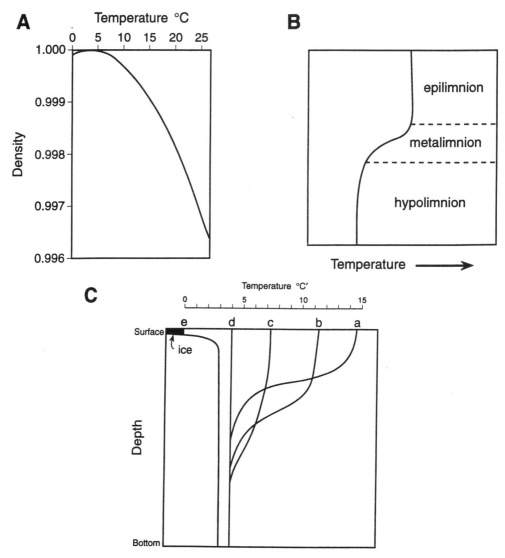

FIG. 13.2. Physical limnology. (A) Density of fresh water as a function of temperature. (B) Thermally stratified water column. (C) Seasonal variation in hypothetical thermal profiles ranging from mid-summer (a) to winter (e).

13.2.1.2. Sediment stratification

Water density is also a function of suspended sediment concentration. Lakes being fed by meltwater streams with high sediment loads (>1 g l^{-1}) may develop a density stratification that is a function of a gradual increase of suspended sediment concentration with depth. As the sediment settles it is being continuously replaced by new inflow during

the melt season (Fig. 13.3b). The density profile due to sediment is more gradual than the temperature profile (Fig. 13.3a). However, the magnitude of density due to sediment may be significantly higher than temperature effects. Table 13.4 lists the published maximum recorded concentration of suspended particulate matter in the waters of glacial lakes (Gilbert and Desloges, 1987a). Gustavson (1975a, b) measured nearly 200 mg l^{-1} at the surface

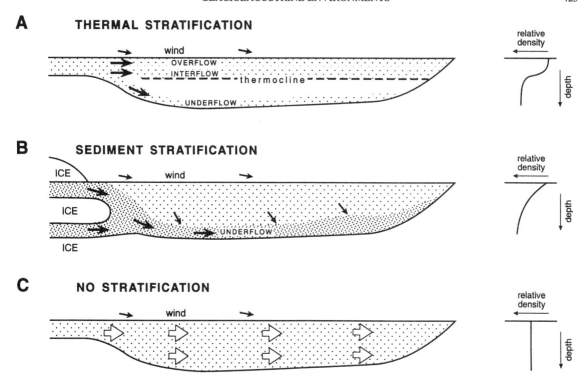

FIG. 13.3. Density stratification in glacier-fed lakes. (A) Thermal stratification, most common in distal lakes. (B) Sediment stratification, most common in ice-contact lakes. (C) No stratification; typical of river-lakes.

TABLE 13.4. Maximum recorded concentration of suspended particulate matter in the waters of glacial lakes (modified from Gilbert and Desloges, 1987a).

Lake	Concentration (mg/l)	Source
Tasikutaaq Lake, Baffin Island	5.7	Lemmen (1984)
Hector Lake, Alberta	>8	Smith et al. (1982)
Bow Lake, Alberta	15	Smith et al. (1982)
Lower Waterfowl Lake, Alberta	17.2	Smith et al. (1982)
Ape Lake, British Columbia	28.7	Gilbert and Desloges (1987a)
Garibaldi Lake, British Columbia	30	Mathews (1956)
Lillooet Lake, British Columbia	49	Gilbert (1975)
Lake Wakatipu, New Zealand	57	Pickrill and Irwin (1983)
Stewart Lakes, Baffin Island*	109	Gilbert et al. (1985)
Cascade Lake, Washington *	110	Campbell (1973)
Hazard Lake, Yukon *	122	Liverman (1987)
Unnamed, Purcell Mountains, B.C.	140	Weirich (1985)
Sunwapta Lake, Alberta	380	Gilbert and Shaw (1981)
Malaspina Lake, Alaska*	723	Gustavson (1975)

*Ice-Contact Lakes.

of Malaspina Lake (an ice-marginal lake) and 800 mg l⁻¹ at mid-depth of the 80 m deep lake. In Ape Lake, British Columbia, an ice-marginal lake (dammed by ice) drained and refilled frequently during a study by Gilbert and Desloges (1987a). Concentrations in Ape Lake ranged from a maximum of 28.7 mg l⁻¹ prior to draining to 1248 mg l⁻¹ after draining.

13.2.2. Chemical and Biological Limnology

13.2.2.1. Chemical processes

Dissolved materials are transported to lakes with mineral detritus but total volume is low. Chemical stratification is generally rare except in coastal ice-contact lakes that have or have had a connection to the sea where dense saline water may be trapped in the bottom water (Hattersley-Smith and Serson, 1964; Harris, 1976). Authigenic carbonate and sodic minerals that typify other lakes (Allen and Collinson, 1986) are rare in glacial-fed lake deposits. Manganese nodules have been noted by Gilbert and Desloges (1987a) and concentrations of trace metals by Pharo and Carmack (1979) and Renberg and Segerstrom (1981). Iron-rich laminations are associated with incomplete circulation (meromixis) and low sedimentation rates (Anthony, 1977). In general, thermal and sediment stratification dominate glacier-fed systems and the importance of chemically precipitated sediments is often masked by detrital sediment. Little is known of the chemical processes involved in glacial-fed lakes.

13.2.2.2. Biological processes

By their nature, glacier-fed lakes are inhospitable and thus biological productivity is relatively low. However, both carbonate and silicious micro-organisms occur in the water column. Fecal pellets produced by organisms aid in binding individual particles together into large flocs and thus enhance sedimentation (Smith and Syvitski, 1982). Trace fossils in the bottom sediment reveal that other forms of life such as worms, gastropods and some flying insects spend at least a portion of their life cycle in the lake (Gibbard and Stuart, 1974; Gibbard and Dreimanis, 1978). However, in general, biological

processes similar to chemical processes are often masked by physical processes.

13.2.3. Sediment Transport and Deposition

Several detailed studies of modern glacier-fed lakes have provided insight into the physical processes related to sediment dispersal that occur when a meltwater stream enters a lake (Mathews, 1956; Nydegger, 1967; Gilbert, 1973; Gustavson, 1975; Hamblin and Carmack, 1978; Carmack et al., 1979; Pharo and Carmack, 1979; Wright and Nydegger, 1980; Pickrill and Irwin, 1982; Smith et al., 1982; Dominik et al., 1983; Weirich, 1984, 1985; Østrem and Olson 1987). Although the specifics of each scenario vary from lake to lake, the overall theme is the same. Short-term fluctuations (hours and days) in meltwater inflow are characteristic and the nature of the lake stratification, which may also vary with time, affects the dispersal of sediment in the lake. These temporal fluctuations in stream and lake properties control the various rhythms of sedimentation that are common to glaciolacustrine deposits. The most important factors affecting sedimentation in glacial-fed lakes are listed in Table 13.2.

As fluvially-transported bedload and suspended sediment load enter the lake, the relative densities of lake and meltwater, as well as the nature of lake stratification, dictate the transport path (Fig. 13.3). If the density of inflowing water is less than the bottom water (hypolimnion) the stream moves as a plume at a level determined by its density, yielding overflows (hypopycnal) at the surface, interflows at any level below the surface, and underflows (hyperpycnal) along the bottom (Bates, 1953; Hamblin and Carmack, 1978)(Fig. 13.3a).

In thermally-stratified lakes, some of the fine suspended sediment is dispersed in the epilimnion by overflows or interflows (Mathews, 1956; Gilbert, 1975; Smith et al., 1982). The rest of the fine sediment is transported by high density turbid flows which 'dive' and move as quasi-continuous underflows onto the lake bottom (Gustavson, 1975; Weirich, 1984). These density underflows are initially driven by the momentum of the inflowing stream which may reach 0.15 ms⁻¹ (Nydegger, 1967; Smith,

1978; Hamblin and Carmack, 1978) and later by gravity as flow continues down the slope of the lake basin. Gravel and coarse sand drop out of the flows early and accumulate in shallow water. Medium to fine sand and silt are deposited by the underflows as they continue into the lake. Fine silt and clay are kept in suspension until flow drops below ≈0.1 ms^{-1}.

In sediment-stratified lakes, glacial meltwater flows entering at the top, middle or bottom of the lake are affected by gravity and move downward and away from the ice until momentum is lost (Fig. 13.3b). Sedimentation is focused in topographic lows because sediment is transported at the base of the hypolimnion. Equal density mixing (homopycnal flow) occurs when the inflowing river water is of equal density to that of the lake and the lake is either unstratified or weakly stratified (Bates, 1953) (Fig. 13.3c). Rather than following a specific level within the lake, the flow expands in three dimensions and the river and lake water gradually mix as the sediment moves by advection throughout the water column.

Winds are generated almost continuously in the vicinity of glaciers. Cold air associated with the ice moves toward warmer areas producing 'katabatic' winds which blow across the surface of lakes setting up epilimnial currents that move slowly away from the ice. Water motion in lakes (including wind-driven flows) are affected by Coriolis forces of the Earth rotation such that flow is directed to the right in the northern hemisphere (Hamblin and Carmack, 1978; Smith *et al.*, 1982) and to the left in the southern hemisphere (Irwin and Pickrill, 1982). These directed currents play an important role in the distribution of sediment on the lake bottom (Smith *et al.*, 1982; Smith and Ashley, 1985).

The extent and longevity of ice cover which normally develops on the lake during the winter months affects the development of thermal stratification within the water column and inhibits wind stress on the lake surface. These factors affect the physical processes of sediment dispersal. Consequently, the amount of time that the lake is ice free each year is important (Table 13.2).

Icebergs are pieces of the glacier that have broken off (calved) and are free to float and drift about the lake under the influence of the wind and/or lake currents (Plate 13.1a). As the density of ice is ≈0.92 g ml^{-1}, nine-tenths of a berg is below the surface. Icebergs commonly disturb (scour and erode) sediment in the shallow parts of the lake (Menzies, 1995b, Chapter 4). Glacial debris is released from the bergs as they melt creating lense-shaped coarse deposits called 'berg dumps' or singular dropstones which are highly visible if found in otherwise fine-grained glaciolacustrine deposit (Thomas and Connell, 1985). Icebergs commonly collect in shallow water, particularly at the lake outlet, and sedimentation is concentrated as debris is released from the ice during melting. Relatively large landforms may accumulate by this process (Gilbert and Desloges, 1987b; McManus and Duck, 1988).

Mass movement is another key mechanism for moving sediment within the lake (Fulton and Pullen, 1969). Glacial debris deposited at the lake margin by fluvial or colluvial processes may become unstable (particularly on steep slopes) and slump, slide or flow into deeper water (Gilbert, 1972; Smith, 1978; Pharo and Carmack, 1979; Pickrill and Irwin, 1983; Siegenthaler and Sturm, 1991). Slump terraces are common on steeper slopes reflecting rotational failures. Sliding may shift deposits to a less steep slope building compressional ridges at the base (Nemec, 1990). The mass movement processes commonly trigger short-lived (minutes) flows of turbid water (Andresen and Bjerrum, 1967; Hampton, 1972; Thomas, 1984a). Called "surge currents" by Smith and Ashley (1985), they are identical to what has been identified as "turbidity currents" in the marine environment (Chapters 14; Menzies, 1995b, Chapter 5). However, these surges are short lived compared to the quasi-continuous density currents that are produced by meltwater inflow that may last for days, weeks or even months. Although both types of density currents are a function of high turbidity, they operate on different time scales and produce different types of deposits. These deposits will be described in detail later in the chapter.

13.3. ICE-CONTACT GLACIER-FED LAKES

The margins of ice-contact lakes continually change as glaciers advance or retreat and as ice melts and sediment is released. Most ice-contact lakes are polymictic and are sediment stratified rather than thermally stratified. Few generalizations can be made about the deposits of these lakes except that sedimentation patterns are highly variable in proximal areas. The high variability of sediment influx to the proximal zone is damped with distance from the ice front and as a consequence the lake bottom sub-environment is typified by steady and relatively low sedimentation rates. The discussion, here, focuses on river-lakes and relatively large semi-permanent ice-marginal lakes (which can be dammed by either topography or ice) as these are the types most likely to leave a sedimentary record (Fig. 13.1).

13.3.1. River-Lakes

River-lakes are short-lived bodies of water that result from partially impounded drainage near the ice front (Fig. 13.1). They are shallow (few metres) slow moving lakes with a low residence time and thus lack stratification. Most are likely to be characterized by equal-density mixing (Fig. 13.3c). Sediments tend to be fine sands and silts because coarse clasts are left as lag and finer material gets carried through the lake (e.g. Matanuska Glacier, Ashley, *unpublished data*; Smith, 1990). Deposits are thin tabular bodies composed of poorly stratified to massive sandy silts that are interbedded with a variety of ice-contact deposits.

13.3.2. Semi-permanent Lakes; Proximal Sub-Environments

The ice proximal environment can be divided into four distinct sub-environments: (a) ice-contact delta, (b) subaqueous fan, (c) ice-cliff margin, and (d) submerged ice ramp (Fig. 13.4). Each is characterized by high sedimentation rates and its own suite of depositional processes and deposits (Table 13.5). The ice-proximal deposits are commonly deformed by seasonal ice marginal fluctuations (Donnelly and Harris, 1989).

The proximal depositional settings of lakes where water lies in direct contact with the ice can be subdivided into those dominated by meltwater discharge (deltas and subaqueous fans) and those dominated by gravity-driven mass-movement processes (ice-cliff margins and submerged ice-ramps).

13.3.2.1. Ice-contact delta

Deltas are landforms built into the lake from fluvially transported detritus disgorged at or near the lake surface. Deltas may build at a single point reflecting a single, stable meltwater stream as the sediment source or deltas may form at several locations along the lake margin suggesting multiple discharge points or one stream that moved frequently. Some deltas build at active ice margins and others on stagnating ice margins. Deltas formed at active ice-margins are fed from subglacial or englacial tunnels where sedimentation eventually builds the deposit to the lake surface. Deltas formed at stagnating ice margins are fed by meltwater streams that originate in active ice but end up flowing over the surface of dead ice and into the lake. Sediment influx occurs mainly during the melt season and deposition generally results in a fan-shape landform because distributaries commonly splay at the mouth of the meltwater stream supplying the sediment. Sediment is transported and deposited along a portion of the delta until the distributaries become inefficient and the flow seeks a steeper gradient and switches to another portion of the delta. Generally only one portion of the delta is active at any time.

Deltas consist of both subaerial and subaqueous portions. The uppermost portion (topset beds) of the delta are deposited by fluvial processes and tend to be thin because vertical aggradation is limited by the local base (lake) level (Fig. 13.5). The bottom contact of the topset sediments is controlled by the depth of fluvial scour in the river channels feeding the delta. The top of the fluvial beds is determined by elevation of overbank flooding. Thus, given a stable lake level, the fluvial topset beds will be about the same thickness as the yearly vertical range of the meltwater streams building the delta.

The classic 'Gilbert' delta tends to form in higher energy, coarse-grained systems; the foreset beds have steep dips (up to 33°)(Fig. 13.5a). Sand and larger clasts which were deposited by the meltwater stream at the margin of the lake move downslope by a variety of mass-movement processes, such as avalanching, creep, slump, debris flows and grain flows (Nemec, 1990). Larger clasts are not imbricated and tend to lie parallel to bedding, indicating that they slid

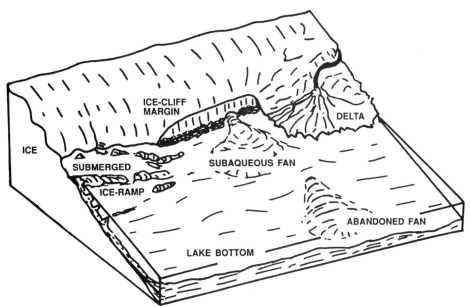

FIG. 13.4. Ice-contact lake depositional environments.

downslope individually rather than being transported as bedload in a density current. Massive to parallel-bedded gravity layers commonly alternate with open work gravel and coarse sand lenses (Martini, 1990). Fining upward and coarsening upward gravel sequences are also common (Cohen, 1979; Clemmensen and Houmark-Nielsen, 1981; Thomas, 1984a).

Deltas with shallowly dipping foresets tend to form in lower energy, fine-grained systems (Fig. 13.5b). Sedimentation occurs from bottom hugging quasi-continuous underflows creating small fan-shaped lobes on the delta front (Jopling and Walker, 1968; Gustavson et al., 1975; Leckie and McCann, 1982). The upper portions of the delta front commonly reveal evidence of rapid sedimentation and dewatering: slumps, ball-and-pillow structures, roll-up structures, pipes and dish structures (Shaw, 1975b, 1977c; Cohen, 1979). In addition, ice-contact deltas typically have interbeds of diamicts and ice-rafted debris and usually have evidence of post-depositional collapse from the melting of underlying or adjacent ice. The glacier functions as a temporary support for the glaciofluvial sediment. When the ice melts either during or following sedimentation the bedding is deformed by gravity-driven processes. The

removal of support produces a distinctive landform that has a coarse-grained steep, 'ice-contact face' on the proximal side and gentle, depositional slope on the lake side. Ice-contact deltas are often called 'heads of outwash', because they represent the source of the sediment supply for the fluvial outwash environment.

On the middle to distal delta front, lobe-shaped sediment bodies composed of climbing-ripple sequences and drape lamination are constructed (Jopling and Walker, 1968; Gustavson et al., 1975; Cohen, 1979). The distal delta front deposits merge imperceptibly with low angle to horizontal bottomset beds which consist mainly of drape lamination built up from waning density underflows (Gustavson et al., 1975; Kelly and Martini 1986).

Clays and fine silt accumulate on the delta foreset beds at locations and at times when current activity at the bed is low. Thus, clay dispersed in the water column may settle on areas of the delta removed from points of inflow (distributary channels) during the non-melt season (fall, winter and early spring). These clay-rich horizons punctuate the deltaic sediments with which may be (but not necessarily be) an annual rhythm.

TABLE 13.5. Ice-contact Glacier-fed Lacustrine Subenvironments and Associated Lithofacies (from Ashley, 1988).

Subenvironments	Processes	Deposits
Delta coarse-grained	mass-movement (creep, slump, debris flows, grain flows), surge currents	Well-sorted coarse sediments. Parallel bedded, steeply dipping foresets (20°-33°), normal and inverse graded beds of sand and gravel. Clasts parallel to bedding planes.
fine-grained	rapid deposition from density underflows	Well-sorted medium to fine sediments. Sets of climbing ripple sequences, draped lamination occurring as low angle <20° foresets of sand and silt. Deformational structures and dewatering structures common.
Subaqueous Fan	deposition from a high velocity expanding jet	Proximal to distal open work gravel, chaotic bedding, convolute lamination, dewatering structures, climbing ripple sequences and drape lamination.
Ice-cliff margin	mass-movement, rock fall, debris slides, calving, squeezed debris at base	Poorly sorted, interbedded mass flow deposits, lacustrine mud, and dropstones.
Submerged ice-ramp	mass-movement, debris flow, creep, slumps, surge currents	Poorly sorted, stacked diamicts interbedded with sand beds, mud drapes.
Lake bottom	deposition from overflows, interflows and underflows and occasional ice bergs	Fine-grained rhythmic, parallel bedded; occasional dropstones and biogenic structures.

(a)

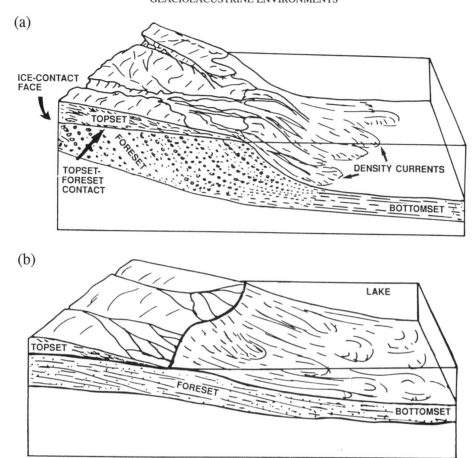

(b)

FIG. 13.5. Ice-contact delta depositional environment. (a) Coarse-grained delta, with high-angle foresets. Density underflows can be generated by inflowing meltwater (quasi-continuous currents) or delta foreset slumps (surge currents). (b) Fine-grained delta, with low-angle foresets that can form in the distal portion of an ice-contact delta or more typically where the delta is physically separated from the ice by an outwash stream..

13.3.2.2. Subaqueous fan

Glaciofluvial detritus carried to the lake via tunnels near or at the base of the ice form subaqueous fan-shaped deposits well below the surface of the lake (Figs 13.4 and 13.6a) (Rust and Romanelli, 1975; Rust, 1977; Gustavson and Boothroyd, 1987; Donnelly and Harris, 1989). The fact that vertical position (elevation) of fans is unrelated to lake level clearly distinguishes them from deltas where the landform surface (topsets) is controlled directly by lake level (i.e. the topset–foreset contact forms at the lake surface). However, both ice-contact deltas and subaqueous fans mark ice-front positions. The individual landforms commonly coalesce forming a

ridge parallel to the ice margin. These features have only recently been recognized as moraines (Fyfe, 1990; Sharpe and Cowan, 1990).

Flows, under a hydraulic head, enter the lake at a point source (Fig. 13.6b) creating a deposit cored with coarse open-work gravel. Rust (1977) found the long axes of the larger clasts were parallel to flow in subaqueous fan deposits as opposed to the perpendicular orientation to flow typical of fluvial deposits. Chaotic bedding with large clasts dispersed in sand and massive 'quick' sands and dewatering structures occur in the proximal fan area representing rapid deposition from the high discharge flows entering the lake (Cheel and Rust, 1986). Mid-fan sediment contains mass flow deposits, ball-and-

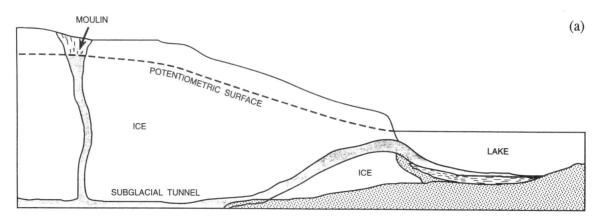

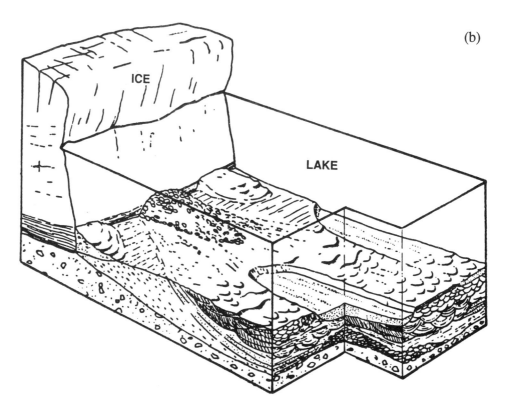

FIG. 13.6. Subaqueous fan depositional environment. (a) Schematic cross-section showing relationship of fan to glacier hydrologic system (after Gustavson and Boothroyd, 1987; reprinted with permission of the authors). (b) Depositional model for fan (after Shaw, 1985; reprinted with permission from *SEPM*.).

pillow structures, dune and ripple crossbedding, flame structures, etc. (Rust, 1977; Postma *et al.*, 1984). Distal fan deposits likely contain climbing ripple sequences and drape laminations which eventually become indistinguishable from lake bottom sediments (Donnelly and Harris, 1989). Fans are abandoned during glacial recession and are later blanketed with fine-grained suspension deposits.

13.3.2.3. Ice-cliff margin

The ice-cliff sub-environment develops when lake water is relatively deep adjacent to the ice (Figs 13.4 and 13.7a). Frequently, calving icebergs maintain the cliff face and thus ice-cliffed margins are likely to occur only on active glacier termini where ice is replenished. Sediment accumulates at the base as an apron, with a source from some glacial debris falling from above or more likely from sediments squeezed out or shoved forward during ice-marginal fluctuations (Barnett and Holdsworth, 1974). The ice cliff debris aprons are composed of coarse clasts and diamicts in the form of ridges oriented parallel to the local ice margin. They are likely to show evidence of glacial tectonization, particularly the squeezing of diamict from beneath the glacier. These ridges would often be considered push or rogen moraines (Chapter 10). The ridges consist of interbedded mass flow deposits, re-sedimented diamict, lacustrine mud and

ice rafted debris, such as isolated dropstones and berg dump.

13.3.2.4. Submerged ice-ramp

The ice-ramp sub-environment occurs only at lake margins of stagnating glaciers where water partially covers melting ice (Figs 13.4 and 13.7b). Basal debris and supraglacial debris elevated and concentrated by repeated over-riding of active ice over dead ice at the terminus is released by melting. The ice margin, as well as ice-cored debris islands and peninsulas in the lake, melts and episodically sheds debris (Evenson *et al.*, 1977; Lawson, 1982). The sediment distribution processes are gravity-driven for the most part, except for wind-driven surface currents which circulate fine suspended sediment. Mass movement processes of creep, slumps, debris flows and surge currents produce a complex lithofacies association of massive to bedded diamicts, massive to cross-bedded sands

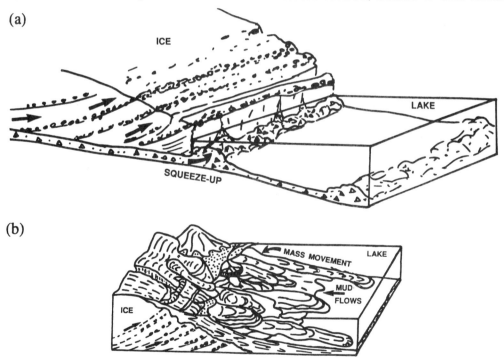

(a)

(b)

FIG. 13.7. (a) Ice-cliff depositional environment. (b) Submerged ice-ramp depositional environment. Sediment is derived from diamicts concentrated at the ice terminus and moves by mass movement processes as the ice margin melts and collapses (after Evenson *et al.* 1977; reprinted from Subaquatic flow tills: a new interpretation for genesis of some laminated till deposits, by E.B. Evenson, A. Dreimanis and J.W. Newsome, *Boreas*, 1977, **6**, 115–133, by permission of the Scandinavian University Press).

and mud drapes (Eyles and Eyles, 1983a; Eyles *et al.*, 1987a). The characteristically poorly sorted, weakly bedded sediments, flame structures and rip-up clasts indicate deposition by a variety of subaqueous flow processes (Table 13.5).

13.3.3. Lake Bottom Sub-Environments

The lake basin depositional setting is the largest glacier-fed lacustrine sub-environment in terms of area (Fig. 13.4). It represents sites in the lake located at a distance sufficiently removed from the points of sediment influx such that highly variable flows entering the lake are damped and the effects of gravity-driven processes are minimized. Sedimentation is more regular and the deposits are finer-grained than in proximal sites (Ashley, 1975; Gustavson, 1975; Smith and Ashley, 1985). Rhythmic parallel laminated sediments are characteristic of lake bottom deposits (Plate 13.2a). Multiple graded laminae (3–5 mm thick) are common and likely represent temporal pulses in the inflow (Plate 13.2b).

Limnological currents, influenced by wind stress on the lake surface, density stratification of the water column and Coriolis forces, determine the transport path of the suspended fraction (silts and clay). Although these fine particles continuously settle out, a large portion remains trapped in the epilimnion until thermal stratification is destroyed during fall cooling (Fig. 13.2c). In fall, the lake water is overturned bringing surface water closer to the lake bottom. Because it is at the end of the melt season the inflowing water is clearer (i.e. less dense) than the deeper lake water and thus overflows on the surface. The net result of the overturn is to clear the lake of most of the suspended sediment. These fines end up as a drape on the lake bottom and become part of the annual deposit. Whatever fine material remains in the water column continues to settle out until the next melt season. There is a relatively sharp contact between the coarser sediments deposited prior to and during the overturn compared to the finer sediments deposited after the overturn. Under ideal conditions one coarse–fine couplet represents an annual deposit and is termed a varve (Plate 13.2a). However, if sediment enters the lake periodically throughout the

year (including winter), the annual signature may become confused and obliterated (Shaw, 1975b; Shaw *et al.*, 1978). Individual, temporally isolated surge-current events, due to an increase in meltwater from the glacier or a mass movement event within the lake itself, may produce an additional coarse–fine couplet. This surge-current deposit can be distinguished from the annual coarse–fine couplet because it is likely to have a gradational contact between the coarse and fine components because of the rapid and continuous settling of sediment out of the flow. It is wiser to use the more general descriptive term rhythmite to describe repetitious bedding rather than the more specific temporal term, varve, unless the deposits are demonstrably annual.

The identification of rhythmically bedded deposits as annual (i.e. varves) provides a means of unravelling the history of the lake in terms of short-term glacier-related processes (i.e. ablation and run-off) and long-term climatic change record (Section 13.5). A variety of techniques have been used to demonstrate the annual nature of the rhythmites: (1) sediment budget considerations in modern lakes— comparing total annual influx with calculated net accumulation represented by the rhythmite (Østrem and Olsen, 1987; Weirich, 1985); (2) isotopic dating using Cs^{137}, Pb^{210} or C^{14} (Ashley, 1979; Coard *et al.*, 1983; Leonard, 1986a); (3) occurrence of seasonally generated microfossils (Simola, 1977; Lemmen *et al.*, 1988; Lotter, 1989); or (4) sedimentary structures (Smith and Ashley, 1985). Below is a list of characteristics that can be used to interpret which rhythmites are annual and not a deposit from a single sedimentation event (river flood, slump etc.):

(1) Although the couplet may, in general, fine from bottom to top, little to no fining occurs in the coarse layer because the layer represents an accumulation over a long time period (weeks to months) from a variety of processes;

(2) There is a relatively sharp contact between coarse and fine portions of the couplet which suggests a break in sedimentation, rather continuous settling. The break may represent the autumn overturn of the water column;

(3) The fine layer fines upwards reflecting a seasonally-controlled diminishing sediment supply in the lake and clearing of the water column during winter months;

PLATE 13.2. (a) Rhythmites, glacial Lake Hitchcock, Massachusetts, U.S.A. (photo by J. Hartshorn). (b) Couplets. Winter clay layers indicated with arrows. Summer silty-fine sand deposits (between the clays) consist of multiple graded laminae.

(4) Trace fossils (*lebenspuren*) on bedding planes both within the coarse layer and at the top of the fine layer suggest several episodes of low to zero sedimentation;

(5) Diatom microfossils occur at the same position within each rhythmite reflect a seasonal bloom (spring?/summer?);

(6) At any one site within the lake the fine layer thickness, which represents slow sedimentation for the same amount of time each year, remains consistent from year to year, whereas the coarse layer thickness varies reflecting the seasonal-to-season variability of rapid sedimentation within the overall frame work of the regular annual cycle.

13.3.4. Jökulhlaups

High discharge events where meltwater dammed beneath or at the margin of the glacier is released suddenly are called jökulhlaups (an Icelandic term) (Chapters 6 and 12). These glacial outburst floods may occur several times within a season, regularly once a year (Clement, 1984; Liverman, 1987) or at several year intervals (Björnsson, 1974; Gilbert and Desloges, 1987a). They are accompanied by dramatic erosional and depositional effects. These catastrophic events can be triggered by, for example, flotation of an ice dam by rising water level (Clarke, 1982); rapid enlargement of tunnels (Mathews, 1973); accelerated melting by volcanic heat sources; or rupturing of ice dam by earthquakes. An empirical relationship was found to exist between peak discharge (Q) of the outburst flood and the initial lake volume (V), such that $Q=0.0075V^{0.667}$ (Clague and Mathews, 1973; Clarke *et al.*, 1986; Mathews and Clague, 1993) (Chapter 6).

The draining of the lake during a single jökulhlaup may be irregular with several episodes of fluctuating discharge superimposed on the main discharge curve presumably reflecting the blocking and clearing of flow paths within the glacier (Sturm *et al.*, 1987). Erosion takes place near to the source of the jökulhlaup and deposition may occur in the ice-dammed lake, if any water remains (Gilbert and Desloges, 1987a; Liverman, 1987), on proglacial fluvial plains (Maizels, 1976) and in distal lakes downstream (Schmok and Clarke, 1989; Waitt, 1985).

Large coarse-grained bedforms occur in proximal areas and coarse erratics are commonly strewn along the river by icebergs transported by the flood. Distal deposits are characterized by a wide-spread blanket of rapidly deposited sediment. Thick coarse–fine couplets (up to a few metres in thickness) may be found tens of kilometres from the burst dam. The couplets are composed of bedded to massive sand overlain by massive silt and clay which represent an entire jökulhlaup or perhaps only a single pulse within an outburst flood (Chapter 12).

13.4. DISTAL GLACIER-FED LAKES

Many lakes fed with meltwater are separated from the glacier by a fluvial outwash plain (sandur). Because of the fluvial 'buffer' separating distal lakes and the source of sediment and water (the glacier), distal lake deposits tend to be fine grained and well sorted in comparison with deposits of ice contact lakes. Distal lakes occur in structural basins or in valleys dammed by drift or bedrock and thus tend to be semi-permanent. The distance from the ice results in the reduction of the density of the meltwater streams by warming and by decreasing the suspended sediment concentration either by deposition or by dilution by non-glacial tributaries carrying low sediment load. Depending upon longevity, distal lakes may evolve gradually from a system supplied directly with glacial detritus into one dominated by meteoric runoff and reworked sediment from ice-free drainage basins. The most important change in parameters through time is the reduction in total annual sediment load to the lake.

13.4.1. Proximal Sub-Environments

Deposition in proximal areas of lakes separated from the glacier occurs primarily in deltas with a minor amount of deposition on the lake margin in between the deltas.

13.4.1.1. Deltas

In general, the process of formation of ice-contact deltas and distal delta s are similar (Section 13.3.2.1). The differences are related to the physical separation of the lake water body from the vagaries of a melting glacier. In distal lake deltas, large clasts and icebergs are trapped in the intermediate fluvial system (Fig. 13.8). The suspended load is decreased and the meltwater warms up as it moves over the sandur. Because distal lakes are not in contact with the glacier they are likely to be thermally stratified with a well-developed thermocline (Fig. 13.3a) as opposed to ice-contact lakes which are sediment stratified (Fig. 13.3b).

The clast sizes composing the deltas vary with the setting and can range from pebble-cobble gravels in 'Gilbert' deltas with steep foresets (30–33°) (Fig. 13.5a) to sands and silts in low-angle foresets (5–15°)(Fig. 13.5b). However, because the deltas in distal lakes are at a distance from the source, they tend to be finer-grained (sands to clay) than ice-contact deltas and dominated by climbing-ripple sequences (Plates 13.3a, b). Many of the fines discharged into the lake are trapped above the thermocline and transported across the lake aided partially by the momentum of the infilling meltwater. Wind stress helps the movement within the epilimnion. Flows gradually slow down and sediment

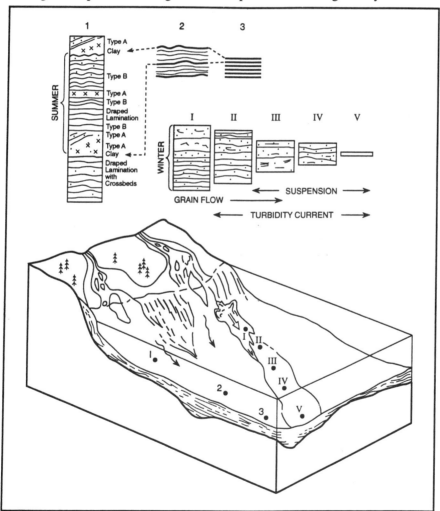

FIG. 13.8. Distal glacier-fed lake depositional environment. Sediments become thinner and finer grained from proximal (1) to distal (3) deposits. Dashed line brackets one years accumulation.

PLATE 13.3. (a) Ripple-drift sequences typical of fine-grained low-angle deltas. A sequence (representing waning flow) consists of Type 'A' ripples at base which grade to Type 'B' and culminates with drape lamination. Scale is 30 cm.

settles through the metalimnion decreasing flow density. Thus, spatial distribution of a 'summer' sediment layer shows a proximal-distal thinning and fining trend, as well as a 'piling-up' of sediment on one side of the lake due to Coriolis effects (Sturm and Matter, 1978; Smith, 1978; Smith *et al.*, 1982). Deposition from the epilimnion occurs as a drape distributed at all elevations in the lake.

The upper delta foreset deposits are commonly composed of rhythmically bedded (up to a metre thick) fining upward sand and gravel beds. These deposits are probably not seasonal but represent episodes of sedimentation during the melt season. Two common causes are (1) high discharge events, or (2) changes in points of sediment influx due to relocation of distributary channels in the topsets.

The deposits on the mid-delta foresets are finer grained (mainly sand), and foreset dips are less steep (<10°) than upper-delta foresets (Fig. 13.4 and Plate 13.3b). Density underflows carrying turbid mixtures of sand, silt and clay frequently deposit climbing ripple-drift sequences capped by silt drapes under waning flow (Jopling and Walker, 1968; Aario, 1972; Gustavson *et al.*, 1975; Shaw, 1975b; Cohen, 1979;

PLATE 13.3. (b) Shallowly dipping sandy foreset beds composed of climbing ripple sequences (Glacial Lake Hitchcock, Massachusetts, U.S.A.). Scale is 30 cm.

Clemmensen and Houmark-Nielsen, 1981; Stone and Force, 1983; Thomas, 1984a, b).

Climbing ripples were classified by Jopling and Walker (1968) into two end-member types: Type A, where stoss-side laminae of ripples are eroded, and Type B where stoss-side laminae are preserved (Fig. 13.9). Type A and Type B are end-members in a continuum of ripple types representing a shift during a waning flow from rapidly migrating ripples with low vertical aggradation rates to slowly migrating ripples with high vertical aggradation rates. The draped lamination capping the sequence is formed when ripple migration ceases and the form is blanketed by a drape of sediment from suspension. A series of flume experiments with varying time-velocity curves simulating waning flow produced climbing ripple–drift sequences similar to natural ones (Ashley *et al.*, 1982). The flume runs with the most 'natural' appearing sequences had vertical accretion rates of 0.3 mhr⁻¹. Thus, most individual

sequences deposited in natural settings likely accumulated within 1–2 hr.

The climbing–ripple sequences are composed mainly of fine sand (3.0–3.5ø) and are commonly repeated. The mid-delta foresets are characterized by lobes marking point sources of sediment (delta distributary channels) which are likely to shift frequently. Thus, the sandy mid-delta foresets are a 3-dimensional mosaic of overlapping lobes (Stone and Force, 1983). Thin, continuous clay drapes presumed to be time-stratigraphic horizons (winter clay) occasionally interrupt the stacked climbing ripple sequences (Gustavson *et al.*, 1975). The flume experiments that suggest a sequence represents a short period of sedimentation (few hours) and field evidence that indicate several sequences may be sandwiched between persistent clay layers both imply that their rhythmicity is due to repeated flow events in one season. An individual sequence is not an annual deposit. Thus, the mid-delta foresets are a site of rapid

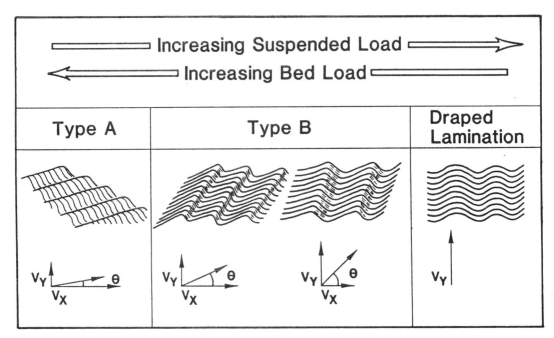

FIG. 13.9. The ripple types reflect a changing proportion of suspended and bed load. V_x = ripple migration velocity; V_y = aggradation rate; θ = angle of climb of ripple crossbedding (from Ashley *et al.*, 1982; reprinted with permission from the International Association of Sedimentologists).

sedimentation receiving deposition from both episodic surge-currents possibly generated by slumping of upper delta foresets or from riverine flood events and quasi-continuous underflows (the subaqueous continuation of meltwater streams) (Fig. 13.8). Because of the high sedimentation rate in this proximal site, the two types of underflow deposits are likely to be interbedded and may be difficult to differentiate. However, some studies in modern lakes comparing the long-term sediment record with historical river inflow data have been able to distinguish the isolated surge-current deposit generated from either a mass movement event or river flood from the background of seasonally-controlled sedimentation (Gilbert, 1975; Sturm and Matter, 1978; Østrem and Olsen, 1987; Siegenthaler and Sturm, 1991).

13.4.1.2. Lake margin

Shoreline sedimentation in between deltas is restricted to colluvium washed from adjacent terrain and sediment dispersed within the epilimnion. The uppermost deposits, beaches, form a narrow bench of coarse, sorted sediment, whereas the shallow shoreface deposits tend to be homogeneous mud and silt (Fig. 13.10) (Sturm and Matter, 1978; N.D. Smith, *unpublished data*). This lack of stratification in deposits accumulating in shallow water may reflect: (1) bioturbation, (2) wave disturbance or (3) mixing of sediment by wind-generated currents.

13.4.2. Lake Bottom Sub-Environments

Rhythmites deposited in the lake basin are the distal equivalents of the lower delta foreset rhythmites and the sedimentary sequences sandwiched between persistent clay layers found in the mid-foreset deposits. These equivalent sedimentary units are shown as 3, 2 and 1 on Fig. 13.8. The sand–silt layer changes gradationally from a complex unit on the mid-delta foreset consisting of climbing ripple drift sequences to lower delta foresets (rippled and multi-laminated beds) to lake bottom (parallel multi-laminated beds). Slump-generated underflows may

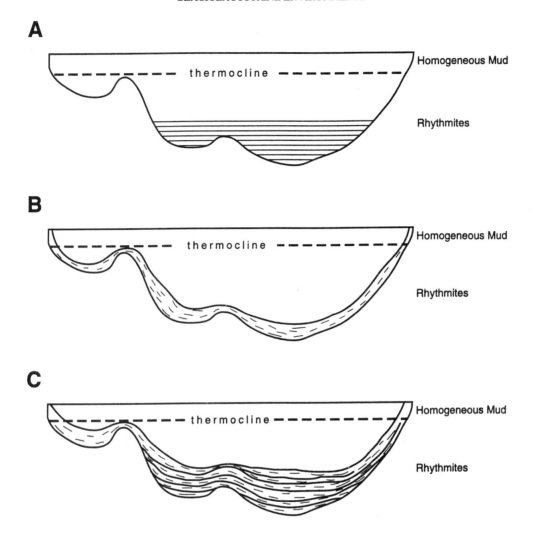

FIG. 13.10. Spatial variation in lake-bottom environment deposits with different dispersal mechanisms. (A) Lakes dominated by underflow. (B) Lakes dominated by overflow/interflow. (C) Lakes which have a combination of all types of flow.

complicate the rhythmicity of lake bottom sediments.

Lake bottom deposits at sites removed (up to a few kilometres) from the point of sediment influx tend to be thin (<1cm) rhythmites consisting of few multiple laminations. The coarser laminae are deposited from summer underflows whereas underflow–interflows deposit laterally equivalent silts and clays along the lake margin and over topographic highs. A thin fining-upward clay layer representing sedimentation during the rest of the year evenly blankets the more variable summer layer. The contact between the two layers of the couplet is relatively sharp which results from alternation of dispersal mechanisms transporting sediment to the site. The silt is transported by bottom-hugging underflows whose dispersal patterns are controlled by topography; thus, total silt thickness varies with topography throughout the basin. The finer layer blankets the coarser sediment, accumulating rapidly following the autumn overturn and then gradually during the rest of the year.

13.5. STRATIGRAPHY AND LANDFORMS

13.5.1. Spatial and Temporal Variations

Glacial-fed lakes exist both during the advancing and retreating phases of a glacier but, during ice advance, proglacial deposits are commonly overridden and incorporated into the base of the ice. Consequently, most glacial lake records are of deglaciation. One landform associated with glacier-fed lakes (deltas) consists of coarse-grained proximal deposits which are perched above surrounding topography. Gilbert's (1890) classic study of the deltas of glacial Lake Bonneville revealed that the topset–foreset contact in the deltas marks the elevation of the lake surface at the time of deposition. Lacustrine beach deposits (coarse, sorted sediment) are commonly found at the same elevation as the topset-foreset contact thus supporting the interpretation of the former lake level (Flint, 1971). Other landforms (deltas or fans) lie as amorphous mounds in valley bottoms (Edwards, 1986). Fans are often blanketed with younger lake bottom sediments and thus not visible as landforms but only show up in cores or borings in the deposits (Sharpe and Cowan, 1990). Extensive flat plains which are usually dissected by younger drainage systems are the landforms produced in the lake bottom environment.

The spatial distribution of sediments in glacier-fed lakes is a function of the mechanisms of sediment distribution. Underflows are controlled by gravity and the deposits are thus constrained to topographic lows (Fig. 13.10a). Sediments dispersed by overflows/interflows blanket topography (Fig. 13.10b). A combination of processes accentuates the thickening in lows and thinning over highs (Fig. 13.10c). Rhythmites are typical below the thermocline and homogeneous muds above.

Depending upon the regional topography (mountains, rolling hills or plains) the details of the deglacial deposits will vary; however, in general, the deposits consist of a series of horizontally stacked morphosequences which stratigraphically onlap and become younger in the direction of ice retreat (Fig. 13.11). A morphosequence is a suite of deposits that are related to a single ice-front position. A morphosequence is generally deposited during a short time period (few hundred years) and is graded to a common local base level (Koteff and Pessl, 1981). A typical glaciolacustrine morphosequence consists of a proximal/distal, coarse-to-fine suite of deposits.

Sedimentation in glacier-fed lakes is dependent on the proximity to ice. As the glacier melts, the sediment source moves and at any given site the sedimentation rate decreases or essentially drops to zero with time. The glacial water and sediment source is replaced by meteoric water and reworked upland deposits. The majority of lakes eventually drain leaving the deposits exposed to subaerial processes.

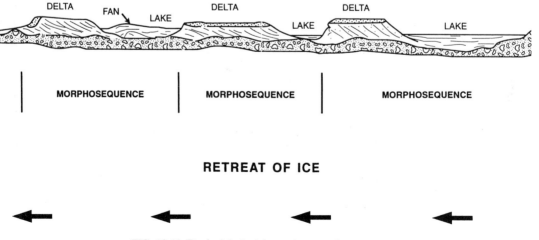

FIG. 13.11. Deglacial glaciolacustrine morphosequences.

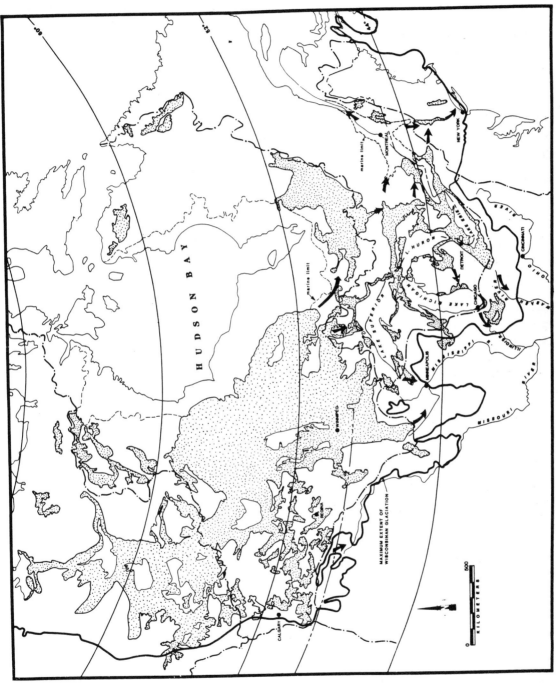

FIG. 13.12. Area covered by proglacial lacustrine sediment (stippled) deposited during the last retreat of the Laurentide Ice Sheet (from Teller, 1987; reprinted with permission of the author). Major avenues of overflow into the Mississippi River and Atlantic Ocean basins shown by arrows.

13.5.2. Sediment Record

Most of the glacial sediment carried to lakes remains in the basin, except for a minor fine-grained portion that may be transported through the lake by epilimnial currents. Lake basins are sediment 'sinks'. These sediments and the organic remains deposited simultaneously with them provide an excellent opportunity for recording a long-term (hundreds to thousands of years) uninterrupted environmental record (Lowe and Walker, 1984) (Menzies, 1995b, Chapters 8 and 14). Sediments of the lakes that once fringed the great ice sheets provide a tangible record from the now extinct water bodies (Teller, 1987) (Fig. 13.12).

Modern alpine glacial lake deposits provide a direct link from past to present by revealing changes in sedimentation, rates, lithic composition and organic content through time (Fulton and Pullen, 1969; Gilbert, 1975; Shaw, 1975b; Sturm and Matter, 1978; Leonard, 1986a, b; Donnelly and Harris, 1989). The stratigraphic record of glacier-fed lakes is the best continental source of continuous record of climate change spanning from the Late Glacial through the Holocene and up to the present day influences of humans (Renberg and Segerstrom, 1981; Leonard, 1986b). The parameters commonly used in reconstructing geochronology are varve thickness (Smith, 1981; Pickrill and Irwin, 1983), diatom (Lemmen et al., 1988) or pollen records, paleomagnetic record (Thompson and Kelts, 1974; Sandgren et al., 1988; Ridge et al., 1990), isotopic dating (Cs^{137}, Pb^{210} and C^{14}) (Ashley, 1979) and tephrochronology (dating of volcanic ash deposits) (Leonard, 1986b), as well as stable isotope records (Lister, 1988) (Menzies, 1995b, Chapter 14).

A several thousand-year varve chronology was developed in Sweden by measuring the absolute thickness of individual varves in sequence at hundreds of localities and correlating the sequences between localities (De Geer, 1912). The underlying assumption in developing the chronology was that the rhythmites were varves and that couplet thickness was controlled by regional climatic factors, temperature and precipitation, and not local factors. The chronology was later corroborated by C^{14} dating. Antevs (1922) developed and successfully used a similar varve chronology for eastern North America (Ridge and Larsen, 1990). The apparent faithful recording of environment change in the continuous sedimentation records of glacial-fed lakes makes glaciolacustrine deposits invaluable sources of information for reconstructing the past.

13.6. CLOSING REMARKS

Concerns about future global warming have raised questions pertaining to the interaction of earth systems (hydrosphere, atmosphere, lithosphere, biosphere and cryosphere). Large complex computer models (Global Climatic Models) have been developed to examine the nature of these interactions and analyze how climate is affected. The only test of the validity and reliability of these model results, aside from waiting patiently, is the history of past climatic change as recorded on the geologic record. Deep sea sediments and continental lake and loess sediments are likely to be the best long-term records (cf. Chapter 2).

It is essential that new research concentrate upon developing techniques in order to interpret continental climates (temperature and moisture) from the sedimentary record. Similarly, improved techniques are required in dating (Menzies, 1995b, Chapter 14) and correlating deposits from widely separated basins (Menzies, 1995b, Chapter 8). Paleoclimatic studies are likely to employ geochemistry and biochemistry to interpret the record left by climate-sensitive organisms. New geophysical methods need to be explored to help unravel the finer scale details of the paleomagnetic record (1–10 years) imprinted upon long-term records (>100 years). The past can hopefully be used to help predict the future.

Chapter 14

MODERN GLACIOMARINE ENVIRONMENTS

R. Powell and E. Domack

14.1. INTRODUCTION

Glaciomarine sediment contains the most detailed and continuous direct record of Late Cenozoic glacial fluctuations and often the first direct indications of a glacial period. In addition, many of the Earth's pre-Cenozoic glacigenic sedimentary rocks are interpreted as glaciomarine (Hambrey and Harland, 1981; Anderson, 1983). Knowledge of glaciomarine environments and processes that occur within them is, therefore, invaluable.

Glaciomarine (glacimarine, glacial marine) environments may be defined as marine environments in sufficient proximity to glacial ice that a glacial signature can be detected within the sediments. Some or all of the sediment is released from grounded or floating glacial or sea ice (cf. Table 14.1). The environment has a plethora of processes and sediment sources due to the combined action of ice (glacial and sea), water (fresh and sea), wind and biological activity.

Knowledge of glaciomarine processes has advanced in the past decade. However, large gaps still remain in understanding these environments and many interesting aspects remain unsolved. Sedimentation processes in cool-temperate, boreal and, to a lesser degree, polar-tundra climatic regions are perhaps best documented whilst processes in polar climatic regions are probably least documented.

This chapter will review present knowledge of sedimentological processes in modern glaciomarine environments. To do so, the glaciological and climatological conditions pertinent to glaciomarine environments will be discussed from the glacier to its margin in seawater and beyond to those related environments distal from the glacier. Non-glacial processes that contribute sediment to glaciomarine environments and modify previously deposited sediment will be considered. An evaluation will be made concerning rates of sedimentation and fluxes of sediment through glaciomarine depositional systems. These aspects are crucial in considering the nature of the sediment record left by glaciers, that ultimately are controlled by advances and retreats of marine-ending glaciers. Finally, glacier fluctuations are discussed in relation to the packages of sediment within a glacial record.

14.1.1. Glaciation and Climate

Glaciation at sea level depends on latitudinal and regional climatic factors such as winter and summer temperatures, insolation, snowfall and cloudiness, and a wide range of climatic conditions which, when

TABLE 14.1.

Climate Region	Summer Temperatures	Glacial Temperature	Equilibrium Line Elevations
Polar Ice Cap	<0°C	<0°C, except at base under special conditions	At calving line or seaward*
Polar Tundra (Subpolar)	<10°C	<0°C	At calving line or 100s m above s.l.
Boreal	>10°C (1 to 3 mo.)	At and below pressure melting	100s m above s.l.
Temperate Oceanic	>10°C (4 to 7 mo.)	At pressure melting throughout	100s to 1,000s m above s.l.

*The potential equilibrium line elevation below present sea level can be calculated, for instance see Robin (1988).

Climate regions and summer temperature characteristics are taken from: Trewartha, G.T., 1968, An introduction to weather climate, McGraw-Hill.

After Anderson and Domack (1991).

combined with relief, are capable of maintaining glaciers at sea level (Table 14.2) (Chapter 4). Most glaciers of coastal Antarctica, for example, are in the Polar Ice Cap category where mean summer temperatures are less than 0°C and ice temperature is subzero. Surface ablation is limited mainly to sublimation; therefore, even though accumulation averages <2 cm year^{-1} on the surface interior of the ice sheet, glaciers can be maintained at sea level. Meltwater and terrestrial vegetation are limited or lacking under these conditions. Unglaciated areas are cold, polar deserts with little atmospheric moisture and rapid evaporation. Thicker portions of the ice sheet and its major drainage systems are melting or melting/freezing at their base. Sea ice is extensive and can be strongly seasonal such as surrounding Antarctica, or persistent as in the Arctic Ocean.

The transition from Polar Ice Cap to Polar-Tundra or Subpolar climatic conditions occurs where mean summer temperatures are >0°C. These conditions occur along the western side of the Antarctic Peninsula, sub-Antarctic islands, coastal Greenland, arctic Canada, Iceland, Svalbard, Nova Semlya and Severnaya Zemla. Some summer melting occurs and glacier equilibrium lines are at or just above sea level, but terrestrial vegetation is limited. Glaciers, under these climatic conditions, experience surface melting during the summer but freezing temperatures remain throughout most of the glacier.

Boreal climates have one to three months with mean temperatures >10°C and, except for summers, are rather dry. These climatic conditions support large areas of mountain glaciers but only a few reach sea level, for example, along the Kenai Peninsula, north-central Gulf of Alaska. Glaciers experience surface melting, but vegetation is shrubs and lowland forest.

The mildest climate supporting glaciation at sea level is Temperate Oceanic, occurring along the eastern Pacific mountainous coasts of North and South America as far as latitude 57°N in southeastern Alaska and northern British Columbia and 47°S in western Patagonia. Mean temperatures for the four- to seven-month summers are >10°C, temperature minima are 0–2°C, and locally, rainfall occurs year-round. Extremes in ablation are balanced by very high snow accumulation, internal ice is temperate, glaciers

TABLE 14.2. Sedimentation rates of recent Antarctic glacio-marine deposits of the continental shelf

Location	Sediment Type	Rate (mm/yr)	Method
East Antarctica			
Mertz-Ninnis Trough, [1] (George V Coast)	diatom mud, ooze	3.1-3.4 3.0	[14]C [210]Pb
Prydz Bay, [2]	diatom ooze	1.5-0.67	[14]C
	silty clay (pebbly mud)	2.1	[14]C
Ross Sea			
Terra Nova Bay [3]	diatom mud	0.05-0.20	[14]C
Antarctic Peninsula			
Andvard Bay, [7]	diatom, pebbly mud	2.0-5.0	[14]C
Hughes Bay, [7]	diatom, pebbly mud	2.8	[210]Pb
Brialmont Cove, [7]	sandy mud	1.7-49.8	[14]C
Marguerite Bay, [5]	diatom mud	0.3 2.0	[14]C [210]Pb
Gerlache Strait, [5,6]	diatom mud	2.7 2.7	[210]Pb [14]C
Bransfield Strait, [5]	diatom mud	1.8	[210]Pb
South Shetland Islands			
Admiralty Bay [7]	sandy, pebbly mud	4.7-33.2	[14]C

References:
1. Domack *et al.* (1989).
2. Domack *et al.* (1991).
3. Krissick (1988).
4. Nittrouer (1988). personal communication.
5. DeMaster *et al.* (1988).
6. Venkatesan and Kaplan (1987).
7. Domack (unpublished NSF report)

are melting or melting/freezing at their base and equilibrium lines are well above sea level. It is the presence of coastal mountains that allows glaciers to thrive under temperate oceanic conditions. These coastlines are biotically productive with dense ground vegetation and even forested glacier termini (e.g. Malaspina Glacier).

14.2. SUBGLACIAL PROCESSES

Generally, transportation and deposition of englacial and subglacial debris by marine-ending glaciers is similar to terrestrial glaciers (Chapter 8; Menzies, 1995b, Chapter 2). However, some differences may occur due to the fast and extensional flow behind grounding lines of major marine-ending glaciers. Knowledge on this question is poor and conjectural. Lodgement of basal debris may occur in the form of till sheets that may be discontinuous. However, because of fast basal ice flow and high basal water pressures, deforming beds may be common. Where subglacial water is in excess, most subglacial sediment appears to be flushed out by meltwater streams (Powell and Alley, *in preparation*). The presence of deforming beds and large volumes of subglacial water may produce different types of depositional processes at grounding lines as will be discussed in Section 14.5.

14.3. MARINE-ENDING TERMINI

Glaciers that terminate in seas or oceans may, in the final few 100 m or kilometres, enter the body of water as (1) a floating terminus, or (2) a grounding tidewater cliff (Plate 14.1). These different modes of terminating in water result in: variations in how sediment arrives at the point of delivery from the ice to the water body; the effect of marine conditions such as temperature, currents, and salinity on sedimentological and glaciological processes; the transient interaction between marine and glacial environments; and the specific sedimentological processes ongoing within the proximal zone of the ice, the immediate proglacial subaquatic zone and in more distal aquatic areas.

14.3.1. Floating Termini

14.3.1.1. Formation and maintenance

Floating termini form when glacial flow lines converge and flow is unconfined due to very low frictional forces between glacier and sea. Commonly the annual 0°C isotherm is at sea level, but even if not, the ice mass is cold (polar). The termini form a continuum from confined floating glacier-tongues, to confined ice shelves, to unconfined floating glacier-tongues, to fringing ice shelves, to large embayment ice shelves (Anderson and Domack, 1991; Fig. 14.1). At present 57% of the Antarctic coastline can be delineated as floating termini, 38% is tidewater termini and 5% is sediment/rock coast (Drewry *et al.*,

1982). Most marine-ending glaciers elsewhere have tidewater termini, although some in the arctic terminate as fringing ice shelves or floating glacier-tongues.

Having a large horizontal surface area at low elevation (<100 m) means that floating termini are susceptible to rapid surface ablation should equilibrium lines rise (cf. Mercer, 1970). This condition probably places a climatic limit for ice shelves at about a mean annual temperature of –8°C and mean summer temperature of –2°C (Swithinbank *et al.*, 1988). Because the temperature of ocean water is often at or above 0°C under floating ice, basal ablation of these ice masses may be just as important as surface ablation (Jacobs, 1989). Nearly all of the mass loss from the George VI ice shelf (about 4.5 to 1.8 x 10^3 kg m^{-2} year^{-1}) is by basal melting aided by warm ocean waters (Bishop and Walton, 1981; Potter *et al.*, 1984).

14.3.1.2. Debris transport and release

Sediment deposited under floating termini comes from glacial sources and marine currents. Basal debris is melted out from floating termini within a few kilometres of a grounding line (Drewry and Cooper, 1981; Jacobs *et al.*, 1979). However, englacial debris entrained by converging outlet glaciers along the coast and then buried by snowfall, can be released from the base by bottom melting in the distal zones of a floating terminus (Powell, 1984). These influences of debris release can be seen in the confined George VI and fringing Larsen Ice Shelves in Antarctica (Domack, 1990). Debris is probably contributed in similar styles to the Ross Ice Shelf, Antarctica, where several large outlet glaciers have grounding lines located up-valley from where they join the ice shelf (cf. Swithinbank *et al.*, 1988).

Release of debris from beneath floating termini depends on oceanic temperature and pressure gradients which are controlled by water depth and circulation (Doake, 1976). For example, water flowing toward a grounding line under a floating terminus flows down a pressure gradient and can potentially melt the basal ice (Foldvik and Kvinge, 1974), whereas outflowing water moving up the pressure gradient will probably freeze to the base

PLATE 14.1. Examples of two types of glacier termini ending in the sea. (A) A floating glacier-tongue of an outlet glacier from the East Antarctic Ice Sheet producing icebergs in the Ross Sea, in which there are also remnants of seasonal sea ice. (B) A grounded vertical cliff of the Margerie Glacier, Glacier Bay, Alaska, with its bergs and a tourist craft in Tarr Inlet.

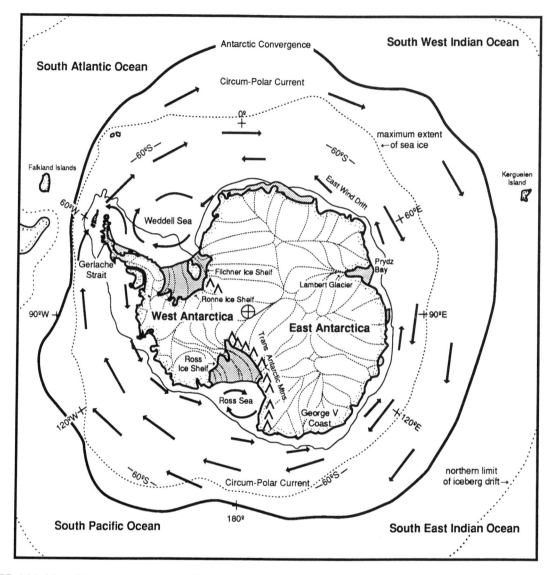

FIG. 14.1. Map of Antarctica and surrounding oceans showing coastal regions mentioned in text and illustrations. Dotted lines on continent are ice flow directions (after Drewry, 1983) and arrows in ocean are current directions. Dashed line in ocean is maximum limit of seasonal pack ice while dotted line is northern extent of drifting icebergs. Solid line in ocean is the approximate location of the southern polar front or Antarctic Convergence. Major ice shelves are indicated by the shaded areas.

(Robin, 1979). The ensuing pattern of debris release may vary amongst floating termini, especially embayment-type ice shelves, and may unpredictably change through the history of a floating terminus depending on oceanic circulation conditions.

There is growing indirect evidence from Antarctica indicating that the grounding lines of large ice shelf systems are dynamic settings with regard to sediment transport and delivery to the marine environment. An early intuitive model by Drewry and Cooper (1981) recognized the role that grounding-line cavities should play in the circulation of basal melt-out and suspended debris. Recent geophysical investigations along the Siple Coast of West

Antarctica have demonstrated that grounding lines of ice streams (fast flowing currents in an ice shelf) may occupy distinct zones of subglacial debris transport in the form of deformable, water saturated diamicton (Alley *et al.*, 1986). 'Till deltas' are one possible product of ice stream/ice shelf interaction (Alley *et al.*, 1989b, Fig. 5) (Section 14.4). The presence of thin subglacial water films is also recognized within the system but the transport of suspended particulates out into the sub-ice shelf or floating glacier-tongue environment remains a major point of future investigation (Barrett, P.J. *et al.*, 1988). Results from the first studies of these processes are described in Section 14.6.4.

Shallow water environments, especially those bordering ice cliff or piedmont settings, are greatly affected by currents of wind or tidal origin. Low siliciclastic sediment supply at the calving line allows such currents to effectively sort and transport sediment into broad aprons of sandy sediments rich in carbonate bioclastic detritus.

14.3.1.3. Ice shelf pumping

Studies of ice shelf grounding-line environments, including the Ronne Ice Shelf, have demonstrated significant vertical movement of the shelf indicating that the grounding-line may in fact undergo periodic uplift and set-down under the influence of tides (Stephenson and Doake, 1982; Kobarg, 1988). The effect within any grounding-line cavities would be similar to a bellows with the expulsion of water and suspended sediment out into the marine setting during falling tide or ice shelf let down (Talbot and Von Brunn, 1987). Unfortunately this pulsing process is difficult to observe directly but when combined with debris rain-out from undermelt, it should produce laminae and deformation structures within sediment such as diamicton. In many cases a disrupted diamicton mélange is produced (Talbot and Von Brunn, 1987).

Indications of how tidal pumping may influence sediment transport are provided by studies of partially-floating termini along the Antarctic Peninsula. Oceanographic data demonstrate that, within the water column, mid-water fine-structure consists of distinct interflows of turbid water adjacent

to the glacial termini in deep water (Fig. 14.2). A mechanism similar to tidal pumping is implicated in the formation of these features (Domack and Williams, 1990; Fig. 14.3). Suspended particulates sampled from these features include quartz silt grains with a fresh appearance and even very fine-grained sand (Plate 14.2).

14.3.1.4. Calving styles and processes

Calving (sometimes termed berging) is a rather infrequent occurrence for ice shelves; the mean annual iceberg production in Antarctica is about 2800 to 3040 km^3 (Ørheim, 1985), 60 to 80% of which comes from embayment ice shelves and 15 to 25% from ice streams and outlet glaciers (Drewry and Cooper, 1981). Calving mechanisms are not fully understood but possible causes can be grouped into those related to: glacial flow (creep-failure with extensional flow, rift zone development, surface and bottom crevasses); a combination of glacial and oceanic processes (Reeh-type calving from force imbalances on the ice face (Reeh, 1968)); bending and flexure by tides and cyclic vibration (Holdsworth and Glynn, 1978); impacts of large icebergs; and oceanic processes (storm and tsunami wave energies and ocean currents) (Kristensen, 1983).

Calving processes from floating termini produce the largest icebergs which can survive long transport distances. For instance, one recent calving event from the Ross Ice Shelf, Antarctica, produced a tabular iceberg which was 154 x 35 km^2 and it drifted across 2000 km of the Ross Sea during a 22 month period (Keys *et al.*, 1990). It contained 1200 km^3 of ice which is nearly 50% of the net annual accumulation of the Antarctic Ice Sheet (Giovinetti and Bentley, 1985). The actual calving event is relatively passive and has no influence on sea floor sediment that can be several hundred meters below (Plate 14.3).

14.3.2. Tidewater Termini

14.3.2.1. Formation and maintenance

Glaciers that end as tidewater termini occur in every climatic setting (a cool-temperate example is shown in Fig. 14.4). Glaciers of temperate ice do not

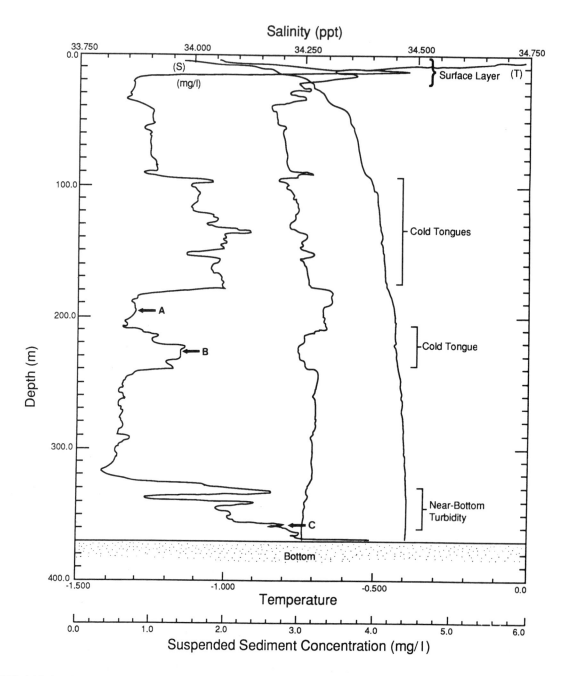

FIG. 14.2. Vertical profile of temperature (T), salinity (S) and suspended particulate matter (mg l⁻¹) within an Antarctic (subpolar) fjord. Location of SEM images (Plate 14.2) of suspended particulate matter are also shown (A,B,C). Note significant suspended particulate matter within mid-water layers (cold tongues) and comparison to surface layer and near bottom turbidity (modified after Domack *et al., in press*).

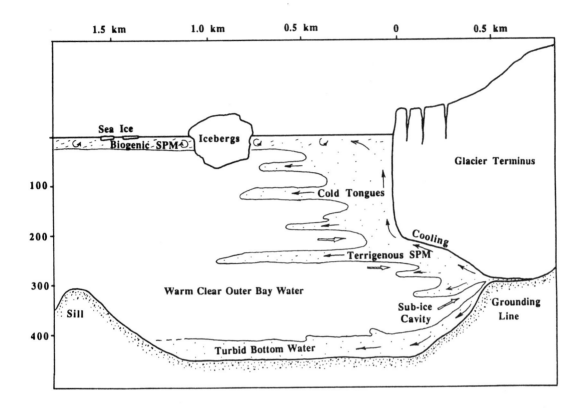

FIG. 14.3. Model for cold tongue (interflow) formation in polar fjord settings as based upon observations in Antarctic fjords (Domack and Williams, 1990; reprinted with permission of the authors and the American Geophysical Union). Circulation is driven by storm or tidal surges within the basal cavities of tidewater termini or ice shelves.

have the tensile strength to float as intact 'slabs' (Powell, 1980), while those of cold ice apparently do not have sufficient flow velocities to maintain a floating terminus (Plate 14.4).

Glaciers form a vertical cliff once they reach sea level because the glacier calves faster than it melts. As the ratio of melting to calving increases a more parabolic surface profile is produced. Under these later circumstances a cliff will remain grounded unless the conditions required for floating are met.

Maintaining a tidewater terminus appears to be primarily a function of the water depth in which the grounded cliff ends (Brown et al., 1982). The ELA of glaciers with these termini is not as sensitive as those with floating termini. If the 0° isotherm is at sea level,

then a cliff is maintained relatively easily in the sea. Where the ELA is at high altitude a large supply of winter snow is required to create sufficient ice flux to maintain a tidewater terminus (Section 14.9).

14.3.2.2. Debris transport and release

Downwasting of a glacier surface produces supraglacial debris by melt-out, especially on valley and outlet glaciers. Most of that debris is from supraglacial sources and, for marine termini, it often falls into the many crevasses caused by related extensional flow. Little englacial and no basal debris is moved to the surface by ice flow at the terminus, because flow lines, under these terminus conditions,

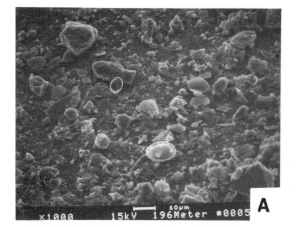

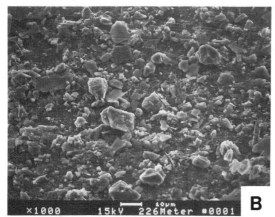

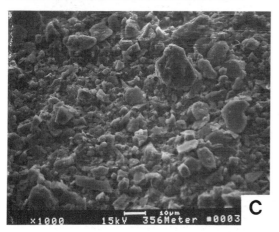

PLATE 14.2. Scanning Electron Microscope photographs of suspended particulate matter within oceanographic features shown in Fig. 14.2. Note dominance of siliciclastic particles in all samples especially sample B from mid-water cold tongue. Greater biogenic (diatom) constituents are observed in sample from less turbid water at 196 m (A).

remain near-horizontal. Englacial debris derived from up-glacier tributaries is often exposed in cross-section on the cliff face in near-vertical layers which may be deformed by local ice flow. All basal debris being horizontal remains near the sea floor and is only exposed above sea-level in shallow water depths.

Melt-out/Fallout. A tidewater terminus is melted by solar heat, stream discharges and sea. In general, melting by air is much slower than calving and other melting rates (Powell and Molnia, 1989). Meltwater streams rising up the cliff can melt small vertical runnels or larger channels that produce 'caves' at sea level (Powell, 1980; Gilbert, 1983; Syvitski, 1989). This process can enhance melt-out of debris and assists in creating calving embayments around the efflux (Sikonia and Post, 1980).

Melt-out from the face produces rockfall and debris grainfall. Dumps of supraglacial debris are produced from the surface and crevasse-fills during calving. Debris flows and slurries slide down the face but disaggregate due to the steep angle of the face before entering the sea (Powell, 1983a). All supraglacial contributions to grounding-line systems are minor except perhaps locally where glaciers have abundant supraglacial debris. Melt-out till similar to that from terrestrial glaciers is difficult to produce in these situations due to fast flow and non-stagnant conditions. Some melt-out deposits may form on the up-glacier side of morainal banks or where glacial 'toes' become buried by glaciomarine sediment.

Squeeze/Push. Sediment that is pushed by marine-ending glaciers is generally preserved by seasonal fluctuations of termini that are undergoing net retreat (e.g. Hoppe, 1959; Smith, 1982; Boulton, 1986b). Because marine-ending glaciers advance across sediment with a high water content, the probability of ejection out from a grounding line by squeeze and liquefaction processes is likely (e.g. Andrews and Smithson, 1966; Powell, 1984; Smith, 1990). Squeeze-up into basal crevasses is also strongly implied by the distinctive sea floor pattern observed after surges (Solheim and Pfirman, 1985) (Menzies, 1995b, Chapter 4).

Mass Movements. Sliding, slumping and sediment gravity flowage are common events at tidewater termini of temperate and subpolar glaciers. These processes are common because of high sedimentation rates, iceberg calving impacts and glacier-push of

PLATE 14.3. Fluvial discharges from temperate tidewater termini in Glacier Bay, Alaska, which of all the glacial sources, contribute the largest volume of sediment to the sea. An upwelling at the sea surface is an expression of the meltwater jet discharging from under Grand Pacific Glacier at about 75 m depth. Gulls are feeding on shrimp that live at the ice face and which are being forced up to the surface by the upwelling (A). An example of a smaller, englacial stream discharging at sea level from Muir Glacier. The diameter of the conduit is about 4 m (B).

FIG. 14.4. Three-dimensional sketch showing the major environments for glaciomarine sedimentation in a cool-temperate climate, based on the present north-northwest Gulf of Alaska (not true orthographic scale). Numbered features are as follows: 1 — Tidewater terminus (see Figs 14.7a and b for details), 2 — Side drainage with gravel pocket beach and talus fan on fjord floor, 3 — Bergstone mud on glacial fjord floor, 4 — Sandur plain in marine outwash fjord, 5 — Sediment gravity flow channels in marine outwash mud, 6 — Entrance sill capped with morainal banks and grounding-line fan, 7 — Continuation of outwash mud, 8 — Rocky shore with gravel pocket beaches, 9 — Tectonically uplifted banks of older rocks exposed by winnowing on the shelf, 10 — Extensive tidal mudflat in estuary, 11 — Recurved and cuspate spits from alongshore transport, 12 — Marine outwash mud on the continental shelf, 13 — Relict sand deposits, 14 — Thick temperate rain forest of spruce/hemlock with muskeg swamps, 15 — Raised marine terrace, 16 — Eolian dunes fed from spit at wave-dominated delta, 17 — Large slide/slump areas on the continental shelf, 18 — Ice-dammed lake with sublacustrine grounding-line fan and lake laminites with iceberg-rafted debris, 19 — Terminal moraine and gravelly beach, 20 — Sandur/delta system with small moraines and gravelly beach, 21 — Modern littoral sand, 22 — Bank on continental shelf of relict glacier-marginal sediment, currently being winnowed, 23 — Tidewater terminus just at sea level with short-headed stream, fan deltas, 24 — River-dominated estuary with stunted barrier island system offshore, 25 — Sea valley on continental shelf being infilled with marine outwash mud (after Powell and Molnia, 1989; reprinted by permission of Elsevier Science Publishers BV, Amsterdam).

grounding-line sediment accumulations (grounding-line depositional systems) such as morainal banks, grounding-line fans and ice-contact deltas (Powell, 1981a, 1990). Small events with short (within 1–2 km) runout distances are common and may occur at intervals of days, hours or even continuously (Powell and Molnia, 1989; Powell, 1990; Phillips *et al.*, 1991). Larger events occur more sporadically but may remove large volumes of sediment from a grounding-line and transport it tens of kilometres away from the glacier (Powell, 1991, *unpublished data*).

Sediment accumulated against a grounding line can collapse back toward the glacier in small events as basal icebergs calve, as well as in quite large volumes when the grounding line retreats from the site (Powell, 1981a, 1991).

Hydrographic Processes. Waves, tides and ocean currents may affect melting of tidewater termini producing distinctive sediments. Melting of ice cliffs by currents flowing toward them and then upwelling to the surface is governed by the relationship:

$$M_l = 6.74 \quad 10^{-6} \quad U_w^{0.8} \quad \Delta T \quad L_{ic}^{-0.2} \quad (14.1)$$

PLATE 14.4. A calving event from the grounded, 55 m high tidewater cliff of Lamplugh Glacier, Glacier Bay, Alaska. The slab is photographed about half way through its rotating fall (A) and then a large splash and wave is produced on impact (B).

where M_1 is the melting rate, U_w is water velocity, ΔT is temperature difference between ice and water and L_{ic} is the characteristic length of the face (Weeks and Campbell, 1973). When currents flow parallel with the face, mean melting rate (M_2) is:

$$M_2 = \frac{1}{h_{sw}} \int_0^{h_w} 3.7 \ 10^{-2} \ (T_\infty - T_s)^{1.5} \left(\frac{x}{k}\right)^{0.25} dx \quad (14.2)$$

where h_{sw} is height of submerged tidewater cliff wall, T_∞ is far field ambient water temperature, T_s is surface water temperature, x is the coordinate in water flow direction and k is a scaling factor (Huppert, 1980; Huppert and Joseberger, 1980). In general, $M_2 < M_1$.

Commonly, waves, tides and ocean currents are second-order controls on grounding-line sedimentation except where large storm events occur on the continental shelf, or where the volume of sediment supplied from the glacier is low as in polar settings, or in shallow water where the processes can rework much of the sediment.

Fluvial Discharges. Fluvial discharges from tidewater termini are greatest from temperate glaciers. Such discharges dominate sediment production in temperate, and to a lesser extent subpolar, glaciers. Streams enter the sea from englacial and subglacial conduits; the latter position providing most of the sediment. Discharge is so high in the melt season that it forms turbulent jets where momentum forces in the flow are stronger than the buoyancy forces of the freshwater entering denser saltwater (Plate 14.5)(Syvitski, 1989).

For slower flows such as during early rising and late falling discharges or from colder glaciers, the effluent is a forced plume where buoyancy dominates momentum forces. That is, when

$$R_n = Q \ B_f^{0.5} \ M_f^{-1.25} \leq 0.56 \text{ flow is a jet}, \quad (14.3)$$

$$> 0.56 \text{ flow is a plume}; \quad (14.4)$$

where R_n is the Richardson number, Q is discharge or volume flux, B_f is buoyancy flux equivalent to $g'Q$ and M_f is momentum flux or $\rho Q U_{mo}$; where $g' = g(\rho_{ad} - \rho_o)/\rho_o$, ρ_{ad} is ambient fluid density, ρ_o the initial discharge density and U_{mo} the initial near-centreline velocity (List, 1982; Powell, 1990). The angle at which the jet issues into the sea can be determined by

the densimetric Froude number (F)

$$F = U_m^2 (g'D_c)^{-1} \quad (14.5)$$

where U_m is mean centreline velocity at the conduit mouth and D_c is conduit diameter (Abraham, 1965). As $|F| \to 0$ the jet is vertical, as $|F| \to \infty$ the jet follows the axis of the conduit and as $F < 0$ when $\rho_{ad} < \rho_o$, a continuous underflow may form (Syvitski, 1989). Underflows have been inferred (e.g. Mackiewicz *et al.*, 1984) and may be enhanced by hyperconcentrating sediment as particles rapidly fall out of transport when the flow detaches from the sea floor (Powell, 1990), but these remain unconfirmed.

Eventually the jet transforms to a plume as it rises to sea level. Occasionally the flow may cause 'boiling' or, at slightly higher discharges, 'fountaining' at sea level above the upwelling (Plate 14.3). The plume spreads as a barotropic flow away from the discharge at a level of neutral buoyancy, which in sea water is commonly at sea level, that is, it forms an overflow or hypopycnal flow. It is from this jet/plume system that laminated sediments (cyclopsams and cyclopels) can form (Mackiewicz *et al.*, 1984; Cowan and Powell, 1990).

14.3.2.3. Calving styles and processes

Tidewater termini differ from floating termini in that the calving line coincides with the grounding-line. Calving from ice cliffs occurs by three processes: (1) spalling of seracs by fracturing above sea level and the iceberg falls into the sea, (2) large sheets shear off the glacier face and the iceberg sinks vertically or topples forward into the sea (Plate 14.4), and (3) detachment below sea level and then the iceberg rises to the surface. While Muir Glacier, Glacier Bay, Alaska was retreating at about 2 km year^{-1}, about 10^6 m^3 day^{-1} of ice was calved (Powell, 1983b). For a fast flowing (3 km year^{-1}) glacier with a stationary tidewater terminus 2 km wide ending in about 100 m of water, about 1.7×10^6 m^3 day^{-1} of ice must be calved.

Subaerial calving occurs by fracture propagation (Iken, 1977; Hughes, 1992b) and if icebergs do not shatter on impact they can travel down through the seawater column. Some blocks can either impact

PLATE 14.5. A berg produced by a submarine calving from the McBride Glacier tidewater cliff in the background, Glacier Bay, Alaska. These types of bergs provide excellent exposures of sections of glaciers otherwise unobtainable. Here, the sole of the glacier with subglacial till is exposed on the left side of the berg and all the basal debris layers to the right are in a stratigraphic-up sense. All of this debris will produce iceberg-rafted debris.

upon the sea floor, or come sufficiently close, so that their preceding pressure wave redistributes sediment (Powell, 1985). Surface waves caused by the impact generally can not influence the sea floor initially because it is too deep (Powell, 1980); however, at shorelines the waves can cause erosion and transport icebergs above the high tide line (Syvitski, 1984).

Icebergs from submarine calving are often termed bergy bits and commonly contain englacial and basal debris (Powell and Molnia, 1989). These icebergs can disturb bottom sediment as they detach, and some are seen with sea floor sediment on top of them. They are also major sources of iceberg-rafted debris (IBRD) from tidewater termini (Gottler and Powell, 1990) (Plate 14.5).

14.4. TYPES OF GROUNDING-LINES

Grounding-line systems are defined as sedimentary depositional systems (*sensu* Fisher and Brown, 1972) formed at grounding-lines of glaciers ending in large bodies of water, such as large lakes or seas (Powell, 1988). Each system is made from one or several sedimentary environments and can be distinguished by its processes of formation, geometry, facies associations and internal architecture.

Recent studies of glaciomarine sediments have shown there are several different varieties of grounding-line systems that include: (1) grounding-line fans (Powell, 1990) and glacier-contact deltas (Thomas, 1984a; McCabe and Eyles, 1988), (2) morainal banks (Powell, 1981), and (3) grounding-line wedges ('till deltas' of Alley *et al.*, 1987b) (Figs 14.5, 14.6, 14.7 and Plate 14.6). Understanding how to recognize each type by lithofacies analysis and being able to infer processes of formation are important for inferring the potential stability of grounding-lines (cf. Powell, 1991) and for interpreting paleoglaciological and paleo-climatological conditions from a sedimentary record.

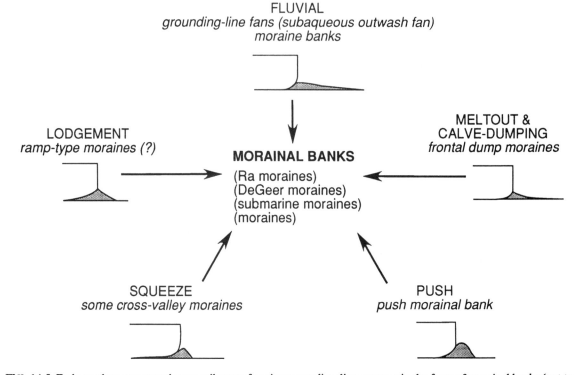

FIG. 14.5. End-member processes that contribute to forming grounding-line systems in the form of morainal banks (not to scale). Beneath each process are terms that have been used to describe the deposit in the literature; here it is recognized that morainal banks can form by the end-member process or any combination of those processes. Alternative terms for morainal banks in the literature are also listed.

Grounding-line fans are point-source depocentres with a fan geometry (*sensu* Mitchum *et al.*, 1977), formed at conduit mouths where meltwater streams discharge subaquatically (Fig. 14.6). They are composed of subaquatic outwash (deposited from the discharge while in contact with the floor of the water body) and a variety of sediment gravity flow and suspension settling deposits. They have also been named transverse eskers and deltas (Glückert, 1975, 1977), beaded eskers (Banerjee and McDonald, 1975), subaqueous outwash fans (Rust and Romanelli, 1975), subaqueous esker deltas (Thomas, 1984b) and glacier-contact fans (Boulton, 1986b) (Menzies, 1995b, Chapter 2).

Glacier-contact deltas are commonly formed by short-headed streams, (cf. Flores, 1975) and are typified by processes, facies associations and the architectural arrangements of fan deltas with glacial influences such as ice-rafting and pushing (cf. Syvitski and Skei, 1983; McCabe and Eyles, 1988; Nemec and Steel, 1988). These fans often develop from grounding-line fans building up to sea level (Glückert, 1975, 1977; Powell, 1990).

Delta processes are grouped into bedload dumping, hemipelagic sedimentation, by-pass and diffusion of sediment (Syvitski *et al.*, 1988). As a river enters the sea its competence is dramatically reduced and bedload is deposited and dumped rapidly near the river mouth. The suspended load can continue into the sea in a hypopycnal flow of fresher, less dense, water flowing over saline water. Particles suspended in this plume eventually settle out and deposit hemipelagic sediment (Section 3.5.1). Meanwhile, some processes on the sea floor act to remove sediment totally from the nearshore delta area. Such processes as low density turbidity currents and large-scale slides behave in this manner. Other bottom processes act to 'smear' bottom sediment

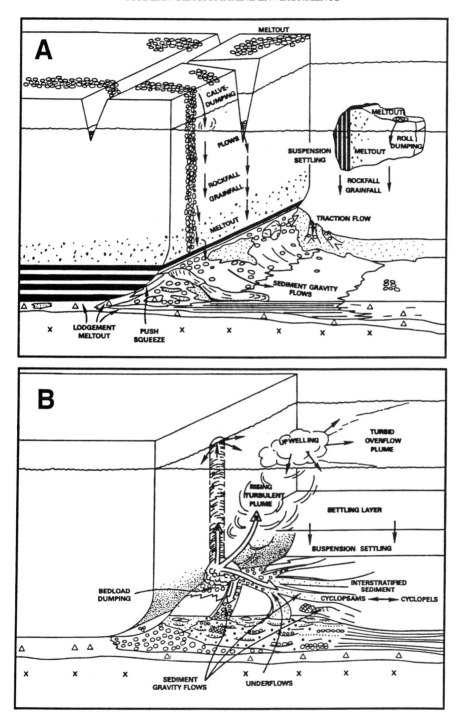

FIG. 14.6. Processes and lithofacies associations contributing to grounding-line systems at tidewater termini of temperate glaciers (not to scale). (A) A morainal bank, (B) submarine outwash of a grounding-line fan and generation of cyclopsams and cyclopels (after Powell and Molnia, 1989; reprinted by permission of Elsevier Science Publishers BV, Amsterdam).

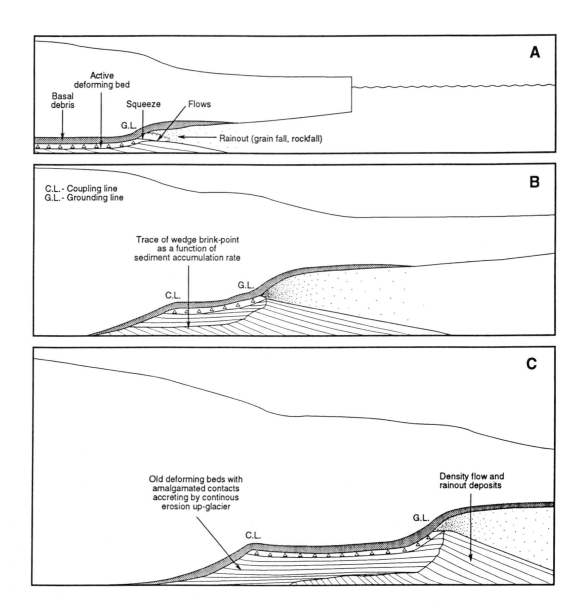

FIG. 14.7. A possible mechanism for formation of grounding-line wedges and how they can expand with glacial advance to form sediment sheets that terminate as wedges. An ice stream feeding an ice shelf has basal debris that is feeding a subglacial deforming bed (A). At the grounding line the deforming bed is squeezed out to form sediment gravity flows on an inclined surface. The wedge of sediment produced is also contributed to by grain fall and rock fall from under-melt of the ice shelf. The volume of that rain-out sediment decreases with distance from the grounding line and also contributes to the thinning of sediment distally. As the sediment builds to meet the base of the ice shelf, the grounding line advances as does the coupling line (B). Upglacier from the coupling line, erosion can occur. Between the coupling line and the grounding line, the deforming bed aggrades by being added to from basal debris, but because transmitted sheer decreases with depth lower sediment in the deforming bed is deposited. With continued advance, the distance between the grounding line and coupling line increases producing a deposit with a sheet-wedge geometry (C). Predictably, the deposit is composed of sediment gravity flow and rain-out facies topped by deformed-bed deposit with amalgamated contacts (not to scale).

PLATE 14.6. View of a grounding line at a tidewater terminus of a temperate Alaskan glacier. The bottom left-hand segment is sorted glaciomarine sediment of gravels up to boulder size draped with a coating of mud from meltwater streams and pebbles from meltout of icebergs and the glacier cliff. The rest of the photograph is a wall of clear glacier ice with bubbles and pebbles within. This was the first observation made of a grounding line, which in the photograph is about 15 cm long.

away into deeper water, such as tidal or wave action, creeps or slumps and perhaps hyperconcentrated sediment underflows from the river. By using simple density contrasts it is easily shown that underflows directly from rivers can occur only if their sediment concentrations are higher than about 30 g l^{-1} (Gilbert, 1982; Powell, 1983).

Morainal banks are made of various glacier-contact lithofacies (Figs 14.5, 14.6 and 14.7) that include chaotic mixtures of diamicton, gravel, rubble, sand and mud facies formed by rock and grain fall processes due to calve-dumping and melt-out (Powell, 1981; Powell and Molnia, 1989), lodgement (Holtedahl, 1960; Glückert, 1975, 1977; Smith, 1982) and squeeze/push processes from under the glacier (e.g. Andrews, 1963; Smith, 1990).

The deposits have a bank geometry (following the terminology of Mitchum *et al.*, 1977) and are analogous to terrestrial end moraines, but are formed subaquatically. Banks created mainly by pushing are push-morainal banks (cf. Powell, 1981; Smith, 1982; Eyles and Eyles, 1984; Boulton, 1986b). Morainal banks also include the ramp-type sublacustrine moraines that form more by subglacial lodgement processes (Holdsworth, 1973; Barnett and Holdsworth, 1974), frontal-dump moraines from melt-out at the terminus and dumping of supraglacial debris during the calving process (Powell, 1981; Syvitski and Praeg, 1989). Other banks are thought to form by squeezing out of subglacial sediment beyond the grounding-line to produce landforms like the cross-valley moraines (Andrews and Smithson, 1966) (cf. Chapter 10). Other forms of morainal banks may have their entire lengths made of facies similar to grounding-line fans (moraine banks) (Leavitt and Perkins, 1935), indicating subglacial discharge of meltwater in either sheet flows or rapidly migrating conduits across a grounding-line. These types of banks confirm a continuum from lone grounding-line fans (that can form beaded eskers) (Banerjee and

McDonald, 1975), to interstratified grounding-line fan and glacier-contact facies to solely glacier-contact facies.

Morainal bank forms are documented in modern process studies in lakes and the sea (e.g. Holdsworth, 1973; Barnett and Holdsworth, 1974; Powell, 1981; 1990; Powell and Molnia, 1989). They are described from older deposits in fjords (Vorren, 1973; Elverhøi et al., 1983; Gilbert, 1985; Syvitski, 1989; Holtedahl, 1989) and continental shelves (Andersen, 1968; Blake, 1977; Matishov, 1978; Wong and Christoffel, 1981; Knebel and Scanlon, 1985; Oldale, 1985; Barnes, 1987; Josenhans et al., 1988; Anderson and Bartek, 1991) and from Pleistocene deposits presently exposed on land (e.g. McCabe, 1986; Ingólfsson, 1987). The latter types have been variously termed moraine banks (Leavitt and Perkins, 1935), DeGeer moraines (e.g. Hoppe, 1959; Smith, 1982), washboard moraines (Elson, 1969), some of the cross-valley moraines of Andrews and Smithson (1966), Ra moraines (e.g. Holtedahl, 1960), submarine moraines or simply, moraines (terminal, end, lateral) (e.g. McCabe et al., 1984; Oldale, 1985; Holtedahl, 1989).

'Till deltas' were defined as wedges of sediment made up of subglacial till overlying dipping strata inferred to be mainly sediment gravity flow deposits (Bentley et al., 1988; Alley et al., 1989b; Blankenship et al., 1989; Kamb and Engelhardt, 1989; Scherer, 1991). They are thought to be forming today beneath the Ross Ice Shelf based on seismic evidence and theoretical considerations. Here the term 'grounding-line wedge' is used rather than 'till delta' to avoid the problem of defining all of the sediment as till, and to avoid an association with sea level that could be inferred from the term 'delta' (Powell and Alley, in preparation).

Grounding-line wedges are thought to form mainly from deforming subglacial till that reaches the grounding-line (Fig. 14.7). There, sediment redistribution by gravity-flow processes produces deposits with dips as low as 1° or less, because the deforming till should contain abundant fine-grained matrix (cf. Mitchum et al., 1977; Sangree and Widmier, 1977; Lawson, 1979a, p. 41; Hampton et al., 1987). Grounding-line advance across these deposits will cause deforming subglacial till to

overlie them disconformably. The up-glacier limit of this disconformity is called the coupling line and coincides with a change in surface slope of the glacier, up-flow from the grounding-line. The deforming layer may thicken downglacier across the coupling line, causing the coupling line to be dynamically significant to the glacier. Grounding-line wedges are less well known than the morainal bank varieties because the one modern example so far postulated (Alley et al., 1987b) is still under investigation (Blankenship et al., 1989; Engelhardt et al., 1990). However, an older deposit on the Antarctic continental shelf has the possibility of being a relict Pleistocene example (cf. Karl et al., 1987, Fig.4).

Dreimanis (1987) identified an older, smaller-scale deposit that may be equivalent to grounding-line wedges. This feature, which he termed a "subaquatic till tongue", occurs in Wisconsinan deposits at the base of a cliff bordering Lake Erie, Ontario. It is described as "a diamicton tongue interpreted as...till...which was deposited, partly by lodgement, under the edge of the Erie lobe, which terminated in a deep proglacial lake, and partly by extrusion and gravity flows of dilated lodgement till..." (Dreimanis, 1987, p. 23).

Powell and Alley (in preparation) suggest that larger-scale deposits independently termed 'till tongues' by King and Fader (1986) also may be older deposits equivalent to the 'till delta' of Alley et al. (1989b) and the 'subaquatic till tongue' of Dreimanis (1987). The 'till tongues' of King and Fader (1986) were defined as wedge-shaped deposits of till interbedded with stratified glacimarine sediment (King et al., 1987). They were mainly described using high-resolution seismic stratigraphy and were attributed to deposition at a fluctuating grounding-line of an ice shelf, by different processes than attributed here to grounding-line wedges.

The apparent major control over the two end-member types of grounding-line systems, morainal banks and grounding-line wedges, is subglacial meltwater. With abundant, free-flowing water morainal banks appear to form; in contrast, grounding-line wedges appear to form with smaller volumes of more confined water and deforming beds.

Using these simplistic limitations, Powell and Alley (in preparation) suggest that deposits such as 'till tongues' may be formed mainly by polar or subpolar glaciers flowing on deforming beds; the glaciers may or may not end as floating termini. Alternatively, morainal banks are most likely formed by temperate or subpolar glaciers where supraglacial water contributes to subglacial water in channelized or sheet flows, and the glacial termini are most commonly tidewater cliffs.

14.5. PROGLACIAL AND PARAGLACIAL ENVIRONMENTS

Proglacial and paraglacial environments lie beyond the glacier proximal settings of grounding-line systems and subglacial ice shelf systems. The term paraglacial was used by Church and Ryder (1972) to define non-glacial processes that are directly conditioned by glaciation and it is used here to include marine systems that receive glacigenic sediment but are not in direct contact with glaciers nor icebergs (Gilbert, 1983; Powell and Molnia, 1989). In many instances processes in proglacial and paraglacial settings are similar. However, the presence of glacial ice sometimes creates distinctive processes in proglacial settings.

14.5.1. Hemipelagic Suspension Settling

Sediment plumes from subglacial and marginal streams are the transportation agents for most glaciomarine sediment in front of temperate and subpolar marine-ending glaciers. The overflow originating from fresh water sources at tidewater termini has been modelled as a barotropic flow originating as a buoyant jet (Calabrese and Syvitski, 1987, Figs 1 and 2). Velocities in different zones are:
for plug flow:

$$U_{x,y} = (Q_o b^{-1}_o H^{-1}_o) \exp [-(y + 0.1x - 0.5 b_o)^2 (2\sigma^2)^{-1}]$$
for $x \le 5.2 b_o$, $y > b_o - 0.2x$ (14.6)

for established flow:
$$U_{x,y} = 2.28 (Q_o b^{-1}_o H^{-1}_o)(b_o x^{-1})^{0.5} \exp [-y^2 (2\sigma^2)^{-1}]$$
for $5.2 b_o < x \le x_b$ (14.7)

for constrained flow:
$$U_{x,y} = 2.28 (Q_o b^{-1}_o H^{-1}_o)(b_o x^{-1}_o)_{0.5} \exp [-y^2 (2\sigma^2)^{-1}]$$
for $x > x_b$ (14.8)

where $U_{x,y}$ is longitudinal velocity, b_o is stream channel width, H_o is channel depth at the stream mouth, Q_o is stream discharge, x is down-fjord distance, y is fjord width distance and $\sigma = 0.108x$ is the standard deviation of the Gaussian distribution for velocity across the plume width.

Sedimentation rates $(Z_h(x,y_o))$ beneath the respective parts of the plume are predicted by the relationships (Syvitski et al., 1988):

$$Z_h(x,y_o) = Z_o \exp [-(\lambda_{rs} U^{-1}_o)x] \quad \text{for } x \le 5.2 b_o \quad (14.9)$$

$$Z_h(x,y_o) = Z_o \exp \{-\lambda_{rs} [(1.76 b_o U^{-1}_o) + 0.29 x^{1.5}$$
$$(U_o b_o 0.5)^{-1}]\} \qquad \text{for } 5.2 b_o < x \le x_b \quad (14.10)$$

$$Z_h(x,y_o) = Z_o \exp \{-\lambda_{rs} [(1.76 b_o U^{-1}_o) - (0.15 x_b^{1.5}$$
$$U^{-1}_o b_o^{-0.5})]\} \qquad \text{for } x > x_b \quad (14.11)$$

where $Z_o = \lambda_{rs} Q_s(U_o b_o)^{-1}$ is the rate of sedimentation at the stream mouth, λ_{rs} is the removal rate constant for the suspended particles, x_b is the distance from the stream mouth where the plume becomes constrained by basin geometry, U_o is the longitudinal velocity component at the stream mouth and Q_s is the suspended load carried by the stream. Under these conditions Calabrese and Syvitski (1987) predicted an exponential decrease in settling rates down flow from a submarine discharge as was later documented by Cowan and Powell (1991) in Alaska.

Proximal marine outwash occurs as a traction facies deposited along the run-out distance of the jet from the submarine efflux of a subglacial stream (Fig. 14.8). Sometimes the jet detachment will be at the glacier face but with larger discharges the zone may be away from the face. At the detachment zone small pebbles to sand sizes appear to fall out and are virtually 'dumped' (Powell, 1990). Particles as coarse as medium sand can remain in the plume as it rises to the surface and sand can be transported by the plume as much as 1 km from a face (Cowan et al., 1988).

Sediment is released from overflow plumes episodically and the areal position of a plume changes due to tides and surface wind sheer (Fig. 14.6). Cowan and Powell (1990) suggest that particles are

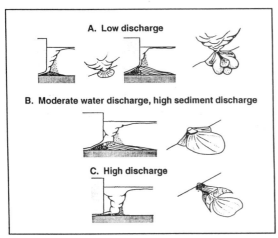

FIG. 14.8. Alternative fan types with different discharges (not to scale). (A) At low discharges the fan grows progressively with coarse traction deposits dumped at the efflux and most deposits from settling or slumping/sliding and sediment gravity flows. (B) At higher sediment discharges the horizontal jet remains in traction to produce rapid sedimentation deposits in sheets or crude scour-and-fill structures. Settling and sediment gravity flow deposits make up more distal deposits. (C) At high discharge the horizontal jet can include a migrating barchanoid bar at the detachment zone (after Powell, 1990; reprinted with permission of the Geological Society of London).

released from a plume during ebb tidal stage when internal turbulence within the plume is smallest. The released particles then fall as a turbid layer through the water column. Coarser particles of sand/silt fall faster than flocs of mud and the result is laminites. Another control on the paths of plumes is the coriolis effect (Gilbert, 1983) that can lead to an increase in sediment accumulation rates at particular sites by 2 to 3 times in high latitude fjords (Syvitski, 1989). Oceanic currents such as those moving along continental shelves are important for mud transport (Carlson *et al.*, 1975), as are eddies both in distal areas and near a glacier face.

Settling of finer particles from the plume can be enhanced by three processes: flocculation, agglomeration and pelletization (Syvitski *et al.*, 1987). Flocculation occurs by electrostatic binding of particles in brackish solutions (salinity > 3‰); agglomeration occurs due to attachment of mineral grains to organic detritus; and pelletization is the result of zooplankton egestation of particles as fecal

pellets. These particle accumulations apparently fall more slowly than sand/silt grains once released from an overflow to allow coarse and fine grained laminae to be produced (Cowan, 1988).

The rising plumes from submarine discharges can have sufficient sediment concentration that they reach a level of neutral buoyancy within the brackish water column and form an interflow (Cowan *et al.*, 1988). Increased sediment concentration may originate from rainstorm events or sudden release of sediment stored subglacially by channel migration if there is a change of head or gradient such as during submarine iceberg calving. Interflows have also been documented from delta fronts in subpolar regimes (Gilbert, 1982). Cold tongues as described from three Antarctic fjords may represent a type of interflow which forms adjacent to subglacial cavities (Domack and Williams, 1990). These processes transport significant quantities of siliciclastic particulates in suspension which may contribute to laminites within inner fjord basins (Domack, 1990) (Plate 14.6).

During winter, these discharges can shut down completely, especially from subpolar glaciers. In cool-temperate climates subglacial and some marginal streams can keep flowing, though with decreased discharge and sediment concentrations. The resulting fjord water is 45 to 170% more saline in winter and has about 96% lower sediment concentrations (Cowan, 1988).

14.5.2. Biological Production

Biological activity adds primary sedimentary particles to glaciomarine environments and they can be a major component of glaciomarine sediments where siliciclastic production is low. Organisms also speed settling rates of siliciclastic particles by pelletization and mix bottom sediment by bioturbation.

An important aspect of polar glaciomarine environments is the abundance of biogenic facies. The primary reason for this facies, particularly in the ice-proximal environment, is the limited amount of siliciclastic sediment supply to the sea. High supply rates of siliciclastic sediment dilutes any biogenic sediment produced in warm glaciomarine settings; but by contrast, bioclastic carbonate and siliceous ooze are

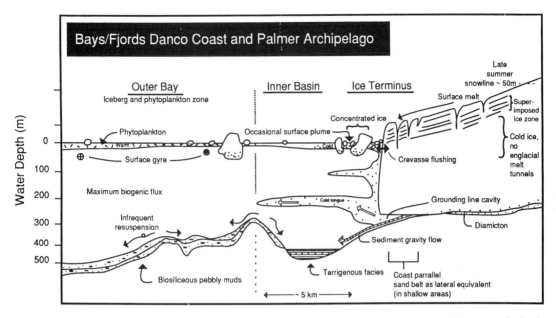

FIG. 14.9. Facies relationships representative of fjords found along the northwestern side of the Antarctic Peninsula, Gerlache Strait. The dominant source of siliciclastic sediment includes, ice-rafted debris, cold tongues (<−0.5°C), and sediment gravity flows. All of these processes are most active in ice-proximal settings near the terminus and within the inner basins. The absence of abundant meltwater prevents the formation of surface overflows and hence, restricts siliciclastic sedimentation to within the inner basins. Phytoplankton (diatoms) are most productive within the warm (>0°C) surface waters of the outer bays where iceberg and sea-ice rafting also takes place. These two processes lead to biosiliceous pebbly muds within the outer basins.

the dominant facies found in Antarctica today. Bioclastic carbonates occur as shell coquina which fringes tidewater cliff margins of the East Antarctic Ice Sheet and as localized deposits of coquina on shelf banks, such as those found in the Ross Sea (Tavianni *et al.*, 1993) and near Prydz Bay (Quilty, 1985). These sediments contain bryozoans barnacle plates, echinoids, a variety of molluscs and abundant and diverse calcareous foraminifera fauna. In more restricted settings, such as fjords along the Ingrid Christianson Coast, carbonate bioherms are actually forming at depths of 10 to 50 m. Here the largest tube worm (*Surpula narconensis*) reefs in the world are forming (Kirkwood and Burton, 1988). Absence of siliciclastic loading and a high rate of biogenic input into the fjord system were considered the key variables in allowing these carbonate reefs to form less than 10 km from the world's largest ice sheet. Other favourable conditions are a restricted entrance and strong tidal currents in Ellis Fjord that prevent icebergs and anchor ice from scouring the sea floor.

Biogenic siliceous muds and oozes are the dominant deposit within deep water basins along the Antarctic continental shelf, including deep fjords along the Peninsula (Fig. 14.9). At depths >500 m, icebergs are prevented from scouring biogenic accumulations and mixing the sea floor sediment and, as a result, diatomaceous particulates accumulate from vertical settling and dilute, sediment gravity-flow deposition. Both processes occur but the primary one is settling of phytoplankton from the pulse of productivity during austral summers. The production of organic carbon and opaline silica by diatoms in Antarctic waters has been studied for a number of years (Smith and Sakshaug, 1990). The present consensus is that productivity is highly variable both spatially and temporally. Factors which lead to higher productivity, however, are not thought to be linked to nutrient supply. Instead, physical stability of the surface layer which is formed as sea ice melts in the austral spring, is the leading variable in controlling production. Where winds and waves are at a minimum as the sea ice is melting, high rates of primary production occur (962 mgC m^{-2} day^{-1}) (Smith and Sakshaug, 1990). Hence, production is usually greatest in nearshore settings, particularly fjord or coastal embayments along the Antarctic Peninsula (3.62 gC m^{-2} day^{-1}) (Burkholder and Mandelli, 1965). The transport of this material to the bottom ultimately depends upon a number of steps as discussed by Dunbar *et al.* (1989). Oxidation and bacterial consumption takes place within the water column and reduces the C/Si ratio of suspended particulate matter (SPM) (Dunbar *et al.*, 1989). Usually resuspension of SPM takes place following deposition and this may be stimulated by periodic increases in bottom current intensity, such as occur during the winter in McMurdo Sound and the Ross Sea. Direct settling to the bottom is also important as shown by the preservation of depositional laminae of seasonal character from basins surrounding the East Antarctic margin (Domack, 1988, Fig. 4a).

In Polar-Tundra, Boreal and Cool-Temperate climates, aquatic biological activity is high, but successions are dominated by siliciclastic sediment due to higher sediment yields from glaciers. Primary productivity is not much different than in Antarctica (Smith and Sakshaug, 1990) and it is thought to be high due to high nutrient levels and well-oxygenated waters at tidewater termini caused by upwelling of deep water (cf. Apollonio, 1973). Annual or biannual phytoplankton blooms occur in fjords in Spitsbergen and Alaska and may produce annual layers in siliciclastic sequences (Sharma, 1979; Elverhøi *et al.*, 1980). In the Arctic and Boreal fjords of Baffin Island and Spitsbergen, a significant proportion of sediment transported to the sea is biogenic; and there is also biological productivity within the fjords. This organic detritus is a major energy source for benthic fauna which, in turn, play an important role in mixing, sorting and binding sediment particles, and in determining the sedimentary chemical environment (Syvitski *et al.*, 1987). Fjords open to the ocean can receive influxes of planktic forms (Elverhøi *et al.*, 1980). However, faunal diversity is commonly low (e.g. Elverhøi *et al.*, 1980) and total productivity is generally controlled by phytoplankton production and subsequent grazing and predation (Syvitski *et al.*,

1987). Phytoplankton blooms commonly occur during spring, but they are delayed until later summer at higher latitudes.

Phytoplankton are generally the largest source of particulate organic carbon in these fjords. Phytoplankton commonly fall to the sea floor in aggregates called 'marine snow' or as fecal pellets from grazing zooplankton (Syvitski *et al.*, 1987). Much of this organic carbon is consumed on its way to the sea floor, but organic carbon commonly comprises most of the carbon in bottom sediment (Syvitski *et al.*, 1990). However, due to high siliciclastic sedimentation, organic carbon is less than 0.1% of bottom sediment near tidewater termini and increases in proportion down-fjord (Syvitski *et al.*, 1990).

Although productivity is high in the Cool, Temperate Gulf of Alaska, population size is controlled by grazing of macro-zooplankton (Sambrotto and Lorenzen, 1987). Zooplankton, in biannual blooms, are uniformly distributed along the Gulf of Alaska due to strong coastal currents and the population is dominated by copepods (Cooney, 1987). Although biological diversity is low, close to Temperate tidewater termini, biological activity is high even though siliclastic sediment accumulation rates are >10 m year^{-1} (Simenstad and Powell, 1990). However, predicting a fossil record from modern settings for biofacies analysis is still not possible because of lack of data. The foraminiferal record is not well known but studies indicate relationships of benthic ecology of foraminifera and molluscs to substrate character and water depth in Alaska (Echols and Armentrout, 1980; Hickman and Nesbitt, 1980; Quinterno *et al.*, 1980; Lagoe *et al.*, 1989).

Macrofaunal communities in cool-temperate to polar-tundra climates are dominated by infaunal populations in soup- and soft-grounds and epifaunal communities on firm- and hard-grounds (e.g. Dale *et al.*, 1989; Aitken, 1990). Common fauna include polychaetes, pelecypods, gastropods, pectens, octocorals, bryozoans, hydrozoans, tunicates, porifera, echinoderms and fucoid algae (Hoskin and Nelson, 1969, 1971; Sharma, 1979; Dale *et al.*, 1989; Aitken, 1990).

14.5.3. Ice Rafting

Ice-rafted debris (IRD) originates from floating termini such as ice shelves as well as icebergs and sea ice; ice-shelf rafted (ISRD), iceberg rafted (IBRD) and sea-ice rafted debris (SIRD) will be used to specifically identify the debris source (cf. Powell, 1984). The glacial character of IRD is important to recognize because sea ice can raft debris in areas where glaciers are absent on land (e.g. Gilbert, 1990). Generally, coarser particles such as gravel and sand sizes are most often recognized as being ice rafted, although contributions of finer particles can be significant (Gilbert, 1990; Powell, 1991). Resulting sediments include discrete lonestones that may be recognized as dropstones if they deform strata (Thomas and Connell, 1985), or as outsized-clasts when their diameter is larger than the thickness of strata in which they are embedded, or in the form of clusters of particles. Other features of IRD are till pellets and mud pellets that were frozen debris encased in glacial ice (Ovenshine, 1970; Powell, 1983b).

14.5.4. Ice-Shelf Rafting

Ice-shelf rafting occurs when basal and englacial sediment is transported beyond a grounding-line by ice deforming into a floating glacier-tongue or an ice-shelf. Direct sampling and observation of this process is very difficult since the floating ice is commonly too thick to allow direct access near the grounding-line.

Debris transport and release are discussed in Section 14.3.2.2. Most particles are released by melt-out and descend to the seafloor in rock falls, grain falls or suspension settling of individual particles. There is a significant lack of particle clusters as can occur when icebergs roll (Section 14.5.5). Because layers at higher elevations within the ice move farther from the grounding-line than lower layers, as happens when you slide the top of a deck of cards, englacial debris encased high above the ice sole can be transported farther than basal debris. Consequently, on the sea floor, ISRD may show a change in provenance as well as shape (from more rounded to more angular) with distance from the grounding-line (Powell, 1984). Deposits from these processes have been termed shelfstone muds or shelfstone diamictons depending on their texture (Powell, 1984).

These remote grounding-lines were observed for the first time recently, by using a submersible, remotely operated vehicle (ROV) at a glacier in Antarctica (Powell et al., 1992). At this glacier, the process of sedimentation sea-ward of the grounding-line involved only direct undermelt of the floating glacier-tongue to produce shelfstone diamicton as glaciomarine sediment overlying subglacial (lodgement?) till (Plate 14.8). The till had ridge-and-furrow morphology and streamlined features indicative of subglacial deposition that revealed little modification after the glacier had detached from the bed. The till was immediately draped by the shelfstone diamicton. Basal debris thinned from 20 to 10 m over a down-flow distance of 1.8 km where the glacier remained in contact with the sea floor, giving an approximate rate of till deposition land-ward of the grounding-line of 0.9 cm year^{-1}. Sea-ward of the grounding-line, all basal debris had been melted out by seawater within about 2 km, giving a rate of glaciomarine sedimentation of 1.8 cm year^{-1} (Powell et al., 1992).

14.5.5. Iceberg Rafting

Rafting of glacial debris by icebergs can provide a significant volume of glaciomarine sediment and IBRD and can be used to indicate glaciation when mixed with marine sediment of a non-glacial origin. The distribution of icebergs in the circum-Antarctic and North Atlantic (Figs 14.1 and 14.10) clearly demonstrate the far-travelled character of IBRD. Hence, while local climatic conditions may not be suitable for glaciation, IBRD can be deposited with 'warm water' sediments when icebergs are transported quickly by ocean currents. Fourier shapes of silt particles can distinguish IBRD from glaciofluvial quartz grains (Cai et al., 1992) and this technique could be useful for distinguishing iceberg-rafted silt from hemipelagic muds accumulated from other sources on continental shelves.

Melt-out, sediment gravity flows and dumping are the main mechanisms of debris release from icebergs, and they contribute particles of clay to boulder size to the sea floor by rock fall, grain fall and suspension

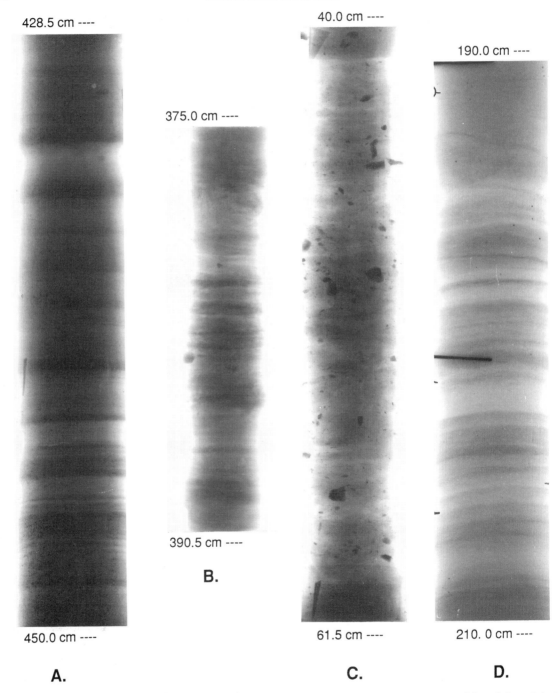

428.5 cm ----

375.0 cm ----

40.0 cm ----

190.0 cm ----

390.5 cm ----

B.

450.0 cm ----

61.5 cm ----

210. 0 cm ----

A.   **C.**   **D.**

PLATE 14.7. X-radiographs of typical Antarctic glaciomarine sediment. Sediment cores are arranged from left to right in order to demonstrate end-member and transitional compositions of siliciclastic (left) and biogenic(right) sediment. Core A (428.5–450.0 cm) is rhythmically laminated silty mud with minor ice-rafted debris from the inner basin of a fjord some 2 km from the glacier terminus; Gerlache Strait, Antarctic Peninsula. Core B (375.0–390.5 cm) is randomly bedded, sandy to silty mud with variable ice-rafted debris at the same position as core A. Core C (40.0–61.5 cm) is siliceous (diatom) pebbly mud with significant ice-rafted debris and is a good example of two-component mixing that occurs some 19 km from the glaciated coast of the Antarctic Peninsula, Gerlache Strait. Core D (190.0–210.0 cm) is a laminated biosiliceous (diatom) ooze with minor ice-rafted debris, at some 35 km from the glaciated coast of George V Land.

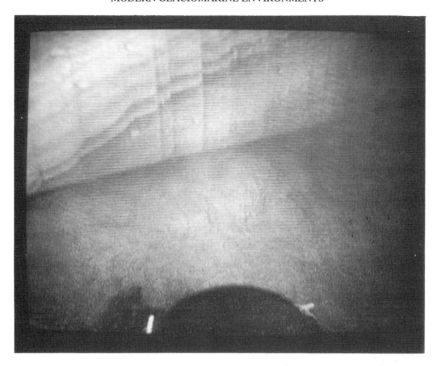

PLATE 14.8. View of a grounding line at a floating glacier-tongue of a polar outlet glacier from the East Antarctic Ice Sheet. Sediment in the foreground is a glaciomarine shelfstone diamicton formed by rainout of debris from melting out of the base of the floating tongue. The glaciomarine diamicton is draped over a stiff subglacial diamicton. In the background, beyond the grounding line which cuts diagonally across the photograph, the glacier is exposed as a vertical cliff because, here, a crevasse extended up from the base of the floating tongue. Basal debris layers dip to the left toward the base of the glacier and extend up to 20 m above the base. Vertical lines in the glacier face are runnels melted by water rising up the face in a similar way to those described for temperate termini (Powell and Molnia, 1989; reprinted by permission of Elsevier Science Publishers BV, Amsterdam). View is 3 to 4 m across.

settling. The ultimate volume of IBRD depends on the size and numbers of icebergs and also their residence times (horizontal velocity), debris concentrations, melting rates and rolling rates. Because melting rates in air are slower than those in water, icebergs eventually get top-heavy and roll. Submarine melting rates vary around and under the iceberg depending on the surface attitudes and currents (Weeks and Campbell, 1973; Gade, 1979; Russell-Head, 1980; Budd et al., 1980; Huppert, 1980). Iceberg rolling is more common for smaller icebergs than for larger tabular icebergs because the latter have a more uniform distribution of mass and they are thus more stable in waves and wind (Weeks and Mellor, 1978). Predicting rates of ice rafting is very difficult, but Drewry and Cooper (1981) and Dowdeswell and Murray (1990) have modelled

hypothetical debris-release histories for icebergs which may vary from one situation to the next.

Potentially, IBRD contributions vary with climatic and glacial regime. Large tabular icebergs produced in Greenland and Antarctica can travel to quite low latitudes (30°N, 40°S) (Figs 14.1 and 14.10) but those produced by large ice shelves contain little debris since the debris is preferentially released from the ice shelf near the grounding-line. However, reports from Antarctica of large icebergs with very thick debris layers are not uncommon. Those with higher debris contents are mainly derived from shorter floating termini and floating glacier-tongues, especially those formed by outlet glaciers (Anderson et al., 1980a; Drewry, 1986). Most debris from Antarctic icebergs is thought to be released on the continental shelf due to long residence times and recirculating drift paths of

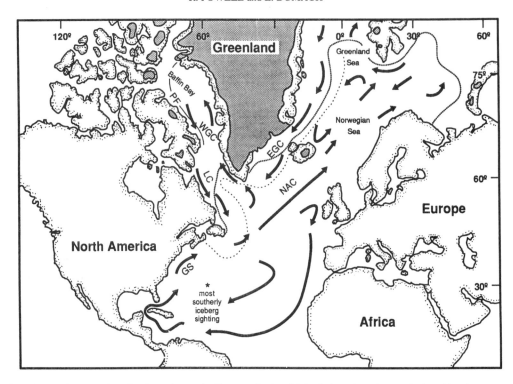

FIG. 14.10. Map of the North Atlantic region and adjacent glaciated (shaded area) and non-glaciated land masses. Dashed line in ocean is maximum limit of seasonal pack ice while dotted line is southern extent of drifting icebergs. Solid line in northern Baffin Bay is the location of the polar front. EGC = East Greenland Current, GS = Gulf Stream, WGC = West Greenland Current, LC = Labrador Current, NAC = North Atlantic Current.

icebergs offshore (Ørheim, 1980; Vinje, 1980; Keys, 1984). Occasionally, icebergs collect in restricted areas because sea ice prevents them drifting after they calve. These iceberg tongues have tabular icebergs separated by lower-lying fast ice. In the North Atlantic large icebergs from Greenland are transported far south by the Labrador Current and can cause problems to shipping lanes (e.g. the 'Titanic' disaster).

Tidewater termini produce smaller icebergs than floating termini and with iceberg transport lengths of only a few tens of kilometres in cool-temperate climates. In boreal and polar-tundra climates these distances can be much longer (Robe, 1980). Because of the rapid melting rate and the coincidence of calving and grounding-lines for tidewater termini, roll-dump structures are common. These structures can be used to infer this environment rather than a floating terminus at a grounding-line since dumping

of large amounts of debris en masse is not likely to occur from the base of an ice shelf (Powell, 1984, 1991).

In iceberg zones of temperate termini, icebergs may be concentrated by various processes (Section 14.5.6); but even when iceberg flux through the ice proximal area is high, the area may remain totally covered with icebergs if their production rate is sufficiently high. High glacier flow speeds and calving speeds can be maintained even under cool-temperate climates, especially during rapid grounding-line retreat. In some situations, rates of IBRD production may be predicted to be high near a tidewater cliff and in others, rates could increase slightly farther away. The resulting deposit on the sea floor depends on the relative rates of IBRD and mud or biogenic production; in other words the problem can be modelled as two component mixing.

14.5.6. Two Component Mixing

'Two component mixing' refers to the admixture of sediment particles derived from processes which are sufficiently distinct that they can be recognized by the character of the particles they contribute to a deposit. The two most prevalent in the glaciomarine environment are ice rafting and various types of current derived sediment, such as that produced by vertical settling from suspension in low density turbidity flows and bottom currents. The coarser (gravel and sand) ice-rafted component can usually be recognized by poor sorting of the glacially transported debris. However, ice-rafted silt and clay is more difficult to discern from hemipelagic particles in bottom sediment. Current-derived components can be expected to have undergone some sorting during the transport and depositional processes and can often be distinguished by stratification and by modal sizes within the sediment. It is possible to distinguish sediment from different sources by separating particle-size populations using detailed size analysis (e.g. Piper, 1976; Kelly, 1986). Shapes of quartz silt particles can be used to distinguish IBRD from silt from other sources (Cai *et al.*, 1992) (Menzies, 1995b, Chapter 13). Depending upon the relative contribution of the sediment components, various sediments are produced ranging from pebbly muds (i.e. compound glaciomarine sediments (Anderson *et al.*, 1980b) or bergstone muds (Powell, 1984)) to diamictons. In temperate and subpolar settings the current-derived component is often supplied by meltwater discharges along or adjacent to glacier termini or deltaic systems.

The manner in which texture of bottom sediment changes can be qualitatively inferred by superimposing a curve of rates of IBRD production onto those for settling of sediment from overflow plumes (Fig. 14.11). At McBride Glacier, Alaska, icebergs melt within a few kilometres of a grounding-line and thus IBRD production is 'high' (Powell, 1991). IBRD volumes become significant with respect to plume deposits beyond approximately one kilometre of the grounding-line (Fig. 14.11) and a bergstone diamicton may accumulate beyond a more proximal bergstone mud zone. That yield includes all sizes of IBRD but, when only the gravel component

of IBRD is considered, the IBRD percentage in bottom sediment is much lower (Fig. 14.11). Therefore, the lonestones which are easily observed in outcrops or cores, are only a small proportion of IBRD that a unit may actually contain.

In most temperate and subpolar regimes fluvial sediment production will dominate iceberg rafting and bergstone muds accumulate. If icebergs are concentrated for any reason, then the proportion of IBRD can increase to produce a bergstone diamicton interfingering with bergstone mud. Circumstances in which a bergstone diamicton may form are, for example, when: (1) meltwater discharge is low due to a cold glacial regime, (2) seasonal discharge decreases but iceberg rafting continues, (3) there is a very high, continuous flux of icebergs in proximal areas, and (4) icebergs occur in a continuous flux in more distal areas where rates of accumulation of marine rock flour have decreased exponentially. Icebergs may be concentrated or have a high continuous flux in situations such as: (1) in constricted basins (e.g. trapped by sills) (Plate 14.9), (2) in oceanic eddies, (3) during catastrophic retreats, (4) during storms blowing icebergs onshore, and (5) during periods when sea ice freezes-in high concentrations of icebergs. Rates of accumulation of the two sediment source-components will probably differ in fjords and on continental shelves because the latter are more open systems.

In polar environments a third, biogenic, component is involved. Because of the limited role of meltwater sedimentation, dropstone diamictons are found in a variety of situations such as near ice shelf grounding-lines and bays with restricted circulation. In general, sediment accumulation rates again decrease from the termini of valley glaciers. This allows for the accumulation of biosiliceous pebbly muds, mixtures of ice rafted and biogenic sediment derived from phytoplankton (Fig. 14.19).

In some areas, coarse IBRD and bioclastic debris are concentrated without a fine-grain component. These deposits have been termed residual glaciomarine sediments (Anderson *et al.*, 1980b) and occur in nearshore areas or shallow banks where marine currents are strong enough to stop fine-grained sediment from accumulating and bioturbation brings finer sediment to the sediment surface where it is removed by bottom currents.

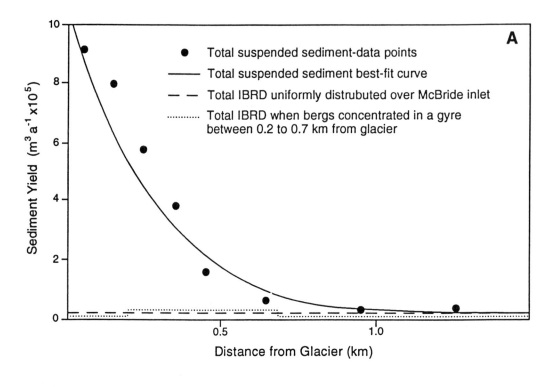

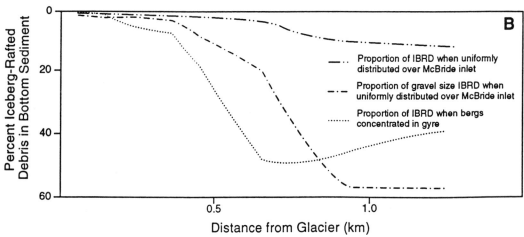

FIG. 14.11. Sediment yields (A) and proportions IBRD in bottom sediment (B) with distance from the tidewater terminus of McBride Glacier, Alaska. In the top plot (A), fields of marine rock flour that settles from suspension are from Cowan and Powell (1991), and IBRD yields are data from Gottler and Powell (1990). In the bottom plot (B), proportions of IBRD in bottom sediment are presented when (1) the total IBRD is distributed uniformly over McBride Inlet, (2) the gravel fraction (taken as the 15% average) of IBRD is distributed uniformly over McBride Inlet, and (3) the total IBRD is concentrated by bergs being in a gyre between 0.2 to 0.7 km from the glacier terminus (after Powell, 1991). (Reprinted individual figures by permission of the author.)

PLATE 14.9. Calving bay of the Columbia Glacier, Alaska (July, 1989). The bay is choked with recently calved icebergs which are prevented from drifting out of the bay by a shallow sill. The concentration of floating icebergs covers 100% of the bay's surface and results in conditions that would be the closest thing to a temperate ice shelf. Water depths within this iceberg-covered bay are up to 380 m deep (Krimmel and Meier, 1989). The dotted line indicates the water line against the grounded tidewater cliff.

In the absence of significant meltwater, most siliciclastic sediment is introduced either as IRD or as sediment gravity flows that are generated by instabilities associated with grounding-line processes (calving, ploughing, etc.). Because of the influence of dilute sediment gravity flows, basin topography plays a major role in the distribution of sediment facies within Antarctic fjords. Silled basins retain siliciclastic sediments within ice-proximal areas and prevent dilution of biogenic detritus within the outer fjord basins (Plate 14.8). Fjords which lack sills contain a more transitional facies gradient between ice-proximal siliciclastic sediment and distal biogenic and ice-rafted deposits (Fig. 14.9).

14.5.7. Sea-Ice Rafting

Formation of sea ice most commonly occurs when air temperatures drop below the freezing point of marine or brackish water. Its formation expels salt and produces brines that form gravity flows that sink down through the water column due to their density and cold temperatures (Lewis and Perkin, 1982). Sea ice can also form under cool-temperate climates where during winter in protected areas heavy snow falls produce a slush of ice crystals in surface water which then freezes as temperatures drop. One problem still remaining is to distinguish SIRD deposited under glacial conditions from that deposited where sea ice forms without glaciers on land (e.g. Reimnitz and Kempema, 1988; Gilbert, 1990).

Sea ice is important to glaciomarine sedimentation because it (1) damps waves (especially during winter storms), (2) changes water column structure, (3) influences biotic productivity, (4) traps icebergs, (5) transports sediment, (6) scours and turbates bottom sediment, (7) deforms shoreline sediment, and (8) reduces calving rates. Sediment can be incorporated into sea ice by (1) freezing of an ice foot, (2) anchor-ice lifting, (3) trapping of sediment during frazil ice/slush ice formation, (4) bottom erosion, (5) stream wash overs, (6) wave wash overs, (7) offshore aeolian transport, and (8) rock falls and slides (see Table 14.1) (e.g. Gilbert, 1983; Reimnitz et al., 1987; Kempema et al., 1989).

SIRD is being increasingly considered as an important component of glaciomarine sediments in distal areas or under colder conditions where other siliciclastic input is low (Clarke et al., 1980; Clarke and Hanson, 1983; Barnes and Reimnitz, 1982; Barnes et al., 1982a; Barrett et al., 1983; Reimnitz and Kempema, 1988; Elverhøi et al., 1989). Sea-ice rafting is important in Arctic seas because of the shallow depths of broad continental shelves and a coastal environment that includes deltaic and wave-dominated systems. Near deltaic systems sea ice has coarse debris incorporated from the sea floor during pressure ridge accretion while waves supply a great deal of fine detritus that is incorporated during freeze-up. Much of the sediment in Arctic sea ice is fine-grained and derived from the littoral zone. Concentrations vary between about 0.05 to 3000 mg l^{-1} (Campbell and Collin, 1958; Barnes et al., 1982a; Elverhøi et al., 1989) but few studies have been conducted. Conditions of debris transport and release are similar to those discussed for icebergs; however, in general, in any one climatic regime icebergs tend to survive longer than sea ice. Antarctic sea ice can be the dominant source of poorly sorted, coarse debris within fjords along the Antarctic Peninsula (Fig. 14.9). Here, rock falls onto fast-ice are common from the rugged cliffs and valley sides during the austral spring. Break-up into pack ice results in the transport of the debris out into fjords during the late spring and early summer. In areas of supercooled bottom water, anchor ice formation can pluck biogenic and coarse detritus from the sea floor. This debris is frozen to the bottom of overlying ice shelves or pack ice and is later rafted further out to sea (Kellogg and Kellogg, 1988). All of the shallow coastal zone surrounding Antarctica (from 0–10 m water depth) has been plucked clean by anchor ice formation (Picken, 1985). The exceptions to this are areas of strong tidal currents. Aeolian transport of fine-sand onto pack and even shelf ice is also significant, especially adjacent to the Dry Valleys along the southwest margin of the Ross Sea (Barrett et al., 1983; Bartek and Anderson, 1991).

14.5.8. Transitional Environments

Transitional environments between terrestrial and fully marine environments mainly include beach, delta, estuarine and tidal flat systems. In some areas these systems are quite important in producing a sedimentary record as they are large depocentres. They can also provide important data for inferring climatic conditions, for example, large deltas are not produced in polar climates but can be significant in cool-temperate and boreal climates. The presence of ice in coastal settings can produce a distinctive record. For example, when grounding-lines of tidewater cliffs are directly at sea level, very gravelly, poorly sorted beaches are formed (Znachko-Yavorskiy, 1964; Sugden and John, 1976); icebergs and sea ice leave deposits such as boulder lines, and deform intertidal sediment into gouges, wallows and push ridges (Barnes, 1982; Barnes et al., 1982a; Reimnitz and Kempema, 1982; Dionne, 1987; Dionne and Brodeur, 1988).

Beaches. In cool-temperate climates, beach sediments range from gravelly to sandy and include all of the features of non-glaciomarine beaches with evidence of ice action superimposed (e.g. Molnia and Wheeler, 1978; Hayes and Michel, 1989). Beach ridge systems with associated aeolian dunes and various forms of spits are common, as are barrier island systems associated with large deltas.

In contrast, beach sediments in more extreme climates tend to have particles that are slightly angular and moderate to poorly sorted, probably due to waves being damped by sea ice (e.g. Powell, 1981b). These beaches have superimposed ice-push ridges and evidence of periglacial action (Nichols, 1961, 1968; Dubrovin, 1979; Gregory et al., 1984). Antarctic beaches are rare and consequently are often sites for penguin rookeries. Because of the penguin activity, old raised beaches have sorted gravels (penguin nests) and penguin bones and mummified carcasses (Whitehouse et al., 1989).

Deltas and estuaries. In order that deltas form, fluvial discharges have to be moderately large. In general, streams in polar ice cap climates do not have high discharges (typically <2 m^3 s^{-1}) nor do they carry high sediment loads (Chinn and Oliver, 1983; Howard-Williams and Vincent, 1986; Gallagher and Burton, 1988; Mosley, 1988). Small glaciofluvial systems that do exist along portions of the Dry Valleys are ephemeral and have sediment yields of 5.9 t km^{-2} year^{-1}; this is two orders of magnitude less than glacial outwash systems of the Arctic alpine regions (Mosley, 1988). The Antarctic fluvial systems have extensive algal mats and filaments and suspended sediment loads typically <10 mg l^{-1} (Howard-Williams and Vincent, 1986). Consequently, with low fluvial sediment delivery, deltas in these conditions are minor and texturally deltaic beach and fluvial sediments are all similar (Powell, 1981b). Certainly, estuaries, lagoons and marshes are not formed.

Large deltas such as the Yukon Delta (Dupre and Thompson, 1979) can form in slightly warmer climates, and the full range of deltas, estuaries, lagoons and marshes form in temperate coastal climates (Hayes and Michel, 1989; Powell and Molnia, 1989). In these latter conditions, fluvial discharges and sediment loads may be very high; for example, peaks of 7×10^3 m^3 s^{-1} and a total sediment production of 1.07×10^{11} kg year^{-1} are reported for the Copper River, Alaska (Reimnitz, 1966). Sedimentary processes and facies at the river mouth and on the delta front are similar to non-glacial situations except for the influence of fluctuations in discharges and indications of ice action; deltas are generally fan deltas (n.b. Syvitski and Skei, 1983; Hayes and Michel, 1989). These deltas can also develop from grounding-line fans once they build to sea level (Powell, 1990). One characteristic that may distinguish fjords headed by grounding-line fans from those headed by deltas is turbidity current channels (Carlson *et al.*, 1989). Both systems have slides/slumps and sediment gravity flows (cf. Prior and Bornhold, 1989; Powell and Molnia, 1989), but only once the gradient of the foreslope increases to that of a delta can turbidity currents build sufficient energy to be erosive on the fjord floor and cut major channels (Carlson *et al.*, 1989). Sediment transport in the buoyant overflow plume has been modelled by Syvitski *et al.* (1988) (Sections 3.4 and 3.5.1).

Grounding-line fans may aggrade and prograde to form ice-contact deltas and, in the transition, when the delta plain is fully intertidal, channels change location during tidal cycles (Powell and Molnia, 1989). Large volumes of sediment may be removed from the delta by this process to produce a pulse of coarser sediment to the sea floor with ebbing tide. A more mature delta can produce laminites with a spring tidal periodicity by tidal draw-down funnelling coarser sediment over the delta brink at low tide (Smith *et al.*, 1990).

Tidal flats. Tidal flats are not common in polar ice cap climates today, mainly because of narrow, high gradient, coast lines in Antarctica. Where low gradient intertidal areas do exist, they are typically the sites of boulder pavements, depending on the onshore movement of floating ice (Hansom, 1983). In warmer climatic conditions, tidal flats can be significant systems with sedimentary characteristics similar to non-glacial tidal flats except for indications of icebergs and sea ice (Bartsch-Winkler and Ovenshine, 1984; Bartsch-Winkler and Schnell, 1984). Some indications of ice are quite distinctive, for example, isolated boulders and boulder barricades, ice wallows and other keel marks and striations and pavements (Dionne, 1985; Aitken *et al.*, 1988; Aitken and Gilbert, 1989).

14.6. OTHER NON-GLACIAL AND MODIFYING PROCESSES

Some non-glacial processes have already been discussed such as sea ice and biological productivity. However, there are several other aspects that deserve to be mentioned.

14.6.1. Hydrographic Processes

In addition to controlling ice melting rates, iceberg paths (Figs 14.1 and 14.10) and paths of turbid freshwater plumes, oceanic currents are important in modifying and redepositing sediments on the sea floor. Hydrography of continental shelves and shallow seas is very site specific in regard to details of circulation patterns, current strengths and whether the environment is dominated by tides or storm waves. Some generalities can be made.

Along-shelf currents commonly transport sediment at velocities that reach several tens of centimetres per second. Those are often geostrophic flows following salinity gradients established by high freshwater influx at the cool-temperate coast and are enhanced by net wave propagation (Powell and Molnia, 1989). These along-shelf currents are capable of transporting sediment to low areas on the shelf. If the lows are glacially-eroded troughs that cross the shelf, then sediment can be subsequently transported offshore to deeper water (tunnel valleys, Menzies, 1995b, Chapter 2). This enhances hemiplagic sedimentation off glaciated coasts.

Dense water can be produced nearshore during winter cooling and, when combined with wind-driven down-welling, can form offshore, gravitationally convective flows. This effect is enhanced by sea ice formation. Alternatively, during summer, strong katabatic winds, flowing from glaciers offshore, cool surface water as well as cause upwelling of deeper water along the coast.

Under Polar Ice Cap climates, the overriding factor of continental shelf hydrography is heat exchange between the atmosphere–ocean and ice–ocean, especially with sea ice formation, but locally with ice shelves. The net result is off-shelf transport and downward salt flux (Jacobs, 1989; Foldvik and Gammelsrod, 1988). These processes drive vertical convection and 'ventilated' dense water across continental shelves and eventually form the ocean bottom waters (Coachman and Aagard, 1974; Jacobs, 1989) which create most of the turnover in deep areas of modern oceans.

Storms are common in high latitudes and waves so propagated may reach over 40 m high. Waves cause sediment erosion and major reworking episodes to alter the original glaciomarine character. They also promulgate mass failure of sediment on shelves. Wave effects are damped by sea ice or high concentrations of icebergs. In fact, storms are one way of concentrating icebergs in local areas, especially at the heads of fjords and in bays.

Tidal currents produce local high frequency fluctuations in near-bottom sediment concentrations that dominate some shelves. Like waves, tidal currents rework glaciomarine sediment, but their deposits will differ.

Shallow water environments of Polar Ice Cap climates are greatly affected by currents of wind or tidal origin, especially bordering ice piedmont settings. Low siliciclastic sediment supply at the calving line allows such currents to effectively sort and transport sediment into broad aprons of sand rich in bioclastic detritus.

Fjords have complex circulation patterns that are strongly influenced by bottom topography. If fjords do not have entrance sills, then waves and tides on a continental shelf propagate into fjords and can be major influences in shallow fjords. Sills significantly influence fjord circulation because deep water renewal is controlled by sill height. Deep water renewal occurs by gravity currents as the density of water outside the sill becomes greater than bottom water of the fjord basin and so flows down into, the basin. Velocities range from 20 to 70 cm s^{-1} (Syvitski et al., 1987). These flows stop anoxia and transport sediment out of, or into a basin (Winters et al., 1985). Tidal currents flowing over sills make sill surfaces sites of erosion and perhaps high biologic activity. Sills form internal waves at the base of the fjord surface layer that may propagate long distances into the fjord and effect the turbid overflow plume. Tidal currents may also scour and redeposit sediment from fjord walls down to the basin floor (Syvitski, 1989) and prevent anchor ice formation. This stratification breaks down during summer as icebergs and pack ice are blown into and out of fjords by winds and storms.

A common characteristic of fjords is to have a maximum homogeneity of water masses during winters by low freshwater input, thermohaline convection due to surface cooling and/or sea ice formation and deep-water renewal (Syvitski et al., 1987). Major stratification within a fjord water column is generated in summer and autumn by surface freshwater flowing seaward as a barotropic flow from streams either at deltas or tidewater termini. Katabatic winds enhance this flow. To balance the hypopycnal outflow, an intermediate extrabasinal water layer either from the continental shelf or the next basin downfjord flows in as a baroclinic current. In polar fjords along the Antarctic Peninsula, stratification is greatest during the spring as sea ice melts and contributes to a low salinity surface layer.

14.6.2. Sediment Gravity Flows

Sites for generating sediment mass movements are depocentres such as grounding-lines or deltas where high sedimentation rates form unstable slopes that collapse. Sediment gravity flows from these depocentres may have short run-out distances because of either significant bottom relief produced by glacial erosion especially in fjords that trap flows, or the continental shelf sloping toward the continent due to either glacial loading (Anderson *et al.*, 1983) or preferential glacial erosion which also traps flows, or a low slope of the continental shelf on which flows rapidly slow. However, events other than high sedimentation rates, such as storm waves, biogenic gas and earthquakes may result in sediment instability and consequent mass movements in subaquatic proglacial areas. Side-entry inputs are important additional sources in fjords.

Processes of sediment redistribution in proglacial regimes are the same as in non-glacial marine settings and large volumes of sediment can be involved (e.g. Kurtz and Anderson, 1980; Wright and Anderson, 1982; Wright *et al.*, 1983; Vorren *et al.*, 1989). Significant features specific to glaciomarine settings are, first, the possibility of producing mass movement deposits on continental shelves on a regular and perhaps rapid basis due to the glacier being on the shelf. In contrast, non-glacial shelves have their sediment sources usually some distance from sites of deposition except during sea level changes. Second, during glacial maxima, glaciers may often extend to the edge of the continental shelf so that all sediment is released directly down the continental slope to form thick wedges of mass flow deposits with glacial signatures such as striated clasts (Hill, 1984; King and Fader, 1986; Piper and Sparkes, 1987; Bonifay and Piper, 1988; Stoker, 1988; Vorren *et al.*, 1989) (Menzies, 1995b, Chapter 5).

14.6.3. Iceberg and Ice Keel Scouring and Turbation

Plough and furrow marks are common on high latitude shelves in northern (Fader *et al.*, 1982; Reimnitz and Kempema, 1982; Vorren *et al.*, 1983; Barnes *et al.*, 1984) and southern hemispheres (Barnes, 1987; Barnes and Lein, 1988). In fact, turbation of bottom sediments can be so intense that primary stratigraphy and biotic communities may be destroyed (Elverhøi, 1984; Gallardo, 1987) and/or remoulded into diamictons (Vorren *et al.*, 1983). Gouges and wallows can also be produced in intertidal areas (cf. Reimnitz and Kempema, 1982; Hansom, 1983) (Menzies, 1995b, Chapter 4).

The depth of iceberg/sea floor interaction in the Antarctic is dependent upon the thickness of tabular icebergs. Studies suggest that the draft of Antarctic icebergs is limited to about 400 m with some non-tabular icebergs as great as 500 m. A variety of iceberg scours, prods and gouges are revealed by side-scan sonar studies on portions of the Antarctic shelf as deep as 500 m (Barnes and Lein, 1988). The effect of this process is to produce an iceberg turbate sediment from relict deposits which lie above 500 m on the shelf. Such deposits would most likely consist of poorly stratified diamictons with convoluted and deformed interbeds of sand or siliceous mud. Fossils would be represented by mixtures of recent and relict assemblages (Menzies, 1995b, Chapters 4 and 5).

14.6.4. Production and Accumulation of Organic Carbon

The content and character of organic carbon in sediments are perhaps one of the more important differences between polar and temperate glaciomarine facies, particularly within mud-dominated facies. Recent studies within shelf and fjord basins in Antarctica have demonstrated that total organic carbon contents within biosiliceous muds can be as high as 3–4% of the total sediment weight (Dunbar, 1988) and up to 6% of the mud fraction (Domack, 1988). Typically, the organic carbon values range between 1–1.5%. More important, however, is the fact that almost all of the carbon is of autochthonous marine (algal) origin. The preserved flux of relatively unstable organic matter in circum-Antarctic basins ranges from 22.8 to 6.45 gC m^{-2} year^{-1} which compares to 4.2 gC m^{-2} year^{-1} in Baffin Island fjords (Syvitski *et al.*, 1990).

The higher Antarctic rate is related to three factors. First, the polar climate inhibits siliciclastic sedimentation and limits dilution of the organic matter. Second, locally high productivity contributes to organic matter loading to the bottom. Third, the cold bottom water temperatures ($<-1.5°C$) limit bacterial metabolism within and along the sediment–water interface. Although bottom waters are fully oxygenated within Antarctic shelf basins, subsurface conditions within the sediment are anoxic, further limiting consumption of the organic matter. Associated authigenic minerals include abundant framboidal pyrite and, less commonly, calcium hexahydrite (ikaite) (Suess *et al.*, 1982).

The biosiliceous mud/ooze facies is widespread along the Antarctic continental shelf within basins >500 m water depth. This has led several authors to suggest that the preservation of both organic carbon and opaline silica within these sediments has an effect upon global budgets for carbon and silica (Ledford-Hoffman *et al.*, 1986; DeMaster *et al.*, 1988; Dunbar, 1988).

In contrast, temperate and subpolar glaciomarine muds are typically poor in organic carbon, with typical values of <1% (Andrews, 1987d; Syvitski *et al.*, 1990). When Total Organic Carbon (TOC) content exceeds 1%, the organic matter is invariably detrital or reworked, being derived from stable refractory sources such as bedrock and terrestrial vegetation. The reasons for this are related to climate and high rates of siliciclastic sedimentation associated with meltwater input. Restricted productivity due to limited sea ice may also be a contributory factor to low indigenous carbon contents. Warmer water temperatures would also allow for higher rates of bacterial metabolism.

14.6.5. Bioturbation

Infaunal organisms disturb sediment and progressively destroy sedimentary structures with increasing intensity away from a glacier (Gilbert, 1982; Dale *et al.*, 1989). Bioturbation may also allow winnowing of finer sediment (Singer and Anderson, 1984) and may homogenize sediment into a diamicton. The best potential for preserving fine sedimentary structures is where sedimentation rates are high and infaunal populations are consequently absent or low.

14.6.6. Aeolian Sources

Offshore transport of sediment by wind can be locally important in high latitude polar ice cap and polar-tundra climates (e.g. Gilbert, 1982; Barrett *et al.*, 1983). Fine sand and silt is transported to the sea directly, or via sea ice. In more temperate climates, wind also transports sediment offshore (Post, 1976; Molnia, 1990) but it is heavily diluted by sediment from other sources.

14.7. SEDIMENTATION RATES AND FLUXES

Potential siliciclastic sediment yields from glaciated basins decrease with decreasing size of the glaciated basin assuming the following conditions: (1) uniform subglacial sediment/rock types, (2) constant subglacial debris conditions (erosion-transportation-release) during glacial shrinkage, and (3) minimal sediment storage (probably true for marine-ending glaciers except during glacial minimum conditions where extensive land areas are exposed in a drainage basin). A logarithmic decrease in sediment yield with decreasing drainage basin area has been documented (Fig. 14.12) for the retreat of the temperate Muir Glacier, Alaska (Powell, 1991). The sediment yield curve is considered to be a continuous function reflecting rates of glacial erosion and flow.

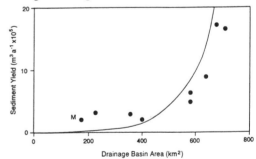

FIG. 14.12. Relation between sediment yields and drainage basin area for a temperate marine-ending glacier. Sediment yields in ice-contact basins (fjord floor basins in which a glacier-terminates) are estimated from seismic profiles from Molnia *et al.* (1984) averaged over the period of time Muir Glacier, Alaska, terminated in or at the head of each basin. Drainage basin areas of Muir Glacier, as determined from Brown *et al.* (1982), become smaller during glacial retreat. Curve is a best fit logarithmic function. M is the yield from present McBride Glacier determined from modern sedimentation rates and is consistent with yields estimated for Muir Glacier from seismic profiles (after Powell, 1991; reprinted with permission of the author).

Rates of glacial sediment accumulation on the sea floor depend on the processes of debris release from the glacier (Fig. 14.13). Accumulation rates are a function of: debris content of the glacier and rate of meltout by sea water; subglacial squeeze/push; sediment discharge from basal/subglacial streams; hemipelagic/pelagic suspension settling from meltwater discharges; debris fall from floating ice; marine hydrographic processes for length of transport and washing and winnowing of fine particles; and sea floor processes involved with mass movement. Of these, sedimentation rates from suspension settling are easiest to measure; very few estimates of other processes have been made. An estimate of relative significance of these processes in a temperate situation is that melting of basal debris produces 0.4% of total sediment yield, stream bedload plus mass-flow 30.6%, and suspension settling 69% (Powell and Molnia, 1989). Estimates of annual increases in volume of grounding-line systems in these temperate conditions are of the order of 10^6 to 10^7 m^3 year^{-1} (Powell and Molnia, 1989; Powell, 1990, 1991).

As expected, rates of sedimentation of hemipelagic siliciclastic glacial sediment logarithmically decrease away from grounding lines (Elverhøi et al., 1980, 1989; Andrews et al., 1985; Görlich, 1986; Andrews, 1987d; Syvitski, 1989; Vorren et al., 1989; Cowan and Powell, 1991; Domack et al., 1991b). The low production of siliciclastic sediment in polar ice cap climates means that sedimentation rates become more a function of biosiliceous productivity and dissolution rates. Surface productivity, in turn, is controlled by sea ice fluctuations and subsequent surface layer mixing.

Table 14.3 lists some recent estimates of sedimentation rates for modern Antarctic glaciomarine sediments. Though far from representative of all depositional environments, especially ice proximal settings, the data do give some basis for comparison to other glaciomarine sequences from cool temperate zones. Sediment accumulation rates vary from 3.4 to 0.5 mm year^{-1} over two orders of magnitude. Most rates range between 0.3 to 3.0 mm year^{-1}. Because most of these sediments are biosiliceous muds or oozes the rates of

siliciclastic sediment accumulation are approximately 70 to 50% of those listed in Table 14.2. The overall low sedimentation rates, however, emphasize the low rate of siliciclastic sediment supply and the absence of extensive fluvial drainage systems.

Another way to measure sedimentation is by determining preserved particle flux which is calculated using sediment bulk density and ^{14}C (long-term) or ^{210}Pb (short-term) chronologies (Table 14.3). Results in the Antarctic demonstrate an order of magnitude variation between 0.471 kg m^{-2} year^{-1} (Amery Trough, Prydz Bay) to 2.285 kg m^{-2} year^{-1} (Andvord Bay, Gerlache Strait, Antarctic Peninsula). In the Ross Sea, the total siliciclastic particle flux ranges from 0.619 kg m^{-2} year^{-1} (Sulzburger Bay) to 0.741 kg m^{-2} year^{-1} (southwest Ross Sea) (Ledford-Hoffman et al., 1986). These results compare to total particle flux values in the Arctic of 0.060 kg m^{-2} year^{-1} (southeast Greenland Shelf), 0.60–1.30 kg m^{-2} year^{-1} (Baffin Island fjords) to 0.40 kg m^{-2} year^{-1} (Baffin Island, shelf trough) (Andrews and Syvitski, 1991).

Total particle flux in the temperate-oceanic fjords of Alaska are several orders of magnitude greater than those of subpolar and polar climates mentioned above (Table 14.4). The results of Powell and others are taken from sediment traps and, therefore, would be expected to represent upper limits to sedimentation when compared to preserved flux as determined by radiogenic isotopic studies of sediment cores. Regardless, the contrasts are extreme and demonstrate the proficiency by which temperate glacial regimes supply sediment into the marine realm.

In order to obtain a reliable comparison of sedimentation rates in different climatic regimes, Cowan and Powell (1991) restricted the data set to those studies that obtained measurements from settling tubes in modern fjord settings and close to grounding lines (Table 14.5). The results show a dramatic 3 to 4 orders of magnitude difference between Temperate and Polar settings, and 1 to 2 orders of magnitude difference between both Temperate and Subpolar settings, as well as Subpolar and Polar settings.

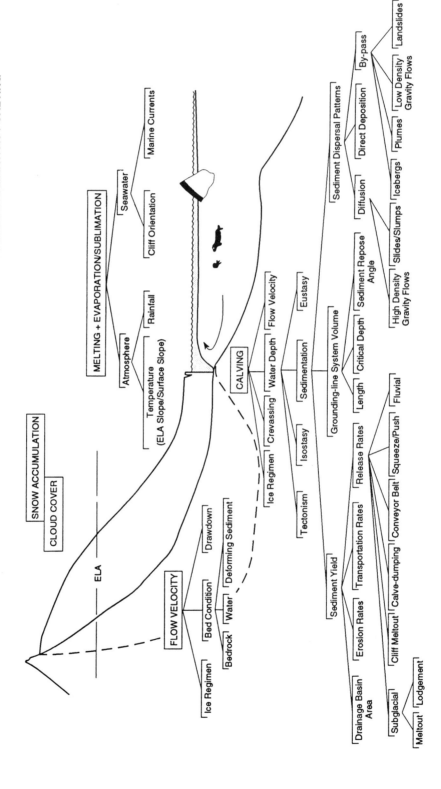

FIG. 14.13. Factors controlling mass balance of glaciers, with tidewater termini emphasizing sedimentological controls at grounding lines (not to scale). Many factors are dependent variables, but feedback loops are not shown for simplicity (after Powell, 1991; reprinted with permission of the author).

TABLE 14.3. Sediment Flux in Glacio-marine Environments

Region Setting	Distance from Glacier (km)	Total Flux (kg m^{-2} a^{-1})	Terrigenous Flux (kg m^{-2} a^{-1})	Biogenic Flux (kg m^{-2} a^{-1})
West Antarctica				
Peninsula Fjords*	5 km	2.285	n.a.	n.a.
	16 km	1.219	n.a.	n.a.
Ross Sea**				
SW Basin	n.a.	1.010	0.741	0.268
Sulzburger Bay	n.a.	0.633	0.619	0.013
East Antarctica				
Amery Trough***	70 km	0.471	0.118	0.353
Mertz-Ninnis Trough#	50 km	1.089	0.033	1.056
South East Greenland@				
Shelf Trough	n.a.	0.060-0.080	n.a.	n.a.
Baffin Island, Canada@				
Fjord	n.a.	0.600-1.300	n.a.	n.a.
Shelf Trough	n.a.	0.400	n.a.	n.a.
Shelf	n.a.	<0.050	n.a.	n.a.
Alaska, Glacier Bay†				
Fjord	0	490 x 10^6		n.a.
	.06	430 x 10^6		n.a.
	.1	180 x 10^6		n.a.
	.8	80 x 10^6		n.a.
	1.0	28 x 10^6		n.a.
	2.75	14 x 10^6		n.a.
	4.5	7.4 x 10^6		n.a.
	8.2	3.7 x 10^6		n.a.
	11	2.2 x 10^6		n.a.
	14.5	0.17 x 10^6		n.a.

*Domack *et al.* (in preparation)
**Ledford-Hoffman *et al.* (1986)
***Domack *et al.* (1991)
#Domack *et al.* (1989)
@Andrews and Syvitski (1991)
†Cai and Powell (unpublished data)

TABLE 14.4. Comparison of Holocene ice-proximal sedimentation rates in fjords or restricted basins (after Cowan and Powell, 1991).

Location	Rate	Distance from Grounding Line	Climatic Classification	Reference
McBride Inlet: Southeast Alaska (tidewater terminus)	200-2000 cm/a	within 1 km	temperate	Cowan and Powell, 1991
Hornsund: West Spitsbergen (tidewater terinus)	10-25 cm/a	within 10 km	subpolar	Görlich, 1986
Kongsfjorden: West Spitsbergen (tidewater terminus)	5-10 cm/a	within 10 km	subpolar	Elverhφi et al., 1983
Mackay Glacier: East Antarctica (floating glacier-tongue)	1.8 cm/a	within 2 km	polar	Powell et al., 1992
Prydz Bay: East Antarctica (inferred sub-ice-shelf or floating glacier-tongue)	0.21 cm/a	within 10 km	polar	Domack et al., 1991

14.8. ADVANCE AND RETREAT OF MARINE-ENDING GLACIERS

Stability of marine-ending glaciers has been undergoing investigation since Mercer (1978) suggested that potential climatic warming (from greenhouse gases) may cause the West Antarctic Ice Sheet to disintegrate rapidly and cause a rise in global sea level.

While stability of marine tidewater termini, ice shelves and terrestrial glacial termini is ultimately a product of mass balance, other variables influence stability and hence the advance and retreat of marine termini (Fig. 14.22). Furthermore, some of these variables involve feedback mechanisms that are more complex in marine systems than in terrestrial systems. As with terrestrial termini, glacial mass balance for marine-ending termini is a function of snow accumulation rate, ablation rate, and glacial flow velocity. Considerations of accumulation rates are the same as for terrestrial systems (Chapter 4).

Non-climatic responses of marine termini have been noted for Pleistocene ice sheets (Mercer, 1968;

Hillaire-Marcel et al., 1981). Retreat of some temperate tidewater termini has been documented historically, and some, after reaching a position of maximum retreat, have stabilized or readvanced. Readvances have been attributed to: (1) decrease in size of the ablation area (Mercer, 1961; Post, 1975; Mayo, 1988), (2) increased precipitation at high altitude following a rise in the equilibrium line altitude (ELA) (cf. Field, 1947, 1979; Mercer, 1961; Goldthwait et al., 1963), and (3) accumulation of sediment at the grounding line that decreases water depth and increases terminus stability (Goldthwait et al., 1963; Post, 1975; Powell, 1984, 1991; Mann, 1986; Mayo, 1988; Warren and Hulton, 1990; Warren and Glasser, 1992).

Marine-ending glaciers differ from terrestrial glaciers in that they have longitudinal profiles near their termini that are concave-up rather than convex-up (Mercer, 1961). Once retreat is initiated by a rise in the equilibrium line, the effect can be cumulative because of the near-horizontal glacial profile which continues to lower by the 'draw-down' effect associated with fast flow (e.g. Hughes, 1973, 1983, 1987c).

Water depth at the grounding line is also considered important because of the potential calving instability at tidewater termini (Brown *et al.*, 1982) and rapid migration of the grounding line of floating termini (Thomas and Bentley, 1978). Changes in water depth can be affected by isostasy, eustasy and sediment accumulation/glacial erosion. In theory, a glacier could initiate its own retreat from a maximum advance by crustal loading as the glacier increased in thickness. If the advance is local and independent of global cooling, then a concomitant eustatic lowering of sea level may not occur. Once initiated, retreat may proceed because rebound is not sufficiently rapid. When rebound had decreased water depth glacial advance could then ensue.

Global cooling and glaciation could allow glaciers to advance onto continental shelves simply by eustatic lowering of sea level. If deglaciation of a sufficiently large ice sheet occurs and global sea level rises, then that can force marine-ending glaciers elsewhere to retreat.

The timing of eustatic and isostatic movements is important in the generation of packages of glaciomarine sediments because such movements are forcing functions that initiate or terminate packages (Powell, 1991). They also have the potential of creating their own feedback loops for glacial fluctuations. However, they do not necessarily influence the detailed sedimentary history within packages, which can be more a function of local and regional sedimentary processes (Menzies, 1995b, Chapter 11).

High latitude continental shelf and slope systems are relatively poorly explored but they potentially hold the key to documenting linkages between major ice sheet fluctuations and global climate changes inferred from other records such as deep-sea cores and ice cores. Some models have been suggested to formulate ideas for interpretations of sediment packages with multiple glaciations over high latitude continental shelves (Andrews, 1990; Boulton, 1990; Anderson *et al.*, 1991; Henrich, 1991; Powell, 1991; Syvitski, 1991; Bartek *et al.*, 1991). However, sedimentary systems tracts and their bounding surfaces need to be mapped in detail by high resolution seismic reflection surveys and sediment coring and drilling need to be performed in order to really test the models and ice sheet–climate linkages.

On a large scale, both glacial isostasy and eustatic sea level influence water depth at termini. Eustatic sea-level fluctuations are the largest global influence on local water depth and may be in-phase or out-of-phase with local advances and retreats of marine termini. The crustal depression phase of isostasy is a more significant effect on marine terminus stability than the rebound phase because rebound and its consequent decrease in water depth commonly appear to lag retreat.

It is the effect of rising eustatic sea level that has been suggested to initiate instability in a grounding-line/ice shelf system that results in rapid disintegration of a marine ice sheet, specifically, the West Antarctic Ice Sheet (Hughes, 1973; Thomas and Bentley, 1978). However, models that place more emphasis on longitudinal stresses and basal sliding indicate that marine ice sheets behave less catastrophically and are more stable to climatic and sea level forcings than the earlier models indicate (Alley and Whillans, 1984; Van der Veen, 1985, 1987a, b). The stability of floating portions of ice masses, that is, ice shelves and floating glacier-tongues, is also controlled by calving processes and those secondary factors which influence calving, such as sea ice, tides and storms. Recent studies suggest that climate warming has induced catastrophic break-up of the Wordie Ice Shelf by increased meltwater and hydrostatic pressure within surface crevasses (Doake and Vaughan, 1991).

On a smaller scale, water depth at a terminus is a function of sea-floor topography that can be modified by either glacial erosion (e.g. over-deepening) that creates deep water to force retreat, or glacial deposition (e.g. grounding-line systems of morainal banks, grounding-line fans, ice-contact deltas) at tidewater termini which causes shoaling (Powell, 1991) and decreases calving rate to enhance glacial stability (cf. Meier and Post, 1987). Although as yet unknown, there must be a critical water depth beyond which each tidewater terminus is unstable (e.g. Hughes, 1992b). Sediment accumulation may allow grounding-lines of floating termini to advance due to grounding-line wedge progradation as discussed above, although the relative change in water depth compared with tidewater termini is small due to the geometries at the grounding-lines.

Recent evidence from various sources points to instability of ice sheets on different time scales. The Laurentide Ice Sheet may have collapsed several times over 10 ka intervals to produce ice-rafted debris layers in the North Atlantic (Heinrich, 1988; Bond *et al.*, 1992; Broecker *et al.*, 1992; Andrews and Tedesco, 1992). The surface of the Greenland Ice Sheet appears to have experienced irregular temperature changes (interstadial episodes) over very short time intervals (500 to 2000 years) (Johnsen *et al.*, 1992). A new model of the West Antarctic Ice Sheet indicates that it could be self-regulated by the rates of erosion and transport of subglacial sediment over periods of the order of 100 ka (MacAyeal, 1992). On a longer time-scale, the West Antarctic Ice Sheet may have been absent about 2 Ma ago with open ocean in the area where it is today (Scherer, 1991). Likewise, even the East Antarctic Ice Sheet, which has been considered to have been stable for tens of millions of years, may have been absent as recently as 3 Ma ago (Webb *et al.*, 1984; Barrett *et al.*, 1992), although that is in debate (Marchant *et al.*, 1993).

The questions arise as to the linkages between ice sheets and climate: Which drives what? Our best long-term record of climatic change comes from deep-sea sediment cores and the Milankovitch theory has become the established model, using these data, to explain climatic change during the periods of time when Earth has been in 'icehouse' periods (Fischer, 1981) through its history. At present, this assertion may need re-evaluation, especially in light of the above information in combination with new data from caves (Winograd *et al.*, 1992) that tend to indicate that Milankovitch forcing is not the only process operating (Chapter 2). One way of testing these ideas is to document the glaciomarine record on high latitude continental shelves and slopes (Imbrie *et al.*, 1993).

Chapter 15

GLACIAL CRUSHING, WEATHERING AND DIAGENETIC HISTORIES OF QUARTZ GRAINS INFERRED FROM SCANNING ELECTRON MICROSCOPY

W. C. Mahaney

15.1. INTRODUCTION

While considerable information is available for glacial-crushing features on quartz particles in tills and diamictons (Krinsley and Doornkamp, 1973; Krinsley and Takahashi, 1962; Hamilton and Krinsley, 1967; Krinsley and Donahue, 1968; Bull, 1981) (see Menzies, 1995b, Chapter 13, and on ice thickness and glacial transport vs. the degree of particle damage (Mahaney et al., 1988b, 1989; Mahaney, 1990a, b; Tison et al., 1993), relatively little is known of the extent to which paleoenvironments can be reconstructed using microtopographical information on glacial grains. In any till, whether for example of lodgement or meltout facies, from supraglacial drift, or from end moraines, a considerable number of quartz fragments are overprinted with various degrees of fresh crushing patterns. In some cases two or more stages of overprinting are indicated (Mahaney, 1990a), separated by weathering features that may result from previous interglacial stages. On rare occasions it is possible to observe on quartz grains, emplaced during the Quaternary, etching features that may be related to ameliorative climatic conditions during the Upper Neogene. This is the case even in areas that today have very severe climates such as the Antarctic, Ellesmere Island and Spitsbergen.

Crushing and dissolution events may be discerned on individual quartz grains by observing relative degrees of crushing freshness affected by different degrees of dissolution. To attain this information higher magnifications (800x–1500x and greater) are necessary to determine the precise nature and character of etching that has occurred. Admittedly, two or three stages of quartz dissolution could result from differential weathering over the same length of time or possibly from imperfections in the quartz lattice. In most cases a judgment must be made as to the relative differences in etching across the particle to determine if the evidence warrants an interpretation of different weathering events. Certainly grains with variable degrees of fresh and old (highly etched) crushing features would have had to be exposed to subaerial weathering at some point prior to glacial transport. Often because several different stages of crushing and weathering are noted, it is sometimes possible to make the assertion that two or more stages of glacial transport separated by weathering events can be perceived. Pedogenic weathering is considered to produce deep etching patterns, whereas mild or soft weathering quite possibly produces minor quartz dissolution during diagenesis (Mahaney, 1990b; Mahaney and Rutter, 1992).

The objective of this chapter is to examine photomicrographic evidence from scanning electron microscopy (described in the previous chapter) of a variety of different till samples from several localities. These photomicrographs provide a wealth of microtextural information, and when they are combined with the elemental chemistry from energy-dispersive spectrometry (EDS), it becomes possible to deduce a considerable amount of information about paleoweathering history, origin and composition of clay coatings, and neoformation of clay minerals (Mahaney et al., 1988a) and to gain insight into the importance of using EDS in microtextural investigations (Vortisch et al., 1987). Thus, quartz particles in glacial sediments from the last glacial maximum (Wisconsinan–Weichselian–Würm) often provide considerable insight into weathering during the previous interglacial (Sangamonian–Eemian) and previous glaciations (Illinoian–Saale–Riss). In some instances it has proved possible to obtain information on pre-Illinoian crushing/weathering events and/or on pre-Quaternary weathering.

15.2. THE SEDIMENTS

Scanning Electron Microscopy photo-micrographic and EDS data from several localities worldwide have been selected to provide representative information on paleoenvironmental reconstruction. These representative examples come from both alpine and continental glaciated areas and contain a variety of different kinds of tills. Examples of glacial grains in severe climatic zones such as Spitsbergen, Ellesmere Island and Antarctica were chosen to illustrate the range of paleoclimates and the different crushing effects of cirque and valley glaciers through to polar ice of increasing thickness. Still other examples from alpine areas in the European Alps and in the Rocky Mountains were singled out to show that paleoenvironmental extremes in the mountain areas were of high magnitude, while glacial crushing processes operated only when the ice achieved a thickness of >200 m (presumably with lower porewater in the underlying till). The presence of porewater in ice above the pressure melting point

may facilitate fracturing and abrasion of quartz by allowing particles to come into contact with one another when shear stress increases in ice of >200 m thickness (Mahaney et al., 1988b).

The term till, used in this chapter, refers to dominantly non-sorted, non-stratified drift that is unconsolidated, unweathered material deposited by a glacier with little or no reworking by meltwater. It is a heterogeneous mix of clay, silt, sand, pebbles, cobbles and boulders varying widely in size and shape. Diamict (Harland et al., 1966) is an all-inclusive term that includes diamicton and diamictite. Diamicton (Flint et al., 1960) is the non-lithified equivalent of diamictite. Diamictite is an all-inclusive non-genetic term first used by Flint et al. (1960) for non-sorted, non-calcareous, sedimentary rocks containing a wide range of particle sizes, including rocks with muddy matrices (e.g. tillites and pebbly mudstones). For all practical purposes tills and underlying surface drift lacking landform relationships are diamictons.

Because glacial microtextures are sometimes considered to result from mechanical weathering of source rocks, samples from supraglacial drift in the Adishy Glacier area, Caucasus Mountains, Russia have been included. The microtopography of these grains likely results from mechanical weathering of rocks from the surrounding cirque walls and/or from mass wasting transport from the cirque walls to the glacier surface, followed by transport along the glacier surface. Alternatively, some grains may have been fractured during aeolian transport up-valley by anabatic winds.

The examples discussed herein are from various sand fractions of the samples collected. Depending upon the distance of transport, quartz may be better preserved in the fine and very fine (63–250 μm) fractions. For long-distance transport (several hundred kilometres or more), quartz is usually more prevalent in these finer fractions, presumably as a result of attrition of coarser particles.

In mountainous volcanic terrain (e.g. East African mountains) the presence of high amounts of quartz could be taken to indicate long-distance transport from surrounding basement rocks, which opens up the possibility of fracturing from collision during aeolian transport.

In the discussions which follow, examples of tills from diverse locations and origins will be described and their surface features interpreted. It is important to be aware that the interpretation of these features is not definitive or conclusive but judgemental and speculative based upon experience and a comprehension of the mechanics of the wear processes involved and invoked to create the microtopographic features observed on the grains. Therefore, the interpretations must be viewed within the context of this particular stage in our understanding of what goes on beneath, within, on and proximal to ice masses.

15.2.1. Antarctica

Tills (and diamictons) from the Antarctic, examined with the SEM, show complex histories and indicate that in addition to extensive crushing, sediments were exposed to rather extreme weathering processes during the Upper Neogene. The samples shown here are from the Transantarctic Mountains, but similar examples are known from Mt. Discovery, Cockburn Island and from the Meseta Formation (glaciomarine sediment of Eocene age) (Mahaney, *unpublished*).

Till 480L (Plate 15.1) from the Vestfold Hills, near Davis Station (68° 30´S), is in an ice-free area ~200 km^2 in Princess Elizabeth Land (Campbell and Claridge, 1987). Deglaciated for at least the last 8000 years, the Vestfold Hills till was deposited by the retreating East Antarctic Ice Sheet (G.C.C. Claridge, *pers. commun.*, 1988) and could contain quartz grains crushed under thick ice and after long transport (several hundred kilometres). Indeed, the grain shown in Plate 15.1 is not only extensively crushed, it is also highly weathered in places. Crushing gave rise to the production of step-features (middle Plate 15.1) that appear fresher and lack dissolution etching features. Thus, it provides an excellent example of overprinted grains transported by continental ice (cf. Mahaney, 1990a).

Till 42 (Plate 15.2) from Inexpressible Island, near Cape Adare (170°E), was deposited by glaciers issuing through the Transantarctic Mountains. The fracture lines, mainly low frequency conchoidal (right) and high frequency subparallel linear (left), appear to have been affected both by soft weathering or abrasion (left) and/or dissolution etching (right). These differing microtextures appear to result from intensive crushing (right), followed by mild dissolution etching. The area to the left was later subjected to intensive crushing that produced high-frequency fresh fracture lines.

Till 84, from the Roberts Massif, Shackleton Glacier, Transantarctic Mountains (82°S), may be of early Miocene age (Sirius Formation) or younger (Claridge, *pers. commun.*, 1989). Two stages of crushing are evident on this grain (Plate 15.3). The edges (lower right and upper left) are intensively crushed and weathered. The front of the grain is also intensively crushed but less intensively weathered by dissolution etching. The palaeoenvironmental interpretation is that the intensively weathered zone probably resulted from an interglacial period sometime in the lower Miocene, or possibly even during the Oligocene Epoch.

Another grain (Plate 15.4) from Till 42 (Plate 15.2) shows two crushing stages apparently separated by intensive dissolution etching. The oldest crushing stage occurs on the edges (lower and upper left) that are also intensively affected by quartz dissolution. On this fragment the younger crushing features are less intense than in Plate 15.2.

Till 84 (Plate 15.5) from the Roberts Massif, yields many grains with pronounced crushing and weathering features. Two older crushing areas separated by degrees of weathering intensity (more intense middle and less intense lower edge) contrast with a fresher zone that shows pronounced high frequency crushing (top) and troughs (right and in enlargement). These troughs, which could be deep corrasion features typical of thick continental ice, are to the order of 8–10 μm across and 4–5 μm deep. They occur neither on grains transported by alpine ice (<200 m) nor on grains moved by expanded mountain ice (~500 m) (approximate ice thickness estimates are derived from moraine positions, valley depths and distance of ice flow in most cases; in others from published sources based upon ice thickness models or profiles).

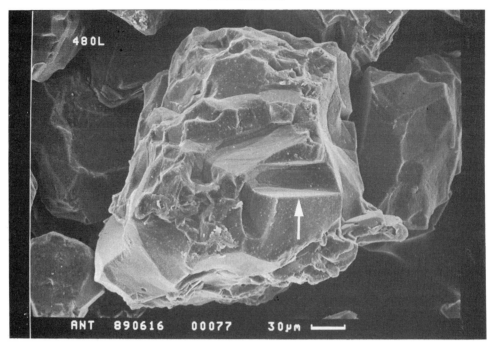

PLATE 15.1. Till 480L from the Vestfold Hills, Antarctica. The edges of this particle were crushed first and then weathered. The main facet of the grain was crushed last giving a series of step-features (arrow) that are fresh. This till, deposited by the retreating inland ice, is typical of recycled grains that are extensively overprinted.

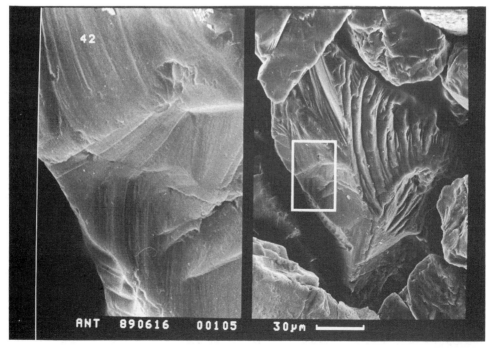

PLATE 15.2. Till 42 from Inexpressible Island, Antarctica, deposited by the Nansen Ice Sheet formed from the coalescing of glaciers issuing through the Transantarctic Mountains. This grain (440x) carries both subparallel linear and conchoidal fractures — high frequency type to left and low frequency type to right. The fractures to the left were softened (enlargement 2200x) by weathering presumably during some previous interglaciation. The conchoidal fractures to the right are slightly affected by dissolution etching which may result from diagenesis.

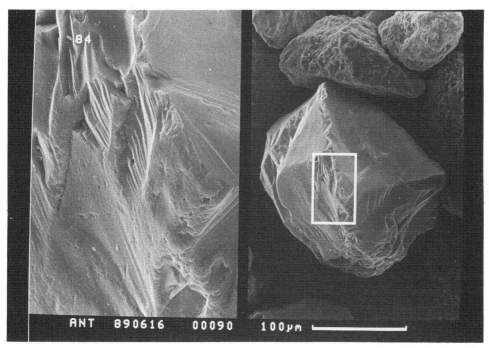

PLATE 15.3. Till 84 from the Roberts Massif, Shackleton Glacier, Antarctica. This till is considered to be of early Miocene age. The lower edges (260x) appear crushed and weathered by extensive dissolution. The remainder of the clast has been soft weathered (enlargement 1300x) possibly during diagenesis.

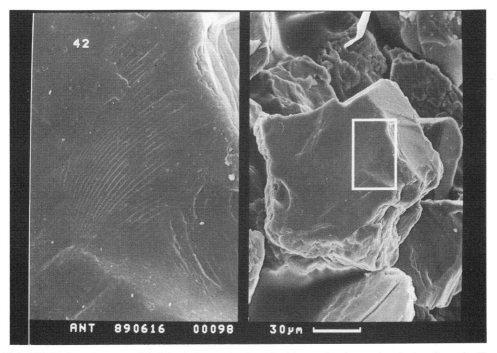

PLATE 15.4. Till 42 from Inexpressible Island, Antarctic, deposited by coalescing outlet glaciers from the Nansen Ice Sheet. Two stages of crushing separated by weathering are identified on this particle (440x). The oldest crushing stage occurs on the bottom and left edges all extensively affected by dissolution etching. Enlargement shows conchoidal fractures at 2200x.

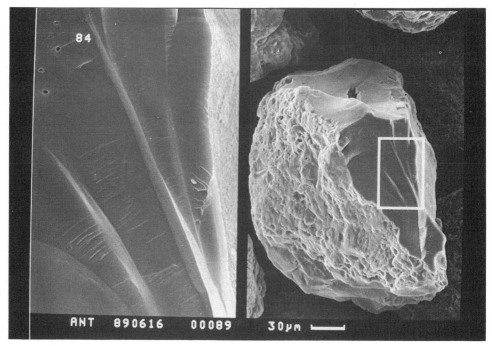

PLATE 15.5. Till 84 from the Roberts Massif, Shackleton Glacier, may be as old as early Miocene. This grain (310x) is extensively crushed and weathered. Two older crushing events appear to have been followed by weathering (soft on lower edge and more intensive on the middle with dissolution etching). Deep low frequency crushing affects attest to the most recent crushing (right and top of grain). Enlargement (1550x) shows minor dissolution etching on most recently crushed area of the particle.

15.2.2. Ellesmere Island, Arctic Canada

Till EI10 from Ellesmere Island, Cape Herschel Plateau (78°36′N; 74°41′30′′W), was sampled at an elevation of 200 m. The till formed a thin veneer of moraine on a glacially polished Precambrian granitic outcrop. Some quartz sand in this till showed regrowth of quartz crystals which could be from Paleozoic outcrops ~35 km to the northwest (W. Blake, *pers. commun.*, 1989). A variety of grains showing different degrees of angularity and roundness were found in this till (Plate 15.6). Most rounded grains (right) carried V-shaped percussion cracks indicating meltwater movement. Most angular grains showed possible evidence for at least two stages of crushing separated by weathering. The grain shown in Plate 15.6 carried fresh crushing features (middle to bottom) and older crushing modified by dissolution etching (top). The degree of dissolution effects, given the high latitude, was unexpected even for Quaternary interglacial weathering. However,

quartz weathered in Quaternary moraines on Spitsbergen (82°N) showed very intensive weathering (Mahaney, *unpublished*).

A second till (EI12) from Ellesmere Island shows quartz with multiple-crushing stages separated by fairly intense dissolution etching (Plate 15.7). On this grain the older crushed surface with dissolution etching probably predates the last glaciation.

15.2.3. Spitsbergen

Till L3 (Plate 15.8), from Spitsbergen, shows a fresh, clean fractured surface with subparallel fracture lines (right) and older linear fractures (bottom and upper left) that are intensively weathered by chemical dissolution. This is quite unexpected given the high latitude but shows the probable weathering power of a previous interglacial climate (Sangamon/Eemian). Close examination of the older intensively weathered surface (low centre) leads to the conclusion that intense pedogenic weathering is probably responsible for the intricate chemical etching.

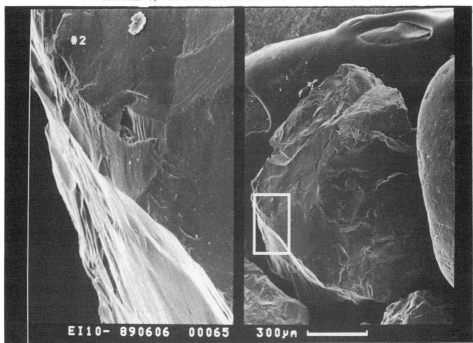

PLATE 15.6. Angular quartz grain (left) from Ellesmere Island showing extensive crushing that is fresher (bottom left and enlargement) and older and weathered (top right). Grain to right is rounded and shows mainly fresh V-shaped percussion cracks from water transport.

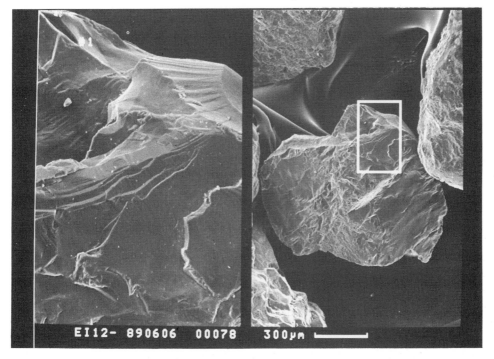

PLATE 15.7. Grain (50x) from till EI12 (Ellesmere Island) showing older crushed surface (centre) and younger surface (right). Enlargement (250x) shows mainly low frequency fractures.

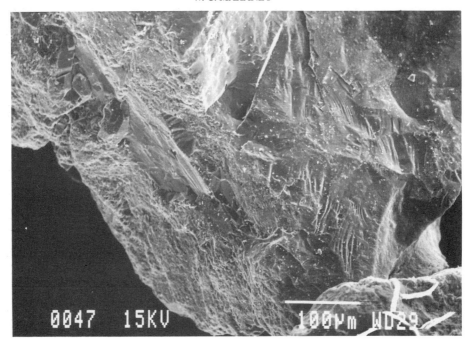

PLATE 15.8. Till L3 showing multiple crushing episodes and multiple weathering preserved *in situ* (centre to bottom) as well as fresh surfaces (right centre).

Till K2, (Plate 15.9), from Spitsbergen, is fairly extensively crushed but shows well-preserved relicts of probable previous crushing and modification by weathering (upper left). The remainder of the crushing consists of isolated subparallel linear fractures presumably consistent with thick valley glaciers (~500 m) (e.g. 10–30% of the surface is freshly crushed but many older crushing features affected by dissolution etching are preserved).

15.2.4. Southern Ontario, Canada

Sample H144 from the Rouge River Valley (Mahaney, 1990a) is from the lower and upper Sunnybrook diamicton of early Wisconsinan (Würm) age. Considerable controversy has been generated over the origin of this sediment as either rainout in a glacial lake (Eyles and Eyles, 1983a) or as lodgement till from grounded ice (Hicock and Dreimanis, 1989; Mahaney, 1990a). The grain shown in Plate 15.10 is from the lowest beds of the Sunnybrook diamicton. This grain is intensively fractured with fresh lines to the left and weathered ones to the right. Because the dissolution etching is so intense (to the right), weathering most probably occurred during a pre-Wisconsinan interglacial climate (Sangamon?). The fresh crushed face to the left is very clean with an absence of adhering particles and no dissolution etching. It is possible that such grains were deposited directly by the ice without any mixing in a slurry of clay and silt with minor sand and pebble/cobble content (i.e. glacial deposition). Almost all grains in the Sunnybrook diamicton have about the same crushing intensity and many have similar complex weathering histories.

Sample 9i (Plate 15.11), located near the top of the Sunnybrook diamicton, displays intensive crushing with probable pre-weathered (weathered prior to transport) surfaces (arrows). The most likely interpretation is that pre-weathering occurred during the Sangamon Interglacial Stage (Eemian). Once again the relative absence of adhering particles supports the opinion of direct deposition by ice. The retention of old pre-weathered surfaces also possibly attests to the preserving condition of ice when other particles are not available for crushing to occur.

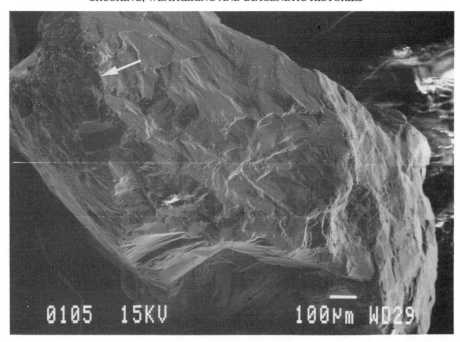

PLATE 15.9. Till K2 (70x) shows moderate crushing effects in places with an older crushed and weathered zone (arrow upper left).

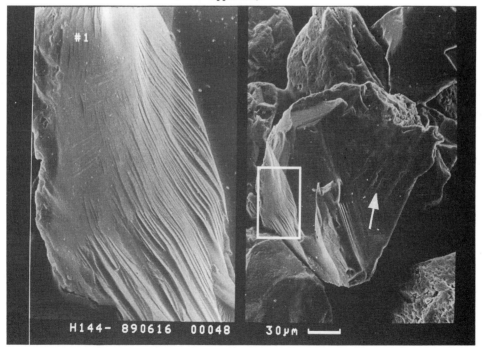

PLATE 15.10. Glacially-crushed quartz particle (300x) in sample 1 of the Sunnybrook diamicton (H144 section) of the Rouge River Valley, south-central Ontario, showing two stages of crushing separated by weathering. The older crushed and weathered area is in the upper right (arrow). The enlargement (1500x) shows fresh, mainly subparallel crushing features unaffected by dissolution etching.

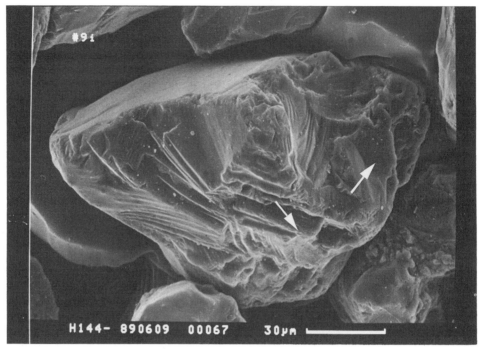

PLATE 15.11. Sample 9i (740x) in the H144 section (Sunnybrook diamicton) with extensive crushing features (see Mahaney, 1990a). Older crushed area (arrows—right and bottom) contains step-like features and low frequency crushing that has been subjected to dissolution etching. Area to left is fresh.

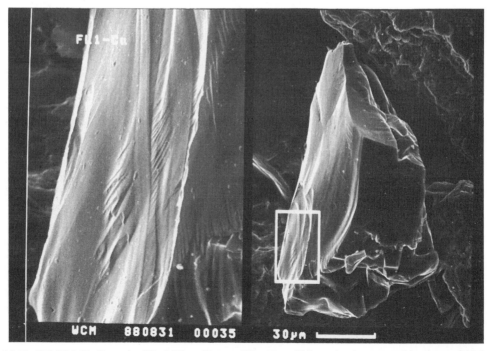

PLATE 15.12. Till FL1-Cu from the Wind River Range, Wyoming. This grain shows combinations of high and low frequency crushing (left and top) that occurred during the Pinedale Glaciation (= Würm, Wisconsinan). The crushed area to the lower right is probably of Bull Lake age (Riss, Illinoian), the weathering effects being produced during the Bull Lake–Pinedale Interglaciation (= Sangamon Stage or Eemian).

Because many such grains are found in the Sunnybrook diamicton, it is possible the ice was relatively free of clastic debris (i.e. clean ice). On the other hand, because all grains were subjected to mild sonication, it is possible that some clay was removed during laboratory preparation.

15.2.5. Wind River Mountains, Wyoming, U.S.A.

Till FL1 (Plates 15.12 and 15.13) of the early stade (<100,000 years) of the Pinedale Glaciation (Mahaney, 1978) contains clasts deposited under expanded mountain ice (Mahaney *et al.*, 1988b) with thicknesses of >500 m and transport distances of ~40 km. Because the climate is dry, it is unlikely to expect significant diagenetic effects from porewater solutions (warmer and wetter climate may produce enough porewater to achieve some weathering effects). The grain shown in Plate 15.12 appears to have variable degrees of crushing separated by weathering. The bottom centre is the oldest with the

most extreme weathering. The crushed area to the bottom left (shown in enlargement) contains softer weathering with minor dissolution (to left) and fresher crushing without dissolution effects to the right.

A second quartz clast (Plate 15.13) shows relatively fresh crushing and few adhering particles (front and lower edge). The left edge and top of the grain are crushed and intensively weathered; possibly these are relicts from weathering events of pre-Pinedale age (e.g. Sangamon) that survived transport after entrainment by Pinedale ice.

Older till (Plates 15.14 and 15.15) of Bull Lake age (Illinoian age, Isotope Stage 6) was deposited by ice presumed to have greater thickness and transport distances 5 km greater than Pinedale ice. Like the younger Pinedale clasts, these grains are crushed over 50% of the clast surface and pre-weathered (weathered during some previous episode of subaerial exposure) features are usually well preserved. The grain in Plate 15.14 probably shows three stages of

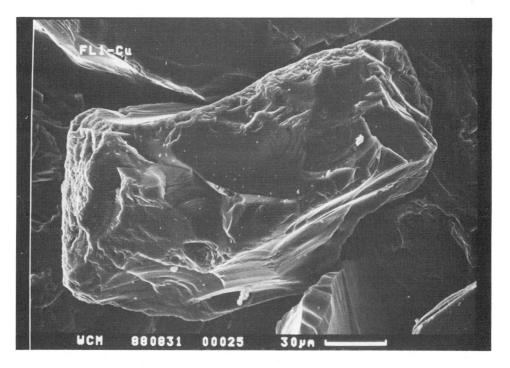

PLATE 15.13. Quartz grain showing an older crushed surface with dissolution etching (left) and fresh crushed surfaces (front and bottom).

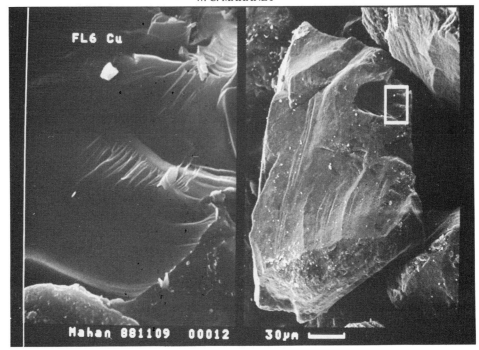

PLATE 15.14. Till FL6 from the early Bull Lake stade (Riss, Illinoian) of the Wind River Range. Three stages of crushing and weathering are evident on this particle. The oldest surface (lower) also has the greatest accumulation of adhering particles. The middle and upper left portion of the grain has fresher crushed surfaces and few adhering particles. The freshest surfaces (upper right) show mainly low frequency crushing, no dissolution etching and no adhering particles.

crushing and is separated by two stages of weathering. The youngest surface shown in enlargement may have been protected from diagenetic processes in a 30 μm wide cavity. The middle area, with a few adhering particles, may have been subjected to soft weathering during the Bull Lake–Pinedale Interglaciation (Sangamon; Isotope Stage 5, >100,000 years). The older surface at the bottom may have been weathered during a pre-Bull Lake interglaciation (~0.5 Ma).

A second sample from till FL6 (Plate 15.15) shows a grain with three stages of crushing. The youngest surface (to left) is a fracture face that probably resulted from mechanical release (physical weathering) of source rocks (Mahaney, 1990b). The area to the right appears to have been subject to variable degrees of dissolution etching and coating of fracture lines. The coating of $CaCO_3$ serves to obscure the fracture lines and could have originated during drier times (i.e. Pinedale Glaciation or the Holocene). The upper area is less thickly coated with

$CaCO_3$ but was probably crushed during the Bull Lake Glaciation.

Till TB18 (Plate 15.16) is from later Pinedale till (Late Glacial age) at an altitude of 3100 m in the Wind River Range (Mahaney, 1978). This clast is so intensively crushed and weathered that it probably has a pre-Pinedale history that is very complex. The upper surface consists of low frequency crushing intensively affected by quartz dissolution. From the degree of weathering it must have at least a Sangamon age (>100,000 years). The remainder of the grain has relatively fresher crushing with only minor dissolution effects that could have a Pinedale interstadial origin. There is nothing to indicate any crushing during the later Pinedale which, given the low ice thickness (~100 m) and short transport distance (<2 km), is not unexpected (Mahaney et al., 1988b, 1989). This grain would appear to be typical of many particles emplaced by thin ice (<200 m) during this time in the Rocky Mountains.

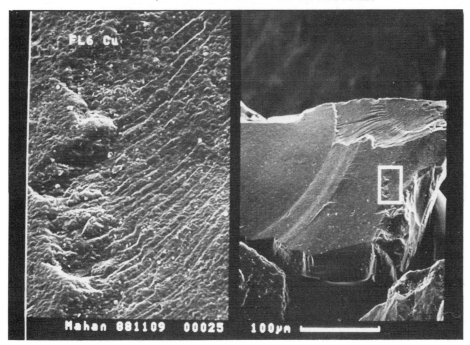

PLATE 15.15. This particle shows a fairly fresh fracture face (left) with little dissolution etching, and glacial crushing features of the low frequency type (middle and top) affected by moderate dissolution weathering. The right edge has the same degree of weathering. Area of enlargement has the oldest degree of crushing and greatest modification by weathering.

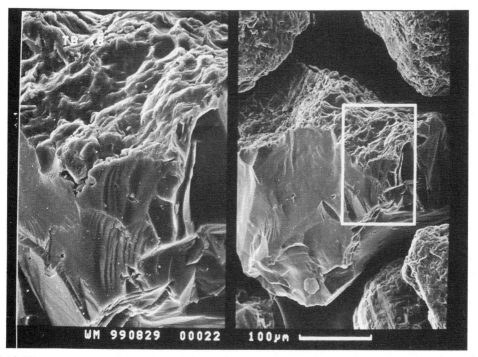

PLATE 15.16. Till TB18, Wind River mountains, Wyoming, of later Pinedale (late glacial) age. The oldest surface (top) is crushed and extensively weathered. The youngest and freshest surface (front) has been affected by dissolution etching and probably predates the Pinedale Glaciation. This is a good example of a recycled grain that inherited its main microtopographic features both from the Pinedale Glaciation and pre-Pinedale weathering.

15.2.6. Austrian Alps

Samples (Plates 15.17 and 15.18) of till and glaciofluvial sediment from a deposit (<50 years) in front of the Horn Kees terminus, in the Zillertal Alps, Austria, show glacial-crushing effects that possibly predate Neoglaciation. A clast from glaciofluvial sediment appears to consist of multiple crushing stages separated by weathering phases. The bottom of the grain is more highly weathered and shows a higher percentage of adhering particles. The upper portion of the grain is fresher, but probably dates from the Würm Glaciation.

Till sample HK6 shows a compound grain with older crushing and extensive dissolution etching (bottom), flanked by fresher crushing with some dissolution effects (top). The enlargement shows some dissolution etching and adhering particles on this younger surface.

15.2.7. Virunga Mountains, Rwanda

Section KAR1 in the Virunga Mountains of Central Africa is a representative Late Quaternary sequence of weathered basalt, aeolian, volcaniclastic (tuff), fluvial and slope wash sediments emplaced over the last 56,000 years (Mahaney 1990b). Glacially-crushed quartz (Plate 15.19) is found in the aeolian and fluvial deposits in this section that date from the last glacial maximum. The older beds have no quartz clasts and the younger horizons have only reworked quartz from aeolian deposits upslope. Possibly these glacial grains were delivered by aeolian processes from the Ruwenzori Mountains (ca. 125 km to the north) when the paleowinds were stronger and/or when the Afroalpine zone was at its maximum extent during the Last Glacial Maximum. Because these quartz particles were likely affected mainly by Würm interstadial climates (cool and dry), they would appear to have been pre-weathered probably during the last interglaciation (Mahaney, 1989).

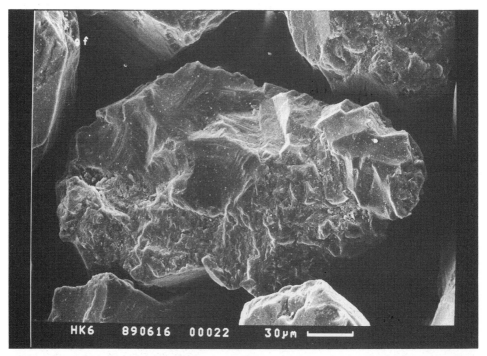

PLATE 15.17. Glaciofluvial sediment (430x) HK6, of Neoglacial age, Zillertal Alps, Austria, showing mainly older crushed and weathered surfaces. The bottom of the grain is more severely weathered and may predate the Würm Glaciation.

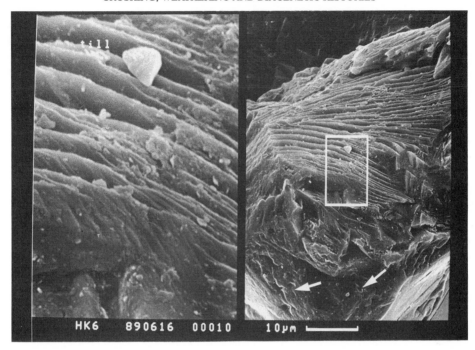

PLATE 15.18. Extensively crushed grain (1520x) showing two stages of weathering that predate Neoglaciation in the Alps. The lower part of the particle is more highly weathered and shows some altered V-shaped percussion features (arrows). The middle and upper portions have been affected by dissolution etching and contain many adhering particles that survived transport during Neoglaciation. Enlargement is 7600x.

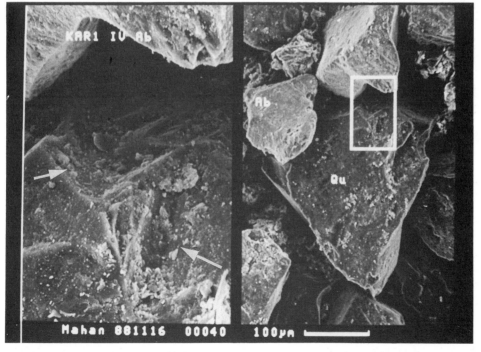

PLATE 15.19. Subsample of horizon IVAb in the KAR1 section, Virunga Mountains, Central Africa. Grains shown here include albite (Ab) and quartz (Qu). Both particles are subangular with rather dull edges. The quartz grain has been crushed (arrows), weathered, and delivered by aeolian transport to the Virunga Mountains (see Mahaney, 1990, for a discussion of the source).

15.2.8. Mount Kenya, Kenya

Tills on Mount Kenya are quartz poor, but all quartz grains are crushed to some degree depending upon ice thickness and distance of transport. The stereopair shown in Plate 15.20 includes isolated linear and conchoidal fractures that are affected by minor dissolution etching. This till was subaerially exposed for ~250,000 years prior to burial by aeolian sediments during the Teleki Glaciation (Illinoian). Particles on the grain are mainly metahalloysite with minor gibbsite (see Vortisch *et al.* (1987) and Mahaney (1990b) for a discussion of the palaeoclimatic significance of relative ratios of these two weathering products).

Two samples from lahars on Mount Kenya, Plates 15.21 and 15.22, are of particular importance as they shed light on glaciation predating the Olduvai subchron (>1.9 Ma). These lahar deposits were emplaced prior to the Gorges Glaciation. The grain

from sample GOR4c shows fracture features coated with MnO_2 and Fe_2O_3 which would appear to indicate glacial crushing followed by diagenetic weathering. The grain from sample HIN7 shows a fresh surface with minor crushing that appears to have remained unaffected by diagenesis following emplacement. Both of these grains possibly suggest that glaciation occurred sometime after 2.4 Ma and prior to the emplacement of the lahars (~2.0 Ma). An alternative hypothesis is that volcanic-mass wasting processes produced the crushing microtextures.

15.2.9. Caucasus Mountains, Russia

Samples of supraglacial drift (Plates 15.23 and 15.24) from the Adishy Glacier in the Caucasus Mountains were studied to determine the effects of mechanical weathering on grains from the surrounding cirque walls that had not been subjected to glacial crushing (Mahaney *et al.*, 1991). As shown

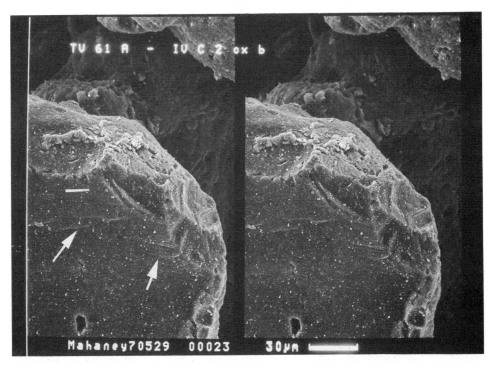

PLATE 15.20. Till in section TV61A, Mount Kenya, of Naro Moru age (isotope stage 8). This till, deposited during the early Brunhes (<730,000 ka) contains little quartz; however, all grains show some crushing (arrows) and weathering. Adhering particles are mainly metahalloysite; minor coatings of gibbsite at top of grain.

PLATE 15.21. Quartz in a lahar deposit (GOR4c) on Mount Kenya of early Quaternary age. Crushing is the low frequency type as shown on enlargement and coating is $MnO_2 + Fe_2O_3$.

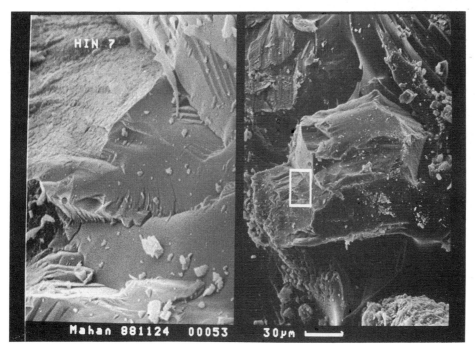

PLATE 15.22. Freshly crushed quartz in lahar HIN7 with fewer adhering particles and weathered surface in upper left of the enlargement. Both quartz grains are considered to represent glacial crushing of particles prior to the Olduvai subchron (>1.87Ma).

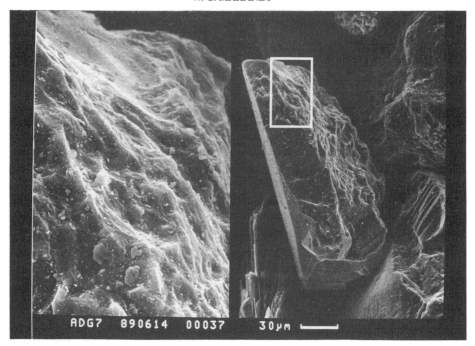

PLATE 15.23. Glacially-crushed grain from the Adishy Glacier, central Caucasus Mountains, Russia. This grain (360x), collected from supraglacial drift probably was recycled by anabatic winds from the lower valley (enlargement 1800x).

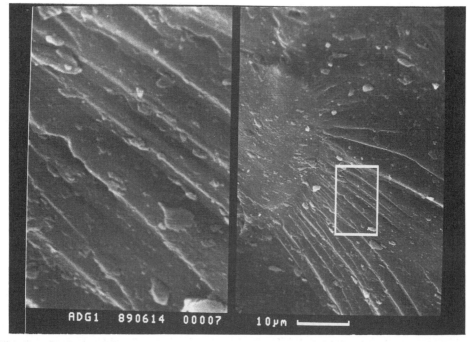

PLATE 15.24. Fracture face (left) and radiating fracture lines from mechanically cleaved grain (1510x) in the Adishy Cirque, Caucasus Mountains. Enlargement (7550x) shows soft chemical weathering that presumably occurred after mechanical release from source rocks.

in Plates 15.23 and 15.24, some grains (Plate 15.23) show relatively pronounced crushing intensively modified by weathering. Conceivably these grains may have been reworked by anabatic winds from older outcrops in the lower valley. Still other grains (Plate 15.24) showed pronounced fracture faces (left) with radiating fracture lines that may possibly result from mechanical release from source rocks. The degree to which many quartz grains were intensively affected by dissolution proved unexpected given the high altitude and severity of the climate (Mean Annual Temperature, MAT estimated at –3°C) at latitude 40°N.

15.3. MICROTEXTURES VS. ICE THICKNESS

The degree to which different microtextures can be correlated to differing ice thicknesses is shown in Fig. 15.1. Almost all these microtextures noted in Fig. 15.1 can be observed on till, diamicton and diamictite samples. However, the alpine tills with ice thicknesses <200 m (which includes most middle and low latitude Holocene ice) lack most fracture and abrasion features, which is presumably related to reduced cryostatic pressure and shear stress. Compared with expanded mountain ice (>500 m) and continental ice with thicknesses of >1000 m, alpine ice seems only to be capable of meltwater release that sometimes produces V-shaped percussion cracks. Overall relief is lower than on grains emplaced by thicker ice masses. Preweathered surfaces, weathered crushing surfaces, adhering particles and precipitation features are less common on grains from alpine ice, presumably a response to higher altitude and lower weathering rates.

Expanded mountain (ca. 500 m in thickness) and continental glaciers (>1000 m in thickness) bring thicker ice and increased transport distances that produce progressively greater fracturing and abrasion. Overall subparallel linear and conchoidal fractures, as well as curved and straight grooves, increase with greater ice thickness. Most interestingly, sharp angular features increase from valley glaciers to continental ice sheets, becoming most pronounced on particles emplaced by the latter (Mahaney et al., 1988b). Crescentic gouges and arc-shaped steps occur with greater frequency on grains

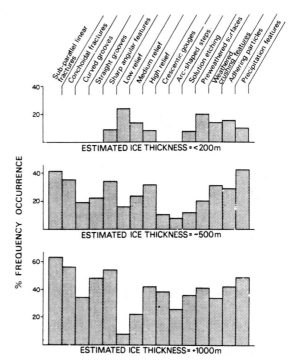

FIG. 15.1. Variations of different microtextures on tills emplaced by glaciers with thicknesses ranging from <200 to >1000 m. Percent frequency of occurrence of microtextures from samples taken from thin cirque glaciers to continental ice sheets as documented from summary statistics from 300 samples. With the exception of low relief, note the progressive increase in nearly all microtextures with increasing ice thickness.

emplaced by continental ice, although they also occur on quartz fragments emplaced by expanded mountain ice.

Despite the theory that increasing slope in alpine ice leads to increasing shear stress (Clark, 1989), the evidence (Mahaney et al., 1988a, b, 1989) suggests that Alpine ice (<200 m thickness) is incapable of quartz fracturing and abrasion.

15.4. CONCLUSIONS

The degree to which glacial-crushing microtextures and their modification by pedogenic weathering and diagenesis can be used to reconstruct the palaeoenvironmental histories of complex quartz grains has been discussed and assessed. In every sample suite studied (which includes tills from many major mountain and continental locales), there are grains present that yield important palaeoenvironmental information. In many cases this involves an interpretation based on the percent of the grain surface affected by crushing and subsequent weathering which may range in intensity from what would appear to be deep dissolution etching to softer weathered forms without noticeable dissolution cavities. Softer 'forms of weathering', in some cases, could be the result of minor abrasion during some subsequent transport of the clast.

The decision to subdivide individual grains into two or more stages of crushing is subject to judicious interpretation and, of course, error. It is possible, in some cases, that different weathering stages might be the result of differential weathering effects acting over the same time period but with different intensities. Therefore, it is essential to examine at least 25 grains per sample in order to insure representative results and to obtain a range of microtextures within one till population (see Vortisch *et al.*, 1987, for a discussion of grain analyses).

Due to their heterogeneous composition, tills contain a wide range of particles with numerous degrees of glacial-crushing features. Some tills even have quartz particles that show only fracture faces, but without fracture lines. Others have rounded grains of meltwater or aeolian origin, but lack deep conchoidal and linear fractures that are the hallmark of transport by ice. Still others have a multitude of grains with preweathered features preserved even after long-distance transport. This diversity of features is valid even with continental glaciers, such as in the Antarctic, where intensively modified quartz grains either from active pedogenesis or from diagenetic effects might be expected to occur. As a result of the preserving features of ice it is fortunate such grains survive glacial transport to provide important information on palaeoenvironmental reconstruction.

REFERENCES

Aario, R. (1972). Associations of bed forms and paleocurrent patterns in an esker delta, Haapajärvi, Finland. *Ann. Acad. Scientiarum Fennicae, ser. A*, pt. III, no. 111, 55 pp.

Aario, R. (1977). Classification and terminology of morainic landforms in Finland. *Boreas*, **6**, 87-100.

Aarseth, I. and Mangerud, J. (1974). Younger Dryas end moraines between Hardangerfjorden and Sognefjorden, Western Norway. *Boreas*, **3**, 3-22.

Aber, J.S. (1985). Definition and model for Kansan glaciation. *Institute for Tertiary-Quaternary Studies-TER-QUA Symposium Series*, **1**, 53-60.

Aber, J.S. (1988a). Ice-shoved hills of Saskatchewan compared with Mississippi delta mudlumps-Implications for glaciotectonic models. *In*: Croot, D.G. (ed.), *Glaciotectonics: Forms and Processes*, pp. 1-9. A.A. Balkema, Rotterdam.

Aber, J.S. (1988b). Bibliography of glaciotectonic references. *In*: Croot, D.G. (ed.), *Glaciotectonics: Forms and Processes*, pp. 195-210. A.A. Balkema, Rotterdam.

Aber, J.S., Croot, D.G. and Fenton, M.M. (eds) (1989). *Glaciotectonic Landforms and Structures*. Kluwer Academic, Boston, Dordrecht, 200 pp.

Abraham, G. (1965). Horizontal jets in stagnant fluid of other density. *Journal of the Hydrological Division, Proceedings of the American Society of Civil Engineers*, **91**(Hy4), 139-154.

Acaroglu, E.R. and Graf, W.H. (1968). Sediment transport in conveyance systems - Part 2. *Bulletin of the International Association of Scientific Hydrology*, **13**, 123-135.

Addison, K. (1981). The contribution of discontinuous rock-mass failure to glacier erosion. *Annals of Glaciology*, **2**, 3-10.

Adhemar, J.A. (1842). Révolutions de la mer. Privately published, Paris.

Agassiz, L. (1840). Études sur les glaciers. Privately published, Neuchâtel, 346 pp.

Aguirre, E. and Pasini, G. (1984). Proposal of the ICS Working Group on the Pliocene/Pleistocene boundary concerning the definition of the Pliocene/Pleistocene boundary stratotype: Draft manuscript, 6 pp.

Aguirre, E. and Pasini, G. (1985). The Pliocene-Pleistocene boundary. *Episodes*, **8**, 116-120.

Ahlmann, H.W. (1948). Glaciological research on the North Atlantic coasts. *Royal Geographical Society, Research Series No. 1*, 83 pp.

Aitken, A.E. (1990). Fossilization potential of Arctic fjord and continental shelf macrofaunas. *In*: Dowdeswell, J.A. and Scourse, J.D. (eds), *Glacimarine Environments: Processes and Sediments. Geological Society of Canada, Special Publication*, **53**, pp. 155-176.

Aitken, A.E. and Gilbert, R. (1989). Holocene nearshore environments and sealevel history in Pangnirtung Fiord, Baffin Island, N.W.T., Canada. *Arctic and Alpine Research*, **21**, 24-44.

Aitken, A.E., Risk, M.J. and Howard, J.D. (1988). Animal-sediment relationships on subarctic intertidal flat, Pangnirtung Fiord, Baffin Island, Canada. *Journal of Sedimentary Petrology*, **58**, 969-978.

Alden, W.C. (1911). Radiation in glacial flow as a factor in drumlin formation. *Bulletin of the Geological Society of America*, **22**, 733-734.

Alean, P., Braun, S., Iken, A. Schram, K. and Zwosta, G. (eds) (1986). Hydraulic Effects at the Glacier Bed and Related Phenomena. International Workshop 16-19 September, 1985, Interlaken, Switzerland. *Mitteilungen der Versuchsanstalt für Wasserbau, Hydrologie und Glaziologie*, No. 90, 148 pp.

Algus, M. (1986). The development of coastal bluffs in a permafrost environment, Kivitoo Peninsula, Baffin Island, Canada. Unpublished Ph.D. Thesis, McGill University, Montreal.

Alkire, B.D. and Andersland, O.B. (1973). The effect of confining pressure on the mechanical properties of sand-ice materials. *Journal of Glaciology*, **12**, 321-325.

Allen, J.R.L. (1965). A review of the origin and characteristics of Recent alluvial sediments. *Sedimentology*, **5**, 89-191.

Allen, J.R.L. (1982). *Sedimentary Structures: Their Character and Physical Basis*. 2 Volumes. Elsevier, Amsterdam, 611 pp. and 679 pp.

Allen, J.R.L. (1985). *Principles of Physical Sedimentology*. George Allen & Unwin, London, 272 pp.

Allen, P. (1984). *Field Guide to the Gipping and Waveney Valleys, Suffolk, May, 1982*. Quaternary Research Association, Cambridge, 116 pp.

Allen, P., Cheshire, D.A. and Whiteman, C.A. (1991). The tills of southern East Anglia. *In*: Ehlers, J., Gibbard, P.L. and Rose, J. (eds), *Glacial Deposits in Great Britain and Ireland*, pp. 255-278. A. A. Balkema, Rotterdam.

Allen, P.A. and Collinson, J.D. (1986). Lakes. *In*: Reading, H.G. (ed.), *Sedimentary Environments and Facies*, 2nd edition, pp. 63-94. Blackwell Scientific, Oxford.

Alley, R.B. (1988). Fabrics in polar ice sheets: Development and prediction. *Science*, **240**, 493-495.

Alley, R.B. (1989a). Water-pressure coupling of sliding and bed deformation: I. Water system. *Journal of Glaciology*, **35**, 108-118.

Alley, R.B. (1989b). Water-pressure coupling of sliding and bed deformation: II. Velocity-depth profiles. *Journal of Glaciology*, **35**, 119-129.

Alley, R.B. (1991). Deforming-bed origin for southern Laurentide till sheets? *Journal of Glaciology*, **37**, 67-76.

Alley, R.B. (1992). Flow-law hypothesis for ice-sheet modelling. *Journal of Glaciology*, **38**, 245-256.

Alley, R.B. and Whillans, I.M. (1984). Response of the East Antarctica ice sheet to sea-level rise. *Journal of Geophysical Research*, **89**, 6487-6493.

Alley, R.B., Blankenship, D.D., Bentley, C.R. and Rooney, S.T. (1986). Deformation of till beneath ice stream B, West Antarctica. *Nature*, **322**, 57-59.

Alley, R.B., Blankenship, D.D., Bentley, C.R. and Rooney, S.T. (1987a). Continuous till deformation beneath ice sheets. *International Association of Hydrological Sciences*, Publication No. 170, 81-91.

Alley, R.B., Blankenship, D.D., Bentley, C.R. and Rooney, S.T. (1987b). Till beneath ice stream B. 3. Till deformation: Evidence and implications. *Journal of Geophysical Research*, **92B**, 8921-8929.

Alley, R.B., Blankenship, D.D., Rooney, S.T. and Bentley C.R. (1987c). Till beneath ice stream B, 4. A coupled ice-till flow model. *Journal of Geophysical Research*, **92**, 8931-8940.

Alley, R.B., Blankenship, D.D., Rooney, S.T. and

Bentley, C.R. (1989a). Water-pressure coupling of sliding and bed deformation: III. Application to Ice Stream B, Antarctica. *Journal of Glaciology*, **35**, 130-139.

Alley, R.B., Blankenship, D.D., Rooney, S.T. and Bentley, C.R. (1989b). Sedimentation beneath ice shelves; The view from ice stream B. *In*: Powell, R.D. and Elverhøi, A. (eds), *Modern Glaciomarine Environments; Glacial and Marine Controls of Modern Lithofacies and Biofacies*. *Marine Geology*, **85**, 101-120.

Alley, R.B., Blankenship, D.D., Rooney, S.T. and Bentley, C.R. (1989c)(Abstract). The sedimentary signature of deforming glacier beds in the Ross Embayment, Antarctica. *In*: Young, N.W. (ed.), *Proceedings of the Symposium on Ice Dynamics*. *Annals of Glaciology*, **12**, 210-211.

Alley, R.B., Lawson, D.E., Evenson, E.B. and Strasser, J.C. (*in preparation*). Basal ice accretion from water flowing beneath overdeepened glaciers. *Journal of Glaciology*.

Alley, R.B., Meese, D.A., Shuman, C.A., Gow, A.J., Taylor, K.C., Grootes, P.M., White, J.W.C. Ram, M., Waddington, E.D., Mayewski, P.A. and Zielinski, G.A. (1993). Abrupt increase in Greenland snow accumulation at the end of the Younger Dryas event. *Nature,* **362**, 527-529.

Åmark, M. (1986). Clastic dikes formed beneath an active glacier. *Geologiska Föreningens i Stockholm Förhandlingar*, **108**, 13-20.

Ambach, W. and Kuhn, M. (1985). The shift of equilibrium-line altitude on the Greenland ice sheet following climatic changes. *In*: *Glaciers, Ice Sheets, and Sea Level: Effect of a CO_2-Induced Climatic Change*. Report of a workshop held in Seattle, Washington, September, 1984. United States Department of Energy, Washington, D.C., pp. 255-257.

Ambach, W., Ambach, E. and Tributsch, W. (1991). Unusual discoveries of corpses immersed in glacier ice after fatal accidents - glaciological aspects. *Journal of Glaciology*, **37**, 185-186.

Ambach, W., Blumthaler, M. and Kirchlechner, P. (1981). Application of the gravity flow theory to the percolation of melt water through firn. *Journal of Glaciology*, **27**, 67-75.

Andersen, B.G. (1968). Glacial geology of western Troms, North Norway. *Norges Geologiske Undersøkelse*, **256**, 160 pp.

Anderson, D.L. (1982). Hotspots, polar wander,

Mesozoic convection and the geoid. *Nature*, **297**, 391-393.

Anderson, E.A. (1976). A point energy and mass balance model of a snow cover. Silver Spring, Maryland. *National Oceanic and Atmospheric Administration, United States Department of Commerce, Technical Memorandum NWS-19.*

Anderson, G.S. (1970). Hydrologic reconnaissance of the Tanana Basin, central Alaska. *United States Geological Survey Hydrological Investigations Atlases, HA-319.*

Anderson, J.B. (1983). Ancient glacial-marine deposits: Their spatial and temporal distribution. *In*: Molnia, B.F. (ed.), *Glacial-Marine Sedimentation*, pp. 3-92. Plenum Press, New York.

Anderson, J.B. and Ashley, G.M. (1991). Glacial Marine Sedimentation: Paleoclimatic significance. *Geological Society of America Special Paper*, 261, 232 pp.

Anderson, J.B. and Bartek, L.R. (1991). Cenozoic glacial history of the Ross Sea revealed by intermediate resolution seismic reflection data combined with drill site information. *American Geophysical Union Antarctic Research Series*, 56, 231-264.

Anderson, J.B. and Domack, E.W. (1991). The glacial marine environment: Subglacial and climatic setting. Foreword. *In*: Anderson, J.B. and Ashley, G.M. (eds), *Glacial Marine Sedimentation: Paleoclimatic significance. Geological Society of America Special Paper*, 261.

Anderson, J.B., Domack, E.W. and Kurtz, D.D. (1980a). Observations of sediment-laden icebergs in Antarctic waters: Implications to glacial erosion and transport. *Journal of Glaciology*, **25**, 387-396.

Anderson, J.B., Kurtz, D.D., Domack E.W. and Balshaw, K.M. (1980b). Glacial and glacial marine sediments of the Antarctic continental shelf. *Journal of Geology*, **88**, 399-414.

Anderson, J.B., Brake, C.F., Domack, E., Myers, N. and Wright, R. (1983). Development of a polar glacial-marine sedimentation model from Antarctic Quaternary deposits and glaciological information. *In*: Molnia, B.F. (ed.), *Glacial-Marine Sedimentation*, pp. 233-264. Plenum Press, New York.

Anderson, J.B., Brake, C.F. and Myers, N.C. (1984). Sedimentation in the Ross Sea continental shelf, Antarctica. *Marine Geology*, **57**, 295-333.

Anderson, J.B., Kennedy, D.S., Smith, M.J. and Domack, E.W. (1991). Sedimentary facies associated with Antarctica's floating ice masses. *In*: Anderson, J.B. and Ashley, G.M. (eds), *Glacial Marine*

Sedimentation: Paleoclimatic Significance. Geological Society of America Special Paper, 261, 1-26.

Anderson, L.W. (1978). Cirque glacier erosion rates and characteristics of Neoglacial tills, Pangnirtung Fjord area, Baffin Island, North West Territories, Canada. *Arctic and Alpine Research*, **10**, 749-760.

Anderson, R.S., Hallet, B., Walder, J.S. and Aubry, B.F. (1982). Observations in a cavity beneath Grinnell Glacier. *Earth Surface Processes and Landforms*, **7**, 63-70.

Andreasen, J.-O. (1985). Seasonal surface-velocity variations on a sub-polar glacier in West Greenland. *Journal of Glaciology*, **31**, 319-323.

Andresen, A. and Bjerrum, L. (1967). Slides in subaqueous slopes in loose sand and silt. *In*: Richards, A.F. (ed.), *Marine Geotechnique*, pp. 221-239. University of Illinois Press, Chicago.

Andrews, J.T. (1963). Cross-valley moraines of the Rimrock and Isotoq valleys, Baffin Island, N.W.T.: A descriptive analysis. *Geographical Bulletin*, **19**, 49-77.

Andrews, J.T. (1971). Englacial debris in glaciers. *Journal of Glaciology*, **10**, 410.

Andrews, J.T. (1972a). Glacier power, mass balances, velocities and erosion potential. *Zeitschrift für Geomorphologie*, **NF13**, 1-17.

Andrews, J.T. (1972b). Englacial debris in glaciers. *Journal of Glaciology*, **11**, 155.

Andrews, J.T. (1975). *Glacial Systems: An Approach to Glaciers and their Environments*. Duxbury Press, North Sciutate, MA.

Andrews, J.T. (1985). *Quaternary Environments of Eastern Canada Arctic, Baffin Bay and West Greenland*. Allen and Unwin, Boston, 774 pp.

Andrews, J.T. (1987a). Postface: The Laurentide Ice Sheet: Research problems. *Géographie Physique et Quaternaire*, **41**, 315-318.

Andrews, J.T. (1987b). Glaciation and sea-level: A case study. *In*: Devoy, R.J.N. (ed.), *Sea Surface Studies: A Global View*, pp. 95-126. Croom Helm, London, New York, Sydney.

Andrews, J.T. (1987c). The late Wisconsin glaciation and deglaciation of the Laurentide ice sheet. *In*: Ruddiman, W.F. and Wright, H.E., Jr (eds), *North America and Adjacent Oceans During the Last Deglaciation*, pp. 13-37. Geological Society of America, Boulder, Colorado, The Geology of North America, K-3.

Andrews, J.T. (1987d). Late Quaternary marine sediment accumulation in fiord-shelf-deep-sea transects, Baffin Island to Baffin Bay. *Quaternary Science Reviews*, **6**,

231-243.

Andrews, J.T. (1990). Fiord to deep sea sediment transfers along the northeastern Canadian continental margin: Models and data. *Géographie Physique et Quaternaire*, **44**, 55-70.

Andrews, J.T. and Matsch, C.L. (1983). *Glacial Marine Sediments and Sedimentation: An Annotated Bibliography*. Bibliography No. 11. Geo Abstracts, Norwich, 227 pp.

Andrews, J.T. and Miller, G.H. (1972). Chemical weathering of tills and surficial deposits in east Baffin Island, N.W.T. Canada. *In*: Adams, W.P. and Helleiner, F.M. (eds), *International Geography 1972*, pp. 5-7. v.1, Paper 0102, University of Toronto Press, Toronto.

Andrews, J.T. and Smithson, B.B. (1966). Till fabrics of the cross-valley moraines north-central Baffin Island, North West Territories, Canada. *Bulletin of the Geological Society of America*, **77**, 271-290.

Andrews, J.T. and Syvitski, J.P.M. (1991). Sediment fluxes along high latitude glaciated continental margins: Northeast Canada and eastern Greenland. *In*: Maybeck, M. and Hay, W. (eds), *Global Sedimentary Fluxes*. NRC/NAS Press.

Andrews, J.T. and Tedesco, K. (1992). Detrital carbonate-rich sediments, northwestern Labrador Sea: Implications for ice-sheet dynamics and iceberg rafting (Heinrich) events in the North Atlantic. *Geology*, **20**, 1087-1090.

Andrews, J.T., Jull, A.J.T., Donahue, D.J., Short, S.K. and Osterman, L.E. (1985). Sedimentation rates in Baffin Island fiord cores from comparative radiocarbon dates. *Canadian Journal of Earth Sciences*, **22**, 1827-1834.

Anonymous (1985). Glaciers, Ice Sheets and Sea Level: Effect of CO_2-induced climatic change. Report of a workshop held in Seattle, Washington, September 1984. United States Department of Energy, Washington, D.C., pp. 241-247.

Antevs, E. (1922). The recession of the last ice sheet in New England. *American Geographical Society Research Series*, No. 11, 120 pp.

Anthony, R.S. (1977). Iron-rich rhythmically laminated sediments in Lake of the Clouds, northeastern Minnesota. *Limnology and Oceanography*, **22**, 45-54.

Apollonio, S. (1973). Glaciers and nutrients in Arctic seas. *Science*, **180**, 491-493.

Archard, J.F. (1953). Contact and rubbing of flat surfaces. *Journal of Applied Physics*, **24**, 981-988.

Arkhipov, S.A., Bespaly, V.G., Faustova, M.A.,

Glushkova, O.Y., Isaeva, L.I. and Velichko, A.A. (1986). Ice-sheet reconstruction. *In*: Šibrava, V., Bowen, D.Q. and Richmond, G.M. (eds), *Quaternary Glaciations in the Northern Hemisphere. Quaternary Science Reviews*, **5**, 475-483.

Arnborg, L. (1955). Hydrology of the glacial river Austurfljót, Chapter 7, the Hoffellssandur - a glacial outwash plain. *Geografiska Annaler*, **37**, 185-201.

Armstrong, T.E., Roberts, B. and Swithinbank, C. (1973). *Illustrated Glossary of Snow And Ice*, 2nd edn. Scott Polar Research Institute, Cambridge, 60 pp.

Ashley, G.M. (1975). Rhythmic sedimentation in glacial Lake Hitchcock, Massachusetts-Connecticut. *In*: Jopling, A.V. and McDonald, B.C. (eds), *Glaciofluvial and Glaciolacustrine Sedimentation. Society of Economic Paleontologists and Mineralogists, Special Publication No. 23*, pp. 304-320.

Ashley, G.M. (1979). Sedimentology of a tidal lake, Pitt Lake, British Columbia, Canada. *In*: Schlüchter, Ch. (ed.), *Moraines and Varves: Origin\Genesis\Classification*, pp. 327-345. A.A. Balkema, Rotterdam.

Ashley, G.M. (1988). Classification of glaciolacustrine sediments. *In*: Goldthwait, R.P. and Matsch, C.L. (eds), *Genetic Classification of Glacigenic Deposits*, pp. 243-260. A.A. Balkema, Rotterdam.

Ashley, G.M., Southard, J.B. and Boothroyd, J.C. (1982). Deposition of climbing-ripple beds: A flume simulation. *Sedimentology*, **29**, 67-79.

Ashley, G.M., Shaw, J. and Smith, N.D. (1985). *Glacial Sedimentary Environments*. Society of Economic Paleontologists and Mineralogists, Short Course No. 16, 246 pp.

Ashmore, P.E. (1982). Laboratory modelling of gravel braided stream morphology. *Earth Surface Processes and Landforms*, **7**, 201-225.

Ashton, W.S. (1984). Variability of flood estimates based on record length for selected Alaskan rivers. *In*: *Alaska's Water: A Cultural Resource*, pp. 77-91. Fairbanks, University of Alaska, Institute of Water Resources, Report IWR-106.

Ashworth, P.J. and Ferguson, R.I. (1986). Interrelationships of channel processes, changes and sediments in a proglacial braided river. *Geografiska Annaler*, **68A**, 361-371.

Atkinson, B.K. (1984). Subcritical crack growth in geological materials. *Journal of Geophysical Research*, **89B**, 4077-4114.

Atkinson, B.K. (1987). *Fracture Mechanics of Rock*. Academic Press, London, 534 pp.

Atkinson, B.K. and Meredith, P.G. (1987). The theory of subcritical crack growth with applications to minerals and rocks. *In*: Atkinson, B.K. (ed.), *Fracture Mechanics of Rock*, pp. 111-166. Academic Press, London.

Atkinson, J.H. and Bransby, P.L. (1978). *The Mechanics of Soils - An Introduction to Critical State Soil Mechanics*. McGraw Hill Book Company, New York.

Attewell, P.B. and Farmer, I.W. (1975). *Principles of Engineering Geology*. Chapman and Hall, New York, 1045 pp.

Avsyuk, G.A. (1955). Temperaturnoe Sostoianie Lednikov. *Iszvestiia Akademii Nauk SSR, Seriia Geograficheskaia*, **1**, 14-31.

Aylsworth, J.M. and Shilts, W.W. (1989). Bedforms of the Keewatin Ice Sheet, Canada. *Sedimentary Geology*, **62**, 407-428.

Baden-Powell, D.F.W. (1948). The chalky boulder clays of Norfolk and Suffolk. *Geological Magazine*, **85**, 279-296.

Bader, H. (1963). Theory of densification of dry snow on high polar glaciers, II. *In*: Kingery, W.D. (ed.), *Ice and Snow; Properties, Processes and Applications*, pp. 351-376. MIT Press, Cambridge, MA.

Bagnold, R.A. (1954). Experiments on a gravity-free dispersion of large solid spheres in a Newtonian fluid under shear. *Proceedings of the Royal Society of London*, **225A**, 49-63.

Baker, C.A. (1977). Quaternary stratigraphy and environments in the Upper Cam Valley. Unpublished Ph.D. Thesis, University of London.

Baker, D., EscherVetter, H., Moser, H., Oerter, H. and Reinwarth, O. (1982). A glacier discharge model based on results from field studies of energy balance, water storage, and flow. In Hydrological Aspects of Alpine and High Mountain Areas (Proceedings of Exeter Symposium, July 1982). *International Association of Hydrological Sciences Publication No. 138*, pp. 103-112.

Baker, H.W., Jr (1976). Environmental sensitivity of submicroscopic surface textures on quartz sand grains - A statistical evaluation. *Journal of Sedimentary Petrology*, **46**, 871-880.

Baker, R.W. (1978). The influence of ice-crystal size on creep. *Journal of Glaciology*, **21**, 485-497.

Baker, V.R. (1973). Paleohydrology and sedimentology of Lake Missoula flooding in eastern Washington. *Geological Society of America Special Paper 144*, 79 pp.

Baker, V.R., Kochel, R.C. and Patton, P.C. (1988). *Flood Geomorphology*. John Wiley, New York, 503 pp.

Banerjee, I. and McDonald, B.C. (1975). Nature of esker sedimentation. *In*: Jopling, A.V. and McDonald, B.C. (eds), *Glaciofluvial and Glaciolacustrine Sedimentation. Society of Economic Paleontologists and Mineralogists, Special Publication 23*, pp. 132-154.

Banham, P.H. (1970). Notes on Norfolk Sections. *In*: Boulton, G.S. (ed.), *Quaternary Research Association Field Guide*. Norwich.

Banham, P.H. (1975). Glacitectonic structures: A general discussion with particular reference to the contorted drift of Norfolk. *In*: Wright, A.E. and Moseley, F. (eds), *Ice Ages: Ancient and Modern*, pp. 69-94. Geological Journal Special Issue 6. Seel House Press, Liverpool.

Banham, P.H. (1977). Glacitectonites in till stratigraphy. *Boreas*, **6**, 101-105.

Banham, P.H. (1988a). Thin-skinned glaciotectonic structures. *In*: Croot, D.G. (ed.), *Glaciotectonics: Forms and Processes*, pp. 21-25. A.A. Balkema, Rotterdam.

Banham, P.H. (1988b). Polyphase glaciotectonic deformation in the Contorted Drift of Norfolk. *In*: Croot, D.G. (ed.), *Glaciotectonics: Forms and Processes*, pp. 27-32. A.A. Balkema, Rotterdam.

Baranowski, S. (1970). The origin of fluted moraine at the fronts of contemporary glaciers. *Geografiska Annaler*, **52A**, 68-75.

Barker, P.F. *et al.* (1987). Ocean drilling program: Glacial history of Antarctica. *Nature*, **328**, 115-116.

Barkov, N.I. (1970). Preliminary results of ice drilling at Vostok Station. *Informatsionnyi Biulleten*, **80**, 58-60.

Barnes, H.L. (1956). Cavitation as a geological agent. *American Journal of Science*, **254**, 493-505.

Barnes, P.W. (1982). Marine ice-pushed boulder ridge, Beaufort Sea, Alaska. *Arctic*, **35**, 312-316.

Barnes, P.W. (1987). Morphologic studies of the Wilkes Land Continental Shelf, Antarctica; glacial and iceberg effects. *In*: Eittreim, S.L. and Hampton, M.A. (eds), *The Antarctic Continental Margin: Geology and Geophysics of Offshore Wilkes Land*, pp. 175-194. Circum-Pacific Council for Energy and Mineral Resources, Earth Science Series, 5A, Houston.

Barnes, P.W. and Lien, R. (1988). Icebergs rework shelf sediments to 500 m off Antarctica. *Geology*, **16**, 1130-1133.

Barnes, P.W. and Reimnitz, E. (1982). Sediment content of nearshore fast-ice, fall 1980 Beaufort Sea, Alaska. *Open File Report United States Geological Survey*,

716, 18 pp.

Barnes P.W., Reimnitz, E. and Fox, D. (1982a). Ice rafting of fine-grained sediment, a sorting and transport mechanism, Beaufort Sea, Alaska. *Journal of Sedimentary Petrology*, **52**, 493-502.

Barnes, P.W., Reiss, T. and Reimnitz, E. (1982b). Sediment content of nearshore fast ice, fall 1980, Beaufort Sea, Alaska. *United States Geological Survey, Open-File Report, 82-716*, 19 pp.

Barnes, P.W., Rearic, D.M. and Reimnitz, E. (1984). Ice gouging characteristics and processes. *In*: Barnes, P.W., Schell, D.M. and Reimnitz, E. (eds), *The Alaskan Beaufort Sea - Ecosystems and Environments*, pp. 185-212. Academic Press, Orlando.

Barnes, P.W., Asbury, J.L., Rearic, D.M. and Ross, C.R. (1987). Ice erosion of a sea-floor knickpoint at the inner edge of the stamukhi zone, Beaufort Sea, Alaska. *Marine Geology*, **76**, 207-222.

Barnett, D.M. and Holdsworth, G. (1974). Origin, morphology, and chronology of sublacustrine moraines, Generator Lake, Baffin Island, Northwest Territories, Canada. *Canadian Journal of Earth Sciences*, **11**, 380-408.

Barrett, P.J. (1980). The shape of rock particles, a critical review. *Sedimentology*, **27**, 291-303.

Barrett, P.J. (1981). Late Cenozoic glaciomarine sediments of the Ross Sea, Antarctica. *In*: Hambrey, M.J. and Harland, W.B. (eds), *Earth's Pre-Pleistocene Glacial Record*, pp. 208-211. Cambridge University Press, Cambridge.

Barrett, P.J., Pyne, A.R. and Ward, B.L. (1983). Modern sedimentation in McMurdo Sound, Antarctica. *In*: Oliver, R.L., James, P.R. and Jago, J.B. (eds), *Antarctic Earth Science*, pp. 550-554. Australian Academy of Science, Canberra, Cambridge University Press, Cambridge.

Barrett, P.J., Elston, D.P., Harwood, D.M., McKelvey B.C. and Webb, P.-N. (1987). Mid-Cenozoic Record of glaciation and sea-level change on the margin of the Victoria Land basin, Antarctica. *Geology*, **15**, 634-637.

Barrett, P.J., Pyne, A.R. and MacPherson, A.J. (1988). Sedimentation in a polar coastal setting, Antarctica. *Geological Society of America, Abstracts with Programs*, **20**, 84-85.

Barrett, P.J., Hambrey, M.J., Harwood, D.M. Pyne, A.R. and Webb, P.N. (1989). General synthesis. *In*: Barrett, P.J. (ed.), *Antarctic Cenozoic History of the CIROS-1 Drillhole*, pp. 241-251. Bulletin in the Miscellaneous Series of the New Zealand Department of Scientific

and Industrial Research, No. 245.

Barrett, P.J., Adams, C.J., McIntosh, W.C., Swisher III, C.C. and Wilson, G.S. (1992). Geochronological evidence supporting Antarctic deglaciation three million years ago. *Nature*, **359**, 816-818.

Barron, E. (1985). Explanations of the Tertiary global cooling trend. *Palaeogeography, Palaeoclimatology, Palaeoecology*, **50**, 45-61.

Barron, J.A., Larsen, B. and Baldauf, C. (1991). Evidence for late Eocene to early Oligocene Antarctic glaciation and observations on late Neogene glacial history of Antarctica: Results from Leg 119. *In*: Barron, J.A., Larsen, B., Baldauf, J.G., Alibert, C., Berkowitz, S., Caulet, J.-P., Chambers, S.R., Cooper, A.K., Cranston, R.E., Dorn, W.U., Ehrmann, W.U., Fox, R.D., Fryxell, G.A., Hambrey, M.J., Huber, B.T., Jenkins, C.J., Kang, S.-H., Keating, B.H., Mehl, K.W., Noh, I., Ollier, G., Pittenger, A., Sakai, H., Schroder, C.J., Solheim, A., Stockwell, D.A., Thierstein, H.R., Tocher, B., Turner, B.R., Wei, W., Mazzullo, E.K. and Stewart, N.J. (eds), *Proceedings of the Ocean Drilling Program, Kerguelen Plateau-Prydz Basin*, pp. 869-891. Proceedings of the Ocean Drilling Program, Scientific Results, 119.

Barry, R.G. (1985). Snow cover, sea ice, and permafrost. *In: Glaciers, Ice Sheets and Sea Level: Effect of a CO_2-induced Climatic Change*. Report of a workshop held in Seattle, Washington, September 1984. United States Department of Energy, Washington, DC, 330 pp.

Bartek, L.R. and Anderson, J.B. (1991). Facies distribution resulting from sedimentation under polar interglacial climatic conditions within a high-latitude marginal basin, McMurdo Sound, Antarctica. *In*: Anderson, J.B. and Ashley, G.M. (eds), *Glacial Marine Sedimentation: Paleoclimatic Significance*, pp. 27-49. Geological Society of America, Special Paper, 261.

Bartek, L.R., Vail, R.R., Anderson, J.B., Emmet, P.A. and Wu, S. (1991). Effect of Cenozoic ice sheet fluctuations in Antarctica on the stratigraphic signature of the Neogene. *Journal of Geophysical Research*, **96**, 6753-6778.

Bartsch-Winkler, S. and Ovenshine, A.T. (1984). Macrotidal subarctic environment of Turnagain and Knik Arms, upper Cook Inlet, Alaska: Sedimentology of the intertidal zone. *Journal of Sedimentary Petrology*, **54**, 1221-1238.

Bartsch-Winkler, S. and Schell, H.R. (1984). Bedding types in Holocene tidal channel sequences, Knik Arm,

Upper Cook Inlet, Alaska. *Journal of Sedimentary Petrology*, **54**, 1239-1250.

Bates, C.C. (1953). Rational theory of delta formation. *American Association of Petroleum Geologists Bulletin*, **37**, 2119-2162.

Beecroft, I. (1983). Sediment transport during an outburst from Glacier de Tsidjiore Nouve, Switzerland, 16-19 June 1981. *Journal of Glaciology*, **29**, 185-190.

Beget, J.E. (1986). Modelling the influence of till rheology on the flow and profile of the Lake Michigan lobe, southern Laurentide ice sheet, U.S.A. *Journal of Glaciology*, **32**, 235-241.

Beget, J.E. (1987). Low profile of the northwest Laurentide ice sheet. *Arctic and Alpine Research*, **19**, 81-88.

Behrendt, J.C. and Cooper, A. (1991). Evidence of rapid Cenozoic uplift of the shoulder escarpment of the Cenozoic West Antarctic rift system and a speculation on possible climatic forcing. *Geology*, **19**, 315-319.

Behrens, H., Bergmann, N., Moser, H., Rauert, W., Stichler, W., Ambach, W., Eisner H. and Pessl, K. (1971). Study of the discharge of alpine glaciers by means of environmental isotopes and dye tracers. *Zeitschrift für Gletscherkunde und Glazialgeologie*, **7**, 79-102.

Behrens, H., Oerter, H. and Reinwarth, O. (1982). Results of tracer tests with fluorescent dyes on the Vernagtferner from 1974 to 1982. *Zeitschrift für Gletscherkunde und Glazialgeologie*, **18**, 65-83.

Belknap, D.F., Kelley, J.T. and Shipp, R.C. (1986). Quaternary stratigraphy of representative Maine estuaries: Initial examination by high resolution seismic reflection profiling. *In*: Fitzgerald, D. and Rosen, P. (eds), *A Treatise on Glaciated Coasts*, pp. 177-207. Academic Press, New York.

Bell, M. and Walker, M.J.C. (1992). *Late Quaternary Environmental Change: Physical and Human Perspectives*. Longman, London, 273 pp.

Bell, R.T. (1970). The Hurwitz Group - A prototype for deposition on metastable cratons. *In*: Baer, A.J. (ed.), Basins and geosynclines of the Canadian Shield. *Geological Survey of Canada, Paper 70-40*, pp. 159-169.

Bender, J.A. and Gow, A.J. (1961). Deep drilling in Antarctica. *International Association of Scientific Hydrology*, **55**, 132-141.

Benson, C.S. (1961). Stratigraphic studies in the snow and firn of the Greenland Ice Sheet. *Folia Geographica Danica*, **9**, 13-37.

Benson, C.S., Harrison, W., Gosink, J., Bowling, S.,

Mayo, L. and Trabant, D. (1986). *Workshop on Alaskan Hydrology: Problems Related to Glacierized Basins*. University of Alaska Geophysical Institute Report UAG-R-306.

Bentley, C.R. (1984). The Ross Ice Shelf Geophysical and Glaciological Survey (RIGGS): Introduction and summary of the measurements performed. *Antarctic Research Series*, **42**, 1-20.

Bentley, C.R., Anandakrishnan, S. and Rooney, S.T. (1988). Seismic studies on the Siple Coast, 1987-1988. *Antarctic Journal of the United States*, **23**, 54-55.

Berg, M.W. van den and Beets, D.J. (1987). Saalian glacial deposits and morphology in The Netherlands. *In*: van der Meer, J.J.M. (ed.), *Tills and Glaciotectonics*, pp. 235-251. A.A. Balkema, Rotterdam.

Berger, A.L. (1977). Long-term variation of the earth's orbital elements. *Celestial Mechanics*, **15**, 53-74.

Berger, A.L. (1978). Long-term variations of caloric insolation resulting from the Earth's orbital elements. *Quaternary Research*, **9**, 139-167.

Berger, A.L. (1981a). Spectrum of climatic variations and possible causes. *In*: Berger, A.L. (ed.), *Climatic Variations and Variability: Facts and Theories*, pp. 411-432. D. Reidel, Dordrecht.

Berger, A.L. (1981b). The astronomical theory of paleoclimates. *In*: Berger, A.L. (ed.), *Climatic Variations and Variability: Facts and Theories*, pp. 501-525. D. Reidel, Dordrecht.

Berger, A.L., Imbrie, J., Hays, J.D, Kukla, G.J. and Saltzman, B. (1984). *Milankovitch and Climate, Part I*. D. Reidel, Dordrecht, 895 pp.

Berger, A.L., Gallée, H., Fichefet, T., Marsiat, I. and Tricot, C. (1990). Testing the astronomical theory with a coupled climate-ice-sheet model. *Palaeogeography, Palaeoclimatology, Palaeoecology (Global and Planetary Change Section)*, **89**, 125-141.

Berger, W.H. (1973). Deep-sea carbonates: Pleistocene dissolution cycles. *Journal of Foraminiferal Research*, **3**, 187-195.

Berger, W.H. (1990). The Younger Dryas cold spell - A quest for causes. *Palaeogeography, Palaeoclimatology, Palaeoecology (Global and Planetary Change Section)*, **89**, 219-237.

Berger, W.H. and Labeyrie, L.D. (1987). *Abrupt Climatic Change*. D. Reidel, Dordrecht, 425 pp.

Berger, W.H., Burke, S. and Vincent, E. (1987). Glacial-Holocene transition: Climate pulsations and sporadic shutdown of NAOW production. *In*: Berger, W.H. and

Labeyrie, L.D. (eds), *Abrupt Climatic Change; Evidence and Implications*, pp. 279-297. D. Reidel, Dordrecht.

Berggren, W.A., Burckle, L.H., Cita, M.B., Cook, H.B.S., Funnell, B.M., Gartner, S., Hays, J.D., Kennett, J.P., Opdyke, N.D., Pastouret, L., Shackleton, N.J. and Takayanagi, Y. (1980). Towards a Quaternary time scale. *Quaternary Research*, **13**, 277-302.

Berger, W., Stauffer, B. and Oeschger, H. (1978a). Past atmospheric composition and climate, gas parameters measured on ice cores. *Nature*, **276**, 53-55.

Berner, W., Stauffer, B. and Oeschger, H. (1978b). Dynamic glacier flow model and the production of internal meltwater. *Zeitschrift für Gletscherkunde und Glazialgeologie*, **13**, 209-217.

Bernhardi, A. (1832). Wie Kamen die aus dem Norden Stammenden Felsbruchstücke und Geschiebe, welche man in Norddeutschland und den benachbarten Ländern findet, an ihre genenwärtigen Fundorte? *Jahrbuch für Mineralogie, Geognosie und Petrefaktenkunde*, **3**, 257-267.

Berthé, D., Choukroune, P. and Jegouzo, P. (1979). Orthogneiss, mylonite and non coaxial deformation of granites: The example of the South Armorican shear zone. *Journal of Structural Geology*, **1**, 31-42.

Berthelsen, A. (1978). The methodology of kineto-stratigraphy as applied to glacial geology. *Bulletin of the Geological Society of Denmark, Special Issue*, **27**, 25-38.

Bezinge, A. (1978). Torrents glaciaires, hydrologie et charriages d'alluvions. *Sonderdruck aus dem Jahrbuch der Schweizerischen Naturforschenden Gesellschaft, Wissenschaftlicher, Teil 1978*, 152-169.

Bezinge, A. (1981). Glacial meltwater streams, hydrology and sediment transport: The case of the Grande Dixence hydroelectricity scheme (in French). Birkhauser. (*See also translation as Bezinge, 1987.*)

Bezinge, A. (1982). Grande Dixence et son hydrology. *Separatdruck aus Mitteilung, Nr. 3 der Landeshydrologie*, Bern, Switzerland.

Bezinge, A. (1987). Glacial meltwater streams, hydrology and sediment transport: The case of the Grande Dixence hydroelectricity scheme. *In*: Gurnell, A.M. and Clark, M.J. (eds), *Glacio-Fluvial Sediment Transfer - An Alpine Perspective*, pp. 473-498. John Wiley, Chichester.

Bezinge, A. and Schafer, F. (1968). Storage pumps and glacial waters. The British Hydromechanics Research Association, Information Department T-1019.

(Translation of original paper: Pompes d'Accumulation et Eaux Glaciers. *Bull. Tech. Suisse Romande*, **49**(20), 282-290 (October 1968).)

Bezinge, A., Clark, M.J., Gurnell, A.M. and Warburton, J. (1989). The management of sediment transported by glacial melt-water streams and its significance for the estimation of sediment yield. *Annals of Glaciology*, **13**, 1-5.

Bidgood, D.E.T. and Harland, W.B. (1961). Palaeomagnetism in some East Greenland sedimentary rocks. *Nature*, **189**, 633-634.

Biju-Duval, B., Deynoux, M. and Rognon, P. (1981). Late Ordovician tillites of the central Sahara. *In*: Hambrey, M.J. and Harland, W.B. (eds), *Earth's Pre-Pleistocene Glacial Record*, pp. 99-107. Cambridge University Press, Cambridge.

Billard, A. (1985). Quaternary Chronologies around the Alps. *In*: Mahaney, W.C. (ed.), *Correlation of Quaternary Chronologies*, pp. 177-189. Geo Books, Norwich.

Billard, A. and Orombelli, G. (1986). Quaternary glaciations in the French and Italian piedmonts of the Alps. *In*: Šibrava, V., Bowen, D.Q. and Richmond, G.M. (eds), *Quaternary Glaciations in the Northern Hemisphere. Quaternary Science Reviews*, **5,** 407-411.

Bindschadler, R. (1983). The importance of pressurized subglacial water in separation and sliding at the glacier bed. *Journal of Glaciology*, **29**, 3-19.

Bindschadler, R., Harrison, W.D., Raymond, C.F. and Crosson, R. (1977). Geometry and dynamics of a surge-type glacier. *Journal of Glaciology*, **18**, 181-194.

Bindschadler, R.A., Roberts, E.P. and MacAyeal, D.R. (1989). Distribution of net mass balance in the vicinity of Crary Ice Rise, Antarctica. *Journal of Glaciology*, **35**, 370-377.

Biot, M.A. (1961). Theory of folding of stratified viscoelastic media and its implications in tectonics and orogenesis. *Geological Society of America Bulletin*, **72**, 1595-1620.

Birchfield, G.E. and Weertman, J. (1983). Topography, albedo-temperature feedback and climate sensitivity. *Science*, **219**, 284-285.

Birnie, R.V. (1992). Estimates of slope retreat around the Ivory Glacier catchment, South Island, New Zealand. *New Zealand Geographer*, **43**, 37-42.

Bishop, J.F. and Walton, J.W.L. (1981). Bottom melting under the George VI ice shelf, Antarctica. *Journal of Glaciology*, **27**, 429-447.

Bjerklie, D. and Carlson, R. (1986). Estimation of glacial

meltwater hydrographs. *In*: Kane, D. (ed.), *Proceedings of Cold Regions Hydrology Symposium*, pp. 345-352. American Water Research Association.

Björnsson, H. (1974). Explanation of jökulhlaups from Grímsvötn, Vatnajökull, Iceland. *Jökull*, **24**, 1-26.

Björnsson, H. (1976). Marginal and supraglacial lakes in Iceland. *Jökull*, **26**, 40-51.

Björnsson, H. (1988). Hydrology of ice caps in volcanic regions. *Vísindafélag Íslendinga, Societas Scientarium Islandica*, 45, Reykjavík. 139 pp. + volume of maps.

Björnsson, H. (1992). Jökulhlaups in Iceland: Prediction, characteristics and simulation. *Annals of Glaciology*, **16**, 95-106.

Björnsson, H. and Kristmannsdóttir, H. (1984). The Grímsvötn Geothermal Area, Vatnajökull, Iceland. *Jökull*, **34**, 25-48.

Blake, W., Jr (1970). Studies of glacial history in Arctic Canada. I. Pumice, radiocarbon dates, and differential postglacial uplift in the eastern Queen Elizabeth Islands. *Canadian Journal of Earth Sciences*, **7**, 634-664.

Blake, W., Jr (1977). Iceberg concentrations as an indicator of submarine moraines, eastern Queen Elizabeth Islands, District of Franklin. *Geological Survey of Canada, Paper 77-1b*, pp. 281-286.

Blanford, W.T., Blanford, H.F. and Theobald, W. (1859). On the geological structure and relations of the Talcheer Coal Field, in the District of Cuttack. *India Geological Survey, Memoir 1*, pp. 33-89.

Blankenship, D.D., Bentley, C.R., Rooney, S.T. and Alley, R.B. (1986). Seismic measurements reveal a saturated, porous layer beneath an active Antarctic Ice Stream. *Nature*, **322**, 54-57.

Blankenship, D.D., Bentley, C.R., Rooney, S.T. and Alley, R.B. (1987). Till beneath ice stream B, 1. properties derived from seismic travel times. *Journal of Geophysical Research*, **92**, 8903-8911.

Blankenship, D.D., Rooney, S.T., Alley, R.B. and Bentley, C.R. (1989). Seismic evidence for a thin basal layer at a second location on ice stream B, Antarctica. (Abstract), *Annals of Glaciology*, **12**, 200.

Blatter, H. (1987). On the thermal regime of an Arctic valley glacier: A study of White Glacier, Axel Heiberg Island, N.W.T. Canada. *Journal of Glaciology*, **33**, 200-211.

Blatter, H. and Hutter, K. (1991). Polythermal conditions in Arctic glaciers. *Journal of Glaciology*, **37**, 261-269.

Blatter, H. and Kappenberger, G. (1988). Mass balance and thermal regime of Laika Ice Cap, Coburg Island. N.W.T. Canada. *Journal of Glaciology*, **34**, 102-110.

Bloom, A.L., Broecker, W.S., Chappell, J.M.A., Matthews, R.K. and Mesolella, K.J. (1974). Quaternary sea level fluctuations on a tectonic coast: New ^{230}Th/^{234}U dates from the Huon Peninsula, New Guinea. *Quaternary Research*, **4**, 185-205.

Blown, I. and Church, M. (1985). Catastrophic lake drainage within the Homathko River basin, British Columbia. *Canadian Geotechnical Journal*, **22**, 551-563.

Bluck, B.J. (1974). Structure and directional properties of some valley sandur deposits in southern Iceland. *Sedimentology*, **21**, 533-554.

Bluck, B.J. (1979). Structure of coarse grained braided stream alluvium. *Transactions of the Royal Society of Edinburgh*, **70**, 181-221.

Bluck, B.J. (1982). Texture of gravel bars in braided streams. *In*: Hey, R.D., Bathurst, J.C. and Thorne, C.R. (eds), *Gravel-Bed Rivers; Fluvial Processes, Engineering and Management*, pp. 339-355. John Wiley, Chichester.

Bluck, B.J. (1987). Bed forms and clast size changes in gravel-bed rivers. *In*: Richards, K.S. (ed.), *River Channels: Environment and Process*, pp. 159-178. Institute of British Geography, Special Publication, 18.

Bluemle, J.P. and Clayton, L. (1984). Large-scale glacial thrusting and related processes in North Dakota. *Boreas*, **13**, 279-299.

Boellstorff, J. (1978). North American Pleistocene stages reconsidered in light of probable Pliocene-Pleistocene continental glaciation. *Science*, **202**, 305-307.

Bogen, J. (1986). Erosjonsprosesser og Sediment transport: Norske Vassdrag. (Erosion and sediment transport in Norwegian rivers) (English summary.) *Norsk Hyrologisk Komite, Oslo, Rapport Nr. 20*.

Bogen, J. (1989). Glacial sediment production and development of hydro-electric power in glacierized areas. *Annals of Glaciology*, **13**, 6-11.

Bond, G., Broecker, W., Johnsen, S., McManus, J., Labeyrie, L., Jouzel, J. and Bonani, G. (1993). Correlations between climate records from North Atlantic sediments and Greenland ice. *Nature*, **364**, 407-412.

Bond, G., Heinrich, H., Broecker, W., Labeyrie, L., McManus, J., Andrews, J., Huon, S., Jantschik, R., Clasen, S., Simet, C., Tedesco, K., Klas, M., Bonani, G. and Ivy, S. (1992). Evidence of massive discharges of icebergs into the North Atlantic Ocean during the last glacial period. *Nature*, **360**, 245-249.

Bonifay, D. and Piper, D.J.W. (1988). Probable Late

Wisconsinan ice margin on the upper continental slope off St. Pierre Bank, Eastern Canada. *Canadian Journal of Earth Sciences*, **25**, 853-865.

Booth, D.B. (1987). Timing and processes of deglaciation along the southern margin of the Cordilleran ice sheet. *In*: Ruddiman, W.F. and Wright, H.E., Jr (eds), *North America and Adjacent Oceans During the Last Deglaciation*. Geological Society of America, Boulder, Colorado. The Geology of North America, K-3, pp. 71-90.

Boothroyd, J.C. (1972). Coarse-grained sedimentation on a braided outwash fan, northeast Gulf of Alaska. Technical Report No.6 - Coastal Research Division. University of South Carolina, 127 pp.

Boothroyd, J.C. and Ashley, G.M. (1975). Processes, bar morphology and sedimentary structures on braided outwash fans, northeastern gulf of Alaska. *In*: Jopling, A.V. and MacDonald, B.C. (eds), *Glaciofluvial and Glaciolacustrine Sedimentation*, pp. 193-222. Society of Economic Paleontologists and Mineralogists, Special Publication 23.

Boothroyd, J.C. and Nummedal, D. (1978). Proglacial braided outwash: A model for humid alluvial-fan deposits. *In*: Miall, A.D. (ed.), *Fluvial Sedimentology*, pp. 641-668. Canadian Society of Petroleum Geologists, Memoir 5, Calgary, Alberta, Canada.

Borland, W.M. (1961). Sediment transport of glacier-fed streams in Alaska. *Journal of Geophysical Research*, **66**, 3347-3350.

Boswell, P.G.H. (1916). The petrology of the North Sea Drift and Upper Glacial brick-earths in East Anglia. *Proceedings of the Geologists' Association*, **27**, 79-98.

Bouchard, M.A. (1989). Subglacial landforms and deposits in central and northern Québec, Canada, with emphasis on Rögen moraines. *Sedimentary Geology*, **62**, 293-308.

Boulton, G.S. (1967). The development of a complex supraglacial moraine at the margin of Sørbreen, Ny Friesland, Vestspitsbergen. *Journal of Glaciology*, **6**, 717-735.

Boulton, G.S. (1968). Flow tills and some related deposits on some Vestspitsbergen glaciers. *Journal of Glaciology*, **7**, 391-412.

Boulton, G.S. (1970a). On the deposition of subglacial and melt-out tills at the margins of certain Svalbard glaciers. *Journal of Glaciology*, **9**, 231-245.

Boulton, G.S. (1970b). On the origin and transport of englacial debris in Svalbard glaciers. *Journal of Glaciology*, **9**, 213-229.

Boulton, G.S. (1971a). Till genesis and fabric in Svalbard, Spitsbergen. *In*: Goldthwait, R.P. (ed.), *Till: A Symposium*, pp. 41-72. Ohio State University Press, Columbus.

Boulton, G.S. (1971b). Englacial debris in glaciers: Reply to the comments of Dr J.T. Andrews. *Journal of Glaciology*, **10**, 410-411.

Boulton, G.S. (1972a). The role of thermal regime in glacial sedimentation. *In*: Price, R.J. and Sugden, D.E. (eds), *Polar Geomorphology*, pp. 1-19. Institute of British Geographers Special Publication 4.

Boulton, G.S. (1972b). Modern Arctic glaciers as depositional models for former ice sheets. *Quarterly Journal of the Geological Society of London*, **128**, 361-393.

Boulton, G.S. (1972c). Englacial debris in glaciers: Reply to the comments of Dr J.T. Andrews. *Journal of Glaciology*, **11**, 155-156.

Boulton, G.S. (1974). Processes and patterns of glacial erosion. *In*: Coates, D.R. (ed.), *Glacial Geomorphology*, pp. 41-87. Binghampton, New York.

Boulton, G.S. (1975). Processes and patterns of subglacial sedimentation: A theoretical approach. *In*: Wright, A.E. and Moseley, F. (eds), *Ice Ages: Ancient and Modern*, pp. 7-42. Geological Journal, Special Issue, 6. Seel House Press, Liverpool.

Boulton, G.S. (1976). The origin of glacially fluted surfaces: Observations and theory. *Journal of Glaciology*, **17**, 287-309.

Boulton, G.S. (1978). Boulder shapes and grain-size distributions of debris as indicators of transport paths through a glacier and till genesis. *Sedimentology*, **25**, 773-799.

Boulton, G.S. (1979). Processes of glacier erosion on different substrata. *Journal of Glaciology*, **23**, 15-37.

Boulton, G.S. (1980). Genesis and classification of glacial sediments. *In*: Stankowski, W. (ed.), *Tills and Glacigene Deposits*, pp. 15-18. Uniwersytet im. Adama Mickiewicza, Poznań. Seria Geografia, No. 20.

Boulton, G.S. (1982). Subglacial processes and the development of glacial bedforms. *In*: Davidson-Arnott, R., Nickling, W. and Fahey, B.D. (eds), *Research in Glacial, Glacio-Fluvial and Glacio-Lacustrine Systems*, pp. 1-31. 6th Guelph Symposium on Geomorphology. Geo Books, Norwich.

Boulton, G.S. (1983). Debris and isotopic sequences in the basal layers of polar ice sheets. *In*: Robin, G. deQ (ed.), *The Climatic Record in Polar Ice Sheets*, pp. 83-89. Cambridge University Press, Cambridge.

Boulton, G.S. (1986a). Geophysics - A paradigm shift in

glaciology. *Nature*, **322,** 18.

Boulton, G.S. (1986b). Push-moraines and glacier-contact fans in marine and terrestrial environments. *Sedimentology*, **33,** 677-698.

Boulton, G.S. (1987a). A theory of drumlin formation by subglacial sediment deformation. *In*: Menzies, J. and Rose, J. (eds), *Drumlin Symposium*, pp. 25-80. A.A. Balkema, Rotterdam.

Boulton, G.S. (1987b). Progress in glacial geology during the last fifty years. *Journal of Glaciology, Special Issue*, 25-32.

Boulton, G.S. (1990). Sedimentary and sea level changes during glacial cycles and their control on glacimarine facies architecture. *In*: Dowdeswell, J.A. and Scourse, J.D. (eds), *Glacimarine Environments: Processes and Sediments*, pp. 15-52. Geological Society of London, Special Publication No. 53.

Boulton, G.S. and Clark, C.D. (1990a). A highly mobile Laurentide ice sheet revealed by satellite images of glacial lineations. *Nature*, **346,** 813-817.

Boulton, G.S. and Clark, C.D. (1990b). The Laurentide Ice Sheet through the last glacial cycle: Drift lineations as a key to the dynamic behaviour of former ice sheets. *Transactions of the Royal Society of Edinburgh, Earth Sciences*, **81,** 327-347.

Boulton, G.S. and Dent, D.L. (1974). The nature and rates of post-depositional changes in recently deposited till from south-east Iceland. *Geografiska Annaler*, **56A,** 121-134.

Boulton, G.S. and Deynoux, M. (1981). Sedimentation in glacial environments and the identification of tills and tillites in ancient sedimentary sequences. *Precambrian Research*, **15,** 397-422.

Boulton, G.S. and Eyles, N. (1979). Sedimentation by valley glaciers; A model and genetic classification. *In*: Schlüchter, Ch. (ed.), *Moraines and Varves: Origin/Genesis/Classification*, pp. 11-23. A.A. Balkema, Rotterdam.

Boulton, G.S. and Hindmarsh, R.C.A. (1987). Sediment deformation beneath glaciers: Rheology and geological consequences. *Journal of Geophysical Research*, **92B,** 9059-9082.

Boulton, G.S. and Jones, A.S. (1979). Stability of temperate ice caps and ice sheets resting on beds of deformable sediment. *Journal of Glaciology*, **24,** 29-43.

Boulton, G.S. and Spring, U. (1986). Isotopic fractionation at the base of polar and sub-polar glaciers. *Journal of Glaciology*, **32,** 475-485.

Boulton, G.S., Dent, D.L. and Morris, E.M. (1974).

Subglacial shearing and crushing, and the role of water pressures in tills from south-east Iceland. *Geografiska Annaler*, **56A,** 134-145.

Boulton, G.S., Jones, A.S, Clayton, K.M. and Kenning, M.J. (1977). A British Ice-sheet model and patterns of glacial erosion and deposition in Britain. *In*: Shotton, F.W. (ed.), *British Quaternary Studies: Recent Advances*, pp. 231-246. Clarendon Press, Oxford.

Boulton, G.S., Morris, E.M., Armstrong, A.A. and Thomas, A. (1979). Direct measurement of stress at the base of a glacier. *Journal of Glaciology*, **22,** 3-24.

Boulton, G.S., Smith, G.D. and Morland, L.W. (1984). The reconstruction of former ice sheets and their mass balance characteristics using a non-linearly viscous flow model. *Journal of Glaciology*, **30,** 140-152.

Boulton, G.S., Smith, G.S., Jones, A.S. and Newsome, J. (1985). Glacial geology and glaciology of the last mid-latitude ice sheets. *Journal of Geological Society of London*, **142,** 447-474.

Boulton, G.S., van der Meer, J.J.M., Ruegg, G.H.J., Beets, D.J., Riezebos, P.A., Castel, I.I.M., Hart, J.K., Quinn, I., Thornton, M. and van der Wateren, F.M. (1989). Preliminary report on the Glacitecs 84' expedition to Spitsbergen. Internal Report of the Fysisch-Geografisch en Bodemkundig Laboratorium. University of Amsterdam.

Bouvet, T. (1958). Contribution a l'etude du phenomena d'erosion por Frottement dans la domaine des turbines hydrauliques. *Bulletin Tech. Suisse Romande*, **84**(3).

Bowen, D.Q. (1978). *Quaternary Geology; A Stratigraphic Framework for Multidisciplinary Work*. Pergamon Press, Oxford, 221 pp.

Bowen, D.Q., Rose, J. McCabe, A.M. and Sutherland, D.G. (1986a). Correlation of Quaternary glaciations in England, Ireland, Scotland and Wales. *In*: Šibrava, V., Bowen, D.Q. and Richmond, G.M. (eds), *Quaternary Glaciations in the Northern Hemisphere. Quaternary Science Reviews*, **5,** 299-340.

Bowen, D.Q., Richmond, G.M., Fullerton, D.S., Šibrava, V., Fulton, R.J. and Velichko, A.A. (1986b). Correlation of Quaternary Glaciations in the northern hemisphere. *In*: Šibrava, V., Bowen, D.Q. and Richmond, G.M. (eds), *Quaternary Glaciations in the Northern Hemisphere. Quaternary Science Reviews*, **5,** 509-510 and Chart 1.

Bowen, D.Q., Sykes, G.A. and Turner, C. (1988). Correlation of marine events and glaciations on the northeast Atlantic margin; discussion. *In*: Shackleton, N.J., West, R.G. and Bowen, D.Q. (eds), *The Past Three Million Years: Evolution of Climatic Variability in the North Atlantic Region*, pp. 209-225. Cambridge University Press, Cambridge.

Bozhinskiy, A.N., Krass, M.S. and Popovnin, V.V. (1986). Role of debris cover in the thermal physics of glaciers. *Journal of Glaciology*, **32**, 255-266.

Bradley, R.S. (1985). *Quaternary Paleoclimatology, Methods of Paleoclimatic Reconstruction.* Allen and Unwin, London, 472 pp.

Bradley, R.S., Diaz, H.F., Eischeid, J.K., Jones, P.D., Kelly, P.M. and Goodess, C.M. (1987). Precipitation fluctuations over northern hemisphere land areas since the mid-19th century. *Science*, **237**, 171-175.

Bradley, W.C., Fahnestock, R.K. and Rowekamp, E.T. (1972). Coarse sediment transport by flood flows on Knik river, Alaska. *Geological Society of America Bulletin*, **83**, 1261-1284.

Bradshaw, R. and Thompson, R. (1985). The use of magnetic measurements to investigate the mineralogy of Icelandic lake sediments and to study catchment processes. *Boreas*, **14**, 203-215.

Braithwaite, R.J. (1980). Regional modeling of ablation in West Greenland. *Grønland Geologie Undersøgelse, Rapport*, 98.

Braithwaite, R.J. (1982). A simple model of runoff from ungauged basins in West Greenland. *Grønlands Geologiske Undersøgelse, Rapport*, 111.

Braithwaite, R.J. (1984). Hydrological modeling in Greenland in connection with hydropower. *Grønland Geologie Undersøgelse. Rapport*, **120**, 90-94.

Braithwaite, R.J. and Olesen, O.B. 1988. Effect of glaciers on annual run-off, Johan Dahl Land, south Greenland. *Journal of Glaciology*, **34**, 200-207.

Brand, G., Pohjola, V. and Hooke, R.LeB. (1987). Evidence for a till layer beneath Storglaciären, Sweden, based on electrical resistivity measurements. *Journal of Glaciology*, **33**, 311-314.

Bray, C.J. and Karig, D.E. (1985). Porosity of sediments in accretionary prisms and some implications for dewatering processes. *Journal of Geophysical Research*, **90**, 768-778.

Brayshaw, A.C., Frostick, L.E. and Reid, I. (1983). The hydrodynamics of particle clusters and sediment entrainment in coarse alluvial channels. *Sedimentology*, **30**, 137-143.

Breitkopf, J.H. (1988). Iron formations related to mafic volcanism and ensialic rifting in the southern margin zone of the Damara orogen, Namibia. *Precambrian Research*, **38**, 111-130.

Bretz, J.H. (1969). The Lake Missoula floods and the channelled scabland. *Journal of Geology*, **77**, 505-543.

Brice, J.C. (1964). Channel patterns and terraces of the Loup Rivers in Nebraska. *United States Geological Survey, Professional Paper 422-D*, 41 pp.

Broccoli, A.J. and Manabe, S. (1987). The influence of continental ice, atmospheric CO_2, and land albedo on the climate of the last glacial maximum. *Climate Dynamics*, **1**, 87-99.

Brodzikowski, K. and Van Loon, A.J. (1980). Sedimentary deformations in Saalian glaciolimnic deposits, near Włostów (Żary area, Western Poland). *Geologie en Mijnbouw*, **59**, 251-272.

Brodzikowski, K. and Van Loon, A.J. (1983). Sedimentology and deformational history of unconsolidated Quaternary sediments of the Jaroszów zone (Sudetic Foreland). *Geologica Sudetica*, **18**, 123-195.

Brodzikowski, K. and Van Loon, A.J. (1985). Penecontemporaneous non-tectonic brecciation of unconsolidated silts and muds. *Sedimentary Geology*, **41**, 269-282.

Brodzikowski, K. and Van Loon, A.J. (1987). A systematic classification of glacial and periglacial environments, facies and deposits. *Earth Science Reviews*, **24**, 297-381.

Brodzikowski, K. and Van Loon, A.J. (1991). *Glacigenic Sediments. Developments in Sedimentology, 49.* Elsevier Science, Amsterdam, 674 pp.

Broecker, W.S. (1975). Climatic change: Are we on the brink of a pronounced global warming? *Science*, **189**, 460-463.

Broecker, W.S. (1986). Oxygen isotope constraints on surface ocean temperatures. *Quaternary Research*, **26**, 121-134.

Broecker, W.S. (1992a). Global warming on trial. *Natural History*, **4**, 6-14.

Broecker, W.S. (1992b). Climate cycles; upset for Milankovitch theory. *Nature*, **359**, 779-780.

Broecker, W.S. and Denton, G.H. (1989). The role of ocean-atmosphere reorganizations in glacial cycles. *Geochimica et Cosmochimica Acta*, **53**, 2465-2501.

Broecker, W.S. and Denton, G.H. (1990a). What drives glacial cycles? *Scientific American*, **262**(1), 49-56.

Broecker, W.S. and Denton, G.H. (1990b). The role of ocean-atmosphere reorganizations in glacial cycles. *Quaternary Science Reviews*, **9**, 305-341.

Broecker, W.S. and Peng, T.-H. (1982). *Tracers in the Sea.* Eldigio Press, Palisades, 690 pp.

Broecker, W.S. and Peng, T.-H. (1989). The cause of the glacial to interglacial atmospheric CO_2 change: A polar alkalinity hypothesis. *Global Biogeochemical*

Cycles, **3**, 215-239.

Broecker, W.S. and van Donk, J. (1970). Insolation changes, ice volumes and the O^{18} record in deep-sea cores. *Review of Geophysics and Space Physics*, **8**, 169-197.

Broecker, W.S., Peteet, D.M. and Rind, D. (1985). Does the ocean-atmosphere system have more than one stable mode of operation? *Nature*, **315**, 21-26.

Broecker, W.S., Andree, M., Wölfli, W., Oeschger, H., Bonani, G., Kennett, J.P. and Peteet, D. (1988). The chronology of the last deglaciation: Implications to the cause of the Younger Dryas event. *Paleoceanography*, **3**, 1-19.

Broecker, W.S., Kennett, J.P., Flower, B.P., Teller, J.T., Trumbore, S., Bonani, G. and Wölfli, W. (1989). Routing of meltwater from the Laurentide Ice Sheet during the Younger Dryas Cold Episode. *Nature*, **341**, 318-321.

Broecker, W.S., Bond, G., Klas, M., Bonani, G. and Wölfli, W. (1990). A salt oscillator in the glacial Atlantic? 1. The Concept. *Paleoceanography*, **5**, 469-477.

Broecker, W.S., Bond, G., McManus, J., Klar, M. and Clark, E. (1992). Origin of the North Atlantic's Heinrich events. *Climate Dynamics*, **6**, 265-273.

Brown, C.S., Meier, M.F. and Post, A. (1982). Calving speed of Alaska tidewater glaciers, with application to Columbia Glacier. *United States Geological Survey, Professional Paper, 1258-C*, 13 pp.

Brown, N.E., Hallet, B. and Booth, D.B. (1987). Rapid soft bed sliding of the Puget Glacial Lobe. *Journal of Geophysical Research*, **92**, 8985-8997.

Brugman, M.M. (1985). The effects of a volcanic eruption on Shoestring Glacier, Mt. St. Helens. Unpublished Ph.D. Thesis (part I), California Institute of Technology.

Brugman, M.M. (1986). Water flow through a surging glacier; Variegated Glacier, Alaska. (Abstract). *In*: Alean, P., Braun, S., Iken, A., Schram, K. and Zwosta, G. (eds), *Hydraulic Effects at the Glacier Bed and Related Phenomena*. International Workshop 16-19 September, 1985, Interlaken, Switzerland. Mitteilungen der Versuchsanstalt für Wasserbau, Hydrologie und Glaziologie, No. 90, pp. 32.

Bryan, F. (1986). High-latitude salinity effects and interhemispheric thermohaline circulations. *Nature*, **323**, 301-304.

Bryson, R.A. and Goodman, B.M. (1980). Volcanic activity and climatic changes. *Science*, **207**, 1041-1044.

Budd, W.F. (1966). The dynamics of the Amery Ice Shelf. *Journal of Glaciology*, **6**, 335-358.

Budd, W.F. (1970a). Ice flow over bedrock perturbations. *Journal of Glaciology*, **9**, 29-48.

Budd, W.F. (1970b). The longitudinal stress and strain-rate gradients in ice masses. *Journal of Glaciology*, **9**, 19-27.

Budd, W.F. (1975). A first simple model for periodically self-surging glaciers. *Journal of Glaciology*, **14**, 3-21.

Budd, W.F. and Jacka, T.H. (1989). A review of ice rheology for ice sheet modelling. *Cold Regions Science and Technology*, **16**, 107-144.

Budd, W.F. and Jenssen, D. (1987). Numerical modelling of the large-scale basal water flux under the West Antarctic Ice Sheet. *In*: Van der Veen, C.J. and Oerlemans, J. (eds), *Dynamics of the West Antarctic Ice Sheet*, pp. 293-320. D. Reidel, Dordrecht.

Budd, W.F. and McInnes, B.J. (1978). The periodically surging Medvezhi Glacier matched with a general ice flow model. *Data of Glaciological Studies, Chronicle, Discussion, Publication 32*, Academy of Sciences of the U.S.S.R., 147-161 and 247-260.

Budd, W.F. and Smith, I.N. (1981). The growth and retreat of ice sheets in response to orbital radiation changes. *International Association of Hydrological Sciences*, No. 131, pp. 369-409.

Budd, W.F. and Young, N.W. (1979). Results from the I.A.G.P. flow-line study inland of Casey, Wilkes Land, Antarctica. *Journal of Glaciology*, **24**, 89-101.

Budd, W.F., Jensen, D. and Radok, U. (1971). Derived Physical Characteristics of the Antarctic Ice Sheet. Australian National Antarctic Research Expedition, Interim Report, Ser. A., Glaciology, Publication No. 120, 178 pp.

Budd, W.F., Keage, P.L. and Blundy, N.A. (1979). Empirical studies of ice sliding. *Journal of Glaciology*, **23**, 157-170.

Budd, W.F., Jacka, T.H. and Morgan, V.I. (1980). Antarctic iceberg melt rates derived from size distributions and movement rates. *Annals of Glaciology*, **1**, 103-112.

Budd, W.F., Corry, M.J. and Jacka, T.H. (1982). Results of the Amery Ice Shelf Project. *Annals of Glaciology*, **3**, 36-41.

Budd, W.F., McInnes, B.J., Jenssen, D. and Smith, I.N. (1987). Modelling the response of the West Antarctic Ice Sheet to a climatic warming. *In*: Van der Veen, C.J. and Oerlemans, J. (eds), *Dynamics of the West Antarctic Ice Sheet*, pp. 321-358. D. Reidel, Dordrecht.

Bull, P.A. (1981). Environmental reconstruction by electron microscopy. *Progress in Physical Geography*, **5**, 368-397.

Burbank, D.W. and Li, J. (1985). Age and paleoclimatic significance of loess of Lanzhou, North China. *Nature*, **316**, 429-431.

Burke, M.J. (1969). The Forth valley: An ice-moulded lowland. *Transactions of the Institute of British Geographers*, **48**, 51-59.

Burkholder, P.R. and Mandelli, E.F. (1965). Carbon assimilation of marine phytoplankton in Antarctica. *Proceedings of the National Academy of Sciences*, **54**, 437-444.

Burkimsher, M. (1983). Investigations of glacier hydrological systems using dye tracer techniques: Observations at Pasterzengletscher, Austria. *Journal of Glaciology*, **29**, 403-416.

Burrows, R.L. (1980). Cross-section, velocity, and bedload data at two erosion sites on the Tanana River near Fairbanks, Alaska, 1979. *United States Geological Survey Open-File Report 80-699*.

Cai, J., Powell, R.D. and Loubere, P. (1992). Source components of glacimarine sediments defined by Fourier shapes of quartz silt particles. *Geological Society of America, Abstracts with Programs*, **24**, 84.

Calabrese, E.A. and Syvitski, J.P.M. (1987). Modelling the growth of a growth prograding delta: Numerics, sensitivity, program code, and users guide. *Geological Survey Canada, Open File 1624*, 64 pp.

Calvert, S.E. and Veevers, J.J. (1962). Minor structures of unconsolidated marine sediments revealed by X-radiography. *Sedimentology*, **1**, 287-295.

Campbell, I.B. and Claridge, G.G.C. (1987). *Antarctica: Soils, Weathering Processes and Environment*. Developments in Soil Science 16, Elsevier, Amsterdam, 368 pp.

Campbell, N.J. and Collin, A.E. (1958). The discoloration of Fox Basin ice. *Journal Fisheries Research Board of Canada*, **15**, 1175-1188.

Campbell, W.J. (1973). Structure and inferred circulation of South Cascade Lake, Washington, U.S.A. *International Association of Scientific Hydrology*, Publication 95, 259-262.

Cant, D.J. (1978). Development of a facies model for sandy braided river sedimentation: Comparison of the South Saskatchewan River and the Battery Point Formation. *In*: Miall, A.D. (ed.), *Fluvial Sedimentology. Canadian Society of Petroleum Geologists, Memoir 5*, Calgary, Alberta, Canada, pp. 627-639.

Cant, D.J. and Walker, R.G. (1978). Fluvial processes and facies sequences in the sandy braided South Saskatchewan River, Canada. *Sedimentology*, **25**, 625-648.

Caputo, M.V. and Crowell, J.C. (1985). Migration of glacial centers across Gondwana during Paleozoic Era. *Geological Society of America Bulletin*, **96**, 1020-1036.

Carbon Dioxide Res. Conf. (CO$_2$RC) (1983). Carbon Dioxide Research Conference: CO$_2$, science and consensus, Proceedings. Berkley Springs, West Virginia, U.S.A.

Carey, S.W. and Ahmad, N. (1961). Glacial marine sedimentation. *In*: Raasch, G.O. (ed.), *Geology of the Arctic*, v.II, pp. 865-894. University of Toronto Press, Toronto.

Carlson, P.R., Conomos, T.J., Janda, R.J. and Peterson, D.H. (1975). Principle sources and dispersal patterns of suspended particulate matter in nearshore surface waters of the northeast Pacific Ocean. *National Technological Information Service*, *E75-10266*, 145 pp.

Carlson, P.R., Powell, R.D. and Rearic, D.M. (1989). Turbidity-current channels in Queen Inlet, Glacier Bay, Alaska. *Canadian Journal of Earth Sciences*, **26**, 807-820.

Carmack, E.C., Gray, C.B.J., Pharo, C.H. and Daley, R.J. (1979). Importance of lake-river interaction on the physical limnology of the Kamloops Lake-Thompson River system. *Limnology and Oceanography*, **24**, 634-644.

Carol, H. (1945). Über einen Versuch, den Gletschergrund mittels Einstiegs durch ein Strudelloch zu erreichen. *Die Alpen*, **21**, 180-184.

Carol, H. (1947). The formation of roches moutonnées. *Journal of Glaciology*, **1**, 57-59.

Carrara, P.E. (1975). The ice-cored moraines of Akudnirmuit Glacier, Cumberland Peninsula, Baffin Island, N.W.T., Canada. *Arctic and Alpine Research*, **7**, 61-67.

Carrara, P.E. and McGimsey, R.G. (1981). The Late-Neoglacial histories of the Agassiz and Jackson Glaciers, Glacier National Park, Montana. *Arctic and Alpine Research*, **13**, 183-196.

Carrara, P.E. and McGimsey, R.G. (1988). Map showing distribution of moraines and extent of glaciers from the mid-19th century to 1979 in the Mount Jackson area, Glacier National Park, Montana. *United States Geological Survey, Miscellaneous Investigations Series, Report I-1508-C*, map scale 1:24,000.

Carrozzi, A.V. (1967). *Studies on Glaciers Preceded by the Discourse of Neuchâtel by Louis Agassiz*. Hafner, New York, 213 pp.

Carruthers, R.G. (1953). *Glacial Drifts and the Undermelt Theory*. Harold Hill, Newcastle upon Tyne, 42 pp.

Carter, R.W.G. (1992). Sea-level: Past, present and future. *In*: Gray, J.M. (ed.), *Applications of Quaternary Research, Quaternary Proceedings 2*, pp. 111-132. Quaternary Research Association, Cambridge.

Cassell, E.A. and Pangburn, T. (1991). Interactive modelling of cold regions watersheds with SSARR. *In*: Prowse, T.D. and Ommanney, C.S.L. (eds), *Northern Hydrology, Selected Perspectives, Proceedings of Northern Hydrology Symposium, July 1990, Saskatoon*, pp. 363-378. National Hydrology Research Institute, Symposium No. 6.

Cassidy, W.A. and Rancitelli, L.A. (1982). Antarctic Meteorites. *American Scientist*, **70**, 156-164.

Catt, J.A. (1988a). Soils of the Plio-Pleistocene: Do they distinguish types of interglacial? *Philosophical Transactions Royal Society London B*, **318**, 539-557.

Catt, J.A. (1988b). *Quaternary Geology for Scientists and Engineers*. John Wiley, Halsted Press, New York, 340 pp.

Cepek, A.G. (1986). Quaternary stratigraphy of the German Democratic Republic. *In*: Šibrava, V., Bowen, D.Q. and Richmond, G.M. (eds), *Quaternary Glaciations in the Northern Hemisphere. Quaternary Science Reviews*, **5**, 359-364.

Chacho, E.F., Jr (1986). The variability of summer runoff from glaciated basins in Alaska. In Workshop on Alaskan Hydrology. *In*: Benson, C. *et al.* (eds), *Problems Related to Glacierized Basins*. University of Alaska, Geophysical Institute, Report UAG-R-306, 70 pp.

Chamberlin, T.C. (1878). The extent and significance of the Wisconsin kettle moraines. *Wisconsin Academy of Science Transactions*, **4**, 201-234.

Chamberlin, T.C. (1888). The rock-scourings of the great ice invasions. *US Geological Survey, Annual Report 7*, pp. 147-248.

Chamberlin, T.C. (1894). Glacial phenomena of North America. *In*: Geikie, J. (ed.), *The Great Ice Age*, pp. 724-775. 3rd edition. Appleton, New York.

Chamberlin, T.C. (1895). Classification of American glacial deposits. *Journal of Geology*, **3**, 270-277.

Chamberlin, T.C. (1896). Nomenclature of glacial formation. *Journal of Geology*, **4**, 872-876.

Chamberlin, T.C. (1899). An attempt to frame a working hypothesis of the cause of glacial periods on an atmospheric basis. *Journal of Geology*, **7**, 545-584, 667-685, 751-787.

Chambers, R. (1853). On glacial phenomena in Scotland and parts of England. *Edinburgh New Philosophical Journal*, **54**, 229-281.

Chandler, F.W., Young, G.M. and Wood, J. (1969). Diaspore in Early Proterozoic quartzites (Lorrain Formation) of Ontario. *Canadian Journal of Earth Sciences*, **6**, 337-340.

Chapman, D.L. (1982). Daily stream flow statistics of Alaskan streams. Anchorage. *National Weather Service, NOAA Technical Memorandum NWS AR-35*.

Chappell, J. (1983). A revised sea-level record for the last 300,000 years from Papua, New Guinea. *Search*, **14**, 99-101.

Chappell, J. and Shackleton, N.J. (1986). Oxygen isotopes and sea level. *Nature*, **324**, 137-140.

Chappell, J. and Thom, B.G. (1978). Termination of last interglacial episode and the Wilson Antarctic surge hypothesis. *Nature*, **272**, 809-810.

Chapple, W.M. (1978). Mechanics of thin-skinned fold-and-thrust belts. *Geological Society of America Bulletin*, **89**, 1189-1198.

Charlesworth, J.K. (1957). *The Quaternary Era, With Special References to its Glaciation*. Edward Arnold, London, 1700 pp.

Charpentier, J. de (1841). *Essai sur les glaciers et sur le terrain erratique du bassin du Rhône*. Marc Ducloix, Lausanne, 363 pp.

Cheel, R.J. and Rust, B.R. (1982). Coarse grained facies of glaciomarine deposits near Ottawa, Canada. *In*: Davidson-Arnott, R., Nickling, W. and Fahey, B.D. (eds), *Research in Glacial, Glacio-fluvial and Glacio-lacustrine Systems*, pp. 279-295, 6th Guelph Symposium on Geomorphology. Geo Books, Norwich.

Cheel, R.J. and Rust, B.R. (1986). A sequence of soft-sediment deformation (dewatering) structures in late Quaternary subaqueous outwash near Ottawa, Canada. *Sedimentary Geology*, **47**, 77-93.

Chen, J. and Ohmura, A. (1990a) On the influence of Alpine glaciers on runoff. *In*: Lang, H. and Musy, A. (eds), *Hydrology of Mountainous Regions I*. International Association of Hydrological Sciences Publication No. 193, pp. 117-125.

Chen, J. and Ohmura, A. (1990b) Estimation of alpine glacier water resources and their change since the 1870s. *In*: Lang, H. and Musy, A. (eds), *Hydrology of Mountainous Regions I*. International Association of

Hydrological Sciences Publication No. 193, pp. 127-135.

Chen, J.H., Curran, H.A., White, B. and Wasserburg, G.J. (1991). Precise Chronology of the last interglacial period: ^{234}U-^{230}Th data from fossil coral reefs in the Bahamas. *Geological Society of America Bulletin*, **103**, 82-97.

Cheng, D.C.-H. and Richmond, R.A. (1978). Some observations on the rheological behaviour of dense suspensions. *Rheologica Acta*, **17**, 446-453.

Chernova, L.P. (1981). Influence of mass balance and run-off on relief-forming activity of mountain glaciers. *Annals of Glaciology*, **2**, 69-70.

Cheshire, D.A. (1983). Till lithology in Hertfordshire and West Essex. *In*: Rose, J. (ed.), *Diversion of the Thames Field Guide*, pp. 50-59. Quaternary Research Association, Cambridge.

Chester, F.M. and Logan, J.M. (1987). Composite planar fabric of gouge from the Punchbowl Fault, California. *Journal of Structural Geology*, **9**, 621-634.

Chinn, T.J.H. (1989). Single folds at the margins of dry-based glaciers as indicators of a glacial advance. *Annals of Glaciology*, **12**, 23-30.

Chinn, T.J.H. and Dillon, A. (1987). Observations on a debris-covered polar glacier "Whisky Glacier"' James Ross Island, Antarctic Peninsula, Antarctica. *Journal of Glaciology*, **33**, 300-310.

Chinn, T.J. and Oliver, A.K.C. (1983). Hydrology and glaciology, Dry Valleys, Antarctica. *Ministry of Works and Development, Christchurch, New Zealand. Annual Report for 1979-1980*. Report No WS808, 49 pp.

Chong, T.S., Christiansen, E.G. and Baer, A.D. (1971). Rheology of concentrated suspensions. *Journal of Applied Polymer Science*, **15**, 2007-2021.

Chorley, R.J., Dunn, A.J. and Beckinsale, R.P. (1964). *The History of the Study of Landforms or the Development of Geomorphology, Volume 1. Geomorphology before Davis*. John Wiley, New York, 678 pp.

Choukroune, P., Gapais, D. and Merle, O. (1987). Shear criteria and structural symmetry. *Journal of Structural Geology*, **9**, 525-530.

Christiansen, E.A. (1971). Geology and groundwater resources of the Melville Area (62K, L) Saskatchewan. *Saskatchewan Research Council, Geology Division, Map No. 12*.

Christiansen, E.A. and Whitaker, S.H. (1976). Glacial thrusting of drift and bedrock. *In*: Legget, R.F. (ed.), *Glacial Till, An Inter-disciplinary Study*, pp. 121-130. Royal Society of Canada, Ottawa, Special Publication No.12.

Chumakov, N.M. and Elston, D.P. (1989). The paradox of Late Proterozoic glaciations at low latitudes. *Episodes*, **12**, 115-120.

Church, M. (1972). Baffin Island Sandurs: A study of Arctic fluvial processes. *Geological Survey of Canada, Bulletin 216*, 280 pp.

Church, M. and Gilbert, R. (1975). Proglacial fluvial and lacustrine environments. *In*: Jopling, A.V. and McDonald, B.C. (eds), *Glaciofluvial and Glaciolacustrine Sedimentation. Society of Economical Paleontologists and Mineralogists, Special Publication No. 23*, Tulsa, pp. 22-100.

Church, M. and Jones, D. (1982). Channel bars in gravel-bed rivers. *In*: Hey, R.D., Bathurst, J.C. and Thorne, C.R. (eds), *Gravel-Bed Rivers; Fluvial processes, Engineering and management*, pp. 291-338. John Wiley, Chichester.

Church, M. and Ryder, J.M. (1972). Paraglacial sedimentation: A consideration of fluvial processes conditioned by glaciation. *Geological Society of America Bulletin*, **83**, 3059-3072.

Churski, Z. (1973). Hydrographic features on the proglacial area of Skeidarárjökull. *Geographia Polonica*, **26**, 209-254.

Clague, J.J. and Mathews, W.H. (1973). The magnitude of jökulhlaups. *Journal of Glaciology*, **12**, 501-504.

Clapperton, C.M. (1975). The debris content of surging glaciers in Svålbard and Iceland. *Journal of Glaciology*, **14**, 395-406.

Clapperton, C.M. (1990a). Quaternary Glaciations in the Southern Hemisphere. *Quaternary Science Reviews*, **9**, 121-304.

Clapperton, C.M. (1990b). Quaternary glaciations in the southern hemisphere: An overview. *Quaternary Science Reviews*, **9**, 299-304.

Clapperton, C.M. (1993). *Quaternary Geology and Geomorphology of South America*. Elsevier, Amsterdam, 796 pp.

Clapperton, C.M. and Sudgen, D.E. (1990). Late Cenozoic glacial history of the Ross Embayment, Antarctica. *Quaternary Science Reviews*, **9**, 253-272.

Clark, C.D. (1993). Mega-scale glacial lineations and cross-cutting ice-flow landforms. *Earth Surface Processes and Landforms*, **18**, 1-29.

Clark, M.J. (1987). The alpine sediment system: A context for glacio-fluvial processes. *In*: Gurnell, A.M. and Clark, M.J. (eds), *Glaciofluvial Sediment Transfer: An Alpine Perspective*, pp. 9-31. John Wiley, Chichester.

Clark, P.U. (1989). Relative differences between glacially crushed quartz transported by mountain and continental ice - some examples from North America and East Africa: Discussion. *American Journal of Science*, **289**, 1195-1198. [reply by Mahaney *et al.*, 1198-1205].

Clarke, G.K.C. (1976). Thermal regulation of glacier surging. *Journal of Glaciology*, **16**, 231-250.

Clarke, G.K.C. (1982). Glacier outburst floods from "Hazard Lake", Yukon Territory, and the problems of flood magnitude prediction. *Journal of Glaciology*, **28**, 3-21.

Clarke, G.K.C. (1986). Surge Mechanisms - Discussion. *In*: Alean, P., Braun, S., Iken, A., Schram, K. and Zwosta, G. (eds), *Hydraulic Effects at the Glacier Bed and Related Phenomena*. International Workshop, 16-19 September 1985. Interlaken, Switzerland. Mitteilungen der Versuchsanstalt für Wasserbau, Hydrologie und Glaziologie, **90**, pp. 133-140.

Clarke, G.K.C. (1987a). A short history of scientific investigations on glaciers. *Journal of Glaciology*, Special Issue, 4-24.

Clarke, G.K.C. (1987b). Subglacial till: A physical framework for its properties and processes. *Journal of Geophysical Research*, **92B**, 9023-9036.

Clarke, G.K.C. (1987c). Fast Glacier flow: Ice streams, surging, and tidewater glaciers. *Journal of Geophysical Research*, **92B**, 8835-8841.

Clarke, G.K.C. (1991). Length, width and slope influences on glacier surging. *Journal of Glaciology*, **37**, 236-246.

Clarke, G.K.C. and Blake, E.W. (1991). Geometric and thermal evolution of a surge-type glacier in its quiescent state: Trapridge Glacier, Yukon Territory, Canada, 1969-89. *Journal of Glaciology*, **37**, 158-169.

Clarke, G.K.C. and Collins, S.G. (1984). The 1981-1982 surge of Hazard Glacier, Yukon Territory. *Canadian Journal of Earth Sciences*, **21**, 297-304.

Clarke, G.K.C. and Jarvis, G.T. (1976). Post-surge temperatures in Steele Glacier, Yukon Territory, Canada. *Journal of Glaciology*, **16**, 261-268.

Clarke, G.K.C. and Mathews, W.H. (1981). Estimates of the magnitude of glacier outburst floods from Lake Donjek, Yukon Territory, Canada. *Canadian Journal of Earth Sciences*, **18**, 1452-1463.

Clarke, G.K.C. and Waldron, D.A. (1984). Simulation of the August 1979 sudden discharge of glacier-dammed Flood Lake, British Columbia. *Canadian Journal of Earth Sciences*, **21**, 502-504.

Clarke, G.K.C., Nitsan, U. and Paterson, W.S.B. (1977).

Strain heating and creep instability in glaciers and ice sheets. *Reviews of Geophysics and Space Physics*, **15**, 235-247.

Clarke, G.K.C., Collins, S.G. and Thompson, D.E. (1984). Flow, thermal structure, and subglacial conditions of a surge-type glacier. *Canadian Journal of Earth Sciences*, **21**, 232-240.

Clarke, G.K.C., Mathews, W.H. and Pack, R.T. (1986a). Comment on "Outburst floods from Glacial Lake Missoula". *Quaternary Research*, **25**, 136-138.

Clarke, G.K.C., Meldrum, R.D. and Collins, S.G. (1986b). Measuring glacier-motion fluctuations using a computer-controlled survey system. *Canadian Journal of Earth Sciences*, **23**, 727-733.

Clarke, G.K.C., Schmock, J.P., Ommanney, C.S.L. and Collins, S.G. (1986c). Characteristics of surge-type glaciers. *Journal of Geophysical Research*, **91**, 7165-7180.

Clarke, T.S., Johnson, D. and Harrison, W.D. (1986a). Some aspects of glacier hydrology in the upper Susitna and Maclaren river basins, Alaska. *In*: Kane, D.L. (ed.), *Proceedings, Cold Regions Hydrology Symposium*, Fairbanks. American Water Resources Association, pp. 329-337.

Clarke, T.S., Harrison, W.D. and Johnson, D. (1986b) Glacier mass balances and runoff in the Upper Susitna and McClaren River basins, 1981-1983. *In*: Benson, C. *et al.* (eds), *Workshop on Alaskan Hydrology: Problems Related to Glacierized Basins*. University of Alaska, Geophysical Institute, Report UAG-R-306, pp. A-74.

Clayton, L. (1964). Karst topography on stagnant glaciers. *Journal of Glaciology*, **5**, 107-112.

Clayton, L. and Attig, J.W. (1987). Drainage of Lake Wisconsin near the end of the Wisconsin Glaciation. *In*: Mayer, L. and Nash, D. (eds), *Catastrophic Flooding*, pp. 139-153. The Binghampton Symposia in Geomorphology International Series, No. 18, Allen and Unwin, Boston.

Clayton, L. and Attig, J.W. (1990). The age of landforms on the Baraboo Hills, Wisconsin. *In*: Canadian Quaternary Association - American Quaternary Association; First Joint Meeting, University of Waterloo, 16 pp.

Clayton, L. and Moran, S.R. (1974). A glacial process-form model. *In*: Coates, D.R. (ed.), *Glacial geomorphology*, pp. 89-119. Publications in Geomorphology, State University of New York, Binghampton, N.Y.

Clayton, L., Teller, J.T. and Attig, J.W. (1985). Surging

of the southwestern part of the Laurentide Ice Sheet. *Boreas*, **14**, 235-241.

Clayton, L., Attig, J.W. and Mickelson, D.M. (1987). Glacier-bed conditions in Wisconsin during late Wisconsin time. *International Union for Quaternary Research*, 3rd International Congress, July 1987. Programme with Abstracts, 145 pp.

Clayton, L., Mickelson, D.M. and Attig, J.W. (1989). Evidence against pervasively deformed bed material beneath rapidly moving lobes of the southern Laurentide Ice Sheet. *Sedimentary Geology*, **62**(2/4), 203-208.

Clement, P. (1984). The drainage of a marginal ice-dammed lake at Nordbogletscher, Johan Dahl Land, South Greenland: *Arctic and Alpine Research* **16**, 209-216.

Clemmensen, L.B. and Houmark-Nielsen, M. (1981). Sedimentary features of a Weichselian glaciolacustrine delta. *Boreas*, **10**, 229-245.

CLIMAP Project Members, (1976). The surface of the ice-age earth. *Science*, **191**, 1131-1137.

CLIMAP Project Members (1981). Seasonal reconstructions of the Earth's surface at the last glacial maximum. *Geological Society of America, Map and Chart Series MC-36*.

Clough, J.W. (1974). RISP radio-echo soundings. *Antarctic Journal of the United States*, **9**, 159.

Coachman, L.K. and Aagaard, K. (1974). Physical oceanography of Arctic and subarctic seas. *In*: Herman, Y. (ed.), *Marine Geology and Oceanography of the Arctic Seas*, pp. 1-72. Springer, New York.

Coard, M.A., Cousen, S.M., Cuttler, A.H., Dean, H.J., Dearing, J.A., Eglinton, T.I., Greaves, A.M., Lacey, K.P., O'Sullivan, P.E. and Pickering, D.A. (1983). Paleolimnological studies of annually-laminated sediments in Loe Pool, Cornwall, U.K. *Hydrobiologia*, **103**, 185-191.

Coats, R.P. (1981). Late Proterozoic (Adelaidean) tillites of the Adelaide Geosyncline. *In*: Hambrey, M.J. and Harland, W.B. (eds), *Earth's Pre-Pleistocene Glacial Record*, pp. 537-548. Cambridge University Press, Cambridge.

Cobbold, P.R. and Quinquis, H. (1980). Development of sheath folds in shear regimes. *Journal of Structural Geology*, **2**, 119-126.

Coffin, J.H. and Ashton, W.S. (1986). Suspended sediment budget of a glacier-fed lake Eklutna Lake, Alaska. *In*: Kane, D.L. (ed.), *Proceedings, Cold Regions Hydrology Symposium, Fairbanks*. American Water Resources Association, pp. 501-508.

Cohen, J.M. (1979). Deltaic sedimentation in glacial Lake Blessington, County Wicklow, Ireland. *In*: Schlüchter, Ch. (ed.), *Moraines and Varves: Origin\Genesis\Classification*, pp. 357-367. A.A. Balkema, Rotterdam.

COHMAP Members, (1988). Climatic changes of the last 18,000 years; observations and model simulations. *Science*, **241**, 1043-1052.

Colbeck, S.C. (1971). A theory of water percolation in snow. *Journal of Glaciology*, **11**(63), 369-385.

Colbeck, S.C. (1973a). Effects of stratigraphic layers on water flow through snow. *US Cold Regions Research and Engineering Laboratory, Research Report 311*.

Colbeck, S.C. (1973b). Theory of metamorphism of wet snow. *US Cold Regions Research and Engineering Laboratory, Research Report 313*.

Colbeck, S.C. (1974a). Water flow through snow overlying an impermeable boundary. *Water Resources Research*, **10**, 119-123.

Colbeck, S.C. (1974b). The capillary effects on water percolation in homogeneous snow. *Journal of Glaciology*, **13**, 85-97.

Colbeck, S.C. (1980). *Dynamics of Snow and Ice Masses*. Academic Press, New York, 468 pp.

Colbeck, S.C. (1982). An overview of seasonal snow metamorphism. *Review of Geophysical Space Physics*, **20**, 45-61.

Colbeck, S.C. and Gow, A.J. (1979). The margin of the Greenland ice sheet at Isua. *Journal of Glaciology*, **24**, 155-165.

Colbeck, S.C. and Parssinen, N. (1978). Regelation and the deformation of wet snow. *Journal of Glaciology*, **21**, 639-650.

Coleman, A.P. (1926). *Ice Ages, Recent and Ancient*. Macmillan, New York, 296 pp.

Coleman, J.M. (1969). Brahmaputra River: Channel processes and sedimentation. *Sedimentary Geology*, **3**, 129-239.

Collins, D.N. (1978). Hydrology of an alpine glacier as indicated by the chemical composition of meltwater. *Zeitschrift für Gletscherkunde und Glazialgeologie*, **13**, 219-238.

Collins, D.N. (1979a). Sediment concentration in melt waters as an indicator of erosion processes beneath an Alpine glacier. *Journal of Glaciology*, **23**, 247-257.

Collins, D.N. (1979b). Hydrochemistry of meltwaters draining from an alpine glacier. *Arctic and Alpine Research*, **11**, 307-324.

Collins, D.N. (1979c). Quantitative determination of the subglacial hydrology of two Alpine glaciers. *Journal* of Glaciology, **23**, 347-362.

Collins, D.N. (1981). Seasonal variation of solute concentration in melt waters draining from an Alpine glacier. *Annals of Glaciology*, **2**, 11-16.

Collins, D.N. (1982). Water storage in an alpine glacier. *In*: Glen, J.W. (ed.), *Hydrological Aspects of Alpine and High-Mountain Areas*. Proceedings of the Exeter Symposium, July 1982. International Association of Hydrological Sciences, Publication 138, pp. 113-122.

Collins, D.N. (1983). Solute yield from a glacierized high mountain basin. *In*: Webb, B.W. (ed.), *Dissolved Loads of Rivers and Surface Water Quantity/Quality Relationships*. Proceedings of the Hamburg Symposium, August 1983, International Association of Hydrological Sciences Publication No. 141, pp. 41-49.

Collins, D.N. (1984). Hydrometeorological conditions, mass balance and runoff from Alpine glaciers. *In*: Oerlemans, J. (ed.),. *Glacier Fluctuations and Climate*, pp. 235-260. Kluwer Academic, Dordrecht.

Collins, D.N. (1985). Climatic variation and runoff from Alpine glaciers. *Zeitschrift für Gletscherkunde und Glazialgeologie*, **20(1984)**, 127-145.

Collins, D.N. (1986). Characteristics of meltwaters draining from the portal of an Alpine glacier during the emptying of a marginal ice-dammed lake. *Materialy Glyatsiologicheskikh Issledovaniy, Khronika, Obsuz-hdeniya*, **58**, 114-122 and 224-232.

Collins, D.N. (1987). Climatic fluctuations and runoff from glacierized Alpine basins. *In*: Solomon, S.J., Beran, M. and Hogg, W. (eds), *Influence of Climate Change and Climate Variability on the Hydrologic Regime and Water Resources*, pp. 77-89, Proceedings of Vancouver Symposium, August 1987. International Association of Hydrological Sciences Publication No. 168.

Collins, D.N. (1988). Suspended sediment and solute delivery to meltwaters beneath an alpine glacier. *Mitteilungen der Versuchsanstalt für Wasserbau, Hydrologie und Glaziologie*, **94**, 147-161.

Collins, D.N. (1989). Seasonal development of subglacial drainage and suspended sediment delivery to melt waters beneath an Alpine glacier. *Annals of Glaciology*, **13**, 45-50.

Collins, D.N. (1990). Seasonal and annual variations of suspended sediment transport in meltwater draining from an alpine glacier. *In*: Lang, H. and Musy, A. (eds), *Hydrology in Mountainous Regions I*. International Association of Hydrological Sciences Publication 193, pp. 439-446.

Collins, D.N. and Taylor, D.P. (1990). Variability of runoff from partly glacierized Alpine basins. *In*: Lang, H. and Musy, A. (eds), *Hydrology in Mountainous Regions I*. International Association of Hydrological Sciences Publication 193, pp. 365-372.

Collins, K. (1983). Scanning electron microscopy of engineering soils. *Geoderma*, **30**, 243-252.

Collinson, J.D. (1970). Bedforms of the Tana River, Norway. *Geografiska Annaler*, **52A**, 31-56.

Collomb, E. (1847). *Preuves de l'existence d'anciens glaciers dans les vallés des Vosges*. Victor Masson, Paris, 246 pp.

Cooney, R.T. (1987). Zooplankton. *In*: Hood, D.W. and Zimmerman, S.T. (eds), *The Gulf of Alaska*, pp. 285-303. Nat. Ocean. Atmos. Adm., United States Department of Commerce.

Corbel, J. (1959). Vitesse de l'erosion. *Zeitschrift für Geomorphologie*, **3**, 1-28.

Costa, J.E. (1988a). Rheologic, geomorphic and sedimentologic differentiation of water floods, hyperconcentrated flows, and debris flows. *In*: Baker, V.R., Kochel, R.C. and Patton, P.C. (eds), *Flood Geomorphology*, pp. 113-122. Wiley Interscience, New York.

Costa, J.E. (1988b). Floods from dam failures. *In*: Baker, V.R., Kochel, R.C. and Patton, P.C. (eds), *Flood Geomorphology*, pp. 439-463. Wiley Interscience, New York.

Court, A. (1957). The classification of glaciers. *Journal of Glaciology*, **3**, 3-7.

Cowan, D.S. (1982). Deformation of partly dewatered and consolidated Franciscan sediments near Piedras Blancas Point, California. *In*: Leggett, J.K. (ed.), *Trench Forearc Geology*. Geological Society of London Special Publication, No. 10, pp. 439-457.

Cowan, D.S. and Silling, R.M. (1978). A dynamic, scaled model of accretion at trenches and its implications for the tectonic evolution of subduction complexes. *Journal of Geophysical Research*, **83**, 5389-5396.

Cowan, E.A. (1988). Sediment transport and deposition in a temperate glacial fjord, Glacier Bay, Alaska. Unpublished Ph.D. Thesis, Northern Illinois University, 432 pp.

Cowan, E.A. and Powell, R.D. (1990). Suspended sediment transport and deposition of cyclically interlaminated sediment in a temperate glacial fjord, Alaska, U.S.A. *In*: Dowdeswell, J.A. and Scourse, J.D. (eds), *Glacimarine Environments: Processes and Sediments*. Geological Society of London, Special Publication No. 53, pp. 75-89.

Cowan, E.A. and Powell, R.D. (1991). Ice-proximal

sediment accumulation rates in a temperate glacial fjord, southeastern Alaska. *In*: Anderson, J.B. and Ashley, G.M. (eds), *Glacial Marine Sedimentation: Paleoclimatic Significance*. , Special Paper, 261, pp. 61-73.

Cowan, E.A., Powell, R.D. and Smith, N.D. (1988). Rainstorm-induced event sedimentation at the tidewater front of a temperate glacier. *Geology*, **16**, 409-412.

Crary, A.P. (1966). Mechanism for fiord formation indicated by studies of an ice-covered inlet. , *Bulletin*, **77**, 911-930.

Croll, J. (1864). On the physical cause of the change of climate during geological epochs. *Philosophical Magazine*, **28**, 121-137.

Croot, D.G. (1981). (Abstract) Depositional landforming processes at the snouts of five surging glaciers in Vestspitsbergen. *Annals of Glaciology*, **2**, 115.

Croot, D.G. (1987). Glacio-tectonic structures: A mesoscale model of thin-skinned thrust sheets? *Journal of Structural Geology*, **9**, 797-808.

Croot, D.G. (ed.), (1988a). *Glaciotectonics, Forms and Processes*. A.A. Balkema, Rotterdam, 212 pp.

Croot, D.G. (1988b). Morphological, structural and mechanical analysis of neoglacial ice-pushed ridges in Iceland. *In*: Croot, D.G. (ed.). *Glaciotectonics, Forms and Processes*, pp. 33-47. A.A. Balkema, Rotterdam.

Crosby, I.B. (1945). Glacial erosion and the buried Wyoming Valley of Pennsylvania. , *Bulletin*, **56**, 389-400.

Crowell, J.C. (1978). Gondwanan glaciation, cyclothems, continental positioning and climate change. *American Journal of Science*, **278**, 1345-1372.

Crowell, J.C. (1981). Early Paleozoic glaciation and Gondwana drift. *In*: McElhinny, M.W. and Valencio, D.A. (eds), *Paleoreconstruction of the Continents*. American Geophysical Union and , Geodynamics Series, 2, pp. 45-49.

Crowell, J.C. (1982). Continental glaciation through geologic times. *In*: *National Research Council, Climate in Earth History*, pp. 77-82. National Academy Press, Washington.

Crowell, J.C. (1983). The recognition of ancient glaciations. *In*: Medaris, L.G., Jr, Byers, C.W., Mickelson, D.M. and Shanks, W.C. (eds), *Proterozoic Geology; Selected papers from an International Proterozoic Symposium*. , Memoir 161, pp. 289-297.

Crowell, J.C. and Frakes, L.A. (1970). Phanerozoic glaciation and the causes of ice ages. *American Journal of Science*, **268**, 193-224.

Crowell, J.C., Rocha-Campos, A.C. and Suárez-Sorulo, R. (1980). Silurian glaciation in central South America. *In*: Cresswell, M.M. and Vela, P. (eds), *Gondwana Five: Selected Papers and Abstracts of Papers Presented at the Fifth International Gondwana Symposium*, pp. 105-110. A.A. Balkema, Rotterdam.

Crowley, T.J., Mengel, J.G. and Short, D.A. (1987). Gondwanaland's seasonal cycle. *Nature*, **329**, 803-807.

Cudlip, W. and McIntyre, N.F. (1987). Seasat altimeter observations of an Antarctic "lake". *Annals of Glaciology*, **9**, 55-59.

Cunningham, F. (1990). *James Forbes, Pioneer Glaciologist*. Scottish Academic Press, Edinburgh, 329 pp.

Dahl, R. (1965). Plastically sculptured detail forms on rock surfaces in northern Nordland, Norway. *Geografiska Annaler*, **47**, 83-140.

Dahl-Jensen, D. (1985). Determination of the flow properties at Dye 3, south Greenland, by bore-hole-tilting measurements and perturbation modelling. *Journal of Glaciology*, **31**, 92-98.

Dahl-Jensen, D. and Johnsen, S.J. (1986). Paleotemperatures still exist in the Greenland ice sheet. *Nature*, **320**, 250-252.

Dahlen, F.A., Suppe, J. and Davis, D. (1984). Mechanics of fold-and-thrust belts and accretionary wedges: Cohesive Coulomb theory. *Journal of Geophysical Research*, **89(B12)**, 10,087-10,101.

Dahlstrom, C.D.A. (1970). Structural geology in the eastern margin of the Canadian Rocky Mountains. *Bulletin of Canadian Petroleum Geology*, **18**, 332-406.

Dale, J.E., Aitken, A.E., Gilbert, R. and Risk, M.J. (1989). Macrofauna of Canadian Arctic fjords. *Marine Geology*, **85**, 331-358.

Dalrymple, G.B. (1964). Potassium-argon dates of three Pleistocene interglacial basalt flows from the Sierra Nevada, California. , *Bulletin*, **75**, 753-758.

Dansgaard, W. (1981). Paleo-climatic studies on ice cores. *In*: Berger, A.L. (ed.), *Climatic Variations and Variability: Facts and Theories*, pp. 193-206. D. Reidel, Dordrecht.

Dansgaard, W. (1987). Ice core evidence of abrupt climatic changes. *In*: Berger, W.H. and Labeyrie, L.D. (eds), *Abrupt Climatic Change; Evidence and Implications*, pp. 223-233. D. Reidel, Dordrecht.

Dansgaard, W. and Oeschger, H. (1989). Past environmental long-term records from the Arctic. *In*: Oeschger, H. and Langway, C.C., Jr (eds), *The Environmental Record in Glaciers and Ice Sheets*, pp.

287-318. John Wiley, New York.

Dansgaard, W. and Tauber, H. (1969). Glacier oxygen-18 content and Pleistocene ocean temperatures. *Science*, **166**, 499-502.

Dansgaard, W., Johnsen, S.J., Møller, J. and Langway, C.C., Jr (1969). One thousand centuries of climatic record from Camp Century on the Greenland Ice Sheet. *Science*, **166**, 377-381.

Dansgaard, W., Johnsen, S.J., Clausen, H.B. and Langway, C.C., Jr (1971). Climatic record revealed by the camp century ice core. *In*: Turekian, K.K. (ed.), *The Late Cenozoic Glacial Ages*, pp. 37-56. Yale University Press, New Haven, Connecticut.

Dansgaard, W., Johnsen, S.J., Clausen, H.B. and Gundestrup, N. (1973). Stable isotope glaciology. *Meddelelser om Grønland*, **197(2)**, 53 pp.

Dansgaard, W., Johnsen, S.J. and Reeh, N. (1975). Climatic changes, norsemen and modern man. *Nature*, **255**, 24-28.

Dansgaard, W., Clausen, H.B., Gundestrup, N., Hammer, C.U., Johnsen, S.F., Kristinsdóttir, P.M. and Reeh, N. (1982). A new Greenland deep ice core. *Science*, **218**, 1273-1277.

Dansgaard, W., Johnsen, S.J., Clausen, H.B., Dahl-Jensen, D., Gundestrup, N., Hammer, C.U. and Oeschger, H. (1984). North Atlantic climatic oscillations revealed by deep Greenland ice cores. *In*: Hansen, J.E. and Takahashi, T. (eds), Climate Processes and Climate Sensitivity, *Geophysical Monograph*, **29**, 288-298.

Dansgaard, W., White, J.W.C. and Johnsen, S.J. (1989). The abrupt termination of the Younger Dryas Climate event. *Nature*, **339**, 532-534.

Dansgaard, W., Johnsen, S.J., Clausen, A.B., Dahl-Jensen, D., Gundestrup, N.S., Hammer, C.U., Avidberg, C.S., Steffensen, J.P., Sveinbjörnsdóttir, A.E., Jouzel, J. and Bond, G. (1993). Evidence for general instability of past climate from a 250-kyr ice-core record. *Nature*, **364**, 218-220.

Davidson-Arnott, R., Nickling, W. and Fahey, B.D. (1982). *Research in Glacial, Glacio-Fluvial and Glacio-Lacustrine Systems*, 6th Guelph Symposium on Geomorphology. Geo Books, Norwich, 318 pp.

Davis, D., Suppe, J. and Dahlen, F.A. (1983). Mechanics of fold-and-thrust belts and accretionary wedges. *Journal of Geophysical Research*, **88**, 1153-1172.

Davis, P.T. and Osborn, G. (1988). Holocene Glacier Fluctuations. *Quaternary Science Reviews*, **7**, 113-242.

Davis, W.M. (1899). Geographical cycle. *Geographical Journal*, **14**, 481-504.

De Geer, G. (1912). A geochronology of the last 12,000 years. *11th International Geological Congress, Stockholm, 1910 Compte Rendu*, **1**, 241-258.

De Geer, G. (1921). Correlation of late glacial clay varves in North America with the Swedish time scale. *Geologiska Föreningen i Stockholm Förhandlingar*, **43**, 70-73.

De Jong, C. (1992). Thresholds for channel change on two contrasting pro-glacial river fans, West Greenland. *Geografiska Annaler*, **74A**, 1-12.

De Master, D.J., Harden, S.L. and Nittrouer, C.A. (1988). Rates of Holocene sedimentation on the western continental shelf of the Antarctic Peninsula. *Antarctic Journal of the United States, Annual Review Issue*, **23**, 94-96.

De Mulder, F.J. and Hageman, B.P. (1989). *Applied Quaternary Research*. A.A. Balkema, Rotterdam, 185 pp.

Deichmann, N., Ansorge, J. and Röthlisberger, H. (1979). (Abstract) Observations of glacier seismicity on Unteraargletscher. *Journal of Glaciology*, **23**, 409.

Demorest, M. (1938). Ice flowage as revealed by glacial striae. *Journal of Geology*, **46**, 700-725.

Demorest, M. (1943). Ice sheets. *Geological Society of America Bulletin*, **54**, 363-399.

Denton, G.H. and Hughes, T.J. (eds), (1981a). *The Last Great Ice Sheets*. John Wiley, New York, p. 484.

Denton, G.H. and Hughes, T.J. (1981b). The Arctic Ice Sheet: An outrageous hypothesis. *In*: Denton, G.H. and Hughes, T.J. (eds), *The Last Great Ice Sheets*, pp. 440-467. John Wiley, New York.

Denton, G.H. and Hughes, T.J. (1983). Milankovitch theory of ice ages; hypothesis of ice-sheet linkage between regional insolation and global climate. *Quaternary Research*, **20**, 125-144.

Denton, G.H. and Karlén, W. (1973). Holocene climate variations - their pattern and possible cause. *Quaternary Research*, **3**, 155-205.

Denton, G.H., Hughes, T.J. and Karlén, W. (1986). Global ice-sheet system interlocked by sea level. *Quaternary Research*, **26**, 3-26.

Denton, G.H., Bockheim, J.G., Wilson, S.C. and Stuiver, M. (1989). Late Wisconsin and early Holocene glacial history, inner Ross Embayment, Antarctica. *Quaternary Research*, **31**, 151-182.

Derbyshire, E., McGown, A. and Radwan, A. (1976). 'Total' fabric of some till landforms. *Earth Surface Processes*, **1**, 17-26.

Desnoyers, J. (1829). Observations sur un ensemble de dépôts marins plus récens que les terrains tertiaires du bassin de la Seine, et constituant une formation géologique distincte; précedées d'un aperçu de la non-simultanéité des bassins tertiaires. *Annales des Sciences Naturelles*, **16**, 171-214, 402-491.

Dewart, G. (1966). Moulins on Kaskawulsh Glacier, Yukon Territory. *Journal of Glaciology*, **6**, 320-321.

Deynoux, M. and Trompette, R. (1981). Late Ordovician tillites of the Taoudeni Basin, West Africa. *In*: Hambrey, M.J. and Harland, W.B. (eds), *Earth's Pre-Pleistocene Glacial Record*, pp. 89-96. Cambridge University Press, Cambridge.

Dhonau, T.J. and Dhonau, N.B. (1963). Glacial structures on the north Norfolk Coast. *Proceedings of the Geologists' Association*, **74**, 433-439.

Dionne, J.-C. (1985). Formes, figures et faciès sédimentaires glaciels des estrans vaseux de régions froides. *Palaeogeography, Palaeoclimatology, Palaeoecology*, **51**, 415-451.

Dionne, J.-C. (1987). Characteristic features of modern tidal flats in cold regions. *In*: de Boer, P.L. (ed.), *Tide-Influenced Sedimentary Environments and Facies,* pp. 301-332. D. Reidel, Dordrecht.

Dionne, J.-C. and Brodeur, D. (1988). Erosion des plates-formes rocheuses littorales par affouillement glaciel. *Zeitschrift für Geomorphologie*, **32**, 101-115.

Dionne, J.-C. and Shilts, W.W. (1974). A Pleistocene clastic dike, Upper Chaudière Valley, Québec. *Canadian Journal of Earth Sciences*, **11**, 1594-1605.

Doake, C.S.M. (1976). Thermodynamics of the interaction between ice shelves and the sea. *Polar Record*, **18**, 37-41.

Doake, C.S.M. (1987). Some aspects of the flow of the Ronne Ice Shelf. *In*: Van der Veen, C.J. and Oerlemans, J. (eds), *Dynamics of the West Antarctic Ice Sheet*, pp. 75-98. D. Reidel, Dordrecht.

Doake, C.S.M. and Vaughan, D.G. (1991). Rapid disintegration of the Wordie Ice Shelf in response to atmospheric warming. *Nature*, **350**, 328-330.

Dodonov, A.E. (1981). Stratigraphy of the Upper Pliocene - Quaternary deposits of Tajikistan (Soviet Central Asia). *Acta Geologica Academia Scientiarum Hungarica*, **22**, 63-73.

Dodonov, A.E. (1984). Stratigraphy and correlation of the Upper Pliocene - Quaternary deposits of central Asia. *In*: Pécsi, M. (ed.), *Lithology and Stratigraphy of Loess and Paleosols*, pp. 201-211. Geographical Research Institute, Hungarian Academy of Sciences, Budapest.

Doeglas, D.J. (1962). The structure of sedimentary deposits of braided rivers. *Sedimentology*, **1**, 167-190.

Dolgushin, L.D. and Osipova, G.B. (1975). Glacier surges and the problem of their forecasting. *International Associations of Hydrological Sciences, Publication No. 104*, 292-304.

Domack, E.W. (1983). Facies of Late Pleistocene glacial-marine sediments on Whidbey Island, Washington: an isostatic glacial-marine sequence. *In*: Molnia, B.F. (ed.), *Glacial-Marine Sedimentation*, pp. 535-570. Plenum Press Publishing, New York.

Domack, E.W. (1987). Preliminary stratigraphy for a portion of the Wilkes Land continental shelf, Antarctica: Evidence from till provenance. *In*: Eittreim, S.L. and Hampton, M.A. (eds), *The Antarctic Continental Margin: Geology and Geophysics of Offshore Wilkes Land, Circum-Pacific Council on Energy and Mineral Resources. Earth Science Series 5A*, pp. 195-203. Houston, Texas.

Domack, E.W. (1988). Biogenic facies in the Antarctic glacimarine environment: basis for a polar glacimarine summary. *Palaeogeography, Palaeoclimatology, Palaeoecology*, **63**, 357-372.

Domack, E.W. (1990). Laminated terrigenous sediments from the Antarctic Peninsula: the role of subglacial and marine processes. *In*: Dowdeswell, J.A. and Scourse, J.D. (eds), *Glaciomarine Environments: Processes and Sediments*, pp. 91-103. Geological Society of London Special Publication, No. 53.

Domack, E.W. and Domack, C.R. (1991). *Cenozoic Glaciation: The marine record established by Ocean Drilling*. Joint Oceanographic Institutions, Washington, DC, 49 pp.

Domack, E.W. and Lawson, D.E. (1985). Pebble fabric in an ice-rafted diamicton. *Journal of Geology*, **93**, 577-591.

Domack, E.W. and Williams, C.R. (1990). Fine structure and suspended sediment transport in three Antarctic fjords. *American Geophysical Union, Antarctic Research Series*, **50**, 71-89.

Domack, E.W., Jull, A.J.T. and Donahue, D.J. (1991a). Holocene chronology for the unconsolidated sediments at Hole 740A: Prydz Bay, East Antarctica. *In*: Barron, J.A., Larsen, B., Baldauf, J.G., Alibert, C., Berkowitz, S., Caulet, J.-P., Chambers, S.R., Cooper, A.K., Cranston, R.E., Dorn, W.U., Ehrmann, W.U., Fox, R.D., Fryxell, G.A., Hambrey, M.J., Huber, B.T., Jenkins, C.J., Kang, S.-H., Keating, B.H., Mehl, K.W., Noh, I., Ollier, G., Pittenger, A., Sakai, H., Schroder, C.J., Solheim, A., Stockwell, D.A., Thierstein, H.R., Tocher, B., Turner, B.R., Wei, W., Mazzullo, E.K. and Stewart, N.J. (eds), Proceedings of the Ocean

Drilling Program, Kerguelen Plateau-Prydz Basin. *Proceedings of the Ocean Drilling Program, Scientific Results,* **119**, 747-750.

Domack, E.W., Jull, A.J.T. and Nakao, S. (1991b). Advance of East-Antarctic outlet glaciers during the Hypsithermal: Implications for the volume state of the Antarctic Ice Sheet under global warming. *Geology,* **19**, 1059-1062.

Dominik, J., Burrus, D. and Vernet, J.-P. (1983). A preliminary investigation of the Rhône River plume in eastern Lake Geneva. *Journal of Sedimentary Petrology,* **53**, 159-163.

Donn, W.L. and Shaw, D.M. (1977). Model of climate evolution based on continental drift and polar wandering. *Geological Society of America Bulletin,* **88**, 390-396.

Donnelly, R. and Harris, C. (1989). Sedimentology and origin of deposits from a small ice-dammed lake, Leirbreen, Norway. *Sedimentology,* **36**, 581-600.

Dowdeswell, J.A. (1982). Supraglacial re-sedimentation from melt-water streams onto snow overlying glacier ice, Sylgjujökull, west Vatnajökull, Iceland. *Journal of Glaciology,* **28**, 365-375.

Dowdeswell, J.A. (1989). On the nature of Svalbard ice bergs. *Journal of Glaciology,* **35**, 224-234.

Dowdeswell, J.A. and Murray, T. (1990). Modelling rates of sedimentation from icebergs. *In:* Dowdeswell, J.A. and Scourse, J.D. (eds), *Glaciomarine Environments: Processes and Sediments,* pp. 121-137. *Geological Society of London Special Publication,* No. 53.

Dowdeswell, J.A. and Scourse, J.D. (eds) (1990). *Glaciomarine Environments: Processes and Sediments. Geological Society of London Special Publication,* No. 53, 423 pp.

Dowdeswell, J.A. and Sharp, M. (1986). Characterization of pebble fabrics in modern terrestrial glacigenic sediments. *Sedimentology,* **33**, 699-710.

Dowdeswell, J.A., Hamilton, G.S and Hagen, J.O. (1991). The duration of the active phase on surge-type glaciers: Contrasts between Svalbard and other regions. *Journal of Glaciology,* **37**, 388-400.

Drake, L.D. (1972). Mechanisms of clast attrition in basal till. *Geological Society of America Bulletin,* **83**, 2159-2166.

Drake, L.D. and Shreve, R.L. (1973). Pressure melting and regelation of ice by round wires. *Proceedings of the Royal Society of London, Series A,* **332**, 51-83.

Dredge, L.A. and Cowan, W.R. (1989). Quaternary geology of the southwestern Canadian Shield. *In:* Fulton, R.G. (ed.), *Quaternary Geology of Canada and Greenland Geological Survey of Canada,* Geology of Canada, No. 1, pp. 214-235.

Dredge, L.A. and Grant, D.R. (1987). Glacial deformation of bedrock and sediment, Magdalen Islands and Nova Scotia, Canada: Evidence for a regional grounded ice sheet. *In:* van der Meer, J.J.M. (ed.), *Tills and Glaciotectonics,* pp. 183-195. A.A. Balkema, Rotterdam.

Dreimanis, A. (1982). Two origins of the stratified Catfish Creek Till at Plum Point, Ontario, Canada. *Boreas,* **11**, 173-180.

Dreimanis, A. (1987). Genetic complexity of a subaquatic till tongue at Port Talbot, Ontario, Canada. *In:* Kujansuu, R. and Saarnisto, M. (eds), *INQUA Report on Activities 1982-1987,* pp. 68-78. ETH, Zurich.

Dreimanis, A. (1988). Tills: their genetic terminology and classification. *In:* Goldthwait, R.P. and Matsch, C.L. (eds), *Genetic Classification of Glacigenic Deposits,* pp. 17-83. A.A. Balkema, Rotterdam.

Dreimanis, A. (1991). The Laurentide Ice Sheet during the last glaciation: A review and some current reinterpretations along its southern margin. *In:* Frenzel, B. (ed.), Klimageschichtliche Probleme der Letzten 130,000 Jahre. *Paläklima Forschung,* **1**, 267-291.

Dreimanis, A. (1993). Small to medium-sized glacitectonic structures in till and in its substratum and their comparison with mass movement structures. *Quaternary International,* **18,** 69-79.

Dreimanis, A. and Lundqvist, J. (1984). What should be called till? *Striae,* **20**, 5-10.

Dreimanis, A. and Schlüchter, Ch. (1985). Field criteria for the recognition of till or tillite. *Palaeogeography, Palaeoclimatology, Palaeoecology,* **51**, 7-14.

Dreimanis, A., and Vagners, U.J. (1971). Bimodal distribution of rock and mineral fragments in basal tills. *In:* Goldthwait, R.P. (ed.), *Till; A Symposium,* pp. 237-250. Ohio State University Press, Columbus, Ohio.

Dreimanis, A. and Vagners, U.J. (1972). The effect of lithology upon texture of till. *In:* Yatsu, E. and Falconer, A. (eds), *Research Methods in Pleistocene Geomorphology,* pp. 66-82. Proceedings of the 2nd Guelph Symposium on Geomorphology, 1971. Geo Abstracts, Norwich.

Drewry, D.J. (1972). A quantitative assessment of dirt-cone dynamics. *Journal of Glaciology,* **11**, 431-446.

Drewry, D.J. (1981). Radio echo sounding of ice masses: Principles and applications. *In:* Cracknell, A.P. (ed.),

Remote Sensing in Meteorology, Oceanography and Hydrology, pp. 270-284. Ellis Horwood, Chichester.

Drewry, D.J. (1983a). *Antarctica: Glaciological and Geophysical Folio.* Scott Polar Research Institute, Cambridge University, Cambridge, 9 map sheets.

Drewry, D.J. (1983b). Antarctic ice sheet: aspects of current configuration and flow. *In*: Gardner, R. and Scoging, H. (eds), *Mega-Geomorphology,* pp. 18-38. Clarendon Press, Oxford.

Drewry, D.J. (1986). *Glacial Geologic Processes.* Edward Arnold, London, 276 pp.

Drewry, D.J. and Cooper, A.P.R. (1981). Processes and models of Antarctic glaciomarine sedimentation. *Annals of Glaciology,* **2**, 117-122.

Drewry, D.J., Jordan, S.R. and Jankowski, E. (1982). Measured properties of the Antarctic ice sheet: surface configuration, ice thickness, volume and bedrock characteristics. *Annals of Glaciology,* **3**, 83-91.

Driedger, C.L. and Fountain, A.G. (1989). Glacier outburst floods at Mount Ranier, Washington. *Annals of Glaciology,* **13**, 51-55.

Driscoll, F.G. (1980). Wastage of the Klutlan ice-cored moraines, Yukon Territory, Canada. *Quaternary Research,* **14**, 31-49.

Drozdowski, E. (1977). Ablation till and related indicatory forms at the margins of Vestspitsbergen glaciers. *Boreas,* **6**, 107-114.

Drozdowski, E. (1987). Surge moraines. *In*: Gardiner, V. (ed.), *International Geomorphology 1986,* pp. 675-692. Proceedings of the First International Conference on Geomorphology, Part 2. John Wiley, Chichester.

Drygalski, E. von, (1897). *Grönland-Expedition der Gesellschaft für Erdkunde zu Berlin.* Berlin.

Du Toit, A.L. (1921). The Carboniferous glaciation of South Africa. *Geological Society of South Africa Transactions,* **24**, 188-277.

Dubrovin, L.I. (1979). Major types of Antarctic ice shores. *Polar Geography,* **3**, 69-75.

Dugmore, A. (1989). Icelandic volcanic ash in Scotland. *Scottish Geographical Magazine,* **105**, 168-172.

Dunbar, R.B. (1988). Holocene sedimentation along the coast of southern Victoria Land. *Antarctic Journal of the United States,* Annual Review Issue, **23**, 110-112.

Dunbar R.B., Leventer, A.R. and Stochton, W.L. (1989). Biogenic sedimentation in McMurdo Sound, Antarctica. *Marine Geology,* **85**, 155-180.

Dunne, T., Price, A.G. and Colbeck, S.C. (1976). The generation of runoff from subarctic snowpacks. *Water Resources Research,* **12**(4), 677-685.

Duplessy, J.-C. (1981). Oxygen isotope studies and Quaternary marine climates. *In*: Berger, A.L. (ed.), *Climatic Variations and Variability: Facts and Theories,* pp. 181-192. D. Reidel, Dordrecht.

Duplessy, J.-C., Lalou, C. and Vinot, A.C. (1970). Differential isotopic fractionation in benthic foraminifera and paleotemperatures reassessed. *Science,* **168**, 250-251.

Dupre, W.R. and Thompson, R. (1979). The Yukon Delta: a model for deltaic sedimentation in an ice-dominated environment. *Proceedings of the 9th Offshore Technology Conference, Houston, Texas,* **3436**, 657-664.

Duval, P. (1976). Lois du fluage transitoire ou permanent de la glace polycristalline pour divers états de contrainte. *Annals de Géophysique,* **32**, 335-350.

Duval, P. (1977). The role of water content on the creep rate of polycrystalline ice. *International Association of Hydrological Sciences,* **118**, 29-33.

Duval, P. (1978). Anelastic behaviour of polycrystalline ice. *Journal of Glaciology,* **21**, 621-628.

Duval, P. (1979). Creep and recrystallization of polycrystalline ice. *Bulletin de Minéralogie,* **102**, 80-85.

Dyke, A.S. and Prest, V.K. (1987). Late Wisconsinan and Holocene history of the Laurentide Ice Sheet. *Géographie Physique et Quaternaire,* **41**, 237-263.

Dyurgerov, M.B. (1986). Computations of mass balance in glacier systems. (In Russian). *Materialy Glyatsiologicheskikh Issledovaniy,* **57**, 8-15.

Easterbrook, D.J. (1986). Stratigraphy and chronology of Quaternary deposits of the Puget Lowland and Olympic Mountains of Washington and the Cascade Mountains of Washington and Oregon. *In*: Šibrava, V., Bowen, D.Q. and Richmond, G.M. (eds), *Quaternary Glaciations in the Northern Hemisphere. Quaternary Science Reviews,* **5**, 145-159.

Eberl, B. (1930). *Die Eiszeitfolge im nördlichen Alpenvorlande.* Benno Filser, Augsburg, 427 pp.

Echelmeyer, K.A. and Harrison, W.D. (1989). (Abstract) Dynamics of Jakobshavns ice stream, west Greenland. *Annals of Glaciology,* **12**, 201.

Echelmeyer, K.A. and Harrison, W.D. (1990). Jakobshavns Isbræ, West Greenland: seasonal variations in velocity - or lack thereof. *Journal of Glaciology,* **36**, 82-88.

Echelmeyer, K.A. and Kamb, B. (1986). Stress-gradient coupling in glacier flow: II. Longitudinal averaging in the flow response to small perturbations in ice thickness and surface slope. *Journal of Glaciology,* **32**, 285-298.

Echelmeyer, K.A. and Wang, Z.-X. (1987). Direct observation of basal sliding and deformation of basal drift at sub-freezing temperatures. *Journal of Glaciology*, **33**, 83-98.

Echols, R.J. and Armentrout, J.M. (1980). Holocene foraminiferal distribution patterns on the shelf and slope, Yakataga-Yakutat area northern Gulf of Alaska. *In*: Field, M.E., Bouma, A.H., Colburn, I.P., Douglas, R.G. and Ingle, J.C. (eds), *Proceedings of the Quaternary Depositional Environments of the Pacific Coast,* pp. 281-303. Society of Economic Paleontologist and Mineralogists, Bakersfield, California.

Eddy. (1977). Historical evidence for the existence of the solar cycle. *In*: White, O.R. (ed.), *The Solar Output and its Variation*, pp. 22-25. Associated University Press, Boulder, Colorado.

Edwards, M. (1986). Glacial Environments. *In*: Reading, H.G. (ed.), *Sedimentary Environments and Facies*, pp. 445-470. 2nd edition. Blackwell Scientific Publishers, Oxford.

Ehlers, J. (1983). *Glacial Deposits in North-West Europe.* A.A. Balkema, Rotterdam, 470 pp.

Ehlers, J., Meyer, K.-D. and Stephan, H.-J. (1984). The pre-Weichselian glaciations of North-West Europe. *Quaternary Science Reviews*, **3**, 1-40.

Ehlers, J., Gibbard, P.L. and Rose, J. (eds) (1991). *Glacial Deposits in Great Britain and Ireland*. A. A. Balkema, Rotterdam, 580 pp.

Ehlers, J., Gibbard, P.L. and Whiteman, C.A. (1992). Elsterzeitliche (Anglian) eisdynamik in East Anglia. *Eiszeitalter und Gegenwart*, **42**, 80-93.

Ehrlich, R. and Davies, D.K. (1968). Sedimentological indices of transport direction, distance, and process intensity in glacio-fluvial sediments. *Journal of Sedimentary Petrology*, **38**, 1166-1170.

Einarsson, T. and Albertsson, K.J. (1988). The glacial history of Iceland during the past three million years. *In*: Shackleton, N.J., West, R.G. and Bowen, D.Q. (eds), *The Past Three Million Years: Evolution of Climatic Variability in the North Atlantic Region*, pp. 227-234. Cambridge University Press, Cambridge.

Eisbacher, G.H. (1985). Late Proterozoic rifting, glacial sedimentation and sedimentary cycles in the light of Windermere deposition, western Canada. *Palaeogeography, Palaeoclimatology, Palaeoecology*, **51**, 231-254.

Eisbacher, G.H. and Clague, J.J. (1984). Destructive mass movements in high mountains - Hazard and management. *Geological Survey of Canada Paper 84-16*, pp. 1-230.

Eissmann, L. (1987). Lagerungsstörungen im Lockergebirge - Exogene und endogene Tektonik im Lockergebirge des nördlichen Mitteleuropa. *Geophysik und Geologie*, Serie 3, Band 3, 7-77.

Elfström, A. and Rossbacher, L. (1985). Erosional remnants in the Båldakatj area, Lapland, northern Sweden. *Geografiska Annaler*, **67A**, 167-176.

Elliott, D. (1976). The motion of thrust sheets. *Journal of Geophysical Research*, **81**, 949-963.

Elliott, R.E. (1965). A classification of subaqueous sedimentary structures based on rheological and kinematical parameters. *Sedimentology*, **5**, 193-209.

Elliston, G.R. (1973). Water movement through the Gornergletscher. *International Association of Hydrological Sciences*, **95**, 79-84.

Ellsaesser, H.W., MacCracken, M.C., Walton, J.J. and Grotch, S.L. (1986). Global climatic trends as revealed by the recorded data. *Reviews in Geophysics*, **24**, 745-792.

Elson, J.A. (1961). Geology of tills. Proceedings of the Fourteenth Soil Mechanics Conference, 13 and 14 October, 1960. *National Research Council of Canada, Associate Committee on Soil and Snow Mechanics, Technical Memorandum No. 69*, 5-17.

Elson, J.A. (1969). Late Quaternary marine submergence of Quebec. *La Revue de Géographique de Montréal*, **23**, 247-258.

Elverhøi, A. (1984). Glacigenic and associated marine sediments in the Weddell Sea, fjords of Spitsbergen and the Barents Sea: a review. *Marine Geology*, **57**, 53-88.

Elverhøi, A., Liestøl, O. and Nagy, T. (1980). Glacial erosion, sedimentation and microfauna in the inner part of Kongsfjorden, Spitsbergen. *Norsk Polarinstitut Skrifter*, **172**, 33-61.

Elverhøi, A., Lønni, Ø. and Seland, R. (1983). Glaciomarine sedimentation in a modern fjord environment, Spitsbergen. *Polar Research*, **1**, 127-149.

Elverhøi, A., Pfirman, S., Solheim, A. and Larsen, B.B. (1989). Glaciomarine sedimentation on epicontinental seas - exemplified by the northern Barents Sea. *Marine Geology*, **85**, 225-250.

Embleton, B.J.J. and Williams, G.E. (1986). Low Palaeolatitude of deposition for late Precambrian periglacial varvites in South Australia: Implications for palaeoclimatology. *Earth and Planetary Science Letters*, **79**, 419-430.

Emiliani, C. (1955). Pleistocene temperatures. *Journal of Geology*, **63**, 538-578.

Emiliani, C. (1966). Palaeotemperature analysis of Caribbean cores P6304-8 and P6304-9 and a generalized temperature curve for the past 425,000 years. *Journal of Geology*, **74**, 109-126.

Emiliani, C. and Shackleton, N.J. (1974). The Brunhes epoch: Isotopic paleotemperatures and geochronology. *Science*, **183**, 511-514.

Engelhardt, H. (1978). Water in glaciers: observations and theory of the behaviour of water levels in boreholes. *Zeitschrift für Gletscherkunde und Glazialgeologie*, **14**, 35-60.

Engelhardt, H.F. (1986). Surge Mechanisms - Discussion. *In*: Alean, P., Braun, S., Iken, A., Schram, K. and Zwosta, G. (eds), *Hydraulic Effects at the Glacier Bed and Related Phenomena*, pp. 133-140. International Workshop, 16-19 September 1985, Interlaken, Switzerland. Mitteilungen der Versuchsanstalt für Wasserbau, Hydrologie und Glaziologie, **90**.

Engelhardt, H.F., Harrison, W.D. and Kamb, B. (1978). Basal sliding and conditions at the glacier bed as revealed by bore-hole photography. *Journal of Glaciology*, **20**, 469-508.

Engelhardt, H.F., Humphrey, N.F., Kamb, B. and Fahnestock, M. (1990). Physical conditions at the base of a fast moving Antarctic ice stream. *Science*, **248**, 247-249.

Ethridge, F.G. and Schumm, S.A. (1978). Reconstructing paleochannel morphologic and flow characteristics: Methodology, limitations, and assessment. *In*: Miall, A.D. (ed.), *Fluvial Sedimentology*, pp. 703-721. Canadian Society of Petroleum Geologists, Memoir 5, Calgary, Alberta, Canada.

Evans, D.J.A. (1989). Apron entrainment at the margins of sub-polar glaciers, north-west Ellesmere Island, Canadian High Arctic. *Journal of Glaciology*, **35**, 317-324.

Evenson, E.B. and Clinch, J.M. (1987). Debris transport mechanisms at active alpine glacier margins: Alaskan case studies. *In*: Kujansuu, R. and Saarnisto, M. (eds), INQUA Till Symposium, Finland 1985. *Geological Survey of Finland Special Paper, 3*, pp. 111-136.

Evenson, E.B., Dreimanis, A. and Newsome, J.W. (1977). Subaquatic flow tills: A new interpretation for the genesis of some laminated till deposits. *Boreas*, **6**, 115-133.

Evenson, E.B., Pasquini, T.A., Stewart, R.A., and Stephens, G. (1979). Systematic provenance investigations in areas of alpine glaciation: Applications to glacial geology and mineral exploration. *In*: Schlüchter, Ch. (ed.), *Moraines and Varves: Origin/Genesis/Classification*, pp. 25-42. A.A. Balkema, Rotterdam.

Evenson, E.B., Schlüchter, Ch. and Rabassa, J. (1983). *Tills and Related Deposits: Genesis, Petrology, Application, Stratigraphy*. A.A. Balkema, Rotterdam, 454 pp.

Ewing, M. and Donn, W.L. (1956). A theory of Ice Ages. *Science*, **123**, 1061-1066.

Eybergen, F.A. (1987). Glacier snout dynamics and contemporary push moraine formation at the Turtmannglacier, Wallis, Switzerland. *In*: van der Meer, J.J.M. (ed.), *Tills and Glaciotectonics*, pp. 217-231. A.A. Balkema, Rotterdam.

Eyles, C.H. and Eyles, N. (1983). Sedimentation in a large lake: a reinterpretation of the late Pleistocene stratigraphy at Scarborough Bluffs, Ontario, Canada. *Geology*, **11**, 146-152.

Eyles, C.H. and Eyles, N. (1984). Glaciomarine sediments of the Isle of Man as a key to Late Pleistocene stratigraphic investigations in the Irish Sea Basin. *Geology*, **12**, 359-364.

Eyles, C.H. and Eyles, N. (1989). The upper Cenozoic White River "tillites" of southern Alaska: Subaerial slope and fan-delta deposits in a strike-slip setting. *Geological Society of America Bulletin*, **101**, 1091-1102.

Eyles, C.H., Eyles, N. and Miall, A.D. (1985). Models of glaciomarine sedimentation and their application to the interpretation of ancient glacial sequences. *In*: Deynoux, M. (ed.), Glacial Record, Proceedings of the Till Mauretania '83 Symposium. *Palaeogeography, Palaeoclimatology, Palaeoecology*, **51**, 15-84.

Eyles, N. (1976). Morphology and development of medial moraines: comments on the paper by R.J. Small and M.J. Clark. *Journal of Glaciology*, **17**, 161-162.

Eyles, N. (1979). Facies of supraglacial sedimentation on Icelandic and Alpine temperate glaciers. *Canadian Journal of Earth Sciences*, **16**, 1341-1361.

Eyles, N. (1983). *Glacial Geology: An Introduction for Engineers and Earth Scientists*. Pergamon Press, Oxford, 409 pp.

Eyles, N. (1993). Earth's glacial record and its tectonic setting. *Earth Science Reviews*, **35**(1/2), 1-248.

Eyles, N. and Clark, B.M. (1988). Storm-influenced deltas and ice scouring in a late Pleistocene glacial lake. *Geological Society of America Bulletin*, **100**, 793-809.

Eyles, N. and Menzies, J. (1983). The subglacial landsystem. *In*: Eyles, N. (ed.), *Glacial Geology*, pp.

19-67. Pergamon Press, Oxford.

Eyles, N. and Miall, A.D. (1984). Glacial Facies. *In*: Walker, R.G. (ed.), *Facies models*, pp. 15-38. Geoscience Canada Reprint Series 1, Second Edition.

Eyles, N. and Rogerson, R.J. (1977). Glacier movement, ice structures, and medial moraine form at a glacier confluence, Berendon Glacier, British Columbia, Canada. *Canadian Journal of Earth Sciences*, **14**, 2807-2816.

Eyles, N. and Rogerson, R.J. (1978). A framework for the investigation of medial moraine formation: Austerdalsbreen, Norway, and Berendon Glacier, British Columbia, Canada. *Journal of Glaciology*, **20**, 99-113.

Eyles, N., Sladen, J.A. and Gilroy, S. (1982). A depositional model for stratigraphic complexes and facies superimposition in lodgement tills. *Boreas*, **11**, 317-333.

Eyles, N., Eyles, C.H. and Miall, A.D. (1983). Lithofacies types and vertical profile models; an alternative approach to the description and environmental interpretation of glacial diamict and diamictite sequences. *Sedimentology*, **30**, 393-410.

Eyles, N., Clark, B.M. and J.J. Clague. (1987). Coarse-grained sediment gravity flow facies in a large supraglacial lake. *Sedimentology*, **34**, 193-216.

Eyles, N., Eyles, C.H. and McCabe, A.M. (1989). Sedimentation in an ice-contact subaqueous setting: The Mid-Pleistocene "North Sea Drifts" of Norfolk, U.K. *Quaternary Science Reviews*, **8**, 57-74.

Fader, G.B., King, L.H. and Josenhans, H.W. (1982). Superficial geology of the Laurentia Channel and the western Grand Banks of Newfoundland. *Geological Survey of Canada, Paper 81-22*, p. 37.

Fahnestock, M. and Humphrey, N.F. (1988). Borehole water level measurements, Columbia Glacier, AK. *Ice*, **86**, 25-26.

Fahnestock, R.K. (1963). Morphology and hydrology of a glacial stream - White River, Mount Rainier, Washington. *United States Geological Survey, Professional Paper, 422-A*, p. 70.

Fahnestock, R.K. (1969). Morphology of the Slims River. *In*: Bushnell, V.C. and Ragle, R.H. (eds), *Icefield Ranges Research Project, Scientific Results, Volume 1*, pp. 161-172. American Geographical Society, New York, and Arctic Institute of North America, Montreal.

Fahnestock, R.K. and Bradley, W.C. (1973). Knik and Matanuska rivers, Alaska: A contrast in braiding. *In*: Morisawa, M. (ed.), *Fluvial Geomorphology*, pp. 220-250. Allen and Unwin, London.

Fairbanks, R.G. (1989). A 17,000-year glacio-eustatic sea level record: influence of glacial melting rates on the Younger Dryas event and deep-ocean circulation. *Nature*, **342**, 637-642.

Fairbanks, R.G. (1990). The age and origin of the "Younger Dryas climatic event" in Greenland ice cores. *Paleoceanography*, **5**, 937-948.

Fairchild, I.J. and Hambrey, M.J. (1984). The Vendian succession of northeastern Spitsbergen: petrogenesis of a dolomite-tillite association. *Precambrian Research*, **26**, 111-167.

Fastook, J.L. (1984). West Antarctica, the sea-level controlled marine instability: past and future. *In*: Hansen, J.E. and Takahashi, T. (eds), *Climate Processes and Climate Sensitivity*, pp. 275-287. American Geophysical Union, Geophysical Monograph No. 29, Washington, D.C.

Fastook, J.L. (1985). Ice shelves and ice streams: three modelling experiments. *In*: *Glaciers, Ice Sheets, and Sea Level: Effect of a CO_2-Induced Climatic Change*, pp. 279-300. Report of a workshop held in Seattle, Washington, September, 1984. United States Department of Energy, Washington, DC.

Fastook, J.L. (1987). The finite-element method applied to a time-dependent flowband model. *In*: Van der Veen, C.J. and Oerlemans, J. (eds), *Dynamics of the West Antarctic Ice Sheet*, pp. 203-221. D. Reidel, Dordrecht.

Fastook, J.L. and Chapman, J.E. (1989). A map-plane finite-element model: Three modelling experiments. *Journal of Glaciology*, **35**, 48-52.

Fastook, J.L. and Hughes, T.J. (1988). A geomorphic method for reconstructing paleo ice sheets, Part II: Glaciology. *In*: Kite, J.S., Lowell, T.V. and Thompson, W.B. (eds), *Contributions to the Quaternary Geology of Northern Maine and Adjacent Canada*. Maine Geological Survey, Bulletin 37, 19-34.

Fastook, J.L. and Hughes, T.J. (*in press*). The changing distribution of crustal ice loads during the last global glaciation cycle. Kluwer Academic Publishers.

Feeser, V. (1988). On the mechanics of glaciotectonic contortion of clays. *In*: Croot, D.G. (ed.), *Glaciotectonics: Forms and Processes*, pp. 63-76. A.A. Balkema, Rotterdam.

Fenn, C.R. (1983). Proglacial streamflow series: measurement, analysis and interpretation. Unpublished Ph.D. Thesis, University of Southampton, England.

Fenn, C.R. (1987). Sediment transfer processes in Alpine Glacier basins. *In*: Gurnell, A.M. and Clark, M.J.

(eds), *Glacio-Fluvial Sediment Transfer - An Alpine Perspective,* pp. 59-85. John Wiley, Chichester.

Fenn, C.R. (1989). Quantifying the errors involved in transferring suspended sediment rating equations across ablation seasons. *Annals of Glaciology,* **13**, 64-68.

Fenn, C.R. and Gomez, B. (1989). Particle size analysis of the sediment suspended in a proglacial stream: glacier de Tsidjiore Nouve, Switzerland. *Hydrological Processes,* **3**, 123-135.

Fenn, C.R. and Gurnell, A.M. (1987). Proglacial channel processes. *In*: Gurnell, A.M. and Clark, M.J. (eds), *Glacio-Fluvial Sediment Transfer - An Alpine Perspective,* pp. 423-498. John Wiley, Chichester.

Fenn, C.R., Gurnell, A.M. and Beecroft, I. (1985). An evaluation of the use of suspended sediment rating curves for the prediction of suspended sediment concentration in a proglacial stream. *Geografiska Annaler,* **67A**, 71-82.

Ferguson, R.I. (1973). Sinuosity of supraglacial streams. *Geological Society of America Bulletin,* **84**, 251-256.

Ferguson, R.I. (1984). Sediment load of the Hunza River. *In*: Miller, K.J. (ed.), *The International Karakoram Project volume 2,* pp. 581-598. Cambridge University Press, Cambridge.

Ferguson, R.I. (1985). Runoff from glacierized mountains: a model for annual variation and its forecasting. *Water Resources Research,* **21**, 702-708.

Fernlund, J.M.R. (1988). The Halland coastal moraines: Are they end moraines or glaciotectonic ridges? *In*: Croot, D.G. (ed.), *Glaciotectonics: Forms and Processes,* pp. 77-90. A.A. Balkema, Rotterdam.

Fischer, A.G. (1981). Climatic oscillations in the biosphere. *In*: Nitecki, M.H. (ed.), *Biotic Crises in Ecological and Evolutionary Time,* pp. 103-131. Academic Press, New York.

Fisher, D.A. (1973). Subglacial leakage of Summit Lake, British Columbia by dye determinations. *International Association of Hydrological Sciences,* **95**, 111-116.

Fisher, D.A. and Koerner, R.M. (1986). On the special rheological properties of ancient microparticle-laden Northern Hemisphere ice as derived from bore-hole and core measurements. *Journal of Glaciology,* **32**, 501-510.

Fisher, D.A., Koerner, R.M., Paterson, W.S.B, Dansgaard, W., Gundestrup, N. and Reeh, N. (1983). Effect of wind scouring on climatic records from ice-core oxygen-isotope profiles. *Nature,* **301**, 205-209.

Fisher, D.A., Reeh, N. and Langley, K. (1985). Objective reconstructions of the Late Wisconsinan Laurentide

Ice Sheet and the significance of deformable beds. *Géographie Physique et Quaternaire,* **39**, 229-238.

Fisher, W.L. and Brown, L.F., Jr (1972). *Clastic Depositional Systems, a Genetic Approach to Facies Analysis,* p. 211. University of Texas, Bureau of Economic Geology, Austin.

Flint, R.F. (1943). Growth of the North American ice sheet during the Wisconsin age. *Geological Society of America Bulletin,* **54**, 325-362.

Flint R.F. (1971). *Glacial and Quaternary Geology.* John Wiley, New York and London, 892 pp.

Flint, R.F., Colton, R.B., Goldthwait, R.P. and Willman, H.B. (1959). *Glacial Map of the United States East of the Rocky Mountains. Geological Society of America,* Boulder, Colorado, Scale 1:1,750,000.

Flint, R.F., Sanders, J.E. and Rodgers, J. (1960). Diamictite, a substitute term for symmictite. *Geological Society America, Bulletin,* **71**, 1809.

Flohn, H. (1974). Background of a geophysical model of the initiation of the next glaciation. *Quaternary Research,* **4**, 385-404.

Flohn, H. and Fantechi, R. (1984). *The Climate of Europe: Past, Present and Future.* D. Reidel, Dordrecht, 356 pp.

Flores, R.M. (1975). Short-headed stream delta: Model for Pennsylvanian Haymond Formation, West Texas. *The American Association of Petroleum Geologists Bulletin,* **59**, 2288-2301.

Foldvik, A. and Gammelsrød, T. (1988). Notes on Southern Ocean hydrography, sea-ice and bottom water formation. *Palaeogeography, Palaeoclimatology, Palaeoecology,* **67**, 3-17.

Foldvik, A. and Kvinge, T. (1974). Conditional instability of sea water at the freezing point. *Deep Sea Research,* **21**, 169-174.

Folk, R.L. (1971). Genesis of longitudinal and Oghurd dunes elucidated by rolling upon grease. *Geological Society of America Bulletin,* **88**, 3461-3468.

Folk, R.L. and Ward, W.C. (1957). Brazos River bar: A study in the significance of grain size parameters. *Journal of Sedimentary Petrology,* **27**, 3-26.

Folland, C.K., Karl, T.R. and Vinnikov, K.Y.A. (1990). Observed climate variations and change. *In*: Houghton, J.T., Jenkins, G.J. and Ephraums, J.J. (eds), *Climate Change - The IPCC Scientific Assessment,* pp. 194-237. Cambridge University Press, Cambridge.

Follmer, L.R. (1983). Sangamon and Wisconsinan pedogenesis in the midwestern United States. *In*: Porter, S.C. (ed.), *Late-Quaternary Environments of the United States, Volume 1 - The Late Pleistocene,* pp. 138-144. Longman, London.

Forbes, E. (1846). On the connection between the distribution of the existing fauna and flora of the British Isles, and the geological changes which have affected their area, especially during the epoch of the Northern Drift. *Great Britain Geological Survey, Memoir 1*, pp. 336-432.

Forbes, J. (1842). Account of his recent observations on glaciers. *Edinburgh New Philosophical Journal*, **33**, 338-352.

Forbes, J. (1843). *Travels through the Alps of Savoy.* Edinburgh.

Fountain, A.G. (1989). The storage of water in, and hydraulic characteristics of the firn of south Cascade Glacier, Washington State, U.S.A. *Annals of Glaciology*, **13**, 69-75.

Fountain, A.G. (1991). Shape, length and flow conditions of conduits at South Cascade Glacier, Washington State, USA. *In*: Prowse, T.L. and Ommaney, C.S.L. (eds), *Northern Hydrology, Selected Perspectives,* pp. 89-104. Proceedings, Northern Hydrology Symposium, Saskatoon, Saskatchewan, 10-12 July 1990. National Hydrology Research Institute Symposium No. 6.

Fountain, A.G. (1992). Subglacial water flow inferred from stream measurements at South Cascade Glacier, Washington, U.S.A. *Journal of Glaciology*, **38**, 51-64.

Fountain, A.G. and Tangborn, W.V. (1985a). The effect of glaciers on streamflow variations. *Water Resources Research*, **21**, 579-586.

Fountain, A.G. and Tangborn, W.V. (1985b). Overview of contemporary techniques. *In*: Young, G.J. (ed.), *Techniques for Prediction of Runoff from Glacierized Areas,* pp. 27-41. International Association of Hydrological Sciences Publication 149.

Fowler, A.C. (1979a). A mathematical approach to the theory of glacier sliding. *Journal of Glaciology*, **23**, 131-141.

Fowler, A.C. (1979b). The use of a rational model in the mathematical analysis of a polythermal glacier. *Journal of Glaciology*, **24**, 443-456.

Fowler, A.C. (1980). The existence of multiple steady states in the flow of large ice masses. *Journal of Glaciology*, **25**, 183-184.

Fowler, A.C. (1984). On the transport of moisture in polythermal glaciers. *Geophysical and Astrophysical Fluid Dynamics*, **28**, 99-140.

Fowler, A.C. (1986). A sliding law for glaciers of constant viscosity in the presence of subglacial cavitation. *Proceedings of the Royal Society of London*, **A407,** 147-170.

Fowler, A.C. (1987a). Sliding with cavity formation. *Journal of Glaciology*, **33**, 255-267.

Fowler, A.C. (1987b). A theory of glacier surges. *Journal of Geophysical Research*, **92B**, 9111-9120.

Fowler, A.C. (1989). A mathematical analysis of glacier surges. *SIAM Journal of Applied Mathematics*, **49**, 246-263.

Fowler, A.C. and Larson, D.A. (1978). On the flow of polythermal glaciers. I. Model and preliminary analysis. *Proceedings of the Royal Society of London*, **A363,** 217-242.

Frakes, L.A. (1986). Mesozoic-Cenozoic climatic history and causes of the glaciation. *In*: Hsü, K.J. (ed.), *Mesozoic and Cenozoic Oceans*, pp. 33-48. Geodynamic Series, 15. American Geophysical Union, Washington.

Francis, J.R.D. (1973). Experiments on the motion of solitary grains along the bed of a water stream. *Proceedings of the Royal Society of London*, **332A**, 443-471.

Freeze, R.A. and Cheery, J.A. (1979). *Groundwater.* Prentice Hall, Englewood Cliffs, 604 pp.

French, H.M. (1976). *The Periglacial Environment.* Longmans, London, 309 pp.

French, H.M. and Goździk, J.S. (1988). Pleistocene epigenetic and syngenetic frost fissures, Bełchatów, Poland. *Canadian Journal of Earth Sciences*, **25**, 2017-2027.

Frenzel, B. and Bludau, W. (1987). On the duration of the interglacial to glacial transition at the end of the Eemian Interglacial (deep-sea stage 5E): Botanical and sedimentological evidence. *In*: Berger, W.H. and Labeyrie, L.D. (eds), *Abrupt Climatic Change; Evidence and Implications*, pp. 151-162. D. Reidel, Dordrecht.

Frolich, R.M., Mantripp, D.R., Vaughan, D.G. and Doake, C.S.M. (1987). Force balance of Rutford Ice Stream, Antarctica. *International Association of Hydrological Sciences Publication 170*, pp. 323-331. Symposium at Vancouver - The Physical Basis of Ice Sheet Modelling.

Frye, J.C., Glass, H.D. and Willman, H.B. (1968). Mineral zonation of Woodfordian loesses of Illinois. *Illinois Geological Survey Circular*, **427**, 44 pp.

Fuji, N. (1988). Palaeovegetation and palaeoclimate changes around Lake Biwa, Japan during the last ca. 3 million years. *Quaternary Science Reviews*, **7**, 21-28.

Fuji, N. and Horowitz, A. (1989). Brunhes epoch paleoclimates of Japan and Israel. *Palaeogeography,*

Palaeoclimatology, Palaeoecology, **72**, 79-88.

Fullerton, D.S. (1986a). Stratigraphy and correlation of glacial deposits from Indiana to New York and New Jersey. *In*: Šibrava, V., Bowen, D.Q. and Richmond, G.M. (eds), *Quaternary Glaciations in the Northern Hemisphere. Quaternary Science Reviews,* **5**, 23-37.

Fullerton, D.S. (1986b). Chronology and correlation of glacial deposits in the Sierra Nevada, California. *In*: Šibrava, V., Bowen, D.Q. and Richmond, G.M. (eds), *Quaternary Glaciations in the Northern Hemisphere. Quaternary Science Reviews,* **5**, 161-169.

Fullerton, D.S. and Richmond, G.M. (1986). Comparison of the marine oxygen isotope record, the eustatic sea level record, and the chronology of glaciation in the United States of America. *In*: V. Šibrava, V., Bowen, D.Q. and Richmond, G.M. (eds), *Quaternary Glaciations in the Northern Hemisphere. Quaternary Science Reviews,* **5**, 197-200.

Fulton, R.J. (1986). Quaternary stratigraphy of Canada. *In*: Šibrava, V., Bowen, D.Q. and Richmond, G.M. (eds), *Quaternary Glaciation in the Northern Hemisphere. Quaternary Science Reviews,* **5**, 207-209.

Fulton, R.J. (1989). Quaternary Geology of Canada and Greenland. Geological Survey of Canada, *Geology of Canada,* No.1, 839 pp.

Fulton, R.J. and Andrews, J.T. (1987). The Laurentide Ice Sheet. *Géographie Physique et Quaternaire,* **41**, 179-318.

Fulton, R.J. and Prest, V.K. (1987). Introduction: The Laurentide Ice Sheet and its significance. *Géographie Physique et Quaternaire,* **41**, 181-186.

Fulton, R.J. and Pullen, M.J.L.T. (1969). Sedimentation in Upper Arrow Lake, British Columbia. *Canadian Journal of Earth Sciences,* **6**, 785-791.

Fulton, R.J., Fenton, M.M. and Rutter, N.W. (1986). Summary of Quaternary stratigraphy and history, Western Canada. *In*: Šibrava, V., Bowen, D.Q. and Richmond, G.M. (eds), *Quaternary Glaciations in the Northern Hemisphere. Quaternary Science Reviews,* **5**, 229-241.

Funder, S. (co-ordinator) (1989). Chapter 13. Quaternary geology of the ice-free areas and adjacent shelves of Greenland. *In*: Fulton, R.J. (ed.), *Quaternary Geology of Canada and Greenland. Geological Survey of Canada, Geology of Canada, No. 1,* pp. 741-792.

Funder, S. and Petersen, K.S. (1980). Glacitectonic deformations in East Greenland. *Bulletin of the Geological Society of Denmark,* **28**, 115-122.

Furbish, D.J. and Andrews, J.T. (1984). The use of hypsometry to indicate long-term stability and response of valley glaciers to changes in mass transfer. *Journal of Glaciology,* **30**, 199-211.

Fyfe, G. (1990). The effect of water depth on ice-proximal glaciolacustrine sedimentation: Salpausselkä I, southern Finland. *Boreas,* **19**, 147-164.

Gade, H.G. (1979). Melting of ice in sea water: A primitive model with application to the Antarctic ice shelf and icebergs. *Journal of Physical Oceanography,* **9**, 189-198.

Gale, S.J. and Hoare, P.G. (1986). Blakeney and Salthouse. *In*: West, R.G. and Whiteman, C.A. (eds), *The Nar Valley and North Norfolk Field Guide,* pp. 78-95. Quaternary Research Association, Cambridge.

Gallagher, J.B. and Burton, H.R. (1988). Seasonal mixing of Ellis fjord, Vestfold Hills, East Antarctica. *Estuarine, Coastal and Shelf Science,* **27**, 363-380.

Gallardo, V.A. (1987). The sublittoral macrofauna benthos of the Antarctic shelf. *Environmental International,* **13**, 71-81.

Galon, R. (1973). Geomorphological and geological analysis of the proglacial area of Skeidarárjökull, central section. *Geographia Polonica,* **26**, 15-56.

Gans, W. de, de Groot, T. and Zwaan, H. (1987). The Amsterdam basin, a case study of a glacial basin in The Netherlands. *In*: van der Meer, J.J.M. (ed.), *Tills and Glaciotectonics,* pp. 205-216. A.A. Balkema, Rotterdam.

Geikie, A. (1863). On the phenomena of the glacial drift of Scotland. *Transactions of the Geological Society of Glasgow,* **1**, 1-190.

Geikie, J. (1984). *The Great Ice Age.* 3rd Edtion.

Genthon, C., Barnola, J.M., Raynaud, D., Lorius, C., Jouzel, J., Barkov, N.I., Korotkevich, Y.S. and Kotlyakov, V.M. (1987). Vostok ice core: climatic response to CO_2 and orbital forcing changes over the last climatic cycle. *Nature,* **329**, 414-418.

Geotimes (1991). U.S., Soviet scientists study paleoclimate history. *Geotimes,* **36**(4), 6.

Gérard, J.-C. and Pecker, J.C. (1990). Modelling the climatic response to solar variability. *In*: Pecker, J.C. (ed.), The Earth's climate and variability of the Sun over recent millennia; Geophysical, astronomical and archaeological aspects. *Philosophical Transactions of the Royal Society of London,* **330A**, 561-574.

Gerdel, R.W. (1945). The dynamics of liquid water in deep snow-packs. *Transactions of the American Geophysical Union,* **26**, 83-90.

Gerdel, R.W. (1954). The transmission of water through snow. *Transactions of the American Geophysical Union,* **35**, 475-485.

Gibbard, P.L. (1980). The origin of stratified Catfish Creek Till by basal melting. *Boreas*, **9**, 71-85.

Gibbard, P.L. and Dreimanis, A. (1978). Trace fossils from late Pleistocene glacial lake sediments in southwestern Ontario, Canada. *Canadian Journal of Earth Sciences*, **15**, 1967-1976.

Gibbard, P.L. and Stuart, A.J. (1974). Trace fossils from proglacial lake sediments. *Boreas*, **3**, 69-74.

Gijssel, K. van. (1987). A lithistratigraphic and glaciotectonic reconstruction of the Lamstedt Moraine, Lower Saxony (FRG). *In*: van der Meer, J.J.M. (ed.), *Tills and Glaciotectonics*, pp. 145-155. A.A. Balkema, Rotterdam.

Gilbert, G.K. (1890). Lake Bonneville. *United States Geological Survey Monograph*, **1**, 438 pp.

Gilbert, G.K. (1903). *Glaciers and Glaciation of Alaska*, p. 231. Harriman Alaska Series, Vol. 3. Doubleday, Page and Company, New York.

Gilbert, G.K. (1906). Crescentic gouges on glaciated surfaces. *Geological Society of America Bulletin*, **17**, 303-316.

Gilbert, G.K. (1910) Harriman Alaska Expedition. *Glaciers and Glaciation*. Volume 3.

Gilbert, R. (1971). Observations on ice-dammed Summit Lake, British Columbia, Canada. *Journal of Glaciology*, **10**, 351-356.

Gilbert, R. (1972). Observations on sedimentation at Lillooet delta, British Columbia. *In*: Slaymaker, H.O. and McPherson, H.J. (eds), *Mountain Geomorphology: Geomorphological Processes in the Canadian Cordillera*, pp. 187-194. B.C. Geographical Series, No.14. Tantalus Press, Vancouver.

Gilbert, R. (1973). Processes of underflow and sediment transport in a British Columbia mountain lake. *Proceedings of the 9th Canadian Hydrology Symposium*, pp. 493-507. National Research Council Canada.

Gilbert, R. (1975). Sedimentation in Lillooet Lake, British Columbia. *Canadian Journal of Earth Sciences*, **12**, 1697-1711.

Gilbert, R. (1982). Contemporary sedimentary environments on Baffin Island, N.W.T., Canada: Glaciomarine processes in fjords of eastern Cumberland Peninsula. *Arctic and Alpine Research*, **14**, 1-12.

Gilbert, R. (1983). Sedimentary processes of Canadian Arctic fjords. *Sedimentary Geology*, **36**, 147-175.

Gilbert, R. (1985). Quaternary glaciomarine sedimentation interpreted from seismic surveys of fiords on Baffin Island, N.W.T. *Arctic*, **38**, 271-280.

Gilbert, R. (1990). Rafting in glaciomarine environments. *In*: Dowdeswell, J.A. and Scourse, J.D. (eds), Glaciomarine Environments: Processes and Sediments, pp. 105-120. *Geological Society of London, Special Publication No. 53*.

Gilbert, R. and Desloges, J.R. (1987a). Sediments of ice-dammed, self-draining Ape Lake, British Columbia. *Canadian Journal of Earth Sciences*, **24**, 1735-1747.

Gilbert, R. and Desloges, J.R. (1987b). A delta built by ice rafting in outflow from a glacial lake. *Geografiska Annaler*, **69A**(3-4), 375-378.

Gilbert, R. and Shaw, J. (1981). Sedimentation in proglacial Sunwapta Lake, Alberta. *Canadian Journal of Earth Sciences*, **18**, 81-93.

Gilbert, R., Syvitski, J.P.M. and Taylor, R.B. (1985). Reconnaissance study of proglacial Stewart Lakes, Baffin Island, District of Franklin. *In*: Current Research, Part A. *Geological Survey of Canada, Paper 85-1A*, 505-510.

Gilliland, R.L. (1989). Solar evolution. *Palaeogeography, Palaeoclimatology, Palaeoecology (Global and Planetary Change Section)*, **75**, 35-55.

Gilpin, R.R. (1979). A model of the "liquid-like" layer between ice and a substrate with applications to wire regelation and particle migration. *Journal of Colloid and Interface Science*, **68**, 235-251.

Gilpin, R.R. (1980). Wire regelation at low temperatures. *Journal of Colloid and Interface Science*, **77**, 435-448.

Giovinetti, M.B. and Bentley, C.R. (1985). Surface balance in ice drainage systems of Antarctica. *Antarctic Journal of the United States*, **20**, 6-13.

Gjessing, J. (1965). On 'plastic scouring' and 'subglacial erosion'. *Norsk Geografisk Tidsskrift*, **20**, 1-37.

Gleick, P.H. (1987). Regional hydrologic consequences of increases in atmospheric CO_2 and other trace gases. *Climatic Change*, **10**, 137-161.

Glen, J.W. (1954). The stability of ice-dammed lakes and other water-filled holes in glaciers. *Journal of Glaciology*, **2**, 316-318.

Glen, J.W. (1955). The creep of polycrystalline ice. *Proceedings of the Royal Society of London*, **A228**, 519-538.

Glen, J.W. (1975). The mechanics of ice. *In*: Johnson, T.C. (ed.), *Cold Regions Science and Engineering*, Monograph 11-C2b. CRREL, Hanover, N.H.

Glen, J.W. (1982). Hydrological aspects of alpine and high-mountain areas. *In*: Glen, J.W. (ed.), Proceedings of the Exeter Symposium, July, 1982. *IAHS Publication 138*, 350 pp.

Glen, J.W., Adie, R.J. and Johnson, D.M. (1973).

Symposium on the Hydrology of Glaciers. Cambridge, September 1969. *International Association of Scientific Hydrology Publication No. 95*, 262 pp.

Glückert, G. (1975). The second Salpausselkä at Karkkila, southern Finland. *Geological Society of Finland Bulletin*, **47**, 45-53.

Glückert, G. (1977). On the Salpausselkä ice-marginal formations in southern Finland. *Zeitschrift für Geomorphologie, Supplementband*, **27**, 79-88.

Goldthwait, R.P. (1951). Development of end moraines in east-central Baffin Island. *Journal of Geology*, **59**, 567-577.

Goldthwait, R.P. (1963). Dating the Little Ice Age in Glacier Bay, Alaska. *Contribution No. 31 of the Institute of Polar Studies*. The Ohio State University, Columbus, Ohio.

Goldthwait, R.P. (1971). *Till: A Symposium*. Ohio State University Press, Columbus, 402 pp.

Goldthwait, R.P. (1979). Giant grooves made by concentrated basal ice streams. *Journal of Glaciology*, **23**, 297-307.

Goldthwait, R.P. and Matsch, C.L. (1988). *Genetic Classification of Glacigenic Deposits*. A.A. Balkema, Rotterdam, 294 pp.

Goldthwait, R.P., McKeller, I.C. and Cronk, C. (1963). Fluctuations of Crillon Glacier system, southeast Alaska. *International Association of Scientific Hydrologists Bulletin*, **8**, 62-74.

Golubev, G.N. (1973). Analysis of the run-off and flow routing for a mountain glacier basin. *In*: Symposium on the Hydrology of Glaciers, Proceedings of the Cambridge Symposium, 7-13 September 1969. *International Association of Scientific Hydrology Publication No. 95*, pp. 41-50.

Gomez, B. (1987). Bedload. *In*: Gurnell, A.M. and Clark, M.J. (eds), *Glaciofluvial Sediment Transfer: An Alpine Perspective*, pp. 355-376. John Wiley, Chichester.

Gomez, B. and Small, R.J. (1983). Genesis of englacial debris within the Lower Glacier de Tsidjiore Nouve, Valais, Switzerland, as revealed by scanning electron microscopy. *Geografiska Annaler*, **65A**, 45-51.

Gomez, B. and Small, R.J. (1986). Medial moraines of the Haut Glacier d'Arolla, Valais, Switzerland: Debris supply and implications for moraine formation. *Journal of Glaciology*, **31**, 303-307.

Goodchild, J.G. (1875). The glacial phenomena of the Eden Valley and the western part of the Yorkshire Dale district. *Quarterly Journal of the Geological Society of London*, **31**, 55-99.

Goodman, D.J., King, G.C.P., Millar, D.H.M. and Robin, G.DeQ. (1979). Pressure-melting effects in basal ice of temperate glaciers: laboratory studies and field observations under Glacier d'Argentière. *Journal of Glaciology*, **23**, 259-271.

Goodwin, I.D. (1988). The nature and origin of a jökulhlaup near Casey Station, Antarctica. *Journal of Glaciology*, **34**, 95-101.

Gordon, J.E. and Birnie, R.V. (1986). Production and transfer of subaerially generated rock debris and resulting landforms on South Georgia: an introductory perspective. *British Antarctic Survey Bulletin*, **72**, 25-46.

Gordon, J.E., Darling, W.G., Whalley, W.B and Gellatly, A.F. (1988). δD-$\delta^{18}O$ relationships and the thermal history of basal ice near the margins of two glaciers in Lyngen, north Norway. *Journal of Glaciology*, **34**, 265-268.

Görlich, K. (1986). Glacimarine sedimentation of muds in Hornsund Fjord, Spitsbergen. *Annales Societatis Geologorum Poloniae*, **56**, 433-477.

Goss, E., Mayewski, P.A. and Lyons, W.B. (1985). Examination of selected microparticles from the Sentik Glacier core, Ladakh Himalaya, India. *Journal of Glaciology*, **31**, 196-197.

Gottler, P.F. and Powell, R.D. (1990). Iceberg-rafted debris in McBride Inlet, Glacier Bay, Alaska. *In*: Milner, A.M. and Wood, J.D. (eds), *Proceedings of the Second Glacier Bay Science Symposium*, pp. 56-61. United States National Park Service, Anchorage.

Gough, D.O. (1981). Solar interior structure and luminosity variations. *Solar Physics*, **74**, 21-34.

Goughnour, R.R. and Andersland, O.B. (1968). Mechanical properties of a sand-ice system. *Journal of the Soil Mechanics Foundation Division. Proceedings of the American Society of Civil Engineers*, **94**, 923-950.

Gow, A.J. (1963). Results of measurements in the 309 meter borehole at Byrd Station, Antarctica. *Journal of Glaciology*, **4**, 771-784.

Gow, A.J. (1975). Time-temperature dependence of sintering in perennial isothermal snowpacks. *International Association of Hydrological Sciences*, **114**, 25-41.

Gow, A.J. and Williamson, T. (1976). Rheological implications of the internal structure and crystal fabrics of the West Antarctic ice sheet as revealed by deep core drilling at Byrd Station. *Geological Society of America Bulletin*, **87**, 1665-1677.

Gow, A.J., Ueda, H.T. and Garfield, D.E. (1968). Antarctic ice sheet: Preliminary results of first core hole to bedrock. *Science*, **161**, 1011-1013.

Gow, A.J., Epstein, S. and Sheehy, W. (1979). On the origin of stratified debris in ice cores from the bottom of the Antarctic ice sheet. *Journal of Glaciology*, **23**, 185-192.

Goździk, J. and Mycielska-Dowgiałło, E. (1982). Badanie wpływu niektórych procesów geologicznych na przekształcenie powierzchni ziarn kwarcowych. *Przeglad Geograficzny*, **54**, 219-241.

Grant, U.S. and Higgins, D.F. (1913). Coastal Glaciers of Prince William Sound and Kenai Peninsula. *United States Geological Survey Bulletin*, **526**, 75 pp.

Gravenor, C.P. and Gostin, V.A. (1979). Mechanisms to explain the loss of heavy minerals from the upper Palaeozoic tillites of South Africa and Australia and the late Precambrian tillites of Australia. *Sedimentology*, **26**, 707-717.

Greenland Ice-Core Project (GRIP) Members (1993). Climate instability during the last interglacial period recorded in the GRIP ice core. *Nature*, **364**, 203-207.

Greuell, W. and Oerlemans, J. (1986). Sensitivity studies with a mass balance model including temperature profile calculations inside the glacier. *Zeitschrift für Gletscherkunde und Glazialgeologie*, **22**, 101-124.

Gregory, M.R., Kirk, R.M. and Mabin, M.C.G. (1984). Shore types of Victoria Land, Ross Dependency, Antarctica. *New Zealand Antarctic Record*, **5**, 23-40.

Gribbin, J. and Lamb, H.H. (1978). Climatic change in historical times. *In*: Gribbin, J. (ed.), *Climatic Change*, pp. 68-82. Cambridge University Press, Cambridge.

Gribbon, P.W.F. (1979). Cryoconite holes on Sermikavsak, west Greenland. *Journal of Glaciology*, **22**, 177-181.

Griffith, A.A. (1924). The theory of rupture. *Proceedings of the 1st International Congress of Applied Mechanics (Delft)*, pp. 55-63.

Griffiths, G.A. (1980). Hydraulic geometry relationships of some New Zealand gravel bed rivers. *Journal of Hydrology (New Zealand)*, **19**, 106-118.

Gripp, K. (1929). Glaziologische und geologische Ergebnisse der Hamburgischen Spitsbergen Expedition 1927. *Abhandlungen der Naturwissenschaftlichen Verein Hamburg*, **22**, 147-249.

Gripp, K. (1975). Eisrandstudien ausgehend von Sermeq, SW-Grönland. *Meddelser om Grönland*, **195**, 1-19.

Groots, P.M., Stuiver, M., White, J.W.G., Johnsen, S. and Jouzel, J. (1993). Comparison of oxygen isotope records from GISP2 and GRIP Greenland ice cores. *Nature*, **366**, 552-554.

Grove, J.M. (1960). A study of Nesigiuv-Breen. *In*:

Lewis, W.W. (ed.), Norwegian Cirque Glaciers. *Royal Geographical Society Research, Series 4*, pp. 69-82.

Grove, J.M. (1988). *The Little Ice Age*. Methuen, London, 498 pp.

Grube, F., Christensen, S., Vollmer, T., Duphorn, K., Klostermann, J. and Menke, B. (1986). Glaciations in north west Germany. *In*: Šibrava, V., Bowen, D.Q. and Richmond, G.M. (eds), *Quaternary Glaciations in the Northern Hemisphere. Quaternary Science Reviews*, **5**, 347-358.

Gudmundsson, G. (1970). Short-term variations of a glacier-fed river. *Tellus*, **22**, 341-353.

Gurnell, A.M. (1982). The dynamics of suspended sediment concentration in a proglacial stream. *In*: Glen, J.W. (ed.), Hydrological Aspects of Alpine and High Mountain Areas, *Proceedings of the Exeter Symposium, July 1982. International Association of Hydrological Sciences Publication No. 138*, pp. 319-330.

Gurnell, A.M. (1987a). Suspended sediment. *In*: Gurnell, A.M. and Clark, M.J. (eds), *Glaciofluvial Sediment Transfer: An Alpine Perspective*, pp. 305-354. John Wiley, Chichester.

Gurnell, A.M. (1987b). Fluvial sediment yield from Alpine, glacierized catchments. *In*: Gurnell, A.M. and Clark, M.J. (eds), *Glaciofluvial Sediment Transfer: An Alpine Perspective*, pp. 415-420. John Wiley, Chichester.

Gurnell, A.M. and Clark, M.J. (eds), (1987). *Glacio-Fluvial Sediment Transfer: An Alpine Perspective*. John Wiley, Chichester, 524 pp.

Gurnell, A.M. and Fenn, C.R. (1984a). Box-Jenkins transfer function models applied to suspended sediment concentration-discharge relationships in a proglacial stream. *Arctic and Alpine Research*, **16**, 93-106.

Gurnell, A.M. and Fenn, C.R. (1984b). Flow separation, sediment source areas and suspended sediment transport in a pro-glacial stream. *In*: Schick, A.P. (ed.),. Channel Processes: Water, Sediment, Catchment Controls. *Catena, Supplement 5*, pp. 109-119.

Gurnell, A.M. and Fenn, C.R. (1985). Spatial and temporal variations in electrical conductivity in a proglacial stream system. *Journal of Glaciology*, **31**, 108-114.

Gurnell, A.M. and Warburton, J. (1990). The significance of suspended sediment pulses for estimating suspended sediment load and identifying suspended sediment sources in Alpine glacier basins. *In*: Lang,

H. and Musy, A. (eds), Hydrology in Mountainous Regions I, pp. 463-470. *International Association of Hydrological Sciences Publication 193.*

Gurnell, A.M., Warburton, J. and Clark, M.J. (1988). A comparison of the sediment transport and yield characteristics of two adjacent glacier basins, Val d'Herens, Switzerland. *In*: Bordas, M.P. and Walling, D.E. (eds), Sediment Budgets, pp. 431-441. *International Association of Hydrological Sciences Publication No. 174.*

Gustavson, T.C. (1974). Sedimentation on gravel outwash fans, Malaspina glacier foreland, Alaska. *Journal of Sedimentary Petrology*, **44**, 374-389.

Gustavson, T.C. (1975a). Bathymetry and sediment distribution in proglacial Malaspina Lake, Alaska. *Journal of Sedimentary Petrology*, **45**, 450-461.

Gustavson, T.C. (1975b). Sedimentation and physical limnology in proglacial Malaspina Lake, southeastern Alaska. *In*: Jopling, A.V. and McDonald, B.C. (eds), *Glaciofluvial and glaciolacustrine sedimentation. Society of Economic Paleontologists and Mineralogists Special Publication No. 23*, pp. 249-263.

Gustavson, T.C. and Boothroyd, J.C. (1982). Subglacial fluvial erosion: a major source of stratified drift, Malaspina glacier, Alaska. *In*: Davidson-Arnott, R., Nickling, W. and Fahey, B.D. (eds), *Glacial, Glaciofluvial and Glaciolacustrine Systems*, pp. 93-116. Proceedings 6th Guelph Symposium on Geomorphology, 1980. Geo Books, Norwich.

Gustavson, T.C. and Boothroyd, J.C. (1987). A depositional model for outwash, sediment sources, and hydrologic characteristics, Malaspina Glacier, Alaska: A modern analog of the southeastern margin of the Laurentide Ice Sheet. *Geological Society of America Bulletin*, **99**, 187-200.

Gustavson, T.C., Ashley, G.M. and Boothroyd, J.C. (1975). Depositional sequences in glaciolacustrine deltas. *In*: Jopling, A.V. and McDonald, B.C. (eds), *Glaciofluvial and Glaciolacustrine Sedimentation. Society of Economic Paleontologists and Mineralogists, Special Publication No. 23*, Tulsa, OK, pp. 264-280.

Guymon, G.L. (1974). Regional sediment yield analysis of Alaska streams. *American Society of Civil Engineers, Journal of the Hydraulics Division*, **100**, 41-51.

Habbe, K.-A. (1989). On the origin of the drumlins of the South German Alpine Foreland. *Sedimentary Geology*, **62**, 357-369.

Habbe, K.-A. (1990). Permafrost in hochletzkaltzeitlichen glazifluvialen Sedimenten - und was daraus folgt. *Quartär*, **41/42**, 14.

Haeberli, W. (1979). Holocene push-moraines in alpine permafrost. *Geografiska Annaler*, **61**, 43-48.

Haeberli, W. (1981). Ice motion on deformable sediments. *Journal of Glaciology*, **27**, 365-366.

Haeberli, W. (1983). Frequency and characteristics of glacier floods in the Swiss Alps. *Annals of Glaciology*, **4**, 85-90.

Haefeli, R. (1963). Note on the history of the Steingletscher lake. *Bulletin of the International Association of Scientific Hydrology*, **8**, 123-125.

Haefeli, R. (1970). Changes in the behaviour of the Unteraargletscher in the last 125 years. *Journal of Glaciology*, **9**, 195-212.

Hagen, J.O., Wold, B., Liestol, O., Østrem, G. and Sollid, J.L. (1983). Subglacial processes at Bondhusbreen, Norway: preliminary results. *Annals of Glaciology*, **4**, 91-98.

Haldorsen, S. (1981). Grain-size distribution of subglacial till and its relation to glacial crushing and abrasion. *Boreas*, **10**, 91-105.

Hallberg, G.R. (1980). Pleistocene stratigraphy in east-central Iowa. *Iowa Geological Survey Technical Information Series*, **10**, 168 pp.

Hallberg, G.R. (1986). Pre-Wisconsin glacial stratigraphy of the central plains region in Iowa, Nebraska, Kansas, and Missouri. *In*: Šibrava, V., Bowen, D.Q. and Richmond, G.M. (eds), *Quaternary Glaciation in the Northern Hemisphere. Quaternary Science Reviews*, **5**, 11-15.

Hallet, B. (1974). Deposits formed by subglacial precipitation of $CaCO_3$. *Bulletin of the Geological Society of America*, **85**, 1003-1015.

Hallet, B. (1976). Deposits formed by subglacial precipitation of $CaCO_3$. *Geological Society of America Bulletin*, **87**, 1003-1015.

Hallet, B. (1979). A theoretical model of glacial abrasion. *Journal of Glaciology*, **23**, 39-50.

Hallet, B. (1981). Glacial abrasion and sliding: Their dependence on the debris concentration in basal ice. *Annals of Glaciology*, **2**, 23-28.

Hallet, B. (*in preparation*) Model of bedrock abrasion with debris entrainment from the bed: I. Local perspective and abrasion patterns. *Journal of Glaciology.*

Hallet, B. (*in preparation*) Model of bedrock abrasion with debris entrainment from the bed: II. Rates of abrasion, quarrying, and other results. *Journal of*

Glaciology.

Hallet, B. and Anderson, M. (1980). Detailed glacial geomorphology of a proglacial bedrock area at Castleguard glacier, Alberta, Canada. *Zeitschrift für Gletscherkunde und Glazialgeologie*, **16**, 171-184.

Hallet, B., Lorrain, R. and Souchez, R. (1978). The composition of basal ice from a glacier sliding over limestones. *Geological Society of America Bulletin*, **89**, 314-320.

Hamblin, P.F. and Carmack, E.C. (1978). River-induced currents in a fjord lake. *Journal of Geophysical Research*, **83**, 885-899.

Hambrey, M.J. (1976). Debris, bubble, and crystal fabric characteristics of foliated glacier ice, Charles Rabots Bre, Okstindan, Norway. *Arctic and Alpine Research*, **8**, 49-60.

Hambrey, M.J. (1979). Origin of foliation in glaciers: comments on a paper by R. L. Hooke and P.J. Hudleston. *Journal of Glaciology*, **22**, 556-559.

Hambrey, M.J. (1983). Correlation of Late Proterozoic tillites in the North Atlantic region and Europe. *Geological Magazine*, **120**, 209-320.

Hambrey, M.J. (1984). Sedimentary processes and buried ice phenomena in the pro-glacial areas of Spitsbergen glaciers. *Journal of Glaciology*, **30**, 116-119.

Hambrey, M.J. and Alean, J. (1992). *Glaciers.* Cambridge University Press, Cambridge, 208 pp.

Hambrey, M.J. and Harland, W.B. (1979). Analysis of pre-Pleistocene glacigenic rocks: Aims and problems. *In*: Schlüchter, Ch. (ed.), *Moraines and Varves: Origin/Genesis/Classification,* pp. 271-275. A.A. Balkema, Rotterdam.

Hambrey, M.J. and Harland, W.B. (eds), (1981a). *Earth's Pre-Pleistocene Glacial Record*, p. 1004. Cambridge University Press, Cambridge.

Hambrey, M.J. and Harland, W.B. (1981b). Summary of Earth's pre-Pleistocene glacial record. *In*: Hambrey, M.J. and Harland, W.B. (eds), *Earth's Pre-Pleistocene Glacial Record,* pp. 943-954. Cambridge University Press, Cambridge.

Hambrey, M.J. and Milnes, A.G.. (1975). Boudinage in glacier ice - Some examples. *Journal of Glaciology*, **14**, 383-393.

Hambrey, M.J. and Müller, F. (1978). Structures and ice deformation in the White Glacier, Axel Heiberg Island, Northwest Territories, Canada. *Journal of Glaciology*, **20**, 41-66.

Hammer, C.U. (1989). Dating by physical and chemical seasonal variations and reference horizons. *In*: Oeschger, H. and Langway, C.C., Jr (eds), *The*

Environmental Record in Glaciers and Ice Sheets, pp. 99-121. John Wiley, New York.

Hammer, C.U., Clausen, H.B. and Dansgaard, W. (1980). Greenland ice sheet evidence of post-glacial volcanism and its climatic impact. *Nature*, **288**, 230-235.

Hammer, C.U., Clausen, H.B., Dansgaard, W., Neftel, A., Kristinsdóttir, P. and Johnson, E. (1985). Continuous impurity analysis along the Dye 3 deep core. *In*: Langway, C.C., Jr, Oeschger, H. and Dansgaard, W. (eds), Greenland Ice Core: Geophysics, Geochemistry and the Environment. American Geophysical Union. *Geophysical Monograph*, **33**, 90-94.

Hammer, C.U., Clausen, H.B. and Tauber, H. (1986). Ice-core dating of the Pleistocene/Holocene boundary applied to a calibration of the ^{14}C time scale. *Radiocarbon*, **28**, 284-291.

Hammer, K.M. and Smith, N.D. (1983). Sediment production and transport in a proglacial stream: Hilda Glacier, Alberta, Canada. *Boreas*, **12**, 91-106.

Hampton, M.A. (1972). The role of subaqueous debris flow in generating turbidity currents. *Journal of Sedimentary Petrology*, **42**, 775-793.

Hampton, M.A., Kravitz, J.H. and Leupke, G. (1987). Geology of sediment cores from the George V continental margin, Antarctica. *In*: Eittreim, S.L. and Hampton, M.A. (eds), *The Antarctic Continental Margin: Geology and Geophysics of Offshore Wilkes Land,* pp. 151-174. Circum-Pacific Council on Energy and Mineral Resources, Earth Science Series 5A, Houston, Texas.

Hance, J.H. (1937). The recent advance of the Black Rapids Glacier, Alaska. *Journal of Geology*, **45**, 775-783.

Hanmer, S. and Passchier, C. (1991). Shear-sense indicators: A review. *Geological Survey of Canada, Special Paper, 90-17*, 72 pp.

Hansom, J.D. (1983). Ice-formed intertidal boulder pavements in the sub-Antarctic. *Journal of Sedimentary Petrology*, **53**, 135-145.

Hantz, D. and Lliboutry, L. (1983). Waterways, ice permeability at depth, and water pressures at Glacier d'Argentière, French Alps. *Journal of Glaciology*, **29**, 227-239.

Harbor, J.M. (1992). Numerical modeling of the development of U-shaped valleys by glacial erosion. *Geological Society of America Bulletin*, **104**, 1364-1375.

Harbor, J.M., Hallet, B. and Raymond, C.F. (1988). A numerical model of landform development by glacial erosion. *Nature*, **333**, 347-349.

Harland, W.B. (1965). Critical evidence for a great Infra-Cambrian glaciation. *Geologische Rundschau*, **54**, 45-61.

Harland, W.B. (1981). The Late Archaean(?) Witwatersrand conglomerates, South Africa. *In*: Hambrey, M.J. and Harland, W.B. (eds), *Earth's Pre-Pleistocene Glacial Record*, pp. 185-187. Cambridge University Press, Cambridge.

Harland, W.B. and Bidgood, D.E.T. (1959). Palaeomagnetism in some Norwegian sparagmites and the late Pre-cambrian ice age. *Nature*, **184**, 1860-1862.

Harland, W.B. and Herod, K.N. (1975). Glaciations through time. *In*: Wright, A.E. and Moseley, F. (eds), *Ice Ages: Ancient and Modern*, pp. 189-216. Geological Journal Special Issue 6. Seel House Press, Liverpool.

Harland, W.B., Herod, K.N, and Krinsley, D. (1966). The definition and identification of tills and tillites. *Earth Science Review*, **2**, 225-256.

Harmer, F.W. (1928). The distribution of erratics and drift. *Proceedings of the Yorkshire Geological Society*, **21**, 79-150.

Harms, J.C. and Fahnestock, R.K. (1965). Stratification, bed forms, and flow phenomena (with an example from the Rio Grande). *Society of Economic Paleontologists and Mineralogists, Special Publication 12*, pp. 84-115.

Harms, J.C., Southard, J.B., Spearing, D.R. and Walker, R.G. (1975). Depositional environments as interpreted from primary sedimentary structures and stratification sequences. *Society of Economic Paleontologists and Mineralogists, Short Course No. 2*, 161 pp.

Harris, C. and Bothamley, K. (1984). Englacial deltaic sediments as evidence for basal freezing and marginal shearing, Leirbreen, southern Norway. *Journal of Glaciology*, **30**, 30-34.

Harris, P.W.V. (1976). The seasonal temperature-salinity structure of a glacial lake: Jökulsárlón, south-east Iceland. *Geografiska Annaler*, **58A**, 329-336.

Harrison, A.E. (1964). Ice surges on the Muldrow Glacier, Alaska. *Journal of Glaciology*, **5**, 365-368.

Harrison, P.W. (1957). A clay-till fabric: its character and origin. *Journal of Geology*, **65**, 275-308.

Harrison, W.D. (1986). Glaciers and sediment. *In*: Benson, C., Harrison, W.D., Gosink, J., Bowling, S., Mayo, L. and Trabant, D. (eds), *Workshop on Alaskan Hydrology: Problems Related to Glacierized Basins*, chapter 4, pp. 53-69. University of Alaska

Geophysical Institute Report UAG-R-306.

Harrison, W.D., Mayo, L.R. and Trabant, D.C. (1975). Temperature measurements on Black Rapids Glacier, Alaska, 1973. *In*: Weller, G. and Bowling, S.A. (eds), *Climate of the Arctic*, pp. 350-352. Geophysical Institute, University of Alaska, Fairbanks, Alaska, U.S.A.

Harrison, W.D, Kamb, B. and Engelhardt, H.F. (1986). Morphology and motion at the bed of a surge-type glacier. *In*: Alean, P., Braun, S., Iken, A., Schram, K. and Zwosta, G. (eds), *Hydraulic Effects at the Glacier Bed and Related Phenomena*, pp. 55-56. International Workshop, 16-19 September, 1985, Interlaken, Switzerland. Mitteilungen der Versuchsanstalt für Wasserbau, Hydrologie und Glaziologie, No. 90.

Harrison, W.D., Raymond, C.F., Sturm, M., Echelmeyer, K.A. and Humphrey, N.F. (1989). (Abstract) First results from automated motion and stream measurements from Fels and Black Rapids Glaciers, Alaska. *Annals of Glaciology*, **12**, 203.

Harrison, W.D., Drage, B.T., Bredthauer, S.R., Johnson, D., Schoch, D. and Follett, A.B. (1983). Reconnaissance of the glaciers of the Susitna River Basin in connection with proposed hydroelectric development. *Annals of Glaciology*, **4**, 99-104.

Harrold, P.E. and Burrows, R.L. (1983). Sediment transport in the Tanana River near Fairbanks, Alaska, 1982. *United States Geological Survey Water Resources Investigations Report 83-4213*.

Hart, J.K. (1987). The genesis of the North East Norfolk drift. Unpublished Ph.D. Thesis, University of East Anglia, U.K.

Hart, J.K. (1990). Proglacial glaciotectonic deformation and the origin of the Cromer Ridge push moraine complex, North Norfolk, England. *Boreas*, **19**, 165-180.

Hart, J.K. and Boulton, G.S. (1991). The interrelation of glaciotectonic and glaciodepositional processes within the glacial environment. *Quaternary Science Reviews*, **10**, 335-350.

Hart, J.K., Hindmarsh, R.C.A. and Boulton, G.S. (1990). Styles of subglacial glaciotectonic deformation within the context of the Anglian ice-sheet. *Earth Surface Processes and Landforms*, **15**, 227-241.

Hart, M.H. (1978). The evolution of the atmosphere of the Earth. *Icarus*, **33**, 23-39.

Hartshorn, J.H. (1958). Flowtill in southeastern Massachusetts. *Geological Society of America Bulletin*, **69**, 477-482.

Harwood, D.M. and Webb, P.-N. (1990). Early Pliocene

deglaciation of the Antarctic ice sheet and late Pliocene onset of bipolar glaciation. *Eos*, **71**, 538-539.

Harwood, D.M., Scherer, R.P. and Webb, P.-N. (1989). Multiple Miocene marine productivity events in West Antarctica as recorded in upper Miocene sediments beneath the Ross Ice Shelf (Site J-9). *Marine Micropaleontology*, **15**, 91-115.

Hattersley-Smith, G. (1964). Rapid advance of glacier in northern Ellesmere Island. *Nature*, **201**, 176.

Hattersley-Smith, G., and Serson, H. (1964). Stratified water of a glacial lake in northern Ellesmere Island. *Arctic*, **17**, 108-111.

Hayes, M.O. and Michel, J. (1989). Glacial geology and geomorphology of North America; Volume 1, Modern clastic depositional systems of south-central Alaska. *28th International Geological Congress. Field Trip Guidebook T101*. p. 42. American Geophysical Union.

Hays, J.D., Imbrie, J. and Shackleton, N.J. (1976). Variations in the earth's orbit: Pacemaker of the ice ages. *Science*, **194**, 1121-1132.

Hecht, A.D., Dansgaard, W., Eddy, J.A., Johnsen, S.J., Lange, M.A., Langway, C.C., Jr, Lorius, C., McElroy, M.B., Oeschger, H., Raisbeck, G. and Schlosser, P. (1989). Group report; long-term ice core records and global environmental changes. *In*: Oeschger, H. and Langway, C.C., Jr (eds), *The Environmental Record in Glaciers and Ice Sheets*, pp. 379-388. John Wiley, New York.

Hein, F.J. and Walker, R.G. (1977). Bar evolution and development of stratification in the gravelly, braided, Kicking Horse River, British Columbia. *Canadian Journal of Earth Sciences*, **14**, 562-570.

Heinrich, H. (1988). Origin and consequences of cyclic ice rafting in the Northeast Atlantic ocean during the past 130,000 years. *Quaternary Research*, **29**, 142-152.

Herman, Y., Osmond, J.K. and Somayajulu, B.L.K. (1989). Late Neogene Arctic paleoceanography, micropaleontology, stable isotopes and chronology. *In*: Herman, Y. (ed.), *The Arctic Seas: Climatology, Oceanography, Geology, and Biology*, pp. 581-655. Van Nostrand Reinhold, New York.

Herron, M.M., Herron, S.L. and Langway, C.C., Jr (1981). Climatic signal of ice melt features in southern Greenland. *Nature*, **293**, 389-391.

Herron, M.M. and Langway, C.C., Jr (1985). Chloride, nitrate and sulfate in the Dye 3 and Camp Century, Greenland, ice cores. *In*: Langway, C.C., Jr, Oeschger, H. and Dansgaard, W. (eds), Greenland Ice Core: Geophysics, Geochemistry and the Environment.

American Geophysical Union, Geophysical Monograph, **33**, 77-84.

Herron, S. and Langway, C.C., Jr (1979). The debris-laden ice at the bottom of the Greenland ice sheet. *Journal of Glaciology*, **23**, 193-207.

Herron, S. and Langway, C.C., Jr (1980). Firn densification: an empirical model. *Journal of Glaciology*, **25**, 373-385.

Herron, S. and Langway, C.C., Jr (1982). A comparison of ice fabrics and textures at Camp Century, Greenland and Byrd Station, Antarctica. *Annals of Glaciology*, **3**, 118-124.

Herron, S., Langway, C.C., Jr and Brugger, K.A. (1985). Ultrasonic velocities and crystalline anisotropy in the ice core from Dye 3, Greenland. *In*: Langway, C.C., Jr, Oeschger, H. and Dansgaard, W. (eds), Greenland Ice Core: Geophysics, Geochemistry, and the Environment. *American Geophysical Union Geophysical Monograph*, **33**, 23-31.

Herterich, K. (1987). On the flow within the transition zone between ice sheet and ice shelf. *In*: Van der Veen, C.J. and Oerlemans, J. (eds), *Dynamics of the West Antarctic Ice Sheet*, pp. 185-202. D. Reidel, Dordrecht.

Heusser, C.J. (1989). Climate and chronology of Antarctica and adjacent South America over the past 30,000 yrs. *Palaeogeography, Palaeoclimatology, Palaeoecology*, **76**, 31-37.

Hewitt, K. (1982). Natural dams and outburst floods of the Karakoram Himalaya. *In*: Glen, J.W. (ed.), *Hydrological Aspects of Alpine and High-Mountain Areas*, pp. 259-269. Proceedings of the Exeter Symposium, July 1982, International Association of Hydrological Sciences, No. 138.

Hewitt, K., Wake, C.P., Young, G.J. and David, C. (1989). Hydrological investigations at Biafo Glacier, Karakorum Range, Himalaya: an important source of water for the Indus River. *Annals of Glaciology*, **13**, 103-108.

Hickman, C.S. and Nesbitt, E.A. (1980). Holocene mollusk distribution patterns in the northern Gulf of Alaska. *In*: Field, M.E., Bouma, A.H., Colburn, I.P., Douglas, R.G. and Ingle, J.C. (eds), *Proceedings of the Quaternary Depositional Environments of the Pacific Coast,* pp. 305-312. Society of Economic Paleontologists and Mineralogists, Bakersfield, California.

Hicks, D.M., McSaveney, M.J. and Chinn, T.J.H. (1990). Sedimentation in proglacial Ivory Lake, Southern Alps, New Zealand. *Arctic and Alpine Research*, **22**,

26-42.

Hicock, S.R. (1990). Genetic till prism. *Geology*, **18**, 517-519.

Hicock, S.R. (1992). Lobal interactions and rheologic superposition in subglacial till near Bradtville, Ontario, Canada. *Boreas*, **21**, 73-88.

Hicock, S.R. and Dreimanis, A. (1989). Sunnybrook drift indicates a grounded early Wisconsin glacier in the Lake Ontario basin. *Geology*, **17**, 169-172.

Hill, P.A. and Billings, D. (1985). Glacial striae and crescentic marks in Nepean Sandstone near Ottawa. *Geoscience Canada*, **12**, 105-108.

Hillaire-Marcel, C., Occhietti, S. and Vincent, J.-S. (1981). Sakami Moraine, Quebec; A 500-km-long moraine without climatic control. *Geology*, **9**, 210-214.

Hindmarsh, R.C.A., Boulton, G.S. and Hutter, K. (1989). Modes of operation of thermo-mechanically coupled ice sheets. *Annals of Glaciology*, **12**, 57-69.

Hirano, M. and Aniya, M. (1988). A rational explanation of cross-profile morphology for glacial valleys and of glacial valley development. *Earth Surface Processes and Landforms*, **13**, 707-716.

Hitchcock, E. (1841). First Anniversary Address before the Association of American Geologists. *American Journal of Science*, **41**, 232-275.

Hjulström, F. (1952). The geomorphology of the alluvial outwash plains (sandurs) of Iceland and the mechanics of braided rivers. *International Geographic Union 17th Congress Proceedings, Washington*, pp. 337-342.

Hobbs, B.E., Means, B.W.D. and Williams, P.F. (1976). *An Outline of Structural Geology*. John Wiley, New York, 571 pp.

Hobbs, P.V. (1974). *Ice Physics*. Clarendon Press, Oxford, 837 pp.

Hodell, D.A., Williams, D.F. and Kennett, J.P. (1985). Late Pliocene reorganization of deep vertical water-mass structure in the western South Atlantic: Faunal and isotopic evidence. *Geological Society of America Bulletin*, **96**, 495-503.

Hodge, S.M. (1974). Variations in the sliding of a temperate glacier. *Journal of Glaciology*, **13**, 349-369.

Hodge, S.M. (1976). Direct measurement of basal water pressures: A pilot study. *Journal of Glaciology*, **16**, 205-218.

Hodge, S.M. (1979). Direct measurement of basal water pressures: Progress and problems. *Journal of Glaciology*, **23**, 309-319.

Hoffman, P.F. (1989). Speculations on Laurentia's first gigayear (2.0 to 1.0 Ga). *Geology*, **17**, 135-138.

Högbom, I. (1923). Ancient inland dunes of northern and middle Europe. *Geografiska Annaler*, **5**, 113-243.

Hoinkes, H. and Rudolph, R. (1962). Variations in the mass balance of Hintereisferner (Oetztal Alps), 1952-1961, and their relation to variations of climatic elements. *International Association of Scientific Hydrology Publications No. 58*, pp. 16-28.

Holdsworth, G. (1973). Ice deformation and moraine formation at the margin of an ice cap adjacent to a proglacial lake. *In*: Fahey, B.D. and Thompson, R.D. (eds), *Research in Polar and Alpine Geomorphology*, pp. 187-199. Third Guelph Symposium, 1973. Geo Abstracts, Norwich.

Holdsworth, G. (1974). Meserve Glacier, Wright Valley, Antarctica: Part I, Basal processes. *Institute of Polar Studies Report No. 37*. Ohio State University, 104 pp.

Holdsworth, G. and Glynn, J.E. (1978). Iceberg calving from floating glaciers by a vibrating mechanism. *Nature*, **274**, 464-466.

Hollin, J.T. (1962). On the glacial history of Antarctica. *Journal of Glaciology*, **4**, 173-195.

Hollin, J.T. (1964). Origin of ice ages: An ice shelf theory for Pleistocene glaciation. *Nature*, **202**, 1099-1100.

Hollin, J.T. (1965). Wilson's theory of ice ages. *Nature*, **208**, 12-16.

Hollin, J.T. (1969). Ice-sheet surges and the geological record. *Canadian Journal of Earth Sciences*, **6**, 903-910.

Hollin, J.T. (1970). Is the Antarctic Ice Sheet growing thicker? *International Association of Hydrological Sciences*, **86**, 363-374.

Hollin, J.T. (1972). Interglacial climates and Antarctic ice surges. *Quaternary Research*, **2**, 401-408.

Holmes, A. (1965). *Principles of Physical Geology*, Nelson, Edinburgh, 532 pp.

Holmes, C.D. (1941). Till fabric. *Geological Society of America Bulletin*, **52**, 1299-1354.

Holmes, C.D. (1952). Drift dispersion in west-central New York. *Geological Society of America Bulletin*, **63**, 993-1010.

Holmes, C.D. (1960). Evolution of till-stone shapes, central New York. *Geological Society of America Bulletin*, **71**, 1645-1660.

Holmlund, P. (1986). Geometry of the near-surface drainage system on Storglaciären, Sweden. *In*: Alean, P., Braun, S., Iken, A., Schram, K. and Zwosta, G. (eds), Hydraulic Effects at the Glacier Bed and Related Phenomena. International Workshop 16-19 September, 1985, Interlaken, Switzerland.

Mitteilungen der Versuchsanstalt für Wasserbau, Hydrologie und Glaziologie, No. 90, pp. 57-58.

Holmlund, P. (1988a). Is the longitudinal profile of Storglaciären, northern Sweden, in balance with the present climate? *Journal of Glaciology*, **34**, 269-273.

Holmlund, P. (1988b). Internal geometry and evolution of moulins, Storglaciären, Sweden. *Journal of Glaciology*, **34**, 242-248.

Holmlund, P. and Hooke, R. LeB. (1983). High water-pressure events in moulins, Storglaciären, Sweden. *Geografiska Annaler*, **65A**, 19-25.

Holtedahl, H. (1989). Submarine end moraines and associated deposits off the south coast of Norway. *Marine Geology*, **88**, 23-48.

Holtedahl, O. (1960). Geology of Norway. *Norges Geologiske Undersøkelse*, **208**, 540 pp.

Hooghiemstra, H. (1984). Vegetational and climatic history of the high plain of Bogotá, Columbia: a continuous record of the last 3.5 million years. Ph.D. Thesis, University of Amsterdam, 368 pp. (also in Dissertatione Botanicae, 79. J. Cramer, Vaduz.)

Hooghiemstra, H. (1989). Quaternary and Upper-Pliocene glaciations and forest development in the tropical Andes: Evidence from a long high-resolution pollen record from the sedimentary basin of Bogotá, Columbia. *Palaeogeography, Palaeoclimatology, Palaeoecology*, **72**, 11-26.

Hooke, R. LeB. (1968). Comments on "The formation of shear moraines; an example from South Victoria Land, Antarctica". *Journal of Glaciology*, **7**, 351-352.

Hooke, R. LeB. (1977). Basal temperatures in polar ice sheets: A qualitative review. *Quaternary Research*, **7**, 1-13.

Hooke, R. LeB. (1984). On the role of mechanical energy in maintaining subglacial water conduits at atmospheric pressure. *Journal of Glaciology*, **30**, 180-187.

Hooke, R. LeB. (1989). (Abstract) Water flow through a glacier situated in an overdeepening: cause or consequence of a till layer at the bed. *Annals of Glaciology*, **12**, 213.

Hooke, R. LeB. (1991). Positive feedbacks associated with erosion of glacial cirques and overdeepenings. *Geological Society of America Bulletin*, **103**, 1104-1108.

Hooke, R. LeB. and Hudleston, P.J. (1980). Ice fabrics in a vertical flow plane, Barnes Ice Cap, Canada. *Journal of Glaciology*, **25**, 195-214.

Hooke, R. LeB., Dahlin, B.B. and Kauper, M.T. (1972). Creep of ice containing dispersed fine sands. *Journal*

of Glaciology, **11**, 327-336.

Hooke, R.LeB., Raymond, C.F., Hotchkiss, R.L. and Gustafson, R.J. (1979). Calculations of velocity and temperature in a polar glacier using the finite-element method. *Journal of Glaciology*, **24**, 131-146.

Hooke, R. LeB., Brzozowski, J. and Bronge, C. (1983). Seasonal variations in surface velocity, Storglaciären, Sweden. *Geografiska Annaler*, **65A**, 263-277.

Hooke, R. LeB., Wold, B. and Hagen, J.O. (1985). Subglacial hydrology and sediment transport at Bondhusbreen, southwest Norway. *Geological Society of America Bulletin*, **96**, 388-397.

Hooke, R. LeB., Holmlund, P. and Iverson, N.R. (1987). Extrusion flow demonstrated by bore-hole deformation measurements over a riegel, Storglaciären, Sweden. *Journal of Glaciology*, **33**, 72-78.

Hooke, R. LeB., Miller, S.B. and Kohler, J. (1988). Character of the englacial and subglacial drainage systems in the upper part of the ablation area of Storglaciären, Sweden. *Journal of Glaciology*, **34**, 228-231.

Hooke, R.LeB., Calla, P., Holmlund, P., Nilsson, M. and Stroeven, A. (1989). A 3 year record of seasonal variations in surface velocity, Storglaciären, Sweden. *Journal of Glaciology*, **35**, 235-247.

Hooke, R.LeB., Laumann, T. and Kohler, J. (1990). Subglacial water pressures and the shape of subglacial conduits. *Journal of Glaciology*, **36**, 67-71.

Hopkins, D.M. (1975). Time-stratigraphic nomenclature for the Holocene Epoch. *Geology*, **3**, 10.

Hoppe, G. (1948). Isrecessionen från Norrbottens Kustland i belysning av de glaciala formelementen. *Geographica*, **No. 20**, 112 pp.

Hoppe, G. (1959). Glacial morphology and inland ice recession in northern Sweden. *Geografiska Annaler*, **41**, 193-212.

Hoppe, G. and Schytt, V. (1953). Some observations on fluted moraine surfaces. *Geografiska Annaler*, **35**, 105-115.

Horowitz, A. (1989). Continuous pollen diagrams for the last 3.5 m.y. from Israel: Vegetation, climate and correlation with oxygen isotope record. *Palaeogeography, Palaeoclimatology, Palaeoecology*, **72**, 63-78.

Horsfield, W.T. (1977). An experimental approach to basement controlled faulting. *Geologie en Mijnbouw*, **56**, 363-370.

Hoskin, C.M. and Nelson, R.V., Jr (1969). Modern marine carbonate sediment, Alexander Archipelago, Alaska. *Journal of Sedimentary Petrology*, **39**, 581-

590.

Hoskin, C.M. and Nelson, R.V., Jr (1971). Size modes in biogenic carbonate sediment, southeastern Alaska. *Journal of Sedimentary Petrology*, **41**, 1026-1037.

Houghton, J.T., Jenkins, G.J. and Ephraums, J.J. (1990). *Climate Change, the I.P.C.C. Scientific Assessment*, 365 pp.

Houghton, R.A. and Woodwell, G.M. (1989). Global climatic change. *Scientific American*, **260**(4), 36-44.

Houmark-Nielsen, M. and Bertelsen, A. (1981). Kineto-stratigraphic evaluation and presentation of glacial-stratigraphic data, with examples from northern Samsö, Denmark. *Boreas*, **10**, 411-422.

Houston, R.S., Lanthier, L.R., Karlstrom, K.K. and Sylvester, G. (1981). Early Proterozoic diamictite of southern Wyoming. *In*: Hambrey, M.J. and Harland, W.B. (eds), *Earth's Pre-Pleistocene Glacial Record*, pp. 795-799. Cambridge University Press, Cambridge.

Hovan, S.A., Rea, D.K., Pisias, N.G. and Shackleton, N.J. (1989). A direct link between the China loess and marine $\delta^{18}O$ records: Aeolian flux to the north Pacific. *Nature*, **340**, 296-298.

Howard-Williams, C. and Vincent, W.F. (1986). The Alph river ecosystem: A major freshwater environment in southern Victoria Land. *New Zealand Antarctic Record*, **7**, 21-33.

Huang, Maohuan (1990). On the temperature distribution of glaciers in China. *Journal of Glaciology*, **36**, 210-216.

Hubbard, B. (1991). Freezing-rate effects on the physical characteristics of basal ice formed by net adfreezing. *Journal of Glaciology*, **37**, 339-347.

Hubbard, B. and Sharp, M. (1989). Basal ice formation and deformation: A review. *Progress in Physical Geography*, **13**, 529-558.

Hubbert, M.K. (1951) Mechanical basis for certain familiar geologic structures. *Geological Society of America Bulletin*, **62**, 355-372.

Hubbert, M.K. and Rubey, W.W. (1959). Role of fluid pressure in mechanics of overthrust faulting. *Geological Society of America Bulletin*, **70**, 115-166.

Hudleston, P.J. (1976). Recumbent folding in the base of the Barnes Ice Cap, Baffin Island, Northwest Territories, Canada. *Geological Society of America Bulletin*, **87**, 1684-1692.

Hudleston, P.J. (1977). Progressive deformation and development of fabric across zones of shear in glacial ice. *In*: Saxena, S.K. and Bhattacharji, S. (eds), *Energetics of Geological Processes*, pp. 121-150. Springer, New York.

Hudleston, P.J. and Hooke, R. LeB. (1980). Cumulative deformation in the Barnes Ice Cap and implications for the development of foliation. *Tectonophysics*, **66**, 127-146.

Hughes, O.L. (1987). Quaternary geology. *In*: Morison, S.R. and Smith, C.A.S. (eds), *XII INQUA Congress Guidebook to Quaternary Research in Yukon*, pp. 12-16. National Research Council of Canada, Ottawa.

Hughes, T.J. (1970). Convection in the Antarctic ice sheet leading to a surge of the ice sheet and possibly to a new ice age. *Science*, **170**, 630-633.

Hughes, T.J. (1971). Structural glaciology of Meserve Glacier, Phase 3. *Antarctic Journal of the United States*, **6**, 127-128.

Hughes, T.J. (1973). Is the West Antarctic ice sheet disintegrating? *Journal of Geophysical Research*, **78**, 7884-7910.

Hughes, T.J. (1975). The west Antarctic ice sheet: Instability, disintegration and initiation of ice ages. *Reviews of Geophysics and Space Physics*, **13**, 502-526.

Hughes, T.J. (1976). The theory of thermal convection in polar ice sheets. *Journal of Glaciology*, **16**, 41-71.

Hughes, T.J. (1977). West Antarctic ice streams. *Review of Geophysics and Space Physics*, **15**, 1-46.

Hughes, T.J. (1981). Numerical reconstruction of paleo-ice sheets. *In*: Denton, G.H. and Hughes, T.J. (eds), *The Last Great Ice Sheets*, pp. 221-261. Wiley Interscience, New York.

Hughes, T.J. (1982). On the disintegration of ice shelves: The role of thinning. *Annals of Glaciology*, **3**, 146-151.

Hughes, T.J. (1983). The stability of the West Antarctic Ice Sheet: what has happened, what will happen. *In*: *Proceedings of the Carbon Dioxide Research Conference: CO_2, Science and Consensus*. Berkley Springs, West Virginia, pp. 19-23.

Hughes, T.J. (1985). The great Cenozoic ice sheet. *Palaeogeography, Palaeoclimatology, Palaeoecology*, **50**, 9-43.

Hughes, T.J. (1987). Ice dynamics and deglaciation models when ice sheets collapsed. *In*: Ruddiman, W.F. and Wright, H.E., Jr (eds), *North America and Adjacent Oceans During the Last Deglaciation, The Geology of North America*, pp. 183-220. Geological Society of America, Boulder.

Hughes, T.J. (1987). Deluge II and the continent of doom: rising sea level and collapsing Antarctic ice. *Boreas*, **15**, 89-100.

Hughes, T.J. (1989). Calving ice walls. *Annals of*

Glaciology, **12**, 74-80.

Hughes, T.J. (1992a). On the pulling power of ice streams. *Journal of Glaciology*, **38**, 125-151.

Hughes, T.J. (1992b). Theoretical calving rates from glaciers along ice walls grounded in water of variable depths. *Journal of Glaciology*, **38**, 282-294.

Hughes, T.J. and Fastook, J.L. (1988). A geomorphic method for reconstructing paleo ice sheets, Part I: Glacial geology. *In*: Kite, J.S., Lowell, T.V. and Thompson, W.B. (eds), Contributions to the Quaternary Geology of Northern Maine and Adjacent Canada. *Maine Geological Survey, Bulletin*, **37**, 3-17.

Hughes, T.J. and Nakagawa, M. (1989). Bending shear: The rate-controlling mechanism for calving ice walls. *Journal of Glaciology*, **35**, 260-266.

Hughes, T.J., Denton, G.H., Andersen, B.G., Schilling, D.H., Fastook, J.L. and Lingle, C.S. (1981). The last great ice sheets: A global view. *In*: Denton, G.H. and Hughes, T.J. (eds), *The Last Great Ice Sheets*, pp. 263-317. Wiley Interscience, New York.

Hughes, T.J., Denton, G.H. and Fastook, J.L. (1985). The Antarctic Ice Sheet: An analog for Northern Hemisphere paleo-ice sheets? *In*: Woldenberg, M.J. (ed.), *Models in Geomorphology*, pp. 25-72. Allen and Unwin, Boston.

Humlum, O. (1981). Observations on debris in the basal transport zone of Mýrdalsjökull, Iceland. *Annals of Glaciology*, **2**, 71-77.

Humlum, O. (1985). Changes in texture and fabric of particles in glacial traction with distance from source, Mýrdalsjökull, Iceland. *Journal of Glaciology*, **31**, 150-156.

Humphrey, N.F. (1986). Suspended sediment discharge from Variegated Glacier during its pre-surge and surge phases of motion. *In*: Benson, C.S., Harrison, W.D., Gosink, J., Bowling, S., Mayo, L. and Trabant, D. (eds), *Workshop on Alaskan Hydrology: Problems Related to Glacierized Basins*. University of Alaska Geophysical Institute Report UAG-R-306, p. A-78.

Humphrey, N.F. (1987). Coupling between water pressure and basal sliding in a linked-cavity hydraulic system. *In*: Waddington, E.D. and Walder, J.S. (eds), *The Physical Basis of Ice Sheet Modelling*. Proceedings of an International Symposium held during the 19th General Assembly of the International Union of Geodesy and Geophysics at Vancouver, B.C., Canada, 9-22 August 1987, *International Association of Hydrological Sciences Publication No. 170*, pp. 105-119.

Humphrey, N.F., Raymond, C.F. and Harrison, W.D.

(1986). Discharges of turbid water during mini-surges of Variegated Glacier, Alaska, U.S.A. *Journal of Glaciology*, **32**, 195-207.

Huppert, H.E. (1980). The physical processes involved in the melting of icebergs. *Annals of Glaciology*, **1**, 97-101.

Huppert, H.E. and Josberger, E.G. (1980). The melting of ice in cold stratified water. *Journal of Physical Oceanography*, **10**, 953-960.

Huston, M.M., Brigham-Grette, J. and Hopkins, D.M. (1990). Paleogeographic significance of middle Pleistocene glaciomarine deposits on Baldwin Peninsula, Northwest Alaska. *Annals of Glaciology*, **14**, 111-114.

Hutchinson, G.E. (1957). *A Treatise on Limnology. Vol. I: Geography, Physics and Chemistry*. John Wiley, New York, 1015 pp.

Hutter, K. (1982a). Dynamics of glaciers and large ice masses. *Annual Review of Fluid Mechanics*, **14**, 87-130.

Hutter, K. (1982b). Glacier flow. *American Scientist*, **70**, 26-34.

Hutter, K. (1983). *Theoretical Glaciology; Material Science of Ice and the Mechanics of Glacier and Ice Sheets*. D. Reidel, Rotterdam, 510 pp.

Hutter, K. and Engelhardt, H.F. (1988). The use of continuum thermodynamics in the formulation of ice-sheet dynamics. *Annals of Glaciology*, **11**, 46-51.

Hutter, K. and Olunloyo, V.O.S. (1981). Basal stress concentration due to abrupt changes in boundary conditions: A cause for high till concentration at the bottom of a glacier. *Annals of Glaciology*, **2**, 29-33.

Hutter, K., Blatter, J. and Funk, M. (1988). A model computation of moisture content in polythermal glaciers. *Journal of Geophysical Research*, **93**, 12,205-12,214.

Hutton, J. (1795). *Theory of the Earth, 2 Vols*. William Creech, Edinburgh. (Reprinted 1959 in facsimile, Hafner, New York, 203 pp.

Iken, A. (1972). Measurements of water pressure in moulins as part of a movement study of the White glacier, Axel Heiberg Island, Northwest Territories, Canada. *Journal of Glaciology*, **11**, 53-58.

Iken, A. (1974). Velocity fluctuations of an Arctic valley glacier, a study of the White glacier, Axel Heiberg Island, Canadian Arctic Archipelago. *Axel Heiberg Island Research Reports, Glaciology Volume 5*. McGill University, Montreal, 116 pp.

Iken, A. (1977). Movement of a large ice mass before breaking off. *Journal of Glaciology*, **19**, 595-605.

Iken, A. (1978). Variations of surface velocities of some Alpine glaciers measured at intervals of a few hours. Comparison with Arctic glaciers. *Zeitschrift für Gletscherkunde und Glazialgeologie*, **13**, 23-35.

Iken, A. (1981). The effect of the subglacial water pressure on the sliding velocity of a glacier in an idealized numerical model. *Journal of Glaciology*, **27**, 407-421.

Iken, A. and Bindschadler, R.A. (1986). Combined measurements of subglacial water pressure and surface velocity of the Findelengletscher, Switzerland. Conclusions about drainage system and sliding mechanism. *Journal of Glaciology*, **32**, 101-119.

Iken, A., Röthlisberger, H., Flotron, A. and Haeberli, W. (1983). The uplift of Unteraargletscher at the beginning of the melt season - a consequence of water storage at the bed? *Journal of Glaciology*, **29**, 28-47.

Imbrie, J. (1985). A theoretical framework for the Pleistocene ice ages. *Journal of the Geological Society of London*, **142**, 417-432.

Imbrie, J. and Imbrie, K.P. (1979). *Ice Ages: Solving the Mystery*. Enslow, Hillside, N.J., 224 pp.

Imbrie, J. and Imbrie, J.Z. (1980). Modelling the climatic response to orbital variations. *Science*, **207**, 943-953.

Imbrie, J., Mix, A.C. and Martinson, D.G. (1993). Milankovitch theory viewed from Devils Hole. *Nature*, **363**, 531-534.

Imbrie, J., Hays, J.D., Martinson, D.G., McIntyre, A., Mix, A.C., Morley, J.J., Pisias, N.G., Prell, W. and Shackleton, N.J. (1984). The orbital theory of Pleistocene climate: Support from a revised chronology of the marine $\delta^{18}O$ record. *In*: Berger, A., Imbrie, J., Hays, J., Kukla, G. and Saltzman, B. (eds), *Milankovitch and Climate, Part I*, pp. 269-305. D. Reidel, Hingham, MA.

Ingólfsson, O. (1987). The late Weichselian glacial geology of the Melabakkar-Asbakkar coastal cliffs, Borgarfjördeen West Iceland. *Jökull*, **37**, 57-80.

Ingólfsson, O. (1988). Large-scale glaciotectonic deformation of soft sediments: a case study of a late Weichselian sequence in Western Iceland. *In*: Croot, D.G. (ed.), *Glaciotectonics: Forms and Processes*, pp. 101-107. A.A. Balkema, Rotterdam.

Irwin, J. and Pickrill, R.A. (1982). Water temperature and turbidity in a glacial fed lake: Lake Tekapo. *New Zealand Journal of Freshwater Research*, **16**, 189-200.

Iverson, N.R. (1990). Laboratory simulations of glacial abrasion: Comparison with theory. *Journal of Glaciology*, **36**, 304-314.

Iverson, N.R. (1991a). Morphology of glacial striae: Implications for abrasion of glacier beds and fault surfaces. *Geological Society of America Bulletin*, **103**, 1308-1316.

Iverson, N.R. (1991b). Potential effects of subglacial water-pressure fluctuations on quarrying. *Journal of Glaciology*, **37**, 27-36.

Iverson, N.R. (1993). Regelation of ice through debris at glacier beds: Implications for sediment transport. *Geology*, **21**, 557-562.

Iverson, N.R. (*in press*). Regelation of ice through subglacial debris: implications for sediment transport. *Geology*.

Jackson, R.G. II, (1976). Depositional model of point bars in the lower Wabash River. *Journal of Sedimentary Petrology*, **46**, 579-594.

Jackson, R.G. II, (1978). Preliminary evaluation of lithofacies models for meandering alluvial streams. *In*: Miall, A.D. (ed.), *Fluvial Sedimentology*. Canadian Society of Petroleum Geologists, Memoir 5, pp. 543-576. Calgary, Alberta, Canada.

Jacobs, S.S. (1989). Marine controls on modern sedimentation on the Antarctic continental shelf. *Marine Geology*, **85**, 121-153.

Jacobs, S.S., Gordon, A.L. and Ardai, J.L., Jr (1979). Circulation and melting beneath the Ross Ice Shelf. *Science*, **203**, 439-443.

Jacobs, S.S., MacAyeal, D.R. and Ardai, J.L., Jr (1986). The recent advance of the Ross Ice Shelf, Antarctica. *Journal of Glaciology*, **32**, 464-474.

Jaeger, J.C. and Cook, N.G.W. (1976) (2nd Edition, 585 pp.), (1979) (3rd Edition, 593 pp.). *Fundamentals of Rock Mechanics*. 2nd Edition, Chapman and Hall Publishing, Halsted Press Book, London.

Jahns, R.H. (1943). Sheet structure in granites: Its origin and use as measure of glacial erosion in New England. *Journal of Geology*, **51**, 71-98.

Jellinek, H.H.G. (1959). Adhesive properties of ice. *Journal of Colloid Science*, **14**, 268-280.

Jezek, K.C. (1984). A modified theory of bottom crevasses used as a means for measuring the buttressing effect of ice shelves on inland ice sheets. *Journal of Geophysical Research*, **89**, 1925-1931.

Jóhannesson, T., Raymond, C.F. and Waddington, E. (1989). Time-scale for adjustment of glaciers to changes in mass balance. *Journal of Glaciology*, **35**, 355-369.

Johnsen, S.J., Clausen, H.B., Dansgaard, W., Fuhrer, K., Gundestrup, N., Hammer, C.U., Iversen, P., Jouzel, J., Stauffer, B. and Steffensen, J.P. (1992). Irregular glacial interstadials recorded in a new Greenland Ice core. *Nature*, **359**, 311-313.

Johnsen, S.J., Dansgaard, W., Clausen, H.B., Gunderstrup, N., Jouzel, J. and Stauffer, B. (*in press*). Glacial interstadials and climate predictability. *Nature*.

Johnson, C.B. (*unpublished*). Characteristics and mechanics of formation of glacial arcuate abrasion cracks. Ph.D. Thesis, Pennsylvania State University, 1975.

Johnson, P.G. (1991a). Discharge regimes of a glacierized basin, Slims River, Yukon. *In*: Prowse, T.D. and Ommanney, C.S.L. (eds), *Northern Hydrology: Selected Perspectives*, pp. 151-164. Proceedings, Northern Hydrology Symposium, July 1990, Saskatoon. National Hydrology Research Institute Symposium No. 6.

Johnson, P.G. (1991b). Pulses in glacier discharge, indicators of the internal drainage system of glaciers. *In*: Prowse, T.D. and Ommanney, C.S.L. (eds), *Northern Hydrology: Selected Perspectives*, pp. 165-176. Proceedings, Northern Hydrology Symposium, July 1990, Saskatoon. National Hydrology Research Institute Symposium No. 6.

Johnson, R.G. (1982). Brunhes-Matuyama magnetic reversal dated at 790,000 yr B.P. by marine-astronomical correlations. *Quaternary Research*, **17**, 135-147.

Johnson, W.H. (1986). Stratigraphy and correlation of the glacial deposits of the Lake Michigan lobe prior to 14 ka BP. *In*: Šibrava, V., Bowen, D.Q. and Richmond, G.M. (eds), *Quaternary Glaciations in the Northern Hemisphere. Quaternary Science Reviews* **5**, 17-22.

Jones, A.S. (1979). The flow of ice over a till bed. *Journal of Glaciology*, **22**, 393-395.

Jones, D.P., Ricker, N.E., Desloges J.R. and Maxwell, M. (1985). Glacier outburst flood on the Noeck River: the drainage of Ape Lake, British Columbia, October 20, 1984. *Geological Survey of Canada, Open File Report*, **1139**, p. 92.

Jones, M.E. and Preston, R.M.F. (eds) (1987). *Deformation of Sediments and Sedimentary Rocks. Geological Society Special Publication* **No. 29**, 350 pp.

Jonsson, J. (1982). Notes on the Katla volcanoglacial debris flows. *Jökull*, **32**, 61-68.

Jopling, A.V. and Walker, R.G. (1968). Morphology and origin of ripple-drift cross-lamination, with examples from the Pleistocene of Massachusetts. *Journal of Sedimentary Petrology*, **38**, 971-984.

Josenhans, H.W., Zevenhuizen, J. and Klassen, R.A. (1986). The Quaternary geology of the Labrador Shelf. *Canadian Journal of Earth Sciences*, **23**, 1190-1213.

Josenhans, H., Balzer, S., Henderson, P., Nielson, E., Thorliefson, H. and Zevenhuizen, J. (1988). Preliminary seismostratigraphic and geomorphic interpretations of the Quaternary sediments of Hudson Bay. *Geological Survey of Canada, Current Research Paper 88-1B*, pp. 271-286.

Jouzel, J. and Souchez, R.A. (1982). Melting-refreezing at the glacier sole and the isotopic composition of the ice. *Journal of Glaciology*, **28**, 35-42.

Jouzel, J., Lorius, C., Petit, J.R., Genthon, C., Barkov, N.I., Kotlyakov, V.M and Petrov, V.M. (1987). Vostock ice core: A continuous isotope temperature record over the last climatic cycle 160,000 years. *Nature*, **329**, 403-408.

Jouzel, J., Raisbeck, G., Benoist, J.P., Yiou, F., Lorius, C., Raynaud, D., Petit, J.R., Barkov, N.I., Korotkevitch, Y.A. and Kotlyakov, V.M. (1989). A comparison of deep Antarctic ice cores and their implications for climate between 65,000 and 15,000 years ago. *Quaternary Research*, **31**, 135-150.

Jouzel, J., Barkov, N.I., Barnola, J.M., Bender, M., Chappellaz, J., Genthon, C., Kotlyakov, V.M., Lipenkov, V., Lorius, C., Petit, J.R., Raynaud, D., Raisbeck, G., Ritz, C., Sowers, T., Stievenard, M., Yiou, M. and Yiou, P. (1993). Extending the Vostock ice-core record of palaeoclimate to the penultimate glacial period. *Nature*, **364**, 407-412.

Kaldal, I. (1978). The deglaciation of the area north and northeast of Hofsjökull, central Iceland. *Jökull*, **28**, 18-31.

Kälin, M. (1971). The active push-moraine of the Thompson glacier, Axel Heiberg Island, Canadian Arctic Archipelago, Dissertation. Zürich, 68 pp.

Kamb, B. (1970). Sliding motion of glaciers: Theory and observation. *Reviews of Geophysics and Space Physics*, **8**, 673-728.

Kamb, B. (1986). Stress-gradient coupling in glacier flow: III. Exact longitudinal equilibrium equation. *Journal of Glaciology*, **32**, 335-341.

Kamb, B. (1987). Glacier surge mechanism based on linked cavity configuration of the basal water conduit system. *Journal of Geophysical Research*, **92**, 9083-9100.

Kamb, B. and Echelmeyer, K.A. (1986a). Stress-gradient coupling in glacier flow: I. Longitudinal averaging of the influence of ice thickness and surface slope. *Journal of Glaciology*, **32**, 267-284.

Kamb, B. and Echelmeyer, K.A. (1986b). Stress-gradient coupling in glacier flow: IV. Effects of the "T" term. *Journal of Glaciology*, **32**, 342-349.

Kamb, B. and Engelhardt, H.F. (1987). Waves of accelerated motion in a glacier approaching surge: the mini-surges of Variegated Glacier, Alaska, U.S.A. *Journal of Glaciology*, **33**, 27-46.

Kamb, B. and Engelhardt, H.F. (1989). Flow mechanics of West Antarctic Ice Streams: Observations by borehole geophysics. *EOS*, **70**, 1081.

Kamb, B. and LaChapelle, E. (1964). Direct observations of the mechanism of glacier sliding over bedrock. *Journal of Glaciology*, **5**, 159-172.

Kamb, B. and Shreve, R.L. (1963). Texture and fabric of ice at depth in a temperate glacier. (Abstract). *Transactions of the American Geophysical Union*, **49**, 103.

Kamb, B., Raymond, C.F., Harrison, W.D., Engelhardt, H.F., Echelmeyer, K.A., Humphrey, N.F., Brugman, M.M. and Pfeffer, T. (1985). Glacier surge mechanism: 1982-1983 surge of Variegated Glacier, Alaska. *Science*, **227**, 469-479.

Karl, H.A., Reimnitz, E. and Edwards, B.D. (1987). Extent and nature of Ross Sea unconformity in the western Ross Sea, Antarctica. *Circum-Pacific Council on Energy and Mineral Resources, Earth Science Series*, **5B**, 77-92.

Karl, T.R., Tarpley, J.D., Quayle, R.G., Diaz, H.F., Robinson, D.A. and Bradley, R.S. (1989). The recent climate record: What it can and cannot tell us. *Reviews of Geophysics*, **27**, 405-430.

Karlén, W. (1988). Scandinavian glacial and climatic fluctuations during the Holocene. *Quaternary Science Reviews*, **7**, 199-209.

Karrow, P.F. and P.E. Calkin (1985). Quaternary Evolution of the Great Lakes. *Geological Association of Canada Special Publication No. 30*, 258 pp.

Kasser, P. (1959). Der Einfluss von Gletscherrückgang und Gletschervorstoss auf den Wasserhaushalt. *Wasser- und Energiewirtschaft*, **51**, 155-168.

Kasser, P. (1973). Influence of changes in the glacierised area on summer runoff in the Porte du Scex drainage basin of the Rhone. *In: Symposium of the Hydrology on Glaciers, Cambridge, 1969*, pp. 221-225. International Association of Scientific Hydrology, Publication 95.

Kasting, J.F. (1984). Comments on the BLAG model; the carbonate-silicate geochemical sysle and its effect on atmospheric carbon dioxide over the past 100 million years. *American Journal of Science*, **284**, 1175-1182.

Kasting, J.F. (1987). Theoretical constraints on oxygen and carbon dioxide concentrations in the Precambrian atmosphere. *Precambrian Research*, **34**, 205-229.

Kasting, J.F. (1989). Long-term stability of the Earth's climate. *Palaeogeography, Palaeoclimatology, Palaeoecology (Global and Planetary Change Section)*, **75**, 83-95.

Kasting, J.F. (1993). New spin on ancient climate. *Nature*, **364**, 759-760.

Kasting, J.F. and Toon, O.B. (1989). Climate evolution on the terrestrial planets. *In*: Atreya, S.K., Pollack, J.B. and Matthews, M.S. (eds), *Origin and Evolution of Planetary and Satellite Atmospheres,* pp. 423-449. University of Arizona Press, Tucson.

Kasting, J.F., Pollack, J.B. and Crisp, D. (1984). Effects of high CO_2 levels on surface temperature and atmospheric oxidation state of the early Earth. *Journal of Atmospheric Chemistry*, **1**, 403-428.

Kazi, A. and Knill, J.L. (1973). Fissuring in glacial lake clays and tills on the Norfolk coast, United Kingdom. *Engineering Geology*, **7**, 35-48.

Kehew, A.E. (1982). Catastrophic flood hypothesis for the origin of the Souris spillway, Saskatchewan and North Dakota. *Geological Society of America Bulletin*, **93,** 1051-1058.

Kehew, A.E. and Clayton, L. (1983). Late Wisconsinan floods and development of the Souris-Pembina spillway system in Saskatchewan, North Dakota, and Manitoba. *In*: Teller, J.T. and Clayton, L. (eds), *Glacial Lake Agassiz.* Geological Association of Canada Special Paper, 26, pp. 187-209.

Kehew, A.E. and Lord, M.L. (1986). Origin and large-scale erosional features of glacial-lake spillways in the northern Great Plains. *Geological Society of America Bulletin*, **97,** 162-177.

Kehew, A.E. and Lord, M.L. (1987). Glacial-lake outbursts along the mid-continent margins of the Laurentide ice-sheet. *In*: Mayer, L. and Nash, D. (eds), *Catastrophic Flooding*, pp. 95-121. The Binghampton Symposia in Geomorphology: International Series, No. 18, Allen and Unwin, Boston.

Keigwin, L. (1982). Isotopic paleoceanography of the Caribbean and east Pacific: Role of Panama uplift in late Neogene time. *Science*, **217**, 350-353.

Keilhack, K. (1926). *In*: Salomon, W. (ed.), *Das Quartär in Grundzüge der Geologie.* Vol. 2, Stuttgart.

Kellogg, T.B. and Kellogg, D.E. (1988). Antarctic cryogenic sediments: biotic and inorganic facies of ice shelf and marine-based ice sheet environments.

Palaeogeography, Palaeoclimatology, Palaeoecology, **67**, 51-74.

Kelly, D.N. (1986). Modern sedimentation in Ferrar Fjord, Antarctica. Unpublished BSc (Hons) Thesis, Victoria University of Wellington, New Zealand, 96 pp.

Kelly, R.I. and I.P. Martini. (1986). Pleistocene glaciolacustrine deltaic deposits of the Scarborough Formation, Ontario, Canada. *Sedimentary Geology*, **47**, 27-52.

Kempema, E.W., Reimnitz, E. and Barnes, P.W. (1989). Sea ice sediment entrainment and rafting in the Arctic. *Journal of Sedimentary Petrology*, **59**, 308-317.

Kennett, J.P. (1977). Cenozoic evolution of Antarctic glaciation, the Circum-Antarctic Ocean, and their impact on global paleoceanography. *Journal of Geophysical Research*, **82**, 3843-3860.

Kennett, J.P. (1990). The Younger Dryas event: An introduction. *Paleoceanography*, **5**, 891-895.

Kennett, J.P. and Thunell, R.C. (1975). Global increase in Quaternary explosive volcanism. *Science*, **187**, 497-503.

Kerr, R.A. (1990). Global warming continues in 1989. *Science*, **247**, 521.

Kerr, R.A. (1992). A revisionist timetable for the ice ages. *Science*, **258**, 220-221.

Keys, H.J.D., Jacobs, S.S. and Barnett, D. (1990). The calving and drift of iceberg B-9 in the Ross Sea, Antarctica. *Antarctic Science*, **2**, 243-257.

Keys, J.R., (1984). *Antarctic Marine Environments and Offshore Oil*. Commission for the Environment, Wellington, N.Z., 168 pp.

Keys, J.R., Anderton, P.W. and Kyle, P.R. (1977). Tephra and debris layers in the Skelton Neve and Kempe Glacier, South Victoria Land, Antarctica. *New Zealand Journal of Geology and Geophysics*, **20**, 971-1002.

Kick, W. (1989). The decline of the last Little Ice Age in High Asia compared with that in the Alps. *In*: Oerlemans, J. (ed.), *Glacier Fluctuations and Climatic Change*, pp. 129-142. Kluwer Academic Publishers, Dordrecht.

King, C.A.M. and Buckley, J.T. (1968). The analysis of stone size and shape in arctic environments. *Journal of Sedimentary Petrology*, **38**, 200-214.

King, L.H. and Fader, G.B.J. (1986). Wisconsinan glaciation of the Atlantic continental shelf of southeast Canada. *Geological Survey of Canada Bulletin*, **363**, 72.

King, L.H., Rokoengen, K. and Gunleiksrud, T. (1987). Quaternary seismostratigraphy of the mid-Norwegian shelf, 65°-67°31'N - a till tongue stratigraphy. *IKU Publications*, Trondheim, 114, 58 pp.

King, L.H., Rokoengen, K., Fader, G.B.J. and Gunleiksrud, T. (1991). Till-tongue stratigraphy. *Geological Society of America Bulletin*, **103**, 637-659.

Kingery, W.D. (1963). *Ice and Snow; Properties, Processes and Applications*. MIT Press, Cambridge, Massachusetts, 684 pp.

Kirkbride, M.P. (1989). The influence of sediment budget on the geomorphic activity of the Tasman Glacier, Mount Cook National Park, New Zealand. Unpublished Ph.D. thesis, University of Canterbury, New Zealand, 395 pp.

Kirkbride, M.P. and Sugden, D.E. (1992). New Zealand Loses its Top. *Geographical Magazine*, **64**, 30-34.

Kirkwood, J.M. and Burton, H.R. (1988). Macrobenthic species assemblages in Ellis Fjord, Vestfold Hills, Antarctica. *Marine Biology*, **97**, 445-457.

Kite, G.W. and Reid, I.A. (1977). Volumetric change of the Athabasca Glacier over the last 100 years. *Journal of Hydrology*, **32**, 279-294.

Kjeldsen, O. (1981). Materialtransportundersokelser i Norske Bre-elver 1980. Norges Vassdrags-og Elektrisitetsvesen. *Vassdragsdirektoratet Hydrologisk Avdeling Rapport*, **4-81**.

Klassen, R.A. (1982). Glaciotectonic thrust plates, Bylot Island, District of Franklin. *Geological Survey of Canada. Current Research, Part A, Paper 82-1A*, pp. 369-373.

Klimek, K. (1973). Geomorphological and geological analysis of the proglacial area of the Skeidarárjökull. Extreme eastern and extreme western sections. *Geographia Polonica*, **26**, 89-113.

Kluiving, S.J. (1989). Glaciotektoniek in de stuwwal van Itterbeck-Uelsen. M.Sc. thesis, University of Amsterdam, 117 pp. (unpublished).

Kluiving, S.J., Rappol, M. and van der Wateren, F.M. (1991). Till stratigraphy and ice movements in eastern Overijssel, The Netherlands. *Boreas*, **20**, 193-205.

Knebel, H.J. and Scanlon, K.M. (1985). Sedimentary framework of Penobscot Bay, Maine. *Marine Geology*, **65**, 305-324.

Knudsen, N.T. and Theakstone, W.H. (1988). Drainage of the Austre Okstindbreen ice-dammed lake, Okstindan, Norway. *Journal of Glaciology*, **34**, 87-94.

Kobarg, W. (1988). The tide-dependent dynamics of the Ekstrom Ice Shelf, Antarctica. *Reports on Polar Research*, **50**, 164 pp.

Kochel, R.C. (1988). Geomorphic impact of large floods: Review and new perspectives on magnitude and frequency. *In*: Baker, V.R., Kochel, R.C. and Patton, P.C. (eds), *Flood Geomorphology*, pp. 169-187. John Wiley, New York.

Kochel, R.C. and Baker, V.R. (1988). Paleoflood analysis using slackwater deposits. *In*: Baker, V.R., Kochel, R.C. and Patton, P.C. (eds), pp. 357-376. Flood Geomorphology. Wiley Interscience Publications, New York.

Kochel, R.C. and Johnson, R.A. (1984). Geomorphology and sedimentology of humid-temperate alluvial fans, central Virginia. *In*: Koster, E.H. and Steel, R.J. (eds), *Sedimentology of Gravels and Conglomerates*, pp. 109-122. Canadian Society of Petroleum Geologists, Memoir 10.

Koerner, R.M. (1977). Distribution of microparticles in a 299-m core through the Devon Island Ice Cap, Northwest Territories, Canada. *International Association of Hydrological Sciences (IAHS)*, **118**, 371-376.

Koerner, R.M. and Fisher, D.A. (1979). Discontinuous flow, ice texture, and dirt content in the basal layers of the Devon Island ice cap. *Journal of Glaciology*, **23**, 209-221.

Kominz, M.A., Heath, G.R., Ku, T.-L. and Pisias, N.G. (1979). Brunhes time scales and the interpretation of climatic change. *Earth and Planetary Science Letters*, **45**, 394-410.

Konovalov, V.G. (1990). Methods for the computations of onset date and daily hydrograph of the outburst from Meitzbacher Lake, Tien-Shan. *In*: Lang, H. and Musy, A. (eds), Hydrology in Mountainous Regions I. *International Association of Hydrological Sciences Publication 193*, pp. 181-187.

Köppen, W. and Wegener, A. (1924). *Die Klimate der Geologischen Vorzeit*. Gebrüder Borntraeger, Berlin, 266 pp.

Koster, E.H. (1978). Transverse ribs: their characteristics, origin and paleohydraulic significance. *In*: Miall, A.D. (ed.), *Fluvial Sedimentology,* pp. 161-186. Canadian Society of Petroleum Geologists, Memoir 5, Calgary, Alberta, Canada.

Köster, R. (1958). Experimenteller Beitrag zur Mechanik von Stauchungszonen. *Meyniania*, **6**, 60-84.

Koteff, C. and Pessl, F., Jr (1981). Systematic ice retreat in New England. *United States Geological Survey Professional Paper No. 1179,* 20 pp.

Kraus, H. (1966). *Freie und Bedeckte Ablation. Ergebnisse Forschung- Unternehmen Nepal Himalaya,*

3, 203-236. Springer, Berlin.

Krigström, A. (1962). Geomorphological studies of sandur plains and their braided rivers in Iceland. *Geografiska Annaler*, **44**, 328-346.

Krimmel, R.M. and Meier, M.F. (1989). Glacial geology and geomorphology of North America; Volume 1, Glaciers and glaciology of Alaska. *28th International Geological Congress. Field Trip Guidebook T301*, American Geophysical Union, 61 pp.

Krimmel, R.M. and Vaughn, B.H. (1987). Columbia Glacier, Alaska: Changes in velocity 1977-1986. *Journal of Geophysical Research*, **92**, 8961-8968.

Krimmel, R.M., Tangborn, W.V. and Meier, M.F. (1973). Water flow through a temperate glacier. *In*: The Role of Snow and Ice in Hydrology: Properties and Processes of Glaciers. *International Association of Hydrological Sciences Publication No. 107*, 401-416.

Krinsley, D.H. and Donahue, J. (1968). Environmental Interpretation of sand grain surface textures by electron microscopy. *Geological Society of America Bulletin*, **79**, 743-748.

Krinsley, D.H. and Doornkamp, J.C. (1973) *Atlas of Quartz Sand Surface Textures*. Cambridge University Press, Cambridge, 91 pp.

Krinsley, D.H. and Marshall, J.R. (1987). Sand grain textural analysis: An assessment. *In*: Marshall, J.R. (ed.), *Clastic Particles*, pp. 2-15. Van Nostrand Reinhold, New York.

Krinsley, D. and Takahashi, T. (1962a). An application of electron microscopy to geology. *N.Y. Academy of Science Transcripts*, **25(1)**, 3-22.

Krinsley, D. and Takahashi, T. (1962b). Surface textures of sand grains - An application of electron microscopy: glaciation. *Science*, **138**, 1262-1264.

Kristensen, M. (1983). Iceberg calving and deterioration in Antarctica. *Progress in Physical Geography*, **7**, 313-328.

Krüger, J. (1979). Structures and textures in till indicating subglacial deposition. *Boreas*, **8**, 323-340.

Krüger, J. (1984). Clasts with stoss-lee form in lodgement tills: A discussion. *Journal of Glaciology*, **30**, 241-243.

Krüger, J. (1985). Formation of push moraine at the margin of Höfdabrekkujökull, south Iceland. *Geografiska Annaler*, **67A**, 199-212.

Krüger, J. (1987). Relationship of drumlin shape and distribution to drumlin stratigraphy and glacial history, Mýrdalsjökull, Iceland. *In*: Menzies, J. and Rose, J. (eds), *Drumlin Symposium,* pp. 257-266. A.A. Balkema, Rotterdam.

Krüger, J. (1994). Glacial processes, sediments, landforms and stratigraphy in the terminus region of Myrdalsjokull, Iceland. *Folia Geographica Danica*, **21**, 233.

Kuhle, M. (1987). Subtropical mountain- and high-land-glaciation as ice age triggers and the waning of the glacial periods in the Pleistocene. *GeoJournal*, **14**, 393-421.

Kuhn, M. (1981). Climate and glaciers. *In: International Association of Hydrological Sciences Publication No. 131* (Symposium at Canberra 1979 - Sea Level, Ice and Climatic Change), pp. 3-20.

Kuhn, M. (1984). (Abstract) Thirty years of mass balance and climate records on Hintereisferner: A basis for modeling. *Annals of Glaciology*, **5**, 212.

Kuhn, M. (1989). Response of the equilibrium line altitude to climate fluctuations: theory and observations. *In*: Oerlemans, J. (ed.), *Glacier Fluctuations and Climatic Change*, pp. 407-417. Kluwer Academic Publishers, Dordrecht.

Kuhn, M., Markl, G., Kaser, G., Nickus, U., Obleitner, F. and Schneider, H. (1985). Fluctuations of climate and mass balance: Different responses of two adjacent glaciers. *Zeitschrift für Gletscherkunde und Glazialgeologie*, **21**, 409-416.

Kukla, G.J. (1977). Pleistocene land-sea correlations. *Earth-Science Reviews*, **13**, 307-374.

Kukla, G.J. (1987a). Loess stratigraphy in central China. *Quaternary Science Reviews*, **6**, 191-219.

Kukla, G.J. (1987b). Pleistocene climates in China and Europe compared to oxygen isotope record. *Paleoecology of Africa*, **18**, 37-45.

Kukla, G.J. (1989). Long continental records of climate - an introduction. *Palaeogeography, Palaeoclimatology, Palaeoecology*, **72**, 1-9.

Kukla, G. and An, Z.S. (1989). Loess stratigraphy in Central China. *Palaeogeography, Palaeoclimatology, Palaeoecology*, **72**, 203-225.

Kulig, J.J. (1985). A sedimentation model for the deposition of glacigenic deposits in central Alberta. Unpublished M.Sc. thesis, University of Alberta, Edmonton, p. 245.

Kumai, M. (1977). Electron microscope analysis of aerosols in snow and deep ice cores from Greenland. *International Association of Hydrological Sciences*, **118**, 341-350.

Kupsch, W.O. (1962). Ice-thrust ridges in Western Canada. *Journal of Geology*, **70**, 582-594.

Kurtz, D.D. and Anderson, J.B. (1979). Recognition and sedimentologic description of recent debris flow deposits from the Ross and Weddell seas, Antarctica. *Journal of Sedimentary Petrology*, **49**, 1159-1170.

Kutzbach, J.E. and Guetter, P.J. (1986). The influence of changing orbital parameters and surface boundary conditions on climate simulations for the past 18,000 years. *The Journal of Atmospheric Sciences*, **43**, 1726-1759.

LaChapelle, E. (1968). Stress-generated ice crystals in a nearly isothermal two-phase system. *Journal of Glaciology*, **7**, 183-198.

Ladurie, E. Le Roy (1971). *Times of Feast, Times of Famine: A History of Climate Since the Year 1000.* Allen and Unwin, London, 428 pp.

Lagerlund, E. (1987). An alternative Weichselian glaciation model, with special reference to the glacial history of Skåne, South Sweden. *Boreas*, **16**, 433-459.

Lagoe, M.B., Eyles, C.H. and Eyles, N. (1989). Paleoenvironmental significance of foraminiferal biofacies in the glaciomarine Yakataga Formation, Middleton Island, Gulf of Alaska. *Journal of Foraminiferal Research*, **19**, 194-209.

Lai, Zuming (1982). A study on the variation coefficient of annual runoff of the rivers in northwest China. *International Association of Hydrological Sciences, Publication No. 138*, 285-294.

Lamb, H.H. (1977). *Climate: Present, Past and Future. Volume 2 - Climatic History and the Future.* Methuen, London, 835 pp.

Lamb, H.H. (1985). *Climatic History and the Future.* Princeton University Press, Princeton, NJ, 835 pp.

Landscheidt, T. (1987). Long-range forecasts of solar cycles and climate change. *In*: Rampino, M.R., Sanders, J.E., Newman, W.S. and Königsson, L.K. (eds), *Climate; History, Periodicity and Predictability*, pp. 421-445. Van Nostrand Reinhold, New York.

Lang, H. (1968). Relations between glacier runoff and meteorological factors observed on and outside the glacier. *In*: IUGG General Assembly, Berne, *International Association of Scientific Hydrology Publication No. 79*, pp. 429-439.

Lang, H. (1980). Theoretical and practical aspects in the computation of runoff from glacier areas. *In*: Proceedings of the Symposium in Tbilisi 1978, Geophysical Committee, Data of Glaciological Studies. *U.S.S.R. Academy of Sciences, Publication 38*, pp. 187-194.

Lang, H. and Dayer, G. (1985). Switzerland case study: Water supply. *In*: Young, G.J. (ed.), Techniques for Prediction of Runoff from Glacierized Areas. *International Association of Hydrological Sciences*

Publication No. 149, pp. 45-57.

Lang, H., Leibundgut, C. and Festel, E. (1979). Results from tracer experiments on the water flow through the Aletschgletscher. *Zeitschrift für Gletscherkunde und Glazialgeologie*, **15**, 209-218.

Lange, M.A. (1987). Quantitative estimates of the mass flux and ice movement along the ice edges in the eastern and southern Weddell Sea. *In*: van der Veen, C.J. and Oerlemans, J. (eds), *Dynamics of the West Antarctic Ice Sheet*, pp. 57-74. D. Reidel, Dordrecht.

Langham, E.J. (1974). Phase equilibria of veins in polycrystalline ice. *Canadian Journal of Earth Sciences*, **11**, 1280-1287.

Langway, C.C., Jr (1970). Stratigraphic analysis of a deep core from Greenland. *Geological Society of America Special Paper*, *125*, 186 pp.

Larson, G.J. (1978). Meltwater storage in a temperate glacier, Burroughs Glacier, southeast Alaska. Columbus: Ohio State University, *Institute of Polar Studies, Report No. 66*.

LaSalle, P. and Shilts, W.W. (1993). Younger Dryas-age readvance of Laurentide ice into the Champlain Sea. *Boreas*, **22**, 25-37.

Laufer, H. and Sommer, N. (1982). Studies on sediment transport in mountain streams of the eastern Alps. *In*: *Proceedings, 14th Congress on Large Dams*. Commission Internationale des Grandes Barrages, pp. 431-453.

Lavrushin, Y.A. (1971). Dynamische Fazies und Subfazies der Grundmoräne. *Zeitschrift für Angewandte Geologie*, **17**, 337-343.

Lawler, D. (1991). Sediment and solute yield from the Jokulsa a Solheimasandi glacierized river basin, southern Iceland. *In*: Maizels, J.K. and Caseldine, C. (eds), *Environmental Change in Iceland: Past and Present*, pp. 303-332. Kluwer Academic Publishers, Dordrecht.

Lawn, B.R. (1967). Partial cone crack formation in a brittle material loaded with a sliding spherical indenter. *Proceedings of the Royal Society of London, Series A*, **299**, 307-316.

Lawn, B.R. and Wilshaw, R. (1975). Review of indentation fracture: Principles and applications. *Journal of Materials Science*, **10**, 1049-1081.

Lawson, D.E. (1976). Observations on flutings at Spencer Glacier, Alaska. *Arctic and Alpine Research*, **8**, 289-296.

Lawson, D.E. (1979a). Sedimentological analysis of the western terminus region of Matanuska Glacier, Alaska. *US CRREL Report 79-9*, Hanover, New Hampshire, 122 pp.

Lawson, D.E. (1979b). A comparison of the pebble orientations in ice and deposits of the Matanuska glacier, Alaska. *Journal of Geology*, **87**, 629-645.

Lawson, D.E. (1979c). Characteristics and origins of the debris and ice, Matanuska Glacier, Alaska. *Journal of Glaciology*, **23**, 437-438.

Lawson, D.E. (1981a). Sedimentological characteristics and classification of depositional processes and deposits in the glacial environment. *US CRREL Report No. 81-27*.

Lawson, D.E. (1981b). Distinguishing characteristics of diamictons at the margin of the Matanuska glacier, Alaska. *Annals of Glaciology*, **2**, 78-84.

Lawson, D.E. (1982). Mobilization, movement and deposition of active subaerial sediment flows, Matanuska Glacier, Alaska. *Journal of Geology*, **90**, 279-300.

Lawson, D.E. (1986). Observations on hydraulic and thermal conditions at the bed of Matanuska Glacier, Alaska. *In*: Alean, P. *et al.* (eds), Hydraulic Effects at the Glacier Bed and Related Phenomena. *Mitteilungen der Versuchanstalt fur Wasserbau, Hydrologie und Glaziologie*, **90**, 69-71.

Lawson, D.E. (1988a). Glacigenic resedimentation: Classification concepts and application to mass-movement processes and deposits. *In*: Goldthwait, R.P. and Matsch, C.L. (eds), *Genetic Classification of Glacigenic Deposits*, pp. 147-172. A.A. Balkema, Rotterdam.

Lawson, D.E. (1988b). Proxy data on subglacial conditions and processes. *In*: Proceedings of Glaciers and Ice Sheets Symposia. AGU 1988 Fall Meeting, December. *EOS*, **69**(44), 1210.

Lawson, D.E. (1993). Glaciohydrologic and glaciohydraulic effects on runoff and sediment yield in glacierized basins. *US CRREL Monograph 93-2*, 108 pp.

Lawson, D.E. and Kulla, J. (1978). An oxygen isotope investigation of the origin of the basal ice of the Matanuska Glacier, Alaska. *Journal of Geology*, **86**, 673-685.

Lawson, D.E., Arcone, S.A. and Delaney, A.J. (1991). Initial interpretation of short-pulse radar profiles of the Matanuska Glacier, Alaska. *EOS*, **72(44)**, 159.

Lawson, D.E., Evenson, E.B., Strasser, J.C. and Alley, R.B. (*in preparation*) Frazil ice in subglacial conduits and its implications for sediment entrainment by glaciers. *Science*.

Laymon, C.A. (1991). Marine episodes in Hudson Strait

and Hudson Bay, Canada, during the Wisconsin Glaciation. *Quaternary Research*, **35**, 53-62.

Leavitt, H.W. and Perkins, E.H. (1935). Glacial geology of Maine, Volume 2. *Maine Technology Experiment Station Bulletin, 30*, 232 pp.

Leckie, D.A. and McCann, S.B. (1982). Glacio-lacustrine sedimentation on low slope prograding delta. *In*: Davidson-Arnott, R., Nickling, W. and Fahey, B.D. (eds), *Research in Glacial, Glacio-Fluvial and Glacio Lacustrine Systems*, pp. 261-278. 6th Guelph Symposium on Geomorphology. Geo Books, Norwich.

Ledford-Hoffman, P.A., De Master, D.J. and Nittrouer, C.A. (1986). Biogenic-silica accumulation in the Ross Sea and the importance of Antarctic continental-shelf deposits in the marine silica budget. *Geochimica et Cosmochimica Acta*, **50**, 2099-2110.

Lehman, S.J. (1993). Flickers within cycles. *Nature*, **361**, 404-405.

Lehman, S.J., Jones, G.A., Keigwin, L.D., Andersen, E.S., Butenko, G. and Østmo, S.-R. (1991). Initiation of Fennoscandian ice-sheet retreat during the last deglaciation. *Nature*, **349**, 513-516.

Lemmen, D.S. (1984). Sedimentation in a glacially-fed arctic lake: Tasikutaak Lake Cumberland Peninsula, Baffin Island, N.W.T. M.Sc. thesis, Department of Geography, Queens's University, Kingston, Ont.

Lemmen, D.S., Gilbert, R., Smol, J.P. and Hall, R.I. (1988). Holocene sedimentation in glacial Tasikutaaq Lake, Baffin Island. *Canadian Journal of Earth Sciences*, **25**, 810-823.

Leonard, E.M. (1986a). Varve studies at Hector Lake, Alberta, Canada and the relationship between glacial activity and sedimentation. *Quaternary Research*, **25**, 199-214.

Leonard, E.M. (1986b). Use of lacustrine sedimentary sequences as indicators of Holocene glacial history, Banff National Park, Alberta, Canada. *Quaternary Research*, **26**, 218-231.

Leonard, E.M. (1986c). Glaciological and climatic controls on lake sedimentation, Canadian Rocky Mountains. *Zeitschrift für Gletscherkunde und Glazialgeologie*, **21**, 35-42

Leopold, L.B. and Maddock, T., Jr (1953). The hydraulic geometry of stream channels and some physiographic implications. *U.S. Geological Survey, Professional Paper*, **252**, 57 pp.

Leopold, L.B. and Wolman, M.G. (1957). River channel patterns: braided, meandering and straight. *U.S. Geological Survey, Professional Paper*, **282-B**, pp. 39-85.

Letréguilly, A. (1988). Relation between the mass balance of western Canadian mountain glaciers and meteorological data. *Journal of Glaciology*, **34**, 11-18.

Leverett, F. (1898). The weathered zone (Yarmouth) between the Illinoian and Kansan till sheets. *Journal of Geology*, **6**, 238-243.

Lewis, E.L. and Perkin, R.G. (1985). The winter oceanography of McMurdo Sound, Antarctica. *In*: Jacobs, S.S. (ed.), Oceanology of the Antarctic Continental Shelf. *American Geophysical Union, Antarctic Research Series, 43*, pp. 145-165.

Libby, W.F. (1955). *Radiocarbon Dating*, 2nd Edition. University of Chicago Press, Chicago, 175 pp.

Lick, W. (1970). The propagation of disturbances on glaciers. *Journal of Geophysical Research*, **75**, 2189-2197.

Liestøl, Ø. (1955a). Glacier-dammed lakes in Norway. *Norsk Geografiska Tidsskrift*, **15**, 122-149.

Liestøl, Ø. (1955b). Bresjøen ved Østerdalsisen. *Norsk Geologiska Tidsskrift*, **35**, 179.

Lindqvist, P.A., Lai, H.H. and Alm, O. (1984). Indentation fracture development in rock continuously observed with a scanning electron microscope. *International Journal of Rock Mechanics and Mineral Sciences and Geomechanics, Abstract*, **21(4)**, 165-182.

Lindstrom, D.R. (1990). The Eurasian Ice Sheet: Formation and collapse resulting from natural atmospheric CO_2 concentration variations. *Journal of Paleoceanography*, **5**, 207-227.

Lindstrom, D.R. and MacAyeal, D.R. (1987). Environmental constraints on West Antarctic ice-sheet formation. *Journal of Glaciology*, **33**, 346-356.

Lindstrom, D.R., and MacAyeal, D.R. (1989). Scandinavian, Siberian, and Arctic Ocean glaciation: Effect of Holocene atmospheric CO_2 variations. *Science*, **245**, 628-631.

Lingle, C.S. (1985). A model of a polar ice stream, and future sea-level rise due to possible drastic retreat of the west Antarctic ice sheet. *Report of a workshop in Seattle, Glaciers, ice sheets, and sea level: effect of a CO_2-induced climatic change*. United States Department of Energy, pp. 317-330.

Lingle, C.S. and Brown, T.J. (1987). A subglacial aquifer bed model and water pressure dependent basal sliding relationship for a West Antarctic ice stream. *In*: Van der Veen, C.J. and Oerlemans, J. (eds), *Dynamics of the West Antarctic Ice Sheet*, pp. 249-285. D. Reidel, Dordrecht.

Lingle, C.S. and Clark, J.A. (1985). A numerical model of interactions between a marine ice sheet and the solid Earth: Application to a West Antarctic ice stream. *Journal of Geophysical Research*, **90**, 1100-1114.

Lipenkov, V. Ya., Barkov, N.I., Duval, P. and Pimienta, P. (1989). Crystalline texture of the 2083m ice core at Vostok Station, Antarctica. *Journal of Glaciology*, **35**, 392-398.

List, E.J. (1982). Turbulent jets and plumes. *Annual Review of Fluid Mechanics*, **14**, 189-212.

Lister, G.S. (1988). A 15,000-year isotopic record from Lake Zürich of deglaciation and climatic change in Switzerland. *Quaternary Research*, **29**, 129-141.

Liu, Tungsheng, Zhang, S. and Han, J. (1986). Stratigraphy and paleoenvironmental changes in the loess of central China. *In*: Šibrava, V., Bowen, D.Q. and Richmond, E.M. (eds), *Quaternary Glaciations in the Northern Hemisphere. Quaternary Science Review*, **5**, 489-495.

Liu, Tungsheng (1987). *Aspects of Loess Research*. Chinaaw China Ocean Press, Beijing, 251 pp.

Liverman, D.G.E. (1987). Sedimentation in ice-dammed Hazard Lake, Yukon. *Canadian Journal of Earth Sciences*, **24**, 1797-1806.

Lliboutry, L. (1962). L'erosion glaciaire. *International Association of Hydrological Sciences Publication*, **59**, 219-225.

Lliboutry, L. (1964). *Traité de Glaciologie, Tome 1*. Masson, Paris, 427 pp.

Lliboutry, L. (1965). *Traité de Glaciologie, Tome 2*. Masson, Paris, 1040 pp.

Lliboutry, L. (1968). General theory of subglacial cavitation and sliding of temperate glaciers. *Journal of Glaciology*, **7**, 21-58.

Lliboutry, L. (1971). The glacier theory. *Advances in Hydroscience*, **7**, 81-167.

Lliboutry, L. (1974). Multivariate statistical analysis of glacier annual balances. *Journal of Glaciology*, **13**, 371-392.

Lliboutry, L. (1976). Physical processes in temperate glaciers. *Journal of Glaciology*, **16**, 151-158.

Lliboutry, L. (1979). Local friction laws for glaciers: A critical review and new openings. *Journal of Glaciology*, **23**, 67-95.

Lliboutry, L. (1983). Modifications to the theory of intraglacial waterways for the case of subglacial ones. *Journal of Glaciology*, **29**, 216-226.

Lliboutry, L.A. (1986). Discharge of debris by glacier Hatunraju, Cordillera Blanca, Peru. *Journal of Glaciology*, **32**, 133.

Lliboutry, L.A. (1987a). Realistic, yet simple bottom boundary conditions for glaciers and ice sheets. *Journal of Geophysical Research*, **92**, 9101-9109.

Lliboutry, L.A. (1987b). *Very Slow Flows of Solids; Basics of Modeling in Geodynamics and Glaciology*. Martinus Nijhoff Publishers, Dordrecht, 510 pp.

Lliboutry, L.A. (1987c). Sliding of cold ice sheets. *In*: Waddington, E.D. and Walder, J.S. (eds), The Physical Basis of Ice Sheet Modelling. Proceedings of an International Symposium held during the 19th General Assembly of the International Union of Geodesy and Geophysics at Vancouver, B.C., Canada, 9-22 August 1987, *International Association of Hydrological Sciences Publication No. 170*, pp. 131-143.

Lliboutry, L. and Reynaud, L. (1981). "Global dynamics" of a temperate valley glacier, Mer de Glace, and past velocities deduced form Forbes' bands. *Journal of Glaciology*, **27**, 207-226.

Lliboutry, L. and Ritz, C. (1978). Ecoulement permanent d'un fluide visqueux non lineaire (corps de Glen) autour d'une sphere parfaitement lisse. *Annales de Geophysique*, **34**, 133-146.

Lliboutry, L., Arnao, B.M., Pautre, A. and Schneider, B. (1977). Glaciological problems set by the control of dangerous lakes in Cordillera Blanca, Peru. I. Historical failures of morainic dams, their causes and prevention. *Journal of Glaciology*, **18**, 239-254.

Lock, G.S.H. (1990). *The Growth and Decay of Ice*. Cambridge University Press, Cambridge, 434 pp.

Locke, C.W. and Locke III, W.W. (1977). Little Ice Age snow-cover extent and paleoglaciation thresholds: North-Central Baffin Island, N.W.T., Canada. *Arctic and Alpine Research*, **9**, 291-300.

Long, D.G.F. (1981). Glacigenic rocks in the Early Proterozoic Chibougamau Formation of northern Quebec. *In*: Hambrey, M.J. and Harland, W.B. (eds), *Earth's Pre-Pleistocene Glacial Record*, pp. 817-820. Cambridge University Press, Cambridge.

Long, W.E. (1959). Preliminary report of the geology of the central range of the Horlick Mountains, Antarctica. *Ohio State University Research Report*, 825-2 pt., **7**, 23 pp.

Loomis, S.R. (1970). Morphology and ablation processes on glacier ice. *In*: Bushnell, V.C. and Ragle, R.H. (eds), *Icefield Ranges Research Project Scientific Results, Volume 2*, pp. 27-31. American Geographical Society and Arctic Institute of North America, New York, Montreal.

Lorius, C., Jouzel, J., Ritz, C., Merlivat, L., Barkov, N.I., Korotkevich, Y.S. and Kotlyakov, V.M. (1985). A

150,000 year climatic record from Antarctic ice. *Nature*, **316**, 591-596.

Lorius, C., Raisbeck, G., Jouzel, J. and Raynaud, D. (1989). Long-term environmental records from Antarctic ice cores. *In*: Oeschger, H. and Langway, C.C., Jr (eds), *The Environmental Record in Glaciers and Ice Sheets,* pp. 343-361. John Wiley, New York.

Lotter, A.F. (1989). Evidence of annual layering in Holocene sediments of Soppensee, Switzerland. *Aquatic Sciences*, **51**, 19-30.

Loubere, P. and Moss, K. (1986). Late Pliocene climatic change and the onset of Northern Hemisphere glaciation as recorded in the northeast Atlantic Ocean. *Geological Society of America Bulletin*, **97**, 818-824.

Lowe, J.J. and Walker, M.J.C. (1984). *Reconstructing Quaternary Environments.* Longman, London, 389 pp.

Ludwig, K.R., Simmons, K.R., Szabo, B.J., Winograd, I.J., Landwehr, J.M., Riggs, A.C. and Hoffman, R.J. (1992). Mass-spectrometer ^{230}Th-^{234}U-^{238}U dating of Devils Hole Calcite Vein. *Science*, **258**, 284-287.

Lundquist, D. (1982). Modeling of runoff from a glacierized basin. In Hydrological Aspects of Alpine and High Mountain Areas, Proceedings of Exeter Symposium, July 1982. *International Association of Hydrological Sciences Publication No. 138*, pp. 131-136.

Lundqvist, J. (1986). Late Weichselian glaciation and deglaciation in Scandinavia. *In*: Šibrava, V., Bowen, D.Q. and Richmond, G.M. (eds), *Quaternary Glaciations in the Northern Hemisphere, Quaternary Science Review*, **5**, 269-292.

Lunkka, J.P. (1988). Sedimentation and deformation of the North Sea Drift Formation in the Happisburgh area, North Norfolk. *In:* Croot, D.G. (ed.), *Glaciotectonics: Forms and Processes,* pp. 109-122. A.A. Balkema, Rotterdam.

Lyell, C. (1830). *Principles of Geology. Volume 1.* p. 511. J. Murray, London.

Lyell, C. (1839). *Nouveaux éléments de géologie.* p. 648. Pitois-Levrault, Paris.

MacAyeal, D.R. (1987). Ice-shelf backpressure: form drag versus dynamic drag. *In*: Van der Veen, C.J. and Oerlemans, J. (eds), *Dynamics of the West Antarctic Ice Sheet,* pp. 141-160. D. Reidel, Dordrecht.

MacAyeal, D.R. (1989a). Ice-shelf response to ice-stream discharge fluctuations: III. The effects of ice-stream imbalance on the Ross Ice Shelf, Antarctic. *Journal of Glaciology*, **35**, 38-42.

MacAyeal, D.R. (1989b). Large-scale ice flow over a viscous basal sediment: Theory and application to ice stream B, Antarctica. *Journal of Geophysical Research*, **94**, 4071-4087.

MacAyeal, D.R. (1992). Irregular oscillations of the West Antarctic ice sheet. *Nature*, **359**, 29-32.

MacClintock, P. and Dreimanis, A. (1964). Reorientation of till fabric by overriding glacier in the St. Lawrence Valley. *American Journal of Science*, **262**, 133-142.

MacCracken, M.C., Hecht, A.D., Budyko, M.I. and Izrael, Y.A. (1990). *Prospects for Future Climate: A special US/USSR Report on Climate and Climate Change.* Lewis Publishers, Chelsea, Michigan, 270 pp.

Mackay, J.R. and Mathews, W.H. (1960). Deformation of soils by glacier ice and the influence of pore pressures and permafrost. *Royal Society of Canada, Transactions Section IV, 3rd Series*, **54**, 27-36.

Mackay, J.R. and Mathews, W.H. (1964). The role of permafrost in ice-thrusting. *Journal of Geology*, **72**, 378-380.

Mackiewicz, N.E., Powell, R.D., Carlson, P.R. and Molnia, B.F. (1984). Interlaminated ice-proximal glacimarine sediments in Muir Inlet, Alaska. *Marine Geology*, **57**, 113-147.

Madsen, V. (1928). Übersicht über die Geologie von Danmark. *Danmarks Geologi Undersog. Afh., Række 5*, **4**, 219 pp.

Magny, M. (1993). Solar influences on Holocene climatic changes illustrated by correlations between past lake-level fluctuations and the atmospheric ^{14}C record. *Quaternary Research*, **40**, 1-9.

Mahaney, W.C. (1978). Late Quaternary stratigraphy and soils in the Wind River Mountains, western Wyoming, *In*: Mahaney, W.C. (ed.), *Quaternary Soils,* pp. 223-264. Geo Abstracts, Norwich, U.K.

Mahaney, W.C. (1989). Quaternary glacial geology of Mount Kenya, *In*: Mahaney, W.C. (ed.), *Quaternary and Environmental Research on East African Mountains,* pp. 121-140. A.A. Balkema, Rotterdam.

Mahaney, W.C. (1990a). Microfabrics and quartz microstructures confirm glacial origin of Sunnybrook drift in the Lake Ontario Basin. *Geology*, **18**, 145-148.

Mahaney, W.C. (1990b). Glacially-crushed quartz grains in late Quaternary deposits in the Virunga Mountains, Rwanda - indicators of wind transport from the north? *Boreas*, **19**, 81-89.

Mahaney, W.C. and Rutter, N.W. (1992). Relative ages of the moraines of the Dalijia Shan, northwestern China. *Catena*, **19**, 179-191.

Mahaney, W.C., Vortisch, W.B. and Fecher, K. (1988a). Sedimentary petrographic study of tephra, glacial and aeolian grains in a Quaternary lithostratigraphic-paleosol sequence on Mount Kenya, East Africa, *Géographie Physique et Quaternaire*, **42**, 137-146.

Mahaney, W.C., Vortisch, W.B. and Julig, P. (1988b). Relative differences between glacially-crushed quartz transported by mountain and continental ice: Some examples from North America and East Africa. *American Journal of Science*, **288**, 810-826.

Mahaney, W.C., Vortisch, W.B. and Spence, J.R. (1988c). Pollen study and scanning electron microscopy of aeolian grains in a compound paleosol in the Mutonga drainage, Mount Kenya. *Journal of African Earth Sciences*, **7**, 895-902.

Mahaney, W.C., Vortisch, W.B. and Julig, P. (1989). Relative differences between glacially-crushed quartz transported by mountain and continental ice - some examples from North America and East Africa, Reply. *American Journal of Science*, **289**, 1198-1205.

Mahaney, W.C., Vaikmae, R. and Vares, K. (1991). Scanning electron microscopy of quartz grains in supraglacial debris, Adishy Glacier, Caucasus Mountains, U.S.S.R. *Boreas*, **20**, 395-404.

Maher, B.A. and Thompson, R. (1992). Paleoclimatic significance of the mineral magnetic record of the Chinese loess and paleosols. *Quaternary Research*, **37**, 155-170.

Maizels, J.K. (1976). A comparison of present-day and Pleistocene proglacial environments, with particular reference to geomorphology and sedimentology. Unpublished Ph.D. Thesis, University of London, U.K., 675 pp.

Maizels, J.K. (1979). Proglacial aggradation and changes in braided channel patterns during a period of glacier advance: an Alpine example. *Geografiska Annaler*, **61A**, 87-101.

Maizels, J.K. (1983). Proglacial channel systems: change and thresholds for change over long, intermediate and short time-scales. *In*: Collinson, N.J. and Lewin, J. (eds), *Modern and Ancient Fluvial Systems*. Special Publication of the International Association of Sedimentologists, **6**, 251-266. Blackwell, Oxford.

Maizels, J.K. (1986). Modeling of paleohydrologic change during deglaciation. *Géographie Physique et Quaternaire*, **40**, 263-277.

Maizels, J.K. (1987). Large-scale flood deposits associated with the formation of coarse-grained braided terrace sequences. *In*: Ethridge, F.G. (ed.), *Recent Developments in Fluvial Sediments*, Society of Economic Paleontologists and Mineralogists, **39**, pp. 135-148.

Maizels, J.K. (1989a). Sedimentology, paleoflow dynamics and flood history of jökulhlaup deposits: Paleohydrology of Holocene sediment sequences in southern Iceland sandur deposits. *Journal of Sedimentary Petrology*, **59**, 204-223.

Maizels, J.K. (1989b). Sedimentology and palaeohydrology of Holocene flood deposits in front of a jökulhlaup glacier, South Iceland. *In*: Beven, K. and Carling, P. (eds), *Floods: Hydrological, Sedimentological, and Geomorphological Implications*, pp. 239-253. John Wiley, Chichester.

Maizels, J.K. (1991). The origin and evolution of Holocene sandur deposits in areas of jökulhlaup drainage, south Iceland. *In*: Maizels, J.K. and Caseldine, C. (eds), *Environmental Change in Iceland: Past and Present*, pp. 267-302. Kluwer Academic Publishers, Dordrecht.

Maizels, J.K. (1992). Boulder ring structures produced during jökulhlaup flows - Origin and hydraulic significance. *Geografiska Annaler*, **74A**, 21-33.

Maizels, J.K. (*in press*). Lithofacies variations within sandur deposits: the role of runoff regime, flow dynamics and sediment supply characteristics. *Sedimentary Geology*.

Maizels, J.K. and Caseldine, C. (1991). *Environmental Change in Iceland: Past and Present*. Kluwer Academic Publishers, Dordrecht, 332 pp.

Maizels, J.K. and Russell, A.J. (1992). Quaternary perspectives on jökulhlaup prediction. *In*: Gray, J.M. (ed.), *Quaternary Proceedings, Applications of Quaternary Research*. Vol. 2, pp. 133-152. Quaternary Research Association, Cambridge.

Malavieille, J. (1987). Kinematics of compressional and extensional ductile shearing deformation in a metamorphic core complex of the northeastern Basin and Range. *Journal of Structural Geology*, **9(5/6)**, 541-554.

Maltman, A.J. (1987). Shear zones in argillaceous sediments - an experimental study. *In*: Jones, M.E. and Preston, R.M.F. (eds), Deformation of Sediments and Sedimentary Rocks. *Geological Society Special Publication No. 29*, pp. 77-87.

Maltman, A.J. (1988). The importance of shear zones in naturally deformed wet sediments. *Tectonophysics*, **145**, 163-175.

Maltman, A.J. (1994). *Geological Deformation of Sediments*. Chapman & Hall, London, 452 pp.

Manabe, S. and Broccoli, A.J. (1985). The influence of continental ice sheets on the climate of an ice age. *Journal of Geophysical Research*, **90**, 2167-2190.

Manabe, S. and Stouffer, R.J. (1993). Century-scale

effects of increased atmospheric CO_2 on the ocean-atmosphere system. *Nature*, **364**, 215-218.

Mandl, G. and Shippam, G.K. (1981). Mechanical model of thrust sheet gliding and imbrication. *In*: McClay, K.R. and Price, N.J. (eds), *Thrust and Nappe Tectonics*, pp. 79-98. Geological Society of London. Blackwell Scientific Publications, Oxford.

Mandl, G., Jong, L.N.J. de and Maltha, A. (1977). Shear zones in granular material. *Rock Mechanics*, **9**, 95-144.

Mangerud, J. (1987). The Alleröd/Younger Dryas boundary. *In*: Berger, W.H. and Labeyrie, L.D. (eds), *Abrupt Climatic Change; Evidence and Implications*, pp. 163-171. D. Reidel, Dordrecht.

Mangerud, J. (1989). Correlation of the Eemian and the Weichselian with deep sea oxygen isotope stratigraphy. *Quaternary International*, **3/4**, 1-4.

Mannerfelt, C.M. (1949). Marginal drainage channels as indicators of the gradients of Quaternary ice caps. *Geografiska Annaler*, **31**, 194-199.

Maohuan, H., Wang, Zhongxiang, Baolin, C. and Jiankang, H. (1989). Some dynamics studies on Urumqi glacier No.1, Tianshan glaciological station, China. *Annals of Glaciology*, **12**, 70-73.

Marchant, D.R., Swisher III, C.C., Lux, D.R., West, D.P., Jr and Denton, G.H. (1993). Pliocene paleoclimate and East Antarctic Ice-Sheet history from surficial ash deposits. *Science*, **260**, 667-670.

Marcus, M.G. (1960). Periodic drainage of glacier-damned Tulsequah Lake, British Columbia. *Geographical Reviews*, **50**, 89-106.

Marshall, J.R. (1987). *Clastic Particles*. Van Nostrand Reinhold, New York, 346 pp.

Martin, L. (1913). Some features of glaciers and glaciation in College Fjord, Prince William Sound, Alaska. *Zeitschrift Gletscherkunde*, **7**, 289-333.

Martini, I.P. (1990). Pleistocene glacial fan deltas in southern Ontario, Canada. *In*: Colella, A. and Prior, D.B. (eds), Coarse-Grained Deltas. *International Association of Sedimentologists, Special Publication 10*, 281-295.

Martinson, D.G., Pisias, N.G., Hays, J.D., Imbrie, J., Moore, T.C. and Shackleton, N.J. (1987). Age dating and the orbital theory of the ice ages: development of a high-resolution 0 to 300,000-year chronostratigraphy. *Quaternary Research*, **27**, 1-29.

Mase, G.E. (1970). *Theory and Problems of Continuum Mechanics*. Schaum's Outline Series. McGraw-Hill, New York, 221 pp.

Mass, C.F. and Portman, D.A. (1989). Major Volcanic eruptions and climate: A critical evaluation. *Journal of Climate*, **2**, 566-593.

Mathewes, R.W., Heusser, L.E. and Patterson, R.T. (1993). Evidence for a Younger Dryas-like cooling event on the British Columbia coast. *Geology*, **21**, 101-104.

Mathews, W.H. (1956). Physical limnology and sedimentation in a glacial lake. *Geological Society of America Bulletin*, **67**, 537-552.

Mathews, W.H. (1964a). Sediment transport from Athabasca Glacier, Alberta. *International Association of Scientific Hydrology Publication 65*, pp. 155-165.

Mathews, W.H. (1964b). Water pressure under a glacier. *Journal of Glaciology*, **5**, 235-240.

Mathews, W.H. (1973). Record of two jökulhlaups. Union Géodésique et Geophysique Internationale. Association Internationale D'Hydrologie Scientifique Commission de Neiges et Glaces. Symposium on the Hydrology of Glaciers. Cambridge, 7-13 September, 1969. *L'Association Internationale D'Hydrologie Scientifique Publication No. 95*, pp. 99-110.

Mathews, W.H. (1974). Surface profiles of the Laurentide ice sheet in its marginal areas. *Journal of Glaciology*, **13**, 37-43.

Mathews, W.H. and Clague, J.J. (1993). The record of jökulhlaups from Summit Lake, northwestern British Columbia. *Canadian Journal of Earth Sciences*, **30**, 499-508.

Matishov, G.G. (1978). The geomorphology of the sea floor and some distinctive features of glacial morphogenesis of the underwater margins of western Spitsbergen. *Oceanology*, **18**, 168-172.

Matsch, C.L. and Ojakangas, R.W. (1991). Comparisons in depositional style of "polar" and "temperate" glacial ice; Later paleozoic Whiteout Conglomerate (West Antarctica) and late Proterozoic Mineral Fork Formation (Utah). *In*: Anderson, J.B. and Ashley, G.M. (eds), *Glacial Marine Sedimentation; Palaeoclimatic Significance. Geological Society of America*, Special Paper, **261**, pp. 191-206.

Matsch, C.L. and Schneider, A.F. (1986). Stratigraphy and correlation of the glacial deposits of the glacial lobe complex in Minnesota and northwestern Wisconsin. *In*: Šibrava, V., Bowen, D.Q. and Richmond, G.M. (eds), *Quaternary Glaciations in the Northern Hemisphere. Quaternary Science Reviews*, **5**, 59-64.

Matthes, F.E. (1930). Geological history of Yosemite Valley. *U.S. Geological Survey Professional Paper*, **160**, 13 pp.

Matthews, J.A. (1987). Regional variation in the composition of Neoglacial end moraines, Jotunheimen, Norway: an altitudinal gradient in clast roundness and its possible palaeoclimatic significance. *Boreas*, **16**, 173-188.

Matthews, R.K. (1984). *Dynamic Stratigraphy: An Introduction to Sedimentology and Stratigraphy*, 2nd Edition 489 pp. Prentice-Hall, Englewood Cliffs, New Jersey.

Mayer, L. and Nash, D. (1987). Catastrophic Flooding. *The Binghampton Symposia in Geomorphology: International Series, No. 18*. Allen and Unwin, Boston, 410 pp.

Mayewski, P.A., Meeker, L.O., Morrison, M.C., Twicker, M.S., Whitlow, S.I., Ferland, K.K., Meese, D.A., Legrand, M.R. and Steffensen, J.P. (1993). Greenland ice core 'signal' characteristics: An expanded view of climatic change. *Journal of Geophysical Research*, **98**, 12,839-12,847.

Mayo, L.R. (1984). Glacier mass balance and runoff research in the U.S.A.. *Geografiska Annaler, Series A*, **66**, 215-227.

Mayo, L.R. (1988). Advance of Hubbard Glacier and closure of Russell Fiord, Alaska; environmental effects and hazards in the Yakutat area. *U.S. Geological Survey Circular*, **1016**, 4-16.

Mayo, L.R. and Trabant, D.C. (1984). Observed and predicted effects of climatic change on Wolverine Glacier, Southern Alaska. *In*: McBeath, J.H. (ed.), *The Potential Effects of Carbon Dioxide-Induced Climatic Changes in Alaska*. Fairbanks: School of Agriculture and Land Resources Management, University of Alaska, Miscellaneous Publication 83-1.

Mazo, V.L. (1987). Effects of glacial erosion on the flow of ice sheets and the morphology of their beds. *In*: Waddington, E.D. and Walder, J.S. (eds), The Physical Basis of Ice Sheet Modelling. Proceedings of an International Symposium held during the 19th General Assembly of the International Union of Geodesy and Geophysics at Vancouver, B.C., Canada, 9-22 August 1987. *International Association of Hydrological Sciences Publication No. 170*, pp. 145-152.

Mazo, V.L. (1989). Waves on glacier beds. *Journal of Glaciology*, **35**, 179-182.

McCabe, A.M. (1986). Glaciomarine facies deposited by retreating tidewater glaciers: An example from the Late Pleistocene of Northern Ireland. *Journal of Sedimentary Petrology*, **56**, 880-894.

McCabe, A.M. and Eyles, N. (1988). Sedimentology of an ice-contact glaciomarine delta, Carey Valley, Northern Ireland. *Sedimentary Geology*, **59**, 1-14.

McCabe, A.M., Dardis, G.F. and Hanvey, P.M. (1984). Sedimentology of a late Pleistocene submarine-moraine complex, County Down, Northern Ireland. *Journal of Sedimentary Petrology*, **54**, 716-730.

McCall, J.G. (1960). The flow characteristics of a cirque glacier and their effects on glacial structure and cirque formation. *In*: Lewis, W.V. (ed.), *Norwegian Cirque Glaciers. Royal Geographic Society (London), Research Series*, **4**, 39-62.

McCarroll, D., Matthews, J.A. and Shakesby, R.A. (1989). "Striations" produced by catastrophic subglacial drainage of a glacier-dammed lake, Mjølkedalsbreen, southern Norway. *Journal of Glaciology*, **35**, 193-196.

McDonald, B.C. and Banerjee, I. (1971). Sediments and bed forms on a braided outwash plain. *Canadian Journal of Earth Sciences*, **8**, 1282-1301.

McDonald, V.C. and Vincent, J.S. (1972). Fluvial sedimentary structures formed experimentally in a pipe and their implications for interpretation of subglacial sedimentary environments. *Geological Survey of Canada Paper*, **72-27**, 30 pp.

McDougall, I. and Wensink, H. (1966). Paleomagnetism and geochronology of the Pliocene-Pleistocene lavas in Iceland. *Earth and Planetary Science Letters*, **1**, 232-236.

McIntyre, N.F. (1985). The dynamics of ice-sheet outlets. *Journal of Glaciology*, **31**, 99-107.

McManus, J. and Duck, R.W. (1988). Localised enhanced sedimentation from icebergs in a proglacial lake in Briksdal, Norway. *Geografiska Annaler*, **70A**, 215-223.

McMeeking, R.M. and Johnson, R.E. (1985). On the analysis of longitudinal stress in glaciers. *Journal of Glaciology*, **31**, 293-302.

McMeeking, R.M. and Johnson, R.E. (1986). On the mechanics of surging glaciers. *Journal of Glaciology*, **32**, 120-132.

McSaveney, M.J. (1978). Sherman Glacier rock avalanche, Alaska, U.S.A. *In*: Voight, B. (ed.), *Rockslides and Rock Avalanches 1, Natural Phenomena*, pp. 197-258. Elsevier, New York.

McWilliams, M.O. and McElhinny, M.W. (1980). Late Precambrian paleomagnetism of Australia: the Adelaide geosyncline. *Journal of Geology*, **88**, 1-26.

Means, W.D. (1976). *Stress and Strain. Basic Concepts of Continuum Mechanics for Geologists*. Springer, New York, 339 pp.

Meer, J.J.M. van der (1987). Micromorphology of glacial sediments as a tool in genetic varieties of till. *In*: Kujansuu, R. and Saarnisto, M. (eds), INQUA Till Symposium, Finland 1985. *Geological Survey of Finland Special Paper, 3*, pp. 77-89.

Meier, M.F. (1960). Mode of flow of Saskatchewan Glacier, Alberta, Canada. *U.S. Geological Survey Professional Paper 351.*

Meier, M.F. (1961). Mass budget of South Cascade glacier 1957-60. *U.S. Geological Survey Professional Paper 424-B*, pp. 206-211.

Meier, M.F. (1965). Glaciers and climate. *In*: Wright, H.E. and Frey, D.G. (eds), *The Quaternary of the United States*, pp. 795-805. Princeton University Press, Princeton, NJ.

Meier, M.F. (1969). Glaciers and water supply. *Journal of the American Water Works Association*, **61**, 8-12.

Meier, M.F. (1973). Hydraulics and hydrology of glaciers. *In*: The Role of Snow and Ice in Hydrology. Proceedings of the Banff Symposium, September 1972. *International Association of Scientific Hydrology Publication No. 107*, pp. 353-369.

Meier, M.F. (1987). Ice sheet models - an overview. *In*: Waddington, E.D. and Walder, J.S. (eds), The Physical Basis of Ice Sheet Modelling. Proceedings of an International Symposium held during the 19th General Assembly of the International Union of Geodesy and Geophysics at Vancouver, B.C., Canada, 9-22 August 1987, *International Association of Hydrological Sciences Publication No. 170*, pp. 1-4.

Meier, M.F. (1989). (Abstract) Relation between water input, basal water pressure, and sliding of Columbia Glacier, Alaska, U.S.A. *Annals of Glaciology*, **12**, 214-215.

Meier, M.F. and Post, A. (1969). What are glacier surges? *Canadian Journal of Earth Sciences*, **6**, 807-817.

Meier, M.F. and Post, A. (1987). Fast tidewater glaciers. *Journal of Geophysical Research*, **92**, 9051-9058.

Meier, M.F., Kamb, B., Allen, C.R. and Sharp, R.P. (1974). Flow of Blue Glacier, Olympic Mountains, Washington, U.S.A. *Journal of Glaciology*, **13**, 187-212.

Meier, M.F. et al. (1985). *Glaciers, Ice Sheets, and Sea Level: Effect of a CO_2-induced climatic change. Report of a Workshop in Seattle, Washington, Sept. 13-15, 1984.* United States Department of Energy, Washington DC, 330 pp.

Mellor, M. (1963). Oversnow travel: flying. *Polar Notes; Occasional Publication of the Stefansson Collection,*

Dartmouth College Library, No. 5, 36-51.

Mellor, M. and Testa, R. (1969). Effect of temperature on the creep of ice. *Journal of Glaciology*, **8**, 131-145.

Menzies, J. (1981). Freezing fronts and their possible influence upon processes of subglacial erosion and deposition. *Annals of Glaciology*, **2**, 52-56.

Menzies, J. (1987). Towards a general hypothesis on the formation of drumlins. *In*: Menzies, J. and Rose, J. (eds), *Drumlin Symposium*, pp. 9-24. A.A. Balkema, Rotterdam.

Menzies, J. (1989a). Subglacial hydraulic conditions and their possible impact upon subglacial bed formation. *Sedimentary Geology*, **62**, 125-150.

Menzies, J. (1989b). Drumlins - Products of Controlled or Uncontrolled Glaciodynamic Response? *Quaternary Science Reviews*, **8**, 151-158.

Menzies, J. (1990a). Brecciated diamictons from Mohawk Bay, S. Ontario, Canada. *Sedimentology*, **37**, 481-493.

Menzies, J. (1990b). Evidence of cryostatic desiccation processes associated with sand, intraclasts, within diamictons, southern Ontario, Canada. *Canadian Journal of Earth Sciences*, **27**, 684-693.

Menzies, J. (1990c). Sand Intraclasts within a diamicton mélange, southern Niagara Peninsula, Ontario, Canada. *Journal of Quaternary Science*, **5**, 189-206.

Menzies, J. (ed.) (1995). *Past Glacial Environments. Sediment, Forms and Techniques.* Butterworth-Heinneman, Oxford.

Menzies, J. and Habbe, K.A. (1992). A cryogenic wedge within gravels, north of Kempten, Bavaria. *Zeitschrift für Geomorphologie*, **36**, 365-374.

Menzies, J. and Maltman, A.J. (1992). Microstructures in diamictons - evidence of subglacial bed conditions. *Geomorphology*, **6**, 27-40.

Mercer, J.H. (1961). The response of fjord glaciers to changes in the firn limit. *Journal of Glaciology*, **3**, 850-858.

Mercer, J.H. (1968). Antarctic ice and Sangamon sea level. *International Association of Hydrological Sciences*, **79**, 217-225.

Mercer, J.H. (1970). A former ice sheet in the Arctic Ocean? *Palaeogeography, Palaeoclimatology, Palaeoecology*, **8**, 19-27.

Mercer, J.H. (1976). Glacial history of southernmost South America. *Quaternary Research*, **6**, 125-166.

Mercer, J.H. (1978). West Antarctic ice sheet and CO_2 greenhouse effect: A threat of disaster. *Nature*, **271**, 321-325.

Mercer, J.H. (1983). Cenozoic glaciation in the southern

hemisphere. *Annual Review of Earth and Planetary Sciences*, **11**, 99-132.

Mercer, J.H. (1984). Simultaneous climatic change in both hemispheres and similar bipolar interglacial warming: Evidence and implications. *In*: Hansen, J.E. and Takahashi, T. (eds), Climate Processes and Climate Sensitivity. *Geophysical Monograph*, **29**, 307-313.

Mesolella, K.J., Mathews, R.K., Broecker, W.S. and Thurber, D.L. (1969). The astronomical theory of climatic change: Barbados data. *Journal of Geology*, **77**, 250-274.

Meyer, K.-D. (1980). Zur Geologie der Dammer und Fürstenauer Stauchendmoränen (Rehburger Phase des Drenthe-Stadiums). *In*: *Festschrift für Georg Keller*. Georg Keller, Osnabrück.

Meyer, K.-D. (1983). Saalian end moraines in Lower Saxony. *In*: Ehlers, J. (ed.), *Glacial Deposits in North-West Europe,* pp. 335-342. A.A. Balkema, Rotterdam.

Meyer, K.-D. (1987). Ground and end moraines in Lower Saxony. *In*: van der Meer, J.J.M. (ed.), *Tills and Glaciotectonics,* pp. 197-204. A.A. Balkema, Rotterdam.

Miall, A.D. (1977). A review of the braided-river depositional environment. *Earth Science Reviews*, **13**, 1-62.

Miall, A.D. (1978). Lithofacies types and vertical profile models in braided river deposits: A summary. *In*: Miall, A.D. (ed.), Fluvial Sedimentology. *Canadian Society of Petroleum Geologists, Memoir 5*, pp. 597-604. Calgary, Alberta, Canada.

Miall, A.D. (1983). Glaciomarine sedimentation in the Gowganda Formation (Huronian), northern Ontario. *Journal of Sedimentary Petrology*, **53**, 477-491.

Miall, A.D. (1985). Architectural-element analysis: a new method of facies analysis applied to fluvial deposits. *Earth-Science Reviews*, **22**, 261-308.

Michel, B. (1978). *Ice Mechanics*. Les Presses de l'Université Laval, Quebec City, P.Q., 499 pp.

Mickelson, D.M. (1986). Conditions at the Bed along the Southern Margin of the Laurentide Ice Sheet: Speculation Based on the Geologic Record. (Abstract). *EOS*, **67**(44), 946.

Mickelson, D.M., Clayton, L. Fullerton, D.S. and Borns, H.W., Jr (1983). The late Wisconsin glacial record of the Laurentide ice sheet in the United States. *In*: Porter, S.C. (ed.), *The Late Pleistocene. Late Quaternary Environments of the United States, Volume I,* pp. 3-37. University of Minnesota, Minneapolis.

Middleton, G.V. (1970). Experimental studies related to problems of flysch sedimentation. *In*: Lajoie, J. (ed.), Flysch Sedimentology in North America. *Geological Association of Canada Special Paper*, **7**, 253-272.

Milankovitch, M.M. (1930). Mathematische Klimalehre und Astronomische Theorie der Klimaschwankungen. *In*: Köppen, W. and Geiger, R. (eds), *Handbuch der Klimatologie*, I, Teil A. Borntraeger, Berlin.

Milankovitch, M.M. (1941). Kanon der Erdbestrahlung und seine Anwendung auf das Eiszeitenproblem. *Royal Serbian Academy, Special Publication*, **133**, p. 633.

Miller, M.M. (1973). Entropy and the self-regulation of glaciers in Arctic and Alpine regions. *In*: Fahey, B.D. and Thompson, R.D. (eds), *Research in Polar and Alpine Geomorphology: Proceedings of the 3rd Guelph Symposium on Geomorphology, 1973*, pp. 136-158. Geo Abstracts, Norwich.

Mills, H.H. (1977). Textural characteristics of drift from some representative Cordilleran glaciers. *Geological Society of America Bulletin*, **88**, 1135-1143.

Mitchell, G.F., Penny, L.F., Shotten, F.W. and West, R.G. (1973). A correlation of Quaternary deposits in the British Isles. *Geological Society of London Special Report, 4,* 99 pp.

Mitchell, J.F.B. (1989). The "Greenhouse effect" and climate change. *Reviews of Geophysics*, **27**, 115-139.

Mitchell, J.F.B., Senior, C.A. and Ingram, W.J. (1989). CO_2 and climate: A missing feedback? *Nature*, **341**, 132-134.

Mitchell, J.M., Jr (1977). *Carbon dioxide and future climate*, pp. 3-9. *Environmental Service, United States Department of Commerce*.

Mitchum, R.M., Jr, Vail, P.R. and Sangree, J.B. (1977). Seismic stratigraphy and global changes of sea level; part 6. Stratigraphic interpretation of seismic reflection patterns in depositional sequences. *American Association of Petroleum Geologists, Memoir 26,* 117-133.

Mix, A.C. (1987). The oxygen-isotope record of glaciation. *In*: Ruddiman, W.F. and Wright, H.E., Jr (eds), *North America and Adjacent Oceans During the Last Deglaciation. The Geology of North America, Volume K-3*, pp. 111-125. Geological Society of America, Boulder, Colorado.

Mix, A.C. and Ruddiman, W.R. (1984). Oxygen-isotope analyses and Pleistocene ice volumes. *Quaternary Research*, **21**, 1-20.

Möbus, G. and Peterss, K. (1983). Kluftstatische

Spannungsanalyse im Geschiebemergel, Analysis of glaciotectonic structures. *4th Glacitectonic Symposium.* Wyższa Szkoła Inżynierska, pp. 139-153.

Möller, D. and Ritter, B. (1988). Glacial geodetic contributions to the mass balance and dynamic of ice shelves. *Annals of Glaciology*, **11**, 89-94.

Molnia, B.F. (1983). *Glacial-Marine Sedimentation.* Plenum Press, New York, 844 pp.

Molnia, B.F. (1990). Subarctic glacial-marine sedimentation. *In*: Anderson, J.B. and Molnia, B.F. (eds), *Glacial-Marine Sedimentation. Short Course - Geology, 9.* American Geophysical Union, 127 pp.

Mosley, M.P. (1988). Bedload transport and sediment yield in the Onyx river, Antarctica. *Earth Surface Processes and Landforms*, **13**, 51-67.

Müller, F. (1961). Jacobsen-McGill arctic research expedition to Axel Heiberg Island, Queen Elizabeth Islands. *In*: Müller, B.S. (ed.), *Preliminary Reports of 1959-1960. By F. Müller and members of the expedition.* McGill University Press, 219 pp.

Müller, F. (1962). Zonation in the accumulation area of the glaciers of Axel Heiberg Island, N.W.T., Canada. *Journal of Glaciology*, **4**, 302-313.

Müller, F. (1976). On the thermal regime of a high-Arctic valley glacier. *Journal of Glaciology*, **16**, 119-133.

Müller, F. and Iken, A. (1973). Velocity fluctuations and water regime of Arctic valley glaciers. *International Association of Hydrological Sciences*, **95**, 165-182.

Mullins, H.T., Eyles, N. and Hinchey, E.J. (1990). Seismic reflection investigation of Kalamalka Lake: A "fiord lake" on the Interior Plateau of southern British Columbia. *Canadian Journal of Earth Sciences*, **27**, 1225-1235.

Murray, T. and Dowdeswell, J.A. (1992). Water throughflow and the physical effects of deformation on sedimentary glacier beds. *Journal of Geophysical Research*, **97B**, 8993-9002.

Muszynski, I. and Birchfield, G.E. (1987). A coupled marine ice-stream-ice-shelf model. *Journal of Glaciology*, **33**, 3-15.

Naden, P.S. and Brayshaw, A.C. (1987). Small- and medium-scale bedforms in gravel-bed rivers. *In*: Richards, K. (ed.), *River Channels. Environment and Process. Institute of British Geographers, Special Publication, 18*, pp. 249-271.

Nakawo, M. (1979). Supraglacial debris of G2 glacier in Hidden Valley, Mukut Himal, Nepal. *Journal of Glaciology*, **22**, 273-283.

Nakawo, M. and Takahashi, S. (1983). A simplified model for estimating glacier ablation under a debris layer. *In*: Glen, J.W. (ed.), Hydrological aspects of alpine and high-mountain areas. *International Association of Hydrological Sciences, Publication 138*, pp. 137-145.

Nakawo, M. and Young, G.J. (1981). Field experiments to determine the effect of a debris layer on ablation of glacier ice. *Annals of Glaciology*, **2**, 85-91.

Nakawo, M. and Young, G.J. (1982). Estimate of glacier ablation under a debris layer from surface temperature and meteorological variables. *Journal of Glaciology*, **28**, 29-34.

Nance, R.D., Worsley, T.R. and Moody, J.B. (1986). Post-Archean biogeochemical cycles and long-term episodicity in tectonic processes. *Geology*, **14**, 514-518.

Nance, R.D., Worsley, T.R. and Moody, J.B. (1988). The Supercontinent cycle. *Scientific American*, **259**, 72-79.

Neftel, A., Oeschger, H., Schwander, J., Stauffer, B. and Zumbrunn, R. (1982). Ice core sample measurements give atmospheric CO_2 content during the past 40,000 yrs. *Nature*, **295**, 220-223.

Nelson, C.S., Hendy, D.H., Jarrett, G.R. and Cuthbertson, A.M. (1985). Near-synchroneity of New Zealand alpine glaciations and Northern Hemisphere continental glaciations during the past 750 Kyr. *Nature*, **318**, 361-363.

Nemec, W. (1990). Aspects of sediment movement on steep delta slopes. *In*: Colella, A. and Prior, D.B. (eds), *Coarse-Grained Deltas. International Association of Sedimentologists, Special Publication 10*, 29-73.

Nemec, W. and Steel, R.J. (1988). *Fan Deltas: Sedimentology and Tectonic Settings: 1987, Bergen, Norway.* Blackie and Son, Glasgow, 444 pp.

Nesbitt, H.W. and Young, G.M. (1982). Early Proterozoic climates and plate motions inferred from major element chemistry of lutites. *Nature*, **299**, 715-717.

Newitt, D.M., Richardson, J.F., Abbott, M. and Turtle, R.B. (1955). Hydraulic conveying of solids in horizontal pipes. *Transactions of the Institute of Chemical Engineers*, **33**, 93-110.

Nichols, R.L. (1961). Characteristics of beaches formed in polar climates. *American Journal of Science*, **259**, 694-708.

Nichols, R.L. (1968). Coastal geomorphology, McMurdo Sound, Antarctica. *Journal of Glaciology*, **7**, 449-478.

Nickling, W.G. and Bennett, L. (1984). The shear

strength characteristics of frozen coarse granular debris. *Journal of Glaciology*, **30**, 348-357.

Nilsson, J. and Sundblad, B. (1975). The internal drainage of Storglaciären and Isfallsglaciären described by an autoregressive model. *Geografiska Annaler, Series A*, **57**, 73-98.

Nilsson, T. (1983). *The Pleistocene; Geology and Life in the Quaternary Ice Age*. D. Reidel, Dordrecht, 651 pp.

Nishiizumi, K., Elmore, D. and Kubik, P.W. (1989a). Update on terrestrial ages of Antarctic Meteorites. *Earth and Planetary Science Letters*, **93**, 299-313.

Nishiizumi, K., Winterer, E.L., Kohl, C.P., Klein, J., Middleton, R., Lal, D. and Arnold, J.R. (1989b). Cosmic loess particles: discussion. *Geological Society of America Bulletin*, **82**, 2357-2360.

North, F.J. (1943). Centenary of the Glacial Theory. *Geologists' Association, Proceedings*, **54**, 1-28.

Nummedal, D., Hine, A.C. and Boothroyd, J.C. (1987). Holocene evolution of the south-central coast of Iceland. *In*: Fitzgerald, D.M. and Rosen, P.S. (eds), *Glaciated Coasts*, pp. 115-150. Academic Press, San Diego.

Nydegger, P. (1967). Investigation of fine-sediment transport in rivers and lakes; concerning the origin of turbidity layers; and river-induced inflow currents in Lake Brienz and other lakes. *Beitr. Geol. Schweiz. Hydrol. Series*, **16**, 1-92.

Nye, J.F. (1952). The mechanics of glacier flow. *Journal of Glaciology*, **2**, 82-93.

Nye, J.F. (1953). The flow law of ice from measurements in glacier tunnels, laboratory experiments and the Jungfraufirn borehole experiment. *Proceedings of the Royal Society of London*, **219A**, 477-489.

Nye, J.F. (1957). The distribution of stress and velocity in glaciers and ice-sheets. *Proceedings of the Royal Society of London*, **239A**, 113-133.

Nye, J.F. (1958a). A theory of wave formation in glaciers. *International Association of Hydrological Sciences*, **47**, 139-154.

Nye, J.F. (1958b). Surges in glaciers. *Nature*, **181**, 1450-1451.

Nye, J.F. (1960). The response of glaciers and ice-sheets to seasonal and climatic changes. *Proceedings of the Royal Society of London*, **256A**, 559-584.

Nye, J.F. (1963). On the theory of the advance and retreat of glaciers. *Geophysical Journal of the Royal Astronomical Society*, **7**, 431-456.

Nye, J.F. (1965). The frequency response of glaciers. *Journal of Glaciology*, **5**, 567-587.

Nye, J.F. (1969a). The effect of longitudinal stress on the shear stress at the base of an ice sheet. *Journal of Glaciology*, **8**, 207-213.

Nye, J.F. (1969b). A calculation on the sliding of ice over a wavy surface using a Newtonian viscous approximation. *Proceedings of the Royal Society of London, Series A*, **311**, 445-67.

Nye, J.F. (1970). Glacier sliding without cavitation in a linear viscous approximation. *Proceedings of the Royal Society of London*, **315A**, 381-403.

Nye, J.F. (1973). Water at the bed of a glacier. *International Association of Hydrological Sciences*, **95**, 189-194.

Nye, J.F. (1976). Water flow in glaciers: jökulhlaups, tunnels and veins. *Journal of Glaciology*, **17**, 181-207.

Nye, J.F. (1989). The geometry of water veins and nodes in polycrystalline ice. *Journal of Glaciology*, **35**, 17-22.

Nye, J.F. (1990). Interpreting the field evidence of past ice sheets: structural stability and genericity. *Annals of Glaciology*, **14**, 208-210.

Nye, J.F. and Frank, F.C. (1973). Hydrology of the intergranular veins in a temperate glacier. *International Association of Hydrological Sciences*, **95**, 157-161

Nye, J.F. and Mae, S. (1972). The effect of non-hydrostatic stress on intergranular water veins and lenses in ice. *Journal of Glaciology*, **11**, 81-101.

Nye, J.F. and Martin, P.C.S. (1968). Glacial erosion. *International Association of Scientific Hydrology Publication*, **79**, 78-86.

O'Brien, D.K., Wenk, H.-R., Ratschbacher, L. and You, Z. (1987). Preferred orientation of phyllosilicates in phyllonites and ultramylonites. *Journal of Structural Geology*, **9**, 719-730.

Oerlemans, J. (1989). On the response of valley glaciers to climatic change. *In*: Oerlemans. J. (ed.), *Glacier Fluctuations and Climatic Change*, pp. 353-371. Kluwer Publications, Dordrecht.

Oerlemans, J. (1991). The role of ice sheets in the Pleistocene climate. *Norsk Geologisk Tidsskrift*, **71**, 155-161.

Oerlemans, J. and Hoogendoorn, N.C. (1989). Mass-balance gradients and climatic change. *Journal of Glaciology*, **35**, 399-405.

Oerlemans, J. and van der Veen, C.J. (1984). *Ice sheets and climate*. D. Reidel, Dordrecht, 217 pp.

Oerter, H. and Moser, H. (1982). Water storage and drainage within the firn of a temperate glacier (Vernagtferner, Oetztal Alps, Austria). *In*: Glen, J.W. (ed.), Hydrological Aspects of Alpine and High

Mountain Areas, Proceedings of the Exeter Symposium, July 1982. *International Association of Hydrological Sciences Publication No. 138*, pp. 71-81.

Oerter, H., Baker, D., Stichler, W. and Rauert, W. (1985). Isotope studies of ice cores from a temperate alpine glacier (Vernagtferner, Austria) with respect to the meltwater flow. *Annals of Glaciology*, 7, 90-93.

Oescheger, H. and Langway, C.C., Jr (1989). *The Environmental Record in Glaciers and Ice Sheets*. Report of the Dahlem Workshop on The Environmental Record in Glaciers and Ice Sheets, Berlin, 13-18 March 1988. John Wiley, Chichester, 400 pp.

Oescheger, H., Beer, J., Siegenthaler, U., Stauffer, B., Dansgaard, W. and Langway, C.C. (1984). Late glacial climate history from ice cores. *In*: Hansen, J.E. and Takahashi, T. (eds), *Climate Processes and Climate Sensitivity*, pp. 299-306. American Geophysical Union, Geophysical Monograph No. 29, Washington, D.C.

Ogilvie, I.H. (1904). Effect of superglacial debris on the advance and retreat of some Canadian glaciers. *Journal of Geology*, 12, 722-743.

Oldale, R.N. (1985). Upper Wisconsinan submarine end moraines off Cape Ann, Massachusetts. *Quaternary Research*, 24, 187-196.

Oldale, R.N. and O'Hara, C.J. (1984). Glaciotectonic origin of the Massachusetts coastal end moraines and fluctuating late Wisconsinan ice margin. *Geological Society of America Bulletin*, 95, 61-74.

Oldale, R.N., Valentine, P.C., Cronin, T.M., Spiker, E.C., Blackwelder, B.W., Belknap, D.F., Wehmiller, J.F. and Szabo, B.J. (1982). Stratigraphy, structure, absolute age, and paleontology of the upper Pleistocene deposits at Sankaty Head, Nantucket Island, Massachusetts. *Geology*, 10, 246-252.

Ørheim, O. (1980). Physical characteristics and life expectancy of tabular Antarctic icebergs. *Annals of Glaciology*, 1, 11-18.

Ørheim, O. (1985). Iceberg discharge and the mass balance of Antarctica. *In*: *Glaciers, Ice Sheets and Sea Level: Effect of CO₂-induced Climatic Change*. Report of a workshop held in Seattle, Washington, September 1984. United States Department of Energy, Washington, DC, pp. 210-215.

Orr, E.D. and Folk, R.L. (1983). New scents on the chattermark trail: weathering enhances obscure microfractures. *Journal of Sedimentary Petrology*, 53, 121-129.

Östling, M. and Hooke, R. LeB. (1986). Water storage in Storglaciären, Kebnekaise, Sweden. *Geografiska Annaler*, 68A, 279-290.

Østrem, G. (1959). Ice melting under a thin layer of moraine, and the existence of ice cores in moraine ridges. *Geografiska Annaler*, 41A, 228-230.

Østrem, G. (1973). Runoff forecasts for highly glacierized basins. In Role of Snow and Ice in Hydrology, Proceedings of Banff Symposium. *International Association of Hydrological Sciences Publication No. 107*, pp. 1111-1132.

Østrem, G. (1975). Sediment transport in glacial meltwater streams. *In*: Jopling, A.V. and McDonald, B.C. (eds), *Glaciofluvial and Glaciolacustrine Sedimentation*, pp. 101-122. Society of Economic Paleontologists and Mineralogists, Special Publication, 23.

Østrem, G. and Brugman, M.M. (1991). *Glacier Mass-Balance Measurements: a Manual for Field Work*. NHRI Science Report No.4, Ministry of Supply and Services Canada, 224 pp.

Østrem, G. and Olsen, H.C. (1987). Sedimentation in a glacier lake. *Geografiska Annaler*, 69A, 123-138.

Østrem, G. and Wold, B. (1986). Glacierized basins and water power production in Norway. *In*: Benson, C.S., Harrison, W.D., Gosink, J., Bowling, S., Mayo, L. and Trabant, D. (eds), *Workshop on Alaskan Hydrology: Problems Related to Glacierized Basins*, pp. A-5-A-9. University of Alaska Geophysical Institute Report UAG-R-306.

Østrem, G., Bridge, C.W. and Rannie, W.F. (1967). Glacio-hydrology, discharge and sediment transport in the Decade Glacier area, Baffin Island, N.W.T. *Geografiska Annaler*, 49A, 268-282.

Østrem, G., Ziegler, T. and Ekman, S.R. (1973). A study of sediment transport in Norwegian glacial rivers, 1969. Fairbanks. *University of Alaska, Institute of Water Resources, Publication No. IWR-35*.

O'Sullivan, P.E. (1983). Annually-laminated lake sediments and the study of Quaternary environmental changes - a review. *Quaternary Science Reviews*, 1, 245-313.

Oswald, G.K.A. and Robin, G.deQ. (1973). Lakes beneath the Antarctic ice sheet. *Nature*, 245, 251-254.

Ovenshine, A.T. (1970). Observations of iceberg rafting in Glacier Bay, Alaska, and the identification of ancient ice-rafted deposits. *Geological Society of America Bulletin*, 81, 891-894.

Owen, T., Cess, R.D. and Ramanathan, V. (1979). Enhanced CO₂ greenhouse to compensate for reduced solar luminosity on early Earth. *Nature*, 277, 640-642.

Page, N.J. (1981). The Precambrian diamictite below the base of the Stillwater Complex, Montana. *In*: Hambrey, M.J. and Harland, W.B. (eds), *Earth's Pre-Pleistocene Glacial Record*, pp. 821-825. Cambridge University Press, Cambridge.

Palmer, A.C. (1972). A kinematic wave model of glacier surges. *Journal of Glaciology*, **11**, 65-72.

Palmer, A.C., Goodman, D.J., Ashby, M.F., Evans, A.G., Hutchinson, J.W. and Ponter, A.R.S. (1983). Fracture and its role in determining ice forces on offshore structures. *Annals of Glaciology*, **4**, 216-221.

Parks, B. and Madison, R.J. (1984). Estimation of selected flow and water quality characteristics of Alaskan streams. *U.S. Geological Survey, Water Resources Investigations Report 84-4247.*

Partl, R. (1977). *Power From Glaciers. The Hydropower Potential of Greenland's Glacial Waters.* Austria: International Institute for Applied Systems Analysis, RR-77-20.

Paterson, W.S.B. (1961). Movement of the Sefstrøms Gletscher, north-east Greenland. *Journal of Glaciology*, **3**, 845-849.

Paterson, W.S.B. (1969). The Meighen Ice Cap, Arctic Canada: Accumulation, ablation and flow. *Journal of Glaciology*, **8**, 341-352.

Paterson, W.S.B. (1972a). Temperature distribution in the upper layers of the ablation area of Athabasca Glacier, Alberta, Canada. *Journal of Glaciology*, **11**, 31-41.

Paterson, W.S.B. (1972b). Laurentide ice sheet: estimated volumes during late Wisconsin. *Reviews of Geophysics and Space Physics*, **10**, 885-917.

Paterson, W.S.B. (1980). Ice sheets and ice shelves. *In*: Colbeck, S.C. (ed.), *Dynamics of Snow and Ice Masses*, pp. 1-78. Academic Press, New York.

Paterson, W.S.B. (1981). *The Physics of Glaciers*. 2nd Edition. Pergamon Press, Oxford, 380 pp.

Paterson, W.S.B. and Hammer, C.U. (1987). Ice core and other glaciological data. *In*: Ruddiman, W.F. and Wright, H.E., Jr (eds), *North America and Adjacent Oceans During the Last Deglaciation*, pp. 91-109. The Geology of North America, Volume K-3. Geological Society of America, Boulder, Colorado.

Paterson, W.S.B., Koerner, R.M., Fisher, D., Johnsen, S.J., Clausen, H.B., Dansgaard, W., Bucher, P. and Oeschger, H. (1977). An oxygen-isotope climatic record from the Devon Island ice cap, arctic Canada. *Nature*, **266**, 508-511.

Paul, M.A. (1981). *Soil Mechanics in Quaternary Science*. Quaternary Research Association, Edinburgh, 122 pp.

Paul, M.A. and Evans, H. (1974). Observations on the internal structure and origin of some flutes in glaciofluvial sediments, Blomstrandbreen, north-west Spitsbergen. *Journal of Glaciology*, **13**, 393-400.

Paul, M.A. and Eyles, N. (1990). Constraints on the preservation of diamict facies (Melt-out tills) at the margins of stagnant glaciers. *Quaternary Science Reviews*, **9**, 51-69.

Paull, C.K., Hills, S.J., Thierstein, H.R., Bonani, G. and Wölfli, W. (1991). ^{14}C offsets and apparently non-synchronous δ^{18}O stratigraphies between nannofossil and foraminiferal pelagic carbonates. *Quaternary Research*, **35**, 274-290.

Payne, A.J. and Sugden, D.E. (1987). Problems of testing ice sheet models: A British case study. *In*: Waddington, E.D. and Walder, J.S. (eds), *The Physical Basis of Ice Sheet Modelling*, pp. 377-380. Proceedings of an International Symposium held during the 19th General Assembly of the International Union of Geodesy and Geophysics at Vancouver, B.C., Canada, 9-22 August 1987, International Association of Hydrological Sciences Publication No. 170.

Peach, B.N. and Horne, J. (1930) *Chapters in the Geology of Scotland.*

Pecker, J.-C. and Runcorn, S.K. (1990). The Earth's Climate and Variability of the Sun Over Recent Millennia: Geophysical, Astronomical and Archaeological Aspects. *Philosophical Transactions of the Royal Society of London*, **330(A)**, 395-687.

Peel, D.A. (1993). Cold answers to hot issues. *Nature*, **363**, 403-404.

Peltier, W.R. (1982). Dynamics of the Ice Age Earth. *In*: Saltzman, B. (ed.), *Advances in Geophysics*, **24**, 1-146. Academic Press, New York.

Penck, A. and Brückner, E. (1909). *Die Alpen im Eiszeitalter.* Tauchnitz, Leipzig, Bd. 1-3, 1199 pp.

Perrin, R.M.S., Rose, J. and Davies, H. (1979). The distribution, variation and origins of Pre-Devensian tills in eastern England. *Philosophical Transactions of the Royal Society of London*, **287B**, 535-570.

Petit, J.P. (1987). Criteria for the sense of movement on fault surfaces in brittle rocks. *Journal of Structural Geology*, **9**, 597-608.

Petit, J.-R., Briat, M. and Royer, A. (1981). Ice age aerosol content from East Antarctic ice core samples and past wind strength. *Nature*, **293**, 391-394.

Pharo, C.H. and Carmack, E.C. (1979). Sedimentation processes in a short residence-time intermontane lake, Kamloops Lake, British Columbia. *Sedimentology*, **26**, processes in a short residence-time intermontane lake, Kamloops Lake, British Columbia. *Sedimentology*, **26**, 523-541.

Philip, J.R. (1980). Thermal fields during regelation. *Cold Regions Science and Technology*, **3**, 193-203.

Phillips, A.C., Smith, N.D. and Powell, R.D. (1991). Laminated sediments in prodeltaic deposits, Glacier Bay, Alaska. *In*: Anderson, J.B. and Ashley, G.M. (eds), *Glacial marine sedimentation: Paleoclimatic Significance. Geological Society of America, Special Paper, 261*, pp. 51-60.

Picard, K. (1964). Die Stratigraphische Stellung der "Warthe-Eiszeit" in Schleswig-Holstein, Deutschland. *INQUA - 6th International Congress. Warsaw 1961, Report No. 2*, pp. 191-197.

Picken, G.B. (1985). Benthic research in Antarctica: past, present, and future. *In*: Gray, J.S. and Christianson, M.E. (eds), *Marine Biology of Polar Regions and Effects of Stress on Marine Organisms*, pp. 167-184. John Wiley, Chichester.

Pickrill, R.A. and Irwin, J. (1982). Predominant headwater inflow and its control of lake-river interactions in Lake Wakatipu. *New Zealand Journal of Freshwater Research*, **16**, 201-213.

Pickrill, R.A. and Irwin, J. (1983). Sedimentation in a deep glacier-fed lake - Lake Tekapo, New Zealand. *Sedimentology*, **30**, 63-75.

Piper, D.J.W. (1976). The use of ice rafted marine sediments in determining glacial conditions. *La Revue de Géographie de Montréal*, **30**, 207-212.

Piper, D.J.W. and Sparkes, R. (1987). Proglacial sediment instability features on the Scotian slope at 63°W. *Marine Geology*, **76**, 15-31.

Piper, J.D.A. (1978). *Paleomagnetism and the Continental Crust*. Open University Press, Milton Keynes, 434 pp.

Platt, J.P. (1984). Secondary cleavages in ductile shear zones. *Journal of Structural Geology*, **6**, 439-442.

Platt, J.P. (1986). Dynamics of orogenic wedges and the uplift of high-pressure metamorphic rocks. *Geological Society of America Bulletin*, **97**, 1037-1053.

Platt, J.P. and Vissers, R.L.M. (1980). Extensional structures in anisotropic rocks. *Journal of Structural Geology*, **2**, 397-410.

Porter, S.C. (1975). Equilibrium-line altitudes of Late Quaternary glaciers in the Southern Alps, New Zealand. *Quaternary Research*, **5**, 27-47.

Porter, S.C. (1976). Pleistocene glaciation in the southern part of the North Cascade Range, Washington. *Geological Society America, Bulletin*, **87**, 61-75.

Porter, S.C. (1981). Recent glacier variations and volcanic eruptions. *Nature*, **291**, 139-142.

Porter, S.C. (1983). The Late Pleistocene, Volume 1 of Wright, H.E., Jr (ed.), *Late Quaternary Environments of the United States*. Longman, London, 407 pp.

Porter, S.C. (1986). Pattern and forcing of Northern Hemisphere glacier variations during the last millennium. *Quaternary Research*, **26**, 27-48.

Porter, S.C. and Orombelli, G. (1981). Alpine Rockfall Hazards. *American Scientist*, **69**, 67-75.

Posamentier, H.W. (1978). Thoughts on Ogive Formation. *Journal of Glaciology*, **20**, 216-218.

Post, A.S. (1972). Periodic surge origin of folded medial moraines on Bering piedmont glacier, Alaska. *Journal of Glaciology*, **11**, 219-226.

Post, A.S. (1975). A preliminary hydrology and historic terminal changes of Columbia Glacier, Alaska. *U.S. Geological Survey, Hydrological Investigations Atlas*, HA-559.

Post, A. S. (1976). Environmental geology of the central Gulf of Alaska coast. *In*: Williams, R.S., Jr and Carter, W.D. (eds), ERTS-1, A New Window on Our Planet. *U.S. Geological Survey, Professional Paper, 929*, pp. 117-119.

Post, A.S. and LaChapelle, E.R. (1971). *Glacier Ice*. University of Washington Press, Seattle, Washington.

Post, A.S. and Mayo, L.R. (1971). Glacier dammed lakes and outburst floods in Alaska. *United States Geological Survey, Hydrologic Investigations Atlas*, HA-455, 100 pp.

Postma, G., Roep, T.B. and Ruegg, G.H.J. (1983). Sandy-gravelly mass-flow deposits in an ice marginal lake (Saalian, Leuvenumsche Beek Valley, Veluwe, The Netherlands), with emphasis on plug-flow deposits. *Sedimentary Geology*, **34**, 59-82.

Potter, J.R., Paren, J.G. and Loynes, J. (1984). Glaciological and oceanographic calculations of the mass balance and oxygen isotope ratio of a melting ice shelf. *Journal of Glaciology*, **30**, 161-170.

Powell, R.D. (1980). Holocene glacimarine sediment deposition by tidewater glaciers in Glacier Bay, Alaska. Unpublished Ph.D. Thesis, Ohio State University, Columbus, Ohio.

Powell, R.D. (1981a). A model for sedimentation by tidewater glaciers. *Annals of Glaciology*, **2**, 129-134.

Powell, R.D. (1981b). Sedimentation conditions in Taylor Valley, Antarctica, inferred from textural analysis of DVDP cores. *American Geophysical Union Antarctic Research Series*, **33**, 331-349.

Powell, R.D. (1983a). Glacial-marine sedimentation

processes and lithofacies of temperate tidewater glaciers, Glacier Bay, Alaska. *In*: Molnia, B.F. (ed.), *Glacial-Marine Sedimentation,* pp. 185-232. Plenum Press, New York.

Powell, R.D. (1983b). Submarine flow tills at Victoria, British Columbia: discussion. *Canadian Journal of Earth Sciences,* **20**, 509-510.

Powell, R.D. (1984). Glacimarine processes and inductive lithofacies modelling of ice shelf and tidewater glacier sediments based on Quaternary examples. *Marine Geology,* **57**, 1-52.

Powell, R.D. (1985). Iceberg calving and its influence on ice-proximal, subaqueous glacigenic lithofacies. *In*: *Abstracts of the 14th Arctic Workshop on Arctic Land-Sea Interaction*, pp. 101-103. Bedford Institute of Oceanography, Halifax, Canada.

Powell, R.D. (1988). Processes and facies of temperate and subpolar glaciers with tidewater fronts. *Short Course Notes. Geological Society of America*, Denver, 114 pp.

Powell, R.D. (1990). Glacimarine processes at grounding-line fans and their growth to ice-contact deltas. *In*: Dowdeswell, J.A. and Scourse, J.D. (eds), *Glacimarine Environments: Processes and Sediments,* pp. 53-73. Geological Society of London, Special Publication No. 53.

Powell, R.D. (1991). Grounding-line systems as second-order controls on fluctuations of tidewater termini of temperate glaciers. *In*: Anderson, J.B. and Ashley, G.M. (eds), *Glacial Marine Sedimentation; Palaeoclimatic Significance. Geological Society of America Special Paper,* **261**, pp. 75-93.

Powell, R.D. and Alley, R.B. (*in preparation*) Significance of grounding-line systems in glacimarine settings: morainal banks versus grounding-line wedges ("till deltas and till tongues"). *Palaeogeography, Palaeoclimatology, and Palaeoecology.*

Powell, R.D. and Molnia, B.F. (1989). Glacimarine sedimentary processes, facies and morphology of the south-southeast Alaska shelf and fjords. *Marine Geology,* **85**, 359-390.

Powell, R.D., Pyne, A.R., Hunter, L.E. and Rynes, N.R. (1992). Polar versus temperate grounding-line sedimentary systems and marine glacier stability during sea level rise by global warming. *Geological Society of America, Abstract with Programs,* **24**, 348.

Power, J.M. (1985). Canada case study: water supply. *In*: Young, G.J. (ed.), *Techniques for Prediction of Runoff from Glacierized Areas,* pp. 59-71. International Association of Hydrological Sciences Publication No. 149.

Prange, W. (1978). Der letzte weichselzeitliche Gletschervorstoß in Schleswig-Holstein - das Gefüge überfahrener Schmelzwassersande und die Entstehung der Morphologie. *Meyniana,* **30**, 61-75.

Preiss, W.V., (1987). The Adelaide geosyncline - Late Proterozoic stratigraphy, sedimentation, palaeontology and tectonics. *Bulletin of the Geological Survey of South Australia,* **53**, 438 pp.

Prell, W.L., Imbrie, J., Martinson, D.G., Morley, J.J., Pisias, N.G., Shackleton, N.J. and Streeter, H.F. (1986). Graphic correlation of oxygen isotope stratigraphy application to the late Quaternary. *Paleoceanography,* **1**, 137-162.

Prentice, M.L. and Matthews, R.K. (1988). Cenozoic ice-volume history: Development of a composite oxygen isotope record. *Geology,* **16**, 963-966.

Prentice, M.L., Denton, G.H., Lowell, T.V., Conway, H.C. and Heusser, L.E. (1986). Pre-Late Quaternary glaciation of the Beardmore Glacier region, Antarctica. *Antarctic Journal of the United States,* **21**, 95-98.

Prest, V.K. (1984). The late Wisconsinan glacier complex. *In*: Fulton, R.J. (ed.), Quaternary Stratigraphy - a Canadian Contribution to IGCP Project 24. *Geological Survey of Canada, Paper 84-80,* pp. 21-36.

Price, R.A. (1973). Large-scale gravitational flow of supracrustal rocks, Southern Canadian Rockies. *In*: DeJong, K.A. and Scholten, R. (eds), *Gravity & Tectonics,* pp. 491-502. John Wiley, New York.

Price, R.J. (1980). Rates of geomorphological changes in proglacial areas. *In*: Cullingford, R.A., Davidson, D.A. and Lewin, J. (eds), *Timescales in Geomorphology,* pp. 79-93. John Wiley, Chichester.

Prior, D.B. and Bornhold, B.D. (1989). Submarine sedimentation on a developing Holocene fan delta. *Sedimentology,* **36**, 1053-1076.

Quigley, R.M. (1983). Glaciolacustrine and glaciomarine clay deposition; a North American perspective. *In*: Eyles, N. (ed.), *Glacial Geology,* pp. 140-167. Pergamon Press, Oxford.

Quilty, P.G. (1985). Distribution of foraminiferids in sediments of Prydz Bay, Antarctica. *South Australia Department of Mines and Energy, Special Publication,* **5**, 329-340.

Quinterno, P., Carlson, P.R. and Molnia, B.F. (1980). Benthic foraminifers from the eastern Gulf of Alaska. *In*: Field, M.E., Bouma, A.H., Colburn, I.P., Douglas, R.G. and Ingle, J.C. (eds), *Proceedings of the*

Quaternary Depositional Environments of the Pacific Coast, pp. 13-21. Society of Economic Paleontologists and Mineralogists, Bakersfield, California.

R&M (1982). *Alaska Power Authority, Susitna Hydroelectric project*; Task 3-hydrology; subtask 3.07 Closeout Report Reservoir Sedimentation. Report for Acres American, Inc., Buffalo, New York.

Rabassa, J. and Clapperton, C.M. (1990). Quaternary glaciations in the southern Andes. *In*: Clapperton, C.M. (ed.), Quaternary Glaciations in the Southern Hemisphere. *Quaternary Science Reviews*, **9**, 153-174.

Rabinowicz, E. (1965). *Friction and Wear of Materials*. John Wiley, New York, 244 pp.

Radok, U., Barry, R.G., Jensen, D., Keen, R.A., Kiladis, G.N. and McInnes, B.J. (1982). *Climatic and Physical Characteristics of the Greenland Ice Sheet, Parts I and II*. Cooperative Institute for Research in Environmental Sciences. University of Colorado, Boulder, 193 pp. and 4 appendices.

Raffle, J.F. and Greenwood, D.A. (1961). Relations between the rheological characteristics of grouts and their capacity to permeate soil. *Proceedings of the 5th International Conference on Soil Mechanics and Foundation Engineering*, 2, pp. 789-794.

Ragan, D.M. (1969). Structures at the base of an ice fall. *Journal of Geology*, **77**, 647-667.

Rainwater, F.H. and Guy, H.P. (1961). Some observations on the hydrochemistry and sedimentation of the Chamberlain Glacier area, Alaska. *U.S. Geological Survey Professional Paper*, 414-C.

Rampino, M.R. and Self, S. (1982). Historic eruptions on Tambnora (1815), Krakatau (1883) and Agung (1963). *Quaternary Research*, **18**, 127-143.

Ramsay, A.C. (1852). On the superficial accumulations and surface markings of North Wales. *Quarterly Journal of the Geological Society*, **8**, 371-376.

Ramsay, A.C. (1855). On the occurrence of angular, subangular, polished and striated fragments and boulders in the Permian breccia of Shropshire, Worcestershire, & c. *Quarterly Journal of the Geological Society*, **11**, 185-205.

Ramsay, J.G. (1982). Rock ductility and its influence on the development of tectonic structures in mountain belts. *In*: Hsü, K.J. (ed.), *Mountain Building Processes*. Academic Press, London.

Ramsay, J.G. and Huber, M.I. (1983). *The Techniques of Modern Structural Geology. Volume 1*. Academic Press, London, 307 pp.

Ramsay, J.G. and Huber, M.I. (1987). *The Techniques of Modern Structural Geology. Volume 2*. Academic

Press, London, 700 pp.

Rango, A., Salomonson, V.V. and Foster, J.L. (1977). Seasonal streamflow estimation in the Himalayan region employing meteorological satellite snow cover observations. *Water Resources Research*, **13**, 109-112.

Rappol, M. (1985). Clast-fabric strength in tills and debris flows compared for different environments. *Geologie en Mijnbouw*, **64**, 327-332.

Rappol, M. and Stoltenberg, H.M.P. (1985). Compositional variability of Saalian till in The Netherlands and its origin. *Boreas*, **14**, 33-50.

Rastas, J. and Seppälä, M. (1981). Rock jointing and abrasion forms on rouches moutonnées, SW Finland. *Annals of Glaciology*, **2**, 159-163.

Raymo, M.E. (1991). Geochemical evidence supporting T.C. Chamberlin's theory of glaciation. *Geology*, **19**, 344-347.

Raymo, M.E., Ruddiman, W.F. and Froelich, P.N. (1988). Influence of late Cenozoic mountain building on ocean geochemical cycles. *Geology*, **16**, 649-653.

Raymo, M.E., Ruddiman, W.F., Backman, J., Clement, B.M. and Martinson, D.G. (1989). Late Pliocene variation in Northern Hemisphere ice sheets and North Atlantic deep water circulation. *Paleoceanography*, **4**, 413-446.

Raymond, C.F. (1971). Flow in a transverse section of Athabasca Glacier, Alberta, Canada. *Journal of Glaciology*, **10**, 55-84.

Raymond, C.F. (1978). Numerical calculation of glacier flow by finite element methods, final technical report for National Science Foundation Grant DPP74-19075, Unpublished Manuscript, 1978.

Raymond, C.F. (1980). Temperate valley glaciers. *In*: Colbeck, S.C. (ed.), *Dynamics of Snow and Ice Masses*, pp. 79-139. Academic Press, New York.

Raymond, C.F. (1987). How do glaciers surge? A review. *Journal of Geophysical Research*, **92**, 9121-9134.

Raymond, C.F. and Harrison, W.D. (1975). Some observations on the behaviour of the liquid and gas phases in temperate glacier ice. *Journal of Glaciology*, **14**, 213-233.

Raymond, C.F. and Harrison, W.D. (1987). Fit of ice-motion models to observations from Variegated Glacier, Alaska. *In*: Waddington, E.D. and Walder, J.S. (eds), *The Physical Basis of Ice Sheet Modelling*, pp. 153-166. Proceedings of an International Symposium held during the 19th General Assembly of the International Union of Geodesy and Geophysics at Vancouver, B.C., Canada, 9-22 August 1987, International Association of Hydrological Sciences

Publication No. 170.

Raymond, C.F. and Harrison, W.D. (1988). Evolution of Variegated Glacier, Alaska, U.S.A., prior to its surge. *Journal of Glaciology*, **34**, 154-169.

Raymond, C.F. and Malone, S. (1987). Propagating strain anomalies during mini-surges of Variegated Glacier, Alaska, U.S.A. *Journal of Glaciology*, **32**, 178-191.

Raymond, C.F., Jóhannesson, T., Pfeffer, T. and Sharp, M.J. (1987). Propagation of a glacier surge into stagnant ice. *Journal of Geophysical Research*, **92**, 9037-9049.

Raynaud, D., Jouzel, J., Barnola, J.M., Chappellaz, J., Delmas, R.J. and Lorius, J. (1993). The ice record of greenhouse gases. *Science*, **259**, 926-933.

Rea, D.K. and Leinen, M. (1989). *Quaternary Climates: The Ocean Sedimentary Record*. Geological Society of America, Short Course, Annual Meeting, St. Louis, 245 pp.

Reboul, H. (1833). *Géologie de la Période Quaternaire, et Introduction à l'Histoire Ancienne*. Levrault, Paris, 222 pp.

Reed, E.C. and Dreeszen, V.H. (1965). Revision of the classification of the Pleistocene deposits of Nebraska. *Nebraska Geological Survey Bulletin*, **23**, 65 pp.

Reeh, N. (1968). On the calving of ice from floating glaciers and ice shelves. *Journal of Glaciology*, **7**, 215-232.

Reeh, N. (1988). A flow-line model for calculating the surface profile and the velocity, strain-rate, and stress fields in an ice sheet. *Journal of Glaciology*, **34**, 46-54.

Reeh, N. (1990). Past changes in precipitation rate and ice thickness as derived from age-depth profiles in ice-sheets; application to Greenland and Canadian Arctic ice core records, *In*: Bleil, U. and Thiede, J. (eds), *Geological History of the Polar Oceans, Arctic versus Antarctic*, pp. 255-271. Kluwer Academic, Dordrecht.

Reeh, N., Oerter, H., Letréguilly, A., Miller, H. and Hubberten, H.-W. (1991). A new, detailed ice-age oxygen-18 record from the ice-sheet margin in central West Greenland. *Palaeogeography, Palaeoclimatology, Palaeoecology (Global and Planetary Change Section)*, **90**, 373-383.

Rees, A.I. (1968). The production of preferred orientation in a concentrated dispersion of elongated and flattened grains. *Journal of Geology*, **76**, 457-465.

Reheis, M.J. (1975). Source, transportation and deposition of debris on Arapaho Glacier, Front Range, Colorado, U.S.A. *Journal of Glaciology*, **14**, 407-420.

Reid, C. (1882). *The Geology of the Country around Cromer*. Memoir of the Geological Survey, Great Britain.

Reid H.F. (1892). Studies of the Muir Glacier, Alaska. *National Geographic Magazine*, **4**, 19-84.

Reid, H.F. (1896). Glacier Bay and its glaciers. *U.S. Geological Survey, 16th Annual Report, Part 1*, 415-461.

Reid, J.R. (1968). Origin and characteristics of supra-glacial drift, Martin River and Sioux glaciers, Alaska (abstract). *Geological Society of America Special Paper 101*, p. 173.

Reimnitz, E. (1966). Late Quaternary history and sedimentation of the Copper River delta and vicinity, Alaska. Unpublished Ph.D. Thesis, Univ. California, San Diego, CA, 160 pp.

Reimnitz, E. and Kempema, E.W. (1982). Dynamic ice-wallow relief of northern Alaska's nearshore. *Journal of Sedimentary Petrology*, **52**, 451-461.

Reimnitz, E. and Kempema, E.W. (1988). Ice rafting: An indication of glaciation? *Journal of Glaciology*, **34**, 254-255.

Reimnitz, E., Kempema, E.W. and Barnes, P.W. (1987). Anchor ice, seabed freezing, and sediment dynamics in shallow arctic seas. *Journal of Geophysical Research*, **92**, 14671-14678.

Renberg, I. (1981). Formation, structure, and visual appearance of iron-rich varved lake sediments. *Verhandlungen Internationen Vereinigung für Limnologies*, **21**, 94-101.

Renberg, I. and Segerstrom, U. (1981). Application of varved lake sediments in paleoenvironmental studies. *Wahlenbergia*, **7**, 125-133.

Reynaud, L. (1987). The November 1986 survey of the Grand Moulin on the Mer de Glace, Mont Blanc massif, France. *Journal of Glaciology*, **33**, 130-131.

Reynaud, L. (1988). Alpine glacier fluctuations and climatic changes over the last century. *In*: Schnee, Eis and Wasser Alpiner Gletscher. *Mitteilungen der Versuchanstalt für Wasserbau, Hydrologie und Glaziologie*. ETH, Zurich, Nr. 94, pp. 127-146.

Reynaud, L., Vallon, M., Martin, S. and Letréguilly, A. (1984). Spatio temporal distribution of the glacial mass balance in the Alpine, Scandinavian and Tien Shan areas. *Geografiska Annaler*, **66A**, 239-247.

Rice, R.J. (1982). The hydraulic geometry of the lower portion of the Sunwapta River Valley train, Jasper National Park, Alberta. *In*: Davidson-Arnott, R., Nickling, W. and Fahey, B.D. (eds), *Research in Glacial, Glacio-fluvial, and Glacio lacustrine Systems*, pp. 151-173. 6th Guelph Symposium on Geomorphology. Geo Books, Norwich.

Richards, K. (1982). *Rivers; Form and Process in Alluvial Channels*. Methuen, London, 358 pp.

Richards, K. (1984). Some observations on suspended sediment dynamics in Storbregrova, Jotunheimen. *Earth Surface Processes and Landforms*, **9**, 101-112.

Richardson, D. (1968). Glacier outburst floods in the Pacific Northwest. *In*: Geological Survey Research, 1968. *U.S. Geological Survey Professional Paper 600-D*, pp. D79-D86.

Richmond, G.M. and Fullerton, D.S. (1986a). Introduction to Quaternary glaciations in the United States of America. *In*: Šibrava, V., Bowen, D.Q. and Richmond, G.M. (eds), *Quaternary Glaciations in the Northern Hemisphere. Quaternary Science Reviews*, **5**, 3-10.

Richmond, G.M. and Fullerton, D.S. (1986b). Summation of Quaternary glaciations in the United States of America. *In*: Šibrava, V., Bowen, D.Q. and Richmond, G.M. (eds), *Quaternary Glaciations in the Northern Hemisphere. Quaternary Science Reviews*, **5**, 183-196.

Richter, W., Schneider, H. and Wager, R. (1951). Die saaleeiszeitliche Stauchzone von Itterbeck-Uelsen (Grafschaft Bentheim). *Zeitschrift der Deutschen Geologischen Gesellschaft*, **102**, 60-75.

Richthofen, F. von. (1877). *China: Ergebnisse eigener Reisen und darauf gegrundeter Studien*. Vol. 1. Reimer, Berlin, 440 pp.

Ridge, J.C. and Larsen, F.D. (1990). Re-evaluation of Antevs' New England varve chronology and new radiocarbon dates of sediments from glacial Lake Hitchcock. *Geological Society of America Bulletin*, **102**, 889-899.

Ridge, J.C., Brennan, W.J. and Muller, E.H. (1990). The use of paleomagnetic declination to test correlations of late Wisconsinan glaciolacustrine sediments in central New York. *Geological Society of America Bulletin*, **102,** 26-44.

Ridky, R.W. and Bindschadler, R.A. (1990). Reconstruction and dynamics of the Late Wisconsin "Ontario" ice dome in the Finger Lakes region, New York. *Geological Society of America Bulletin*, **102,** 1055-1064.

Riedel, W. (1929). Zur Mechanik geologischer Brucherscheinungen, ein Beitrag zum Problem der Fiederspalten. *Zentralblatt Mineralogischer, Geologischer und Paläontologischer Abteilung*, 354-368.

Rigsby, G.P. (1960). Crystal orientation in glacier and in experimentally deformed ice. *Journal of Glaciology*, **3**, 589-606.

Riley, N.W. (1982). Rock wear by sliding ice. Ph.D. Thesis, University of Newcastle, U.K.

Rind, D. and Peteet, D. (1985). Terrestrial conditions at the last glacial maximum and CLIMAP sea-surface temperature estimates: Are they consistent? *Quaternary Research*, **24**, 1-22.

Ringberg, B., Holland, B. and Miller, U. (1984). Till stratigraphy and provenance of the glacial chalk rafts at Kvangby and Ängdala, southern Sweden. *Striae*, **20**, 79-90.

Rio, D., Sprovieri, R. and Thunell, R. (1991). Pliocene-lower Pleistocene chronostratigraphy: A re-evaluation of Mediterranean type sections. *Geological Society of America Bulletin*, **103**, 1049-1058.

Rist, S. (1955). Skeidarárhlaup 1954. *Jökull*, **5**, 30-36.

Rist, S. (1973). Jökulhlaupaannáll 1971, 1972 og 1973. *Jökull*, **23**, 55-60.

Ritz, C. (1987). Time dependant boundary conditions for calculation of temperature fields in ice sheets. *In*: Waddington, E.D. and Walder, J.S. (eds), *The Physical Basis of Ice Sheet Modelling*, pp. 207-216. Proceedings of an International Symposium held during the 19th General Assembly of the International Union of Geodesy and Geophysics at Vancouver, B.C., Canada, 9-22 August 1987, International Association of Hydrological Sciences Publication No. 170.

Robardet, M. and Doré, F. (1988). The late Ordovician diamictic formations from southwestern Europe: North-Gondwana glaciomarine deposits. *Palaeogeography, Palaeoclimatology, Palaeoecology*, **66**, 19-31.

Robe, R.Q. (1980). Iceberg drift and deterioration. *In*: Colbeck, S.C. (ed.), *Dynamics of Snow and Ice Masses*, pp. 211-259. Academic Press, New York.

Roberts, J.D. (1976). Late Precambrian dolomites, Vendian glaciation, and synchroneity of Vendian glaciations. *Journal of Geology*, **84**, 47-63.

Robin, G. deQ. (1955). Ice movement and temperature distribution in glaciers and ice sheets. *Journal of Glaciology*, **2**, 523-532.

Robin, G. deQ. (1976). Is the basal ice of a temperate glacier at the pressure melting point? *Journal of Glaciology*, **16**, 183-196.

Robin, G. deQ. (1977). Ice Cores and Climatic Change. *Philosophical Transactions of the Royal Society of London*, **B280**, 143-168.

Robin, G. deQ. (1979). Formation, flow, and disintegration of ice shelves. *Journal of Glaciology*, **24**, 259-271.

Robin, G. deQ. (1983). *The Climatic Record In Polar Ice Sheets*. Cambridge University Press, Cambridge, 212 pp.

Robin, G. deQ. (1988). The Antarctic ice sheet, its history and response to sea level and climatic changes over the past 100 million years. *Palaeogeography, Palaeoclimatology, Palaeoecology*, **67**, 31-50.

Robin, G. deQ. and Swithinbank, C. (1987). Fifty years of progress in understanding ice sheets. *Journal of Glaciology, Special Issue*, 33-47.

Robin, G. deQ. and Weertman, J. (1973). Cyclic surging of glaciers. *Journal of Glaciology*, **12**, 3-18.

Robin, G. deQ., Drewry, D.J. and Meldrum, D.T. (1977). International studies of ice sheet and bedrock. *Philosophical Transactions of the Royal Society London*, **279B**, 185-196.

Rocha-Campos, A.C. and Dos Santos, P.R. (1981). The Itararè Subgroup, Aquidauana Group and San Gregório Formation, Paraná basin, southeastern South America. *In*: Hambrey, M.J. and Harland, W.B. (eds), *Earth's Pre-Pleistocene Glacial Record*, pp. 842-852. Cambridge University Press, Cambridge.

Rodine, J.D. and Johnson, A.M. (1976). The ability of debris, heavily freighted with coarse clastic materials, to flow on gentle slopes. *Sedimentology*, **23**, 213-234.

Rogerson, R.J. (1985). Measured re-advance of a debris-covered glacier terminus in the President Range, Yoho National Park, British Columbia, Canada. *Journal of Glaciology*, **31**, 13-17.

Rogerson, R.J., Olsen, M.E. and Branson, D. (1986). Medial moraines and surface melt on glaciers of the Torngat Mountains, northern Labrador, Canada. *Journal of Glaciology*, **32**, 350-354.

Ronnert, L. and Mickelson, D.M. (1992). High porosity of basal till at Burroughs glacier, southeastern Alaska. *Geology*, **20**, 849-852.

Rooney, S.T., Blankenship, D.D., Alley, R.B. and Bentley, C.R. (1986). *Structure and Continuity of a Till Layer Beneath Ice Stream B, West Antarctica*. American Geophysical Union. Whistler Village, B.C., Canada.

Rose, J. (1974). Small scale spatial variability of some sedimentary properties of lodgement and slumped till. *Proceedings of the Geologists' Association*, **85**, 223-237.

Rose, J. and Allen, P. (1977). Middle Pleistocene stratigraphy in south-east Suffolk. *Journal of the*

Geological Society London*, **133**, 83-102.

Röthlisberger, F. (1986). *10,000 Jahre Gletschergeschichte der Erde*. Sauerländer, Aarau, Switzerland, 416 pp.

Röthlisberger, H. (1968). Erosive processes which are likely to accentuate or reduce the bottom relief of valley glaciers. *International Association of Scientific Hydrology Publication*, **79**, 87-97.

Röthlisberger, H. (1972). Water pressure in intra- and subglacial channels. *Journal of Glaciology*, **11**, 177-203.

Röthlisberger, H. (1980). Gletscherbewegung und Wasserabfluss. *Wasser, Energie, Luft*, **72**, 290-294.

Röthlisberger, H. (1981). Eislawinen und Ausbrüche von Gletscherseen. *In*: Kasser, P. (ed.), *Gletscher und Klima -Glaciers et Clima*, pp. 170-212. Jahrbuch der Schweizerischen Naturforschenden Gesellschaft, wissenschaftlicher Teil. Birkhäuser, Basel.

Röthlisberger, H. and Iken, A. (1981). Plucking as an effect of water-pressure variations at the glacier bed. *Annals of Glaciology*, **2**, 57-62.

Röthlisberger, H. and Lang, H. (1987). Glacial hydrology. *In*: Gurnell, A.M. and Clark, M.J. (eds), *Glacio-Fluvial Sediment Transfer - an Alpine Perspective*, pp. 207-284. John Wiley, Chichester.

Röthlisberger, H., Iken, A. and Spring, U. (1979). (Abstract) Piezometric observations of water pressure at the bed of Swiss glaciers. *Journal of Glaciology*, **23**, 429.

Rotnicki, K. (1976). The theoretical basis for and a model of the origin of glaciotectonic deformations. *Quaestiones Geographicae*, **3**, 103-139.

Rotnicki, K. (1991). Retrodiction of palaeodischarges of meandering and sinuous alluvial rivers and its palaeohydroclimatic implications. *In*: Starkel, L., Gregory, K.J. and Thornes, J.B. (eds), *Temperate Palaeohydrology*, pp. 431-471. John Wiley, New York.

Ruddiman, W.F. (1987a). Northern oceans. *In*: Ruddiman, W.F. and Wright, H.E., Jr (eds), *North America and Adjacent Oceans During the Last Deglaciation*, pp. 137-154. The Geology of North America, Volume K-3. Geological Society of America, Boulder, Colorado.

Ruddiman, W.F. (1987b). Synthesis: The Ocean ice-sheet record. *In*: Ruddiman, W.F. and Wright, H.E., Jr (eds), *North America and Adjacent Oceans During the Last Deglaciation*, pp. 463-478. The Geology of North America, Volume K-3. Geological Society of America, Boulder, Colorado.

Ruddiman, W.F. (1991). Notes on long-term forcing of Arctic climate. *Norsk Geologisk Tidsskrift*, **71**, 167-168.

Ruddiman, W.F. and Kutzbach, J.E. (1989). Forcing of late Cenozoic Northern Hemisphere climate by plateau uplift in southern Asia and the American West. *Journal of Geophysical Research*, **94**, 18409-18427.

Ruddiman, W.F. and McIntyre, A. (1981). Oceanic mechanisms for amplification of the 23,000-year ice-volume cycles. *Science*, **212**, 617-627.

Ruddiman, W.F. and McIntyre, A. (1982). Severity and speed of Northern Hemisphere glaciation pulses: The limiting case? *Geological Society of America Bulletin*, **93**, 1273-1279.

Ruddiman, W.F. and McIntyre, A. (1984). Ice-age thermal response and climatic role of the surface Atlantic Ocean, 40° N to 63° N. *Geological Society of America Bulletin*, **95**, 381-396.

Ruddiman, W.F. and Raymo, M.E. (1988). Northern Hemisphere climate régimes during the past 3 Ma: Possible tectonic connections. *In*: Shackleton, N.J., West, R.G. and Bowen, D.Q. (eds), The Past Three Million Years: Evolution of Climatic Variability in the North Atlantic Region. *Philosophical Transactions of the Royal Society of London*, **B 318**, 411-430.

Ruddiman, W.F. and Wright, H.E., Jr (1987). Introduction. *In*: Ruddiman, W.F. and Wright, H.E., Jr (eds), *North America and Adjacent Oceans During the Last Deglaciation,* pp. 1-12. The Geology of North America, Volume K-3. Geological Society of America, Boulder, Colorado.

Ruddiman, W.F., Raymo, M. and McIntyre, A. (1986). Matuyama 41,000-cycles: North Atlantic Ocean and Northern Hemisphere ice sheets. *Earth and Planetary Science Letters*, **80**, 117-129.

Ruddiman, W.F., Raymo, M.E., Martinson, D.G., Clement, B.M. and Backman, J. (1989). Pleistocene Evolution: Northern Hemisphere ice sheets and North Atlantic Ocean. *Paleoceanography*, **4**, 353-412.

Rudolph, R. (1962). Abflusstudien an Gletscherbachen. *Veroffentlichung des Museum Ferdinandeum, Innsbruck*, **41**, 118-266.

Ruegg, G.H.J. (1981a). Sedimentary features and grain size of glaciofluvial and periglacial Pleistocene deposits in The Netherlands and adjacent parts of Western Germany. *Verh. Naturwiss. Verein Hamburg*, **24**, 133-154.

Ruegg, G.H.J. (1981b). Ice-pushed Lower and Middle Pleistocene deposits near Rhenen (Kwintelooijen): sedimentary-structural and lithological/granulometrical investigations. *In*: Ruegg, G.H.J. and Zandstra, J.G. (eds), Geology and Archaeology of Pleistocene Deposits in the Ice-pushed Ridge Near Rhenen and Veenendaal. *Mededelingen Rijks Geologische Dienst*, **35**, 165-177.

Ruhe, R.V. (1969). *Quaternary Landscapes in Iowa*. 255 pp. Iowa State University Press, Ames, Iowa.

Runcorn, S.K. (1990). Closing remarks: geophysical. *In*: Pecker, J.-C. and Runcorn, S.K. (eds), The earth's climate and variability of the sun over recent millennia: geophysical, astronomical and archeological aspects. *Philosophical Transactions of the Royal Society of London*, **330(A)**, 685-687.

Rundle, A. (1985). Braid morphology and the formation of multiple channels, the Rakaia, New Zealand. *Zeitschrift für Geomorphologie, N.F., Supplement-Band*, **55**, 15-37.

Russell, A.J. (1989). A comparison of two recent jökulhlaups from an ice-dammed lake, Søndre Strømfjord, West Greenland. *Journal of Glaciology*, **35**, 157-162.

Russell, A.J. (1991). The geomorphological and sedimentological effects of jökulhlaups. Unpublished Ph.D. Thesis, University of Aberdeen, Aberdeen, 135 pp.

Russell, A.J. (1993). Obstacle marks produced by flow around stranded ice blocks during a glacier outburst flood (Jökulhlaup) in west Greenland. *Sedimentology*, **40**, 1091-1111.

Russell, A.J., Aitken, J.F. and De Jong, C. (1990). Observations on the drainage of an ice-dammed lake in West Greenland. *Journal of Glaciology*, **36**, 72-74.

Russell-Head, D.S. (1980). The melting of-free drifting icebergs. *Annals of Glaciology*, **1**, 119-122.

Russell-Head, D.S. and Budd, W.F. (1979). Ice sheet flow properties derived from bore-hole shear measurements combined with ice core studies. *Journal of Glaciology*, **24**, 117-130.

Rust, B.R. (1972). Structure and process in a braided river. *Sedimentology*, **18**, 221-245.

Rust, B.R. (1975). Fabric and structure in glaciofluvial gravels. *In*: Jopling, A.V. and McDonald, B.C. (eds), *Glaciofluvial and Glaciolacustrine Sedimentation*, pp. 238-248. *Society of Economic Paleontologists and Mineralogists, Special Publication No. 23*.

Rust, B.R. (1977). Mass flow deposits in a Quaternary succession near Ottawa, Canada: diagnostic criteria for subaqueous outwash. *Canadian Journal of Earth Sciences*, **14**, 175-184.

Rust, B.R. (1978). Depositional models for braided

alluvium. *In*: Miall, A.D. (ed.), *Fluvial Sedimentology*, pp. 605-625. Canadian Society of Petroleum Geologists, Memoir 5, Calgary, Alberta, Canada.

Rust, B.R. and Koster, E.H. (1984). Coarse alluvial deposits. *In*: Walker, R.G. (ed.), *Facies Models, Second Edition*, pp. 53-69. Geoscience Canada, Reprint Series 1, Toronto.

Rust, B.R. and Romanelli, R. (1975). Late Quaternary subaqueous outwash deposits near Ottawa, Canada. *In*: Jopling, A.V. and McDonald, B.C. (eds), *Glaciofluvial and Glaciolacustrine Sedimentation*, pp. 177-192. *Society of Economic Paleontologists and Mineralogists, Special Publication No. 23*.

Ruszczyńska-Szenajch, H. (1980). Glacial erosion in contradistinction to glacial tectonics. *In*: Stankowski, W. (ed.), *Tills and Glacigene Deposits*, pp. 71-76. Wydawnictwa Uniwersytetu Im. A. Mickiewicza, Poznań, Seria Geografia, No. 20.

Ruszczyńska-Szenajch, H. (1987). The origin of glacial rafts: detachment, transport, deposition. *Boreas*, **16**, 101-112.

Rzechowski, J. (1986). Pleistocene till stratigraphy in Poland. *In*: Šibrava, V., Bowen, D.Q. and Richmond, G.M. (eds), *Quaternary Glaciations in the Northern Hemisphere. Quaternary Science Reviews*, **5**, 365-372.

Sagan, C. and Mullen, G. (1972). Earth and Mars: evolution of atmospheres and surface temperatures. *Science*, **177**, 52-56.

Sambrotto, R.N. and Lorenzen, C.J. (1987). Phytoplankton and primary production. *In*: Hood, D.W. and Zimmerman, S.T. (eds), *Natl. Ocean. Atmos. Adm., United States Department of Commerce*, pp. 249-282.

Sanderson, T.J.O. (1979). Equilibrium profile of ice shelves. *Journal of Glaciology*, **22**, 435-460.

Sanderson, T.J.O. and Doake, C.S.M. (1979). Is vertical shear in an ice shelf negligible? *Journal of Glaciology*, **22**, 285-292.

Sandford, A.R. (1959). Analytical and experimental study of simple geologic structures. *Geological Society of America Bulletin*, **70**, 19-52.

Sandgren, P., Björck, S., Brunnberg, L. and Kristiansson, J. (1988). Palaeomagnetic records from two varved clay sequences in the Middle Swedish ice marginal zone. *Boreas*, **17**, 215-227.

Sangree, J.B. and Widmier, J.M. (1977). Seismic stratigraphy and global changes of sea level, Part 9; Seismic interpretation of clastic depositional facies. *In*: Payton, C.E. (ed.), Seismic Stratigraphy; Applications to Hydrocarbon Exploration. *American Association of Petroleum Geologists Memoir, 26*, pp. 165-184.

Sarmiento, J.L. (1993). Atmospheric CO_2 stalled. *Nature*, **365**, 697-698.

Sarnthein, M., Winn, K. and Zahn, R. (1987). Paleoproductivity of ocean upwelling and the effect on atmospheric CO_2 and climatic change during deglaciation times. *In*: Berger, W.H. and Labeyrie, L.D. (eds), *Abrupt Climatic Change; Evidence and Implications*, pp. 311-337. D. Reidel, Dordrecht.

Saunderson, H.C. (1977). The sliding bed facies in esker sands and gravels: a criterion for full-pipe (tunnel) flow? *Sedimentology*, **24**, 623-638.

Saunderson, H.C. (1982). Bed form diagrams and the interpretation of eskers. *In*: Davidson-Arnott, R., Nickling, W. and Fahey, B.D. (eds), *Research in Glacial, Glacio-fluvial, and Glacio lacustrine Systems*, pp. 139-150. Proceedings of the 6th Guelph Symposium on Geomorphology, 1980. Geo Books, Norwich.

Saunderson, H.C. and Jopling, A.V. (1980). Palaeohydraulics of a tabular, cross-stratified sand in the Brampton esker, Ontario. *Sedimentary Geology*, **25**, 169-188.

Schack Pedersen, S.A. (1988). Glacitectonite: brecciated sediments and cataclastic sedimentary rocks formed subglacially. *In*: Goldthwait, R.P. and Matsch, C.L. (eds), *Genetic Classification of Glacigenic Deposits*, pp. 89-91. A.A. Balkema, Rotterdam.

Schaefer, I. (1953). Die donaueiszeitlichen Abagerungen an Lech und Wertach. *Geologica Bavarica*, **19**, 13-64.

Schaefer, I. (1968). The succession of fluvioglacial deposits in the northern Alpine foreland. *In*: *Glaciations of the Alps*. International Congress (INQUA USA) 1965, Proceedings, 14: 9-14.

Scherer, R.P. (1991). Quaternary and Tertiary microfossils from beneath Ice Stream B: Evidence for a dynamic West Antarctic Ice Sheet history. *Palaeogeography, Palaeoclimatology, Palaeoecology (Global Planetary Change Section)*, **90**, 395-412.

Schermerhorn, L.J.G. (1974). Late Precambrian mixtites: Glacial and/or nonglacial? *American Journal of Science*, **274**, 673-824.

Schermerhorn, L.J.G. (1983). Late Proterozoic glaciation in the light of CO_2 depletion in the atmosphere. *In*: Medaris, L.G., Jr, Byers, C.W., Mickelson, D.M. and Shanks, W.C. (eds), *Proterozoic geology: selected papers from an international symposium. Geological Society of America, Memoir 161*, pp. 309-315.

Schmok, J.P. and Clarke, G.K.C. (1989). Lacustrine

sedimentary record of ice-dammed Neoglacial Lake Alsek. *Canadian Journal of Earth Sciences*, **26**, 2092-2105.

Scholz, C.H. (1987). Wear and gouge formation in brittle faulting. *Geology*, **15**, 493-495.

Scholz, C.H. and Engelder, J.T. (1976). The role of asperity indentation and ploughing in rock friction-I. *International Journal of Rock Mechanics and Geomechanics Abstracts*, **13**, 149-154.

Schubert, C. and Clapperton, C.M. (1990). Quaternary glaciations in the northern Andes (Venezuela, Columbia and Equador). *In*: Clapperton, C.M. (Ed.), *Quaternary Glaciations in the Southern Hemisphere. Quaternary Science Reviews*, **9**, 123-135.

Schubert, G. and Yuen, D.A. (1982). Initiation of ice ages by creep instability and surging of the East Antarctic ice sheet. *Nature*, **296**, 127-130.

Schumm, S.A. (1960). The shape of alluvial channels in relation to sediment type. *U.S. Geological Survey, Professional Paper 352-B*, 30 pp.

Schumm, S.A. and Lichty, R.W. (1963). Channel widening and flood-plain construction along Cimarron River in southwestern Kansas. *U.S. Geological Survey, Professional Paper 352-D*, pp. 71-88.

Schumm, S.A. and Khan, H.R. (1972). Experimental study of channel patterns. *Geological Society of America Bulletin*, **83**, 1755-1770.

Schwander, J. (1989). The transformation of snow to ice and the occlusion of gasses. *In*: Oeschger, H. and Langway, C.C., Jr (eds), *The Environmental Record in Glaciers and Ice Sheets*, pp. 53-67. Report of the Dahlem Workshop on The Environmental Record in Glaciers and Ice Sheets, Berlin, 13-18 March 1988. John Wiley, Chichester.

Schweizer, J. and Iken, A. (1992). The role of bed separation and friction in sliding over an undeformable bed. *Journal of Glaciology*, **38**, 77-92.

Scofield, J.P. (1986). Flow characteristics of Byrd Glacier, Antarctica. MSc. thesis. University of Maine, Orono, May 1988, 136 pp.

Scotese, C.R., Bambach, R.K., Barton, C., van der Voo, R. and Ziegler, A.M. (1979). Paleozoic base maps. *Journal of Geology*, **87**, 217-277.

Seaberg, S.Z., Seaberg, J.Z., Hooke, R.LeB. and Wiberg, D.W. (1988). Character of the englacial and subglacial drainage system in the lower part of the ablation area of Storglaciären, Sweden, as revealed by dye-trace studies. *Journal of Glaciology*, **34**, 217-227.

Serebryanny, L.R. and Orlov, A.V. (1982). Genesis of marginal moraines in the Caucasus. *Boreas*, **11**,
279-289.

Shabtaie, S. and Bentley, C.R. (1987). West Antarctic ice streams draining into Ross Ice Shelf: configuration and mass balance. *Journal of Geophysical Research*, **92**, 1311-1336.

Shackleton, N.J. (1967). Oxygen isotope analyses and Pleistocene temperatures reassessed. *Nature*, **215**, 15-17.

Shackleton, N.J. (1969). The last interglacial in the marine and terrestrial record. *Proceedings of the Royal Society of London*, **B174**, 135-154.

Shackleton, N.J. (1982). The deep-sea sediment record of climate variability. *Progress in Oceanography*, **11**, 199-218.

Shackleton, N.J. (1987). Oxygen isotopes, ice volume and sea level. *Quaternary Science Reviews*, **6**, 183-190.

Shackleton, N.J. and Opdyke, N.D. (1973). Oxygen isotope and palaeomagnetic stratigraphy of equatorial Pacific core V28-238: oxygen isotope temperatures and ice volumes on a 10^5 year and a 10^6 year scale. *Quaternary Research*, **3**, 39-55.

Shackleton, N.J., Backman, J., Zimmerman, H., Kent, D.V., Hall, M.A., Roberts, D.G., Schnitker, D., Baldauf, J.G., Desprairies, A., Homrighausen, R., Huddlestun, P., Keene, J.B., Kaltenback, A.J. Krumsiek, K.A.O., Morton, A.C., Murray, J.W. and Westberg-Smith, J. (1984). Oxygen isotope calibration of the onset of ice-rafting and history of glaciation in the North Atlantic region. *Nature*, **307**, 620-623.

Shakesby, R.A. (1985). Geomorphological effects of jökulhlaups and ice-dammed lakes, Jotunheimen, Norway. *Norsk Geografisk Tidsskrift*, **39**, 1-16.

Sharma, G.D. (1979). *The Alaskan Shelf: Hydrographic, Sedimentary, and Geochemical Environment*, 498 pp. Springer, New York.

Sharp, M.J. (1982a). Modification of clasts in lodgement tills by glacial erosion. *Journal of Glaciology*, **28**, 475-481.

Sharp, M.J. (1982b). A Comparison of the landforms and sedimentary sequences produced by surging and non-surging glaciers in Iceland. Unpublished Ph.D. Thesis, University of Aberdeen.

Sharp, M.J. (1985a). Sedimentation and stratigraphy at Eyjabakkajökull - an Icelandic surging glacier. *Quaternary Research*, **24**, 268-284.

Sharp, M.J. (1985b). "Crevasse-fill" ridges - a landform type characteristic of surging glaciers? *Geografiska Annaler, Series A*, **67**, 213-220.

Sharp, M.J. (1988). Surging glaciers: behaviour and

mechanism. *Progress in Physical Geography*, **12**, 349-370.

Sharp, M.J. and Gomez, B. (1986). Processes of debris comminution in the glacial environment and implications for quartz sand-grain micromorphology. *Sedimentary Geology*, **46**, 33-47.

Sharp, M.J., Gemmell, J.C. and Tison, J.L. (1989). Structure and stability of the former subglacial drainage system of the Glacier de Tsanfleuron, Switzerland. *Earth Surface Processes and Landforms*, **14**, 119-134.

Sharp, M.J., Lawson, W. and Anderson, R.S. (1988). Tectonic processes in a surge-type glacier. *Journal of Structural Geology*, **10**, 499-515.

Sharp, M.J., Dowdeswell, J.A. and Gemmell, J.C. (1989). Reconstructing past glacier dynamics and erosion from glacial geomorphic evidence: Snowdon, North Wales. *Journal of Quaternary Science*, **4**, 115-130.

Sharp, R.P. (1949). Studies of superglacial debris on valley glaciers. *American Journal of Science*, **247**, 289-315.

Sharpe, D.R. and Cowan, W.R. (1990). Moraine formation in northwestern Ontario: product of subglacial fluvial and glaciolacustrine sedimentation. *Canadian Journal of Earth Sciences*, **27**, 1478-1486.

Sharpe, D.R. and Shaw, J. (1989). Erosion of bedrock by subglacial meltwater, Cantley, Quebec. *Geological Society of America Bulletin*, **101**, 1011-1020.

Shaw, J. (1975a). The formation of glacial flutings. *In*: Suggate, R.P. and Cresswell, M.M. (eds), *Quaternary Studies*, pp. 253-258. The Royal Society of New Zealand, Wellington.

Shaw, J. (1975b). Sedimentary successions in Pleistocene ice-marginal lakes. *In*: Jopling, A.V. and McDonald, B.C. (eds), *Glaciofluvial and Glaciolacustrine Sedimentation*, pp. 281-303. *Society of Economic Paleontologists and Mineralogists, Special Publication No. 23*, Tulsa, OK.

Shaw, J. (1977). Sedimentation in an alpine lake during deglaciation, Okanagan Valley, British Columbia, Canada. *Geografiska Annaler*, **59A**, 221-240.

Shaw, J. (1979). Genesis of the Sveg tills and Rögen moraines of central Sweden: a model of basal melt out. *Boreas*, **8**, 409-426.

Shaw, J. (1980). Drumlins and large-scale flutings related to glacier folds. *Arctic and Alpine Research*, **12**, 287-298.

Shaw, J. (1983). Drumlin formation related to inverted meltwater erosional marks. *Journal of Glaciology*, **29** (**103**), 461-479.

Shaw, J. (1985). Subglacial and ice marginal environments. *In*: Ashley, G.M., Shaw, J. and Smith, N.D. (eds), *Glacial Sedimentary Environments*, pp. 7-84. *Society of Economic Palaeontologists and Mineralogists, Short Course 16*, Tulsa, OK.

Shaw, J. (1988). Subglacial erosion marks, Wilton Creek, Ontario. *Canadian Journal of Earth Sciences*, **25**, 1256-1267.

Shaw, J. and Archer, J. (1978). Winter turbidity current deposits in late Pleistocene glaciolacustrine varves, Okanagan Valley, British Columbia, Canada. *Boreas*, **7**, 123-130.

Shaw, J. and Freschauf, R.C. (1973). A kinematic discussion of the formation of glacial flutings. *Canadian Geographer*, **17**, 19-35.

Shaw, J., Gilbert, R. and Archer, J.J.J. (1978). Proglacial lacustrine sedimentation during winter. *Arctic and Alpine Research*, **10**, 689-699.

Shaw, J., Kvill, D. and Rains, B. (1989). Drumlins and catastrophic subglacial floods. *Sedimentary Geology*, **62**, 177-202.

Shimek, B. (1909). Aftonian sands and gravels in western Iowa. *Geological Society of America Bulletin*, **20**, 399-408.

Shoemaker, E.M. (1986a). Subglacial hydrology for an ice sheet resting on a deformable aquifer. *Journal of Glaciology*, **32**, 20-30.

Shoemaker, E.M. (1986b). Debris-influenced sliding laws and basal debris balance. *Journal of Glaciology*, **32**, 224-231.

Shoemaker, E.M. (1988). On the formulation of basal debris drag for the case of sparse debris. *Journal of Glaciology*, **34**, 259-264.

Shoemaker, E.M. (1990). The ice topography over subglacial lakes. *Cold Regions Science and Technology*, **18**, 323-329.

Shoemaker, E.M. (1991). On the formation of large subglacial lakes. *Canadian Journal of Earth Sciences*, **28**, 1975-1981.

Shoemaker, E.M. (1992). Water sheet outburst floods from the Laurentide Ice Sheet. *Canadian Journal of Earth Sciences*, **29**, 1250-1264.

Shoji, H. and Langway, C.C., Jr (1985). Comparison of mechanical tests on the Dye-3, Greenland ice core and artificial laboratory ice. *Annals of Glaciology*, **6**, 305.

Shotton, F.W. (1965). Normal faulting in British Pleistocene deposits. *Quarterly Journal of the Geological Society of London*, **121**, 419-434.

Shreve, R.L. (1972). Movement of water in glaciers. *Journal of Glaciology*, **11**, 205-214.

Shreve, R.L. (1984). Glacier sliding at subfreezing temperatures. *Journal of Glaciology*, **30**, 341-347.

Shreve, R.L. (1985a). Esker characteristics in terms of glacier physics. Katahdin Esker System, Maine. *Geological Society of America Bulletin*, **96**, 639-646.

Shreve, R.L. (1985b). Late Wisconsin ice-surface profile calculated from esker paths and types, Katahdin esker system, Maine. *Quaternary Research*, **23**, 27-37.

Šibrava, V. (1986). Correlation of European glaciations and their relation to the deep-sea record. *In*: Šibrava, V., Bowen, D.Q. and Richmond, G.M. (eds), *Quaternary Glaciations in the Northern Hemisphere. Quaternary Science Reviews*, **5**, 433-441.

Šibrava, V., Bowen, D.Q. and Richmond, G.M. (eds), (1986). *Quaternary Glaciations in the Northern Hemisphere. Quaternary Science Reviews*, **5**, 510 pp.

Sigbjarnarson, G. (1973). Katla and Askja. *Jökull*, **23**, 45-50.

Sigenthaler C. and Sturm, M. (1991). Slump induced surges and sediment transport in Lake Uri, Switzerland. *International Association for Theoretical and Applied Limnology Congress Proceedings 1989*, **241**, 955-958.

Sikonia, W.G. and Post, A. (1980). Columbia Glacier, Alaska: Recent ice loss and its relationship to seasonal terminal embayments, thinning and glacier flow. *U.S. Geological Survey Hydrologic Investigations Atlas*, HA-619, 3 sheets..

Simenstad, C.A. and Powell, R.D. (1990). Benthic, epibenthic and planktenic invertebrates in ice-proximal glacimarine environs: life in the turbidity lane. *In*: Milner, A.M. and Wood, J.D. (eds), *Proceedings of the Second Glacier Bay Science Symposium*, pp. 120-126. United States National Park Service, Anchorage.

Simola, H. (1977). Diatom succession in the formation of annually-laminated sediments in Lovojärvi, a small eutrophicated lake. *Annals Botanici Fennici*, **14**, 143-148.

Simons, D.B., Richardson, E.V. and Nordin, C.F. (1965). Sedimentary structures generated by flow in alluvial channels. *Society of Economic Paleontologists and Mineralogists, Special Publication, No. 12*, pp. 34-52.

Singer, J.K. and Anderson, J.B. (1984). Use of total grain-size distributions to define bed erosion and transport for poorly sorted sediment undergoing simulated bioturbation. *Marine Geology*, **57**, 335-359.

Sissons, J.B. (1967). *The Evolution of Scotland's Scenery*. Oliver & Boyd, Edinburgh, 259 pp.

Sissons, J.B. (1974). A Late-glacial ice cap in the central Grampians, Scotland. *Transactions of the Institute of British Geographers*, **62**, 95-114.

Sissons, J.B. (1979). The Loch Lomond Stadial in the British Isles. *Nature*, **280**, 199-203.

Sissons, J.B. (1981). Late glacial marine erosion and a jökulhlaup deposit in the Beauly Firth. *Scottish Journal of Geology*, **17**, 7-19.

Slater, G. (1926). Glacial tectonics as reflected in disturbed drift deposits. *Proceedings of the Geologists Association*, **37**, 392-400.

Slater, G. (1943). Exhibit of sections and models illustrating the drifts of East Anglia. *Proceedings of the Geologists' Association*, **54**, 124.

Slatt, R.M. (1971). Texture of ice-cored deposits from ten Alaskan valley glaciers. *Journal of Sedimentary Petrology*, **41**, 828-834.

Sloss, L.L. (1963). Sequences in the cratonic interior of North America. *Geological Society of America Bulletin*, **74**, 93-114.

Small, R.J. (1987a). Englacial and supraglacial sediment: transport and deposition. *In*: Gurnell, A.M. and Clark, M.J. (eds), *Glacio-Fluvial Sediment Transfer - an Alpine Perspective*, pp. 111-145. John Wiley, Chichester.

Small, R.J. (1987b). Moraine sediment budgets. *In*: Gurnell, A.M. and Clark, M.J. (eds), *Glacio-Fluvial Sediment Transfer - an Alpine Perspective*, pp. 165-197. John Wiley, Chichester.

Small, R.J. and Clark, M.J. (1976). Morphology and development of medial moraines: reply to comments by N. Eyles. *Journal of Glaciology*, **17**, 162-164.

Small, R.J., Clark, M.J. and Cawse, T.J.P. (1979). The formation of medial moraines on alpine glaciers. *Journal of Glaciology*, **22**, 43-52.

Smart, C.C. (1990). Comments on: "Character of the englacial and subglacial drainage system in the lower part of the ablation area of Storglaciären, Sweden, as revealed by dye-trace studies". *Journal of Glaciology*, **36**, 126-128.

Smiraglia, C. (1989). The medial moraines of Ghiacciaio dei Forni, Valtellina, Italy: morphology and sedimentology. *Journal of Glaciology*, **35**, 81-84.

Smith, A.G. and Briden, J.C. (1977). *Mesozoic and Cenozoic Paleocontinental Maps*. Cambridge University Press, Cambridge, 63 pp.

Smith, A.G., Hurley, A.M. and Briden, J.C. (1981). *Phanerozoic Paleocontinental World Maps*. Cambridge University Press, London, 102 pp.

Smith, D.G. (1976). Effect of vegetation on lateral migration of anastomosed channels of a glacier meltwater river. *Geological Society of America*

Bulletin, **87,** 857-860.

Smith, D.G. (1983). Anastomosed fluvial deposits: modern examples from Western Canada. *In:* Collinson, J.D. and Lewin, J. (eds), *Modern and Ancient Fluvial Systems,* pp. 155-168. International Association of Sedimentologists, Special Publication, No. 6.

Smith, D.G. and Smith, N.D. (1980). Sedimentation in anastomosed river systems: examples from alluvial valleys near Banff, Alberta. *Journal of Sedimentary Petrology,* **50,** 157-164.

Smith, G.D. and Morland, L.W. (1981). Viscous relations for the steady creep of polycrystalline ice. *Cold Regions Science and Technology,* **5,** 141-150.

Smith, G.W. (1982). End moraines and the pattern of last ice retreat from central and south coastal Maine. *In:* Larson, G.J. and Stone, B.D. (eds), *Late Wisconsinan Glaciation of New England,* pp. 195-209. Kendall-Hunt, Dubuque, Iowa.

Smith, I.N. and Budd, W.F. (1981). The derivation of past climate changes from observed changes in glaciers. *In:* Allison, I. (ed.), *Sea Level, Ice and Climatic Change,* pp. 31-52. Proceedings of the Canberra Symposium, December 1979. IAHS Publication 131.

Smith, N.D. (1970). The braided stream depositional environment: comparison of the Platte River with some Silurian clastic rocks, North-Central Appalachians. *Geological Society of America Bulletin,* **81,** 2993-3014.

Smith, N.D. (1974). Sedimentology and bar formation in the Upper Kicking Horse River, a braided outwash stream. *Journal of Geology,* **82,** 205-223.

Smith, N.D. (1978). Sedimentation processes and patterns in a glacier-fed lake with low sediment input. *Canadian Journal of Earth Sciences,* **15,** 741-756.

Smith, N.D. (1981). The effect of changing sediment supply on sedimentation in a glacier-fed lake. *Arctic and Alpine Research,* **13,** 75-82.

Smith, N.D. (1985). Proglacial Fluvial Environment. *In:* Ashley, G.M., Shaw, J. and Smith, N.D. (eds), *Glacial Sedimentary Environments.* Society of Economic Paleontologists and Mineralogists, Short Course, 16, pp. 85-134.

Smith, N.D. (1990). The effects of glacial surging on sedimentation in a modern ice-contact lake, Alaska. *Geological Society of America Bulletin,* **102,** 1393-1403.

Smith, N.D. and Ashley, G.M. (1985). Proglacial Lacustrine Environment. *In:* Ashley, G.M., Shaw, J. and Smith, N.D. (eds), *Glacial Sedimentary Environments,* pp. 135-216. Society of Economic Paleontologists and Mineralogists, Short Course 16, Tulsa, OK.

Smith, N.D. and Syvitski, J.P.M. (1982). Sedimentation in a glacier-fed lake: The role of pelletization on deposition of fine-grained suspensates. *Journal of Sedimentary Petrology,* **52,** 503-513.

Smith N.D., Vendl, M.A. and Kennedy, S.K. (1982). Comparison of sedimentation regimes in four glacier-fed lakes of western Alberta. *In:* Davidson-Arnott, R., Nickling, W. and Fahey, B.D. (eds), *Research in Glacial, Glacio-fluvial, and Glacio-lacustrine Systems,* pp. 203-238. 6th Guelph Symposium on Geomorphology. Geo Books, Norwich.

Smith, N.D., Phillips, A.C. and Powell, R.D. (1990). Tidal drawdown: A mechanism for producing cyclic sediment laminations in glaciomarine deltas. *Geology,* **18,** 10-13.

Smith, W.O. and Sakshaug, E. (1990). Polar phytoplankton. *In:* Smith, W.O. (ed.), *Polar Oceanography, Part B.* Academic, New York, pp. 477-525.

Sneed, E.D. and Folk, R.L. (1958). Pebbles in the lower Colorado River, Texas; A study in particle morphogenesis. *Journal of Geology,* **66,** 114-150.

Solheim, A. (1988). Glacial geology of the northern Barents Sea, with emphasis on the surge-related, ice-proximal depositional environment. *Norsk Polarinstitut Rapportserie,* **47.**

Solheim, A. and Pfirman, S.L. (1985). Sea-floor morphology outside a grounded surging glacier, Bråsvellbreen, Svalbard. *Marine Geology,* **65,** 127-143.

Sønstegaard, E. (1979). Glaciotectonic deformation structures in unconsolidated sediments at Os, south of Bergen. *Norsk Geologisk Tidsskrift,* **59,** 223-228.

Souchez, R.A. and Lorrain, R.D. (1978). Origin of the basal ice layer from alpine glaciers indicated by its chemistry. *Journal of Glaciology,* **20,** 319-328.

Souchez, R.A. and Lorrain, R.D. (1987). The subglacial sediment system. *In:* Gurnell, A.M. and Clark, M.J. (eds), *Glacio-Fluvial Sediment Transfer - an Alpine Perspective,* pp. 147-164. John Wiley, Chichester.

Souchez, R.A. and Lorrain, R.D. (1991). *Ice Composition and Glacier Dynamics.* Springer, Berlin, 207 pp.

Souchez, R.A., Lorrain, R.D. and Lemmens, M.M. (1973). Refreezing of interstitial water in a subglacial cavity on an alpine glacier as indicated by the chemical composition of ice. *Journal of Glaciology,*

12, 453-459.

Souchez, R., Lorrain, R.D., Tison, J.-L. and Jouzel, J. (1988). Co-isotopic signature of two mechanisms of basal-ice formation in Arctic outlet glaciers. *Annals of Glaciology*, **10**, 163-166.

Southard, J.B., Smith N.D. and Kuhnle, R.A. (1984). Chutes and lobes: newly identified elements of braiding in shallow gravelly streams. *In*: Koster, E.H. and Steel, R.J. (eds), Sedimentology of Gravels and Conglomerates. *Canadian Society of Petroleum Geologists, Memoir 10*, pp. 51-59.

Spencer, A.M. (1971). Late Precambrian glaciation in Scotland. *Memoir of the Geological Society of London*, **6**, 1-48.

Spring, U. (1980). Intraglazialer Wasserabfluss: Theorie und Modellrechnungen. *Mitteilungen der Versuchsanstalt für Wasserbau, Hydrologie und Glaziologie, No. 48*, 197 pp.

Spring, U. (1983). Turbulent flow in intraglacial conduits: Temperature induced instabilities (Abstract). Euromech 171 Mechanics of Glaciers, Interlaken, Switzerland, 19-23 September 1983, 25 pp.

Spring, U. and Hutter, K. (1981a). Numerical studies of jökulhlaups. *Cold Regions Science and Technology*, **4**, 227-244.

Spring, U. and Hutter, K. (1981b). Conduit flow of a fluid through its solid phase and its application to intraglacial channel flow. *International Journal of Engineering Sciences*, **20**, 327-363.

Starkel, L. (1991). Long-distance correlation of fluvial events in the temperate zone. *In*: Starkel, L., Gregory, K.J. and Thornes, J.B. (eds), *Temperate Palaeohydrology: Fluvial Processes in Temperate Zone During the Last 15 000 Years*, pp. 473-495. John Wiley, Chichester.

Stauffer, B., Neftel, A., Oeschger, H. and Schwander, J. (1985). CO_2 concentration in air extracted from Greenland ice samples. *American Geophysical Union Geophysical Monograph*, **33**, 85-89.

Stauffer, B., Lochbronner, E., Oeschger, H. and Schwander, J. (1988). Methane concentration in the glacial atmosphere was only half that of the preindustrial Holocene. *Nature*, **332**, 812-814.

Steiner, J. and Grillmair, E. (1973). Possible galactic causes for periodic and episodic glaciations. *Geological Society of America Bulletin*, **84**, 1003-1018.

Stenborg, T. (1965). Problems concerning winter runoff from glaciers. *Geografiska Annaler*, **47A**, 141-184.

Stenborg, T. (1968). Glacier drainage connected with ice structures. *Geografiska Annaler*, **50A**, 25-53.

Stenborg, T. (1969). Studies on the internal drainage of glaciers. *Geografiska Annaler*, **51A**, 13-41.

Stenborg, T. (1970). Delay of run-off from a glacier basin. *Geografiska Annaler*, **52A**, 1-30.

Stenborg, T. (1973). Some viewpoints on the internal drainage of glaciers. *International Association of Scientific Hydrology, Publication No. 95*, pp. 117-129.

Stephan, H.-J. (1987). Form, composition and origin of drumlins in Schleswig-Holstein. *In*: Menzies, J. and Rose, J. (eds), *Drumlin Symposium*, pp. 335-345. A.A. Balkema, Rotterdam.

Stephens, G.C., Evenson, E.B., Tripp, R.B. and Detra, D. (1983). Active alpine glaciers as a tool for bedrock mapping and mineral exploration: A case study from Trident Glacier, Alaska. *In*: Evenson, E.B., Schlüchter, Ch. and Rabassa, J. (eds), *Tills and Related Deposits: Genesis, Petrology, Application, Stratigraphy*, pp. 195-204. A.A. Balkema, Rotterdam.

Stephenson, S.N. and Doake, C.S.M. (1982). Dynamic behaviour of Rutford ice stream. *Annals of Glaciology*, **3**, 295-299.

Stevens, R.L. (1990). Proximal and distal glacimarine deposits in southwestern Sweden: contrasts in sedimentation. *In*: Dowdeswell, J.A. and Scourse, J.D. (eds), *Glacimarine Environments: Processes and Sediments*, pp. 307-316. Geological Society Special Publication No.53.

Stockmal, G.S. (1983). Modeling of large-scale accretionary wedge deformation. *Journal of Geophysical Research*, **88**, 8271-8287.

Stoker, M.S. (1988). Pleistocene ice-proximal glaciomarine sediments in boreholes from the Hebrides shelf and Wyville-Thomson ridge, NW U.K. continental shelf. *Scottish Journal of Geology*, **24**, 249-262.

Stone, B.D. and Force, E.R. (1983). Coarse-grained glaciolacustrine deltaic sedimentation in ice marginal and ice-channel environments. *Geological Society of America, Abstracts with Programs*, **15**, 126

Strasser, J.C., Lawson, D.E., Evenson, E.B., Gosse, J.C. and Alley, R.B. (1992). Frazil ice growth at the terminus of the Matanuska Glacier, Alaska, and its implications for sediment entrainment in glaciers and ice sheets. *Geological Society of America, Abstracts with Programs*, **24**, 78.

Stuiver, M. and Braziunas, T.F. (1989). Atmospheric ^{14}C and century-scale solar oscillations. *Nature*, **338**, 405-408.

Stuiver, M. and Quay, P.D. (1981). Changes in

atmospheric carbon-14 attributed to a variable sun. *Science*, **207**, 11-19.

Stuiver, M., Heusser C.J. and Yang, I.C. (1978). North American glacial history extended to 75,000 years ago. *Science*, **200**, 16-21.

Stuiver, M., Denton, G.H., Hughes, T.J. and Fastook, J.L. (1981). History of the marine ice sheet in West Antarctica during the last glaciation: A working hypothesis. *In*: Denton, G.H. and Hughes, T.J. (eds), *The Last Great Ice Sheets*, pp. 319-436. Wiley-Interscience, New York.

Stuiver, M., Braziunas, T.F., Becker, B. and Kromer, B. (1991). Climatic, solar, oceanic and geomagnetic influences on late-glacial and Holocene atmospheric $^{14}C/^{12}C$ change. *Quaternary Research*, **35**, 1-24.

Stump, E. and Fitzgerald, P.G. (1992). Episodic uplift of the Transantarctic Mountains. *Geology*, **20**, 161-164.

Stupavsky, M., Symons, D.T.A. and Gravenor, C.P. (1982). Evidence for metamorphic remagnetization of Upper Precambrian tillite in the Dalradian Supergroup of Scotland. *Transactions Royal Society of Edinburgh, Earth Sciences*, **73**, 59-65.

Sturm, M. (1979). Origin and composition of clastic varves. *In*: Schlüchter, Ch. (ed.), *Moraines and Varves: Origin/Genesis/ Classification*, pp. 281-285. A.A. Balkema, Rotterdam.

Sturm, M. (1987). Observations on the distribution and characteristics of potholes on surging glaciers. *Journal of Geophysical Research*, **92**, 9015-9022.

Sturm, M. and Benson, C.S. (1985). A history of jökulhlaups from Strandline Lake, Alaska, U.S.A. *Journal of Glaciology*, **31**, 272-280.

Sturm, M. and Cosgrove, D.M. (1990). An unusual jökulhlaup involving potholes on Black Rapids Glacier, Alaska Range, Alaska, U.S.A. *Journal of Glaciology*, **36**, 125-126.

Sturm, M. and Matter, A. (1978). Turbidites and varves in Lake Brienz (Switzerland): deposition of clastic detritus by density currents. *In*: Matter, A. and Tucker, M.E. (eds), *Modern and Ancient Lake Sediments*, pp. 147-168. International Association of Sedimentologists, Special Publication No. 2.

Sturm, M., Beget, J. and Benson, C. (1987). Observations of jökulhlaups from ice-dammed Strandline Lake, Alaska: implications for paleohydrology. *In*: Mayer, L. and Nash, D. (eds), *Catastrophic Flooding*, pp. 79-94. Allen and Unwin, Boston.

Sturm, M., Hall, D.K., Benson, C.S. and Field, W.O. (1991). Non-climatic control of glacier-terminus fluctuations in the Wrangell and Chugach Mountains, Alaska, U.S.A. *Journal of Glaciology*, **37**, 348-356.

Suess, E., Balzer, W., Hesse, K.-F., Müller, P.J., Ungerer, C.A. and Wefer, G. (1982). Calcium carbonate hexahydrate from organic-rich sediments of the Antarctic Shelf: precursors of glendonites. *Science*, **216**, 1128-1131.

Suess, H.E. (1973). Natural radiocarbon. *Endeavour*, **32**, 34-38.

Sugden, D.E. (1977). Reconstruction of the morphology, dynamics, and thermal characteristics of the Laurentide Ice Sheet at its maximum. *Arctic and Alpine Research*, **9**, 21-47.

Sugden, D.E. (1978). Glacial erosion by the Laurentide Ice Sheet. *Journal of Glaciology*, **20**, 367-391.

Sugden, D.E. (1979). (Abstract) Ice-sheet erosion - a result of maximum conditions? Discussion. *Journal of Glaciology*, **23**, 402-404.

Sugden, D.E. and John, B.S. (1976). *Glaciers and Landscape - A Geomorphological Approach*. Edward Arnold, London, 320 pp.

Sugden, D.E., Clapperton, C.M. and Knight, P.G. (1985). A jökulhlaup near Søndre Strømfjord, west Greenland, and some effects on the ice-sheet margin. *Journal of Glaciology*, **31**, 366-368.

Sugden, D.E., Clapperton, C.M., Gemmell, J.C. and Knight, P.G. (1987). Stable isotopes and debris in basal glacier ice, South Georgia, Southern Ocean. *Journal of Glaciology*, **33**, 324-329.

Sugden, D.E., Glasser, N. and Clapperton, C.M. (1992). Evaluation of large roches moutonnées. *Geografiska Annaler*, **74A**, 253-264.

Sugden, D.E., Knight, P.G., Livesey, N., Lorrain, R.D., Souchez, R.A., Tison, J.-L. and Jouzel, J. (1987a). Evidence for two zones of debris entrainment beneath the Greenland ice sheet. *Nature*, **328**, 238-241.

Sugden, D.E., Clapperton, C.M., Gemmell, J.C. and Knight, P.C. (1987b). Stable isotopes and debris in basal glacier ice, South Georgia, Southern Ocean. *Journal of Glaciology*, **33**, 324-329.

Sundquist, E.T. (1987). Ice core links CO_2 to climate. *Nature*, **329**, 389-390.

Süsstrunk, A. (1951). Sondage du glacier par la méthode sismique. *La Houille Blanche*, 6e An., Special No. A, 309-319.

Sutherland, D.G. (1984). Modern glacier characteristics as a basis for inferring former climates with particular reference to the Loch Lomond Stadial. *Quaternary Science Reviews*, **3**, 291-309.

Sutton, J. (1963). Long-term cycles in the evolution of the continents. *Nature*, **198**, 731-735.

Swinzow, G.K. (1962). Investigation of shear zones in the ice sheet margin, Thule area, Greenland. *Journal of Glaciology*, **4**, 215-229.

Swithinbank, C. (1978). Remote sensing of iceberg thickness. *In*: Husseiny, A.A. (ed.), *Proceedings of International Conference and Workshops on Iceberg Utilization for Fresh Water Production, Weather Modification, and Other Applications, 1st*, pp. 100-107. Iowa State University, Ames, October 2-6. Pergamon Press, New York.

Swithinbank, C. (1988). Antarctica. *In*: Williams, R.S. and Ferrigno, J.G. (eds), Satellite Image Atlas of Glaciers of the World. *U.S. Geological Survey, Professional Paper 1386-B*, 278 pp.

Swithinbank, C.T., Chinn, T.J., Williams, R.S. and Ferrigno, J.G. (1988). Satellite image atlas of glaciers of the world, Antarctica. *U.S. Geological Survey, Professional Paper, 1386-B*.

Syvitski, J.P.M. (1984). Sedimentology of Arctic Fjords Experiment: HU83-O28 data report, Vol. 2, Can. Data Rep. *Hydrog. Ocean Sci.*, **28**, 1130 pp.

Syvitski, J.P.M. (1989). On the deposition of sediment within glacier-influenced fjords: oceanographic controls. *Marine Geology*, **85**, 301-329.

Syvitski, J.P.M. and Praeg, D.B. (1989). Quaternary sedimentation in the St. Lawrence estuary and adjoining areas, eastern Canada: An overview based on high resolution seismo-stratigraphy. *Géographie Physique et Quaternaire*, **43**, 291-310.

Syvitski, J.P.M. and Skei, J.M. (eds), (1983). Sedimentology of Fjords. *Sedimentary Geology*, **36**, 75-339.

Syvitski, J.P.M., Burrell, D.C. and Skei, J.M. (1987). *Fjords: Processes and Products*. Springer, New York, 379 pp.

Syvitski, J.P.M., Smith, J.N., Calabrese, E.A. and Boudreau, B.P. (1988). Basin sedimentation and the growth of prograding deltas. *Journal of Geophysical Research*, **93**, 6895-6908.

Syvitski, J.P.M., LeBlanc, K., William, G. and Cranston, R.E. (1990). The flux and preservation of organic carbon in Baffin Island Fjords. *In*: Dowdeswell, J.A. and Scourse, J.D. (eds), *Glacimarine Environments: Processes and Sediments*, pp. 177-199. *Geological Society of London Special Publication No. 53*.

Talbot, C.J. and Von Brunn, V. (1987). Intrusive and extrusive (micro)melange couplets as distal effects of tidal pumping by a marine ice sheet. *Geological Magazine*, **124**, 513-525.

Tangborn, W.V. (1980a). Two models for estimating climate-glacier relationships in the North Cascades, Washington, U.S.A. *Journal of Glaciology*, **25**, 3-21.

Tangborn, W.V. (1980b). Contribution of glacier runoff to hydroelectric power generation on the Columbia River. In International Symposium on Computation and Prediction of Runoff from Glaciers and Glacierized Areas, Tbilisi, Georgian S.S.R., 3-11 September, 1978. Mater. *Glyatsiologicheskikh Issled. Khronika. Obsuzhdeniay*, **39**, pp. 62-67.

Tangborn, W.V. (1984). Prediction of glacier derived runoff for hydroelectric development. *Geografiska Annaler*, **66A(3)**, 257-265.

Tangborn, W.V. (1986). A seasonal flow forecasting model for the Susitna River, Alaska. *In*: Benson, C.S., Harrison, W.D., Gosink, J., Bowling, S., Mayo, L. and Trabant, D. (eds), *Workshop on Alaskan Hydrology: Problems Related to Glacierized Basins*. University of Alaska Geophysical Institute, Report UAG-R-306, pp. A58-A64.

Tangborn, W.V., Krimmel, R.M. and Meier, M.F. (1975). A comparison of glacier mass balance by glaciological, hydrological and mapping methods, South Cascade Glacier, Washington. *International Association of Hydrological Sciences*, **104**, 185-196.

Tarar, R.N. (1982). Water resources investigations in Pakistan with the help of Landsat imagery snow surveys 1975-1978. *International Association of Scientific Hydrology Publication No. 138*, pp. 177-190.

Tarling, D.H. (1974). A palaeomagnetic study of Eocambrian tillites in Scotland. *Journal of the Geological Society of London*, **130**, 163-170.

Tarr, R.S. and Butler, B.S. (1909). The Yakutat Bay Region, Alaska. *United States Geological Survey, Professional Paper No. 64*.

Tarr, R.S. and Martin, L. (1914). *Alaskan Glacier Studies*. National Geographical Society, Washington, 498 pp.

Taylor, K.C., Lomorey, G.W., Doyle, G.A., Alley, R.B., Grootes, P.M., Mayewski, P.A., White, J.W.C. and Barlow, L.K. (1993). The "flickering" switch of late Pleistocene climate change. *Nature*, **361**, 432-436.

Taylor, S.R. and McLennan, S.M., (1985). *The Continental Crust: its Composition and Evolution*. Blackwell Scientific, Oxford, 312 pp.

Tchalenko, J.S. (1970). Similarities between shear zones of different magnitudes. *Geological Society of America Bulletin*, **81**, 1625-1640.

Tchalenko, J.S. and Ambraseys, N.N. (1970). Structural analysis of the Dasht-e-Bayaz (Iran) earthquake fractures. *Geological Society of America Bulletin,* **81,** 41-60.

Teller, J.T. (1987). Proglacial lakes and the southern margin of the Laurentide Ice Sheet. *In*: Ruddiman, W.F. and Wright, H.E., Jr (eds), *North America and Adjacent Oceans During the Last Deglaciation. The Decade of North American Geology* Geological Society of America, Boulder, CO, K3, pp. 39-69.

Teller, J.T. (1990). Volume and routing of Late-Glacial runoff from the southern Laurentide ice sheet. *Quaternary Research,* **34,** 12-23.

Teller, J. and Clayton, L. (1983). Glacial Lake Agassiz. *Geological Association of Canada Special Paper,* **26,** 451 pp.

Teller, J.T. and Thorliefson, L.H. (1987). Catastrophic flooding into the Great Lakes from Lake Agassiz. *In*: Mayer, L. and Nash, D. (eds), *Catastrophic Flooding. The Binghampton Symposia in Geomorphology,* pp. 121-138. International Series, No. 18, Allen and Unwin, Boston.

Terzaghi, K. (1943). *Theoretical Soil Mechanics.* John Wiley, New York, 510 pp.

Terzaghi, K. and Peck, R.B. (1967). *Soil Mechanics in Engineering Practice.* John Wiley, New York, 729 pp.

Theakstone, W.H. (1967). Basal sliding and movement near the margin of the glacier Østerdalsisen, Norway. *Journal of Glaciology,* **6,** 805-816.

Theakstone, W.H. (1979). Observations within cavities at the bed of the glacier Østerdalsisen, Norway. *Journal of Glaciology,* **23,** 273-281.

Theakstone, W.H. and Knudsen, N.T. (1981). Dye tracer tests of water movement at the glacier Austre Okstindbreen, Norway. *Norsk Geografisk Tidsskrift,* **35,** 21-28.

Thiede, J., Clark, D.L. and Herman, Y. (1990). Late Mesozoic and Cenozoic paleoceanography of the northern polar oceans. *In*: Grantz, A., Johnson, L. and Sweeney, J.F. (eds), *The Arctic Ocean Region.* Geological Society of America, The Geology of North America, Volume 50, pp. 427-458.

Thomas, G.S.P. (1984a). A late Devensian glaciolacustrine fan-delta at Rhosesmor, Clwyd, North Wales. *Geological Journal,* **19,** 125-141.

Thomas, G.S.P. (1984b). Sedimentation of a subaqueous esker-delta at Strabathie, Aberdeenshire. *Scottish Journal of Geology,* **20,** 9-20.

Thomas, G.S.P. and Connell, R.J. (1985). Iceberg drop, dump and grounding structures from Pleistocene glacio-lacustrine sediments, Scotland. *Journal of Sedimentary Petrology,* **55,** 243-249.

Thomas, R.H. (1973). The creep of ice shelves: theory. *Journal of Glaciology,* **12,** 45-53.

Thomas, R.H. (1976). Ice velocities on the Ross Ice Shelf. *Antarctic Journal of the United States,* **11,** 279-281.

Thomas, R.H. (1979). The dynamics of marine ice sheets. *Journal of Glaciology,* **24,** 167-177.

Thomas, R.H. and Bentley, C.R. (1978a). The equilibrium state of the eastern half of the Ross Ice Shelf. *Journal of Glaciology,* **20,** 509-518.

Thomas, R.H. and Bentley, C.R. (1978b). A model for Holocene retreat of the West Antarctic Ice Sheet. *Quaternary Research,* **10,** 150-170.

Thomas, R.H., Sanderson, T.J.P. and Rose, K.E. (1979). Effect of climatic warming on the West Antarctic ice sheet. *Nature,* **277,** 355-358.

Thomas, R.H., MacAyeal, D.R., Eilers, D.H. and Gaylord, D.R. (1984). Glaciological studies on the Ross Ice Shelf, Antarctica, 1973 - 1978. *Antarctic Research Series,* **42,** 21-53.

Thompson, A. (1988). Historical development of the proglacial landforms of Svinafellsjökull and Skaftafellsjökull, Southeast Iceland. *Jökull,* **38,** 17-31.

Thompson, D.E. (1979). Stability of glaciers and ice sheets against flow perturbations. *Journal of Glaciology,* **24,** 427-442.

Thompson, L.G. (1989). Ice-core records with emphasis on the global record of the last 2000 years. *In*: Bradley, R.S. (ed.), *Global Changes of the Past,* pp. 201-224. University Corporation for Atmospheric Research/Office for Interdisciplinary Earth Studies, Boulder, Colorado.

Thompson, L.G. and Mosley-Thompson, E. (1982). Microparticle concentration and size-distribution determinations from the J-9 Core, Ross Ice Shelf. *Antarctic Journal of the United States,* **17,** 83-85.

Thompson, L.G. and Mosley-Thompson, E. (1987). Evidence of abrupt climatic change during the last 1,500 years recorded in ice cores from the tropical Quelccaya Ice Cap, Peru. *In*: Berger, W.H. and Labeyrie, L.D. (eds), *Abrupt Climatic Change; Evidence and Implications,* pp. 99-110. D. Reidel, Dordrecht.

Thompson, L.G., Hamilton, W.L. and Bull, C. (1975). Climatological implications of microparticle concentrations in the ice core from "Byrd" Station, western Antarctica. *Journal of Glaciology,* **14,** 433-444.

Thompson, L.G., Mosley-Thompson, E., Dansgaard, W. and Grootes, P.M. (1986). The Little Ice Age as recorded in the stratigraphy of the tropical Quelccaya Ice Cap. *Science*, **234**, 361-364.

Thompson, R. and Kelts, K. (1974). Holocene sediments and magnetic stratigraphy from Lakes Zug and Zurich, Switzerland. *Sedimentology*, **21**, 577-596.

Thomsen, H.H., Thorning, L. and Olesen, O.B. (1989). Applied glacier research for planning hydro-electric power, Illulissat/Jakobshavn, West Greenland. *Annals of Glaciology*, **13**, 257-261.

Thorarinsson, S. (1939). Hoffellsjökull, its movement and drainage. *Geografiska Annaler*, **21**, 189-215.

Thorarinsson, S. (1953). Some new aspects of the Grímsvötn problem. *Journal of Glaciology*, **2**, 267-274.

Thorarinsson, S. (1957). The jökulhlaups from the Katla area in 1955 compared with other jökulhlaups in Iceland. *Jökull*, **7**, 21-35.

Thorarinsson, S. (1969). Glacier surges in Iceland, with special reference to the surges of Brúarjökull. *Canadian Journal of Earth Sciences*, **6**, 875-882.

Thorarinsson, S. (1974). *Vötnin Stríð. Saga Skeiðarárhlaupa og Grímsvatnagosa.* (In Icelandic) Bökaútgáfa Menningarsjóðs, Reykjavík, 255 pp.

Thorp, P.W. (1986). A mountain icefield of Loch Lomond Stadial age, western Grampians, Scotland. *Boreas*, **15**, 83-97.

Thorp, P.W. (1991). Surface profiles and basal shear stresses of outlet glaciers from a Late-glacial mountain ice field in western Scotland. *Journal of Glaciology*, **37**, 77-88.

Thorson, R.M. (1980). Ice-sheet glaciation of the Puget Lowland, Washington, during the Vashon Stade (Late Pleistocene). *Quaternary Research*, **13**, 303-321.

Tison, J.-L. and Lorrain, R.D. (1987). A mechanism of basal ice-layer formation involving major ice-fabric changes. *Journal of Glaciology*, **33**, 47-50.

Tison, J.-L., Petit, J.R. and Barnola, J.M. (1993). Debris entrainment at the ice-bedrock interface in sub-freezing temperature conditions (Terre Adelie, Antarctica). *Journal of Glaciology*, **39**, 303-315.

Tison, J.-L., Petit, J.R., Barnola, J.M. and Mahaney, W.C. (1993). Debris entrainment at the ice-bedrock interface in sub-freezing temperature conditions (Adelie Land, Antarctica). *Journal of Glaciology*, **39**, 303-315.

Titus, J.G. (1987). The greenhouse effect, rising sea level and society's response. *In*: Devoy, R.J.N. (ed.), *Sea Surface Studies: a Global View*, pp. 499-528. Croom Helm Publishing, London.

Todtmann, E. (1936). Einige Ergebnisse von glazialgeologischen Untersuchungen am Südrand des Vatnajökull auf Island (1931 und 1934). *Zeitschrift der Deutsche Geologische Gesellschaft*, **88**.

Todtmann, E.M. (1960). Gletscherformen auf Island (Vatnajökull). *Universität Hamburg. Abhandlungen aus dem Gebiet der Auslandkunde*, Band 65, Reihe C (Naturwissenschaften), Band 19.

Tómasson, H. (1974). Grímsvatnahlaup 1972. Mechanism and sediment discharge. *Jökull*, **24**, 27-39.

Tómasson, H., Palsson, S. and Ingólfsson, P. (1980). Comparison of sediment load transport in the Skeidara Jökulhlaups in 1972 and 1976. *Jökull*, **30**, 21-33.

Tronov, M.V. (1962). On the role of summer snowfall in glacier variations. *International Association of Scientific Hydrology Publication No. 58*, pp. 262-269.

Tufnell, L. (1984). *Glacier Hazards*. Longman Publishing, London, 97 pp.

Turner, C. and West, R.G. (1968). The subdivision and zonation of interglacial periods. *Eiszeitalter und Gegenwart*, **19**, 93-101.

Tvede, A.M. (1982). Influence of glaciers on the variability of long runoff series. *Proceedings of 5th Northern Research Basin Symposium, Ullmsvang, Norway*, pp. 179-189.

Twenhofel, W.H. (1932). *Treatise on Sedimentation*. 2nd Edition. Williams and Wilkins Publishing, Baltimore, 661 pp.

Tyndall, J. (1864). On the conformation of the Alps. *Philosophical Magazine*, **28**, 225-271.

Unruh, J.R. (1991). The uplift of the Sierra Nevada and implications for late Cenozoic epeirogeny in the western Cordillera. *Geological Society of America Bulletin*, **103**, 1395-1404.

Urey, H.C. (1947). The thermodynamic properties of isotopic substances. *Journal of the Chemical Society*, 562-581.

Vallon, M., Petit, J.-R. and Fabre, B. (1976). Study of an ice core to the bedrock in the accumulation zone of an alpine glacier. *Journal of Glaciology*, **17**, 13-28.

Vallot, J. (1898). Exploration des moulins de la Mer de Glace. *Annales de l'Observatoire du Mont Blanc*, **3**, 185-190

Van Andel, V.H. (1989). Global change - do only the present and the future count? *Terra Nova*, **1**, 236-237.

Van Campo, E., Duplessy, J.C., Prell, W.L., Barratt, N. and Sabatier, R. (1990). Comparison of terrestrial and marine temperature estimates for the past 135 kyr off southeast Africa: a test for GCM Simulations of palaeoclimate. *Nature*, **348**, 209-212.

Van der Veen, C.J. (1985). Response of a marine ice

sheet to changes at the grounding line. *Quaternary Research*, **24**, 257-267.

Van der Veen, C.J. (1987a). Longitudinal stresses and basal sliding: a comparative study, *In*: Van der Veen, C.J. and Oerlemans, J. (eds), *Dynamics of the West Antarctic Ice Sheet*, pp. 223-248. D. Reidel, Dordrecht.

Van der Veen, C.J. (1987b). The West Antarctic Ice Sheet: the need to understand its dynamics. *In*: Van der Veen, C.J. and J. Oerlemans, J. (eds), *Dynamics of the West Antarctic Ice Sheet*, pp. 1-16. D. Reidel, Dordrecht.

Van der Veen, C.J. and Oerlemans, J. (1987). *Dynamics of the West Antarctic Ice Sheet*. D. Reidel, Dordrecht, 368 pp.

Van der Veen, C.J. and Whillans, I.M. (1989). Force budget: I. Theory and numerical methods. *Journal of Glaciology*, **35**, 53-60.

Van der Veen, C.J. and Whillans, I.M. (1990). Flow laws for glacier ice: comparison of numerical predictions and field measurements. *Journal of Glaciology*, **36**, 324-339.

Van der Wiel, A.M. and Wijmstra, T.A. (1987). Palynology of the lower part (76-120m) of the core Tenaghi Philippon II, Middle Pleistocene of Macedonia, Greece. *Review of Palaeobotany and Palynology*, **52**, 73-88.

van Donk, J. (1976). O^{18} record of the Atlantic Ocean for the entire Pleistocene Epoch. *Geological Society of America Memoir 145*, 147-163.

Varnes, D.J. (1978). Slope movement types and processes. *In*: Schuster, R.L. and Krizek, R.J. (eds), *Landslides, Analysis and Control*, pp. 11-33. National Academy of Science, Washington, D.C.

Veevers, J.J. and Powell, C. McA. (1987). Late Paleozoic glacial episodes in Gondwanaland reflected in transgressive-regressive depositional sequences in Euramerica. *Geological Society of America Bulletin*, **98**, 475-487.

Velichko, A.A. and Faustova, M.A. (1986). Glaciations in the East European region of the USSR. *In*: Šibrava, V., Bowen, D.Q. and Richmond, G.M. (eds), *Quaternary Glaciations in the Northern Hemisphere. Quaternary Science Reviews*, **5**, 447-461.

Vere, D.M. and Benn, D.I. (1989). Structure and debris characteristics of medial moraines in Jotunheimen, Norway: implications for moraine classification. *Journal of Glaciology*, **35**, 276-280.

Vernekar, A.D. (1960). Long-period variations of incoming solar radiation. *In*: *Research on the Theory of Climate, 2*. Report of the Travellers Research Center, Inc., Hartford, Connecticut, 289 pp.

Vernekar, A.D. (1972). Long-period global variations of incoming solar radiation. *Meteorological Monographs*, 12(34). American Meteorological Society, Boston, Mass.

Viete, G. (1960). Zur Entstehung der glazigenen Lagerstörungen unter besonderer Berücksichtigung der Flözdeformationen im mitteldeutschen Raum. *Freiberger Forschungshefte*, **C-78**, 1-257.

Vinje, T.E. (1980). Some satellite-tracked iceberg drifts in the Antarctic. *Annals of Glaciology*, **1**, 83-87.

Visser, J.N.J. (1981). The Mid-Precambrian tillite in the Griqualand West and Transvaal Basins, South Africa. *In*: Hambrey, M.J. and Harland, W.B. (eds), *Earth's Pre-Pleistocene Glacial Record*, pp. 180-184. Cambridge University Press, Cambridge.

Visser, J.N.J. (1983). Submarine debris flow deposits from the Upper Carboniferous Dwyka Tillite Formation in the Kalahari Basin, South Africa. *Sedimentology*, **30**, 511-523.

Visser, J.N.J. (1989). The Permo-Carboniferous Dwyka Formation of Southern Africa: Deposition by a predominantly subpolar marine ice sheet. *Palaeography, Palaeoclimatology, Palaeoecology*, **70**, 377-391.

Vivian, R. (1970). Hydrologie et erosion sous-glaciaire. *Révue de Géographie Alpine*, **58**, 241-265.

Vivian, R. (1980). The nature of the ice-rock interface: the results of investigation on 20,000 m^2 of the rock-bed of temperate glaciers. *Journal of Glaciology*, **25**, 267-277.

Vivian, R. and Bocquet, G. (1973). Subglacial cavitation phenomena under the Glacier d'Argentière, Mont Blanc, France. *Journal of Glaciology*, **12**, 439-451.

Vivian, R.A. and Zumstein, J. (1973). Hydrologie sous-glaciaire au Glacier d'Argentière (Mont-Blanc, France). *International Association of Hydrological Sciences*, **95**, 53-64.

Vornberger, P.L. and Whillans, I.M. (1986). Surface features of ice Stream B, Marie Byrd Land, west Antarctica. *Annals of Glaciology*, **8**, 168-170.

Vorren, T.O. (1973). Glacial geology of the area between Jostedalsbreen and Jotunheimen, South Norway. *Norges Geologiske Undersøekelse*, **291**, 46 pp.

Vorren, T.O. (1977). Grain-size distribution and grain-size parameters of different till types on Hardangervidda, south Norway. *Boreas*, **6**, 219-227.

Vorren, T.O., Hald, M., Edvardsen, M. and Lind-Hansen, O.-W. (1983). Glacigenic sediments and sedimentary environments on continental shelves: general principles with a case study from the Norwegian shelf. *In*: Ehlers, J. (ed.), *Glacial Deposits in North-West Europe*, pp. 61-73. A.A. Balkema, Rotterdam.

Vorren, T.O., Lebesbye, E., Andereasson, K. and Larsen, K.B. (1989). Glacigenic sediments on a passive continental margin as exemplified by the Barents Sea. *Marine Geology*, **85**, 251-272.

Vortisch, W.B., Mahaney, W.C. and Fecher, K. (1987). Lithology and weathering in a paleosol sequence on Mount Kenya: East Africa. *Geologica et Paleontologica*, **21**, 245-255.

Waddington, E.D. (1986). Wave ogives. *Journal of Glaciology*, **32**, 325-334.

Waddington, E.D. (1987). Geothermal heat flux beneath ice sheets. *In*: Waddington, E.D. and Walder, J.S. (eds), The Physical Basis of Ice Sheet Modelling. Proceedings of an International Symposium held during the 19th General Assembly of the International Union of Geodesy and Geophysics at Vancouver, B.C., Canada, 9-22 August 1987, *International Association of Hydrological Sciences Publication No. 170*, pp. 217-226.

Waddington, E.D. and Walder J.S. (eds), (1987). The Physical Basis of Ice Sheet Modelling. Proceedings of an International Symposium held during the 19th General Assembly of the International Union of Geodesy and Geophysics at Vancouver, B.C., Canada, 9-22 August 1987, *International Association of Hydrological Sciences Publication No. 170*, 384 pp.

Wahlen, M., Allen, D., Deck, B. and Herchenroder, A. (1991). Initial measurements of CO_2 concentrations 1530 to 1940 AD in air occluded in the GISP2 ice core from Greenland. *Geophysical Research Letters*, **18**, 1457-1460.

Waitt, R.B., Jr (1985). Case for periodic, colossal jökulhlaups from Pleistocene glacial Lake Missoula. *Geological Society of America Bulletin*, **96,** 1271-1286.

Wakahama, G. (1964). Plastic deformation of polycrystalline ice. *Low Temperature Sciences*, **Series A, 22**, 1-24.

Wakahama, G. (1968). The metamorphism of wet snow. International Association of Hydrological Sciences Assembly Bern, 1967. *International Association of Scientific Hydrology Publication Publication No. 79*, pp. 370-379.

Wakahama, G., Kuroiwa, D., Kobayashi, O., Tanuma, K., Endo, Y., Mizuno, Y. and Kobayashi, S. (1973). Observations of permeating water through a glacier body. *Low Temperature Science A*, **31**, 217-219.

Walder, J.S. (1982). Stability of sheet flow of water beneath temperate glaciers and implications for glacier surging. *Journal of Glaciology*, **28**, 273-293.

Walder, J.S. (1986). Hydraulics of subglacial cavities. *Journal of Glaciology*, **32**, 439-445.

Walder, J.S. and Hallet, B. (1979). Geometry of former subglacial water channels and cavities. *Journal of Glaciology*, **23**, 335-346.

Walder, J.S. and Hallet, B. (1985). A theoretical model of the fracture of rock during freezing. *Geological Society of America Bulletin*, **96,** 336-346.

Walder, J.S. and Hallet, B. (1986). The physical basis of frost weathering: Toward a more fundamental and unified perspective. *Arctic and Alpine Research*, **18**, 27-32.

Walling, D.E. (1984). Dissolved loads and their measurement. *In*: Hadley, R.F. and Walling, D.E. (eds), *Erosion and Sediment Yield: Some Methods of Measurement and Modelling*, pp. 111-177. Geo Books, Norwich.

Walters, R.A. and Dunlap, W.W. (1987). Analysis of time series of glacier speed: Columbia Glacier, Alaska. *Journal of Geophysical Research*, **92**, 8969-8975.

Ward, L.G., Stephen, M.F. and Nummedal, D. (1976). Hydraulics and morphology of glacial outwash distributaries, Skeidarársandur, Iceland. *Journal of Sedimentary Petrology*, **46**, 770-777.

Warnke, D.A. and Gram, R. (1969). The study of mineral-grain surfaces by interference microscopy. *Journal of Sedimentary Petrology*, **39**, 1599-1604.

Warren, C.R. and Glasser, N.F. (1992). Contrasting response of south Greenland glaciers to recent climatic change. *Arctic and Alpine Research*, **24**, 124-132.

Warren, C.R. and Hulton, N.R.J. (1990). Topographic and glaciological controls on Holocene ice-sheet margin dynamics, Central West Greenland. *Annals of Glaciology*, **14**, 307-310.

Warren, W.P. (1987). Periglacial periods in Ireland. *In*: Boardman, J. (ed.), *Periglacial Processes and Landforms in Britain and Ireland*, pp. 101-111. Cambridge University Press, Cambridge.

Wateren, D.F.M. van der (1981). Glacial tectonics at the Kwintelooijen sandpit, Rhenen, the Netherlands. *Mededelingen Rijks Geologische Dienst*, **35**, 252-268.

Wateren, D.F.M. van der (1985). A model of glacial tectonics, applied to the ice-pushed ridges in the Central Netherlands. *Bulletin of the Geological*

Society of Denmark, **34**, 55-74.

Wateren, D.F.M. van der (1986). Push moraines and glacial tectonics. *In*: Beets, D.J. (ed.), *Saalian Glacial Deposits and Morphology in the Netherlands and Distribution of Fennoscandian Erratic Indicator Pebbles in the Netherlands. Explanatory Notes*. Rijks Geologische Dienst, Haarlem.

Wateren, D.F.M. van der (1987). Structural geology and sedimentology of the Dammer Berge push moraine, FRG. *In*: van der Meer, J.J.M. (ed.), *Tills and Glaciotectonics*, pp. 157-182. A.A. Balkema, Rotterdam.

Wateren, D.F.M. van der (1992). Structural Geology and Sedimentology of Push Moraines. Processes of soft sediment deformation in a glacial environment and the distribution of glaciotectonic styles. Amsterdam, Ph.D. Thesis, 238 pp.

Wateren, F.M. van der (1994). Proglacial subaquatic outwash fan and delta sediments in push moraines - Indicators of subglacial meltwater activity. *Sedimentary Geology*, **91**, 145-172.

Wateren, D.F.M. van der, Kluiving, S.J., Jiskoot, H. and Verbers, A.L.L.M. (in press). Proglacial subaquatic fan and delta sediments in push moraines - indicators of subglacial meltwater activity in a deformable bed. *Sedimentary Geology*.

Watson, R.A. (1980). Landform development on moraines of the Klutlan Glacier, Yukon Territory, Canada. *Quaternary Research*, **14**, 50-59.

Watson, R.A. and Wright, H.E., Jr (1980). The end of the Pleistocene: A general critique of chronostratigraphic classification. *Boreas*, **9**, 153-163.

Washburn, A.L. (1969). Weathering, frost action and patterned ground in the Mesters Vig District, Northeast Greenland. *Meddr. Grønland*, **176**, 303 pp.

Washburn, A.L. (1973). *Periglacial Processes and Environments*. Edward Arnold, London, 320 pp.

Washburn, A.L. (1979). *Geocryology: a Survey of Periglacial Processes and Environments*. Edward Arnold Publishing, London, 406 pp.

Wayne, W.J. and Corte, A.E. (1983). Multiple glaciations of the Cordon del Plata, Mendoza, Argentina. *Palaeogeography, Palaeoclimatology, Palaeoecology*, **42**, 185-209.

Webb, P.N. and McKelvey, P.C. (1959). Geological investigations in South Victoria Land, Antarctica. *New Zealand Journal of Geology and Geophysics*, **2**, 120-136.

Weber, J.N. (1958). Recent grooving in lake bottom sediments at Great Slave Lake, Northwest Terrray

production rates of ^{10}Be and ^{26}Al in quartz from glacially polished rocks. *Journal of Geophysical Research*, **94**, 17907-17915.Waitt, R.B., Jr (1980). About forty last-glacial Lake Missoula jökulhlaups through Southern Washington. *Journal of Geology*, **88**, 653-679.

Webb, P.N. (1990). The Cenozoic history of Antarctica and its global impact. *Antarctic Science*, **2**, 3-21.

Webb, P.N., Harwood, D.M., McKelvey, B.C., Mercer, J.H. and Stott, L.D. (1984). Cenozoic marine sedimentation and ice-volume variation on the East Antarctic craton. *Geology*, **12**, 287-291.

Webb, W.W. and Hayes, C.E. (1967). Dislocations and plastic deformation of ice. *Philosophical Magazine*, **16**, 909-925.

Weeks, W.F. and Campbell, W.J. (1973). Icebergs as a fresh-water source: an appraisal. *Journal of Glaciology*, **12**, 207-233.

Weeks, W.F. and Mellor, M. (1978). Some elements of iceberg technology. *In*: Husseiny, A.A. (ed.), *Iceberg Utilization*, pp. 45-57. Pergamon Press, Oxford.

Weertman, J. (1957a). Deformation of floating ice shelves. *Journal of Glaciology*, **3**, 38-42

Weertman, J. (1957b). On the sliding of glaciers. *Journal of Glaciology*, **3**, 33-38.

Weertman, J. (1958). Travelling waves on glaciers. *International Association of Hydrological Sciences*, **47**, 162-168.

Weertman, J. (1961a). Equilibrium profile of ice caps. *Journal of Glaciology*, **3**, 953-964.

Weertman, J. (1961b). Mechanism for the formation of inner moraines found near the edge of cold ice caps and ice sheets. *Journal of Glaciology*, **3**, 965-978.

Weertman, J. (1962). Stability of ice-age ice caps. *U.S. Cold Regions Research and Engineering Laboratory Research Report*, **97**, 12 pp.

Weertman, J. (1964a). The theory of glacier sliding. *Journal of Glaciology*, **5**, 287-303.

Weertman, J. (1964b). Rate of growth or shrinkage of non-equilibrium ice sheets. *Journal of Glaciology*, **5**, 145-158.

Weertman, J. (1966). Effect of a basal water layer on the dimensions of ice sheets. *Journal of Glaciology*, **6**, 191-207.

Weertman, J. (1969). Water lubrication mechanism of glacier surges. *Canadian Journal of Earth Sciences*, **6**, 929-942.

Weertman, J. (1972). General theory of water flow at the base of a glacier or ice sheet. *Reviews of Geophysics and Space Physics*, **10**, 287-333.

Weertman, J. (1973). Creep of ice. *In*: Whalley, E., Jones, S.J. and Gold, L.W. (eds), *Physics and Chemistry of Ice*, pp. 320-337. Royal Society of Canada, Ottawa.

Weertman, J. (1979). The unsolved general glacier sliding problem. *Journal of Glaciology*, **23**, 97-115.

Weertman, J. (1980). Bottom crevasses. *Journal of Glaciology*, **25**, 185-188.

Weertman, J. (1986). Basal water and high-pressure basal ice. *Journal of Glaciology*, **32**, 455-463.

Weertman, J. (1987a). Basal water and high-pressure basal ice. *Journal of Glaciology*, **33**, 131.

Weertman, J. (1987b). Impact of the International Glaciological Society on the development of glaciology and its future role. *Journal of Glaciology*, *Special Issue*, **33**, 86-90.

Weertman, J. and Birchfield, G.E. (1982). Subglacial water flow under ice streams and West Antarctic ice-sheet stability. *Annals of Glaciology*, **3**, 316-320.

Weertman, J. and Birchfield, G.E. (1983a). Basal water film, basal water pressure, and velocity of travelling waves on glaciers. *Journal of Glaciology*, **29**, 20-27.

Weertman, J. and Birchfield, G.E. (1983b). Stability of sheet water flow under a glacier. *Journal of Glaciology*, **29**, 374-382.

Wegener, A. (1966). *The Origin of Continents and Oceans*. (Translation of the 1915 original) Dover, New York, 246 pp.

Weidick, A. (1984). Studies of glacier behaviour and glacier mass balance in Greenland - a review. *Geografiska Annaler*, **66A**, 183-195.

Weirich, F.H. (1984). Turbidity currents: monitoring their occurrence and movement with a three-dimensional sensor network. *Science*, **224**, 384-387.

Weirich, F.H. (1985). Sediment budget for a high energy glacial lake. *Geografiska Annaler*, **67A**, 83-99.

Wellman, P. (1982). Surging of Fisher Glacier, eastern Antarctica: evidence from geomorphology. *Journal of Glaciology*, **28**, 23-28.

West, R.G. (1967). The Quaternary of the British Isles, *In*: Rankama, K. (ed.), *The Quaternary. Volume 2*, pp. 1-87. Wiley Interscience, New York.

West, R.G. and Donner, J.J. (1956). The glaciations of East Anglia and the East Midlands: a differentiation based on stone orientation measurements of the tills. *Quarterly Journal of the Geological Society of London*, **112**, 69-91.

West, R.G., Rose, J., Coxon, P., Osmaston, H. and Lamb, H.H. (1988). The record of the Cold Stages. *In*: Shackleton, N.J., West, R.G. and Bowen, D.Q.

(eds), *The Past Three Million Years: Evolution of Climatic Variability in the North Atlantic Region*, pp. 95-112. Cambridge University Press, London.

Westgate, J.A., Stemper, B.A. and Péwé, T.L. (1990). A 3 m.y. record of Pliocene-Pleistocene loess in interior Alaska. *Geology*, **18**, 858-861.

Whalley, W.B. and Langway, C.C., Jr (1980). A scanning electron microscope examination of subglacial quartz grains from Camp Century core, Greenland - a preliminary study. *Journal of Glaciology*, **25**, 125-131.

Whillans, I.M. (1987). Force budget of ice sheets. *In*: Van der Veen, C.J. and Oerlemans, J. (eds), *Dynamics of the West Antarctic Ice Sheet*, pp. 17-36. D. Reidel, Dordrecht.

Whillans, I.M. and Johnsen, S.J. (1983). Longitudinal variations in glacial flow: theory and test using data from the Byrd Station Strain network, Antarctica. *Journal of Glaciology*, **29**, 78-97.

White, S.H., Burrows, S.E., Carreras, J., Shaw, N.D. and Humphreys, F.J. (1980). On mylonites in ductile shear zones. *Journal of Structural Geology*, **2**, 175-187.

Whitehouse, I.E., Chinn, T.J.H., Hofle, H.C. (1989). Radiocarbon dates from raised beaches, Terra Nova Bay, Antarctica. *Geological Journal*, **38**, 321-334.

Whitehouse, I.E. and Griffiths, G.A. (1983). Frequency and hazard of large rock avalanches in the central Southern Alps, New Zealand. *Geology*, **11**, 331-334.

Whiteman, C.A. (1983). Great Waltham. *In*: Rose, J. (ed.), *Diversion of the Thames Field Guide*, pp. 163-169. Quaternary Research Association, Cambridge.

Whiteman, C.A. (1987). Till lithology and genesis near the southern margin of the Anglian ice sheet in Essex, England. *In*: van der Meer, J.J.M. (ed.), *Tills and Glaciotectonics*, pp. 55-66. A.A. Balkema, Rotterdam.

Whiteman, C.A. (1990). Early and Middle Pleistocene stratigraphy in Central Essex, England. Unpublished Ph.D. Thesis, University of London. 771 pp.

Wigley, T.M.L. and Kelly, P.M. (1990). Holocene climatic change, ^{14}C wiggles and variations in solar irradiance. *Philosophical Transactions of the Royal Society of London*, **330(A)**, 547-560.

Wigley, T.M.L., Ingram, M.J. and Farmer, G. (1981). *Climate and History: Studies in Past Climates and Their Impact on Man*. Cambridge University Press, Cambridge, 530 pp.

Wijmstra, T.A. (1969). Palynology of the first 30 m of a 120 m deep section of Northern Greece. *Acta Botan. Neerl.*, **18**, 511-527.

Willett, H.C. (1951). Extrapolation of sunspot climate

relationships. *Journal of Meteorology*, **8**, 1-6.

Willett, H.C. (1987). Climatic responses to variable solar activity; past, present and predicted. *In*: Rampino, M.R., Sanders, J.E., Newman, W.S. and Königsson, L.K. (eds), *Climate; History, Periodicity and Predictability*, pp. 404-414. Van Nostrand Reinhold, New York.

Williams, D.F., Thunell, R.C. Tappa, E., Rio, D. and Raffi, I. (1988). Chronology of the Pleistocene oxygen isotope record: 0-1.88 m.y.B.P. *Palaeogeography, Palaeoclimatology, Palaeoecology*, **64**, 221-240.

Williams, G.E. (1975). Late Precambrian glacial climate and the Earth's obliquity. *Geological Magazine*, **112**, 441-544.

Williams, P.F. and Rust, B.R. (1969). The sedimentology of a braided river. *Journal of Sedimentary Petrology*, **39**, 649-679.

Williams, R.S., Jr, Hall, D.K. and Benson, C.S. (1991). Analysis of glacier facies using satellite techniques. *Journal of Glaciology*, **37**, 120-128.

Willis, I.C., Sharp, M.J. and Richards, K.S. (1990). Configuration of the drainage system of Midtdalsbreen, Norway, as indicated by dye-tracing experiments. *Journal of Glaciology*, **36**, 89-101.

Willman, H.B. and Frye, J.C. (1970). Pleistocene stratigraphy of Illinois. *Illinois State Geological Survey Bulletin* **94**, 204 pp.

Wilson, A.T. (1964). Origin of ice ages: an ice shelf theory for Pleistocene glaciation. *Nature*, **201**, 147-149.

Wilson, K.C. (1965). Derivation of the regime equations from relationships for pressurized flow by use of the principle of minimum energy-degradation rate. *Civil Engineering Report 51*, Queen's University, Kingston, Ontario.

Wilson, K.C. (1969). Comments on "Sediment transport in conveyance systems". *Bulletin of the International Association of Scientific Hydrology*, **14**, 131-134.

Windley, B.F. (1977). *The Evolving Continents*. John Wiley, New York.

Winograd, I.J., Coplen, T.B., Landwehr, J.M., Riggs, A.C., Ludwig, K.R., Szabo, B.J., Kolesar, P.T. and Revesz, K.M. (1992). Continuous 500,000-year climate record from vein calcite in Devils Hole, Nevada. *Science*, **258**, 255-259.

Winters, G.V., Syvitski, J.P.M. and Muillet, L. (1985). Distribution and dynamics of particulate matter in Baffin Island fjords. Abstract, *14th Arctic Workshop*, Dartmouth, pp. 73-77.

Wold, B. and Østrem, G. (1979). Subglacial

constructions and investigations at Bondhusbreen, Norway. *Journal of Glaciology*, **23**, 363-379.

Woldstedt, P. (1954). *Das Eiszeitalter I*. Ferdinand Enke, Stuttgart, 374 pp.

Woldstedt, P. and Duphorn, K. (1974). *Norddeutschland und angrenzende Gebiete im Eiszeitalter*. Koehler, Stuttgart, 500 pp.

Wolfe, J.A. (1985). Distribution of major vegetational types during the Tertiary. *In*: Sundquist, E.T. and Broecker, W.S. (eds), The carbon cycle and atmospheric CO_2: Natural variations Archaean to present. *Geophysical Monograph*, **32**, 357-375.

Wolff, E.W. and Doake, C.S.M. (1986). Implications of the form of the flow law for vertical velocity and age-depth profiles in polar ice. *Journal of Glaciology*, **32**, 366-370.

Wong, H.K. and Christoffel, D.A. (1981). A reconnaissance seismic survey of McMurdo Sound and Terra Nova Bay, Ross Sea. *In*: McGinnis, L.D. (ed.), Dry Valley Drilling Project. *American Geophysical Union, Antarctic Research Series*, **33**, 37-62.

Woodruff, F. and Savin, S.M. (1989). Miocene deepwater oceanography. *Paleoceanography*, **4**, 87-140.

Woodworth-Lynas, C.M.T. (1990). Observed deformation structures beneath relict iceberg scours. *In*: Clark, J.I., Konuk, I, Poorooshasb, F., Whittick, J. and Woodworth-Lynas, C. (eds), *Ice Scouring and the Design of Offshore Pipelines*, Proceedings of an invited workshop, April 18-19th, Calgary, Alberta. Canada Oil and Gas Lands Administration and Centre for Cold Ocean Resources Engineering, pp. 103-125.

Worsley, T.R. and Kidder, D.L. (1991). First-order coupling of paleogeography and CO_2, with global surface temperature and its latitudinal contrast. *Geology*, **19**, 1161-1164.

Worsley, T.R. and Nance, R.D. (1989). Carbon redox and climate control through Earth history, a speculative reconstruction. *Palaeogeography, Palaeoclimatology, Palaeoecology (Global and Planetary Change Section)*, **75**, 259-282.

Worsley, T.R., Nance, R.D. and Moody, J.B. (1984). Global tectonics and eustacy for the past 2 billion years. *Marine Geology*, **58**, 373-400.

Worsley, T.R., Nance, R.D. and Moody, J.B., (1986). Tectonic cycles and the history of the Earth's biochemical and paleoceanographic record. *Paleoceanography*, **1**, 233-263.

Wright, A.E. and Moseley, F. (1975). *Ice Ages: Ancient and Modern*. Geological Journal, Special Issue 6, Seel

House, Liverpool, 320 pp.

Wright, H.E., Jr (1973). Tunnel valleys, glacial surges, and subglacial hydrology of the Superior Lobe, Minnesota. *In*: Black, R.F., Goldthwait, R.P. and Willman, H.B. (eds), The Wisconsinan Stage. *Geological Society of America Memoir No. 136*, pp. 251-276.

Wright, H.E., Jr (1980). Cores of soft lake sediment. *Boreas*, **9**, 107-114.

Wright, R. and Anderson, J.B. (1982). The importance of sediment gravity flow to sediment transport and sorting in a glacial marine environment: Eastern Weddell Sea, Antarctica. *Geological Society of America Bulletin*, **93**, 957-963.

Wright, R., Anderson, J.B. and Fisco, P.P. (1983). Distribution and association of sediment gravity flow deposits and glacial/glacial-marine sediments around the continental margin of Antarctica. *In*: Molnia, B.F. (ed.), *Glacial-Marine Sedimentation*, pp. 265-300. Plenum Press, New York.

Wright, R.F. and Nydegger, P. (1980). Sedimentation of detrital particulate matter in lakes: influence of currents produced by inflowing rivers. *Water Resources Research*, **16**, 597-601.

Wright, W.B. (1937). *The Quaternary Ice Age*. 2nd Edition. MacMillan, London, 478 pp.

Yang, Zhenniang, (1982). Basic characteristics of runoff in glacierized areas of China. *In*: Glen, J.W. (ed.), Hydrological aspects of alpine and high-mountain areas. Proceedings of the Exeter Symposium, July 1982. *IAHS Publication 138*, pp. 295-307.

Yatsu, E. (1966). *Rock Control in Geomorphology*. Sozosha, Tokyo, 135 pp.

Yeo, G.M., (1981). The Late Proterozoic Rapitan glaciation in the northern Cordillera. *In*: Campbell, F.H.A. (ed.), Proterozoic Basins of Canada. *Geological Survey of Canada, Paper 81-10*, pp. 25-46.

Yong, R.N.Y. and Warkentin, B.P. (1975). *Soil Properties and Behaviour,* 449 pp. Elsevier Scientific Publishing Company, Amsterdam.

Young, G.J. (1980). Streamflow formation in a glacier-ized watershed in the Rocky Mountains, Canada. In International Symposium. Computation and Prediction of Runoff from Glaciers and Galcierized Areas, Tbilisi, Georgian S.S.R, 3-11 September 1978. *Obsuzhdeniya, Mater. Glyatsiologicheskikh Issled. Khronika.*, Vol. 39, pp. 55-62.

Young, G.J. (1985). Techniques for the Prediction of

Runoff from Glacierized Areas. *International Association of Hydrological Sciences, No. 149*, 149 pp.

Young, G.J. (1990). Glacier hydrology. *In*: Prowse, T.D. and Ommanney, C.S.L. (eds), *Northern Hydrology: Selected Perspectives*. Proceedings, Northern Hydrology Symposium. July 1990, Saskatoon. National Hydrology Research Institute Symposium No. 6, pp. 35-162.

Young, G.M. (1973). Tillites and aluminous quartzites as possible time markers for middle Precambrian (Aphebian) rocks of North America. *In*: Young, G.M. (ed.), *Huronian Stratigraphy and Sedimentation*. Geological Association of Canada Special Paper, 12, pp. 97-127.

Young, G.M. (1983). Tectono-sedimentary history of Early Proterozoic rocks of the northern Great Lakes region. *In*: Medaris, L.G., Jr (ed.), *Early Proterozoic geology of the Great Lakes region. Geological Society of America, Memoir 160*, pp. 15-32.

Young, G.M. (1991). The geologic record of glaciation: relevance to the climatic history of Earth. *Geoscience Canada*, **18**, 100-108.

Young, G.M. and Gostin, V.A. (1989a). An exceptionally thick upper Proterozoic (Sturtian) glacial succession in the Mount painter area, South Australia. *Geological Society of America Bulletin*, **101,** 834-845.

Young, G.M. and Gostin, V.A. (1989b). Depositional environment and regional stratigraphic significance of the Serle Conglomerate: a Late Proterozoic submarine fan complex, South Australia. *Palaeogeography, Palaeoclimatology, Palaeoecology*, **71**, 237-252.

Young, G.M. and McLennan, S.M. (1981). Early Proterozoic Padlei Formation, Northwest Territories, Canada. *In*: Hambrey, M.J. and Harland, W.B. (eds), *Earth's Pre-Pleistocene Glacial Record*, pp. 790-794. Cambridge University Press, Cambridge.

Young, G.M. and Nesbitt, H.W. (1985). The Gowganda Formation in the southern part of the Huronian outcrop belt, Ontario, Canada: stratigraphy, depositional environments and regional tectonic significance. *Precambrian Research*, **29**, 265-301.

Yuen, D.A. and Schubert, G. (1979). The role of shear heating in the dynamics of large ice masses. *Journal of Glaciology*, **24**, 195-212.

Yuen, D.A., Saari, M.R. and Schubert, G. (1986). Explosive growth of shear-heating instabilities in the down-slope creep of ice sheets. *Journal of Glaciology*, **32**, 314-320.

Zagwijn, W. (1959). Zur stratigraphischen und

pollenanalytischen Gliederung der pliozänen Ablagerungen im Roertal-Graben und Venloer Graben der Niederlande. *Fortschritte in der Geologie von Rheinland und Westfalen*, **4**, 5-26.

Zagwijn, W.H. (1986). The Pleistocene in the Netherlands with special reference to glaciation and terrace formation. *In*: Šibrava, V., Bowen, D.Q. and Richmond, G.M. (eds), Quaternary Glaciations in the Northern Hemisphere. *Quaternary Science Reviews*, **5**, 341-345.

Źielinski, G.A., Mayewski, P.A., Meeker, L.D., Whitlow, S., Twickler, M.S., Morrison, M., Taylor, K., Alley, R., Gow, A.J. and Meese, D. (1993). A continuous record of volcanism (Present - 7000 B.C.) and climatic implications. Unpublished manuscript.

Znachko-Yavorskiy, G.A. (1964). Marine geological observations during the eighth cruise of the diesel electric ship 'OB' (Preliminary Results). *Information Bulletin, Soviet Antarctic Expedition*, **5**, 123-125.

Zolnai, A.I., Price, R.A. and Helmstaedt, H. (1984). Regional cross section of the Southern Province adjacent to Lake Huron, Ontario: implications for the tectonic significance of the Murray Fault Zone. *Canadian Journal of Earth Sciences*, **21**, 447-456.

Zumberge, J.H. (1955). Glacial erosion in tilted rock layers. *Journal of Geology*, **63**, 149-158.

SUBJECT INDEX

Aavatsmarkbreen, Svalbard 120, 133
abandoned proglacial channels 401
ablation 85-86, 93, 108, 146, 147, 148, 149, 167, 197, 204, 215, 218, 262, 265, 273, 279, 289-291, 347, 358, 365, 434, 448
ablation dominant moraines (AD) *see moraines*
ablation rates 96-97, 108, 203, 249, 263, 291
ablation season 110, 291
ablation zone (area) 101, 180, 266, 270, 275, 276, 277, 340, 367
Abramova Glacier, C.I.S. 188
abrasion 241, 242-249, 261, 270, 285, 286, 341
  spatial variation 248, 249, 256
  on stoss surfaces 247-250, 256
abrasion models 246-250
abrasion rates 244, 248, 249, 250, 256, 342, 484
abrasion of a rough bed 247-250
abrasion of a subglacial debris layer 250-251
abrasive wear 243-244, 286
accretionary wedge 328, 329-332
  internal sediments 330
accumulation 85-86, 107, 108, 117, 146, 148, 149, 167, 180, 203, 215, 262, 365, 451
accumulation area ratios (AAR) 291
accumulation rates 84, 96-97, 108, 192, 484
accumulation zone (area) 101, 108, 265, 266, 270, 276, 277, 279, 314, 339
"activity" (ice mass) *see ice mass flux*
"activity index" 113
adfreezing 265, 269, 270, 271, 276, 277
Adhemar, J. 10
Adishy Glacier area, Caucasus Mts., Russia 488, 502, 504
advance sequence of glaciotectonic overprinting (style/model) 335
aeolian debris 265, 269, 277, 401, 476, 480, 488, 500, 501, 502
aeolian reworking 410, 476
aerosols (*see atmospheric gases*) 46, 50, 72, 73, 74, 277
Afghanistan 17
Africa 13, 15, 18, 59, 108, 488, 500, 502, 503

Afroalpine zone 500
Aftonian interglacial 25
Agassiz, L. 9
air bubbles (in glacier ice) 103, 123, 124
Alaska 15, 38, 52, 62, 103, 108, 120, 127, 158, 164, 165, 172, 173, 174, 175, 180, 181, 182, 183, 184, 185, 187, 188, 190, 191, 192, 199, 208, 225, 235, 237, 239, 266, 268, 271, 272, 274, 277, 282, 284, 290, 291, 300, 305, 341, 343, 345, 346, 354, 355, 356, 359, 360, 362, 375, 379, 383, 391, 392, 396, 397, 399, 400, 407, 408, 422, 428, 446, 449, 455, 456, 457, 458, 459, 465, 468, 473, 474, 477, 480, 481, 483
albedo 65, 67, 68, 107
albite 501
Aletschgletsher, Switzerland 367
algal mats 477
Alleröd 36, 48, 49, 50, 62, 112-113, 368
Alberta 140, 165, 202, 325, 329, 369, 371, 396, 397, 425
Alley's 'thin till approximation' 161
Alley's till channels 228, 229
Alpine glacial classification 23-24
Alpine tills 505
Altithermal *see Hypsithermal*
Amazon basin 18
Amazon River, Brazil 371
amber ice 120, 121, 272
America *see United States of America*
Amery Ice Shelf 167, 276, 280
Amery Trough, Prydz Bay 481, 483
anabatic winds 488, 504, 505
anisotropy in sediment rheology 163, 218
anoxia 478, 480
anchor ice 476, 478
Andes 20, 38, 57, 62, 66, 291, 370, 446
Andvord Bay, Antarctica 481
Anglian glaciation 25, 306-308
Antarctica 15, 18, 19, 48, 49, 50, 51, 58, 59-64, 86, 93, 145, 272, 275, 277, 280, 452, 464, 467, 446, 448, 450, 470, 471, 475, 476, 477, 479, 480, 481, 487, 488, 489-492
  circum-Antarctica 469

east Antarctica 149, 447, 467, 468, 483, 484
  Ross Sea 447, 467, 468, 483
  sub-Antarctic Islands 59, 446
  west Antarctica 450, 468, 483
Antarctic Ice Sheet 20, 44, 45, 47, 48, 49, 50, 79, 109,
111, 121, , 124, 133, 147, 165, 167, 171, 172, 179,
215, 270, 282, 289, 451
  East Antarctic Ice Sheet 20, 49, 50, 112, 113, 149,
176, 177, 193, 233, 282, 231, 449, 471, 486, 489
  West Antarctic Ice Sheet 20, 43, 79, 81, 112, 120,
122, 125, 143, 154, 158, 160, 161, 162, 168, 169, 170,
174, 175, 176, 177, 178, 179, 231, 317, 484, 485, 486
Antarctic Convergence (map) 450
Antarctic Peninsula 446, 447, 451, 467, 468, 470, 476,
478, 481
anthropogenic emissions 50, 73
anti-greenhouse effect 17, 65
Ape Lake, British Columbia 425, 426
aphelion 67
Arapaho Glacier, Colorado 283
arboreal pollen 38
Archean 13
  Early Archean 11
Arctic Ocean 446, 476, 481
Argentina 58
artesian conditions see quasi-artesian conditions
Asia 36, 37, 66, 80, 280
asperities on glacier bed 143, 154
Astronomical theory of glaciation 10, 11, 13, 16, 28,
31-35, 36, 48, 63, 65, 66-69, 70, 74, 486
asymmetric isoclinal folds 320
Athabasca Glacier, Alberta 140, 165, 202
Atmospheric gases
  Carbon dioxide 12, 13, 14, 15, 16, 17, 19, 48, 49, 50,
63, 65, 66, 68, 69, 70, 71, 73, 75
  Methane 46, 48, 49, 50, 65, 75
atmospheric dust 49, 63, 65, 121
atmospheric transparency (transmissivity) 71, 109
Australia 15, 18, 59
Austria 36, 108, 109, 200, 322, 329, 368, 500, 501
Austrian Alps 36, 108, 109, 200, 217, 322, 329, 368,
500, 501
authigenic minerals 426, 480
autochthonous sediments 322
autonomous basal cavities (confined) 155
avalanche snow 108
avalanches 174, 201, 265, 269, 273, 283, 339
  ice cliffs 201, 273, 368
  rock 201, 266, 267, 274, 275, 287, 339
Axel Heiberg Island, Canada 174, 325

Ayutor Glaciers, Tien Shan Mts. 282, 283, 290

Badlands (Lake Missoula), U.S.A. 239
backpressure (in ice shelves) 167, 170, 171, 224
backwater zones (see slackwater) 239, 379
bacterial metabolism 480
Baffin Island 90, 121, 128, 131, 268, 282, 283, 284,
361, 368, 370, 372, 379, 392, 393, 396, 407, 468, 479,
481, 483
balance gradient
  net 154
ball & pillow structures 429, 432
Baltbare Glacier, Karakoram 188
Baltic Shield see Fennoscandian Shield
Banham's model of glaciotectonic deformation 296,
298, 308
Barbados 34, 35
Barnes Ice Cap, Baffin Island 121, 128, 282, 282
bar & channel facies models for braided rivers and
glacial outwash 407-410
barotropic flow 458, 465, 478
bars
  braid 380, 387, 399, 408
  boulder 382, 383, 410
  downstream migration 401, 404, 408, 409, 413
  expansion 383
  formation 374, 401
  lateral 408
  linguoid 403, 408, 409
  longitudinal 403, 404, 405, 408, 409, 410
  models 407-408
  pendant 382
  point 403, 408
  residual 382, 413
  spillway expansion 382
  spool 403
  tails 398
  types 389
basal cavity formation see cavities
basal creep instability 113, 119
basal crevasse dykes 194, 195
basal debris 120-121, 123, 134, 135, 143, 155, 177,
195, 246, 261, 268, 277, 284, 339, 340, 342, 348, 459
  debris discharge 269, 277-284, 285 , 347, 350
basal debris transport 179, 194, 270, 271-273, 274,
277, 279, 283, 286, 289, 339, 342, 350
basal hydraulic regime 155, 164
basal ice/bed heat flux 151, 219

basal ice/bed separation
  cavitation 144
  no cavitation 144
basal ice (debris foliated) 302
basal ice friction 113, 117, 175, 197, 219, 220
basal ice thermal zonation 82-90, 118, 215, 276
basal ice velocity 118, 153, 157, 158, 175-176, 194, 205, 217, 225, 226, 228, 231, 247
basal meltwater 90, 93, 117-124, 194, 312
  film 174, 197, 218, 219-220, 227, 228, 353
  pressure 143, 157, 158, 182, 183, 190, 192, 216, 217, 222, 224, 225, 226, 247, 251, 253, 256, 265
basal plane of ice crystal 103
basal sliding (slip) 81, 93, 94, 147, 151-165, 176, 177, 219
  sliding with no separation (cavities) 151, 152-155, 156
  sliding with separation (no cavities) 151, 155-158
  and subglacial sediment deformation 160-165
  velocity 118, 153, 157, 158, 159, 165, 171-176, 177, 194, 205, 217, 225, 226, 228, 231, 247
basal stress
  local stress concentrations 154
  normal pressure distribution over bed 144
  shear stress 92-97, 110, 142, 152, 153, 154, 161, 165, 217, 331
  stress conditions 142-144, 162, 193
basal shear stress and surging 189
basal uncoupling 89-90
basal water layer 94, 143, 152, 157, 193
basal water pressure 143, 157, 158, 182, 183, 190, 192, 193, 194, 216, 217, 222, 224, 225, 226, 247, 251, 253, 256, 265
  pressure constriction 191, 226
basal thermal conditions *see subglacial thermal conditions*
Bas Arolla, Switzerland 357
basin(s) (glaciated) 80, 87, 88
  overdeepened 209, 256, 315
beaches 476
Beaufort Sea, Arctic Canada 112
bed *see subglacial*
bed armour 373, 374
bed interface models 156
bed roughness 152, 193, 225, 254
bedload 373, 378
bedrock fracture *see rock mechanics*
belemnite 30
benthic fauna 468
  faunal diversity 468

benthonic formanifera 28, 30
berg dumps 427, 433
berging *see calving*
bergschründ 251, 256, 340
bergstone 456, 473
Bering Glacier, Alaska 185, 422
Bering Strait 21
beryllium concentrations (^{10}Be) 45, 47, 48, 49, 50, 71
Bezymyanny Glacier, Tien Shan Mts., C.I.S. 188
Bijou Creek type facies model 410, 411, 413
bimodal grain sizes 285, 305, 398, 403
Bingham model 312
Bingham viscoplastic material 159, 160, 164
bioclastic detritus (carbonate) 451, 466, 467, 473
biogenic processes (in lakes) 421, 434
biogenic sediment 265, 454, 466, 468, 470, 472, 473, 475, 476
biological limnology 426
biosiliceous muds & oozes 467, 470, 473, 479, 480, 481
bioturbated sediments 421, 440
bioturbation 36, 440, 480
Black Rapids Glacier, Alaska 185, 188
Blakeney Esker, England 305
blown snow 108
Bluck's model of braided stream alluvial deposition 397, 407, 409
Bluck's model of lateral bar sedimentology 408
Blue Glacier, Washington State 158, 162, 173, 175
Bogotá, Columbia 39, 41
Bölling 50, 62
Bolivia 58
Bondhusbreen, Norway 274, 282, 283
Bossons Glacier, French Alps 376, 377, 383, 390, 391, 394, 395, 396, 397, 399, 400
boudinage 319
boudins 316, 320, 322, 334
boulder barricade 477
boulder bars 382, 410
boulder berms 383
boulder lags (*pavements*) 382, 383, 477
boulder lines 476
boulder rings 414
boulder sheets (*cobble*) 392, 395, 410
boulder tables 290
Boulton's Model of Abrasion 194, 246
boundary conditions 78-99
Bow Lake, Alberta 425
Bow River, Alberta 396
"bow wave" structure in diamicton 297

Brahmaputra River, India 371
braid bar division (development) 380, 386
braiding
  abandoned channels 401
  avulse processes 387
  bar-channel systems 389, 407-408
  bar-top surfaces 387, 399, 403, 408
  Bluck's facies models 397, 407, 408
  "braiding index" 391
  "braiding parameter" 389, 391
  Cant's facies model 410
  channel sinuosity 391, 408
  chute channel development 388
  development 386-387, 390
  "primary" anastomosis 386
  "secondary" anastomosis 387, 388
  splays 387
  "switching" 386
Brazil 15, 18, 371
Breiðamerkurjökull, Iceland 122, 158, 161, 162, 164, 256, 282, 284, 286, 298, 312, 313, 316, 399
Brenva Glacier, Italy 291
British Columbia 90, 105, 108, 215, 425, 426, 446
British Isles 24-25, 58
brittle deformation in rock 243, 286
brittle failure in sediment 194, 310-311, 330
brittle fracture of rock *see rock mechanics*
Brúarjökull, Iceland 185, 188
Brückner, E. 23
Brunhes Paleomagnetic chron 24, 29, 34, 42, 68, 502
Buckland, W. 10
buckle folds 320
"bulldozing" of sediment 309, 330
Bull Lake Glaciation (Illinoian) 496, 497, 498
"bullet boulders" 251
buoyancy 92
Burroughs Glacier 268
buried channels *see subglacial meltwater channels*
Burokurmas Glacier, C.I.S. 188
Byrd Glacier, Antarctica 125, 167
Byrd Ice Core (Byrd Station) 43, 44, 45, 47, 59, 114, 120, 122, 124, 282
Byrs Glacier, C.I.S. 188

C-foliations 319
C-channels *see Clarke channels*
c^{14} *see radiocarbon dating*

Cailleux-Tricart roundness index (R_l) 399, 400
Calabria, Italy 22
calcium carbonate coatings on quartz 498
calcium carbonate dissolution in sea water 28, 29
calcium hexahydrite (ikaite) 480
calcareous foraminifera 28, 30
calcite *see vein calcite*
California 38, 108
calving 96, 170, 176, 276, 309, 433, 454, 456, 457, 461, 462, 475, 476
  instability 485
  pressure wave 459
  Reeh-type 451
  styles & processes 451, 458-459
  submarine calving 459
calving bay 475
calving fronts 84
calving line 451, 458
camera observation
  subglacial 264
  submarine 471
Camp Century Core 43, 44, 45, 51, 115, 121, 122, 123
Camp Milicent ice core 46
Canada 13, 25, 44, 53, 623, 66, 80, 86, 87, 89, 90, 96, 98, 103, 105, 108, 111, 112, 118, 121, 126, 128, 129, 130, 131, 140, 154, 158, 159, 164, 165, 174, 179, 180, 184, 185, 186, 192, 202, 215, 230, 235, 270, 282, 283, 284, 289, 290, 306, 325, 329, 355, 361, 368, 369, 371, 372, 379, 383, 388, 391, 395, 396, 397, 407, 425, 426, 443, 446, 464, 468, 477, 479, 481, 483, 487, 488, 491, 493, 494-497
Canadian Shield 80, 87, 312
Cant's facies model of a sandy braided river 410
Canterbury Plains, New Zealand 375, 378
Canwell Glacier, Alaska 127
Cape Adare, Antarctica 489
Cape Herschel Plateau, Ellesmere Island 492
Carbon dioxide *see Atmospheric gases*
carbon 14 dating *see radiocarbon dating*
carbonates 426
Caribbean Sea 28, 34
Carroll Glacier, Alaska 188
Cascade Lake, Washington State 425
Cascade Mountains, Washington State 51
Casement Glacier, Alaska 120, 173
cataclastic shear zone 312
catchment denudation rate 280, 370
catchment sediment yield 370, 371, 372
Caucasus Mts. 282, 290, 488, 502, 504

Causes of Glaciation 65-73
  atmospheric changes 65, 71
  land/ocean distribution 65, 71
  mega-surges 179
  plate tectonics 12, 15, 65
  solar effects (albedo, radiation, shielding, sunspots)
12, 14, 18, 63, 65, 67, 69, 70
  supercontinentality 12
  explosive vulcanism 64, 66, 71-72
cavitation 241, 283
  marks 382
cavity 152, 155-158, 182, 197, 209, 217, 252, 344,
353
  autonomous basal (confined) 155
  beneath clasts 246
  at grounding line 450, 451, 466
  englacial 202
  in lee of rock step 157, 252, 253, 254, 265
  on sinusoidal bed 144, 255
  orifice types (see linked-cavity systems) 191, 192
  rate of closure 158, 216
  sliding with cavities (separation) see basal sliding
  sliding with no cavities (separation) see basal sliding
  water-filled 151, 157, 254, 255
Cenozoic see Late Cenozoic
central Asia see Asia
chalk banding in till 320, 322
chaotic bedding 431
Charpentier, J. de 9
Chelmsford, England 296
chemical limnology 426
chemical stratification in glacial lakes 426
chemical weathering see corrasion
Chibougamau Formation, Canada 13
Childs Glacier, Alaska 188
China 37, 38, 39, 122, 158, 162, 165, 177, 217, 313,
370
Chile 58
chlorine/beryllium ratios ($^{36}Cl/^{10}Be$) 48
chronosequence 23-28
chute channels 380, 387, 388, 392, 395
C.I.S. (Commonwealth of Independent States) 24, 38,
42, 58, 172, 180, 182, 184, 188 215, 282, 283, 290,
329
cirque 270, 488, 502, 505
  erosion 256
  glaciers 280
Clarke-channels (C-Channels) 219, 221, 227
classic subdivisions of glacial periods 22-28
clast-bed contact 242, 247, 270, 285

clast concentration 247, 248, 249, 288
  clast streamlining close to ice/bed 249, 261, 272,
299
  effective contact force 161, 244-247, 248, 253, 285
clast drag effects 245, 249
clast fabric 137, 164, 298, 399-401
  bar-top 400, 401
  dip 164, 300
clast imbrication 305, 372, 392, 395, 403, 404
clast "ploughing" 160, 162, 163, 164, 194, 299, 348
  Ploughing Index 163
clast velocity in abrasive wear 244, 245, 246, 247
clay coatings 488
clay domains 163
clay minerals 319
  neoformation 488
  oriented 321
CLIMAP (Climate: Long-range Investigation,
Mapping and Prediction) 49, 62, 78
Climate in relation to Glaciomarine conditions
  boreal 446, 468
  polar-ice cap 445, 446, 466, 478, 479, 480, 481
  polar-tundra 445, 446, 468, 473, 480,
  temperate oceanic 445, 446, 466, 468, 473, 479, 481
Climatic Optimum 51, 64, 74
climatic warming 70, 165, 484
climbing ripples 421, 429, 432, 437, 439, 441
clotted ice facies (or dispersed) 116, 120
coccolith 30
cold-based ice conditions (dry-based) 121, 151, 153,
200, 268, 269, 271, 273, 284, 307
cold ice 85, 135
"cold patches" 117, 152, 264, 344
"cold spikes" 137
cold tongues (sea water) 452, 453
"cold wave" 114
collapse dolines (thermokarst) 279
collapse structures (post depositional) (glacial lakes)
429
Colorado 283, 284
Columbia 38, 41
Columbia Glacier, Alaska 158, 164, 174, 175, 475
Columbia Plateau, Washington State 382
comma forms see sichelwannen
comminution see abrasion
compressional ridges 427
compressive flow (ice) 146, -147, 215, 270, 273
concentric (box) folds 325, 327, 328
  Jura style 326, 333
conchoidal fractures 489, 490, 491, 505

conduit (subglacial) tortuosity 193
Congo River 371
conjugate faults 320
conjugate shears 321
continental rifting 13, 14, 15, 21, 22
"Contorted Drifts" 307
contorted surface medial moraines 180
convection hypothesis (cells) (in ice) 136, 137
convergent ice flow 82
copepods 468
Copper river, Alaska 477
coquina (shells) 467
coral growth curves 36
coral terraces 31, 35
Cordilleran Ice Sheet 52, 53, 62, 98
Cordilleran Mountains 446
Coriolis Forces 427, 466
corrasion 340, 489
correlations of Late Cenozoic glaciation - north/south
hemispheres 52-55
cosmic dust 265
Coulomb failure criterion 251, 294, 311, 331
coupling line 464
crack
   density 286
   faces 252
   microcracks 286
   pre-existing 252
   tip growth 207, 209, 252, 254, 259
   water pressure 252, 259
crack growth in rock 207, 209, 243, 244, 251, 252,
254
creep instability and surging 189
crenulations (compressive) 320
crescentic furrows (see sichelwannen) 223, 241
   fractures 243, 245, 249
   gouges 244, 505
Cretaceous 19
Crete Ice Core 52, 72
crevasses 116-117, 124134, 147, 180, 197, 201, 204,
217, 273, 345, 453
crevasse types 132, 170, 194
   basal 194
   geometry 206
   plunge pools 206
   steps 206
critical hydraulic gradient 229
critical obstacle size 94, 141, 154, 163, 220
critical particle size see terminal grade concept
critical volume concentration (c.v.c.) 160, 231

Croll, J. 10, 28
Cromer interglacial 25
Cromer Tills 307
cross beds 392, 404, 408, 413, 421, 433
cross-stratification 402, 410, 433
crushing & shattering of particles 341, 487-506
   crushing intensity 489, 493, 495, 496, 498
   crushing patterns 487
   crushing stages 491, 494, 495, 500
   grain facets 489, 505
   etching features 489, 492, 494, 495, 496, 499, 500,
501, 502
      meltwater action (v-shapes) 492, 493, 501, 505
      step features 489, 490, 496
      weathering intensity 489, 495, 498
crushing experiments on rock 286, 287
cryoconites 290
cryostatic pressures 294
crystalline slip (dislocation climb) 103, 140
currents (in lakes)
   density 427, 431, 438
   quasi-continuous 421, 427, 431, 440
   surge 427, 434, 440
   turbidity 427, 438
   wind driven 421
'cusp catastrophe' 178
cyclopels 458, 461
cyclopsams 458, 461
cylindrical folds 320, 325
Czech Republic 36

Dammer Berge push moraine 319, 324, 325, 326, 327
Danco Coast, Antarctica 467
Dansgaard-Oeschger cycles 49, 50
Darcy's Law 228-229
Darcy-Weisbach Law 209
dating ice cores 47
Davis Station, Antarctica 489
"dead ice" 135
Dead Sea, Israel/Jordan 38, 42
debris (mud) "pockets" or "clots" 121, 123
debris
   entrainment 179, 248, 261, 264, 265, 267, 270, 338,
342, 350
   moraine septa 266, 270, 273, 276, 339
   sources 264-270, 342
   stacking 179, 195
debris aprons 433

debris blebs 421
debris discharge *see basal debris discharge*
debris flows 265, 269, 295, 303, 304, 316, 328, 409, 421, 422, 429
debris grainfall 454, 461, 462, 481
debris/ice concentrations *see ice/debris concentrations*
debris meltout *see meltout*
debris-release mechanism 147, 295, 350
debris-rich ice bands 145, 195, 271, 302, 339, 342
Decade River, Baffin Island 361
décollement
  ice 133
  plane of (in sediment) 298, 313, 328, 329
decoupling (basal ice/bed interface) 143, 145, 178, 179, 228
deep sea cores 28
deep sea records 35, 73
Deep Sea Drilling Project (*DSDP*) 29, 44, 60, 61, 73
deformable bed/hydraulic instability and surging 189, 192, 193
deformable beds & hydraulic systems 226-231
deformation
  deformable bed conditions 81, 82, 87, 94, 101, 122, 135, 143, 147, 158, 163, 164, 165, 175, 177, 178-179, 216, 218, 231, 242, 250, 272, 283, 287, 295, 299, 338, 340, 349, 448, 451, 462
  advective flow of meltwater 227
  ductile 319, 328
  hysteresis effects 160
  in soft sediments 160, 163, 310-313, 329, 330, 349
  passive deformations 332-333
  rate of 312
  subglacial till layer 158, 159, 161, 164, 215, 232, 317, 340, 349, 464
  thickness of debris layer 232
  'thin till approximation' (Alley's) 161
  thixotropic 410
  till deltas 451, 459-460, 462, 464
  viscosity of sediment 160, 163, 226, 310-313, 329, 330, 349
  wedge in front of a glacier 330
deformation and glacier movement 158-165
 'ploughing' 158, 160, 162, 163
deformation structures 298, 313, 320
deltas & estuaries 477 (*see glaciolacustrine*)
De Geer, G. 444
Denmark 36, 320, 322, 323
density mixing (homopycnal) 427, 438
Des Moines Ice Lobe, Laurentide Ice Sheet 90
detached folds in sediment 316

deuterium/hydrogen ratio 30, 46, 47
Devensian glaciation 25, 77, 402, 403, 404
  tills 306-308
Devensian Ice Sheet 193
Devils Hole, Nevada 39, 44, 49, 68, 69, 74
Devon Island Ice Cap, Canada 44, 108, 121, 154, 289
Devonian 12, 16, 18
dewatering structures 429, 432
  ball & pillow 429, 432
  flame 432, 434
  pipes & dish 429
diagenesis of snow *see snow transformation*
diagenetic histories of quartz grains 487-506
diamicton 87, 117, 279, 293-308, 488, 505 (*see till*)
  dilated 163
  glaciolacustrine 421, 433
  glaciomarine 448, 451, 463, 471
  laminations 307
  matrix 307
  mélange 451
  "pods" 307
  quartz grains 487-506
  rims (*kettles*) 380
  shelfstone 469, 471
  shell fragments 308
diamictite *see tillite*
diatom microfossils 436, 444, 454, 467
dilatant behaviour in sediments 159, 161, 162, 250
dilation in shear zone 318
diluvium 10
dirt cones 290
discharge
  (ice) flux 92
  meltwater 165, 180, 198, 200, 204, 213, 217, 228, 231, 252, 353-362, 365, 393, 394
dislocation climb *see crystalline slip*
distal front delta sediments 429
distal lakes 419, 436-442
  model 437
  proximal sub-environments 436
    deltas 436-440
    lake margin 440
  lake bottom sub-environments 441-442
distributary channels (deltas) 429, 440
divergent ice flow 82
Djankat, Caucasus Mts. 282, 290
dome *see ice dome*
Dome C (Circe) Ice Core 43, 44, 45, 58, 59, 59, 60, 61
Donau glaciation 24

Don River, C.I.S. 58
Donjek River, Yukon Terr. 395, 396
Donjek type facies model 409, 411, 413
drag coefficient 245
Drake Passage 21, 66
Drance River, France 356
draped laminations 421, 429, 432, 438, 440
  model 440
drapes 401
dryas octapetala 36
drift deposits 10
drift prospecting *see mineral exploration*
Drift Theory 10
dropstones 421, 427, 433, 469
'drowned bed' (meltwater) 143, 151, 219
drumlin(s) 80, 86, 134, 312, 315, 333, 335, 375
dry bed conditions *see polar ice*
dry snow zone 116
Dry Valleys, Antarctica 477
ductile failure in sediments 310-311, 320, 328
dunes (fluvial) 401, 402, 406, 409
Dwyka Formation 18
Dye 3 Core 43, 44, 45, 49, 51, 288
dye tracers 203, 206, 225
dyke-sill ice flow pattern *see convection hypothesis*
dynamic balance line (DBL) 182, 184
dynamics of ice flow *see ice flow dynamics*
Dyngjujökull, Iceland 188

earthquakes 266
East Africa 488, 500, 502, 503
East Anglia 25, 306-308, 322
East Fork Glacier, Alaska 173
Eem interglacial (Eemian) (*n.b. Ipswichian*) 25, 35, 49, 74, 488, 492, 494, 496
effective stress 143, 159, 161, 177, 226, 310-313
  fluctuations 143
Eistektonik *see glaciotectonics*
Ekalugad River, Canada 372
Ekalugad River sandur 379
Ekström Ice Shelf, Antarctica 147
ELA (equilibrium line altitude) 106-108, 110, 146, 147, 453, 484
Ellesmere Island, Canada 165, 487, 488, 492, 493
Ellesmere Island ice cores 289
Ellis Fjord, Antarctica 467
Elster glaciation 24, 58

Emiliani, C. 28
end moraine 24
Enderby Land, E. Antarctica 112
energy-dispersive spectrometry (EDS) 488
energy flux 110
"energy of glacierization" 113
Engabreen, Norway 371
englacial
  debris 121, 143, 205, 213, 266, 269, 270, 271, 277, 282, 285, 293, 338, 339, 340, 350, 369, 370, 448
  debris units (ellipsoidal, lensate) 270
  debris meltout 151, 201, 300, 315, 454
  debris stratification 300
  debris pathways 143, 205, 275, 338
  debris septa 273, 276, 279, 339
  hydraulic systems 200, 201-209, 213, 215, 217
  stratified 123, 287
  water flow 194, 256
englacial conduit 202, 203, 204, 205, 206, 207, 209, 210, 211, 274-275, 287, 305, 339, 348, 350, 358, 368, 428, 455, 457
  hydraulic pressure 205, 206, 207, 209
  meandering 206
englacial lakes 418
englacial temperatures 114
englacial watertable (*see potentiometric surface*) 203, 206, 213, 432
enhanced plastic flow (of ice) 152, 153
England 296, 298, 305
  eastern England (tills) 306-308
Eocene 489
epifauna 468
epilimnion 420, 421, 426, 434, 437, 440
epitaxial ice growth 344
equilibrium line 96-97, 106, 107-108, 145, 146, 147, 149, 173, 262, 266, 270, 274, 314, 448
equipotential planes of water pressure
  (*see potentiometric surface*) 204, 206, 207, 223
Erdalsbreen, Norway 356, 361, 372
Erebus Glacier Tongue, Antarctica 167
Erie Ice Lobe, Laurentide Ice Sheet 464
erratic trains 277
erosion, rates of 255-256, 342, 350
esker(s) 80, 86, 215, 222, 375, 381
  beaded 460, 463
  subaqueous esker deltas 460
  transverse 460
Esmarck, J. 10
etching patterns 487, 489, 492, 494, 495, 496, 499, 500, 501, 502

Europe 17, 24-25, 36, 38, 50, 57, 63, 80, 112-113, 291, 313, 315, 317
European Pollen Zones 36, 37, 38, 48, 50, 63
eustatic changes in sea and ocean level 12, 20, 34, 36, 59, 75, 78, 80, 165, 168, 171, 175
eye shapes in asymmetrical folds 320, 322
Eyjabakajökull, Iceland 188
extending flow (ice) 133, 146-147, 204, 215, 273
extra-glacier factors 106-108, 111
extraterrestrial sediment 269, 283

Facies coding system classification
  Miall's 411, 412, , 413
  Rust's 416
facies units (litho) in glacier-fed lakes 420, 430
Facies Models
  Bijou Creek type (Miall) 410, 411, 413
  Donjek type (Miall) 409, 411, 413
  GI Lithotype (Rust) 409
  GII Lithotype (Rust) 409
  GIII Lithotype (Rust) 409
  grounding line (submarine) 461
  Maizels' jökulhlaup sandur 414, 415
  Platte type (Miall) 410, 411, 413
  SI Lithotype (Rust) 410
  SII Lithotype (Rust) 410
  Scott type (Miall) 409, 411, 413
  South Saskatchewan (Miall) 410, 411, 413
  Trollheim type (Miall) 409, 411, 413
Fairbanks, Alaska 355
fast ice 472
Fe$_2$O$_3$ coatings 502, 503
Fedchenko Glacier, C.I.S. 283
feedback response of ice mass 108
Fennoscandian Ice Sheet 165, 270, 375
  Scandinavian ice 307
Fennoscandian Shield 87, 112, 312
Filchner Ice Shelf, Antarctica 167
fiords see fjords
Findlngletscher, Switzerland 352
Finland 135
firn 103, 114
Fjallsjökull, Iceland 297, 303, 304
fjords 80, 90, 313, 452, 453, 464, 465, 467, 468, 470, 476, 477, 478, 481
  basins 466, 475, 480
  facies relationships 467

sedimentation rates 484
valley 179
water stratification 478
film lubrication layer see basal water layer
finite-element method 252, 253
firn line 273
first-order glacial landscapes see glacial landscape types
Fisher Glacier, E. Antarctica 193
fission track-tephrochronology 36, 52
flame structures 432
flickers (dust fluctuations) 49
floating glacier termini 174, 176, 448-451, 469, 471
  confined floating glacier-tongues 95, 111, 174, 263, 448, 449, 451, 471
  confined ice shelf see ice shelf
  debris transport & release 448-451
  fringing ice shelf see ice shelf
  large embayment ice shelf see ice shelf
  tidal influence 176, 451
  unconfined floating glacier-tongues 174, 448, 451
floods (see jökulhlaups) 352, 354, 356, 357, 359, 365, 372, 380, 382, 393, 398, 401, 403, 410, 413, 436
  hydrograph 355
Florida 17
flow till see till
fluidization 301
flume experiments 439-440
flute(s) 193, 194, 382
  fluted moraine(s) 261
  fluted terrain 80
fluvial inwash debris 274
folds (in ice) see ice tectonics
foliation in basal ice 302
Forbes bands 136
fore-deepened channels (troughs) 82, 89
foraminifera 28, 30, 36, 467, 468
formation of glacier ice 103-105
Forest Beds, East Anglia 25
fossil fuels 75
Fowler's model of surge activation 193
fracture features on quartz grains
  conchoidal 489, 490, 491, 499, 502, 505
  fracture lines 489, 492, 502, 504, 505
  intensity 494
  subparallel linear 489, 492, 494, 495, 505
  troughs 489
fracture geometry 285
fractures (in ice) see ice tectonics
fragile ice (crystals, spicules) 119-120

France 120, 122, 173, 209, 213, 282, 284, 352, 356, 376, 377, 384, 390, 391, 395, 396, 397, 399, 400
frazil ice 344, 476
  epitaxial ice growth 344
frictional heat 84
frictional wear *see wear processes*
freeze-thaw processes 251, 339
freezing front 294
freezing ice zones 78-99, 152, 344
"freezing-on" hypothesis 120, 121, 122, 205, 342, 344
French Alps 120, 122, 173, 209, 215, 322, 329, 352, 356, 368, 376, 377, 384, 390, 391, 395, 396, 397, 399, 400
friction melting (basal) and surging 189, 219
frontal instability (ice) and surging 189
frost cracking (*shattered debris*) 251, 399
Froude number 392, 393, 458
frozen bed conditions *see subglacial thermal conditions*
frozen sediments 192, 197, 218, 227, 313, 380
"full-pipe flow" 305, 306
  sliding bed concept 306

Ganges River, India 371
Garibaldi Lake, British Columbia 425
gastropods 37
Gauckler-Manning-Stickler equation 209
Gauss-Matuyama paleomagnetic boundary 58
Geilston, Scotland 297
Generator Lake, Baffin Island 128
geochemical processes (subglacial) 154
George V Land, Antarctica 470
George VI Ice Shelf, Antarctica 448
Georgian Bay, Ontario 230
geothermal heat (flux) 84, 110, 113, 117, 143, 175, 197, 217, 219, 246, 247, 299, 368
Gerlache Strait, Antarctica 467, 470, 481
German Alpine Foreland 23, 58
Germany 23-25, 313, 317, 319, 320, 324, 325, 326, 327, 335
gibbsite 502
Gigjukvisl river, Iceland 383, 396
"Gilbert" delta 428, 437
Gipping Valley, Suffolk, England 295, 296, 307
glacial delta 313, 329
glacial deposition 80-99, 293-308
glacial erosion 80-99, 241-259
glacial hydrology 101

glacial landscape types 80, 81
  first-order glacial terrains 86-90
glacial mill *see moulin*
glacial outwash 201, 277, 324, 329, 436
  composite fans 375
  distal bar-heads 397, 406, 407
  distal fan-delta 375
  distal plain-delta 375
  facies 401-405
  grain size 398
  grain size v. gradient 397
  longitudinal profiles 379
  plains 24, 275, 374, 375
  Pleistocene sandur 375
  proximal bar-tails 398, 406, 407
  streams 279
  valley-sandur 386
  "valley-train" deposits 375, 377, 390, 394
  zone of accumulation 375
glacial overdeepening 209, 256, 315
glacial overprinting 315, 333, 335, 487, 490
  crushing patterns 487
  etching patterns 487, 489, 492, 494, 495, 496, 499, 500, 501, 503
  step features 489, 490, 496
  weathering features 487, 490, 495, 502
glacial physics *see glaciology*
glacial polish 249
glacial thrust terrains 315
glaciated basin(s) *see basin*
glacial hydrology 197-239
  meltwater sources 197, 202, 225, 227
  englacial hydraulic systems 200, 201-209, 213, 215, 217
  supraglacial hydraulic systems 198-201, 217
glacial isostasy (glacio-isostasy) *see isostatic readjustment*
Glaciated Terrain-Ice Sheet Model *see GT-IS Model*
glaciated trough(s) *see trough*
Glacier Bay, Alaska 108, 449, 455, 459, 483
"glacier burst" *see jökulhlaups*
glacier sensitivity 105, 106
glacier (ice sheet) surges *see surges*
glacier tectonics *see glaciotectonics*
Glacier d'Argentiére, French Alps 120, 122, 173, 282, 284
Glacier de Tsidjiore Nouve, Swiss Alps 277, 290, 357, 370, 393
glaciofluvial
  deposition 293, 305-306, 328

sediments 307, 417-444, 500
glaciolacustrine environments 165, 417-444
  berg dumps 427, 433
  biological processes 426
  chemical processes 426
  currents 421, 431
  deltas 428-431, 436-440
  diamictons 421
  distal deltas 436
  distal lakes 419, 436-442
  englacial lakes 418
  factors affecting sedimentation 420
  floods 436
  grain flows 437
  ice-cliff margin 430, 433
  ice cover 427
  jökulhlaups 436
  ice-rafting 421, 433
  kettle lakes 418
  lake bottom 430, 434-436, 440-441
  morphosequences 442
  mud drapes 434
  physical limnology 420
  proximal ice-contact lakes 417-419, 422, 427-428,
430
  rip-up clasts 434
  river lakes 41, 419, 428
  sedimentation rates 426, 428
  sediment stratification 424-426, 427
  sediment transport & deposition 426-427, 434
  sediments 307, 313
  squeeze-up processes 433
  stratigraphy 442-444
  subaqueous fan 428, 431-432
  subglacial lakes 231-233, 418
  submerged ice ramp 428, 433-434
  supraglacial lakes 418
  surge currents 427, 431, 434, 440
  suspended particle concentrations 425, 426, 432
  thermal stratification 420-424, 437
  turbidity currents 427, 438
  water body turnover (mixing) *see water body*
turnover
glaciolacustrine facies (litho) units 421, 430
  bioturbated sediments 421, 440
  climbing ripples 421, 429, 432, 437, 439, 441
  coarse-fine couplets 434, 435, 441
  cross-bedding 421, 432, 433
  debris blebs 421
  diamictons 421

draped laminations 421, 429, 432, 438, 440
  model 440
dropstones 421, 427, 433
iron-rich laminations 426
load structures 421
parallel bedding 421, 434, 441
rhythmites 239, 421, 434, 435, 436, 441, 442, 444
tabular bedding 421, 428
varves 434
glaciolacustrine morphosequences 442
  stacked 442
glaciology 101-138
glaciomarine environments 165, 308, 445-486
  aeolian sources 480
  bioturbation 480
  deformed muds 479
  gravity flows 467, 475, 479
  grounding lines 90, 111, 135, 142, 165, 174, 224,
265, 276, 277, 448, 450, 451, 456, 464, 476
  coupling line 464
  ice rafting 469-473
  Intraformational folds 308
  marine-ending termini 448-459
  model 453, 456
  paraglacial environments 465-477
  sedimentation rates 447, 474, 480-484,
  sediments 13, 18, 277, 307, 470, 471
  sediment yields 474, 480
  stratified diamictons 479
  subglacial processes 447-448
  suspended sediments 452, 454, 474
  suspended particulate matter (SPM) 468
  water turnover 478
glaciotectonics
  definitions 310
  deformation 194, 195, 296, 298, 306, 309, 310, 330
  differential loading 195
  ice regimes 313-315
  processes of 309-335
  squeeze-up 143, 195, 433, 461
  structures 149, 195, 307, 313
  styles 317-322
glaciotectonite 310, 311, 322
Glen Roy, Scotland 239
Glen Spean, Scotland 239
Glen's Law 113, 124, 137, 139, 150, 178
global climatic modelling 67-72
global cooling 485
Gorges Glaciation (Kenya) 502
Görnera stream, Switzerland 354, 360, 361

Görnergletscher, Switzerland 204, 352, 354, 355, 360, 361, 362, 371
Görnersee, Switzerland 355
Gondwanaland 12, 16, 17, 20
Graenalon ice dammed lake, Iceland 369
grain fracture 270, 271, 487-506
grain-grain interaction 301, 340, 487-506
grain packing 103
grain size 163, 226, 273, 284, 285-287, 305, 342, 344, 345, 346, 347, 374, 395-398
Grand Moulin, Mer de Glace, French Alps 208
Grand Pacific Glacier, Alaska 455
Grande del Juncal Glacier, Argentina 188
Grande del Nevado Glacier, Argentina 188
Grande Dixence SA, France 352, 356
gravity flows 467, 475, 479
Great Basin, western U.S.A. 39
Great Blakenham, Suffolk, England 296, 306
Great Lakes (Canada/U.S.A.) 52, 62, 90, 103, 165, 174
Greece 38
greenhouse effect 12, 13, 20, 47, 49, 50, 75, 78
Greenland 15, 16, 21, 22, 59-64, 90, 268, 369, 375, 380, 386, 395, 406, 407, 446, 471, 484
Greenland Ice Core Project (GRIP) 39, 44, 46, 47, 48, 49, 50, 74
Greenland Ice Sheet 20, 39, 44, 45, 46, 48, 49, 50, 51, 52, 60, 70, 72, 74, 79, 84, 89, 93, 105, 109, 115, 121, 122, 123, 146, 154, 164, 172, 174, 175, 176, 179, 215, 217, 268, 282, 288, 289, 373, 392, 486
Greenland Ice Sheet Project 2 (GISP2) 44, 47, 48, 49, 50, 72
Greenland-Norwegian Sea 69
Greenland Shelf 481, 483
Grimsvötn, Iceland 230, 235, 236, 239, 368, 369
Grindelwald, Switzerland 173
Grinnell Glacier, Montana 120
Griquatown West basin, South Africa 13
grooves 80, 249, 382
grounding line 90, 111, 135, 165, 168, 174, 224, 265, 276, 277, 448, 451, 456, 469, 485
  lithofacies model 461
  types 459-465
    ice-contact deltas 459-460, 485
    fans (outwash) 459-460, 461, 477, 485
      types 466
    morainal banks 459-460, 461, 463, 464, 485
    wedges (till deltas) 459-460, 462, 464, 479
      model 462
      subaquatic till tongue 464

till tongues 464
GT-IS Model (Glaciated Terrain-Ice Sheet Model) 81-99
Guelph, Ontario 306
Gulf of Alaska 392, 456, 468
Gulf of Bothnia 87
Günz glaciation 24

H-beds 148, 155, 176, 178, 192, 193, 218, 226
  type A 155, 156
  type B 155, 156
Hagafellsjökull eystri, Iceland 188
Hagafellsjökull vestari, Iceland 188
Hallet's Model of Abrasion 194, 246, 256, 257
Hamilton-Niagara Regions, Ontario 284
Happisburgh, England 307
hard (rigid) beds (H-beds) 148, 155, 176, 178, 192, 193, 218, 226
Hatunraju Glacier, Peru 283
Hatherton Glacier, Antarctica 125
Hazard Glacier, Alaska 188
Hazard Lake, Yukon Terr. 235, 425
headwall 90
heat advection in subglacial/englacial meltwater streams 225
heat flux see basal ice/bed heat flux or geothermal heat
'heat pump' effect 152
heat transfer (conduction) 290
Hector Lake, Alberta 425
Heinrich 'events' 50, 51
helicoidal flow (in ice) (cork-screw) 135-136, 137
hemipelagic sediments 460, 469, 478
hemipelagic suspension 465, 481
"herring-bone" till fabrics 137
Hertzian contact 243
high-angle structures (glaciotectonic) 333
high level debris transport 269, 273-274, 283, 285, 286
Highland Ice Field, Baffin Island 130, 131
high resolution seismic surveys 160, 215, 417
Hilda Glacier, Alberta 369, 371
"hill-depression forms" 325
Himalayan Mountains 66, 213, 291, 407
Hintereisferner Glacier, Austria 108, 109
Hochstetter icefall, Tasman Glacier, New Zealand 275
Hoffellsandur, Iceland 379, 393

Holmströmbreen moraine, Svalbard 325, 327, 326, 328, 333, 335
  moraine model 326, 333
Holstein interglacial (*n.b. Hoxnian*) 25, 58, 74
Holocene 22, 35, 44, 47, 48, 49, 50, 51, 59, 62, 64-65, 70-72, 73, 289, 498
Holocene ice-proximal sedimentation rates in fjords 484
"homogenised" diamictons 296, 298, 316, 322
homopycnal flow (density mixing) 427, 438
Hooke's model of channel form 222
horizontal bedding *see planar bedding*
Horn Kees glacier, Austrian Alps 500, 501
Hornsund, west Greenland 484
Hudson Bay 13, 80, 87, 96
hummocky moraine 84, 195
  pitted zone of net basal melting 87
  pitted zone of net basal erosion 89
hummocky outwash (pitted) 375, 381, 410
Humphrey model of subglacial meltwater 228
Huon Peninsula 32
Huronian Supergroup 13
Hurwitz Group 13
Hitchcock, E. 10
Hoxnian interglacial 25
Hudson Bay 53
Hudson Strait 53
Hunza River, Pakistan 371
hydraulic erosion 89
hydraulic gradient 158, 225, 229, 305
hydraulic 'jump' 193
hydraulic roughness 221
hydraulic seal 231, 236
hydro-electric power (HEP) 215, 351, 352, 356
hydrographic processes 477-478, 481
  along-shelf currents 478
  continental shelves 477
  down-ward salt flux 478
  geostrophic flows 478
  off-shelf transport 478
  salinity gradients 478
  tidal currents 478
  upwelling currents 478
hydrology of glaciers *see glacial hydrology*
hydrolysis 252
hydrostatic pressure 93, 218, 222, 224, 225, 235, 237, 245, 252, 253, 254, 305
hyperconcentrated flow 374, 405, 406, 413, 414, 458
hypolimnion 420, 421, 426, 427
hypopycnal 426, 460

Hypsithermal 51, 64, 74
hysteresis effects of slurry rheology 160
hysteresis loop 357, 358, 361, 372

ice adhesion 144
ice/bed interface *see subglacial*
icebergs 170, 309, 427, 437, 458, 473, 477, 479
  distribution 469, 472
  flux 472
  gouges 479
  keel scours 479
  melting rates 471
  prods 479
  rafting 469-472
  rolling rates 471
  scour 427, 479
  tabular 170, 471, 479
iceberg calving *see calving*
iceberg rafted debris (IBRD) 459, 469-472, 473, 474
  production 473
ice-cliff margin 430, 433
  debris aprons 433
ice-contact lakes (proximal) 233, 235, 417-419, 421, 427-428, 430
  model 429
ice-contact delta 428-431, 456, 460
  bottomsets 429, 431
  collapse structures (post depositional) 429
  distal delta sediments 429
  distributary channels 429, 440
  foreset beds 429, 431, 437, 438, 440
  "Gilbert" delta 428
  ice-rafted debris 429, 433
  interbedded diamictons 429, 433
  topset beds 428, 431
  vertical aggradation 428
ice-core thin section 46
ice-cored debris islands & peninsulas 433
ice-cored landforms 291, 304, 359, 362
ice creep (glide) 84, 113, 117, 124, 263, 289
ice crystal(s) (shape, size) 103-105, 117, 134, 139, 140, 150, 270, 271
  node 206
ice dam (failure) 367-368
ice-dammed lakes 180, 225, 233, 235, 356, 368, 369, 436
ice/debris concentrations 121, 150, 195, 268, 271, 272,

279, 288, 289, 342, 345, 346, 347, 350
ice density 90, 103, 263
ice divide 82, 84, 98, 147
ice dome(s) 84, 96
ice dynamics 78-99
ice fabric 105, 117, 134, 135
ice facies types (zones) 116-117, 120, 271, 340, 345, 348
  banded 271, 340
  basal 271, 348
  diffused 271, 340
  dispersed subfacies 271, 272, 340, 346
  englacial 271, 348
  stratified subfacies 271, 272, 279, 340, 347
ice falls (seracs) 174, 201, 458
ice flow dynamics 139-196
  ice sheets 147-165
  ice shelves 166-171
  valley glaciers 165-166
ice flow models 48, 289
ice flowbands 82, 95-96
ice glaze 120
ice hardness coefficient 93
ice keel scouring (see keel marks) 171, 477, 479
  furrow marks 479
  plough marks 479
  remoulded diamictons 479
ice loading see isostatic readjustments
ice lobe 84, 112
  surging 193
ice marginal compressive belt 315, 322-332
ice marginal lakes 233, 235, 368, 383, 418
  river lakes (ice dammed/topographically dammed)
233, 235, 418, 419, 428
  semi-permanent lakes (ice dammed/topographically
dammed) 180, 233, 235, 418, 419, 428
ice marginal streams 274
ice margins 84
  grounded 85, 165, 174, 176, 262, 263
  marine 92, 93, 96-98, 262
  terrestrial 91, 96-98, 262
ice mass flux ("activity") 113
ice mass stability/instability 143
  decoupling 143
ice mass types 102-103, 110, 153
ice movement see glaciology
ice rafting (in lakes) 421, 429, 433
ice rafting (in marine environments) 460, 467, 469-472, 473, 486
  ice rafted debris (IRD) 469, 470, 475

  till & mud pellets 469
ice shelf rafting 469
ice shelf rafted debris (ISRD) 469
rates of ice rafting 471
sea ice rafted debris (SIRD) 469, 475-476
ice ramp (submerged) see submerged ice ramp
ice rheology 121, 134, 139-146, 204, 222, 289,
  anisotropic inclusions 205
  crystal cavities 141
  crystal lattice defects 140
  crystal reorientation 151
  deformation (elastic/plastic) 140, 141, 151, 177, 205
  dislocation climb 103, 140, 141, 150
  enhanced plastic flow 152, 153
  grain boundary melting 151
  grain boundary slip 140
  ice creep curves 141
  ice creep 'states' 139, 141, 150, 153, 166
  ice crystal structure 139
  longitudinal strain rate 146, 147, 152, 169, 170, 182,
185, 204, 270, 273
  non-linear creep 153
  permeability 220
  primary creep 141
  secondary creep 141, 150, 151
  secondary flow patterns 193
  strain softening 141, 152
  stress-strain behaviour 139, 140-142, 150, 151
  syntectonic recrystallization 141
  tertiary creep 141
  viscoelastic response 139
  viscoplastic 146, 149
  yield strength 141, 146, 149
ice rise(s) (rumples) 84, 135, 142, 167, 170, 174
  dynamic drag 170
  form drag 170
ice saddles 84, 96
ice shear zones 273
ice sheet asymmetry 111
ice sheet modelling 77-99, 148-149, 485
  model schools see GT-IS, Main, Melbourne
ice sheet reconstruction see ice sheet models
ice shelf 84, 117, 142, 166-167, 171, 224, 263, 265,
275, 276, 448, 476
  backpressure 167, 170, 171, 224
  bottom crevasses 170
  bottom melting 165, 166
  confined 167-168, 448
    parallel/diverging/converging edges 169-170
    strain rate 170

grounding line 90, 111, 135, 165, 168, 74, 224, 265, 276, 277, 448, 451, 456, 469, 485
    advance/retreat 172, 175
  flow mechanics 166-171
  fringing 448
  ice rises (rumples) 84, 135, 142, 167, 170, 174
  large embayment ice shelf 448, 450
  profiles (advance/retreat) 171
  pulsing 451
  pumping 451
  rafting 469
  response to internal /external influences 172
  stability 168-169, 484
  stress conditions within ice shelves 167, 169-171
  sub-ice shelf 451
  tidal influence 167, 176, 451
    unconfined 92, 167-168, 169, 448
     longitudinal strain rate 169
    undermelt 451
ice side-wall friction 142, 146
ice stagnation kettle lakes (*network*) 418, 419, 423
ice stratigraphy 121, 205
ice-stream interaction moraines (ISI) *see moraines*
Ice Stream B, W. Antarctica 122, 143, 154, 158, 160, 161, 162, 164, 168,
  172, 173, 174, 231, 317
Ice Stream C, W. Antarctica 168, 172, 174, 175, 179
ice streams 80, 89-90, 94-97, 112, 167, 168, 176, 313, 317
  terrestrial 81
  marine 81
ice stream edge drag 133, 174, 175, 451
ice structures 124-137
  ogives 124, 134, 136-137 (*see Forbes bands*)
  stratification 124
ice surface conditions 85-86, 113, 116-117
ice surface slope 90, 141, 207
  low profiles 94, 193
ice tectonics 105, 121, 123, 133, 134-137, 141 270
ice thickness versus microtextures (SEM) 505
ice tongue 84
ice viscosity 153, 263-264
ice velocity 84, 85, 92, 101, 113, 118, 142, 153, 157, 158, 159, 165, 171-179, 194, 217, 226, 228, 231, 247, 484
  balance 171-175, 176, 177
  basal 118, 153, .157, 158, 173, 175-176, 194, 205, 247
  class A - ordinary 178
  class B - fast 178

  class C - surging 178
  diurnal variation 175
  'draw-down' effect 484
  seasonal variation 175
  surface 171-175
  transverse 165
ice wedges 380
icehouse concept 12
Iceland 122, 158, 161, 162, 164, 185, 188, 230, 233, 234, 236, 239, 256, 279, 282, 284, 286, 297, 298, 302, 303, 304, 312, 313, 316, 368, 369, 370, 371, 375, 377, 381, 382, 383, 384, 385, 388, 391, 393, 395, 396, 399, 402, 405, 406, 410, 413, 415
Iceland-Faeroes Ridge 21
ikaite 480
Illinoian glaciation 25, 53, 488, 496, 497, 498, 502
imbrication (of clasts) 305, 372, 392, 395, 403, 404
incompetent strata 310, 330
indentation fracture (hardness) 243, 244
India 10, 13, 371
  Talchir boulder beds 10, 18
Indian Ocean 47
Inexpressible Island, Antarctica 489, 490, 491
Ingrid Christianson Coast, Antarctica 467
inner channel *see chute channel*
"inner moraines" 271
Innuitian Ice Sheet 86
insolation curves 60, 61, 72
insulating effect of supraglacial debris on ablation rate *see supraglacial environments*
interflow 425, 426, 442, 453, 466
interglacial periods 23-28, 112
inter-hemisphere synchroneity 59, 67
inter-island channels 89-90
internal deformation of ice 103, 113, 124, 147, 149-151
International Decade of Ocean Exploration *see CLIMAP project*
International Geological Correlation Program (IGCP) 51
intraglacial fluvial transport 274
inverse grading 374
Iowa 26, 53
Iran 17
iron oxide (Fe_2O_3) coatings 502, 503
iron-rich laminations 426
Isere River, France 356
Isfallsglaciären, Sweden 290
isostatic readjustments (rebound, depression) 75, 81, 86-88, 91, 98, 110, 168, 171, 310, 374, 485

Isotope Stages 29, 31-32, 35, 38, 39, 44, 45, 48, 49, 51, 52, 53, 57, 70, 74
isotopic dating 417, 434, 444, 481
isotopic fractionation 30, 47
Israel 38, 42-43
Italy 291
Itararé Formation, South America 18
Italy 22
Ivory Glacier 282, 283

Jakobshavn Isbræ, west Greenland 164, 172, 174, 175, 176
James River Ice Lobe, Laurentide Ice Sheet 90
Japan 38, 39, 42
joints 251, 254
  joint patterns 251
jökulhlaup facies model 410-414
jökulhlaups 191, 194, 225, 230, 233-239, 367-369, 395, 401, 413, 418, 436
  boulder ring structures 414, 415
  discharges 235, 368, 372, 392, 393, 395
  effects of 239
  hydrographs 235, 368, 369
  outwash 382
  record of jökulhlaups 234, 238
  residual hillocks 415
  sediment transport 235
  sources 235-236
  streamlining 382, 413, 415
  wash-limits 413
jökulhlaup initiation mechanisms 236-239
  englacial/subglacial hydrostatic pressure effects 236-237
  ice-barrier buoyancy 236-237
  plastic deformation 236-237
jökulhlaup sandur facies model 414
Jordan River-Dead Sea area 40-43
Jurassic 19
juvenile water 197

kame(s) 80, 215, 310, 375
Kansan glaciation 25
Karabatkak Glacier, C.I.S. 283
Karakoram Mts. 239, 291
Kaskawalsh Glacier, Yukon Terr. 290
katabatic winds 427, 478

Katla volcano, Iceland 368, 369, 406
keel marks 171, 477, 479
Keewatin District, N.W.T., Canada 87
Kemp Coast, E. Antarctica 112
Kenai Peninsula, Alaska 446
Kenya 108, 502, 503
Kesselwandferner, Austria 109
kettle holes 379, 380, 410
  till-filled 413
kinetic energy 222
kinetic sieving 273, 305
kinematic bulge 180
kinematic wave 101, 111, 113, 174,
kinematic wave and surging 189
kink bands 320
Knik Glacier, Alaska 362
Knik sandur, Alaska 395, 396, 397
Kolka Glacier, C.I.S. 188
Köngsfjorden, west Spitsbergen 484
Kozeny-Carmen relation 231
Krumbein form-roundness 286
Kutiah Glacier, Karakoram 188

laboratory simulations of abrasive wear 245-246
Labrador 90, 112, 165, 270
Labrador Current 472
Labrador Sea 69
lahars 502, 503
Lake Agassiz, Manitoba 368
Lake Baikal, Russia 39
Lake Biwa, Japan 39, 42
Lake Bonneville, Utah 442
lake bottom environments & sediments 434-436
Lake Erie, Canada/U.S.A. 62, 464
Lake Hitchcock, Massachusetts 435, 439
Lake Huron, Canada/U.S.A. 62
Lake Manitoba, Manitoba 90
lake margins 440-441
Lake Missoula, Montana 239, 382
Lake Superior, Canada/U.S.A. 13
Lake Wakatipu, New Zealand 425
Lake Winnipeg, Manitoba 90
"Laminated Diamicton", East Anglia 307
"laminated ice" 120, 121
lamination
  coarse-fine couplets 434, 435, 441
  Cyclopels 458, 461

cyclopsams 458, 461
draped laminations 421, 429, 432, 438, 440
  model 440
iron-rich laminations 426
multiple graded laminae 434, 441
rhythmic parallel 434, 435, 441, 470
ripple-drift cross lamination 402, 403, 408, 438, 439, 440
  sinusoidal ripple cross lamination 402
  varve 434
laminites 466
Lamplugh Glacier, Alaska 457
Lamstedt push moraine, Germany 335
Landsat Imagery 382
landsliding and surging 189
lapse rates 107
Larsen Ice Shelf, Antarctica 448
Last Glacial Maximum 59-64
Late Cenozoic glaciation 19-26, 28, 52-55, 65-66, 73
late summer runoff 365
latent heat
  freezing 114, 152, 222
  melting 114, 152
Late-Pinedale Interglaciation (Sangamon) 496
lattice defects (ice crystals) 140
Laurasia 18
Laurentide Ice Sheet 25, 52, 53, 62, 80, 90, 96, 98, 112, 118, 165, 193, 270, 298, 368, 375, 382, 443, 464, 486
layer parallel shear (M shear) 321
Lawson Type Flows (flow tills)
  Type I 301
  Type II 301
  Type III 301
  Type IV 301
Lead [210] chronologies 481
lebenspuren 436
Lewis River, Canada 355, 372
Lewis River sandur 379, 383, 396
Lierdalen, Norway 258
Lillooet Lake, British Columbia 425
limnology
  biological 426
  chemical 426
  physical 420, 424
linked-cavity mechanism for surging (*Kamb*) 187-192, 194
  orifice development 191, 192
  tunnel instability 191
linked-cavity model 190

linked-cavity system 186, 187-192, 216, 219, 225-226, 227, 348, 351, 352, 353
liquefaction 301, 454
lithostratigraphy 28
Little Ice Age 50, 51, 52, 62, 64-65, 70-72, 74, 291, 324, 377, 399
load structures 332, 421
lodgement process of till deposition *see till*
loess 36-38, 370, 395
Loess Plateau, China 38
loess stratigraphy 37
lonestones 469, 473
longitudinal compression 319
longitudinal extension 319
longitudinal strain rate
  in ice sheets 146, 147, 152, 182, 204, 270, 273
  in ice shelves 169, 170
  in surging glaciers 185
low-angle structures (glaciotectonic) 333
Lower Waterfowl Lake, Alberta 425
Lower Wright Glacier, Antarctica 133
Lowestoft Till 298, 306-308
lubrication (basal) and surging 189
Luochuan, China 38
Lyell, C. 10
Lyngsdalselva, Norway 367, 393, 396

M-beds 148, 176, 178, 193, 218, 226
M-shear (layer parallel) 321
McBride Glacier, Alaska 459, 473, 474, 480
McBride Inlet, Alaska 474, 484
Macedonia, Greece 38, 40
Mackay Glacier, E. Antarctica 484
MacLaren Glacier, Alaska 208
McMurdo Ice Shelf 277, 468
macrofauna 468
magnetic signature *see paleomagnetism*
magnetite (detrital) 38
Maine School of ice sheet modelling 79-81
Maizels' jökulhlaup sandur facies model 414, 415
Makarovbreen, Svalbard 268
Malaspina Glacier, Alaska 423, 447
Malaspina Lake, Alaska 425, 426
Manganese dioxide coatings 502, 503
Manitoba 90, 368
Manning roughness 222
mantle rheology 78
Margerie Glacier, Glacier Bay, Alaska 449

marine-ending glacier termini 448-459, 484-486
  advance & retreat 484-486
  floating termini 176, 448-451
  grounded margin 165, 176, 449
  stability 484
  tidewater termini 174, 176, 224, 451-459, 474, 484
marine ice sheet 80, 84, 89, 168
marine snow *see phytoplankton*
mass balance 92, 96-97, 105-113, 146, 165, 176, 178, 197, 261, 280, 281, 291, 315, 325, 329, 332, 365, 482, 484
mass balance relationship with tidewater termini 165, 482
mass movement into glacial lakes 421, 427, 429, 434
  debris flow 421
  slump terraces 427
  surge current 421, 427, 434, 440
  turbidity currents 427, 438
Massachusetts 256, 435, 439
Matanuska Glacier, Alaska 120, 158, 199, 208, 268, 271, 272, 282, 284, 300, 305, 343, 345, 346, 359, 362, 422, 428
Matanuska sandur, Alaska 395
matrix-supported sediments 305, 374, 404
Matuyama Paleomagnetic Chron 29, 34, 35, 56
Maunder Minimum 70, 71
medial moraine *see moraine*
medial moraine septa *see debris septa*
medieval warm interval 70
Medvezhiy Glacier, Tadzhikstan 172, 180, 182, 184, 188
megaflutes 315, 333
megaripples 407, 408, 410, 413
mega-blocks of sediment 322, 329
Melbourne School of ice sheet modelling 78-79
melted core and surging 189
melting glacial bed conditions *see subglacial thermal bed conditions*
melting ice zones 78-99
meltout 147, 151, 461, 469
  process of release 294
meltwater (*see subglacial meltwater channels*)
  abandoned channels (*proglacial*) 401
  abrasive capacity of streams 351
  bank scour 374
  channels (*see meltwater channels*)
  chemistry 264
  discharge characteristics 165, 180, 198, 200, 204, 213, 217, 228, 353-362, 365, 393, 394
  'drowned bed' 143, 151, 219

  entrainment 350-362
  erosion rate 353
  flowcells 374
  jets (buoyant) (*see plume*) 224, 455, 458, 461, 465
  "live" bed 393
  proglacial environments 180, 365
  sediment concentrations in streams 350
  sources 197, 225, 227
  subglacial drainage network 352
  seasonal fluctuations 198, 200, 201, 204, 213, 217, 365, 366, 367, 372-374
  sediment transport 209, 350-362, 370, 373, 374
  transit time 367
meltwater channels 381
  abandoned channels (*proglacial*) 401
  braided networks (*see braiding*) 376, 386
  bank instability 386
  channel piracy 209
  distal 383
  downstream decline in grain size 396, 397
  "dynamic metastable equilibrium" 393
  flow depth 392
  flow velocity 392-394
  fluvial bedforms 401, 403
  Froude number 392, 393
  gradient 383-393
  hydraulic geometry 209, 394
  hydraulic gradient 209
  large-scale cross beds 392, 410, 413, 433
  megaripples 407, 408, 410, 413
  mobile-bed sediments 398
  morphology & pattern 383-395, 398, 401
  open-channel flow 213
  proximal 383
  rates of channel closure 209
  turbulence 205
  width-depth ratios 383
meltwater systems 197-226
  channel flow 218, 221-225
  discrete (segregated) 197
  drainage routes 218
  englacial systems 200, 201-209, 213, 215, 217
  linked-cavity flow 197, 216, 218, 219, 227
  model of drainage routes 198
  non-steady state channels 213, 220, 221
  seepage via tubes & veins 203, 204, 207
  sheet (film) flow 197, 218, 219-220
    thickness 219, 220
  steady/quasi-steady state 198, 208-213
  steady state channels 208-212

subglacial systems 148, 152, 155, 159, 182, 209, 213-, 312
  supraglacial systems 198-201, 204, 215, 217
  vein system 203-204, 207
    moisture flux 203
Mendenhall outwash, Alaska 396
Mer de Glace, France 173, 208
meromixis 426
Mertz-Ninnis Trough, E. Antarctica 483
Meserve Glacier, Antarctica 145
Meseta Formation, Antarctica 489
Mesozoic 66
metahalloysite 502
metalimnion 420, 421, 438
metamorphoses of snow *see snow transformation*
meteorites 265
Miall's facies model types 411, 413
  Bijou Creek type 410, 411, 413
  coding scheme 412
  Donjek type 409, 411, 413
  Platte type 410, 411, 413
  Scott type 409, 411, 413
  South Saskatchewan type 410, 411, 413
  Trollheim type 409, 411, 413
Michigan 13, 193
micromorphology 321, 334
micro-structures 317
microtextures 488, 489, 505
  versus ice thickness 505
microtopography of grains 487
Midtalsbreen, Norway 207, 226
Mikkaglaciären, Sweden 200
Mississippi river 235
Missouri 26
Missouri river 235
Missouri Coteau 325
Milankovitch, M. 10, 24, 28, 66
Milankovitch Theory (*effect, forcing*) 10, 13, 16, 31-35, 36, 48, 63, 65, 66-69, 70, 73, 486
Miocene 20, 52, 58, 489, 491
Mindel glaciation 24
mineral exploration 277
mineralogy 285
mini-surges 182, 183, 184
mixing (water turnover in lakes) *see water body turnover*
MnO_2 coatings 502, 503
mobile (soft) beds (M-beds) 148, 176, 178, 193, 218, 226
models *see ice sheet and/or ice flow models*

models of glacial abrasion 242
model of medial moraine development 278
Mohr-Coulomb Failure Law 121, 294
moisture flux form ice veins 203
molluscs 467, 468
mono-crystalline ice 139
mono-thermal basal ice conditions 119
monomictic lakes 471
Møns Klint, Denmark 323
monsoon shifts 39
Montana 13, 120, 193, 239, 382
Mont Collon, Switzerland 173
moraine(s) 80, 261, 274, 358, 359, 385, 454, 456
  ablation dominant (AD) 273, 274
  annual 233
  avalanche 273
  beaded 271, 273, 274
  composite ridges 193
  cross-valley 460, 463, 464
  De Geer 460, 464
  dump end 315
  end 24, 315, 329, 464
  fluted *see fluted moraine*
  frontal-dump 463
  hummocky *see hummocky moraine*
  ice-cliff fed 274
  ice-covered 195, 201
  ice-front & boulder 384
  ice-stream interaction (ISI 273, 274
  lateral 201, 261, 339, 341, 377, 464
  marginal 266
  medial 180, 201, 271, 273, 275, 276, 277, 278, 288, 339, 341
  push 309, 313, 315, 316, 317, 319, 324, 325, 326, 327, 328, 329, 332, 433, 460, 463
  ramp-type 460, 463
  Ra 460, 464
  Rogen 433, 460
  submarine 464
  squeeze 460, 461, 463
  supraglacial 181, 205, 263, 277, 278
  terminal 90, 91, 277, 338, 375, 383, 384, 464
  washboard 464
moraine breach 368
morphosequences (glaciolacustrine) 442
morphostratigraphy 24-28
Moscow, Russia 329
moulin (glacial mill) 124, 200, 201, 202, 203, 204, 206, 209, 210, 212, 217, 339, 345
Mount Kenya 108, 502, 503

mountain building 21, 22
Mt. Discovery, Cockburn Island 489
mud pellets 469
"mud walls" 333
Mueller Glacier, New Zealand 274, 279, 286
Muir Glacier, Alaska 455, 458, 480
Muldrow Glacier, Alaska 185, 188
Munich, Germany 23
Mushketov Glacier, Tien Shan Mts., C.I.S. 188
Muzgazy Glacier, C.I.S. 188
mylonites 310, 319
Myrdalsjökull, Iceland 282, 368
Myrsdalssandur, Iceland 382, 388, 402, 406, 413

N-channels *see Nye channels*
Nansen Ice Sheet, Antarctica 490, 491
nappes (in soft sediment) 195, 316, 333
   fold-thrust 326, 327, 328
   gravitational 326, 327
   internal imbrication 326, 328
Naro Moru age (Isotope stage 8) 502
Nebraska 26
Nebraskan glaciation 25
negative gravity anomalies 86
net basal adfreezing of ice 120
net mass balance 108-109, 174, 217
Nevada 39, 44, 48, 49, 68, 69, 74
névé 103, 114, 121
Neogene 487
Neoglacial 261, 324, 500, 501
Netherlands 321, 322, 325, 327, 334, 335
New Guinea 32, 35, 59
New Hampshire 286
New York State 53, 112, 164
New Zealand 17, 59, 62, 90, 99, 266, 274, 279, 282, 283, 286, 288, 290, 291, 370, 375, 378, 382, 407, 425
Newney Green, Essex, England 296
Newtonian fluid flow 154, 311
Newton's Law 90
Nigardsbreen, Norway 127, 216, 355, 358, 360, 361, 372
nival flood period 365
non-coaxial deformation 317
non-dilated sediment 162, 163
non-laminar flow in ice 134-137
non-pervasive deformation 160, 162
non-steady state channels 213, 220, 221
Nordenskjöldbreen, Svalbard 268, 282

Norfolk, England 306-308, 317
North America 62, 80, 107, 217, 298, 313, 315, 317, 325
North American glacial chronology 25-28, 35, 113
North Atlantic Ocean 15, 50, 68-69, 368, 469, 472
North Atlantic Deep Water Flow (NADW) (*Current*) 20, 49, 50, 69-70
North Atlantic DSDP 29, 44, 60, 61, 73
North Dakota 325
North European glacial chronology 24-25
North Sea 25
North Sea Drifts 306-308
Northern hemisphere glacial correlations 54-55
Norway 63, 89, 99, 119, 127, 128, 173, 207, 215, 216, 226, 242, 258, 270, 274, 283, 283, 355, 356, 358, 359, 360, 361, 362, 367, 371, 372, 393, 407
Nova Semlya, Russia 446
nunatak 265, 270, 275
Nye channels (N-channels) 155, 191, 194, 219, 221, 227

obliquity 16, 34
ocean-atmosphere interchanges 69, 71
ocean circulation 450
ogives 124, 134, 136-137 (*see Forbes bands and wave ogives*)
Older Dryas 48, 49, 50
Olduvai subchron 502, 503
Oligocene 20, 58, 489
Omsbreen, Norway 128, 242
Ontario 13, 53, 112, 230, 284, 464, 494-497, 496
opaline silica 467, 480
open-channel flow 213
Ordovician 16, 17-18
orbital periods 34, 65, 67
organic carbon 467, 468, 479-480
   Total Organic carbon (TOC) 480
orifice (*see linked-cavity system*) 190, 191, 192, 225
   step type 191, 192
   wave type 191
osar *see esker*
Østerdalisen, Norway 119, 120, 173
outwash plains *see glacial outwash plains*
Overberg, Sweden 300
overconsolidation 298
overdeepened glacial basins 209, 256, 315, 333, 485
overflow (hypopycnal) 425, 426, 434, 442, 458, 460, 477, 478

Overijessel, The Netherlands 322
"over heating" 149
"overlengthening" of glacier 291
overprinting *see glacial overprinting*
Oxygen Isotope Ratios 28, 30-31, 43, 46, 52, 70, 121, 133
Oxygen Isotope Stratigraphy 28-36, 44, 52, 58, 59, 306

P-forms 209, 223, 229
Pacific Ocean 30, 32
Pakistan 215, 217, 239, 371
Pakitsaq, Greenland 44
paleoenvironments 487
paleohydrologic analysis (variables) 366
paleomagnetism 12, 16, 24, 36, 38, 52, 53, 417, 444
paleosols 25-26, 37-38, 53, 296
paleothermal layers in snow 116
paleothermometer 30
paleowinds 500
Paleozoic 16-19, 492
Palmer Archipelago, Antarctica 467
Panama Isthmus 21, 66
Pangaea 18
paraglacial environments 464-477
parallel bedding 421, 434, 441
parautochthonous sediment 315
particle
  compaction (consolidation) 310
  evolution of shape 286, 287-289
  form ratio 399
  particle to particle contact 161
  roundness (sphericity) 287, 288, 399, 400
  shape 163, 285, 286, 287, 288, 399
  shape maturity/immaturity 288, 289
  shear strength 310
particle size distribution *see grain size distribution*
particle trajectory 262, 270, 276, 280, 287
Patagonia 90, 375, 446
Patagonian Andes Mountains 22, 58
^{210}Pb chronologies 481
Pearlette ash layer 26
pedogenic weathering 487, 492
Pee Dee Formation (PDB) 30
pelagic sedimentation rates 36, 481
Penck, A. 23
pendant bars 382
"penetratively cleaved" or sheared sediments 298, 310

penguin rookeries 476
Peninsula Fjords, W. Antarctica 483
penitentes 290
percolation zone 114, 116
percussion cracks 492
periglacial conditions 37, 476
perihelion 67
permafrost *see frozen sediments*
permeability of sediments 312
Permian 10, 18
  Early Permian 18
  Late Permian 18
  breccias 10
Permo-Carboniferous 12, 16, 65
Peru 58, 165, 283
pervasive deformation 160, 161, 163, 250, 298, 349
Peters Glacier, Alaska 188
Peyto Glacier, Alberta 105
Peyto outwash, Alberta 396, 397
Phanerozoic 12, 16-19
physical constants (ice & water) 214
phytoplankton 467, 468, 473
  fecal pellets 468
  marine snow 468
piedmont glaciers 279
"pillow structures" in sediments 307
Pinedale Glaciation (Wisconsinan) 496, 497, 498, 499
pingo formation 380
pipe-flow of porewater in subglacial sediment 192, 218, 226, 228
pipes & dish structures 429
pitted erosion zone 82, 87-89
pitted moraine *see hummocky moraine*
Place Glacier, British Columbia 105
planar bedding
  cross-bedding 404, 408, 413
  horizontal 403, 404
plane of décollement *see décollement*
planktonic formanifera 28, 30, 32, 43
"plastering on" of glacial debris 299
plastic attenuation hypothesis 137 (*see ogives*)
plastic deformation in ice 272
plastic deformation in rock 243, 286
plastic deformation in sediment 311
Plate tectonics 12, 21, 22, 65
Platte type facies model 409, 411, 413
Pleistocene 23-60, 110, 289, 317, 325, 484
Pleistocene/Holocene boundary 22, 51, 70
Pliocene 28, 34, 38, 53
Pliocene/Pleistocene boundary 22, 56

Ploughing Index 163
"ploughing" at the ice/bed interface 158, 348
"ploughing" of marine sediments 309, 313
plucking *see quarrying*
plume 224, 458, 461, 466
  barotropic flow 458, 465
plunge pools 382
Poland 25, 58, 368
polar ice 93, 105, 121, 148, 175, 176, 203, 204, 215, 218, 233, 263, 265
pollen records 38-43, 56, 444
polycrystalline ice 93, 103, 139, 140, 150, 204
polymictic lakes 421
polymodal grain sizes 285, 398, 403
Polythermal Type A (bed condition) 122
Polythermal Type B (bed condition) 122-123
Polythermal Type C (bed condition) 123-124
Polythermal-based ice conditions 121-124, 147, 215, 264-265, 268, 269, 271, 272, 273, 279, 283
Polythermal subglacial bed types 122-124, 215
polythermal ice bed models 117-124
porewater 151, 159, 164, 228, 231, 488
  pressure 143, 152, 192, 218, 298, 301, 313
  solutions 497
Port Askaig Tillite, Scotland 16
postglacial isostatic rebound *see isostatic readjustment*
potassium/argon dating 52
potentiometric surface (englacial water table) 203, 204, 206, 207, 432
potholes 209, 221, 241, 382
pre-Bull Lake-Pinedale interglacial 498, 499
Precambrian 13, 492
    Precambrian Shield *see Canadian or Fennoscandian Shield*
Pre-Cenozoic glaciations 11-22
precession 34, 65
precipitation of glacier (ice sheet) surface 106
pre-Elster interglacial 25
pre-Illinoian 52
pre-Pinedale *see pre-Bull Lake*
Pre-Pleistocene glacial sediments
  Archean 11, 13
  Permian 10
Pre-Quaternary glacial sediments *see Pre-Pleistocene glacial sediments*
pressure melting 152
pressure melting point (pmp) 84, 151, 152, 153, 158, 177, 203, 204, 263, 272, 343
pressure waves (in ice) 136 *(see ogives)*
pressure waves from calving 459

Princess Elizabeth Land, Antarctica 489
proglacial environments 111, 193, 201, 278, 305, 332, 365-416
  aeolian sediments 401
  anastomosing channel systems 377, 408
  bank materials 370
  bank scour 374
  bar formation (deposition) 374
  braid system development 374, 380, 382, 408
  buried ice 379
  chute channels 380
  delta 332
  distal channels 378
  drainage systems 368, 376, 383-395
  hydraulics 392-393
  ice blocks (*stranded*) 374, 379, 380, 383, 392, 395
  kettles 379, 380, 410, 413
  kettled outwash 383
  lakes (pools) 148, 263, 283, 313, 374, 443
  longitudinal profile 375, 378
  megaripples 407, 408, 410, 413
  permafrost 380
  pingo formation 380
  pitted outwash 375, 381, 383, 410
  sediment pathways 369, 372, 398
  vegetation colonization 380
  "washed" sandur spreads 383
proglacial facies models (*see Bluck, and Cant*) 405-414
  Bijou Creek type (Miall) 410, 411, 413
  Donjek type (Miall) 409, 411, 413
  GI Lithotype (Rust) 409
  GII Lithotype (Rust) 409
  GIII Lithotype (Rust) 409
  Platte type (Miall) 410, 411, 413
  SI Lithotype 410
  SII Lithotype (Rust) 410
  Scott type (Miall) 409, 411, 413
  South Saskatchewan (Miall) 410, 411, 413
  Trollheim type (Miall) 409, 411, 413
proglacial sediment facies 401-416
  boulder 405
  cross beds 392, 413
  cross-lamination 402
  cross-stratification 402
  debris-flow 404
  drapes (*silt-skins*) 401, 438
  falling-limb deposition 401
  fine-grained 401
  gravel 403-405

jökulhlaup 410
mud-flow 404
"pulsed" sediment influx events 401
sand 401
Type 'A' ripple-drift cross-lamination 402, 403, 408, 438, 439,
   model 440
Type 'B' ripple-drift cross-lamination 438, 439
   model 440
Type 'S' sinusoidal ripple lamination 402
prograding sediment units 332
Proterozoic
   Early 12-14
   Middle 14-15
   Late 12, 15-16, 19
provenance (clast) 147, 274, 277-284, 469
proximal-distal facies models for braided rivers and glacial outwash 408-410 proximal ice-contact lakes 233, 235, 417-419, 421, 427-428, 430
   englacial lakes 418
   ice-marginal lakes 233, 235, 418
   subglacial lakes 231-233, 235, 418
   supraglacial lakes 199, 233, 235, 418
Prydz Bay, Antarctica 467, 481, 484
Puget Ice Lobe (Pleistocene), Washington State 162
Puget Sound, Washington State 53
"pulling force" 92, 95
"pulsed" sediment influx events (proglacial) 401
push moraines 309, 313, 315, 316, 317, 319, 324, 325, 326, 327, 328, 329, 332, 433
   annual 325
   internal composition 325
   tectonic style of development 194, 195, 325-328, 433
push ridges (glaciomarine) 476
pyrite (framboidal) 480

Q-beds 124, 148, 178, 218, 226,
quarrying 241, 251-255, 256, 287, 310, 340
   of lee surfaces 250, 251, 252, 254, 256, 259, 265
   rates 257
"quarrying law" 256
quartz grains 486-506
   adhering particles 494, 497, 498, 500, 501, 502, 503, 505
   arc-shaped steps 505
   coatings 498, 502, 503

conchoidal fractures 489.490, 491, 505
corrasion features 489
crushing patterns 489, 494, 495
dissolution 487, 489, 490, 491, 494, 495, 496, 497, 498, 500, 501, 502, 505
   etch features 489, 492, 494, 495, 496, 499, 500, 501, 502
   gouges 505
   grooves 505
   fracture lines 489, 504
   microtextures 488, 489, 505
   regrowth crystals 492
   soft weathering 489, 490, 503, 504
   step features 489, 490, 496
   troughs 489
   V-shaped percussion cracks 492, 493, 501, 505
quasi-artesian conditions (hydraulic systems) 224
quasi-hard/mobile beds (Q-beds) 124, 148 , 178, 218, 226
Quebéc 13, 53
Queen Elizabeth Islands, Canadian Arctic Archipelago 86
Quelccaya Ice Cap 51
quick sands 431

R-channels see Röthlisberger channels
radiocarbon dating 29, 35, 36, 47, 52, 71, 444, 481
rafts see sediment rafts (intraclasts or ice rafts)
raised beaches 87, 474
raised boulder delta 382
Ramsay, W. 10
Ravak Glacier, C.I.S. 188
readvance sequence of glaciotectonic overprinting (style/model) 335
receiving area (in surging glacier) 180, 182
red beds 16
red clay (Late Pliocene) 38
release mechanisms (surges) 189
resedimentation processes (reworking) 295, 302, 303, 305, 345, 362, 433
regelation 85, 87, 90, 117, 121, 140, 147, 152, 177, 245, 246, 249, 264, 271, 289, 338, 344
   Weertman regelation 120, 154, 264
Rehburg line 325, 335
relaxation time see response time
Rendu Glacier, Alaska 188
Renland Ice Core, Greenland 44
reservoir area (in surging glacier) 180, 182, 192

response (relaxation) time (*ice mass adjustment*) 107, 116, 146, 217
  for meltwater channel systems 395
rheology of ice *see ice rheology*
rheology of sediments *see soft sediment deformation*
rhythmites 239, 421, 434, 435, 436, 441, 442, 444, 470
  subglacially deformed 321
Richardson Number 458
Riedel shears 319, 320, 321, 334
riegels *see rock steps*
rigid subglacial bed 299
ring shapes in asymmetrical folds 320
rip-up clasts 434
ripples
  climbing 421, 429, 432, 437, 439, 441
  fluvial 401, 403
  megaripples 407, 408, 410, 413
  types 440
ripple lamination
  Type 'A' ripple-drift cross lamination 402, 403, 408, 438, 439
    model 440
  Type 'B' ripple-drift cross-lamination 438, 439
    model 440
  Type 'S' sinusoidal ripple lamination 402
Riss glaciation 24, 25, 56, 488, 496, 497, 498
river lakes (ice dammed/topographically dammed) 180, 233, 235, 418, 419, 428
Roberts Massif, Antarctica 489, 491
Robin "Heat Pump" Effect 117, 264, 344
roche moutonnée 80, 248, 249, 256
rock dam 368
rock dislodgement 254-255
rock falls 174, 265, 269, 273, 279, 287, 291, 339, 368, 454, 461, 462
rock flour 243, 284
rock fracture 251-254, 261, 270, 340
rock hardness 242
rock mechanics (fracture, microcracking etc.) 207, 241, 242
rock steps 207, 209, 223, 252, 253, 256
rock strength 251
  yield strength 143
Rocky Mountains, Canada/U.S.A. 325, 329, 370, 407, 498
Rogen moraines 433
Ronne Ice Shelf, Antarctica 167, 451
rooted structures in sediment 316, 322
Rose Township, New York State 164

Roslin Glacier, East Greenland 268
Ross Ice Shelf 81, 111, 125, 167, 169, 170, 174, 448, 451, 464
Ross Embayment 58
Ross Sea 58, 59, 447, 449, 467, 468, 481, 483
Rossby waves 66, 70
rotational failures 427
Röthlisberger channels (R-channels) 143, 219, 220, 221, 222, 224, 225, 227, 228, 229
  Röthlisberger's model of channel form 221-222
Rouge river valley, Ontario 494, 495
Russell Glacier, West Greenland 406
Russia 24, 39, 57, 58, 329, 446, 488, 502, 504
Russian Plain (central) 57
Rust's facies model types 416
  coding scheme 416
  GI Lithotype 409
  GII Lithotype 409
  GIII Lithotype 409
  SI Lithotype 410
  SII Lithotype 410
Rutford Ice Stream, Antarctica 172
Ruwenzori Mts., Rwanda 500
Rwanda 500-501

S-foliations 319
Saalian end moraines 329
Saale glaciation 24, 327, 488
Saalian till 321
saddle *see ice saddle*
St. Elias Mountains 62
St. Lawrence 368, 371
sandurs *see glacial outwash*
Sangamon interglacial 25, 35, 49, 53, 488, 492, 494, 496, 497
salinity (*sea water*) 30, 65, 69-70, 117, 276, 426, 452, 460, 466, 478
  downward salt flux 478
  surface layer salinity 478
salt export (*sea water*) 70
sand wedges 296
Saskatchewan 174, 325, 329
Saskatchewan Glacier, Alberta 129
Scandinavian Ice Sheet 24-25, 53, 58, 63, 327
Scandinavian Shield *see Fennoscandian Shield*
Scanning Electron Microscopy (SEM) 454, 487-506
  photomicrography 488
schötter *see glacial outwash plains*

Scotland 16, 77, 99, 239, 297, 402, 403
Scott outwash river, Alaska 392, 396, 399, 400
Scott type facies model 408, 409, 411, 413
Scottish erratics 307
Scottish Late-Devensian Ice Sheet 77
scour marks 382, 392, 403, 415
scoured terrain 80, 86-89, 382,
sea ice 63, 446, 477
sea ice rafted debris (SIRD) 469, 475-476
sea level change see eustatic Change
seasonal pack ice 472, 476, 478
sea surface temperatures (SST) 28, 34, 47, 59, 63, 65, 117, 276
sea water 47
  bottom turbidity 452
  cold tongues 452
  temperature 445, 446, 452
second-order glacial landscapes see glacial landscape types
secondary ice flow 137, 193
sediment anisotropy (rheology) 163, 218
sediment budgets 270
sediment block rafting 273, 322
sediment facies see proglacial sediment facies
sediment faulting & folding 313
sediment flux from glacierized basins 337, 349, 349-350, 351-353, 354
  discharge characteristics 353-362
sediment flux through a subglacial conduit 351, 352
sediment/landform associations 101
sediment modification in transport 285-289
sediment rafts (floes) 194, 313, 322, 328, 329
sediment stratification in glacial lakes 424-426, 427
sediment strength (shear) 294, 310, 316, 331
  strain rate 312
sediment trap arrays 417
sediments in meltwater streams
  bed load 354, 355, 358, 370, 372, 373, 378
  discharge hydrograph 357, 359
  diurnal variations 354, 356, 360, 370, 372
  non-glacial sources 369
  suspended sediment 353-362, 369-374
  seasonal variations 354
  sediment concentrations 354
  total annual yield 354, 358, 370, 371, 372
  transport rating curve 355, 373
  yield 362, 370, 372
sedimentology associated with surging 193-195
  model of sedimentation by surging glaciers 195
SEM see scanning electron microscopy

semi-permanent lakes (ice dammed/topographically dammed) 180, 233, 235, 418, 419, 428
  ice-contact delta 233, 235, 428-431
  subaqueous fan 428, 431-432
  submerged ice ramp 428, 433-434
Sentinel Glacier, British Columbia 105
seracs 174, 201, 458
Severnaya Zemla, Russia 446
Seward Glacier, Alaska 103
Shackelton Glacier, Antarctica 489, 491
"shadowing effect" 248, 249, 256
shear fabric in sediment 318, 319, 322
shear heating (ice) (see "overheating") 149
shear planes (in ice) 121.135
shear planes (in sediment) 318
shear zones (in ice) 121
shear zones (in sediment) 315-322, 322
sheared snouts (by surging glaciers) 180
sheath folds 320, 322
sheet structures in rock 256
shell coquina 467
Sherman Glacier, Alaska 266, 290
Shini-bini Glacier, C.I.S. 188
Shoemaker's Model of Basal Debris 248, 256
shorelines (raised) see raised beaches
Shropshire, England 10
Siberia 18, 39
sichelwannen 223, 230
Sidujökull, Iceland 188
Sierra Nevada, California 108
silicate rocks 252
silicicalastic sediments 451, 453, 466, 467, 468, 470, 475, 476, 478, 479, 480, 481
silicon-oxygen bonds 252
siliceous microfossils 30, 426
siliceous ooze 467, 470
silt drapes 438
"silt-skins" 401
silt transport in meltwater 359
Silurian 17-18
sintering (rounding of snow particles) 103
Siple Coast, W. Antarctica 450
Sirius Formation, Antarctica 58, 489
Skaftafellsjökull, Iceland 383, 384, 385
Skaftavkos ice dammed lake, Iceland 369
Skeidarar river, Iceland 233, 396
Skeidararsandur, Iceland 370, 372, 375, 377, 381, 383, 391, 396, 397
Skoasandur, Iceland 415
slackwater deposits 239, 383

sliding coefficient of ice 93
Slims River sandur, Alaska/Yukon Terr. 379, 383,
391
"slippery spots" 84
Slovakia 36
slump terraces 427
slushflows 339
snow 103, 108, 448
   densification (consolidation) 103, 117
   firnification 265
   transformation 103, 116-117, 149
   snow ablation (deflation) see ablation rates, zone
snow accumulation see accumulation rates, zone
'snow blitz' 175
snowfield 111
snowlines 59, 62
snowpacks (perennial) 339
Snowy Pass Supergroup 13
sodic minerals 426
soft sediment deformation or mobilization see
deformation
Sogndal, Norway 258
soils see paleosols
Solar causes of glaciation
   faint young sun theory 12, 13, 14, 19
solar radiation 105, 109
solar variability 46
solar winds 71
Solheimajökull, Iceland 302, 371, 383, 405
Solheimasandur, Iceland 413
Sonnblick, Austria 109
Sørbreen Glacier, Svalbard 133, 268
Sore Buchananisen, Svalbard 286
Sorge's Law 103
sorting index 398
South Africa 13, 15
South America 15, 18, 20, 38, 39, 51, 58, 59, 60, 62,
375
South Cascade Glacier, Washington State 158
South Saskatchewan type facies model 410, 411, 413
South Shetland Islands 447
South west Africa 15, 18
Southern Alps, New Zealand 62, 291
Southern Atlantic Continental Shelf 57
Southern Atlantic Ocean 60, 179
Southern hemisphere glacial correlations 56-57
spalling 254
SPECMAP 39, 44, 49, 52
spillway expansion bars 382
Spitsbergen see Svalbard

spodosols 37
spring meltwater flows ('spring event') 217, 225, 365
squeeze-up processes 433, 454, 461, 463, 481
Stages (isotope) see isotope stages
stagnant ice 203, 213, 280, 293, 333, 381, 428
Standard Mean Ocean Water (SMOW) 30, 46
steady-state ice sheet models 78-99, 118
steady state meltwater channels 208-212
steady-state temperatures 116
Steele Glacier, Yukon Terr. 235
Stewart Lakes, Baffin Island 425
"sticky spots" or "stick-slip" 84, 117, 152
stillstand 110
Stillwater complex, western U.S.A. 13
streamlined bedforms 315
Storglaciären, Sweden 158, 173, 200, 206, 209, 211,
256
Storsteinfjell, Norway 359
strain ellipses 318-319, 320
strain rate
   normal 124
strain rate of ice 146, 147, 150, 151
   longitudinal strain rate 146, 182, 185
Strandline Lake, Alaska 235, 237
stratified sediments 226, 313
stratotype 22, 24
stream power 374, 378, 392
streamlined terrains 315, 335, 382, 413
stress (glacial) (active/passive) 310, 311
stress-anisotropic fabric see ice fabric
stress corrosion 243, 252
striae 80, 86, 87, 243, 249, 477
   geometry 244
striation(s) see striae
Strontium isotope ratios (^{87}Sr/^{86}Sr) 66
sub-Antarctic Islands 446
subaqueous
   esker deltas 460
   fan 428, 431-432, 460
   flows 434
   chaotic bedding 431
   distal deposits 432
   model 432
   quick sands 431
subglacial environments 193
   bed conditions 78-99, 114, 117-124, 143, 314, 315,
317, 342, 343, 343, 451
      hard (rigid) beds (H-beds) 148, 155, 176, 178, 192,
193, 218, 226
      model of subglacial interfaces 211

mobile (soft) beds (M-beds) 148, 176, 178, 193, 218, 226
  quasi-hard/mobile beds (Q-beds) 124, 148 , 178, 218, 226
  bed permeability 216
  debris 120-121, 123, 134, 135, 143, 163, 213, 246, 261, 266, 268, 277, 284, 339, 340, 342, 348, 459
  frozen conditions 148
  hydrology 78, 82, 148, 155, 175, 178, 183
  ice/bed separation 143, 144, 145, 253, 254, 255, 259, 343
    model 156
  melting rates 156
  thermal conditions 78-99, 101, 113, 114, 117-124, 143, 159, 178, 197, 263, 276, 298, 314
subglacial cavities 143, 152-158, 220, 294
  formation 143, 156, 157, 247
  rate of closure 158
  temperatures 119
subglacial crevasses (basal) 133, 454
subglacial hydraulic systems 148, 152, 155, 159, 182, 187, 209, 213-229, 274-275, 305, 312, 339, 348, 350, 351, 352, 353, 358, 368, 428, 432, 457, 481
  discharge 156, 187-191, 192-193, 194, 207, 217-218, 222, 223, 224, 225, 226, 228, 231
  channel (conduit) flow 24, 143, 155, 216, 219-225, 227, 241
    backpressure 222
    bottom channels 223
    cross-section 221
    channel form 221-222, 224
      Hooke's Model 222
      Röthlisberger's Model 221-222
    channel types 221
    closure rate 228
    C-channels 219, 221, 227
    'gradient' channels 223
    location 223
    maturation & degeneration 223, 227
    N-channels 155, 191, 192, 193, 219, 221, 227
    R-channels 143, 219, 220, 221, 222, 224, 225, 227, 228, 229
    till channels (Alley's) 228, 229
  hydraulic stability between channel/films 223, 224-225
  hydraulic system 'switching' 225
  linked cavity flow 225-226, 227
  sheet flow 143, 152, 155, 174, 209, 212, 213, 219-220, 224, 227, 228
subglacial lakes 231-233, 235, 418

catastrophic flood 233
  hydraulic seal 231, 236
  rate of leakage 236
  temperature 235, 239
  tunnel geometry/length 235, 239
subglacial meltwater 148, 152, 155, 159, 182, 212, 215, 312
  cavities 155, 156
  discharge 156, 187-191, 192-193, 194, 207, 217-218, 222, 223, 224, 225
  channels 24, 143, 155, 216, 219-225, 227, 228
  erosion 241
  film (sheet) 143, 152, 155, 174, 209, 212, 213, 219-220, 224, 227, 228
  hydraulic gradient 158
  hydraulic 'jump' 193
  pressure 155, 165, 182, 183, 190, 192, 216, 217, 218, 222, 224, 225, 226, 249-256, 265
  pressure constrictions 191, 226
  undermelt under surging 194
subglacial sediment/hydraulic sensitivity mechanism for surging (Clarke) 192-193, 194
  pipe-flow 192, 226
subglacial shear conditions
  non-pervasive 148
  pervasive 148
subglacial streamlining processes 299, 312, 315, 335
subglacial till see till
subglacial volcanoes 368, 369
submarine diapirs 313
submarine mud transport 466
  eddies (gyres) 466, 473
submarine outwash fans 459-460, 461
  types 466
submerged ice ramp 428, 433-434
Sub-Antarctic 57, 60
sub-nappe zone in sediment 316, 320
sub-polar ice masses 113-124, 184, 215
subsole drag 295
Sugran Glacier, C.I.S. 188
Sulasandur, Iceland 377
sulphur (sulphuric acid) aerosols 72
Sulzburger Bay, Antarctica 481, 483
summer ablation 108
summer accumulation 108
summer balance 108
summer temperatures 106
Summit Ice Core, Greenland 39, 44, 46, 48, 74
Sun see Solar causes of glaciation

Sunnybrook diamicton, Ontario 494, 495, 496
Sunwapta Lake, Alberta 425
Sunwapta outwash, Alberta 388
super-cavitation and surging 189
Supercontinent cycle 12, 16
Supercontinent fragmentation 14
Supercycle theory 12
supercooled meltwater 204, 235, 344
supraglacial environments 101, 105-113, 143, 193, 197, 305
 debris 201, 213, 263, 266, 269, 270, 273, 274, 275, 279, 287, 289, 291, 293, 295, 302, 338, 345, 350, 369, 370, 453, 488, 504
  entrainment of debris 201, 265-266, 270, 338, 350
  flow till 301, 302, 303, 309, 310, 315, 316, 328
  hydraulic systems 198-201, 204, 215, 217
  insulating effect 197, 290, 291
  meltout 295, 299, 302, 453
  meltwater storage 200
  streams 194, 339, 340
supraglacial lakes 198, 199, 233, 235, 418
surges 113, 143, 145, 149, 156, 172, 175, 178, 179-195, 198, 201, 203, 219, 220, 225, 283, 291, 309, 317, 329, 331-332, 353, 368, 454
 basal cavitation 182
 comparison between the Variegated & Medvezhiy Glaciers 184
 dynamic balance line (DBL) 182, 184
 glacier length and probability of surging 180
 hypotheses of surge behaviour 185-195
  hydraulic disruption 186-193, 194
 longitudinal strain rate 185
 mega-surges 179
 meltwater discharge 184, 187-194
 mini-surges 182, 183, 184
 periodicity 185
 pre-surge phase 182, 185, 194
 release mechanisms 189
 sedimentology associated with surging 193-195
 surface morphology 180-182
 surge/non-surge cycle 182-185
 triggers 182
 undermelt 194
 velocity 183, 184, 186, 187
Susitna Glacier, Alaska 181, 354
Susitna River, Alaska 354
suspended sediment in meltwater streams *see sediments in meltwater streams*
suspended sediment in glacial lakes 421, 425, 426, 432

suspended marine sediment 265, 452, 460
 agglomeration 466
 flocculation 466
 hemipelagic 460, 478
 pelletization 466
 particulate matter (SPM) 468
 rainout 471
 sediment yields 474
Svalbard 16, 120, 133, 199, 203, 205, 268, 282, 286, 313, 317, 325, 326, 327, 328, 333, 335, 446, 468, 484, 487, 488, 492-494, 495
Sveg tills 300
Sweden 158, 165, 200, 206, 209, 211, 256, 290, 300, 444
Swiss Alps 173, 204, 217, 277, 290, 322, 325, 329, , 357, 360, 368, 370, 383, 393, 407
Switzerland 131, 173, 204, 215, 277, 290, 325, 352, 354, 355, 357, 360, 361, 362, 367, 370, 371
syntectonic recrystallization *see ice tectonics*
syntectonic sedimentation 328

tabular bedding 421, 428
Tadzhikistan 38, 172, 180, 182, 184, 188
Talchir Boulder Beds, India 10, 18
Tanana River, Alaska 355
tangential cross-beds 408
Taikutaaq Lake, Baffin Island 425
Tanyrnas Glacier, C.I.S. 188
Tasman Glacier, New Zealand 266, 267, 275, 282, 286, 288, 290, 291
Tasman Seaway 21
Tasmania 59
tectonic activity and surging 189
tectonic lamination 320
tectonic styles *see glaciotectonics*
tectonic transport direction 319
Teigadaljökull, Iceland 188
Tekapo River, New Zealand 382
Teleki Glaciation (Illinoian), East Africa 502
temperate ice 93, 117, 147, 148, 151, 177, 178, 184, 202, 203, 215, 217, 227, 251, 261, 263, 264, 268, 269, 271, 274, 279, 283, 284, 312, 342
Tenaghi-Philippan, Macedonia 38, 40
tension fractures 320
tephra 395
tephrochronology 36, 444
terminal grade concept 284, 285, 286, 287
Termination (*Stages - interglacials*) 33, 35, 44, 69, 70

terrace sequences 374, 375, 383, 410
Tertiary clays 325, 327
"tertiary comminution" 285
thawed bed conditions *see subglacial thermal bed conditions*
thermal classification of ice mass types 102-103
thermal diffusivity 116
thermal stratification of glacial lakes 420-424, 425, 426, 434, 437
  density profile 424
  epilimnion 420, 421, 424, 426, 434, 437, 440
  hypolimnion 420, 421, 424, 426
  metalimnion 420, 421, 424, 438
  thermocline 420, 421, 424, 425, 437, 441, 442
thermal structures of ice masses 113-124
thermocline 420, 421, 424, 425, 441, 442
  interflow 425, 426, 441, 442, 466
  overflow (hypopycnal) 425, 426, 434, 441, 442, 458, 460, 477, 478
  underflow (hyerpycnal) 425, 426, 440, 441, 442, 458, 461
thermohaline circulation (conveyor belt) 69, 70, 478
thermokarst 195, 198, 279, 380
thermoluminescence 36, 52
thermo-mechanical processes 113
three-grain intersections (boundaries) *see meltwater vein systems*
through-valleys 179
tidal flats 477
thin film *see basal water layer*
thin sections *see micromorphology*
thixotropic deformation 410
Thompson Glacier, Axel Heiberg Island 325
thrust planes (in ice) 135
Tibet 17, 38, 66
tidewater glaciers 174, 176, 224
tidewater positions (margins) 84
tidewater termini 174, 176, 224, 451-459, 468, 482, 484
  critical water depth (instability) 485
  debris transport & release 453
  fluvial discharges 455, 458
  hydrographic processes 456-458, 477-478
  mass balance relationships 165, 482
  mass movements 454
  meltout/fallout 454
  meltwater jet 224, 454, 458, 461
  squeeze/push 454, 461, 463
  stability 484
Tien Shan Mountains 282, 290

Tikke Glacier, Alaska 188
till (deposition) 295-305, 488, 500, 505 (*see diamictons*)
  abrasion by till 250
  chalk banding 320, 322
  deformation 295, 298, 312, 315, 317, 322, 333
  deformation till model (Boulton) 316
  deposition 256, 275, 459
  flow 295, 298, 300-305, 316
  flow mechanics (gravity, squeeze) 227, 301, 461
  lodgement 293, 295, 297, 298, 299, 306-308, 448, 461, 464, 487
  meltout 295, 298, 299-300, 306-308, 461, 487
  stratified 300, 307
  sublimation 295, 298, 299
  till dilatancy 250
  till matrix 250, 284, 286
    water-saturated 155, 160, 215, 250, 301, 312, 332
till channels (Alley's) 228, 229
till deltas 451, 459-460, 464
till fabric *see clast fabric*
till pellets 469
till prism 295
till sheets 448
  pushed/stacked 195, 315, 328-329, 333
till types 295
till wedges 320
tillite 13, 20
tilt of Earth's spin 16, 48, 65
time lag *see response time*
time series temperature measurements 417
Tingin sandur, Canada 379, 396
"Titanic disaster" 472
Total Organic Carbon (TOC) 480
topographic inundation 268, 269
trace fossils 426, 434, 435, 436
traction flow 461
traction-zone debris *see basal debris*
Transantarctic Mountains 18, 22, 58, 59, 489
transport distances 261
transport pathways 270-284, 340, 341, 398
transport rates 283
transportation processes 261-292
  englacial conduit 187, 202, 203, 204, 205, 206, 207, 209, 210, 211, 217, 274-275, 287, 305, 339, 348, 350, 358, 368, 428
    ice shelves 275
  subglacial conduit 155, 175, 182, 187, 209, 213, 219, 274-275, 305, 339, 348, 350, 351, 352, 353, 358, 368, 428

transposed foliation in sediments 316, 323, 324, 334

Transvaal basin, South Africa 13

Trapridge Glacier, Yukon Terr. 126, 158, 165, 180, 184, 186, 192

triaxial compression 251

tributary glaciers 111

triple point pressure 222

Triumvirate Glacier, Alaska 237

Trollheim type facies model 409, 411

Trollbergdalsbreen, Norway 355

trough(s) (glaciated) 80, 90, 375

trough cross-stratification 402, 404, 408, 413, 414

tube worm (*Surpula narconensis*) reefs 467

tubes & tubules *see meltwater vein systems*

tundra conditions 37

Tungnárjökull, Iceland 188

tunnel valleys 209, 312, 315, 333, 478

turbulent jets 224, 458, 461

Turkey 17

Turtmann Glacier, Switzerland 325

'Two Component Mixing' glaciomarine environments 73

Tyeen Glacier, Alaska 188

Type 'A' ripple-drift cross lamination 402, 403, 439
model 440

Type 'B' ripple-drift cross-lamination 438, 439
model 440

Type 'S' sinusoidal ripple lamination 402

U-shaped valleys 256, 257, 258

Upstream B *see Ice Stream B*

Ukraine 24

uncoupling *see basal uncoupling*

underflow (hyperpcynal) 425, 426, 440, 442, 458, 461

undeformed foreland sediments 333

Ungava 87, 112, 270

unimodal grain sizes 398

United States of America (*U.S.A.*) 15, 17, 21 26, 28, 38, 39, 44, 52, 53, 62, 68, 69, 74, 90, 98, 103, 108, 112, 120, 127, 158, 162, 164, 165, 172, 173, 174, 175, 180, 181, 182, 183, 184, 185, 187, 188, 190, 191, 192, 199, 208, 225, 235, 237, 239, 256, 266, 268, 271, 272, 274, 277, 282, 283, 284, 290, 291, 300, 305, 325, 341, 343, 345, 346, 354, 355, 356, 359, 360, 362, 375, 382, 383, 391, 392, 395, 396, 397, 399, 400, 407, 408, 422, 423, 425, 426, 428, 435, 439, 442, 443, 446, 449, 455, 456, 457, 458, 459, 465, 468, 473, 474, 475, 477, 480, 481, 483, 496

upwelling of ocean water 478

uranium-series dating 39

uranium/thorium dating 58

Uruguay 18

Urumqi Glacier, China 122, 158, 162, 173, 177

Utah 442

Uversbreen, Svalbard 203

Uzbekistan 38

V-shaped percussion cracks 492, 493, 501, 505

Valley Farm Paleosol 296

valley glaciers 337-363

valley train deposits 375, 377, 390, 394

valley wall frictional drag 133, 174, 175

Variegated Glacier, Alaska 158, 172, 173, 174, 180, 182, 183, 184, 185, 187, 188, 190, 191, 192, 225, 354

varves 434
chronology 444
thickness 444

Vashon Ice Lobe (Pleistocene), Washington State 164

Vatnajökull Ice Cap, Iceland 236, 279, 368, 377

Vatnsdalslon ice dammed lake, Iceland 369

vein calcite 39, 44, 68

Venetz, I. 10

Verngtbach Glacier, Austrian Alps 201

Vernagtgletscher, Switzerland 367

Vesl-Skautbreen, Norway 173

Vestfold Hills, Antarctica 489, 490

Virunga Mts., Rwanda 500-501

viscoplastic exponent of ice 93

viscous heat dissipation 205

volcanic activity 16, 65, 71-72, 395, 413, 502

volcanoes *see subglacial volcanoes*

volcanic terrain 488

Vostock Ice Core 39, 44, 47, 48, 49, 50, 59

Vostock Station 231

Vrica section 22

Waimakariri River, New Zealand 378

wallows (glaciomarine) 476, 477, 479

Walsh Glacier, Alaska 188

Warthe moraine stage 24

Washington State 52, 53, 99, 158, 162, 164, 173, 175, 382, 383, 395, 396, 425, 426

Watana Reservoir, Susitna River, Alaska 354

water body turnover (lakes)

density mixing (homopycnal) 426, 428, 434
dimictic 421
meromixis 426
monmictic 421
polymictic 421
water turnover (oceans) 478
water entrainment & transport *see meltwater*
waterlain sediments 308
Watts Glacier 268, 282
water layer *see basal water layer*
wave ogives *see ogives*
wave trains (in ice) 136 (*see ogives*)
wear 244-259, 285
weathering 53, 66, 251, 269, 339, 487-506, 502, 505
corrasion 489
intensities (*rates*) 13, 14, 15, 66
dissolution 241, 487, 490, 491, 494, 495, 495, 497, 499, 500, 501, 502, 504
etching 491, 492, 494, 495, 496, 499, 500, 501, 502
features 487
frost cracking 251
hydrolysis 252
pedogenic 487, 492
porewater solutions 497
pre-weathering 494, 497, 505
soft weathering 489, 491
wedge *see accretionary wedge or till deltas or grounding line wedge*
"Weertman regelation" film 120, 154, 223, 264
Weishsel glaciation 24, 58, 368
West Fork Glacier, Alaska 188
wet-based ice conditions (warm-based) 119-121, 153
wet bed conditions *see temperate ice*
wet snow zone 116
White Glacier, N.W.T., Canada 111, 174
White River sandur, Washington State 379, 383, 396
wind drift of snow 107
wind mixing in glacial lakes (wind stress) 421, 427, 434, 437
Wind River Mts., Wyoming 496, 497-499
winter ablation 108
winter accumulation 108
winter balance 108

winter snow 124
Wisconsin 53, 193
Wisconsinan glaciation 25-28, 34, 45, 48, 49, 50, 73, 78, 98, 103, 368, 488, 494, 496, 497, 500
Witwatersrand succession, South Africa 13
Wolstonian glaciation 25
Worcestershire 10
Wordie Ice Shelf, Antarctica 485
Wright Valley, Antarctica 145
Würm glaciation 25-28, 34, 48, 49, 73, 78, 98, 103, 368, 488, 494, 496, 497, 500
Wyoming 13, 26, 52, 496

Xian, China 38
Xifeng, China 37
X-radiographs 470

Yana outwash 408
Yarmouth interglacial 25
Yellowstone National Park, Wyoming 26
yield strength of ice 121, 124, 149
yield strength of rock 143
yield strength of sediment 160, 161, 162, 228
young Sun theory *see Solar causes of glaciation*
Younger Dryas 34, 36, 48, 49, 50, 62, 70, 72, 74
Yukon river delta 477
Yukon Territory 15, 62, 126, 158, 159, 165, 179, 180, 184, 185, 186, 192, 235, 290, 379, 383, 391, 395, 396, 425, 477

Zaire 371
Zarovshankiy Glacier, C.I.S. 283
Zermatt, Switzerland 352
Zillertal Alps, Austria 500, 501
zooplankton 466, 468
copepods 468